# ELECTRICAL ENGINEERING

# ELECTRICAL ENGINEERING

## (OBJECTIVE TYPE)

*By*

**S.S. Gupta**

**Nitin Saxena**
*Lecturer*
*Department of Electrical Engineering*
*Moradabad Institute of Technology*
*Moradabad (U.P.)*

UNIVERSITY SCIENCE PRESS
(An Imprint of Laxmi Publications Pvt. Ltd.)
**BANGALORE • CHENNAI • COCHIN • GUWAHATI • HYDERABAD**
**JALANDHAR • KOLKATA • LUCKNOW • MUMBAI • PATNA**
**RANCHI • NEW DELHI**

*Published by :*

UNIVERSITY SCIENCE PRESS

(*An Imprint of Laxmi Publications Pvt. Ltd.*)

113, Golden House, Daryaganj,

New Delhi-110002

*Phone* : 011-43 53 25 00

*Fax* : 011-43 53 25 28

www.laxmipublications.com

info@laxmipublications.com



*Price* : **Rs. 450.00** *Only.*

*Third Edition* : 2011

**OFFICES**

| | | | |
|---|---|---|---|
| ✆ **Bangalore** | 080-26 75 69 30 | ✆ **Jalandhar** | 0181-222 12 72 |
| ✆ **Chennai** | 044-24 34 47 26 | ✆ **Kolkata** | 033-22 27 43 84 |
| ✆ **Cochin** | 0484-237 70 04, 405 13 03 | ✆ **Lucknow** | 0522-220 99 16 |
| ✆ **Guwahati** | 0361-251 36 69, 251 38 81 | ✆ **Mumbai** | 022-24 91 54 15, 24 92 78 69 |
| ✆ **Hyderabad** | 040-24 65 23 33 | ✆ **Ranchi** | 0651-221 47 64 |
| | | ✆ **Patna** | 0612-230 00 97 |

UEE-9603-450-ELECTRICAL ENGG. (OT)-GUP

**C. 3602 /011/06**

*Typeset at* : Goswami Associates, Delhi.

*Printed at* : Ajit Printers

# CONTENTS

# PREFACE

Why another book on electrical engineering? The current market is flooded with books dedicated to this title. Flurring through the pages of the same, the students, especially those sparking in the field as amateurs and newcomers, find some degree of difficulty. This volume is dedicated to target these budding intellectuals. The current work is dedicated to target these needs of students, teachers and professionals who desire to acquire a firm foundation for the area known to generation through ages.

We hope that the present volume will benefit the entire electrical engineering fraternity. Constructive criticisms for improvement of the book in successive editions are welcome. The readers are always welcome to contact the authors for their doubts if any.

**—Authors**

*To*
*SAI BABA*
*my parents, wife,*
*&*
*Son 'Ojas'*

Chapter 1

# RESISTANCE

**Atom.** All matter has electricity in the form of electrons and protons in the atom. The electron is the basic quantity of negative electricity and proton is the basic quantity of positive electricity. Electrons in motion provide an electron current. An electric current is a closed path for electron flow. Potential difference must exist in order that current may flow through a circuit. In the external circuit outside the voltage source, electrons flow from negative terminal toward the positive terminal.

**Particles in the Atom**

| *Particle* | *Location* | *Charge* | *Mass* |
|---|---|---|---|
| Electron | Orbital shells | $0.16 \times 10^{-18}$ coulomb, negative | $9.108 \times 10^{-28}$ gm |
| Proton | Nucleus | $0.16 \times 10^{-18}$ coulomb, positive | $1.672 \times 10^{-24}$ gm |
| Neutron | Nucleus | None | $1.675 \times 10^{-24}$ gm |

The masses of a proton and a neutron are almost equal and the mass of each is approximately 1840 times the mass of electron.

The unit of charge is coulomb. One coulomb of charge is approximately the charge of $6.28 \times 10^{18}$ electrons. If a charge $q$ passes across a conductor in a time $t$ seconds, a current $i$ is said to flow, such that

$$i = \frac{q}{t}$$

The unit of current is coulomb per second and is called the ampere.

Thus, $$\boxed{\text{Ampere} = \frac{\text{Coulomb}}{\text{Second}}}$$

**Various Types of Electric Charges**

| *Type of charge* | *Amount of charge* | *Polarity* | *Application* |
|---|---|---|---|
| Electron | $Q_e - 0.16 \times 10^{-18}$ C | Negative | In wire conductors, vacuum tubes and *N*-type semi-conductors |
| Ion | $Q_e$ or multiples of $Q_e$ | Positive or Negative | In liquids and gases |
| Hole | $Q_e = 0.16 \times 10^{-18}$ C | Positive | In *P*-type semi-conductors |

**Volt** is the electromotive force necessary to cause one ampere of current to flow through a resistance of one ohm.

**Electron volt** is a unit in terms of which the energies of atomic particles are expressed. It is the work done when an electron, whose charge is *e* coulombs, is moved in an electric field through a potential difference of 1 volt against the force (newtons) acuting on the charge.

Thus $$1 \text{ electron volt} = e \text{ joules}$$
$$eV \text{ joules} = V \text{ electron volts.}$$

**Power.** Electrical power $p$ is the product of impressed voltage $v$ and resulting current $i$. Thus
$$p(\text{watts}) = v\ (\text{volts}) \times i(\text{ amperes})$$

If power $p$ is a periodic function of time $t$ with period $T$, then the average power,
$$P = \frac{1}{T}\int_0^T p\,dt.$$

**Energy, W.** As power $p$ is the time rate of energy transfer,
$$p = \frac{dW}{dt}$$
and
$$W = \int_{t_1}^{t_2} p\,dt$$
where W is the energy transferred during the time interval.

**Ohm's Law.** According to this law–"For a given conductor the ratio of the steady direct potential difference $v$ between its ends to the steady direct current $i$ flowing through the conductor, is a constant provided the physical conditions of the conductor *e.g.*, temperature, length, cross-section remain unchanged".

Thus $$V \propto I$$
or
$$V = IR$$
where $R$ is a constant for the conductor called the electrical resistance.

When P is power (watts), $V$ is voltage (volt), $I$ is current (amperes) and $R$ is resistance (ohms), then following relations are valid

| | | |
|---|---|---|
| $P = VI$ | $P = I^2R$ | $P = \frac{V^2}{R}$ |
| $I = \frac{P}{V}$ | $R = \frac{P}{I^2}$ | $R = \frac{V^2}{P}$ |
| $V = \frac{P}{I}$ | $I = \sqrt{\frac{P}{R}}$ | $V = \sqrt{PR}$ |

**Ohm's law does not apply under the following conditions:**

1. Non-linear resistors like vacuum radio valves, semi-conductors, gas-filled tubes, etc.
2. Electrolytes where enormous gases are produced on either electrode.
3. Appliances like metal rectifiers, crystal detectors, etc. in which operation depends on the direction of current.
4. Metals which get heated up due to flow of current through them.

**Resistance** is the property of a substance due to which it opposes the flow of electricity (*i.e.,* electrons) through it.

**Conductors of Electricity.** A conductor of electricity has very low resistance. All metals, acids and salt solutions are good conductors of electricity. Silver, copper and aluminium are the best conductors. Copper and aluminium are used for wire conductors.

Those substances which offer high resistance to flow electricity are known as poor conductors or insulators. Thus poly vinyl chloride (*PVC*), bakelite, mica, glass, rubber, dry wood, etc. are insulators of electricity.

**Laws of Resistance.** The resistance $R$ of a conductor.

(*a*) Varies directly as its length ($L$).

(*b*) Varies inversely as the cross-section ($A$) of the conductor.

(*c*) Depends on the nature of the material.

(*d*) Depends on the temperature of the conductor.

Thus for a conductor of a particular material at any temperature

$$R \propto \frac{L}{A}$$

or

$$R = \rho\frac{L}{A}$$

where ρ is a constant known as specific resistance of resistivity of the material of the conductor. Thus specific resistance of a material is the resistance between the opposite faces of a metre cube of that material.

In SI units the unit of resistivity is ohm-cm.

**Temperature Co-efficient of Resistance.** This factor with symbol ($\alpha$) states how much the resistance changes for a change in temperature. A positive value of $\alpha$ means $R$ increases as temperature increases and a negative value of $\alpha$ means $R$ decreases as temperature increases.

All metals in their pure form, such as copper and tungsten, have a positive temperature co-efficient. The $\alpha$ for tungsten, for example is 0.005. Although $\alpha$ is not exactly constant, an increase in wire resistance caused by a rise in temperature can be calculated approximately from the formula

$$R_t = R_0 + R_0\,(\alpha\,.\,\Delta t)$$

where $R_0$ is the resistance at 20°C, $R_t$ is the higher resistance at higher temperature and $\Delta t$ is the temperature rise above 20°C.

Carbon has a negative temperature co-efficient. In general, $\alpha$ is negative for all semi-conductors, including germanium and silicon. Also all electrolyte solutions, such as sulphuric acid and water, have a negative $\alpha$.

A negative value of $\alpha$ means less resistance at higher temperatures. The negative a has a practical application in the use of carbon thermistors. A thermistor can be connected as a series component to decrease its resistance to compensate for the increased hot resistance of wire conductors.

Metal alloys constant and Manganin have the value of zero for $\alpha$. They can be used for precision wire wound resistors, which do not change resistance when the temperature increases.

**Resistivities and Temperature Co-efficients**

| *Material* | *Resistivity in ohm-metre at 20°C* | *Temperature co-efficient at 20°C* |
|---|---|---|
| Aluminium | $2.8 \times 10^{-8}$ | $40 \times 10^{-4}$ |
| Carbon | $3000 - 7000 \times 10^{-8}$ | $-5 \times 10^{-4}$ |
| Constantan | $49 \times 10^{-8}$ | $0.1 \times 10^{-4}$ |
| Copper | $1.72 \times 10^{-8}$ | $39.3 \times 10^{-4}$ |
| Silver | $1.64 \times 10^{-8}$ | $38 \times 10^{-4}$ |
| Bakelite | $10^{10}$ | |
| Mica | $10^{15}$ | |
| Rubber | $10^{16}$ | |

**Series Circuits.** There is only one current $I$ in a series circuit $I = \dfrac{V_T}{R_T}$, where $V_T$ is the voltage applied across the total series resistance $R_T$. This $I$ is the same in all the series components. The total resistance of a series string is the sum of the individual resistances. The applied voltage $V_T$ equals the sum of the series $IR$ drops.

**Parallel Circuits.** There is only one voltage $V_A$ across all components in parallel. The total line current equals the sum of all the branch currents. For any number of *equal* parallel resistances, $R_T$ is the value of one resistance divided by the number of resistances. An open in one branch results in no current through that branch, but the other branches can have their normal current. A short circuit has zero resistance, resulting in excessive current. When one branch is short circuited, all the parallel paths are inactivated. The entire current flows through the short circuit and bypasses the remaining branches.

**Comparison of Series and Parallel Circuits**

| *Series circuit* | *Parallel circuit* |
|---|---|
| 1. Same current flows through all components | 1. Voltage is the same across all branches. |
| 2. $R_T = R_1 + R_2 + R_3 + ......$ | 2. $G_r = G_1 + G_2 + G_3 + ......$ (G = conductance) |
| 3. $R_T$ must be more than the largest individual $R$. | 3. $R_T$ must be less than the smallest branch $R$. |
| 4. $P_T = P_1 + P_2 + P_3 + ......$ | 4. $P_T = P_1 + P_2 + P_3 + ......$ |
| 5. $V_T = V_1 + V_2 + V_3 + ......$ | 5. $I_t = I_1 + I_2 + I_3 + ......$ |
| 6. Applied voltage is divided into 1 $R$ voltage drops. | 6. Main line current is divided into branch currents |
| 7. Largest $I$ $R$ drop is across the largest series $R$. | 7. The largest branch $I$ is in the smallest parallel $R$. |
| 8. Open in one component causes entire circuit to be open | 8. Open in one branch does not prevent $I$ in other branches. |

**Hot Resistance.** With resistance wire made of Tungsten, Nichrome, Iron or Nickel, there is substantial difference, in the amount of resistance the wire has when hot in normal operation and when cold without its normal load. Thus a cold tungston wire filament of a bulb having resistance of 100 ohms may have hot resistance of 600 to 800 ohms when bulb is switched on.

**Super Conductivity.** The opposite effect of hot resistance is to cool a metal down to very low temperature to reduce its resistance. Near absolute zero at 0°K or – 273°C, some metals abruptly lose practically all their resistance. As an example, the metal tin when cooled by liquid helium becomes super conductive at 3.7°K.

**Resistors.** The two main characteristics of a resistor are its $R$ in ohm and the wattage rating. Resistors are available in a very wide range of $R$ values, from a fraction of an ohm to many mega ohms. The power rating may be as high as several hundred watts or as low 1/10 watt. Power rating of a resistor indicates the maximum wattage the resistor can dissipate without excessive heat. Too much heat can make the resistor burn open.

Commonly used resistors are wire wound resistors and carbon resistors. Both carbon and wire wound resistors can be either fixed or variable. Variale type carbon resistors are used for controls, such as the volume control in a radio or the contrast control in a television. An application for a variable wire wound resistor is to divide the voltage from a power supply.

**Carbon Composition Resistors.** These are made of finely divided carbon or graphite mixed with a powdered insulating material in the proportions needed for the desired values. The resistor element is enclosed in a plastic case for insulation and mechanical strength. Joined to the two ends of the carbon resistance element are metal caps and leads of tinned copper wire for soldering the resistor connections into a circuit. Carbon resistors are available in resistance values 1 ohm to 20 Mega ohms. Their power rating is generally $\frac{1}{10}, \frac{1}{8}, \frac{1}{4}, \frac{1}{2}$, 1 or 2 watts.

**Carbon Film Resistors.** These have a carbon film coating around an insulator like alumina. These are cheaper than hot-moulded carbon-composition resistors.

**Metal Film Resistors.** These have a conductive coating on a glass substrate. These resistances can be used for more accurate values.

**Cermet Film Resistors.** These have a carbon coating fixed into a solid ceremic substrate. These resistors have better stability with heat.

**Resistor Colour Coding.** Carbon resistors are physically smaller in size as such it is difficult to write resistance value. Colour coding of these resistors is used to identify the value and tolerance. Normally these resistors are provided with four bands. Colour bands are printed at one end of the insulating body, which is usually tan. The first band (*A*) indicates the first digit in the numerical value of resistance. The second band (*B*) gives the second digit. The third band is the decimal multiplier, giving the number of zeroes after the two digits. The resulting number is the resistance in ohms.

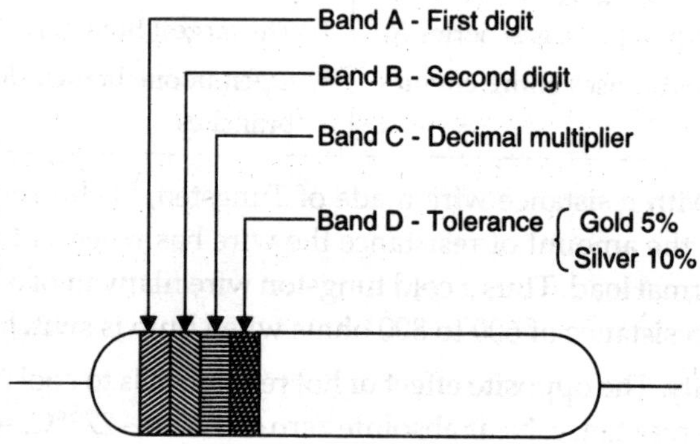


***FIGURE 1.1***

**Colour Code**

| *Colour* | *Value* | *Colour* | *Value* |
|---|---|---|---|
| Black | 0 | Green | 5 |
| Brown | 1 | Blue | 6 |
| Red | 2 | Violet | 7 |
| Orange | 3 | Gray | 8 |
| Yellow | 4 | White | 9 |

**Example:** The colours of the four bands are red, green, orange and gold. The values for these colours are respectively 2, 5, 3 and 5%. The value of the resistance will be 25000 ohms with 5% tolerance.

**Resistors under 10 ohms.** Such resistors have a third stripe of gold or silver which are fractional decimal multipliers. When the third stripe is gold, multiply the first two digits by 0.1. Silver is a multiplier of 0.01. Thus if the stripes are yellow, blue and gold, the resistance will be $46 \times 0.1 = 4.6$ ohms. Gold and silver are fractional multipliers only in the third stripe.

**Tolerance.** Gold and silver stripes in fourth band represent tolerance of ± 5% and ± 10%. When there is no fourth stripe, the tolerance is ± 20 percent. The smaller the tolerance, the more accurate the resistance value and the higher the cost. Some resistors may have a fifth colour band to designate a military failure rate indicating reliability.

**Schematic Symbols for Various Resistors**

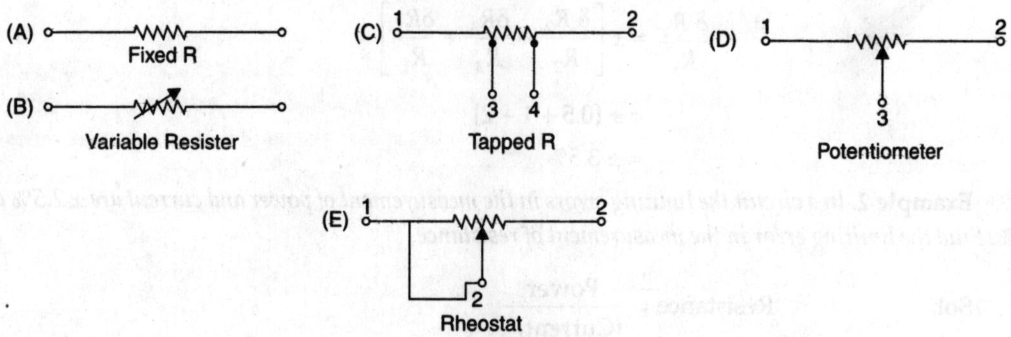


*FIGURE 1.2*

**Limiting Errors.** The accuracy and precision of a resistor depends upon its material used and the workmanship that goes into making the resistor. The choice of resistor for a particular application depends upon the accuracy desired. If only a fair degree of accuracy is desired, it is not economical to use expensive resistors. But a resistance used for an application requiring a high degree of accuracy has to use expensive material and high skill in manufacture is required for that. The manufacturers specify the deviations from the nominal value of a particular resistance. The limits of these deviations from the specified value are defined as **limiting errors** or **guarantee error.** Thus a resistor may have nominal value of 100 ohms with a limiting error of ± 10 ohms. The magnitude of the resistor will be between 90 ohms and 110 ohms.

**Sum of Limiting Errors.** Let us consider two resistors with limiting errors as given below

$$R_1 \pm x\% \quad \text{and} \quad R_2 \pm y\%$$

Now if these resistors are connected in series, the limiting error of the combination will be

$$R_1 + R_2 \pm (x + y)\%.$$

**Example 1.** *The solution for the unknown resistance for a Wheatsone bridge is*

$$R_x = \frac{R_2 R_3}{R_1}$$

*If*

$$R_1 = (100 \pm 0.5\%)\ ohms$$

$$R_2 = (1000 \pm 1\%)\ ohms$$

$$R_3 = (500 \pm 2\%)\ ohms.$$

*Determine the magnitude of the unknown resistance and the limiting error in percent.*

**Sol.** The unknown resistance

$$R_x = \frac{1000 \times 500}{100} = 5000 \text{ ohms}$$

Relative limiting, error of unknown resistance is

$$\frac{\delta R_x}{R_x} = \pm\left[\frac{\delta R_2}{R_2} + \frac{\delta R_3}{R_3} + \frac{\delta R_1}{R_1}\right]$$

$$= \pm [0.5 + 1 + 2]$$

$$= \pm 3.5\%$$

**Example 2.** *In a circuit the limiting errors in the measurement of power and current are ± 2.5% and ± 2.0%. Find the limiting error in the measurement of resistance.*

**Sol.**

$$\text{Resistance} = \frac{\text{Power}}{(\text{Current})^2}$$

$$\frac{\delta R}{R} = \pm\left(\frac{\delta P}{P} + 2\frac{\delta I}{I}\right)$$

$$= \pm (2.5 + 2 \times 2.0)$$

$$= \pm 6.5\%.$$

## OBJECTIVE TYPE QUESTIONS

1. The unit for electrical conductivity is
   (*a*) ohms/metre (*b*) mho/metre (*c*) mho-metre (*d*) mho.
2. The unit of temperature co-efficient of resistance is
   (*a*) mho (*b*) ohm per degree kelvin
   (*d*) mho per degree kelvin (*d*) per degree kelvin.
3. Candela is the unit of
   (*a*) brightness (*b*) colour
   (*c*) luminous intensity (*d*) magnetic flux.
4. Metre has been accurately defined in terms of
   (*a*) diameter of earth (*b*) diameter of atom
   (*c*) wavelength of krypton (*d*) wavelength of platinum.
5. Second has been defined in terms of
   (*a*) time taken by light to travel from sun to earth
   (*b*) revolution of earth on its axis
   (*c*) transition between two hyperfine levels of the ground state of Ceasium—133 atom.
   (*d*) wavelength of krypton.

6. **Which one of the following is a derived SI unit?**
(*a*) Candela (*b*) Kelvin (*c*) Hertz (*d*) Ampere.

7. **Illumination of one lumen per square metre is a**
(*a*) candela (*b*) lux (*c*) tesla (*d*) none of the above.

8. **Resistance: Ohm:**
(*a*) Inductance: Farad (*b*) Capacitance: Henry
(*c*) Conductance: mho (*d*) Lumen: Steradian.

9. **The unit for electric field intensity is**
(*a*) coulomb (*b*) weber/metre (*c*) newton (*d*) newton/coulomb.

10. **1 newton is the same as**
(*a*) $10^5$ ergs (*b*) $10^5$ joules (*c*) $10^5$ dynes (*d*) $10^5$ W.

11. **All of the following are equivalent to watt EXCEPT**
(*a*) amperes × volts (*b*) joule/sec (*c*) amperes/volt (*d*) $(\text{amperes})^2$ ohm.

12. **The value of following is given by**

$$\frac{1000 \text{ kilo amperes} \times 0.1 \text{ micro ampere}}{10 \text{ amperes} \times 10 \text{ milli amperes}}$$

(*a*) 1 (*b*) 10 (*c*) 0.1 (*d*) 0.01

13. **1 coulomb is the same as**
(*a*) $10^9$ esu (*b*) $3 \times 10^9$ esu (*c*) $3 \times 10^6$ esu (*d*) $10^{-6}$ esu.

14. **Magnetic flux is measured in**
(*a*) amperes (*b*) ampere turns
(*c*) ampere turns/cm (*d*) webers.

15. **1 Tesla is the same as**
(*a*) 1 henry/$(\text{metre})^2$ (*b*) 1 farad/volt (*c*) 1 weber/$\text{metre}^2$ (*d*) 1 dyne/$\text{cm}^2$.

16. **A 10 kΩ resistor has to carry a current of 3 milli amperes. Which of the following resistor will be suitable for this?**
(*a*) 10 kΩ, 1/8 W (*b*) 10 kΩ, 1/4 W (*c*) 10 kΩ, 1/2 W (*d*) 10 kΩ, 1 W.

17. **Permeability is expressed in terms of**
(*a*) maxwell (*b*) weber (*c*) henry/metre (*d*) farad/metre.

18. **Which of the following is heaviest?**
(*a*) Electron (*b*) Neutron (*c*) Proton (*d*) All have same mass.

19. **1 atomic mass unit is equal to**
(*a*) $1.66 \times 10^{-24}$ gm (*b*) $6.25 \times 10^{-27}$ gm (*c*) $1.6 \times 10^{-19}$ gm (*d*) $3.33 \times 10^{-35}$ gm.

20. **For which of the following 'ampere second' could be the unit?**
(*a*) Power (*b*) Charge (*c*) Energy (*d*) Reluctance

21. Permittivity is expressed in terms of
(a) Farad metre (b) Farad/metre$^2$ (c) farad (d) farad/metre.

22. One litre of air at NTP weigh nearly
(a) 1.3 milligram (b) 1.3 gram (c) 1.3 decigram (d) 1.3 kilogram

23. 1 angstrom is equal to
(a) $10^{-8}$ millimetre (b) $10^{-8}$ centimetre (c) $10^{-8}$ metre (d) $10^{-8}$ kilometre.

24. International ohm has been defined in terms of the resistance of
(a) a cube of platinum (b) a cube of gold
(c) a cylinder of copper (d) a column of mercury.

25. One newton metre is the same as
(a) one joule (b) ten joules (c) one joule second (d) one watt.

26. One volt is the same as one
(a) watt per ohm (b) watt hour
(c) coulomb per joule (d) joule per coulomb.

27. The unit of electrical energy is
(a) volt-ampere (b) watt (c) joule second (d) kilowatt hour.

28. Joule per second is the same as
(a) volt (b) watt (c) mho (d) none of the above.

29. Ionization current in liquids and gases results from a flow of
(a) protons (b) free electrons
(c) positive or negative ions (d) ions that are lighter in weight than electrons.

30. The charge of an electron is $1.60 \times 10^{-19}$ coulomb. The number of electrons that will have combined charge of 1 coulomb will be
(a) $6.25 \times 10^{16}$ (b) $6.25 \times 10^{18}$ (c) $6.25 \times 10^{20}$ (d) $6.25 \times 10^{22}$.

31. The time variation of the charges $q$ passing through a certain conductor is given by the equation

$$q = 50t^2$$

where $t$ is in seconds.

The equation for instantaneous current $i$ will be
(a) $i = 50t$ (b) $i = 100t$ (c) $i = 200t + C$ (d) $i = 200t$.

32. Reluctance: Permeance:
(a) Voltage: Current (b) Current: Resistance
(c) Resistance: Conductance (d) Ohm: Farad.

33. Out of the following which one is the lightest?
(a) Neutron (b) Proton
(c) Electron (d) All have identical weight.

34. When an electron is removed from an atom it becomes
(a) a neutron (b) a proton (c) a positive ion (d) a negative ion.

35. The ratio of mass of proton to that of electron is nearly

(a) 2 (b) 25 (c) 184 (d) 1840.

36. One micro volt is

(a) 0.001 V (b) 0.0001 V (c) 0.00001 (d) 0.00001 V.

37. An electron is a vacuum tube is observed to loose $1.6 \times 10^{-21}$ joules of energy while moving from point $P$ to point $Q$. The voltage of point $Q$ with respect to point $P$ is

(a) – 0.01 V (b) – 0.1 V (c) + 0.1 V (d) + 1 V.

38. The minimum charge on an ion is

(a) zero (b) equal to charge of an electron

(c) equal to the charge of the number of electrons in an atom

(d) equal to the atomic number of the atom.

39. The number of electrons in the outermost orbit of carbon atom is

(a) 1 (b) 2 (c) 4 (d) 8.

40. Which is the best conductor of electricity?

(a) Carbon (b) Silver (c) Copper (d) Iron.

41. Ten million electrons each of which carries a charge of $-1.6 \times 10^{-19}$ coulomb, are removed from an initially neutral pith ball. The resulting charge on the pith ball will be

(a) $-1.6 \times 10^{-26}$ coulomb (b) $-1.6 \times 10^{-22}$ coulomb

(c) $-1.6 \times 10^{-14}$ coulomb (d) $-1.6 \times 10^{-12}$ coulomb.

42. Which one of the following is semi-conductor?

(a) Copper (b) Silicon (c) Neon (d) Lithium.

43. An element with 14 protons and 14 electrons has the atomic number of

(a) 0 (b) 7 (c) 14 (d) 28.

44. What is the electron in valance shell of an element with atomic number 3?

(a) 0 (b) 1 (c) 2 (d) 3.

45. A battery can supply 11 joules of energy to move 5 coulombs of charge. How much is the voltage of the battery?

(a) 55 V (b) 11 V (c) 5 V (d) 2.2 V.

46. Which of the following statements is false?

(a) An element with atomic number 12 and 12 orbital electrons.

(b) An ion is a charged atom.

(c) A resistance of 600 ohms has a conductance of 6 mho

(d) The force of attraction or repulsion between charges is due to their electric field.

47. The diameter of the nucleus of an atom is of the order of

(a) $10^{-25}$ m (b) $10^{21}$ m (c) $10^{-27}$ m (d) $10^{-14}$ m.

48. A neutral dielectric has added to it $12.5 \times 10^{18}$ electrons. What is the charge in coulombs?

(a) $12.5 \times 10^{18}$ C (b) $12.5 \times 10^{9}$ C (c) $6.25 \times 10^{6}$ C (d) 2 C.

**49.** $12.5 \times 10^{18}$ electrons are added in a dielectric having charge + 2 coulombs. What is the final charge?

(*a*) 4 C (*b*) 2 C (*c*) 1 C (*d*) Zero.

**50.** Which of the following can have positive or negative polarity?

(*a*) Electron (*b*) Ion (*c*) Hole (*d*) All of the above.

**51.** Which of the following exists in *p*-type semi-conductors only?

(*a*) Electron (*b*) Ion (*c*) Hole (*d*) All of the above.

**52.** When two resistors are connected in series

(*a*) they must both have the same resistance value

(*b*) the voltage across each must be the same

(*c*) they must have different resistance values

(*d*) there is only one path for current through both resistances.

**53.** With a 2 ohm resistance and a 4 ohm resistance in series across a 6 *V* battery, the current

(*a*) in the larger resistance is $1\frac{1}{2}$ A (*b*) in the smaller resistance is 3 A

(*c*) in both resistances is 1 A (*d*) in both resistances is 2 A.

**54.** When two resistances are connected in parallel

(*a*) the current through each must be the same

(*b*) the voltage across each must be the same

(*c*) their combined resistance equals the sum of the individual values

(*d*) each must have the same resistance value.

**55.** The resistance of a heating coil required in a 1500 watt kettle used on 200 V mains will be

(*a*) 13.33 Ω (*b*) 26.67 W (*c*) 88.8 Ω (*d*) 168.45 W.

**56.** In Fig. 1.3 if curve *D* represents resistivity for iron, then curve C is likely to represent resistivity for

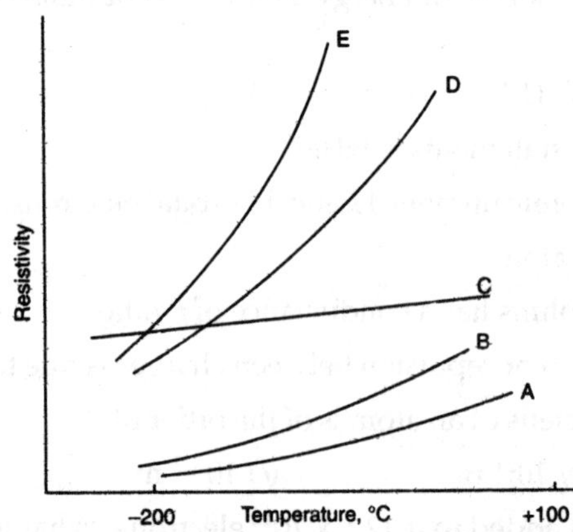


***FIGURE 1.3***

(a) manganin (b) copper (c) lead (d) platinum.

57. Which curve is likely to represent resistivity for copper

(a) curve A or B (b) curve D (c) curve E (d) curve D or E.

58. With eight 10 MΩ resistors connected in parallel across a 10 V source, the main line current will be

(a) 0.8 μA (b) $\frac{1}{8}$ μA (c) 8 μA (d) 80 μA.

59. Ohm's law is not applicable to

(a) carbon resistors (b) circuits with low current densities

(c) high voltage circuits (d) vacuum tubes.

60. Which of the following parallel grouping of resistors will have the highest value of equivalent resistance?

(a) Two 500 Ω resistances (b) Five 10 kΩ resistances

(c) 100 Ω, 200 Ω and 300 Ω resistances (d) Two 5 kΩ and two 2 kΩ resistances.

61. Three conductances in parallel have the values

$$G_1 = 9000\ \mu\ \text{mho}$$
$$G_2 = 7000\ \mu\ \text{mho}$$
$$G_3 = 22000\ \mu\ \text{mho}$$

Total conductance will be

(a) 0.38 μ mho (b) 0.038 mho (c) 0.038 μ mho (d) 0.38 mho.

62. With 10 V applied across ten 50 ohm resistances in parallel, the current through each resistance will be

(a) 20 A (b) 10 A (c) 2 A (d) 0.2 A.

63. With resistances of 500 Ω, 400 Ω, 300 Ω, 200 Ω and 100 Ω in parallel, the equivalent resistance will be

(a) less than 100 Ω (b) more than 1 MΩ (c) about 500 Ω (d) about 1 kΩ.

64. With four resistances connected in parallel, if each dissipates 10 W the total power supplied by the voltage source equals

(a) 400 W (b) 100 W (c) 40 W (d) 10 W.

65. The good conductor of electricity is

(a) Porcelain (b) Mica (c) Rubber (d) None of the above.

66. The resistance of a conductor varies

(a) inversely as its length (b) directly as its cross-section area

(c) directly as resistivity of the material (d) due to all above factors.

67. Irrespective of whether the circuit is ac or dc, the following is most effective in reducing the magnitude of the current

(a) Resistor (b) Capacitor (c) Inductor.

68. Spot the odd one out

(a) Resistance (b) Capacitive reactances

(c) Inductive reactance (d) Conductance.

69. An electron of mass $m$ kg and having a charge $e$ coulombs travels from rest through a potential difference of 100 volts. Its kinetic energy will be

(a) 100 $e$ joules (b) $me$ joules (c) $\frac{me}{100}$ joules (d) $\frac{100}{e}$ joules.

**Questions 70 to 73 refer to Fig. 1.4.**

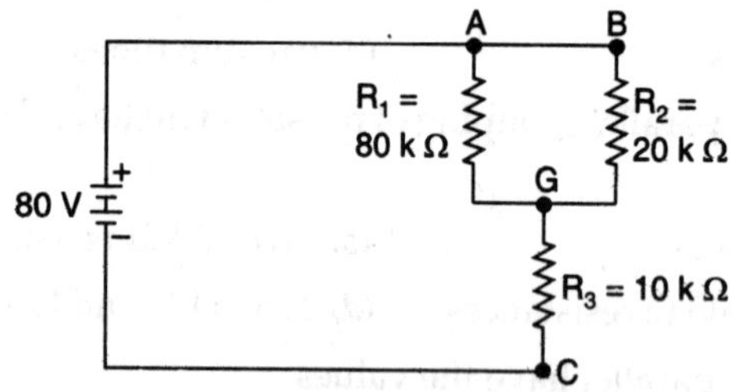


*FIGURE 1.4*

70. The voltage across $R_2$ will be

(a) 20 V (b) 33.3 V (c) 49.23 V (d) 60 V.

71. The current through $R_3$ will be

(a) 4 A (b) 2 A (c) 3 mA (d) 0.615 mA.

72. If the point G is connected to ground the voltage $V$ AG will be

(a) + 20 V (b) + 49.2 V (c) – 20 V (d) – 40 V.

73. The current through $R_2$ with G connected to ground the will be

(a) 2.46 mA (b) 4 mA (c) 2 A (d) 4 A.

74. The filament of an electric bulb is made of

(a) Steel (b) Copper (c) Tungsten (d) Carbon.

75. In a series circuit with unequal resistances

(a) the lowest resistance has the highest voltage drop

(b) the highest resistance has the highest voltage drop

(c) the lowest resistance has the highest current

(d) the highest resistance has the most of current through it.

76. What is the equivalent resistance for the circuit shown in Fig. 1.5

(a) 10.45 Ω (b) 18.75 Ω (c) 104.5 Ω (d) 245 Ω

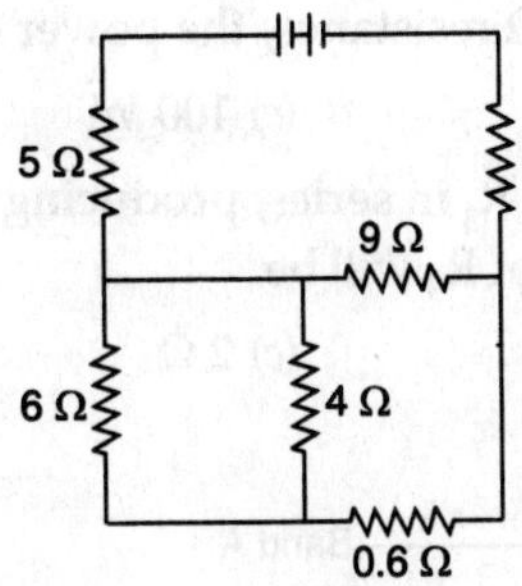


*FIGURE 1.5*

**77.** For the circuit shown in Fig. 1.6 the limiting value of current is

(*a*) 1 A (*b*) 0.1 A (*c*) 0.01 A (*d*) 1 mA.

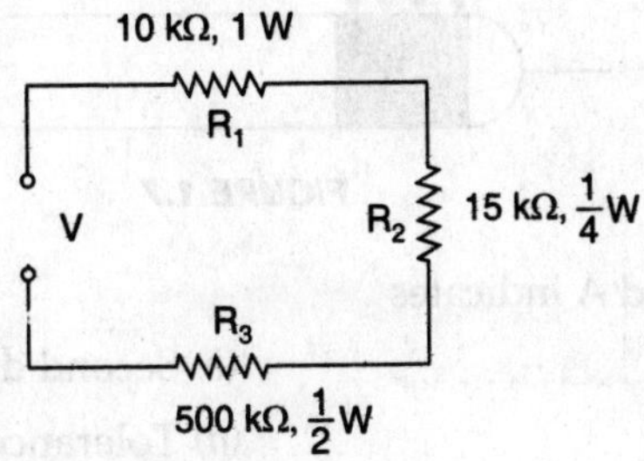


*FIGURE 1.6*

**78.** In a parallel bank with unequal branch resistances

(*a*) the current is highest in the lowest *R* (*b*) the voltage is highest across the lowest *R*

(*c*) the current is highest in the highest *R* (*d*) the current is equal in all branches.

**79.** In which of the following circuits will the voltage source produce the most current?

(*a*) 10 V across 10 Ω resistance (*b*) 10 V across two 10 Ω resistances in series

(*c*) 10 V across two 10 Ω resistances in parallel

(*d*) 1000 V across a 1 MΩ resistance.

**80.** An electric heater draws 1000 watts from a 250 volt source. What power will be drawn by this heater from a 200 V source?

(*a*) 1250 W (*b*) 860 W (*c*) 640 W (*d*) 500 W.

**81.** A circuit allows 1 mA of current to flow with 1 V applied. The conductance of the circuit equals

(*a*) 0.002 Ω (*b*) 0.005 μ mho (*c*) 1 mho (*d*) 1000 μ mho.

**82.** A string of two 1000 Ω resistances is in series with a parallel bank of two 1000 Ω resistances. The total resistance of the series-parallel circuit equals

(*a*) 250 Ω (*b*) 2500 Ω (*c*) 3000 Ω (*d*) 4000 Ω.

**83.** An electric boiler drawn 10 amperes at 200 volts for a period of 5 hours. The quantity of electricity in coulombs which passes through the boiler will be

(*a*) 180,000 coulombs (*b*) 360,000 coulombs (*c*) 10,000 coulombs (*d*) 4000 coulombs.

**84.** With 100 V applied across a 10 kΩ resistance, the power dissipation will be

(*a*) 1 mW (*b*) 1 W (*c*) 100 W (*d*) 1 kW.

**85.** 100 V is applied across $R_1$, $R_2$ and $R_3$ in series, producing 10 A in the series circuit. $R_1$ equals 6 Ω and $R_2$ equals 2 Ω. The value of $R_3$ will be

(*a*) 10 Ω (*b*) 4 (*c*) 2 Ω (*d*) 1 Ω.

**Questions 86 to 89 refer to Fig. 1.7.**

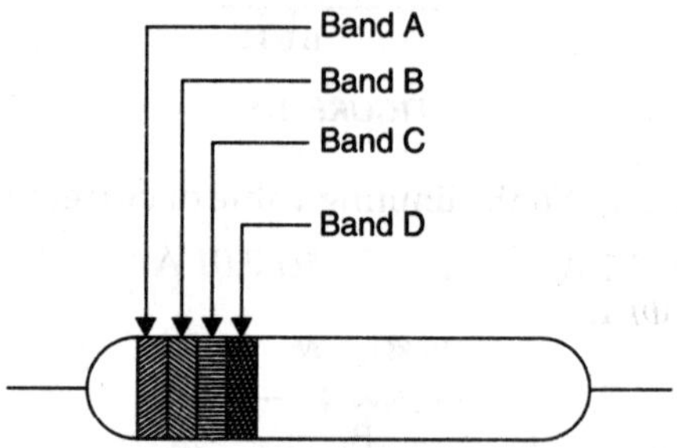


***FIGURE 1.7***

**86.** In resistor colour, band A indicates

(*a*) First digit (*b*) Second digit

(*c*) Decimal multiplier (*d*) Tolerance.

**87.** In resistor colour coding, decimal multiplier is represented by

(*a*) band *A* (*b*) band *B* (*c*) band *C* (*d*) band *D*.

**88.** In resistor colour coding, tolerance is indicated by

(*a*) band *A* (*b*) band *B* (*c*) band *C* (*d*) band *D*.

**89.** In resistance colour coding darker colours are assigned values

(*a*) closer to 1 (*b*) closer to 9

(*c*) in the middle around 5 (*d*) there is no hard and fast rule.

**90.** In resistance colour coding red colour is assigned the value

(*a*) 0 (*b*) 1 (*c*) 2 (*d*) 3.

**91.** In resistance colour coding yellow colour is assigned the value of

(*a*) 2 (*b*) 4 (*c*) 6 (*d*) 8.

**92.** In resistance colour coding the value of 5 is represented by

(*a*) green colour (*b*) brown colour (*c*) orange colour (*d*) violet colour.

**93.** Which of the following colour is assigned value more than 5?

(*a*) Green (*b*) Orange (*c*) Black (*d*) Gray.

**94.** In resistance colour coding white colour represents the value of

(*a*) 9 (*b*) 7 (*c*) 5 (*d*) 3.

**95.** **In tolerance band, silver lining represents the tolerance of**

(a) 1% (b) 5% (c) 10% (d) 20%.

**96.** **A tolerance of 20% in the value of carbon resistor is represented**

(a) by gold band (b) by silver band (c) by black (d) without any band.

**97.** **If a carbon resistor has stripes as yellow, violet and gold. The value of resistor will be**

(a) 4700 Ω ± 5% (b) 470 Ω ± 5% (c) 47 Ω (d) 4.7 Ω.

**98.** **A resistance of 0.47 ohm will be represented by**

(a) yellow, green and white stripes (b) yellow, violet, black and white stripes

(c) yellow, violet and silver stripes (d) more of the above.

**99.** **A carbon resistor having brown, red, orange and silver stripes can have the maximum value of**

(a) 1080 Ω (b) 1200 Ω (c) 12000 Ω (d) 13,200 Ω.

**100.** **What is the multiplier for yellow in the third stripe?**

(a) 10 (b) 100 (c) 1000 (d) 10,000.

**101.** **For a carbon composition resistor colour-coded with yellow, violet, orange and silver stripes from left to right, the resistance and tolerance are**

(a) 740 Ω ± 5% (b) 4700 Ω ± 10% (c) 7400 Ω ± 1% (d) 47 kΩ + 10%.

**102.** **A resistance having rating 10 ohm, 10 W is likely to be a**

(a) wire wound resistor (b) carbon resistor

(c) variable resistor (d) none of the above.

**103.** **A resistance having rating as 1 MΩ, 1 W is likely to be a**

(a) wire wound resistor (b) carbon resistor

(c) any of the above (d) none of the above.

**104.** **Which of the following is typical resistance and power dissipation value for a wire wound resistor?**

(a) 10 Ω 45 W (b) 10,000 Ω, 1 W (c) 1 MΩ, $\frac{1}{4}$ W (d) 500 Ω, $\frac{1}{2}$ W.

**105.** **A resistor is to be connected across a 45 V battery to provide 1 mA of current. The required resistance with a suitable wattage rating will be**

(a) 45000 Ω, $\frac{1}{4}$ W (b) 450 Ω, 2 W (c) 4.5 Ω, 1 W (d) 45 Ω, 10 W.

**106.** **For a carbon-composition resistor colour-coded with green, black, gold and silver stripes from left to right, the resistance and tolerance are**

(a) 500 Ω ± 10% (b) 50 W ± 10% (c) 5 Ω ± 10% (d) 0.5 W ± 10%.

**107.** **Two 2 kΩ, 1 W resistor are connected in parallel. Their combined resistance value and wattage rating is**

(a) 4 kΩ, 2 W (b) 1 kΩ, 2 W (c) 1 kΩ, 4W (d) 1 kΩ, 1 W.

**108.** Which of the following is a preferred resistor value?

(*a*) 47 (*b*) 520 (*c*) 43,000 (*d*) 54,321.

**109.** The stripes on resistors are given below. Which one represents 2.2 MΩ ± 10% resistor

(*a*) Red, red, gold, gold (*b*) Red, red, green, silver

(*c*) Red, green, red, gold (*d*) Red, blue, black, silver.

**110.** What should be the resistance and power rating of a carbon composition resistor to fit the following requirements:

5 V, IR drop with 100 mA current and a safety factor of 2 for power dissipation

(*a*) 50 Ω, 1 W (*b*) 50 kΩ, 1 W (*c*) 5 Ω, $\frac{1}{2}$ W (*d*) 5 Ω, 1 W.

**111.** At room temperature the coil of copper in a certain relay has a resistance of 160 ohms. After normal voltage has been applied to the coil for several hours, the resistance was 180 Ω. If the resistance of copper changes 0.4 represent for each degree centigrade rise in temperature, the temperature rise of the coil was about

(*a*) 10°C (*b*) 31.25°C (*c*) 120°C (*d*) 300°C.

**112.** A thermister has

(*a*) zero temperature co-efficient (*b*) variable temperature co-efficient

(*c*) positive temperature co-efficient (*d*) negative temperature co-efficient.

**113.** A resistance wire of 5 ohms is further drawn so that the diameter reduces to one fifth of its original diameter. The resistance of the drawn wire will be

(*a*) 1 Ω (*b*) 5 Ω (*c*) 25 Ω (*d*) 125 Ω.

**114.** An essential condition for Ohm's law to be applicable to a circuit is that

(*a*) resistance must be at room temperature

(*b*) voltage should be only dc

(*c*) the current should flow in forward direction only

(*d*) the resistance must be uniform.

**115.** In case the resistance of a GLS lamp changes as the voltage across the bulb changes, then it can be concluded that

(*a*) the change is caused by variation in internal inductance of the bulb

(*b*) the bulb contains some gas at significant pressure

(*c*) ionisation of the gas within the bulb causes this change

(*d*) the filament resistance is non-linear.

**116.** If $I$ be the current, $R$ the resistance and $t$ the time, then according to Joule's law, the heat produced will be proportional to

(*a*) $IRt$ (*b*) $I^2 Rt$ (*c*) $I^2R^2t$ (*d*) $I^2R^2t^2$.

**117.** Which of the following is not a standard rating of GLS lamps?

(*a*) 25 W (*b*) 40 W (*c*) 75 W (*d*) 200 W.

**118.** Electric heater wires are generally made of

(a) carbon (b) steel (c) copper (d) nichrome.

**119.** A standard 25 W bulb is installed in series with a 750 W heater. Now if the bulb is replaced by a 60 W bulb

(a) the heater output will reduce (b) the heater output will remain unaltered

(c) the heater output will increase (d) the bulb will not work.

**120.** A nichrome wire used as heater coil has the resistance of 0.5 kΩ/m for a heater of 200 W at 2000 V, the length of the wire required will be

(a) 20 m (b) 40 m (c) 60 m (d) 80 m.

**121.** Which one of the following does not have negative temperature co-efficient?

(a) Rubber (b) Paper (c) Mercury (d) Gutta Parcha.

**122.** Varistors are

(a) carbon resistors (b) insulator

(c) non-linear resistors (d) resistors with zero temperature co-efficient.

**123.** 100 centimetre length of a wire of cross-section $A$ has resistance of $R$ ohms. The resistance of the wire of the same material having length 200 centimetres and cross-sectional area 2 $A$ will be

(a) $4R$ (b) $2R$ (c) $R$ (d) $\frac{R}{4}$.

**124.** A 1500 watt immersion rod takes 10 minutes to heat a bucket of water. Which of the following modification will help in heating the bucket of water in less than ten minutes?

(a) Reducing the supply voltage

(b) Increasing the length of heating element

(c) Reducing the length of heating element

(d) Using element of larger diameter resistance wire.

**125.** A 100 W, 200 V bulb is connected to 100 V supply. Its wattage will be

(a) 400 W (b) 100 W (c) 50 W (d) 25 W.

**126.** The hot resistance of a bulb is 5 times the cold resistance. Accordingly the hot resistance of a 100 W, 200 V bulb will be

(a) 100 Ω (b) 400 Ω (c) 800 Ω (d) 2000 Ω.

**127.** Which resistor will be physically larger in size?

(a) 1 MΩ, $\frac{1}{2}$ W (b) 220 kΩ, 1 W (c) 1 kΩ, $\frac{1}{4}$ W (d) 10 ohm, 50 W.

**128.** The cold resistance of a 100 W, 250 V bulb will be

(a) 100 Ω (b) 125 Ω (c) 625 Ω (d) 1250 Ω.

**129.** Nichrome wire is an alloy of

(*a*) copper and silver (*b*) chromium and vanadium

(*c*) nickel and chromium (*d*) lead and zinc.

**130.** A circuit allows one micro ampere current to flow through it when the voltage applied is one volt. The conductance of the circuit is

(*a*) 1 milli-mho (*b*) 1 μ-mho (*c*) $10^6$ mho (*d*) none of the above.

**131.** The value of supply voltage for 1 kW, 10 Ω load will be

(*a*) 220 V (*b*) 200 V (*c*) 100 V (*d*) 50 V.

**132.** Which of the following material has least value of resistivity?

(*a*) Nichrome (*b*) Copper (*c*) Lead (*d*) Stainless steel.

**133.** 100 resistors of 1 kΩ each are connected in parallel. Their equivalent resistance will be

(*a*) 0.01 Ω (*b*) 0.1 Ω (*c*) 1 Ω (*d*) 10 Ω.

**134.** Three resistors of 15 ohm each are connected to form the sides of a triangle. The equivalent resistance across any two points of the triangle will be

(*a*) 1 Ω (*b*) 10 Ω (*c*) 15 Ω (*d*) 20 Ω.

**135.** Which of the following can have negative temperature co-efficient?

(*a*) Electrolytes (*b*) Liquid metals (*c*) Metallic alloys (*d*) Compounds of silver.

**136.** Four elements having conductances $G_1$, $G_2$, $G_3$ and $G_4$ are connected in parallel. Their combined conductance $G$ is given by

(*a*) $G = G_1 + G_2 + G_3 + G_4$ (*b*) $\frac{1}{G} = \frac{1}{G_1} + \frac{1}{G_2} + \frac{1}{G_3} + \frac{1}{G_4}$

(*c*) $G = \frac{\Sigma G_1 G_2 G_3}{G_1 G_2 G_3 G_4}$ (*d*) $G = \frac{\Sigma G_1 G_2 G_3}{G_1 + G_2 + G_3 + G_4}$.

**137.** The resistance of 100 metre length of copper wire of 24 gauge is '*a*' ohms. The resistance of 100 metres of copper wire of 18 gauge will be

(*a*) more than '*a*' ohms (*b*) '*a*' ohms

(*c*) less than '*a*' ohms.

**138.** A closed switch has a resistance of

(*a*) zero (*b*) infinity (*c*) about 10 Ω (*d*) at least 1 kΩ.

**139.** Which arrangements will give a circuit resistance of 4 ohms?

(*a*) Three resistance of values 2 Ω, 3 Ω and 6 Ω in parallel

(*b*) 2 Ω resistance in series with parallel combination of 3 ohm and 6 ohm resistance

(*c*) 3 ohm resistance in series with parallel combination of 2 ohm and 3 ohm resistance.

(*d*) 6 ohm resistance in series with parallel combination of 2 ohm and 3 ohm resistance.

**140.** Resistors of following values are connected in parallel

5 Ω, 50 Ω, 500 Ω, 5 kΩ, 50 kΩ, 500 kΩ, 500 kΩ, 5 MΩ

Their equivalent resistance will be nearly

(*a*) 4.5 Ω (*b*) 45 Ω (*c*) 45000 Ω (*d*) 4500,000 Ω.

**141.** Insulating materials have the function of

(*a*) storing very high currents (*b*) conducting very large currents

(*c*) preventing a short circuit between conducting wires

(*d*) preventing an open circuit between the voltage source and the load.

**142.** In which of the following case the *IR* drop will be least?

(*a*) 6 μA current through 1 MΩ (*b*) 5 mA current through 1 kΩ

(*c*) 6 A current through 1 Ω (*d*) 60 A current through 0.01 Ω.

**143.** Two resistances $R_1$ and $R_2$ have combined value in series as 4 ohms and in parallel $\frac{15}{6}$ ohm. Their values are

(*a*) 1 Ω and 2 Ω (*b*) 2 Ω and 3 Ω (*c*) 1.5 Ω and 2.5 Ω (*d*) 2.5 Ω and 4.5 Ω.

**144.** Which of the following has negative temperature co-efficient?

(*a*) Constantan (*b*) Carbon (*c*) Manganin (*d*) Tungsten.

**145.** A 5 metre length of 20 gauge copper conductor wire will have a resistance of

(*a*) less than 1 Ω (*b*) around 5 Ω (*c*) 9.45 Ω (*d*) nearly 1 MΩ.

**146.** If a wire conductor of 0.01 Ω resistance is doubled in length, its resistance will become

(*a*) 0.05 Ω (*b*) 0.02 Ω (*c*) 0.005 Ω (*d*) 0.002 Ω.

**147.** The resistivity of copper is nearly

(*a*) $1.7 \times 10^{-4}$ ohm-cm (*b*) $1.7 \times 10^{-6}$ ohm-cm

(*c*) $1.7 \times 10^{-8}$ ohm-cm (*d*) $1.7 \times 10^{-10}$ ohm-cm.

**148.** When the length and diameter of an aluminium conductor are tripled, the resistance will

(*a*) increase by $33\frac{1}{2}$% (*b*) increase by $66\frac{2}{3}$%

(*c*) reduce by $33\frac{1}{3}$% (*d*) reduce by $66\frac{2}{3}$%.

**149.** Three resistors $R_1 = 47$ kΩ, 1 W, $R_2 = 1$ kΩ, $\frac{1}{2}$ W and $R_3 = 0$ Ω, 1 W are connected in parallel. The equivalent resistance will be

(*a*) 47.1 kΩ, $\frac{1}{2}$ W (*b*) 47.1 kΩ, 1 W (*c*) 4.7 kΩ, $\frac{1}{2}$ W (*d*) 0 Ω, 1 W.

**150.** The rating of a fuse wire is always expressed in

(*a*) amperes (*b*) ampere-volts (*c*) kWh (*d*) ampere-hours.

## ANSWERS

| | | | | | | |
|---|---|---|---|---|---|---|
| **1.** (*b*) | **2.** (*d*) | **3.** (*c*) | **4.** (*c*) | **5.** (*c*) | **6.** (*c*) | **7.** (*b*) |
| **8.** (*c*) | **9.** (*d*) | **10.** (*c*) | **11.** (*c*) | **12.** (*a*) | **13.** (*b*) | **14.** (*d*) |
| **15.** (*c*) | **16.** (*a*) | **17.** (*c*) | **18.** (*b*) | **19.** (*a*) | **20.** (*b*) | **21.** (*d*) |
| **22.** (*b*) | **23.** (*b*) | **24.** (*d*) | **25.** (*a*) | **26.** (*d*) | **27.** (*d*) | **28.** (*b*) |
| **29.** (*c*) | **30.** (*b*) | **31.** (*b*) | **32.** (*c*) | **33.** (*c*) | **34.** (*c*) | **35.** (*d*) |
| **36.** (*d*) | **37.** (*a*) | **38.** (*b*) | **39.** (*c*) | **40.** (*c*) | **41.** (*d*) | **42.** (*b*) |
| **43.** (*c*) | **44.** (*b*) | **45.** (*d*) | **46.** (*c*) | **47.** (*d*) | **48.** (*d*) | **49.** (*d*) |
| **50.** (*b*) | **51.** (*c*) | **52.** (*d*) | **53.** (*c*) | **54.** (*b*) | **55.** (*b*) | **56.** (*a*) |
| **57.** (*a*) | **58.** (*c*) | **59.** (*d*) | **60.** (*b*) | **61.** (*b*) | **62.** (*d*) | **63.** (*a*) |
| **64.** (*c*) | **65.** (*d*) | **66.** (*c*) | **67.** (*a*) | **68.** (*d*) | **69.** (*a*) | **70.** (*c*) |
| **71.** (*d*) | **72.** (*b*) | **73.** (*a*) | **74.** (*c*) | **75.** (*b*) | **76.** (*a*) | **77.** (*d*) |
| **78.** (*a*) | **79.** (*c*) | **80.** (*c*) | **81.** (*d*) | **82.** (*b*) | **83.** (*a*) | **84.** (*b*) |
| **85.** (*c*) | **86.** (*a*) | **87.** (*c*) | **88.** (*d*) | **89.** (*a*) | **90.** (*c*) | **91.** (*b*) |
| **92.** (*a*) | **93.** (*d*) | **94.** (*a*) | **95.** (*c*) | **96.** (*d*) | **97.** (*d*) | **98.** (*c*) |
| **99.** (*d*) | **100.** (*d*) | **101.** (*d*) | **102.** (*a*) | **103.** (*b*) | **104.** (*a*) | **105.** (*a*) |
| **106.** (*c*) | **107.** (*b*) | **108.** (*a*) | **109.** (*b*) | **110.** (*a*) | **111.** (*b*) | **112.** (*d*) |
| **113.** (*d*) | **114.** (*d*) | **115.** (*d*) | **116.** (*b*) | **117.** (*c*) | **118.** (*d*) | **119.** (*c*) |
| **120.** (*b*) | **121.** (*c*) | **122.** (*c*) | **123.** (*c*) | **124.** (*c*) | **125.** (*d*) | **126.** (*d*) |
| **127.** (*d*) | **128.** (*c*) | **129.** (*c*) | **130.** (*b*) | **131.** (*c*) | **132.** (*b*) | **133.** (*d*) |
| **134.** (*b*) | **135.** (*a*) | **136.** (*a*) | **137.** (*c*) | **138.** (*a*) | **139.** (*b*) | **140.** (*a*) |
| **141.** (*c*) | **142.** (*d*) | **143.** (*c*) | **144.** (*b*) | **145.** (*a*) | **146.** (*b*) | **147.** (*b*) |
| **148.** (*d*) | **149.** (*d*) | **150.** (*a*). | | | | |

Chapter **2** **DC CIRCUITS**

Many types of circuits have components that are not in series, in parallel, or in series-parallel. Sometimes a circuit may have two or more voltage or current sources. Some of the laws used in solving such circuits are given below:

**Kirchhoff's Current Law (KCL).** The sum of the currents entering a junction is equal to the sum of the currents leaving the junction. If the currents towards a junction are considered positive and those away from the same junction negative, then this states that the algebraic sum of all the currents meeting at a common junction is zero.

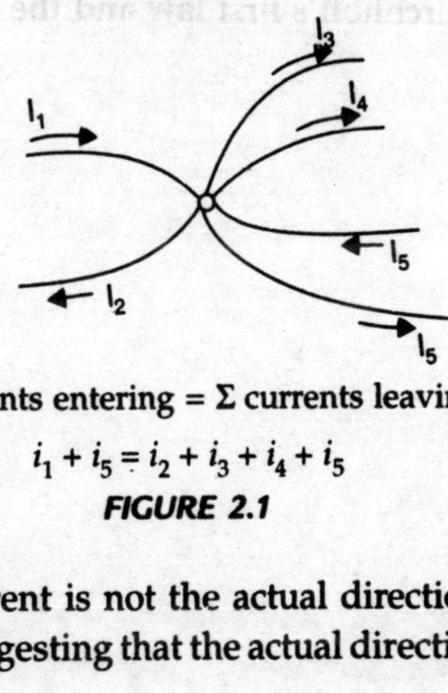


Σ Currents entering = Σ currents leaving

$$i_1 + i_5 = i_2 + i_3 + i_4 + i_5$$

*FIGURE 2.1*

While applying Kirchhoff's law to any network, the direction of flow of current may be assumed either clockwise or anti-clockwise. Whatever the direction of current is assumed, the same should be used throughout the solution of the question. In case the assumed direction of current is not the actual direction, then on solving the question, this current will have negative sign suggesting that the actual direction of flow is reverse of the assumed direction.

**Kirchhoff's Voltage Law (KVL).** The sum of the rises of potential around any closed circuit equals the sum of the drops of potential in that circuit. In other words, the algebraic sum of the potential differences around a closed circuit is zero.

Kirchhoff's laws are applicable both to DC and AC voltages and currents. However, in the case of alternating currents and voltages any emf of self inductance or that existing across a capacitor should be also taken into account.

**Example 1.** *If each side of the cube offers resistance R ohms, then equivalent resistance between points X and Y will be*

(*a*) R (*b*) 12 R (*c*) $\frac{5}{6}$ R (*d*) $\frac{6}{5}$ R.

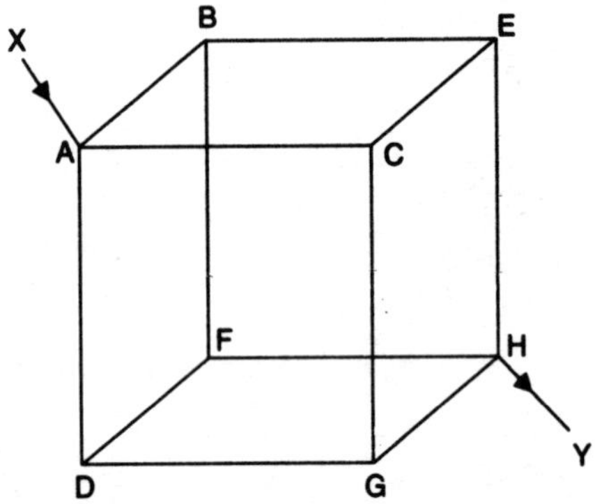


*FIGURE 2.2*

**Sol.** Referring to Fig. 2.3 if *I* is the current entering and leaving the framework, then by Kirchhoff's first law and the symmetry of the arrangement.

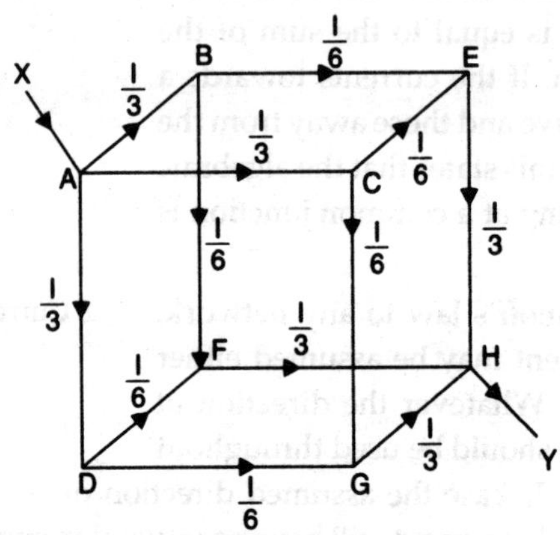


*FIGURE 2.3*

Current in *AB, AC, AD, EH, FH, GH*

$$= \frac{I}{3}$$

Current in *BE, BF, CE, CG, DF, DG*

$$= \frac{I}{6}$$

Hence, P.D. between $X$ and $Y$ = P.D. between $A$ and $B$ + P.D. between $B$ and $E$ + P. D. between $E$ and $H$

$$IReq = \frac{I}{3}R + \frac{I}{6}R + \frac{I}{3}R$$

$$= \frac{5}{6}IR$$

$$\therefore \quad Req = \frac{5}{6}R.$$

**Some Definitions**

**Circuit.** A circuit is a conducting path through which either an electric current flows or is intended to flow.

**Loop.** A closed path is a loop.

**Circuit Parameters.** The various elements like resistance, inductance and capacitance are called circuit parameters. The parameters may be lumped or distributed.

**Mesh.** A mesh is the simplest possible loop. A mesh current is assumed to flow around the mesh without branching.

**Node.** A principal node is a branch point where currents divide or combine.

**Linear Circuit.** A linear circuit is one whose parameters are constant *i.e.,* they do not change with voltage or current.

**Non-linear Circuit.** A circuit in which the parameters change with voltage or current is known as non-linear circuit.

**Unilateral Circuit.** A circuit whose properties or characteristics change with the direction of its operation is known as unilateral circuit. A vacuum tube rectifier is a unilateral circuit because it cannot perform rectification in both directions.

**Bilateral Circuit.** A circuit whose properties or characteristics are the same in either direction, is known as bilateral circuit.

**Electric Network.** A combination of various electric elements, connected in any manner whatever, is known as electric circuit.

**Passive Network.** A circuit which does not contain any source of emf is known as passive network.

**Active Network.** A circuit which contains one or more than one source of emf is known as active network.

**Star-Delta Conversion.** In the analysis of networks, it is often helpful to convert a delta to star or *vice-versa*. The formulas for these transformations are derived from Kirchhoff's laws. In these formulas the letters used as subscripts are $R_A$, $R_B$ and $R_C$ in the delta (Δ). While resistances are numbered $R_1$, $R_2$ and $R_3$ in the star.

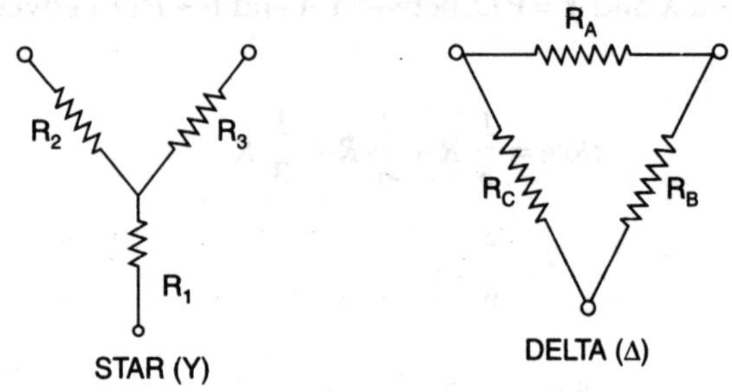


*FIGURE 2.4*

**Conversions of Star to Delta (or T to π)**

$$R_A = \frac{R_1R_2 + R_2R_3 + R_3R_1}{R_1}$$

$$R_B = \frac{R_1R_2 + R_2R_3 + R_3R_1}{R_2}$$

$$R_C = \frac{R_1R_2 + R_2R_3 + R_3R_1}{R_3}$$

In general,

$$R_\Delta = \frac{\Sigma \text{ all cross products in } Y}{\text{Opposite } R \text{ in } Y}$$

**Conversion of Delta (Δ) to Star (Y) or π to T**

$$R_1 = \frac{R_BR_C}{R_A + R_B + R_C}$$

$$R_2 = \frac{R_CR_A}{R_A + R_B + R_C}$$

$$R_3 = \frac{R_AR_B}{R_A + R_B + R_C}$$

In general,

$$R_{(Y)} = \frac{\text{Product of two adjacent } R \text{ in delta}}{\Sigma \text{ all } R \text{ in delta}}$$

**Superposition Theorem.** The theorem extends the use of Ohm's law to circuits that have more than one source. In brief, we can calculate the effect of one source at a time and then super impose the results of all the sources. According to superposition theorem. In a network with two or more sources the current or voltage for any component is the algebraic sum of the effects produced by each source acting separately.

**Conditions for Superposition.** All the components must be linear and bilateral in order to super impose currents and voltages.

Active components such as transistors, semi-conductor diodes, and electron tubes are never bilateral and often are non-linear.

**Thevenin's Theorem.** Any network with two open terminals *A* and *B* can be replaced by a single voltage source $V_{Th}$ in series with a single resistance $R_{Th}$ driving the *AB* terminals. $V_{Th}$ is the voltage produced by the network across terminals *A* and *B*. $R_{Th}$ is the resistance across the open *AB* terminals with all sources killed.

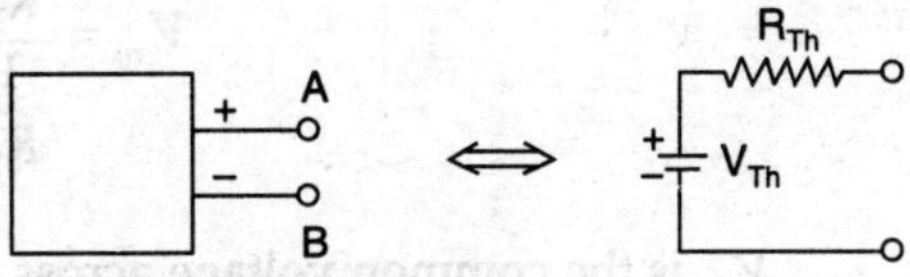

*FIGURE 2.5*

Which way to look at the resistance of a series-parallel circuit depends on where the source is connected. When the source is short-circuited for thevenizing a circuit, the A and B terminals become the reference. Looking back from A and B to calculate $R_{Th}$, the view point becomes reversed from the way $V_{Th}$ is determined.

**Norton's Theorem.** Any two-terminal network can be replaced by a single current source $I_N$ in parallel with a single resistance $R_N$. The value of $I_N$ is the current produced by the network through the short-circuited terminals. $R_N$ is the resistance across the open terminals with all sources killed.

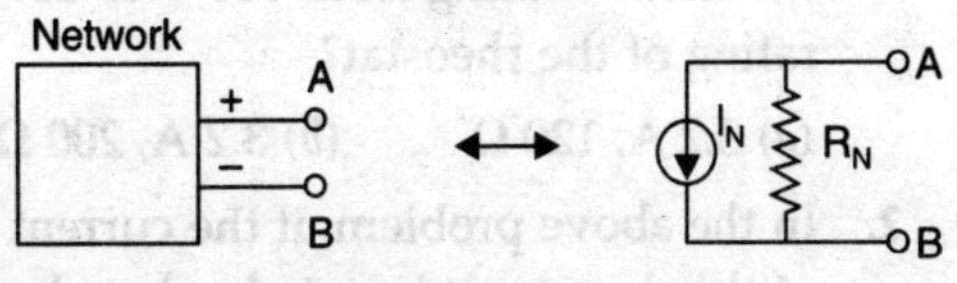

*FIGURE 2.6*

**Thevenin-Norton Conversions.** Thevenin's theorem says that any network can be represented by a voltage source and series resistance, while Norton' theorem says that the same network can be represented by a current source and shunt resistance. The conversion directly from a Thevenin form to a Norton form and *vice-versa* is possible using the following relations:

Thevenin from Norton

$$R_{Th} = R_N$$

$$V_{Th} = I_N \times R_N$$

Norton from Thevenin

$$R_N = R_{Th}$$

$$I_N = V_{Th} \div R_{Th}$$

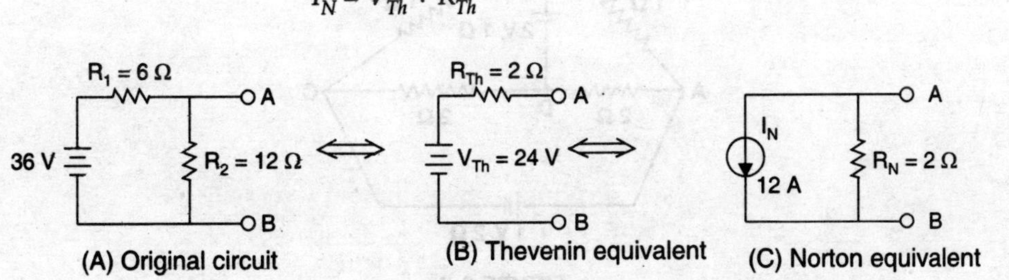

*FIGURE 2.7*

**Millman's Theorem.** The common voltage across parallel branches with different $V$-sources can be determined from the relation

$$V_{xy} = \frac{\frac{V_1}{R_1} + \frac{V_2}{R_2} + \frac{V_3}{R_3}}{\frac{1}{R_1} + \frac{1}{R_2} + \frac{1}{R_3}} \text{ ... etc.}$$

$V_{xy}$ is the common voltage across all branches.

## OBJECTIVE TYPE QUESTIONS

1. The field winding of a dc motor has a resistance of 80 ohms and is to be excited from a 250 volt source. A field rheostat with 20 steps of resistance is to vary the voltage applied to the field winding from 100 V to 250 V. What should be the total resistance and the current rating of the rheostat?

   (*a*) 3.2 A, 120 Ω (*b*) 3.2 A, 200 Ω (*c*) 1.33 A, 200 Ω (*d*) 1.33 A, 120 Ω.

2. In the above problem, if the current is to be decreased in equal increments as each element of the rheostat is inserted, what should be the resistances of the first and last elements?

   (*a*) 14 Ω, 2.3 Ω (*b*) 2.3 Ω, 14 Ω (*c*) 4.6 Ω, 7 Ω (*d*) 7Ω, 4.6 Ω.

**Questions 3 to 5 refer to data given below**

*AB*, *BC*, *CD* and *DA* are resistors of 1, 1, 2 and 2 ohms, respectively in series. Between *A* and *C* is a 1 volt cell of resistance 2 ohms, *A* being positive. Between *B* and *D* is a volt cell of 1 ohm resistance, *B* being positive Fig. 2.8.

3. The current in BA will be

   (*a*) 0.1 A (*b*) 0.2 A (*c*) 0.4 A (*d*) 1 A.

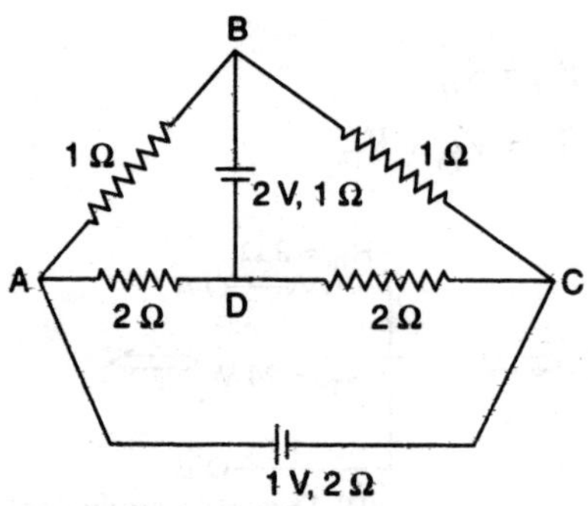


*FIGURE 2.8*

4. The current in $AD$ will be

(a) 0.2 A (b) 0.5 A (c) 0.75 A (d) 1 A.

5. The current in $BC$ will be

(a) 0.2 A (b) 0.4 A (c) 0.5 A (d) 0.6 A.

6. "In a linear, bilateral network having more than one source, the current and voltage in any part of the network can be found by adding algebraically the effect of each source separately."

The above statement is associated with

(a) Superposition theorem (b) Thevenin's theorem

(c) Norton's theorem (d) Millman's theorem.

7. "Any network with two open terminals $a$ and $b$ can be replaced by a single voltage source $V_{Th}$ in series with a single resistance $R_{Th}$ driving the $ab$ terminals."

The above statement is associated with

(a) Superposition theorem (b) Thevenin's theorem

(c) Norton's theorem (d) None of the above.

8. "Any two-terminal network can be replaced by a single current source $I_N$ in parallel with a single resistance $R_N$."

The above statement is associated with

(a) Superposition theorem (b) Thevenin's theorem

(c) Norton's theorem (d) Millman's theorem.

9. Thevenin's equivalent for the circuit shown in Fig. 2.9 will be

(a) $V_{Th} = 15$ V, $R_{Th} = 3\ \Omega$ (b) $V_{Th} = 10$ V, $R_{Th} = 5\ \Omega$

(c) $V_{Th} = 5$ V, $R_{Th} = 7\ \Omega$ (d) $V_{Th} = 3$ V, $R_{Th} = 10\ \Omega$.

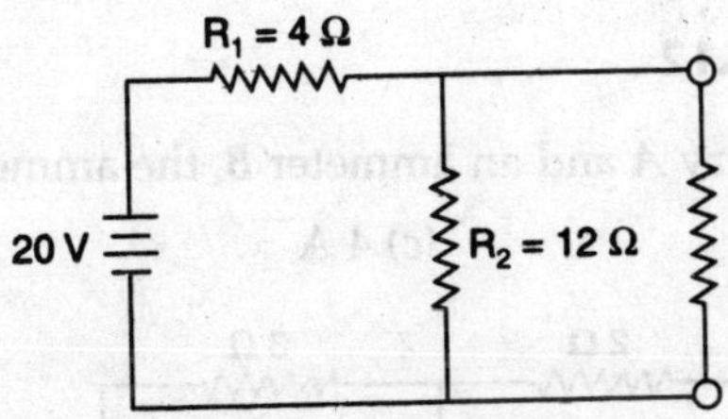


FIGURE 2.9

10. For the circuit shown in Fig. 2.9, what will be $I_L$?

(a) 1 A (b) 2 A (c) 3 A (d) 4 A.

11. For the network shown in Fig. 2.10 the resistance between $E$ and $F$ will be

(a) 5.6 Ω (b) 10.6 Ω (c) 16.6 Ω (d) 21.6 Ω

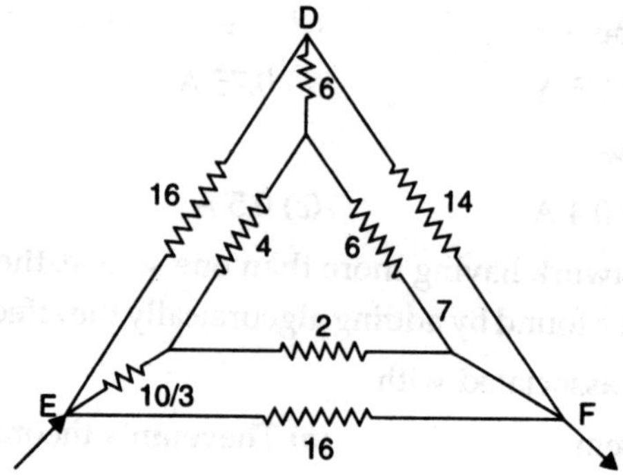


*FIGURE 2.10*

12. The current drawn from 2 volt battery in the network shown in Fig. 2.11 will be

(*a*) 0.39 A (*b*) 0.66 A (*c*) 0.78 A (*d*) 1.32 A.

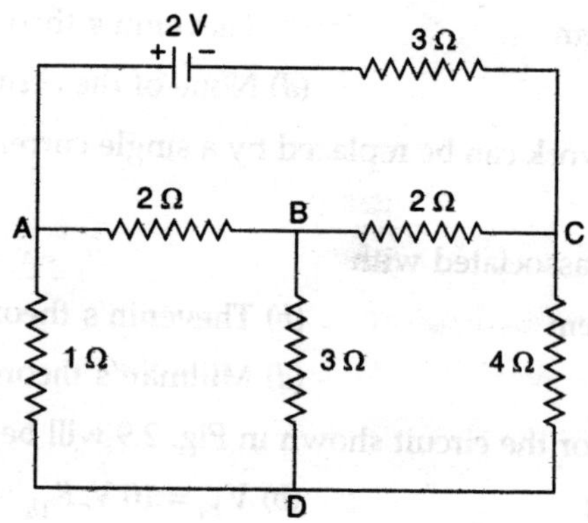


*FIGURE 2.11*

**Questions 13 to 15 refer to Fig. 2.12**

13. In the network shown a battery *A* and an ammeter *B*, the ammeter current will be

(*a*) 2 A (*b*) 3 A (*c*) 4 A (*d*) 6 A.

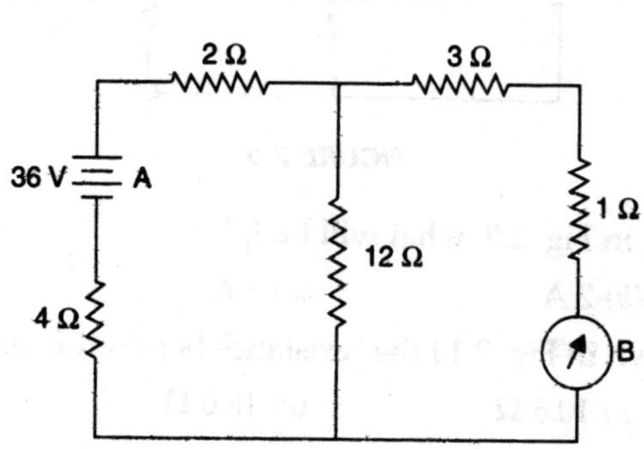


*FIGURE 2.12*

14. If battery and ammeter are replaced with each other in Fig. 2.12, the battery current will be

(a) 2.0 A (b) 3.0 A (c) 4.5 A (d) 5.5.

15. The value of transfer resistance is

(a) 4 Ω (b) 6 Ω (c) 8 Ω (d) 12 Ω

16. Using Thevenin's theorem, the current in branch *CD* of the network shown in Fig. 2.13 will be

(a) 0.7 A (b) 1.7 A (c) 2.7 A (d) 3.7 A.

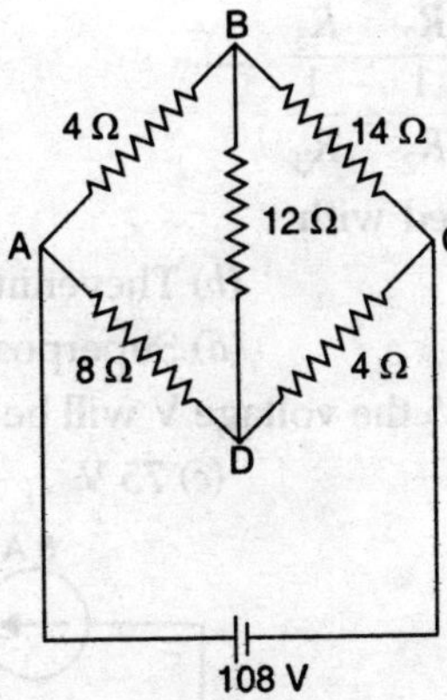


FIGURE 2.13

17. A bridge network *ABCD* has arms *AB*, *BC*, *CD* and *DA* of resistance 1, 1, 2 and 1 ohms respectively. If the detector *AC* has resistance of 1 ohm, determine by star delta transformation the network resistance as viewed from the battery terminals *BD*.

(a) $R_T = 0.78\ \Omega$ (b) $R_T = 1\ \Omega$ (c) $R_T = 1.18\ \Omega$ (d) $R_T = 1.78\ \Omega$.

18. A π network is shown in Fig. 2.14. The equivalent *T* network is shown by

(a) Figure *A* (b) Figure *B* (c) Figure C (d) Figure *D*.

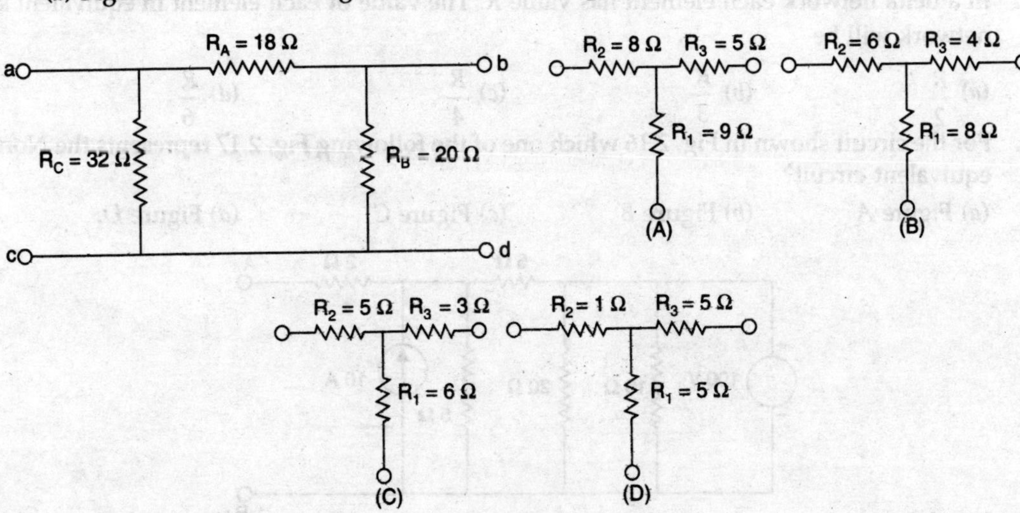


FIGURE 2.14

19. "In any linear network containing bilateral impedances and generators, the ratio of voltage $V$ introduced in one mesh to the current $I$ in any second mesh is the same as the ratio obtained if the positions of $V$ and $I$ are interchanged, other emf being removed."
The above statement is associated with
(*a*) Norton's theorem (*b*) Thevenin's theorem(*c*) Kirchhoff's law (*d*) Reciprocity theorem.

20. The common voltage across parallel branches with different voltage sources can be determined by the relation

$$V = \frac{\frac{V_1}{R_1} + \frac{V_2}{R_2} + \frac{V_3}{R_3}}{\frac{1}{R_1} + \frac{1}{R_2} + \frac{1}{R_3}} \; ......$$

The above statement is associated with
(*a*) Millman's theorem (*b*) Thevenin's theorem
(*c*) Norton's theorem (*d*) Superposition theorem.

21. For the circuit shown in Fig. 2.15 the voltage V will be
(*a*) 55 V (*b*) 65 V (*c*) 75 V (*d*) 85 V.

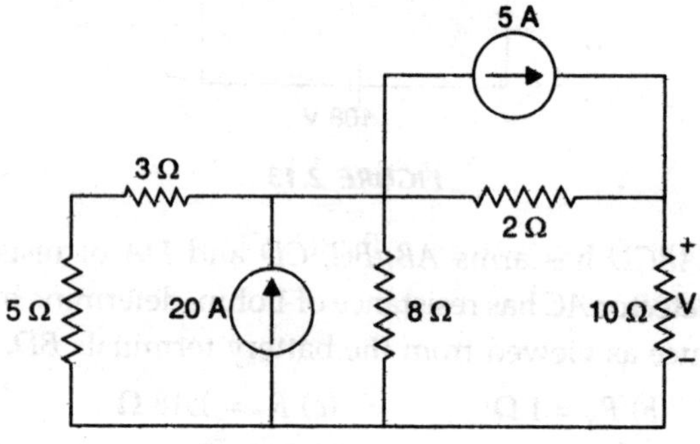


**FIGURE 2.15**

22. In a delta network each element has value $R$. The value of each element in equivalent star network will be
(*a*) $\frac{R}{2}$ (*b*) $\frac{R}{3}$ (*c*) $\frac{R}{4}$ (*d*) $\frac{R}{6}$.

23. For the circuit shown in Fig. 2.16 which one of the following Fig. 2.17 represents the Norton equivalent circuit?
(*a*) Figure *A* (*b*) Figure *B* (*c*) Figure C (*d*) Figure *D*.

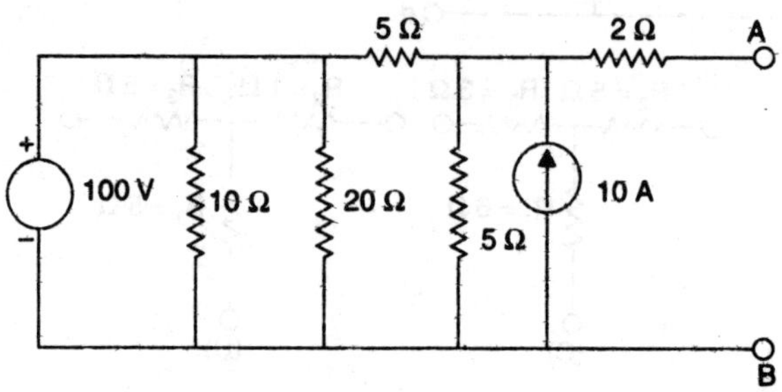


**FIGURE 2.16**

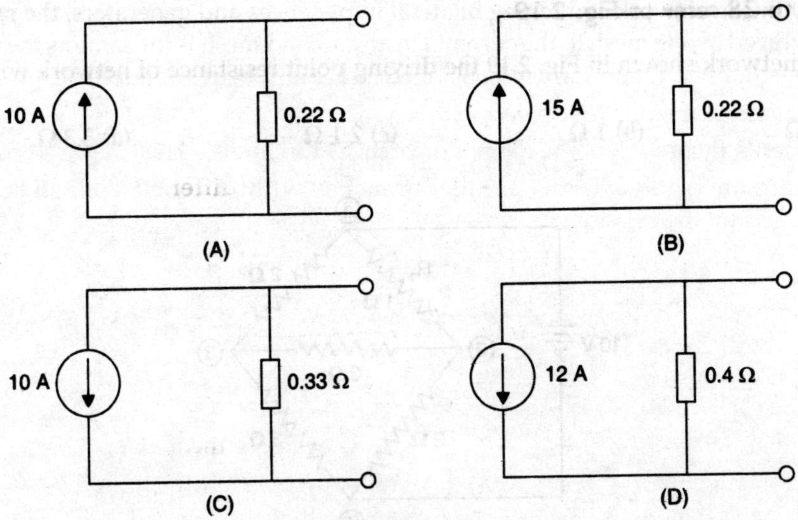


FIGURE 2.17

**24.** In the circuit shown in Fig. 21.8 (*a*) the value of voltage across 500 ohms resistor will be

(*a*) 25.5 V (*b*) 35.5 V (*c*) 49.5 V (*d*) 61.5 V.

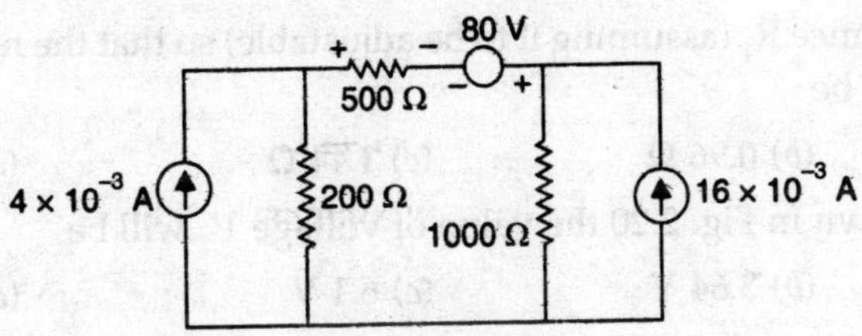


FIGURE 2.18 (*a*)

**25.** In the circuit shown in Fig. 2.18 (*b*) the voltage across the 4 ohm resistor will be

(*a*) $\frac{9}{17}$ A (*b*) $\frac{13}{17}$ A (*c*) $\frac{25}{17}$ A (*d*) $\frac{31}{17}$ A.

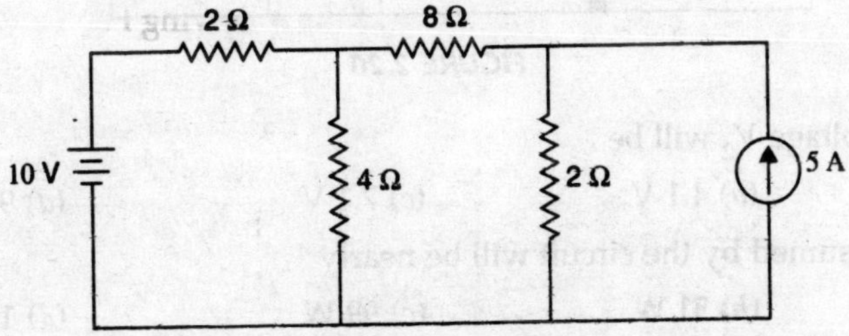


FIGURE 2.18 (*b*)

**Questions 26 to 28 refer to Fig. 2.19**

26. For the network shown in Fig. 2.19 the driving point resistance of network will be

(a) $\frac{30}{17}$ Ω (b) 1 Ω (c) 2.1 Ω (d) 3.3 Ω.

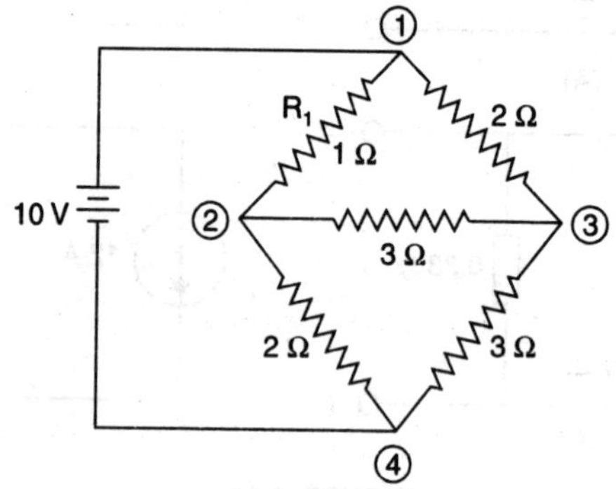


*FIGURE 2.19*

27. The current delivered by the source will be

(a) 1.2 A (b) 2.4 A (c) 3.6 A (d) 4.8 A.

28. The value of resistance $R_1$ (assuming it to be adjustable) so that the nodes (2) and (3) become equipotential, will be

(a) 0.48 Ω (b) 0.96 Ω (c) 1.44 Ω (d) 2.14 Ω

29. For the circuit shown in Fig. 2.20 the value of voltage $V_1$ will be

(a) 2.64 V (b) 3.64 V (c) 6.1 V (d) 9.1 V.

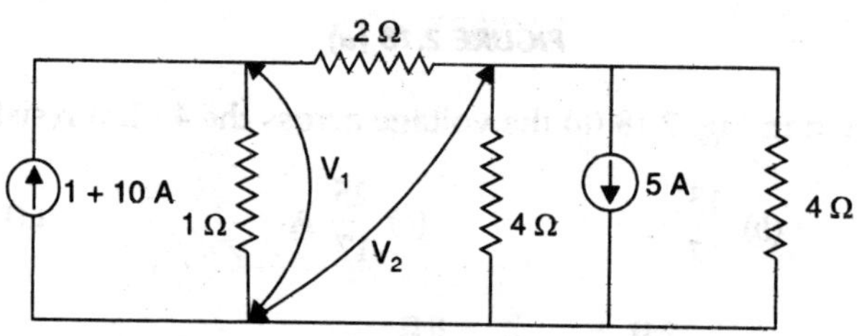


*FIGURE 2.20*

30. The value of voltage $V_2$ will be

(a) 2.1 V (b) 4.1 V (c) 7.1 V (d) 9.1 V.

31. The power consumed by the circuit will be nearly

(a) 11 W (b) 51 W (c) 99 W (d) 119 W.

**Questions 32 and 33 Refer to Fig. 2.21**

32. For the circuit shown in Fig. 2.21 the current through $R_3$ will be

(a) 1.5 A (b) 2 A (c) 2.5 A (d) 2.75 A.

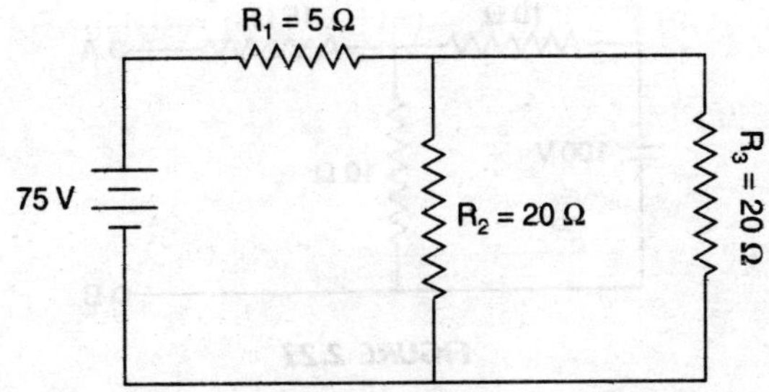


*FIGURE 2.21*

**33. If the value of $R_3$ is increased by 30% the value of current through $R_3$ will be**

(*a*) 2.6 A (*b*) 2.5 A (*c*) 2.2 A (*d*) 2 A.

**34. Kirchhoff's law is applicable to**

(*a*) AC circuits only (*b*) DC circuits only

(*c*) both ac as well as dc circuits (*d*) passive networks only.

**35. Which of the following represents a relation for non-linear resistor?**

(*a*) $v = -5i$ (*b*) $i + 5v = 0$ (*c*) $i = 2v^2$ (*d*) none of the above.

**36. For the circuit shown in Fig. 2.22 the current through 2 ohm resistor will be**

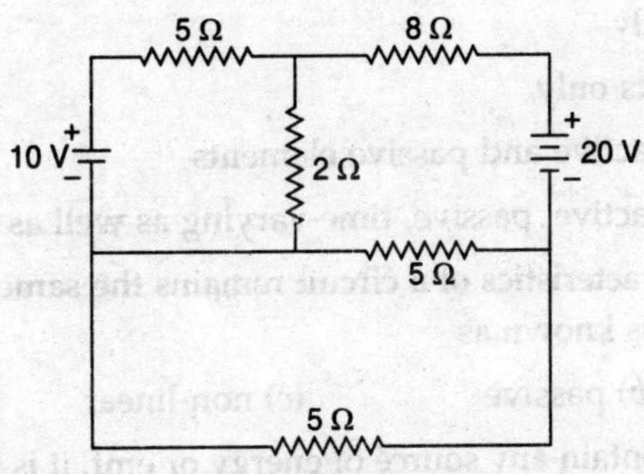


*FIGURE 2.22*

(*a*) 1.68 Ω (*b*) 2.68 Ω (*c*) 3.68 Ω (*d*) 4.68 Ω.

**37. A star network has each element of value *R*. The value of elements of equivalent network will be**

(*a*) $\frac{R}{2}$ (*b*) *R* (*c*) $\frac{3}{2}R$ (*d*) 3 *R*.

**38. Kirchhoff's law is not applicable to circuits with**

(*a*) non-linear resistances (*b*) passive elements

(*c*) distributed parameters (*d*) lumped parameters.

**39. For the network shown in Fig. 2.23 the Norton's equivalent circuit is shown in Fig. 2.24. The value of *R* will be**

(*a*) 5 Ω (*b*) 10 Ω (*c*) 15 Ω (*d*) 20 Ω.

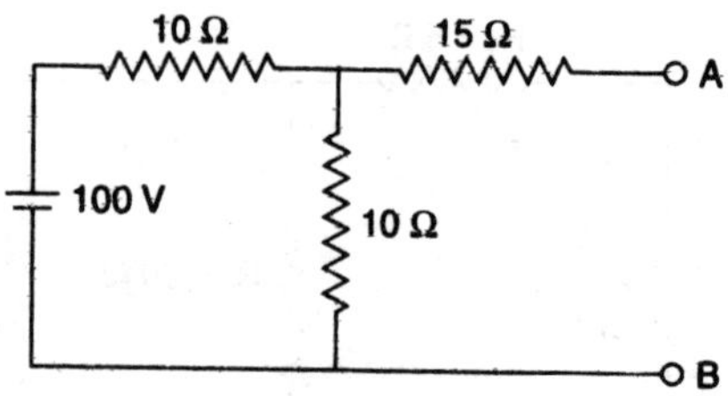


FIGURE 2.23

40. The value of $I$ will be

(a) 2.5 A (b) 3.5 A (c) 4 A (d) 5 A.

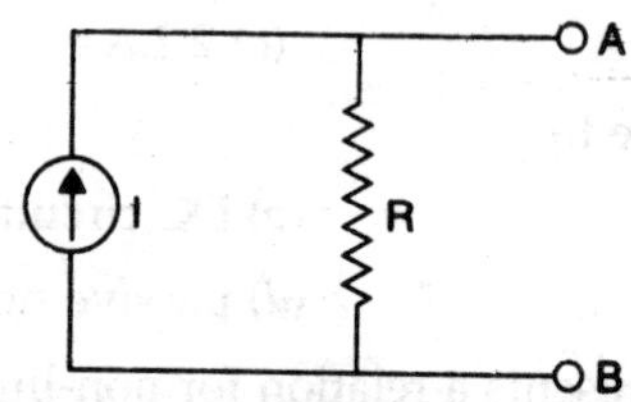


FIGURE 2.24

41. Kirchhoff's voltage law applies to circuits with
   (a) linear elements only
   (b) non-linear elements only
   (c) linear, non-linear, active and passive elements
   (d) linear, non-linear, active, passive, time-varying as well as time-invariant elements.
42. If the function or characteristics of a circuit remains the same whatever the direction of its operation, the circuit is known as
   (a) active (b) passive (c) non-linear (d) bilateral.
43. If a circuit does not contain any source of energy or emf, it is known as
   (a) active network (b) passive network (c) unilateral circuit (d) bilateral circuit.
44. Which of the following statement is false?
   (a) a 10 A source with 10 Ω parallel $R$ will supply 5 A to a 10 Ω load $R_L$
   (b) a node voltage is a voltage between current nodes
   (c) a Thevenin-Norton conversion means converting a voltage source to a current source
   (d) a Thevenin equivalent is a parallel circuit.
45. Six resistor of 8 ohm each are connected to form the figure shown in Fig. 2.25. The resistance between points $X$ and $Y$ will
   (a) 1 Ω (b) 2 Ω (c) 4 Ω (d) 8 Ω.

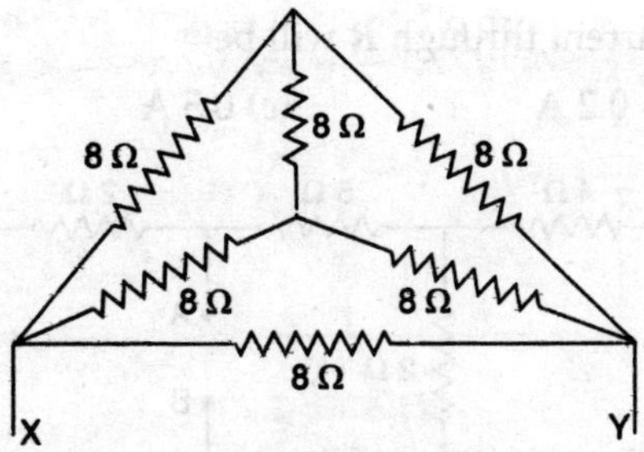


FIGURE 2.25

**Questions 46 to 48 refer to Fig. 2.26**

**46.** The current through 10 Ω resistor will be

(*a*) 0.663 A (*b*) 1 A (*c*) 1.295 A (*d*) 1.895 A.

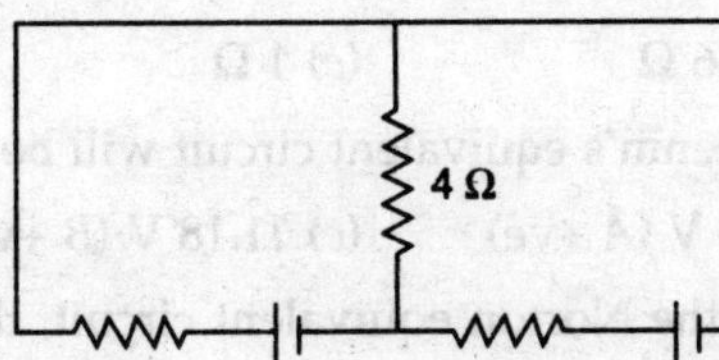


FIGURE 2.26

**47.** The current through 8 ohm resistor will be

(*a*) 0.663 A (*b*) 1.035 A (*c*) 1.63 A (*d*) 1.895 A.

**48.** The current through 4 ohm resistor will be

(*a*) 0.263 A (*b*) 0.526 A (*c*) 0.883 A (*d*) 0.999 A

**Questions 49 and 50 refer to Fig. 2.27**

**49.** If the current through branch *OA* is zero, the value of *R* will be

(*a*) 2 Ω (*b*) 3 Ω (*c*) 4 Ω (*d*) 6 Ω.

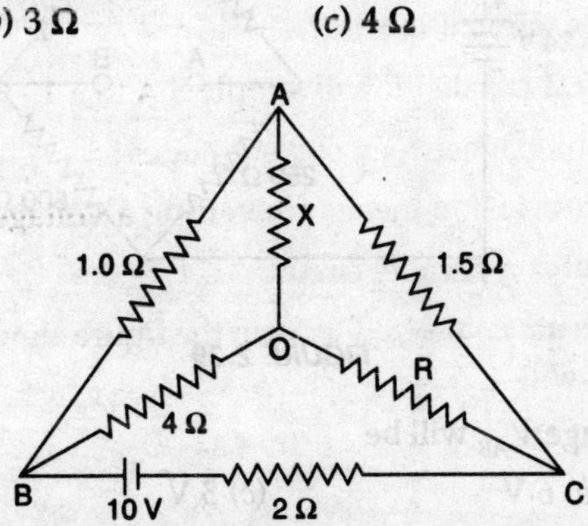


FIGURE 2.27

50. In the above case the current through $R$ will be

(a) 0.1 A (b) 0.2 A (c) 0.5 A (d) 1 A.

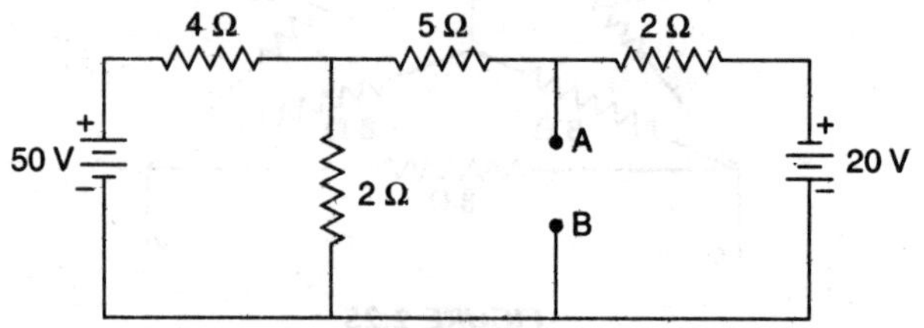


FIGURE 2.28

**Questions 51 to 54 refer to Fig. 2.28**

51. For Thevenin equivalent circuit of the circuit shown in Fig. 2.28 the equivalent Thevenin resistance will be

(a) 0.52 Ω (b) 0.66 Ω (c) 1 Ω (d) 1.52 Ω

52. The source voltage in Thevenin's equivalent circuit will be

(a) 5.8 V (B +ve) (b) 5.8 V (A +ve) (c) 11.18 V (B +ve) (d) 11.18 V (A +ve).

53. For the circuit of Fig. 2.28 the Norton equivalent circuit, the equivalent Norton resistance will be

(a) 1.52 ohms (b) 2.55 ohms (c) 3.55 ohm (d) 4.55 ohms.

54. In the Norton equivalent circuit, the Norton current will be

(a) 1.1 A (b) 2.35 A (c) 7.35 A (d) 10.35 A

**Questions 55 and 56 refer to bridge circuit shown in Fig. 2.29**

55. For Thevenin equivalent circuit of bridge circuit shown in Fig. 2.29 $R_{AB}$ will be

(a) 810 Ω (b) 750 Ω (c) 500 Ω (d) 55.5 Ω

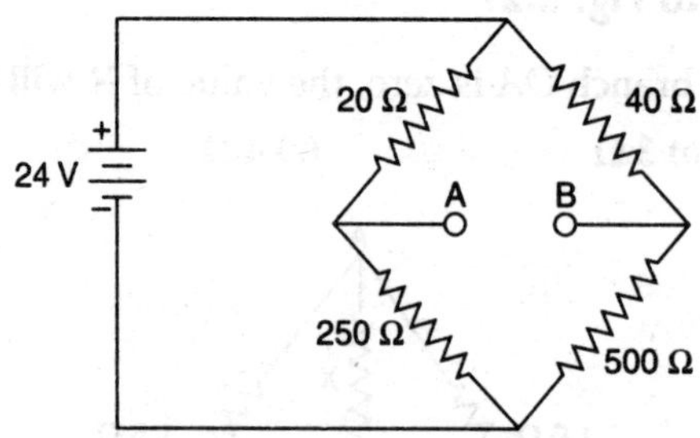


FIGURE 2.29

56. In the above case voltage $V_{AB}$ will be

(a) 12 V (b) 6 V (c) 3 V (d) zero.

57. In the bridge circuit of the 500 ohm resistor is changed to 475 ohms, the Thevenin equivalent voltage will be
(*a*) 0.0863 V (*A* +ve) (*b*) 0.0863 V (*B* +ve) (*c*) 0.163 V (*A* +ve) (*d*) 0.163 V (*B* +ve).

58. In the bridge circuit shown in Fig. 2.30 a galvanometer is connected across *AB* with a resistance of 100 ohms and a sensitivity of 0.5 μA/mm. What will be the deflection of the galvanometer
(*a*) 0.195 m (*b*) 0.088 m (*c*) 0.044 m (*d*) 0.022 m.

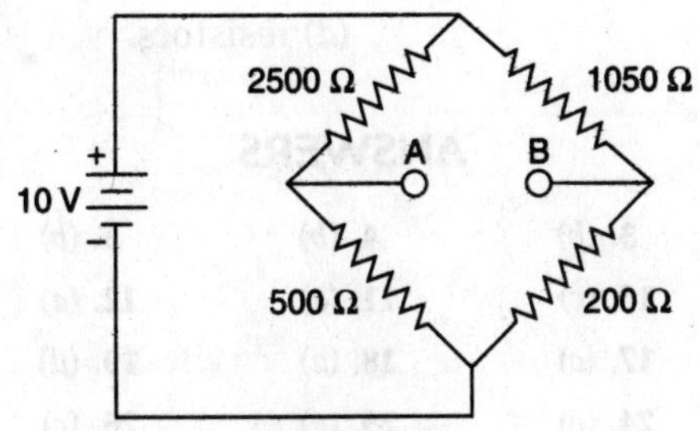


***FIGURE 2.30***

59. For the circuit shown in Fig. 2.31, the equivalent star circuit will have all the following resistance EXCEPT
(*a*) 2.5 Ω (*b*) 5.5 Ω (*c*) 6.25 Ω (*d*) 10.5 Ω.

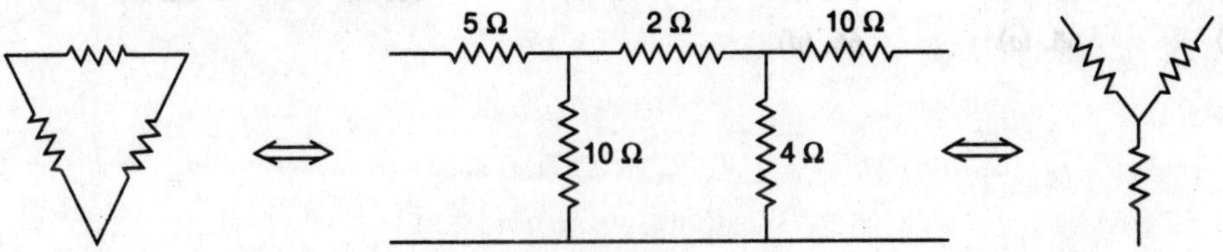


***FIGURE 2.31***

60. For the circuit in Fig. 2.31 in the equivalent delta connected network will have all the following resistance EXCEPT
(*a*) 10.25 Ω (*b*) 17.2 Ω (*c*) 43 Ω (*d*) 53 Ω.

61. A circuit has resistors, capacitors and semi-conductor diodes. The circuit will be known as
(*a*) linear circuit (*b*) bilateral circuit (*c*) non-linear circuit (*d*) none of the above.

62. Which of the following is non-linear circuit parameter?
(*a*) wire wound resistor (*b*) condenser
(*c*) transistor (*d*) inductance.

63. A circuit having same properties in either direction is known as
(*a*) bilateral circuit (*b*) non-linear circuit (*c*) passive network (*d*) none of the above.

64. The circuit parameters may be
I. Active II. Passive III. Linear IV. Non-linear.
Which of the following is valid for a gas diode?
(*a*) I and II (*b*) I and IV (*c*) II and III (*d*) II and IV.

65. A capacitor is generally a

(*a*) linear and bilateral component
(*b*) non-linear and active component
(*c*) bilateral and active component
(*d*) active, passive, linear and non-linear component.

66. Which of the following is linear and bilateral parameter?

(*a*) transistor
(*b*) semi-conductor diodes
(*c*) electron tubes
(*d*) resistors.

## ANSWERS

| | | | | | | |
|---|---|---|---|---|---|---|
| **1.** (*a*) | **2.** (*b*) | **3.** (*b*) | **4.** (*b*) | **5.** (*b*) | **6.** (*a*) | **7.** (*b*) |
| **8.** (*c*) | **9.** (*a*) | **10.** (*c*) | **11.** (*a*) | **12.** (*a*) | **13.** (*b*) | **14.** (*b*) |
| **15.** (*d*) | **16.** (*c*) | **17.** (*c*) | **18.** (*a*) | **19.** (*d*) | **20.** (*a*) | **21.** (*d*) |
| **22.** (*b*) | **23.** (*a*) | **24.** (*c*) | **25.** (*c*) | **26.** (*c*) | **27.** (*d*) | **28.** (*b*) |
| **29.** (*b*) | **30.** (*d*) | **31.** (*d*) | **32.** (*c*) | **33.** (*d*) | **34.** (*c*) | **35.** (*c*) |
| **36.** (*b*) | **37.** (*d*) | **38.** (*c*) | **39.** (*d*) | **40.** (*a*) | **41.** (*d*) | **42.** (*d*) |
| **43.** (*b*) | **44.** (*d*) | **45.** (*c*) | **46.** (*d*) | **47.** (*c*) | **48.** (*a*) | **49.** (*d*) |
| **50.** (*c*) | **51.** (*d*) | **52.** (*c*) | **53.** (*a*) | **54.** (*c*) | **55.** (*d*) | **56.** (*d*) |
| **57.** (*a*) | **58.** (*a*) | **59.** (*b*) | **60.** (*d*) | **61.** (*c*) | **62.** (*c*) | **63.** (*a*) |
| **64.** (*b*) | **65.** (*a*) | **66.** (*d*). | | | | |

# Chapter 3 BATTERIES

**The Voltaic Cell.** When two different conducting materials are immersed in an electrolyte, the chemical reaction of forming a new solution results in the separation of charges. This arrangement for converting chemical energy into electrical energy is a voltaic cell.

The charged conductors are electrodes, serving as the terminals of the cell for connection to an external circuit. Then the potential difference resulting from the separated charges enables the cell to function as a source of applied voltage. The voltage across the cell's terminals forces current to flow in the external circuit. Electrons from the negative terminal of the cell flow through the external circuit and return to the positive terminal. The chemical action in the cell continuously separates charges to maintain the terminal voltage that produces current in the circuit.

**Carbon Zinc Dry Cell.** In this a zinc container acts as the negative electrode and at the same time holds the electrolyte. For positive electrode a carbon rod is used. The electrolyte is the solution of ammonium chloride. The electrolyte is in the form of a paste.

The output voltage is 1.4 to 1.6 volts regardless of the size. The cells may be connected in series for higher voltage. By connecting cells in parallel, current carrying capacity can be increased.

**Lead Acid Wet Cell.** These cells are widely used particularly in automobiles. In this the positive and negative electrodes consist of a group of plates welded to a connecting strap. The plates are immersed in a dilute solution of sulphuric acid for the electrolyte. Each plate is a grid or framework, made of a lead-antimony alloy. In the forming process, the active material in the positive plate is changed to lead peroxide to produce the positive electrode. The negative electrode is spongy lead. During charging the reverse reaction takes place.

$$Pb + PbO_2 + 2H_2SO_4 \underset{\text{Discharge}}{\overset{\text{Charge}}{\rightleftharpoons}} 2PbSO_4 + 2H_2O$$

On discharge, the lead (Pb) and lead peroxide ($PbO_2$) electrodes supply Pb ions that combine with the sulphate ions ($SO_4$) to form lead sulphate ($PbSO_4$) and water ($H_2O$).

The capacity of such batteries is expressed in terms of ampere hours. The typical values are 100 to 300 ampere hours. The rating is usually expressed for 80°F temperature. At higher temperatures, the rating will be reduced. Lower temperatures also reduce the current and voltage output.

The state of charge of battery is usually measured by measuring the specific gravity of the electrolyte. The specific gravity of fully charged battery is 1.280. When the specific gravity is about 1.150 the battery is fully discharged.

**Primary Cells**

| *Type* | *Voltage* | *Remarks* |
|---|---|---|
| Carbon-zinc | 1.5 | Low cost, low current capacity<br>Used for flashlights and toys. |
| Zinc-chloride | 1.5 | Similar to carbon-zinc battery but high current capacity. |
| Manganese alkaline | 1.5 | Hydroxide electrolyte; high current capacity. |
| Silver oxide | 1.5 | Hydroxide electrolyte; miniature button sizes. |
| Lithium | 2.95 | Long life, high cost, high current capacity and energy density. |
| Mercury | 1.35 | Cathode is mercuric oxide. |

**Secondary Cells**

| *Type* | *Voltage* | *Remarks* |
|---|---|---|
| Lead-acid | 2.2 | Wet electrolyte; lowest cost; very low internal resistance; very high current ratings. |
| Nickel-iron | 1.36 | Edison cell; wet hydroxide electrolyte. Industrial uses. |
| Silver-zinc | 1.86 | Rechargeable dry cell, high efficiency. |
| Silver-cadmium | 1.1 | Rechargeable dry cell, high current rating and energy density. |
| Nickel-cadmium | 1.25 | Most common rechargeable dry battery. |
| Gelled electrolyte, lead-acid | 2.1 | Moist electrolyte; sealed unit. |

**Primary Cells.** In a primary cell, the chemical process of forming the solution is not reversible. For instance, zinc can dissolve in ammonium chloride, but the process cannot be reversed to form the zinc electrode from this solution. Flashlight cell is an example of a primary cell.

**Secondary Cells.** In these cells the chemical action is reversible. The electrodes can dissolve in solution with current in one direction, or the current can be reversed to make the solution build up the electrodes. When the electrodes are going into solution, the cell is discharging as the current tends to neutralize the separated charges. Then the cell is used as a voltage source to produce current in a load resistance. During charging the opposite of above occurs. The charging current must be supplied by an external voltage, with the cell then serving just as a load resistance.

Since the secondary cell can be recharged, it is also a storage cell. The most common form of storage cell is the lead-sulphuric acid cell generally used in automobile batteries.

**Solar Cells.** This contains photosensitive silicon cells. The semi-conductor units generate voltage output when exposed to light. Typical output is 0.26 V per cell.

**Internal Resistance.** All batteries have some internal resistance $r_i$. With load current $I_L$, the internal $I_L$, $r_i$ drop reduces the voltage across the output terminals. When $I_L$ makes the terminal voltage drop to one-half the zero load voltage, the external $R_L$ equals the internal $r_i$.

Matching a cell means making the $R_L$ equal to the generator's $r_i$. The result is maximum power delivered to the load from the cell.

A constant voltage source has a very low internal resistance. Output voltage is relatively constant with changing values of load because of the small internal voltage drops.

A constant current source has a very high internal resistance. This determines the constant value of current in the source circuit relatively independent of the load resistance.

## OBJECTIVE TYPE QUESTIONS

1. The output voltage of carbon-zinc dry battery is
   (*a*) 1.1 V (*b*) 1.5 V (*c*) 1.77 V (*d*) 2.2 V.
2. The output of lead acid wet cell is
   (*a*) 1.1 V (*b*) 1.5 V (*c*) 1.77 V (*d*) 2.2 V.
3. Which of the following primary cell has output voltage other than 1.5 V?
   (*a*) carbon-zinc (*b*) zinc-chloride (*c*) silver oxide (*d*) lithium.
4. Which of the following primary cell has the highest voltage?
   (*a*) carbon-zinc (*b*) manganese-alkaline
   (*c*) mercury (*d*) lithium.
5. Which of the following primary cell has the lowest voltage?
   (*a*) mercury (*b*) lithium (*c*) zinc-chloride (*d*) carbon-zinc.
6. Which secondary cell has the lowest output voltage?
   (*a*) lead acid (*b*) silver-cadmium (*c*) silver-zinc (*d*) nickel-iron.
7. Which secondary cell has the highest voltage output?
   (*a*) lead-acid (*b*) nickel-cadmium (*c*) silver-zinc (*d*) silver cadmium.
8. The output voltage of silver-zinc secondary cell is
   (*a*) 1.1 V (*b*) 1.25 V (*c*) 1.86 V (*d*) 2.1 V.
9. The output voltage of silver cadmium secondary cell is
   (*a*) 1.1 V (*b*) 1.36 V (*c*) 7.86 V (*d*) 2.2 V.
10. Which of the following secondary cells are rechargeable type?
    (*a*) nickel-cadmium (*b*) silver-cadmium (*c*) silver-zinc (*d*) all of the above.

11. The active material on the positive plates of a fully-charged lead acid cell is
(*a*) lead peroxide (*b*) lead pentoxide (*c*) red lead (*d*) pure lead.

12. The active material on the negative plates of a fully charged lead acid cell is
(*a*) lead peroxide (*b*) lead pentoxide (*c*) red lead (*d*) pure lead.

13. What will happen if a lead acid battery is kept in discharged condition for a long time?
(*a*) the battery will regain some charge (*b*) the acid and water in electrolyte will dissociate
(*c*) the plates will become sulphated (*d*) the plates will get dissolved in electrolyte.

14. While preparing electrolyte for a lead-acid battery
(*a*) acid is poured into water (*b*) water is poured into acid
(*c*) anyone of the two can be added to the other chemical.

15. A constant-voltage generator has
(*a*) low internal resistance (*b*) high internal resistance
(*c*) minimum efficiency (*d*) minimum current capacity.

16. A lead acid battery discharges at the rate of 8 A for 10 hours. How many coulombs of charge must be put back into the battery to restore the original charge, assuming 100 per cent efficiency?
(*a*) $2.88 \times 10^6$ C (*b*) $2.88 \times 10^5$ C (*c*) $2.88 \times 10^4$ C (*d*) $2.88 \times 10^3$ C.

17. The ampere hour capacity of a lead acid battery depends on
(*a*) density of electrolyte (*b*) size of container
(*c*) number of plates (*d*) all of the above.

18. When electrolyte level in a lead-acid battery is low, the proper remedy is
(*a*) to add some dilute acid (*b*) to add distilled water
(*c*) to shake the battery (*d*) to recharge the battery.

19. The current in a chemical cell is a movement of
(*a*) positive-hole charges (*b*) positive and negative ions
(*c*) positive ions only (*d*) negative ions only.

20. A 6 volt lead-acid battery has an internal resistance of 0.01 Ω.
How much current will flow if the battery has a short circuit?
(*a*) 6 A (*b*) 60 A (*c*) 600 A (*d*) 6000 A.

21. Which of the following battery is used for air-craft?
(*a*) Nickel iron battery (*b*) Lead acid battery
(*c*) Silver oxide battery (*d*) Dry cell battery.

22. Emf of a dry cell is
(*a*) 1 V (*b*) 1.1 V (*c*) 1.5 V (*d*) 2.2 V.

23. Cells are connected in parallel
(*a*) to increase the internal resistance (*b*) to increase the voltage output
(*c*) to increase the efficiency (*d*) to increase the current capacity.

24. Thirty zinc-carbon dry cells are connected in series. The total voltage output will be
(*a*) 1.5 V (*b*) 30 V (*c*) 45 V (*d*) 60 V.

25. Main reason for sedimentation in lead acid batteries is
(*a*) charging at high rate (*b*) discharging at high rate
(*c*) frequent charging and discharging (*d*) extremely slow charging and discharging.

26. Which of the following is used as an electrolyte in a lead acid battery?
(*a*) distilled water (*b*) sulphuric acid
(*c*) hydrochloric acid (*d*) sodium chloride.

27. A battery can supply 110 joules of energy to move 10 coulombs of charge. How much is the voltage of the battery?
(*a*) 1100 V (*b*) 110 V (*c*) 11 V (*d*) 10 V.

28. When cells are connected in series
(*a*) internal resistance decreases (*b*) current capacity increases
(*c*) overall efficiency increases (*d*) voltage output increases.

29. Which of the following is a primary cell?
(*a*) dry cell (*b*) nickel iron cell (*c*) lead acid cell (*d*) none of the above.

30. Which of the following finds application in solar cell?
(*a*) silicon (*b*) lead (*c*) copper (*d*) zinc.

31. Which of the following is a dry storage cell?
(*a*) edison cell (*b*) nickel-cadmium cell
(*c*) mercury cell (*d*) carbon-zinc cell.

32. Which of the following cell has a reversible chemical reaction?
(*a*) carbon-zinc (*b*) lead-acid (*c*) silver oxide (*d*) mercury oxide.

33. Over discharging of lead acid battery may cause
(*a*) sulphation (*b*) sedimentation
(*c*) treeing in the negative plates (*d*) erosion of plate material.

34. Which of the following device is used for the measurement of specific gravity of a lead acid battery?
(*a*) pyrometer (*b*) hydrometer (*c*) lactometer (*d*) electrometer.

35. Which of the following is false?
(*a*) a lead acid cell can be recharged
(*b*) a primary cell has an irreversible chemical reaction
(*c*) a storage cell has a reversible chemical reaction
(*d*) a carbon zinc cell has unlimited shelf life.

36. Which of the following battery is used in automobiles?
(*a*) solar cell (*b*) dry cell battery (*c*) lead-acid battery (*d*) mercury cell.

37. The specific gravity of a fully charged lead-cell is
(*a*) 1.0 (*b*) 1.1 (*c*) 1.22 (*d*) 2.12.
38. During discharge the specific gravity of a lead acid cell should not be allowed to fall below
(*a*) 1.0 (*b*) 1.155 (*c*) 1.55 (*d*) 1.83.
39. Separators are used in lead-acid battery to
(*a*) prevent shorting of plates (*b*) prevent earthing of plates
(*c*) prevent corrosion of plates (*d*) reduce damage to the container.
40. The internal resistance of a lead-acid cell is likely to increase if
(*a*) density of acid goes below 1.15 gm/cc (*b*) gassing of electrolyte occurs
(*c*) water is added to electrolyte (*d*) the battery discharges below 1.8 V.
41. A battery is charged at 5 amperes for 10 hours and discharged at 8 amperes for 8 hours. The ampere hour efficiency of the battery is
(*a*) 80% (*b*) 64% (*c*) 50% (*d*) 40%.
42. The term "trickle charge" is associated with
(*a*) Dry cells (*b*) Silver-zinc batteries
(*c*) Lead acid batteries (*d*) Primary cells only.
43. The state of charge of a battery can be ascertained from
(*a*) colour of battery terminals (*b*) level of electrolyte
(*c*) density of electrolyte (*d*) rate of air bubble formation.
44. The watt-hour efficiency of a lead acid battery is generally around
(*a*) 99% (*b*) 90% (*c*) 75% (*d*) 60%.
45. The ampere hour efficiency of a lead acid battery is generally around
(*a*) 99.9% (*b*) 90–95% (*c*) 80–85% (*d*) 60–70%.
46. In a lead acid battery, the level of electrolyte should be
(*a*) below the level of plates (*b*) upto the level of plates
(*c*) above the level of plates
(*d*) container should be full of electrolyte so that there is no air inside.
47. A generator has a 100 V output on open circuit, which drops to 50 V with a load current of 50 mA and an $R_L$ of 1000 ohms. The internal resistance $r_i$ equals
(*a*) 25 Ω (*b*) 50 Ω (*c*) 100 Ω (*d*) 1000 Ω.
48. The internal resistance of a dry cell is of the order of
(*a*) 20 to 50 ohms (*b*) 2 to 5 ohms (*c*) 0.2 to 0.5 ohms (*d*) 0.02 to 0.05 ohms.
49. Satellite power requirement is provided through
(*a*) lead acid batteries (*b*) dry cells
(*c*) solar cells (*d*) nickel-cadmium cells.

50. Overcharging of lead acid battery is likely to result in
(*a*) sulphation (*b*) sedimentation (*c*) chlorination (*d*) none of the above.

51. Which of the following has lower specific gravity?
(*a*) water (*b*) dilute sulphuric acid
(*c*) concentrated sulphuric acid.

52. The capacity of a lead acid battery is 100 ampere hours at 40°C. If the temperature is lowered to –20°C the capacity will
(*a*) increase sharply (*b*) increase by about 10%
(*c*) remain unaltered (*d*) drop to about 60%.

53. Which electrolyte for lead acid battery will have the lowest freezing point?
(*a*) sp. gr. 1.285 (*b*) sp. gr. 1.20 (*c*) sp. gr. 1.19 (*d*) sp. gr. 1.12.

54. Which Fig. 3.1 represents the charging and discharging cycle of a lead-acid battery?
(*a*) Figure A (*b*) Figure B (*c*) Figure C (*d*) Figure D.

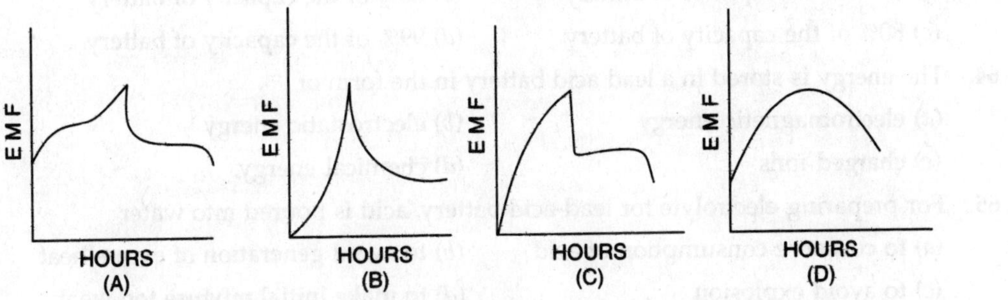


*FIGURE 3.1*

55. "The mass of any substance liberated, deposited or removed at an electrode by a given quantity of electricity is proportional to the chemical equivalent of that substance."
The above statement is associated with
(*a*) Henry's law (*b*) Faraday's law (*c*) Coulomb's law (*d*) Fleming's law.

56. Normal life of an automobile battery is nearly
(*a*) 3 to 6 months (*b*) 6 to 8 months (*c*) 2 to 3 years (*d*) 5 to 10 years.

57. Which among the following constitutes the major load for an automobile battery?
(*a*) Parking lights (*b*) Brake light (*c*) Spark plugs (*d*) Self starter.

58. The container of a lead acid battery must be
(*a*) light (*b*) black in colour (*c*) metallic (*d*) acid proof.

59. The container for lead acid batteries is usually made of
(*a*) plastic (*b*) bakelite
(*c*) moulded hard rubber (*d*) fibre glass.

60. The statement "the mass of the products of an electrolytic process is equal to the quantity of electricity passed through the electrolyte times a factor that varies with the element deposited." The above statement is associated with

(*a*) Coulomb's law (*b*) Gauss's law (*c*) Faraday's law (*d*) none of the above.

61. If the specific gravity of electrolyte in a lead acid battery is 1.20, it can be concluded that

(*a*) the battery has sulphated

(*b*) the battery is damaged and can not be recharged

(*c*) the battery is half charged

(*d*) the battery is fully charged.

62. Under normal charging rate, the charging current should be around

(*a*) 5% of capacity (*b*) 10% of capacity (*c*) 25% of capacity (*d*) 50% of capacity.

63. If a battery is to be charged at a much higher rate as compared to normal charging rate, the charging should be restricted to

(*a*) 50% of the capacity of battery (*b*) 60% of the capacity of battery

(*c*) 80% of the capacity of battery (*d*) 99% of the capacity of battery.

64. The energy is stored in a lead acid battery in the form of

(*a*) electromagnetic energy (*b*) electrostatic energy

(*c*) charged ions (*d*) chemical energy.

65. For preparing electrolyte for lead-acid battery, acid is poured into water

(*a*) to conserve consumption of acid (*b*) to avoid generation of excess heat

(*c*) to avoid explosion (*d*) to make initial mixture too weak.

66. Which of the following can give the best indication regarding the state of charge of a lead-acid battery?

(*a*) specific gravity of electrolyte (*b*) colour of electrolyte

(*c*) open circuit voltage of battery (*d*) weight of the battery.

67. For easy identification the colour of positive plate of a fully charged lead acid battery is marked as

(*a*) white (*b*) black

(*c*) deep chocolate brown (*d*) red.

68. Which of the following factor adversely affects the capacity of a lead acid battery?

(*a*) specific gravity of electrolyte (*b*) temperature of surroundings

(*c*) rate of discharge (*d*) all of the above.

69. When load resistance $R_L$ equals the generator resistance $r_2$, which of the following will be maximum?

(*a*) power in $R_L$ (*b*) current

(*c*) voltage across $R_L$ (*d*) efficiency of the circuit.

70. A generator has an open-circuit emf of 180 V. Its terminal voltage drops to 150 V with an $R_L$ of 5 kΩ. What is the internal resistance?
(*a*) 100 kΩ (*b*) 10 kΩ (*c*) 1 kΩ (*d*) 100 Ω.

71. What is the reasonable value of current that can be continuously drawn from a 120 ampere hour capacity lead-acid battery?
(*a*) 1 A (*b*) 15 A (*c*) 40 A (*d*) 80 A.

72. The common impurity in the electrolyte of a lead acid battery is
(*a*) lead crystals (*b*) chlorine (*c*) iron (*d*) dust particles.

73. Petroleum jelly is applied on the electrical connections to the lead acid battery to
(*a*) prevent short circuiting (*b*) prevent corrosion
(*c*) prevent local heating (*d*) reduce path resistance.

74. When two batteries are to be connected in parallel, it should be ensured that
(*a*) they have identical internal resistance (*b*) they have same make
(*c*) they have same emf (*d*) they have same ampere hour capacity.

75. In a lead acid battery during charging the electrolyte becomes ...... and during discharging it becomes ......
(*a*) stronger ...... weak (*b*) weak ...... stronger
(*c*) stronger ...... stronger (*d*) weak ...... weak.

76. A sulphated battery usually gives
(*a*) low ampere hour capacity (*b*) low voltage
(*c*) low specific gravity (*d*) all of the above.

77. Sulphuric acid is
(*a*) monovalent negative ion (*b*) bivalent negative ion
(*c*) monovalent positive ion (*d*) trivalent positive ion.

78. If a lead acid battery is discharged to below 1.5 V, the damage may be by way of
(*a*) melting of plates (*b*) sulphation
(*c*) discomposition between terminals and plates
(*d*) deposition of sulphates on plates.

79. Which of the following statement is incorrect about lead-acid batteries?
(*a*) the number of plates is always odd
(*b*) the number of positive plates is one less than the number of negative plates
(*c*) the electrolyte is weak sulphuric acid (*d*) none of the above.

80. The typical output of a solar cell is
(*a*) 2.2 V (*b*) 2.0 V (*c*) 1.1 V (*d*) 0.26 V.

## ANSWERS

| | | | | | | |
|---|---|---|---|---|---|---|
| 1. (*b*) | 2. (*d*) | 3. (*d*) | 4. (*d*) | 5. (*a*) | 6. (*b*) | 7. (*a*) |
| 8. (*c*) | 9. (*a*) | 10. (*d*) | 11. (*a*) | 12. (*d*) | 13. (*c*) | 14. (*a*) |

15. (*a*) 16. (*b*) 17. (*c*) 18. (*b*) 19. (*b*) 20. (*c*) 21. (*a*)
22. (*c*) 23. (*d*) 24. (*c*) 25. (*a*) 26. (*b*) 27. (*c*) 28. (*d*)
29. (*a*) 30. (*a*) 31. (*b*) 32. (*b*) 33. (*a*) 34. (*b*) 35. (*d*)
36. (*c*) 37. (*c*) 38. (*b*) 39. (*b*) 40. (*d*) 41. (*b*) 42. (*c*)
43. (*c*) 44. (*c*) 45. (*b*) 46. (*c*) 47. (*d*) 48. (*c*) 49. (*c*)
50. (*b*) 51. (*a*) 52. (*d*) 53. (*a*) 54. (*a*) 55. (*b*) 56. (*c*)
57. (*d*) 58. (*d*) 59. (*c*) 60. (*c*) 61. (*c*) 62. (*b*) 63. (*c*)
64. (*d*) 65. (*b*) 66. (*a*) 67. (*c*) 68. (*d*) 69. (*a*) 70. (*c*)
71. (*c*) 72. (*c*) 73. (*b*) 74. (*c*) 75. (*a*) 76. (*d*) 77. (*b*)
78. (*b*) 79. (*d*) 80. (*d*).

Chapter **4** 

# DC GENERATOR

**A direct current generator is a machine that converts mechanical energy into electrical energy in the form of direct current. The generator works on the principle that when mangetic flux is cut by a conductor or a number of conductors, emf is induced in the conductors and the mechanical power supplied by the prime mover is converted into electrical power.**

**In a DC generator the magnetic field is stationary and the conductors move. The magnetic flux is cut and the voltage is induced in the copper conductors.**

**EMF Equation. The emf induced in the generator is given by**

$$E = \frac{\phi Z N}{60} \times \frac{P}{A} \text{ volts}$$

**where $\phi$ = flux reaching the armature from each path**

**$Z$ = total number of conductors in the armature**

**$N$ = armature speed in r.p.m.**

**$A$ = number of parallel paths through the winding**

**$A$ = 2 for ware-winding**

**$A$ = $P$ for lap-winding**

## Main Parts of DC Generator

**Field Magnets. Consist of a circular yoke with inwardly projecting poles with alternate *N-S* polarity.**

**Yoke. The yoke of a large machine is fabricated from steel. Small machines may have cast iron yoke.**

**Poles.** The poles are generally, laminated and made of sheet steel. Laminated poles are secured to the yoke by a bolt or bolts screwed into a bar passing through the laminations.

When flux entering the armature if $\phi$, the leakage factor $\lambda$ then the total flux through the pole core is $\lambda\phi$. The fluxes carried by various parts of the circuit are

| | |
|---|---|
| Yoke | $\frac{1}{2}\lambda\phi$ |
| Pole | $\lambda\phi$ |
| Air gap | $\phi$ |
| Teeth under one pole | $\phi$ |
| Armature core | $\frac{1}{2}\phi$ |

**Armature.** The armature consists of core and the winding. The core obviously has to be magnetic in order to provide a path of low reluctance to the lines of force. Iron being good conductor of electricity, the rotation of the solid core in a strong magnetic field would result in very heavy eddy currents being set up. To minimise eddy currents the core is made of thin sheet laminations insulated from one another by varnish, thin paper, or even the oxide coating.

The armature also curries winding. The winding may be lap winding or wave winding. The resultant pitch, $y$ is given by

$$y = y_B - y_F \text{ for lap winding}$$
$$y = y_B + y_F \text{ for wave-winding}$$

where $y_B$ = back pitch

$y_F$ = from pitch.

For generating large currents lap winding is used in which coil connections are made from one segment, through the two sides of a coil and hence back to the next adjacent segment, from which similar connection is made to the next coil. Small generators use wave winding. Wave windings do not generate circulatory currents (as in lap winding) due to unbalancing of magnetic reluctances or flux densities.

**Commutator.** The function of commutator is to facilitate the collection of current from the armature. It is a cylindrical structure built up of segments of high-conductivity hard drawn copper, insulated from one another by copper. The segments have $V$-grooves so that they cannot fly out under the action of centrifugal force, and the insulation in these grooves is in the form of conical mecanite rings.

**Brushes.** The brushes, which are almost invariably of some form of carbon, are housed in brush holders. There are several types of holders, the box type probably being the best. The brush is pressed on the commutator by a spring whose tension can be adjusted by placing the small lever in one of the notches. At the top of the brush a flexible copper "pigtail" is mounted, and this conveys the current collected by the brush to the holders. There may be several brushes per spindle, according to the magnitude of the current to be collected.

**Armature Reaction.** Armature reaction is the effect of magnetic field, set up by the armature current, on the distribution of flux under main poles of a generator. Two effects of this magnetic field are (*i*) weakening of the main flux.

(*ii*) cross-magnetisation or distortion of magnetic flux.

**Demagnetising AT per pole** are given by

$$\frac{\theta_m}{360} \times ZI$$

where $\theta_m$ = forward lead angle in mechanical or geometrical or angular degrees

$Z$ = total number of armature conductors

$I$ = current in each armature conductor

$= \frac{I_a}{Z}$ for wave winding

$= \frac{I_a}{P}$ for lap winding

**Cross-magnetising AT per pole** are given by

$$ZI\left[\frac{1}{2P} - \frac{\theta_m}{360}\right]$$

Whenever lead angle is given in electrical degrees, it should be coverted into mechanical degrees by the relation

$$\theta\ (\text{mechanical}) = \frac{\theta\ (\text{electrical})}{\text{pair of poles}}$$

or $$\theta_m = \frac{\theta_e}{P/2} = \frac{2\theta_e}{P}$$

**Compensating Windings.** These are used for large direct current machines subjected to large fluctuations in load. The purpose of providing these windings is to neutralize the cross-magnetising effect ot armature reaction. These windings are embedded in slots in the pole-shoes and the connected in series with armature in such a way that the current in them flows in opposite direction to that flowing in armature conductors directly below the pole shoes.

The number of armature ampere turns per pole for compensating winding.

$$= 0.7 \times \frac{ZI}{2P}$$

$$= 0.7 \times \text{armature amp-turns/pole.}$$

**Inter Poles.** These are small poles located between the main poles. These are would with comparatively few heavy gauge copper wire turns and are connected in series with the armature so that they carry full armature current. Their polarity is the same as that of the pole ahead in the direction of rotation.

The purpose of providing interpoles is two fold:

(*i*) These poles help in sparkless commutation even on overloads. These poles neutralize reactance voltage which is also due to armature current.

(*ii*) These poles neutralize the cross magnetizing effect of armature reaction at all loads.

**Equalizer Rings.** The function of equalizer rings is to avoid unequal distribution of current at the brushes thereby helping to get sparkless commutation.

**Different Kinds of Generators.** Based on method of excitation the generators may be classified as

(*i*) Separately excited generators

(*ii*) Self-excited generators.

In separately excited generators the field magnet excitation is supplied from an external source such as a battery or a separate small generator called an exciter.

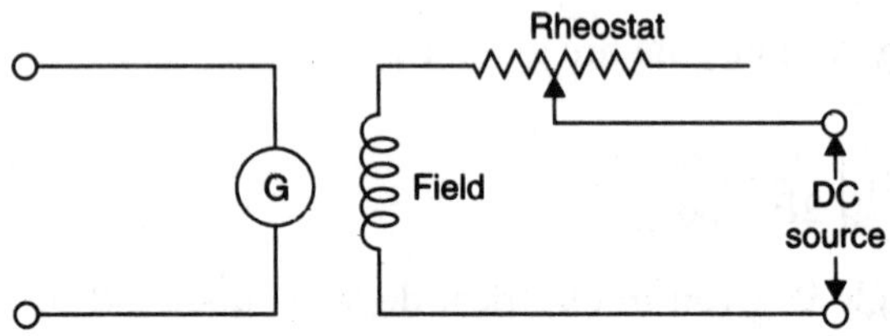


*FIGURE 4.1*

A self-excited generator supplies its own field current to its field magnets. Thus an external source is not needed. There are three type of self-excited generators:

(*i*) **Series Wound Generators.** In this the field coils and the armature coils are in series, hence the same current flows through them as through the external load.

(*ii*) **Shunt Wound Generators.** In this case the field coils are in parallel with the armature coils.

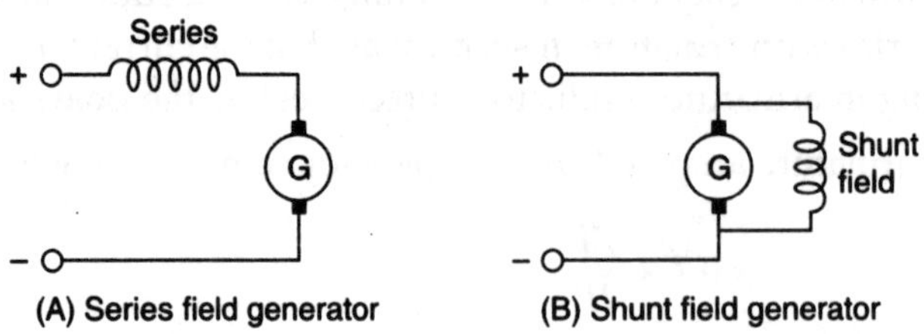


*FIGURE 4.2*

(*iii*) **Compound Wound Generators.** In this the excitation is partly supplied by series field and partly by shunt field. When the shunt field coils are connected across the armature circuit only, it is called short shunt and with the inclusion of the series coils it is called long shunt. However, in this case major portion of the excitation is usually supplied by the shunt field.

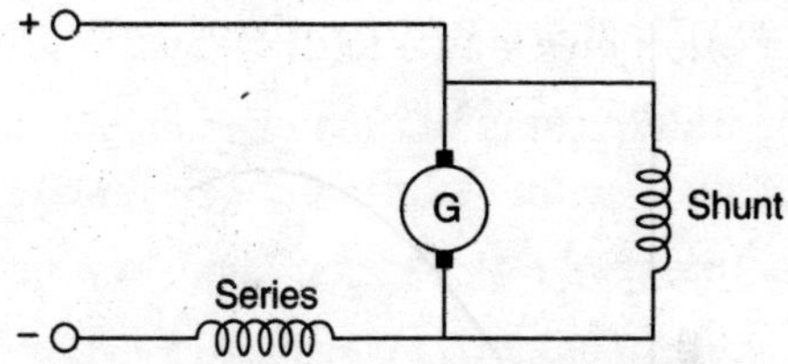

(A) Compound short shunt generator

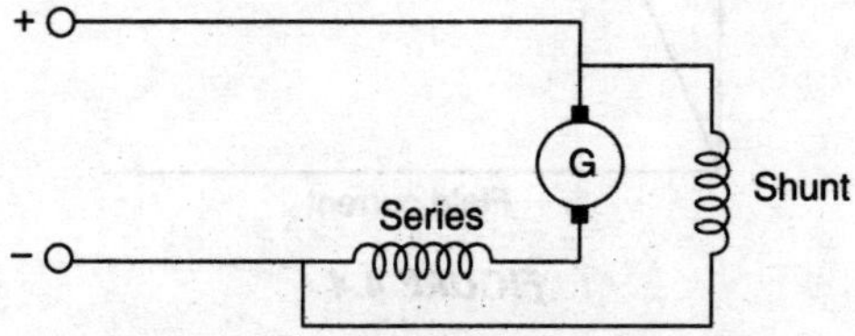

(B) Compound long shunt generator

*FIGURE 4.3*

The compound generators are also called as cumulative and differential. In cumulative compound generators the magnetizing forces of the shunt and the series field ampere-turns are cumulative, *i.e.*, they both tend to set up flux in the air gap in the same direction. However, in case the ampere turns of the shunt winding oppose those of the series winding, the machine is said to be differentially compound-wound generator.

## Characteristics of DC Generators

The following characteristics are generally studied:

(*i*) **Magnetization Characteristic.** This gives the relation between the emf induced in the armature at no-load and the exciting current *i.e.*, field current. It is also called the no load saturation characteristic.

(*ii*) **External and Internal Characteristics.** The variation of terminal voltage and the load current gives external characteristic and the graph between emf generated and the armature current $I_a$ is known as internal characteristic.

(*i*) **Separately Excited Generator.** For a generator

$$E = \frac{\phi ZN}{60}\left(\frac{P}{A}\right) \text{ volts}$$

In case the speed is constant,

$$E = K\phi$$

When $I_f$, the field current increases, the voltage generated $E$ also increases directly while the poles are unsaturated. After saturation the curve deviates from linear relationship as shown in Fig. 4.4.

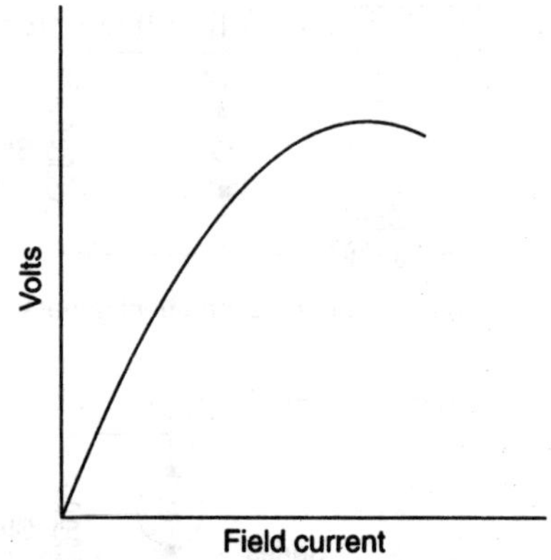


*FIGURE 4.4*

(*ii*) **Load Saturation Curve.** A curve showing relation between the terminal voltage *V* and field current $I_f$ when the generator is loaded, is called load saturation curve. A load saturation curve can be deduced from no-load saturation curve when armature reaction and armature resistance is known. The effect of armature reaction drop and armature drop is shown in Fig. 4.5.

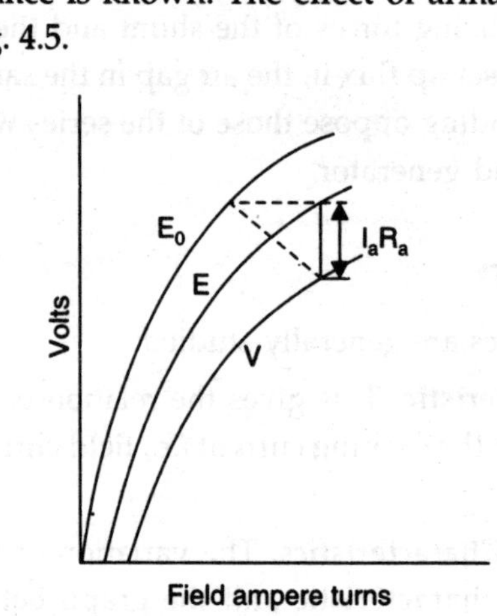


*FIGURE 4.5*

In Fig. 4.6 *OB* represents internal characteristic while *OC* represents external characteristic.

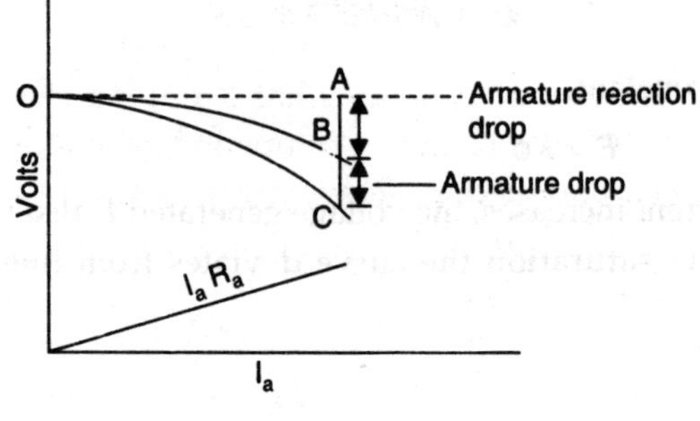


*FIGURE 4.6*

**Shunt Generator.** A shunt generator gives full voltage on noload and the voltage drops as the load on the generator increases. Thus voltage on load is given by

$$V = E - I_a R_a.$$

where $I_a R_a$ is armature drop. The characteristics of a shunt generator are shown in Fig. 4.7.

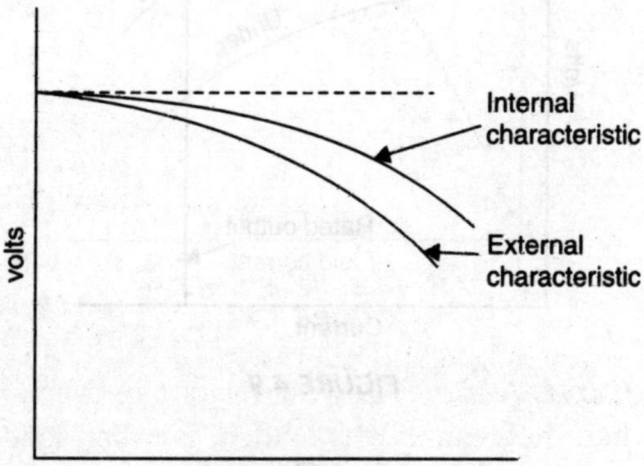


*FIGURE 4.7*

**Series Generator.** In these generators as field windings are in series with the armature, they carry full armature current $I_a$. As $I_a$ is increased, flux and hence generated emf is also increased as shown in Fig. 4.8. *OCC* and internal characteristic curves are shown in Fig. 4.8. This generator has rising voltage characteristic. At higher loads voltage starts decreasing due to excessive demagnetizing effects of armature reaction.

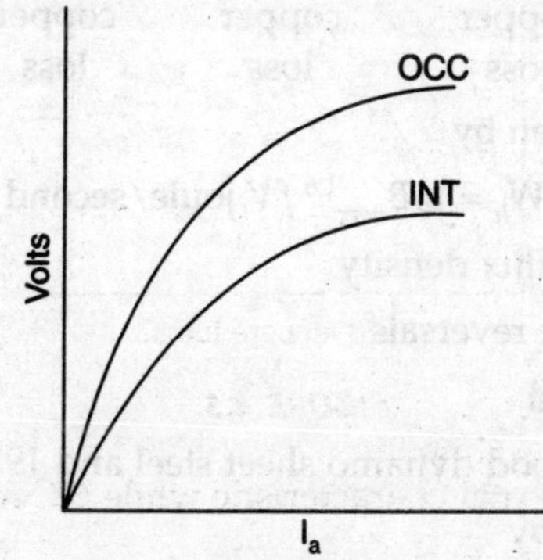


*FIGURE 4.8*

**Compound Generator.** A shunt generator has drooping characteristics while a series generator has rising characteristic. Accordingly a compound generator may be flat compounded, level compounded or under compounded, as shown in Fig. 4.9.

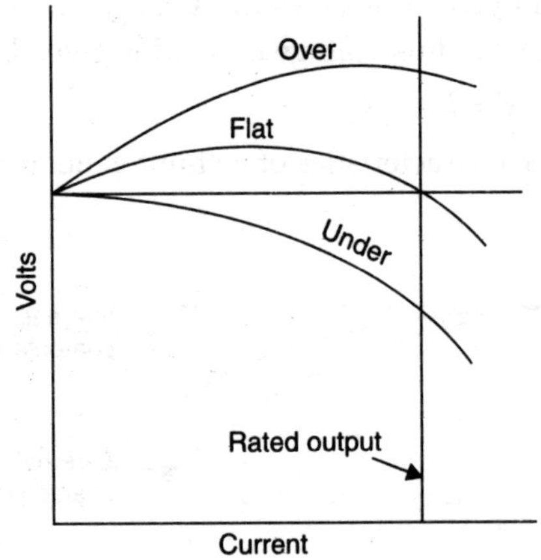


*FIGURE 4.9*

**Losses in a Generator**

Total losses in generator

- Iron losses
  - Hysteresis
  - Eddy current
- Copper losses
  - Armature copper loss
  - Shunt copper loss
  - Series copper loss
- Mech. losses
  - Wind-age
  - Friction

**Hysteresis Loss.** ($W_h$) is given by

$$W_h = \eta\, B_{max}^{1.6} f V \text{ joule/second}$$

where $B_{max}$ = maximum value of flux density

$f$ = number of magnetic reversals

$V$ = volume of iron used

Value of $\eta$ is 502 J/m$^3$ for good dynamo sheet steel and 191 J/m$^3$ for silicon steel.

**Eddy Current** loss is given by

$$W_e = K\,(B_{max})^2 f^2 t^2\, V \text{ watt}$$

where $B_{max}$ = maximum flux density

$f$ = frequency of magnetic reversals

$t$ = thickness of laminations

$V$ = volume of armature core

**Copper Losses.** Armature copper loss = $I_a^2 R_a$

where $R_a$ = resistance of armature and interpoles and series field windings.

Field copper loss = $I_{sh}^2 R_{sh}$

**Stray Losses.** Usually magnetic and mechanical losses are collectively known as Stray Losses.

**Condition for Maximum Efficiency.** For maximum efficiency.

Variable loss = Constant loss.

The variation of efficiency with load is shown in Fig. 4.10.

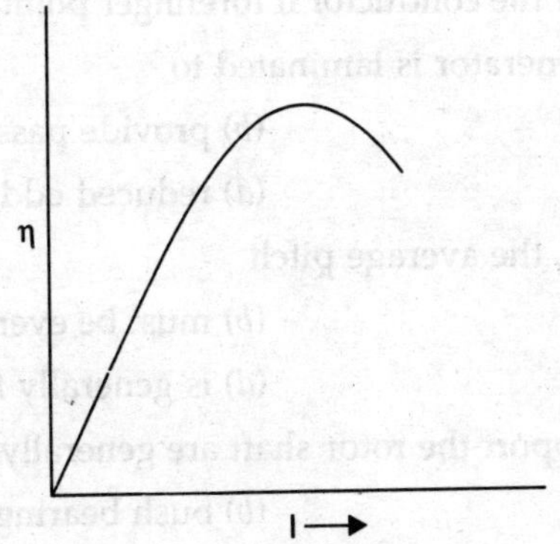


*FIGURE 4.10*

## OBJECTIVE TYPE QUESTIONS

1. The resistance of armature winding depends on
   (*a*) number of conductors (*b*) length of conductor
   (*c*) cross-sectional area of the conductor (*d*) all of the above.
2. The commutator segments are connected to the armature conductors by means of
   (*a*) brazing (*b*) copper lugs (*c*) resistance wires (*d*) insulation pads.
3. In case of a four pole DC generator provided with a two layer lap winding with sixteen coils, the pole pitch will be
   (*a*) 4 (*b*) 8 (*c*) 16 (*d*) 32.
4. In case of a wave wound generator, the average pitch
   (*a*) must be odd (*b*) must be even
   (*c*) may be odd or even (*d*) is generally fractional.
5. Which of the following could be approximately the thickness of laminations of a DC machine ?
   (*a*) 0.005 mm (*b*) 0.05 mm (*c*) 0.5 mm (*d*) 5 mm.
6. Fleming's right hand rule regarding direction of induced emf, correlates
   (*a*) magnetic flux, direction of motion and the direction of emf induced
   (*b*) magnetic flux, direction of current flow and resultant force
   (*c*) magnetic flux, direction of force and direction of motion of conductor
   (*d*) magnetic field strength, induced voltage and current.

7. While applying Fleming's right hand rule to find the direction of induced emf, the thumb points towards
   (*a*) direction of flux
   (*b*) direction of induced emf
   (*c*) direction of motion of the conductor if forefinger points along the lines of flux
   (*d*) direction of motion of the conductor if forefinger points in the direction of induced emf.
8. The armature of a DC generator is laminated to
   (*a*) reduce the bulk
   (*b*) provide passage for cooling air
   (*c*) insulate the core
   (*d*) reduced eddy current losses.
9. In case of wave winding, the average pitch
   (*a*) must be odd
   (*b*) must be even
   (*c*) could be odd or even
   (*d*) is generally fractional.
10. The bearings used to support the rotor shaft are generally
    (*a*) magnetic bearings
    (*b*) bush bearings
    (*c*) needle bearings
    (*d*) ball bearings.
11. The field coils of a DC generator are usually made of
    (*a*) carbon (*b*) cast iron (*c*) copper (*d*) mica.
12. In a commutator
    (*a*) mica is harder than copper
    (*b*) copper is harder than mica
    (*c*) mica and copper are equally hard.
13. In DC generators, the cause of rapid brush wear may be
    (*a*) rough commutator surface
    (*b*) imperfect constant
    (*c*) severe sparking
    (*d*) any of the above
14. For generating large currents on DC generators, which winding is generally preferred?
    (*a*) lap winding
    (*b*) progressive wave winding
    (*c*) retrogressive wave winding
    (*d*) current depends on design.
15. In a shunt generator the voltage build up generally restricted by
    (*a*) speed limitation
    (*b*) armature heating
    (*c*) insulation restrictions
    (*d*) saturation of iron.
16. If a DC generator fails to build up the probable cause could not be
    (*a*) imperfect brush contact
    (*b*) field resistance less than the critical resistance
    (*c*) no residual magnetism in the generator
    (*d*) faulty shunt connections tending to reduce the residual magnetism.
17. In DC generators the pole shoes are fastened to the pole core by
    (*a*) welding (*b*) brazing (*c*) rivets (*d*) counter sunk screws.
18. According to Flemings right hand rule for finding the direction of induced emf, when middle finger points in the direction of induced emf, forefinger will point in the direction of

(*a*) lines of force (*b*) motion of conductor
(*c*) either of the above (*d*) none of the above.

19. In lap winding, the number of brushes is always
(*a*) 2 (*b*) same as the number of poles
(*c*) double the number of poles (*d*) half the number of poles.

20. The purpose of providing dummy coils in a generator is
(*a*) to reduce eddy current losses (*b*) to enhance flux density
(*c*) to amplify voltage (*d*) to provide mechanical balance for the rotor.

21. If $B$ is the flux density, $l$ the length of conductor and $v$ the velocity of conductor, then the induced emf is given by
(*a*) $B.l.v$ (*b*) $B.l.v^2$ (*c*) $B.l^2.v$ (*d*) $B.l^2.v^2$.

22. Core laminations are generally made of
(*a*) cast iron (*b*) carbon (*c*) silicon steel (*d*) stainless steel.

23. In a DC generator if $P$ be the number of poles and $N$ be the rpm of rotor, then the frequency of magnetic reversals will be
(*a*) $\frac{NP}{2}$ (*b*) $\frac{NP}{60}$ (*c*) $\frac{NP}{120}$ (*d*) $\frac{NP}{3000}$.

24. For a DC generator when the number of poles and the number of armature conductors is fixed, then which winding will give the higher emf?
(*a*) lap winding (*b*) wave winding
(*c*) either of (*a*) and (*b*) above (*d*) depends on other features of design.

25. The material for commutator brushes is generally
(*a*) copper (*b*) mica (*c*) carbon (*d*) cast iron.

26. The insulating material used between the commutator segments is normally
(*a*) mica (*b*) graphite (*c*) insulating vanish (*d*) paper.

27. A DC generator is provided with wave winding and $P$ as the number of poles. If $a$ is the degree of multiplicity, then the number of parallel paths will be
(*a*) 2 (*b*) $2a$ (*c*) $aP$ (*d*) $2aP$.

28. A 4 pole DC generator rotates at 1500 rpm. How many times in a second, will the emf induced in each armature conductor change its direction?
(*a*) 50 (*b*) 100 (*c*) 200 (*d*) 400.

**Questions 29 to 31 Refer to Data Given Below:**

A 4 pole generator has 288 surface conductors. The armature is lap wound, and the armature current is 120 amperes. The brushes are advanced 15 space degrees

29. The number of demagnetizing conductors will be
(*a*) 24 (*b*) 48 (*c*) 96 (*d*) 192.

**30.** The number of demagnetizing ampere conductors will be

(*a*) 720 (*b*) 1440 (*c*) 2160 (*d*) 2880.

**31.** The number of cross magnetizing ampere turns will be

(*a*) 720 (*b*) 1440 (*c*) 2880 (*d*) none of the above.

**32.** In a duplex winding for a 4 pole machine, the number of parallel paths will be

(*a*) 2 (*b*) 4 (*c*) 8 (*d*) 32.

**33.** Flashing the field of a DC generator means

(*a*) neutralising residual magnetism (*b*) creating residual magnetism by a DC source

(*c*) making the magnetic losses of force parallel

(*d*) increasing flux density by adding extra turns of windings on poles.

**34.** The emf induced in the armature of a shunt generator is 600 volts. The armature resistance is 0.1 ohm. If the armature current is 200 amperes, the terminal voltage will be

(*a*) 640 V (*b*) 620 V (*c*) 600 V (*d*) 580 V.

**35.** The characteristics of a DC generator are as shown in Fig. 4.11. The ordinate and abscissae of the graph will be respectively

(*a*) terminal voltage and load current (*b*) armature resistance and voltage drop

(*c*) terminal voltage and current (*d*) ampere turns and volts.

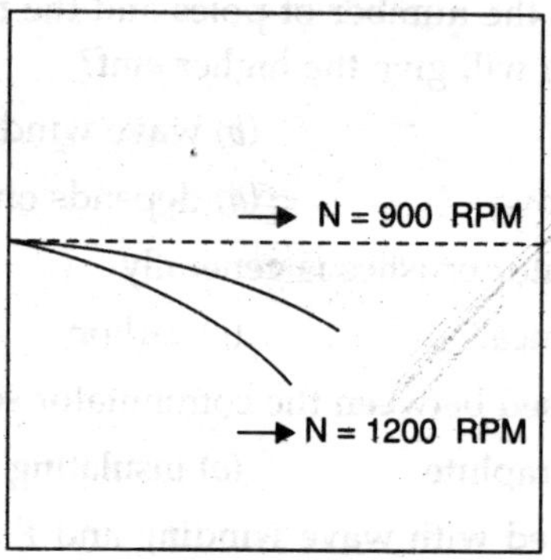


*FIGURE 4.11*

## Questions 36 to 38 Refer to the Data Given Below:

A 20 kW 220 volt shunt generator has an armature resistance of 0.07 ohm and a shunt field resistance of 900 ohms.

**36.** The field loss of the generator is

(*a*) 242 W (*b*) 388 W (*c*) 592 W (*d*) 1080 W.

**37.** The armature loss of the generator is

(*a*) 121 W (*b*) 333 W (*c*) 488 W (*d*) 592 W.

38. Power developed in the armature will be
(*a*) 18.83 kW (*b*) 19.97 kW (*c*) 20 kW (*d*) 20.83 kW.

39. The losses occurring in a DC generator are given below. Which loss in likely to have the least proportion?
(*a*) magnetic losses (*b*) armature copper losses
(*c*) mechanical losses (*d*) field copper losses.

40. The hysteresis loss in a DC mechine least depends on
(*a*) frequency of magnetic reversals (*b*) maximum value of flux density
(*c*) volume and grade of iron (*d*) rate of flow of ventilating air.

41. In a DC generator all of the following could be the effects of iron losses except
(*a*) loss of efficiency (*b*) excessive heating of core
(*c*) increase in terminal voltage (*d*) rise in temperature of ventilating air.

42. The losses occurring in a DC generator are given below. Which loss is likely to have highest proportion at rated load of the generator?
(*a*) hysteresis loss (*b*) field copper loss
(*c*) armature copper loss (*d*) eddy current loss.

43. If '*t*' be the thickness of the laminations, then eddy current loss in a DC generator will vary as
(*a*) $\frac{1}{t}$ (*b*) $\frac{1}{t^2}$ (*c*) $t$ (*d*) $t^2$.

44. Which of the following loss in a DC generator varies significantly with the load current?
(*a*) Field copper loss (*b*) Windage loss
(*c*) Armature copper loss (*d*) None of the above.

45. Which of the following loss in a DC generator will increase rapidly as compared to the remaining losses, when the speed of generator is increased?
(*a*) Friction loss (*b*) Copper loss (*c*) Hysteresis loss (*d*) Eddy current loss.

46. For a properly designed DC generator, the overall efficiency could be of the order of
(*a*) 95% (*b*) 80% (*c*) 60% (*d*) 50%.

47. If $B_{max}$ is the maximum flux density, then eddy current loss will very as
(*a*) $B_{max}$ (*b*) $(B_{max})^2$ (*c*) $(B_{max})^{1.6}$ (*d*) $(B_{max})^{2.4}$

48. The hysteresis loss in a DC generator varies with the frequency of magnetic reversals as
(*a*) $\frac{1}{f}$ (*b*) $f$ (*c*) $f^{1.6}$ (*d*) $f^2$.

49. Which of the following step is likely to result in reduction of hysteresis losses in a DC generator?
(*a*) providing laminations in armature core
(*b*) providing laminations in stator

(*c*) using non-magnetic material for frame.

(*d*) using material of low hysteresis co-efficient for armature core material.

50. The voltage drop for which of the following type of brush can be expected to be the least?

(*a*) carbon brushes (*b*) graphite brushes

(*c*) metal graphite brushes.

**Questions 51 and 52 Refer to Data Given Below:**

A 80 kW DC shunt generator, running at 1000 rpm delivers its full load output at 118 volts. Its losses, expressed as the percentage of full load output, are as under:

| | |
|---|---|
| Core losses | 1.1% |
| Friction and windage losses | 1.3% |
| Total copper losses at full load | 5.7% |
| Shunt field copper losses | 0.25% |

51. The stray losses in watts will be

(*a*) 860 W (*b*) 1920 W (*c*) 2240 W (*d*) 3680 W.

52. Total of variable losses at full load will be

(*a*) 1000 W (*b*) 1920 W (*c*) 3320 W (*d*) 4360 W.

53. Which of the following loss will be least as compared to the remaining?

(*a*) stray losses (*b*) variable losses at full load

(*c*) mechanical losses (*d*) constant losses.

54. Full load efficiency of the generator will be

(*a*) 68.5% (*b*) 73.5% (*c*) 80.8% (*d*) 92.51%.

55. Which loss in a DC generator varies with load?

(*a*) copper loss (*b*) eddy current loss (*c*) hysteresis loss (*d*) windage loss.

56. Which loss in a DC generator does not vary with load as well as flux density?

(*a*) copper loss (*b*) eddy current loss (*c*) hysteresis loss (*d*) windage loss.

57. Which of the following loss in a DC generator is dissipated in the form of heat?

(*a*) mechanical loss (*b*) core loss (*c*) copper loss (*d*) all of the above.

58. Two DC shunt generators, each with armature resistance of 0.02 Ω and field resistance of 50 Ω run in parallel and supply a total current of 1000 amperes to the load circuit. If their emf's are 270 and 265 volts their bus bar voltage will be

(*a*) 270 V (*b*) 267.5 V (*c*) 265 V (*d*) 257.4 V.

59. The net output of the two machines respectively will be

(*a*) 100 kW, 120 kW (*b*) 160.7 kW, 96.7 kW

(*c*) 140.5 kW, 120.3 kW (*d*) 133 kW, 103 kW.

**60.** The emf generated by a shunt wound DC generator is $E$. Now while pole flux remains constant if the speed of the generator is doubled, the emf generated will be

(*a*) $E$ (*b*) $\frac{E}{2}$ (*c*) $2E$ (*d*) slightly less than $E$.

**61.** The essential condition for parallel operation of two DC generators is that they must have

(*a*) same kW rating (*b*) same operating rpm

(*c*) same drooping voltage characteristics (*d*) same percentage regulation.

**62.** In a DC generator the actual flux distribution depends upon

(*a*) shape of the pole shoe (*b*) size of air gap

(*c*) clearance between the tips of the adjacent pole shoes

(*d*) all of the above.

**63.** A DC generator is driven at 300 rpm and the external diameter of the armature is 180 cms while core length is 35 cms. The maximum flux density in the air gap is 10,000 gauss. The maximum value of emf induced in the armature conductors will be

(*a*) 1 V (*b*) 1.9 V (*c*) 9.9 V (*d*) 99 V.

**64.** Two DC generators *A* and *B* have 6 poles each. Generator *A* has wave wound armature while generator *B* has lap wound armature. The ratio of the included emf in generators *A* and *B* will be

(*a*) 1 : 3 (*b*) 2 : 3 (*c*) 3 : 1 (*d*) 3 : 2.

**65.** Which of the following losses are significantly reduced by laminating the core of a DC generator?

(*a*) hysteresis lesses (*b*) eddy current losses

(*c*) copper losses (*d*) windage losses.

**66.** Which curve in Fig. 4.12 represents the terminal voltage (*V*), load current (*I*) characteristics of a shunt generator?

(*a*) curve *A* (*b*) curve *B* (*c*) curve *C* (*d*) curve *D*.

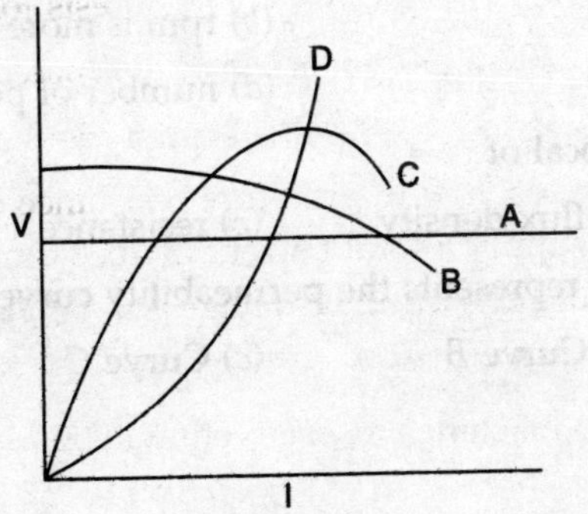


*FIGURE 4.12*

67. In DC generators, lap winding is used for
(*a*) low voltage, low current (*b*) high voltage, hight current
(*c*) low voltage, high current (*d*) high voltage, low current.

68. Which of the following rule can be used to determine the direction of emf induced in the armature conductors of a DC generator?
(*a*) cork screw rule (*b*) flemings left hand rule
(*c*) flemings right hand rule (*d*) none of the above.

69. Which of the following component of a DC generator plays vital role for providing direct current of a DC generator?
(*a*) equaliser rings (*b*) dummy coils (*c*) commutator (*d*) eyebolt.

70. Which of the following generators are preferred for parallel operation?
(*a*) series generators (*b*) shunt generators
(*c*) compound generators (*d*) all of the above.

71. In a DC generator the ripples in the direct emf generated are reduced by
(*a*) using equaliser rings (*b*) using conductor of annealed copper
(*c*) using commutator with large number of segments
(*d*) using carbon brushes of superior quality.

72. The armature core of a DC generator is usually made of
(*a*) cast iron (*b*) silicon steel
(*c*) copper (*d*) non-ferrous material.

73. DC generator generally preferred for charging automobile batteries is
(*a*) shunt generator (*b*) series generator
(*c*) long shunt compound generator (*d*) any of the above.

74. In a DC generator the number of mechanical degrees and electrical degrees will be the same when
(*a*) rpm is less than 300 (*b*) rpm is more than 300
(*c*) number of poles is 2 (*d*) number of poles is 4.

75. Permeance is the reciprocal of
(*a*) reluctance (*b*) flux density (*c*) resistance (*d*) ampere-turns.

76. In Fig. 4.13 which curve represents the permeability curve for iron?
(*a*) Curve *A* (*b*) Curve *B* (*c*) Curve *C* (*d*) Curve *D*.

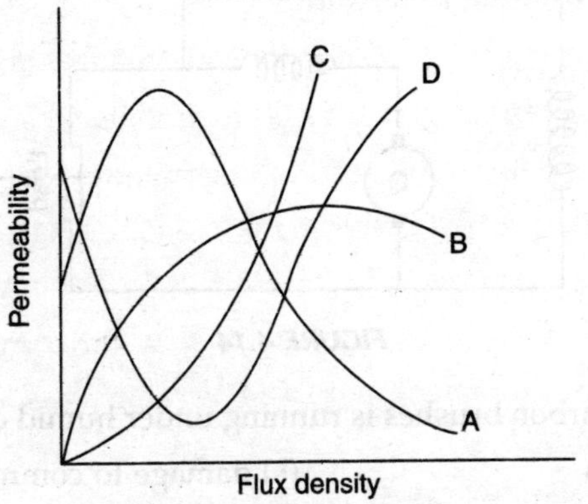


*FIGURE 4.13*

**77.** **In case of DC generators, the armature voltage control is considered as suitable, if the machine is driven**

(*a*) at constant load (*b*) at constant current

(*c*) at constant torque (*d*) at constant *VA*.

**78.** **The total losses in a well designed DC generator of 10 kW will be nearly**

(*a*) 100 W (*b*) 500 W (*c*) 1000 W (*d*) 1500 W.

**79.** **The yoke of a DC generator is usually made of**

(*a*) cast iron (*b*) copper (*c*) stainless steel (*d*) silicon steel.

**80.** **Compensating winding are used in DC generators**

(*a*) primarily to neutralize the magnetizing action of the armature currents

(*b*) mainly to reduce the eddy currents by providing local short circuits

(*c*) to provide path for the circulation of cooling air

(*d*) none of the above.

**81.** **In DC generators the polarity of interpoles**

(*a*) is the same as that of the main pole ahead

(*b*) is the same as that of the immediately preceding pole

(*c*) is opposite to that of the main pole ahead

(*d*) is neutral as these poles do not play part in generating emf.

**82.** **The emf generated in a DC generator is directly proportional to**

(*a*) number of poles (*b*) flux per pole

(*c*) speed of armature (*d*) all of the above.

**83.** **Fig. 4.14 represents a**

(*a*) series generator (*b*) shunt generator

(*c*) short shunt compound generator (*d*) long shunt compound generator.

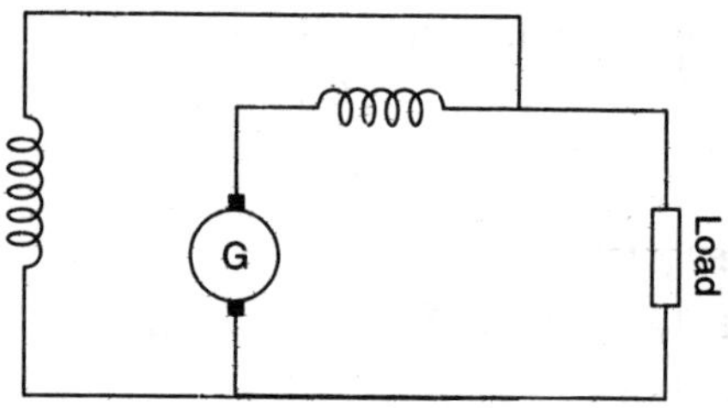


*FIGURE 4.14*

**84.** A DC generator with carbon brushes is running under humid conditions. This will result in

(*a*) sparking at brushes (*b*) damage to commutator segments

(*c*) more brush wear at positive terminal (*d*) more brush wear at negative terminal.

**85.** In a DC generator the magnetic neutral axis coincides with the geometrical neutral axis, when

(*a*) there is no load on the generator (*b*) the generator runs on full load

(*c*) the generator runs on over load (*d*) the generator runs on designed speed.

**86.** In a DC generator the critical resistance refers to the resistance of

(*a*) brushes (*b*) field (*c*) armature (*d*) load.

**87.** In a DC generator, the armature reaction results in

(*a*) magnetisation of interpoles (*b*) demagnetisation of the centres of poles

(*c*) demagnetisation of the leading pole tip and magnetisation of the trailing pole tip.

(*d*) magnetisation of the leading pole tip and demagnetisation of the trailing pole tip.

**88.** In a DC generator in case the brushes are moved so as to bring them in magnetic neutral axis, then there will be

(*a*) demagnetisation (*b*) cross magnetisation

(*c*) cross magnetisation as well as demagnetisation

(*d*) none of the above.

**89.** When to DC generators are running in parallel an equalizer bar is used.

(*a*) to increase the series flux (*b*) to increase the generated emf

(*c*) to reduce the combined effect of armature reaction of both the machines.

(*d*) since the two identical machines will pass approximately equal currents to the load.

**90.** In a DC generator in order to reduce sparking at brushes, the self induced emf in the coil is neutralized by all of the following EXCEPT.

(*a*) interpoles (*b*) dummy coils

(*c*) compensating winding (*d*) shifting of axis of brushes.

**91.** In a DC generator in case the resistance of the field winding is increased, then the output voltage will.

(*a*) increase (*b*) decrease

(*c*) remain unaffected (*d*) fluctuate heavily.

**92.** The condition of maximum efficiency for a DC generator is

(*a*) eddy current losses = stray losses (*b*) hysteresis losses = eddy current losses

(*c*) copper losses = 0 (*d*) variable losses = constant losses.

**93.** An excitor for a turbo generator is a

(*a*) separately excited generator (*b*) shunt generator

(*c*) series generator (*d*) compound generator.

**94.** Which of the following helps in reducing the effect of armature reaction in a DC generator?

(*a*) Interpoles (*b*) Compensating winding

(*c*) Both interpoles as well as compensating winding

(*d*) Interpoles, compensating winding as well as dummy coils.

**95.** With a DC generator which of the following regulation is preferred?

(*a*) 100% regulation (*b*) infinite regulation

(*c*) 50% regulation (*d*) 1% regulation.

**96.** Which generator would you prefer for feeding long DC transmission lines?

(*a*) series generator (*b*) shunt generator

(*c*) overcompound generator (*d*) flat compound generator.

**97.** DC generators are normally designed for maximum efficiency around

(*a*) full load (*b*) rated rpm (*c*) rated voltage (*d*) all of the above.

**98.** In case of a flat compounded generator

(*a*) voltage generated is less than the rated voltage

(*b*) generated voltage is proportional to the load on the generator

(*c*) voltage remains constant irrespective of the load

(*d*) speed varies in proportion to the load on the generator.

**99.** The polarity of a DC generator can be reversed by

(*a*) increasing field current (*b*) reversing field current

(*c*) reversing field current as well as direction of rotation

(*d*) any of the above.

**100.** In DC generators, the brushes on commutator remain in contact with conductors which

(*a*) lie under north pole (*b*) lie under south pole

(*c*) lie under interpolar region (*d*) are farthest from the poles.

**101.** Which of the following generator will have negligible terminal voltage while running on no load?

(*a*) series generator (*b*) shunt generator

(*c*) compound generator (*d*) separately excited generator.

**102.** In Fig. 4.15 which curve represents the characteristics of a over compounded generator?

(*a*) curve *A* (*b*) curve *B* (*c*) curve *C* (*d*) curve *D*.

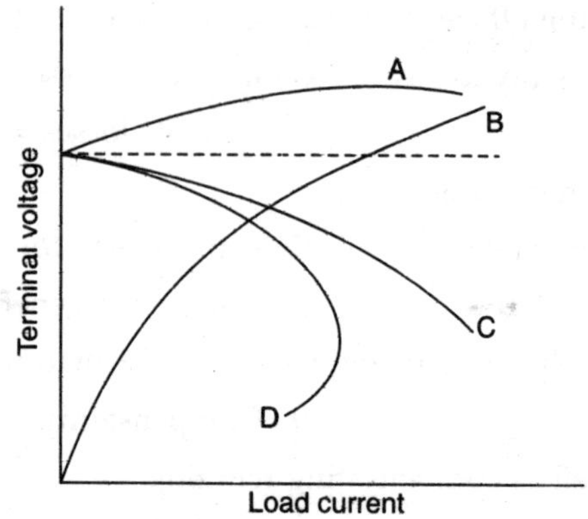


*FIGURE 4.15*

**103.** Curve C is likely to represent the characteristics of a

(*a*) shunt generator (*b*) series generator
(*c*) differentially compounded generator (*d*) under compounded generator.

**104.** Curve *B* is likely to represent the characteristics of a

(*a*) level compounded generator (*b*) under compounded generator
(*c*) shunt generator (*d*) none of the above.

**105.** Which of the following DC generator will be in a position to build up without any residual magnetism in the poles?

(*a*) series generator (*b*) shunt generator
(*c*) compound generator (*d*) none of the above.

**106.** In over compounded generator, full load terminal voltage is

(*a*) almost zero (*b*) less than no load terminal voltage
(*c*) more than no load terminal voltage (*d*) equal to no load teminal voltage.

**107.** In DC generators on no load, the air gap flux distribution in space is

(*a*) sinusoidal (*b*) triangular (*c*) pulsating (*d*) flat topped.

**108.** When $Y_b$ is the back pitch and $Y_f$ is the front pitch, then in case of simplex lap winding

(*a*) $Y_b + Y_f = 2$ (*b*) $Y_b + Y_f = 1$ (*c*) $Y_b + Y_f = -1$ (*d*) $Y_b + Y_f = 0$.

**109.** In a DC generator, the iron losses mainly take place in

(*a*) yoke (*b*) commutator
(*c*) armature conductors (*d*) armature rotor.

**110.** In case of a DC generator with wave winding which one of the following is not valid?

(*a*) Armature has only two parallel paths
(*b*) The minimum number of brushes required is equal to the number of poles
(*c*) Armature winding is a closed circuit
(*d*) Between two adjecent commutator segments the number of coils in series is equal to the number of pairs of poles.

**111.** In a level compounded DC generator, full load terminal voltage is
(*a*) negligibly low (*b*) equal to no load terminal voltage
(*c*) more than no load terminal voltage (*d*) less than no load terminal voltage.

**112.** In a lap wound DC generator having $P$ as number of poles and $Z$ as number of conductors, the maximum number of equalizer rings will be
(*a*) $Z$ (*b*) $P$ (*c*) $\frac{2Z}{P}$ (*d*) $\frac{2P}{Z}$.

**113.** A DC welding generator has
(*a*) lap winding (*b*) wave winding (*c*) duplex winding (*d*) any of the above.

**114.** Unlike alternators, DC generators are installed near the load centres to reduce
(*a*) iron losses (*b*) line losses (*c*) sparking (*d*) corona losses.

**115.** Two generators *A* and *B* running in parallel are supplying power to a common load of 500 kW. Generator *A* has armature resistance equal to half that of *B*. Which of the following statement is correct?
(*a*) both generators will share load equally
(*b*) generator *A* will take load less than the load taken by generator *B*
(*c*) generator *B* will take more load as compared to generator *A*
(*d*) none of the above.

**116.** The terminal voltage of a DC shunt generator drops on load because of all of the following reasons EXCEPT,
(*a*) armature reaction (*b*) armature resistance drop
(*c*) field weaking due to armature reaction and armature resistance drop
(*d*) commutation.

**117.** In a DC generator windage loss varies with speed ($N$) in the proportion of
(*a*) $N$ (*b*) $N^2$ (*c*) $\frac{1}{N}$ (*d*) $\frac{1}{N^2}$.

**118.** Which of the following statement about DC generators is false?
(*a*) equalising bus bars are used with parallel running of DC shunt generators
(*b*) compensating winding in a DC machine helps in commutation
(*c*) in a DC generator interpole winding is connected in series with the armature winding
(*d*) back pitch and front pitch are both odd and approximately equal to the pole pitch.

**119.** In a DC generator
(*a*) external characteristics = internal characteristics + armature reaction
(*b*) internal characteristics = magnetisation characteristics – ohmic drop
(*c*) external characteristics = magnetisation characteristics – ohmic drop – armature reaction
(*d*) magnetisation characteristics = external characteristics.

120. The demagnetising component of armature reaction in DC generator
(*a*) results in sparking trouble (*b*) reduces generated emf
(*c*) increases armature speed (*d*) reduces interpole flux density.

121. Magnetic field in a DC generator is produced by
(*a*) permanent magnets (*b*) electromagnets
(*c*) both (*a*) and (*b*) above (*d*) none of the above.

122. There are two critical resistances for
(*a*) series generator (*b*) shunt generator
(*c*) permanent magnet generator (*d*) separately excited generator.

123. A shunt generator running at 1000 rpm has induced emf as 200 V. If the speed increases to 1200 rpm, the induced emf will be nearly
(*a*) 150 V (*b*) 175 V (*c*) 240 V (*d*) 290 V.

124. A sinusoidal voltage of 5 Hz is applied to the filed of a shunt generator. The armature voltage wave
(*a*) will be zero (*b*) will be of 5 Hz
(*c*) will be of 5 × *N* Hz (*d*) will be of *N*/5 Hz.

125. The number of brushes in commutator depends on
(*a*) amount of current to be collected (*b*) speed of armature
(*c*) type of winding (*d*) voltage.

## ANSWERS

| | | | | | | |
|---|---|---|---|---|---|---|
| 1. (*d*) | 2. (*b*) | 3. (*b*) | 4. (*c*) | 5. (*c*) | 6. (*a*) | 7. (*c*) |
| 8. (*d*) | 9. (*c*) | 10. (*d*) | 11. (*c*) | 12. (*a*) | 13. (*d*) | 14. (*a*) |
| 15. (*d*) | 16. (*b*) | 17. (*d*) | 18. (*a*) | 19. (*b*) | 20. (*d*) | 21. (*a*) |
| 22. (*c*) | 23. (*c*) | 24. (*b*) | 25. (*c*) | 26. (*a*) | 27. (*b*) | 28. (*b*) |
| 29. (*c*) | 30. (*b*) | 31. (*c*) | 32. (*c*) | 33. (*b*) | 34. (*d*) | 35. (*a*) |
| 36. (*a*) | 37. (*d*) | 38. (*d*) | 39. (*c*) | 40. (*d*) | 41. (*c*) | 42. (*c*) |
| 43. (*d*) | 44. (*c*) | 45. (*d*) | 46. (*a*) | 47. (*b*) | 48. (*b*) | 49. (*d*) |
| 50. (*c*) | 51. (*b*) | 52. (*d*) | 53. (*c*) | 54. (*d*) | 55. (*a*) | 56. (*d*) |
| 57. (*d*) | 58. (*b*) | 59. (*b*) | 60. (*c*) | 61. (*c*) | 62. (*d*) | 63. (*c*) |
| 64. (*c*) | 65. (*b*) | 66. (*b*) | 67. (*c*) | 68. (*c*) | 69. (*c*) | 70. (*b*) |
| 71. (*c*) | 72. (*b*) | 73. (*a*) | 74. (*c*) | 75. (*a*) | 76. (*a*) | 77. (*c*) |
| 78. (*b*) | 79. (*a*) | 80. (*a*) | 81. (*a*) | 82. (*d*) | 83. (*d*) | 84. (*d*) |
| 85. (*a*) | 86. (*b*) | 87. (*c*) | 88. (*c*) | 89. (*d*) | 90. (*b*) | 91. (*b*) |
| 92. (*d*) | 93. (*b*) | 94. (*c*) | 95. (*d*) | 96. (*c*) | 97. (*a*) | 98. (*c*) |
| 99. (*b*) | 100. (*c*) | 101. (*a*) | 102. (*a*) | 103. (*b*) | 104. (*c*) | 105. (*d*) |
| 106. (*c*) | 107. (*d*) | 108. (*a*) | 109. (*d*) | 110. (*b*) | 111. (*b*) | 112. (*c*) |
| 113. (*a*) | 114. (*b*) | 115. (*d*) | 116. (*d*) | 117. (*b*) | 118. (*a*) | 119. (*c*) |
| 120. (*b*) | 121. (*c*) | 122. (*b*) | 123. (*c*) | 124. (*b*) | 125. (*a*) | |

# Chapter 5 DC MOTORS

A dc motor is a device that converts electrical energy into mechanical energy. It works on the principle that when a current carrying conductor is placed in a magnetic field, it experiences a magnetic force, whose direction is given by Flamings Left Hand rule. The magnitude of this force *F*, is given by

$$F = BIl \text{ newtons}$$

The construction of a dc generator and a dc motor is basically identical.

**Back EMF.** As the motor armature rotates, the conductors also rotate thereby cutting the flux. This results in emf induced in the conductors, whose direction is in opposition to that of the applied voltage. This emf is known as back emf.

**Voltage Equation.** The voltage equation of a motor is

$$V = E_b + I_a R_a$$

where $V$ is the applied voltage.

**Condition for Maximum Power.** The mechanical power developed by a motor is maximum when back emf is equal to half the applied voltage. No doubt the efficiency under this conditions will be 50%.

**Torque.** The torque developed by the motor is given by

$$T_a = 0.159\, \phi\, Z\, I_a\, (P/A) \text{ Nm}$$

From the above equation it is evident that

$$T_a \propto \phi\, I_a$$

In a series motor, $\phi$ is directly proportional to $I_a$ (before saturation) because field windings carry full armature current

$$\therefore \qquad T_a \propto I_a^2.$$

**Speed Regulation** of a motor is the change in speed when the load on the motor is reduced from rated value to zero, expressed as percent of the rated load speed. Thus

$$\% \text{ speed regulation} = \frac{\text{No load speed} - \text{Full load speed}}{\text{Full load speed}} \times 100$$

**Characteristics.** The characteristic curves for motors represent the relation between following quantities:

(*i*) Torque and armature current. This is also known as electrical characteristic

(*ii*) Speed and armature current

(*iii*) Speed and torque characteristic. This is also known as mechanical characteristic.

For motor characteristics the following relations are useful

$$T_a \propto \phi I_a$$

$$N \propto \frac{E_b}{\phi}$$

## Series Motors

(*i*) **Torque—armature current characteristic.**

In a series motor field winding also carries the armature current so that $\phi \propto I_a$ upto the point of magnetic saturation. Hence

$$T_a \propto \phi I_a$$
$$\propto (I_a)^2$$

Hence $T_a - I_a$ curve is a parabola as shown in Fig. 5.1.

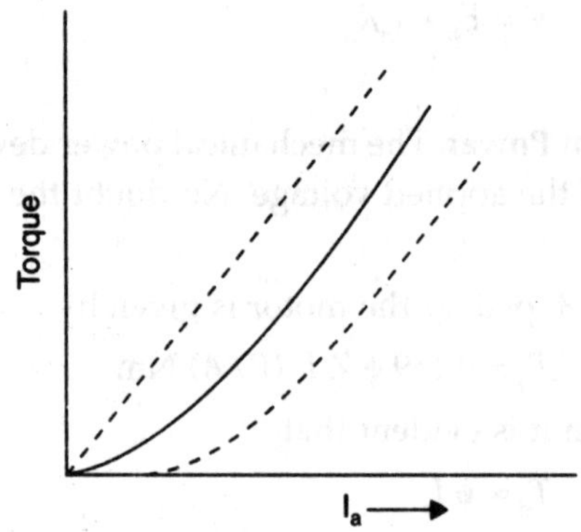


*FIGURE 5.1*

After saturation, $\phi$ becomes almost independent of $I_a$, so that $T_a \propto I_a$ or.iy. In this region the characteristic follows linear relation. This is shown in Fig. 5.1. The shaft torque $T_{sh}$ is less than the armature torque due to stray losses. It is represented by dotted line.

From the above it can be concluded that a prior to magnetic saturation, on heavy loads, a series motor exerts a torque proportional to the square of armature current. Thus when huge starting torque for accelerating heavy masses, such as hoists, electric trains, etc. is required, series motor is ideal.

(*ii*) **Speed** — armature corrent characteristic.

Variations of speed can be seen from the relation

$$N \propto \frac{E_b}{\phi}$$

Variation of back emf for a various load currents is small and hence $E_b$ may be approximately treated as constant. However, as $\phi$ increases, $I_a$ also increases. Thus speed varies inversely as armature current. This is shown in Fig. 5.2.

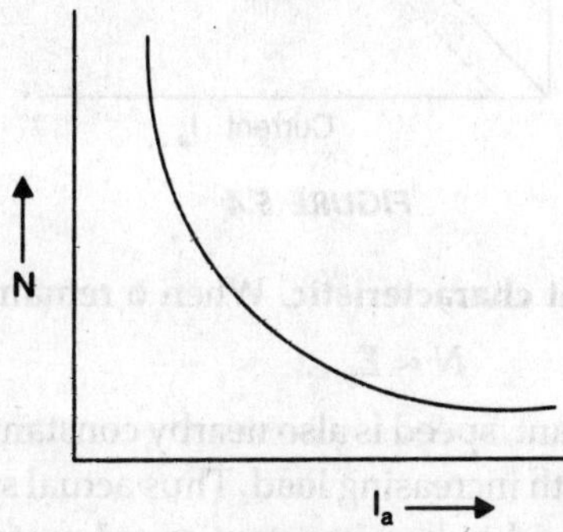


***FIGURE 5.2***

One small $I_a$, the value of $N$ is very high, therefore a series motor should not be started on no load or light loads. Also it can be concluded from Fig. 5.2 that a series motor is a variable speed motor.

(*iii*) **Mechanical Characteristic.** For a series motor when speed is high, torque is low and *vice-versa*. This relation is shown in Fig. 5.3.

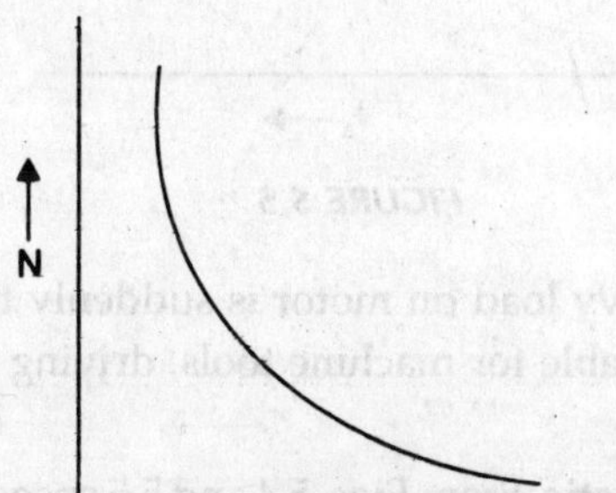


***FIGURE 5.3***

## Shunt Motors

(*i*) **Torque — armature current characteristic.** In shunt motors ϕ can be considered to be practically constant, hence, $T_a \propto I_a$. Thus electrical characteristic is nearly a straight line (at heavy loads ϕ decreases somewhat due to increased armature reaction) Fig. 5.4.

Shaft torque being less armature torque is shown by a dotted straight line. From the above it can be concluded that a shunt motor should not be started on heavy loads.

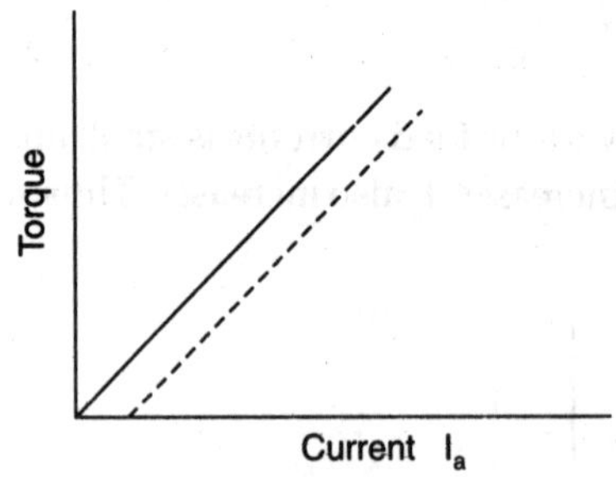


*FIGURE 5.4*

(*ii*) **Speed—armature current characteristic.** When ϕ remains constant,

$$N \propto E_b$$

As $E_b$ is also practically constant, speed is also nearby constant, as shown in Fig. 5.5. However, $E_b$ decreases slightly more than ϕ with increasing load. Thus actual speed curve is slightly drooping. Thus a shunt motor can be considered to be a constant speed motor.

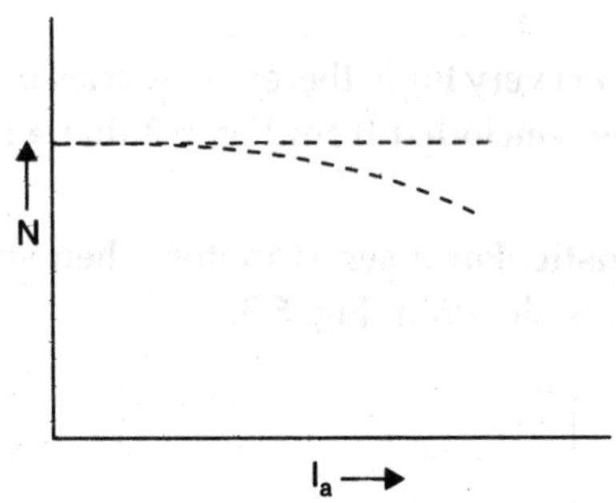


*FIGURE 5.5*

In view of above even if heavy load on motor is suddenly thrown off, the speed variation will be small. These motors are suitable for machine tools, driving shafts for power transmission, etc.

(*iii*) **Speed—torque characteristic.** From Figs. 5.4 and 5.5 speed torque characteristic for shunt motor can be deduced as shown in Fig. 5.6.

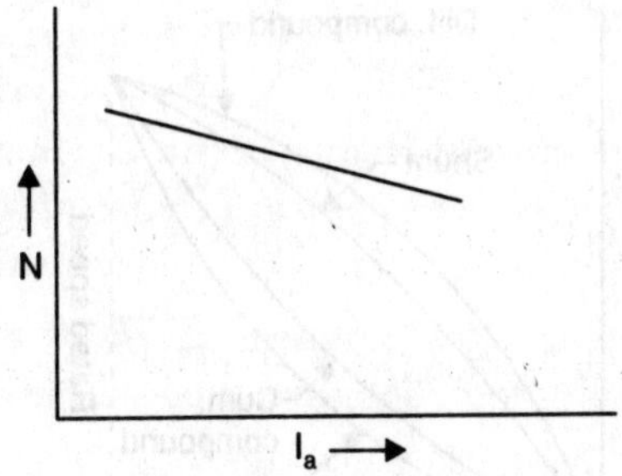


FIGURE 5.6

## Compound Motors

**(i) Cumulatively and Differentially Compound Motors.**

These motors are provided with both series and shunt windings. The motor is said to be cumulatively compound motor when series excitation helps the shunt excitation. However, in case series field opposes the shunt field, the motor is said to be differentially compound.

**Speed**–armature current characteristic of these motors is shown in Fig. 5.7. Cumutatively compound motors are used where series characteristics are desired and also where possibility of throwing off the entire load suddenly, exists. Some of the applications are coal cutting machines, heavy machine tools, etc., Sometimes a fly wheel is also used on motor shafts to take care of fluctuations in load, such as power presses, rolling mills etc.

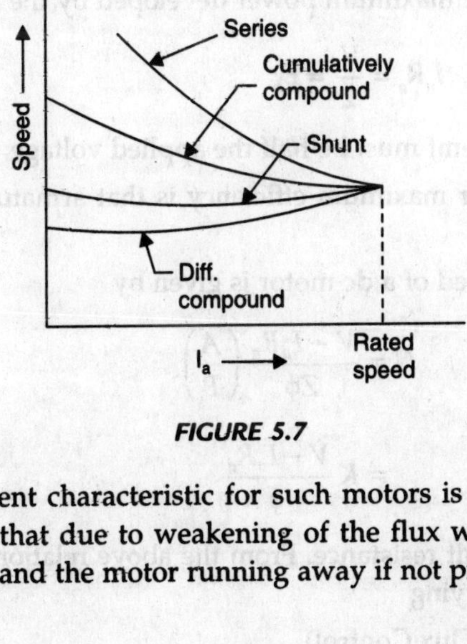


FIGURE 5.7

**Torque**–armature current characteristic for such motors is shown in Fig. 5.8. One of the drawback of such motors is that due to weakening of the flux with increase in load there is a tendency to speed instability and the motor running away if not properly designed.

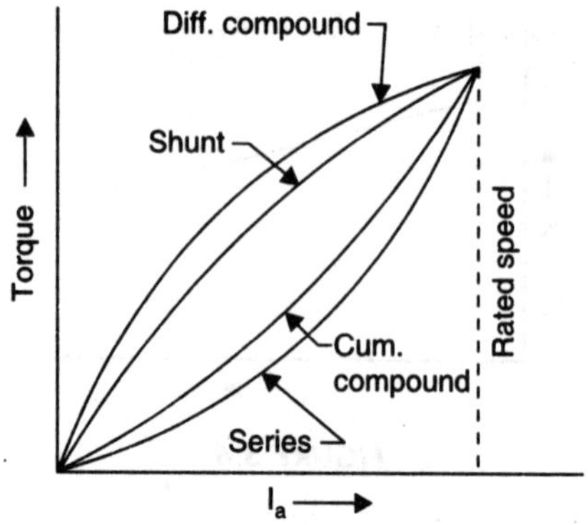


*FIGURE 5.8*

**Losses and Efficiency.** The losses in a dc motor as well as dc generator are identical. Thus losses in a motor are

(*i*) **Copper losses**

(*ii*) **Magnetic losses**

(*a*) Hysteresis loss

(*b*) Eddy current loss

(*iii*) **Mechanical losses**

(*a*) Friction loss at bearings and commutator

(*b*) Air-friction or windage loss of rotating armature

The condition for maximum power developed by the motor is

$$I_a R_a = \frac{V}{2} = E_b$$

That is the back emf must be half the applied voltage.

The condition for maximum efficiency is that armature copper losses are equal to constant losses.

**Speed Control.** The speed of a dc motor is given by

$$N = \frac{V - I_a R_a}{Z\phi}\left(\frac{A}{P}\right)$$

$$= K\frac{V - I_a R_a}{\phi}$$

where $R_a$ is the armature circuit resistance. From the above relation it can be concluded that the speed can be controlled by varying

(*i*) $\phi$ *i.e.*, flux per pole (Flux Control)

(*ii*) $R_a$ *i.e.*, resistance of the armature circuit (Rheostatic Control)

(*iii*) $V$ *i.e.*, applied voltage (Voltage Control)

These methods of speed control apply to series, shunt as well as compound motors. Some methods of speed control are given below.

**Ward Leonard Control**

This method of control is used for shunt motors where wide and very sensitive speed control is desired as in case of elevators, colliery winders, etc. $M_2$ is the motor subjected to speed control. The field of this motor is connected to dc supply mains. The speed variation of motor is possible by applying variable voltage across its armature. This variable voltage is obtained from motor-generator set $M_1 - G$. In this $M_1$ can be dc or ac motor which is mechanically coupled to generator $G$.

As long as $M_1$ runs on constant speed, output of $G$ is fed to motor $M_2$. However, the output of the generator can be varied through field regulator. The voltage generated in $G$ can be reversed through reversing switch $R_s$. Thus direction of rotation of $M_2$ can be reversed also. Thus besides controlling speed of $M_2$ even direction of rotation can be reversed,

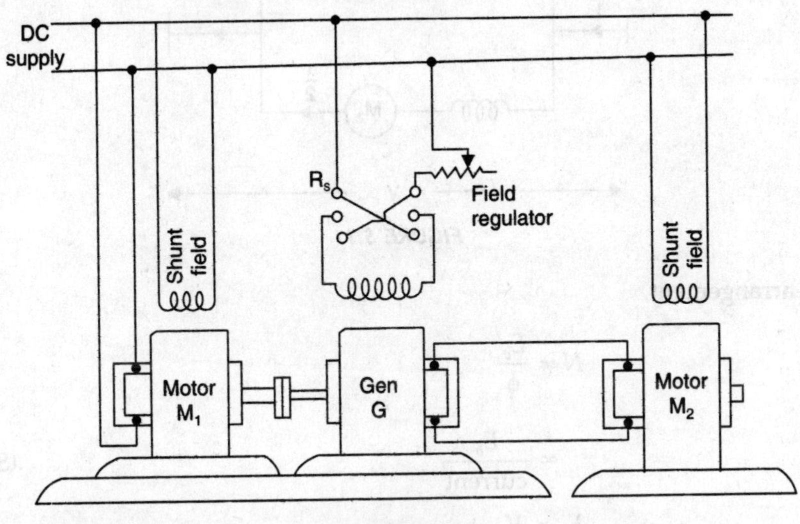


*FIGURE 5.9*

**Series Parallel Control.** This method of speed control is used in electric traction. In this two or more similar mechanically coupled motors are used. At low speeds, the motors are joined in series and for high speeds, they are joined in parallel.

**Series Running.** When motors are connected in series the same current passes through them. However, voltage across each motor is half the applied voltage.

Since
$$\text{Speed} \propto \frac{E_b}{\phi}$$
$$\propto \frac{\frac{V}{2}}{I}$$
$$\propto \frac{V}{2I}$$

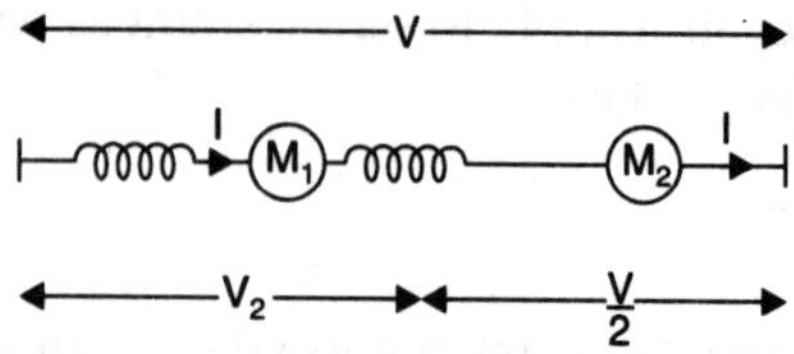


*FIGURE 5.10*

**Parallel Running.** When motors are joined in parallel, the voltage across each motor is $V$, but the current drawn by each motor will be $\frac{I}{2}$.

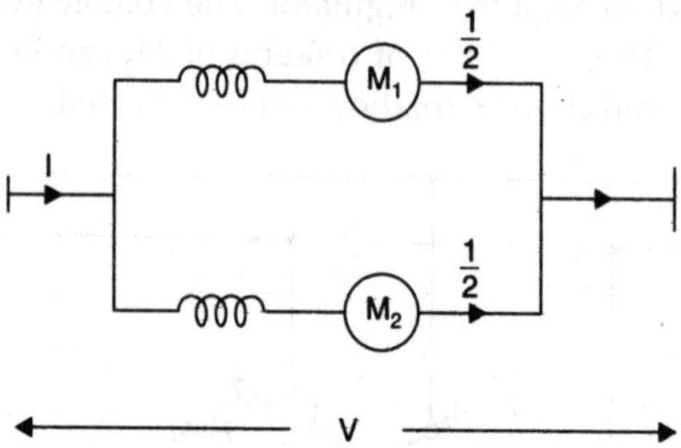


*FIGURE 5.11*

In this arrangement,

$$N \propto \frac{E_b}{\phi}$$

$$\propto \frac{E_b}{\text{current}} \qquad \text{(Series motors)}$$

but $E_b \simeq V$

hence $$N \propto \frac{V}{\frac{I}{2}} \propto \frac{2V}{I}$$

and torque $$T \propto \frac{I^2}{4}.$$

From the above it can be concluded that

(*i*) Speed of motors in series arrangement is one-fourth of the speed of the motors when in parallel.

(*ii*) The torque is four times that produced by motors when in parallel.

**Plugging.** In this method of speed control, the connections of armature terminals are reversed so that motor tends to run in the reverse direction. This tendency of armature is used for braking the motor. This method of speed control is applicable to shunt as well as series motor and is widely used in controlling elevators, rolling mills, printing presses, machine tools etc.

**Regenerative Bracking.** In a dc motor if $E_b$ becomes more than $V$, then it can be used as a generator, there by cutting down its torque. This is possible in hauling loads like cranes, etc. Where load becomes prime mover and drives the motor as a generator. Consequently, direction of $I_a$ and hence of armature torque is reversed and speed falls until $E_b$ becomes less than $V$. Thus during slowing down of motor, the power is returned to the line.

## Testing

(*i*) **Brake test.** Brake test on dc motors is conducted to measure the mechanical output of the motor. In this a pulley is subjected to braking action to measure the torque or horsepower.

(*ii*) **Swinburne's test.** This test is used to measure the losses. This test is applicable to shunt as well as level compounded motors. In this the motor is run on no-load at its rated speed and voltage. Shunt and armature currents are measured. Thus no-load losses can be determined. These consist of

(*a*) iron loss in core (*b*) friction loss

(*c*) windage loss (*d*) armature copper loss.

By measuring armature resistance, armature copper loss can be separated and thereby constant losses of the machine can be evaluated. In this way losses at any load in the machine can be estimated and efficiency of the motor can be calculated.

(*iii*) **Back to back test.** This is also known as regenerative or Hopkinson's test. In this test two identical machines are mechanically coupled and one of them is run as a motor and the other as a generator. The mechanical output of the motor drives the generator and the electrical output of generator is used in supplying the greater part of the input to the motor. The additional input to the set accounts for losses in the set.

(*iv*) **Retardation or Running down test.** This method is applicable to shunt motors (and also generators) and is used for finding stray losses. In this the motor is speeded up slightly beyond its normal speed and then supply is cut off from the armature while keeping the field excited. This results in armature slowing down and its kinetic energy is used to meet the rotational losses *i.e.*, friction, windage and iron losses.

## OBJECTIVE TYPE QUESTIONS

1. Which curve in Fig. 5.12 represents speed-torque characteristics of a dc series motor?

(*a*) curve *A* (*b*) curve *B* (*c*) curve *C* (*d*) curve *D*.

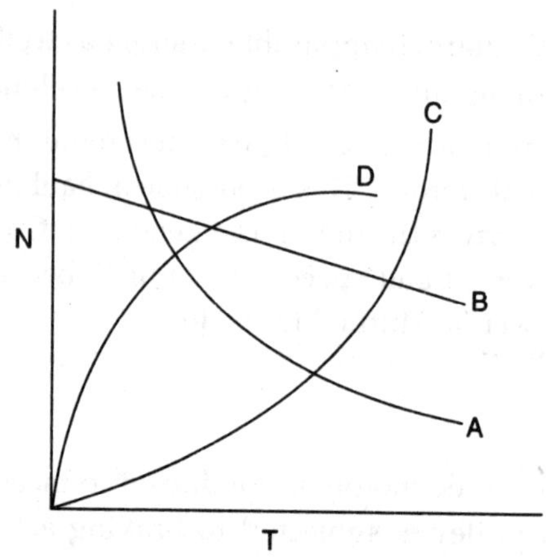


*FIGURE 5.12*

2. No load speed of which of the following motor will be highest?
   (*a*) shunt motor (*b*) series motor
   (*c*) cumulatively compound motor (*d*) differentially compound motor.
3. The direction of rotation of a dc series motor can be changed by
   (*a*) interchanging supply terminals (*b*) interchanging field terminals
   (*c*) either of (*a*) or (*b*) above (*d*) none of the above.
4. In case a dc motor is connected across ac supply
   (*a*) the motor will burn (*b*) the motor will run at rated speed
   (*c*) the motor will run at corresponding synchronous speed
   (*d*) the motor will run at slow speed.
5. Which of the following application requires high starting torque?
   (*a*) lathe machine (*b*) centrifugal pump (*c*) locomotive (*d*) air blower.
6. If a dc motor is to be selected for conveyers, which motor would be preferred?
   (*a*) series motor (*b*) shunt motor
   (*c*) differentially compound motor (*d*) cumulatively compound motor.
7. Which dc motor will be preferred for machine tools?
   (*a*) series motor (*b*) shunt motor
   (*c*) cumulatively compound motor (*d*) differentially compound motor.
8. Which dc motor will be preferred for constant speed line shafting?
   (*a*) cumulatively compound motor (*b*) differentially compound motor
   (*c*) shunt motor (*d*) series motor.
9. Usually wide and sensitive speed control is desired in case of
   (*a*) centrifugal pumps (*b*) elevators
   (*c*) steel rolling mills (*d*) colliery winders.

10. The speed of a motor falls from 1100 rpm at no load to 1050 rpm at rated load. The speed regulation of the motor is

(a) 2.36% (b) 4.76% (c) 6.77% (d) 8.84%.

11. Differentially compound dc motors can find applications requiring

(a) high starting torque (b) low starting torque

(c) variable speed (d) frequent on-off cycles.

12. Which dc motor is preferred for elevators?

(a) shunt motors (b) differentially compound motor

(c) cumulatively compound motor (d) series motor.

13. Which figure represents the efficiency of a dc motor?

(a) Figure A (b) Figure B (c) Figure C (d) Figure D.

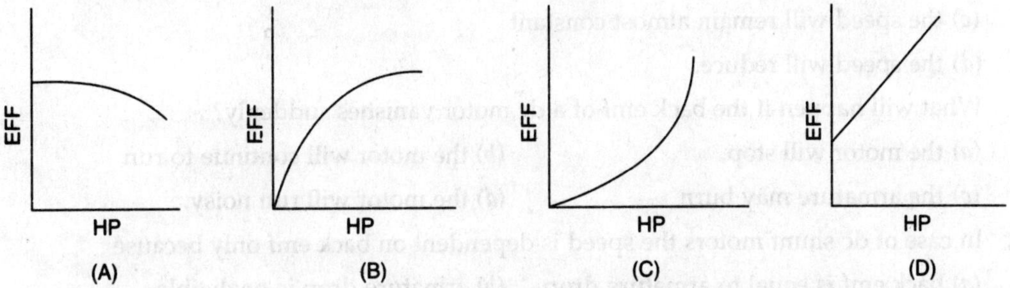


*FIGURE 5.13*

14. A dc motor armature is taking 50 A from the line, it develops torque *T*. The field strength is reduced to 75 per cent of its original value and the current increased to 80 A. The new value of the torque will be

(a) *T* (b) 1.2*T* (c) 0.8*T* (d) 0.64*T*.

15. In a dc motor the armature resistance is 0.1 ohm. When connected across; 110 volt mains the armature takes 20 amperes, and its speed is 1200 rpm. What will be the speed when the armature takes 50 amperes from the same mains with the field increased by 10%?

(a) 939 rpm (b) 978 rpm (c) 1000 rpm (d) 1061 rpm.

16. According to Fleming's left hand rule, when the forefinger points in the direction of the field or flux, the middle finger will point on the direction of

(a) current in the conductor (b) movement of conductor

(c) resultant force on conductor (d) none of the above.

17. If the field of a dc shunt motor gets opened while motor is running

(a) the speed of the motor will be reduced

(b) the armature current will reduce

(c) the motor will attain dangerously high speed

(d) the motor will continue to run at constant speed.

18. Starters are used with dc motors because
(*a*) these motors have high starting torque
(*b*) these motors are not self starting
(*c*) back emf of these motors is zero initially
(*d*) to restrict armature current as there is no back emf while starting.

19. Which of the following dc motor has self loading properties?
(*a*) series motor (*b*) shunt motor
(*c*) cumulatively compound motor (*d*) differentially compound motor.

20. In dc shunt motors as load is reduced
(*a*) the speed will increase abruptly
(*b*) the speed will increase in proportion to reduction in load
(*c*) the speed will remain almost constant
(*d*) the speed will reduce.

21. What will happen if the back emf of a dc motor vanishes suddenly?
(*a*) the motor will stop (*b*) the motor will continue to run
(*c*) the armature may burn (*d*) the motor will run noisy.

22. In case of dc shunt motors the speed is dependent on back emf only because
(*a*) back emf is equal to armature drop (*b*) armature drop is negligible
(*c*) flux is proportional to armature current
(*d*) flux is practically constant in dc shunt motors.

23. A cumulatively compounded long shunt generator when operating as a motor would run as
(*a*) cumulatively compounded long shunt motor
(*b*) cumulatively compounded short shunt motor
(*c*) differentially compounded short shunt motor
(*d*) differentially compounded long shunt motor.

24. In dc motors by controlling field flux
(*a*) speeds above the rated speed can be achieved
(*b*) speeds below the rated speed can be achieved
(*c*) speeds above as well as below rated speed can be achieved
(*d*) speed from zero to infinitely high speeds can be achieved.

25. The armature voltage control of dc motor provides
(*a*) constant torque drive (*b*) constant voltage drive
(*c*) constant current drive (*d*) none of the above.

26. The speed regulation of a dc motor is

(a) $\frac{\text{Lowest speed}}{\text{Highest speed}}$ (b) $\frac{\text{Highest speed} - \text{Lowest speed}}{\text{Average speed}}$

(c) $\frac{\text{No load speed} - \text{Full load speed}}{\text{Full load speed}}$ (d) $\frac{\text{No load speed} - \text{Full load speed}}{\text{No load speed}}$.

27. As there is no back emf at the instant of starting a dc motor, in order to prevent a heavy current from flowing through the armature circuit

(a) a resistance is connected in series with the armature

(b) a resistance is connected parallel to the armature

(c) armature is temporarily open circuited

(d) a high value resistor is connected across the field winding.

28. The purpose of retardation test on dc shunt machines is to find out

(a) stray losses (b) eddy current losses

(c) field copper losses (d) windage losses.

29. The speed of a dc motor is

(a) always constant (b) proportional to load current

(c) directly proportional to back emf

(d) inversely proportional to the product of back emf and flux.

30. The retardation test on the shunt motors is performed to determine

(a) copper losses (b) stray losses (c) field losses (d) windage losses.

31. The speed of a dc shunt motor can be increased by

(a) increasing the resistance in the armature circuit

(b) increasing the resistance in the field circuit

(c) reducing the resistance in the field circuit.

(d) reducing the resistance in the armature circuit.

32. Two dc series motors when connected in series draw a supply current of $I$ amperes and run at $N$ rpm. Now if the motors are connected in parallel and if the motors continue to draw current $I$, then the speed of the motors will be

(a) $N$ (b) $\frac{N}{4}$ (c) $4N$ (d) $2N$.

33. The speed of a series wound dc motor

(a) can be controlled by shunt field regulator

(b) cannot be controlled by diverter

(c) increases as flux increases

(d) increases as armature circuit resistance increases.

34. If $I_a$ be the armature current, then speed of a dc shunt motor is
(*a*) independent of $I_a$ (*b*) proportional to $I_a$
(*c*) varies as $(I_a)^2$ (*d*) varies as $\frac{1}{I_a}$.

35. If $I_a$ be the armature current, the speed of a dc series motor will be
(*a*) independent of $I_a$ (*b*) proportional to $I_a$
(*c*) inversely proportional to $I_a$ (*d*) inversely proportional to $(I_a)^2$.

36. In a dc shunt motor, under the conditions of maximum power, the current in the armature will be
(*a*) almost negligible (*b*) rated full load current
(*c*) less than full load current (*d*) more than full load current.

37. Which dc motor will have least percentage increase of input current, for the same percentage increase in torque?
(*a*) shunt motor (*b*) series motor
(*c*) cumulatively compound motor (*d*) separately excited motor.

38. In case the back emf and the speed of a dc motor are doubled, the torque developed by the motor will
(*a*) remain unchanged (*b*) reduce to one-fourth value
(*c*) increases four folds (*d*) be doubled.

39. At the instant of starting when a dc motor is put on supply, it behaves like
(*a*) a highly resistive circuit (*b*) a low resistance circuit
(*c*) a capacitive circuit (*d*) none of the above.

40. Which of the following test will be suitable for testing two similar dc series motors of large capacity?
(*a*) Swinborne test (*b*) Hopkinson's test (*c*) Field test (*d*) Brake test.

41. Now a days dc motors are widely used in
(*a*) pumping sets (*b*) air compressors (*c*) electric traction (*d*) machine shops.

42. By looking at which part of motor, it can be easily confirmed that a particular motor is dc motor?
(*a*) frame (*b*) shaft (*c*) commutator (*d*) stator.

43. The speed of a dc motor can be varied by varying
(*a*) field current (*b*) applied voltage
(*c*) resistance in series with armature (*d*) any of the above.

44. Two dc series motors are connected in series to produce a torque $T$. Now if the motors are connected in parallel, the torque produced will be
(*a*) $\frac{T}{4}$ (*b*) $\frac{T}{2}$ (*c*) $2T$ (*d*) $4T$.

45. Which one of the following is not necessarily the advantage of dc motors over ac motors?
(*a*) low cost (*b*) wide speed range
(*c*) stability (*d*) high starting torque.

46. For a dc shunt motor if the excitation is changed
(*a*) torque will remain constant
(*b*) torque will change but power will remain constant
(*c*) torque and power both will change
(*d*) torque, power and speed, all will change.

47. In which of the following applications dc series motor is invariably used?
(*a*) starter for a car (*b*) drive for a water pump
(*c*) fan motor (*d*) motor for operation in ac or dc.

48. Which motor has the poorest speed control?
(*a*) differentially compound motor (*b*) cumulatively compound motor
(*c*) shunt motor (*d*) series motor.

49. In dc machines fractional pitch winding is used
(*a*) to improve cooling (*b*) to reduce copper losses
(*c*) to increase the generated emf (*d*) to reduce the sparking.

50. A three point starter is considered as suitable for
(*a*) shunt motors (*b*) shunt as well as compound motors
(*c*) shunt, compound and series motors (*d*) all dc motors.

### Questions 51 to 53 Refer to Fig. 5.14

Fig. 5.14 shows characteristics for different dc motors

51. Which curve represents the characteristics for a dc series motor?
(*a*) curve *A* (*b*) curve *B* (*c*) curve C (*d*) curve *D*.

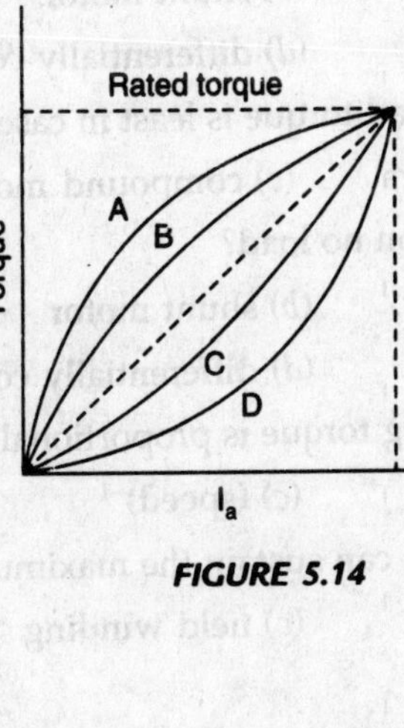


*FIGURE 5.14*

52. Which curve is closer to the characteristics of a shunt motor?
(*a*) curve *A* (*b*) curve *B* (*c*) curve *C* (*d*) curve *D*.
53. Which curve represents the characteristics of differentially compound motor?
(*a*) curve *A* (*b*) curve *B* (*c*) curve *C* (*d*) curve *D*.
54. In a series motor, the shaft torque is less than the armature torque due to
(*a*) hysteresis losses (*b*) eddy current losses
(*c*) stray losses (*d*) all of the above.
55. Which of the following class of insulation can withstand the highest temperature?
(*a*) *F* (*b*) *B* (*c*) *H* (*d*) *C*.
56. Small dc motors upto 5HP usually have
(*a*) 2 poles (*b*) 4 poles (*c*) 6 poles (*d*) 8 poles.
57. Hopkinson's test on dc machines is conducted at
(*a*) no load (*b*) part load (*c*) full load (*d*) over load.
58. In case the conditions for maximum power for a dc motor are established, the efficiency of the motor will be
(*a*) 100% (*b*) around 90%
(*c*) anywhere between 75% and 90% (*d*) less than 50%.
59. During rheostatic braking of dc series motors
(*a*) motor is run as a generator (*b*) motor is reversed in direction
(*c*) motor is run at reduced speed.
60. For which type of dc motor, dynamic braking is generally used?
(*a*) shunt motor (*b*) series motors
(*c*) compound motors (*d*) all of the above.
61. A shearing machine has cyclic load consisting of intermittent light and heavy loads. Which of the following dc motor will be suitable for this purpose?
(*a*) series motor (*b*) shunt motor
(*c*) cumulatively compound motor (*d*) differentially compound motor.
62. The ratio of starting torque to full load torque is least in case of
(*a*) series motors (*b*) shunt motors (*c*) compound motors.
63. Which motor should not be started on no load?
(*a*) series motor (*b*) shunt motor
(*c*) cumulatively compound motor (*d*) differentially compound motor.
64. During rheostatic braking the braking torque is proportional to
(*a*) speed (*b*) $(\text{speed})^2$ (*c*) $(\text{speed})^{-1}$ (*d*) $(\text{speed})^{-2}$.
65. In a dc motor which of the following can sustain the maximum temperature rise?
(*a*) slip rings (*b*) commutator (*c*) field winding (*d*) armature winding.

**66.** Which of the following law (rule) can be used to determine the direction of rotation of a dc motor?

(*a*) Lenz's law (*b*) Faraday's law

(*c*) Coulomb's law (*d*) Fleming 's left hand rule.

**67.** Which of the following load normally needs starting torque more than the rated torque?

(*a*) blowers (*b*) conveyors (*c*) air compressors (*d*) centrifugal pumps.

**68.** Which method of braking is generally used in elevators?

(*a*) plugging (*b*) regenerative braking

(*c*) rheostatic braking (*d*) none of the above.

**69.** The starting resistance of a dc motor is generally

(*a*) low (*b*) around 500 ohms (*c*) 1000 ohms (*d*) infinitely large.

**70.** In case of conductively compensated dc series motors, the compensating winding is provided

(*a*) as separately wound unit (*b*) in parallel with armature winding

(*c*) in series with armature winding (*d*) in parallel with field winding.

**71.** Sparking at the commutator of a dc motor may result in

(*a*) damage to commutator segments (*b*) damage to commutator insulation

(*c*) increased power consumption (*d*) all of the above.

**72.** Ward Leonard control is basically a

(*a*) voltage control method (*b*) field diverter method

(*c*) field control method (*d*) armature resistance control method.

**73.** Which of the following motor is preferred for operation in highly explosive atmosphere?

(*a*) series motor (*b*) shunt motor

(*c*) air motor (*d*) battery operated motor.

**74.** Which of the following test can be used to determine no load losses in a dc shunt motor?

(*a*) Running down test (*b*) Swinburne's test

(*c*) Field test (*d*) Brake test.

**75.** What will happen when the supply terminals of a dc shunt motor are reversed?

(*a*) the motor will burn (*b*) the motor will not run

(*c*) the motor will run as a generator (*d*) the motor will run in reverse direction.

**76.** If the supply voltage for a dc motor is increased, which of the following will decrease?

(*a*) starting torque (*b*) operating speed (*c*) full load current (*d*) all of the above.

**77.** When the speed of a dc motor increases

(*a*) back emf increases and current drawn decreases

(*b*) back emf as well as current drawn both increase

(*c*) back emf as well as current drawn both decrease

(*d*) back emf decreases and current drawn increases.

78. As compared to an induction motor, the air gap in a dc motor is
(*a*) less than 50% (*b*) between 50% and 90%
(*c*) same (*d*) more.

79. For constant torque drive which speed control method is preferred?
(*a*) field control (*b*) armature voltage control
(*c*) shunted armature control (*d*) mechanical loading system.

80. In Ward Leonard control the lower limit of speed is imposed by
(*a*) residual magnetism of the generator (*b*) core losses of motor
(*c*) mechanical losses of motor and generator together
(*d*) all of the above.

81. Field winding of a dc series motor is usually provided with thicker wire
(*a*) to provide large flux (*b*) to reduce the use of insulating materials
(*c*) as it carries large load current (*d*) in order to reduce eddy currents.

82. Which one of the following is not the function of pole shoes in a dc machine?
(*a*) to reduce eddy current losses (*b*) to support the field coils
(*c*) to spread out flux for better uniformity
(*d*) to reduce the reluctance of the magnetic path.

83. In regenerative braking
(*a*) motor energy is dissipated as heat
(*b*) motor energy is dissipated in armature heating
(*c*) motor energy is dissipated in windage losses
(*d*) motor is made to run as a generator.

84. The mechanical power developed by a dc shunt motor will be maximum when the ratio of back emf to applied voltage is
(*a*) 4.0 (*b*) 2.0 (*c*) 1.0 (*d*) 0.5.

85. The condition for maximum power in case of dc motors is
(*a*) back emf = 2 × supply voltage (*b*) back emf = $\frac{1}{2}$ × supply voltage
(*c*) supply voltage = $\frac{1}{2}$ × back emf (*d*) supply voltage = back emf.

86. Regenerative braking on shunt motors is used when
(*a*) the load is variable (*b*) the load is constantly decreasing
(*c*) the load also acts as a braking force (*d*) the load has overhauling characteristics.

87. Which of the following method gives the greatest breaking torque?
(*a*) regenerative braking (*b*) plugging
(*c*) rheostatic braking.

88. For which of the following application a dc motor is preferred over an ac motor?
   (a) low speed operation    (b) high speed operation
   (c) variable speed operation    (d) fixed speed operation.

89. What will happen if the field winding of a dc series motor gets opened up by chance, when the motor is in operation?
   (a) the motor speed will increase tremendously
   (b) the motor will draw large armature current
   (c) the motor winding will burn    (d) none of the above.

90. What will happen in case a 220 V dc series motor is connected to 230 V ac supply?
   (a) the armature winding of motor will burn
   (b) the motor will vibrate violently
   (c) the motor will run with less efficiency and more sparking
   (d) the motor will not run.

91. In dc machines the residual magnetism is of the order of
   (a) 2 to 3 per cent    (b) 10 to 15 per cent    (c) 20 to 25 per cent    (d) 50 to 75 per cent.

92. In class *B* dc motors, the maximum permissible temperature rise in armature winding is
   (a) 25ºC    (b) 50ºC    (c) 75ºC    (d) 100ºC.

93. If $T_a$ be the torque and $I_a$ the armature current for a dc motor, then which of the following relation is valid before saturation?
   (a) $T_a \propto I_a$    (b) $T_a \propto \frac{1}{I_a}$    (c) $T_a \propto I_a^2$    (d) $T_a \propto \frac{1}{(I_a)^2}$.

94. The variation of one of the characteristic of a dc series motor is shown in Fig. 5.15. This characteristic could be
   (a) useful torque    (b) back emf    (c) speed    (d) terminal voltage.

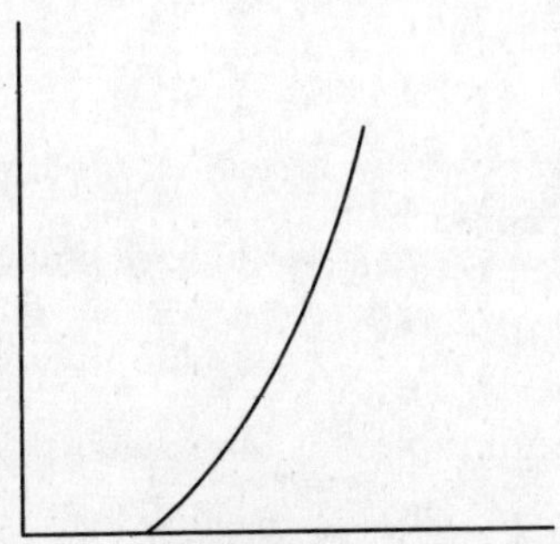


*FIGURE 5.15*

95. The main disadvantage of the Ward Leonard control method is
   (a) high initial cost    (b) high maintenance cost
   (c) low efficiency at light loads    (d) all of the above.

96. Which dc motor is generally preferred for cranes and hoists?
(*a*) series motor (*b*) shunt motor
(*c*) cumulatively compound motor (*d*) differentially compound motor.

97. Which of the following is a class C insulation?
(*a*) cotton tape (*b*) teflon (*c*) mica (*d*) none of the above.

98. A 200 V dc motor will be subjected to high voltage test at
(*a*) 250 V (*b*) 500 V (*c*) 660 V (*d*) 1400 V.

99. A brake test on dc motors is usually restricted to
(*a*) small horsepower motors (*b*) variable speed motors
(*c*) high speed motors (*d*) open frame type motors.

100. Which of the following can be used to control the speed of a dc motor?
(*a*) thermister (*b*) thyristor (*c*) thyratron (*d*) transistor.

## ANSWERS

| | | | | | | |
|---|---|---|---|---|---|---|
| 1. (*a*) | 2. (*b*) | 3. (*c*) | 4. (*a*) | 5. (*c*) | 6. (*a*) | 7. (*b*) |
| 8. (*c*) | 9. (*d*) | 10. (*b*) | 11. (*b*) | 12. (*c*) | 13. (*b*) | 14. (*b*) |
| 15. (*d*) | 16. (*a*) | 17. (*c*) | 18. (*d*) | 19. (*d*) | 20. (*c*) | 21. (*c*) |
| 22. (*d*) | 23. (*d*) | 24. (*a*) | 25. (*a*) | 26. (*c*) | 27. (*a*) | 28. (*a*) |
| 29. (*c*) | 30. (*b*) | 31. (*b*) | 32. (*c*) | 33. (*c*) | 34. (*a*) | 35. (*c*) |
| 36. (*d*) | 37. (*b*) | 38. (*a*) | 39. (*b*) | 40. (*c*) | 41. (*c*) | 42. (*c*) |
| 43. (*d*) | 44. (*a*) | 45. (*a*) | 46. (*b*) | 47. (*a*) | 48. (*d*) | 49. (*d*) |
| 50. (*b*) | 51. (*d*) | 52. (*a*) | 53. (*b*) | 54. (*c*) | 55. (*d*) | 56. (*a*) |
| 57. (*c*) | 58. (*d*) | 59. (*a*) | 60. (*d*) | 61. (*c*) | 62. (*b*) | 63. (*a*) |
| 64. (*a*) | 65. (*c*) | 66. (*d*) | 67. (*b*) | 68. (*a*) | 69. (*a*) | 70. (*c*) |
| 71. (*d*) | 72. (*a*) | 73. (*c*) | 74. (*b*) | 75. (*d*) | 76. (*c*) | 77. (*a*) |
| 78. (*d*) | 79. (*b*) | 80. (*a*) | 81. (*c*) | 82. (*a*) | 83. (*d*) | 84. (*d*) |
| 85. (*b*) | 86. (*c*) | 87. (*b*) | 88. (*c*) | 89. (*d*) | 90. (*c*) | 91. (*a*) |
| 92. (*c*) | 93. (*c*) | 94. (*a*) | 95. (*d*) | 96. (*a*) | 97. (*c*) | 98. (*d*) |
| 99. (*a*) | 100. (*b*). | | | | | |

# Chapter 6 AC FUNDAMENTALS

An alternating quantity is one which acts in alternate positive and negative directions, whose magnitude undergoes a definite series of changes in definite intervals of time, and in which the sequence of changes while negative is identical with the sequence when positive. As shown in Fig. 6.1 the graph repeats after regular time intervals and one repetition is called a complete cycle. The time $T$ of one cycle is called the Periodic time and the number of cycles per second, the frequency.

Hence $$T = \frac{1}{f}$$

$$f = \frac{1}{T}.$$

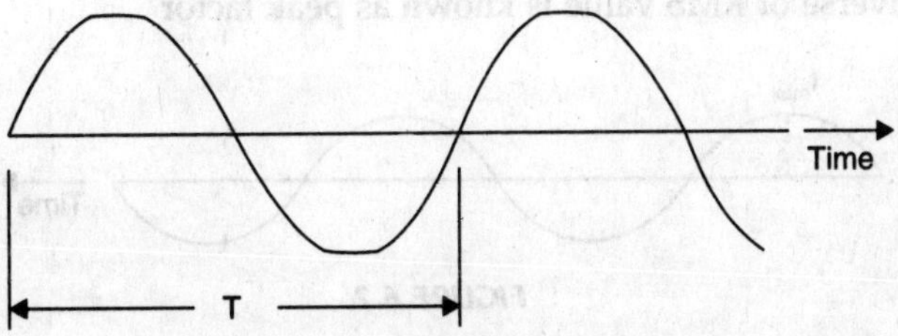


*FIGURE 6.1*

**Effective Values.** The effective value of an alternative emf or current is given by that direct emf or current which, when applied to a given circuit for a given time, produces the same expenditure of energy as when the alternating emf or current is applied to the same circuit for the same time. This is also known as root mean square (RMS) value or virtual, value.

The general function $Y_{(t)}$, with period $T$, has an effective value $Y_{rms}$ given by

$$Y_{rms} = \sqrt{\frac{1}{T}\int_0^T Y(t)^2\,dt}$$

The effective value of the functions $a \sin \omega t$ and $a \cos \omega t$ is $\frac{a}{\sqrt{2}}$.

### RMS Value for Several Sine and Cosine Terms

The function $X(t) = a_0 + (a_1 \cos \omega t + a_2 \cos 2\omega t + ......) + (b_1 \sin \omega t + b_2 \sin 2\omega t + ......)$ has an effective value given by

$$Y_{rms} = \sqrt{a_0^2 + \frac{1}{2}(a_1^2 + a_2^2 + ......) + \frac{1}{2}(b_1^2 + b_2^2 + ......)}$$

Also, if $A$, is the effective value of $a_1 \cos \omega t$, then

$$A_1 = \frac{a_1}{\sqrt{2}},\ A_1^2 = \frac{a_1^2}{2} \text{ and}$$

$$Y_{rms} = \sqrt{a_0^2 + (A_1^2 + A_2^2 + ......) + (B_1^2 + B_2^2 + ......)}$$

**Form Factor.** The ratio of effective value to average value is the form factor $F$ of the waveform. It has use in voltage generation and instrument correction factors.

$$\text{Form Factor} = \frac{Y_{rms}}{Y_{av}} = \sqrt{\frac{\frac{1}{T}\int_0^T Y(t)^2\,dt}{\frac{1}{T}\int_0^T Y(t)\,dt}}$$

**Peak Factor.** Inverse of RMS value is known as peak factor

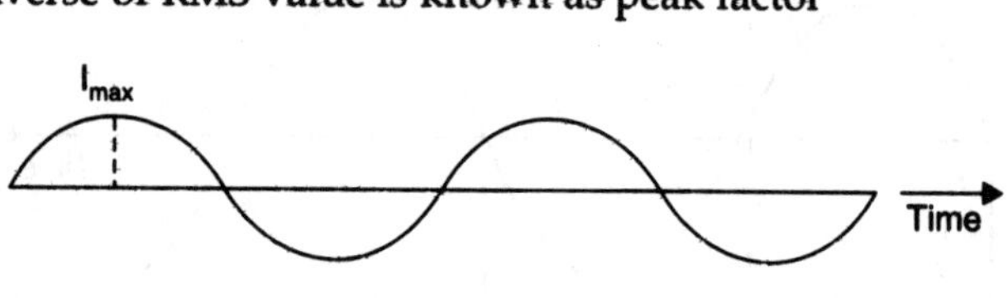


*FIGURE 6.2*

**Sine Wave**

$$\text{RMS Value} = \frac{I_{max}}{\sqrt{2}} = 0.707\,I_{max}$$

$$\text{Mean Value} = \frac{2}{\pi} I_m = 0.636\,I_m$$

$$\text{Form Factor} = \frac{\text{RMS value}}{\text{Mean value}} = \frac{0.707\,I_{max}}{0.636\,I_{max}} = 1.11$$

$$\text{Peak Factor} = \frac{1}{0.707} = 1.41$$

**Half Wave Rectified Sine Wave**

$$\text{RMS Value} = \frac{I_m}{2} = 0.5\ I_m$$

$$\text{Mean Value} = \frac{1}{\pi} I_m = 0.318\ I_m$$

$$\text{Form Factor} = \frac{0.5}{0.318} = 1.57$$

$$\text{Peak Factor} = \frac{1}{0.5} = 2.0$$

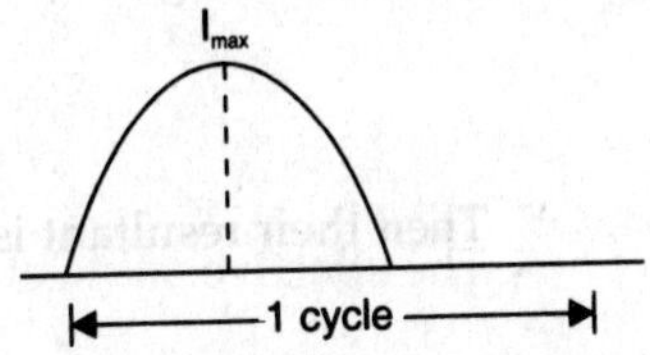


*FIGURE 6.3*

**Full Wave Rectified Sine Wave**

$$\text{RMS Value} = \frac{I_m}{\sqrt{2}} = 0.707\ I_m$$

$$\text{Mean Value} = \frac{2}{\pi} I_m$$

$$= 0.636\ I_m$$

$$\text{Form Factor} = \frac{0.707}{0.636} = 1.11$$

$$\text{Peak Factor} = \frac{1}{0.707} = 1.41.$$

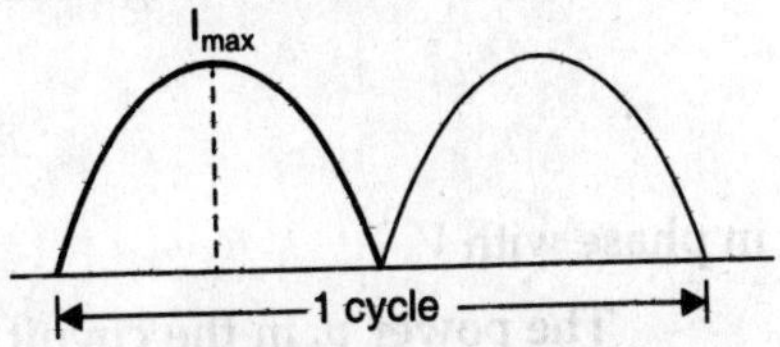


*FIGURE 6.4*

**Rectangular Wave**

RMS Value = $I_m$

Mean Value = $I_m$

Form Factor = 1.0

Peak Factor = 1.0.

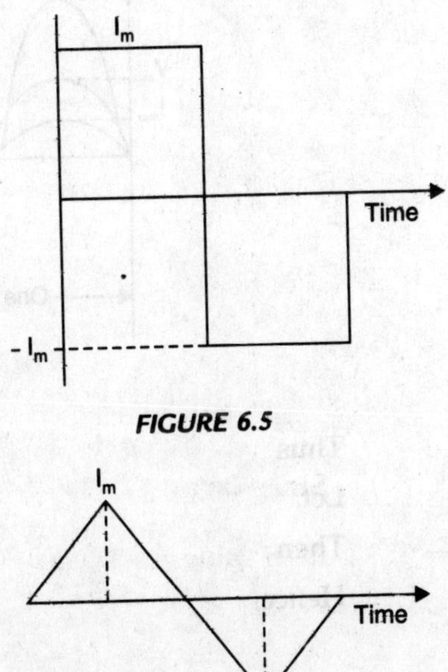


*FIGURE 6.5*

**Triangular Wave**

$$\text{RMS Value} = \frac{I_m}{\sqrt{2}} = 0.578\ I_m$$

$$\text{Mean Value} = \frac{I_m}{2} = 0.5\ I_m$$

$$\text{Form Factor} = \frac{0.578}{0.2} = 1.16$$

$$\text{Peak Factor} = \frac{1}{0.578} = 1.73.$$

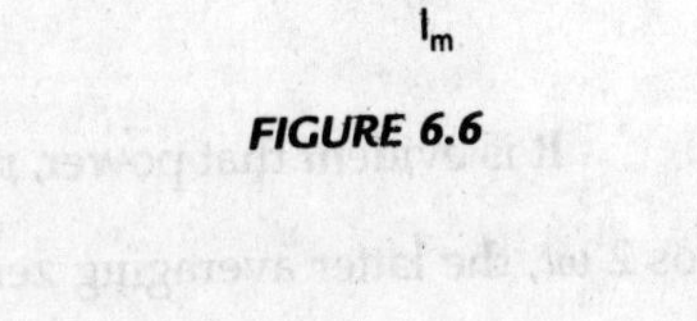


*FIGURE 6.6*

**Addition of Alternating EMFs or Currents.** Let there be two emfs' acting at the same time, their equations being

$$e_1 = E_1 \sin \omega t$$
$$e_2 = E_2 \sin (\omega t + \phi)$$

Then their resultant is given by

$$E^2_{max} = 0.707 \times \sqrt{(E_1^2 + E_2^2 + 2E_1E_2 \cos \phi)}$$

The instantaneous value is given by

$$e = (E_1 + E_2 \cos \phi) \sin \omega t + E_3 \sin \phi \cos \omega t.$$

**Purely Resistive Circuit.** A purely resistive circuit is one possessing neither inductance nor capacitance. Hence, if a current passes through the circuit, no back-emf will be set up by any change in current. The applied voltage $V$ has therefore to overcome the $RI$ drop only as in a direct current circuit. Thus we have, using RMS values

$$I = \frac{V}{R},$$

in phase with $V$.

The power $p$, in the circuit at any instant is the product of the instantaneous voltages and the instantaneous current.

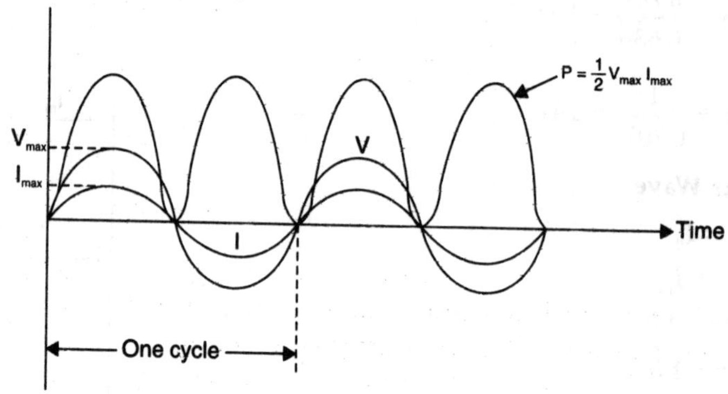


*FIGURE 6.7*

Thus $p = vi$

Let $v = V_{max} \sin \omega t$

Then $i = I_{max} \sin \omega t$

Hence $p = V_{max} I_{max} \sin^2 \omega t$

$$= \frac{1}{2} V_{max} I_{max} (1 - 2 \cos \omega t)$$

It is evident that power, $p$, has a constant part $\frac{1}{2} V_{max} I_{max}$ and fluctuating part $-\frac{1}{2} V_{max} I_{max}$ $\cos 2 \omega t$, the latter averaging zero over a complete cycle.

A wattmeter measures the average of the instantaneous power, $p$, because power is a scalar and not a vector quantity.

Hence average power

$$p = \text{average of } p$$
$$= \text{steady part of } p$$
$$= \frac{1}{2} V_{max} I_{max}$$
$$= \frac{V_{max}}{\sqrt{2}} \cdot \frac{I_{max}}{\sqrt{2}}$$
$$= V \cdot I.$$

Thus power in a purely resistive circuit is given by the product of the rms voltage and the rms current. The fluctuating power has twice the frequency of the voltage or current. This is illustrated by the curve of power on the wave diagram.

**Purely Inductive Circuit.** In purely inductive circuits the applied voltage is equal and opposite to the self induced emf at every instant.

When $$v = V_{max} \sin \omega t$$
$$= L \frac{di}{dt}$$
$$i = \frac{V_{max}}{L_{\omega}} \cos \omega t$$
$$T = \frac{V_{max}}{L.\omega} \sin\left(\omega t - \frac{\pi}{2}\right)$$

Thus current set up by a sinusoidal applied voltage is also sinusoidal and it lags the voltage by $\pi/2$ radians. Also the current is limited by the quantity $L\omega$, called the inductive reactance. Thus for inductive circuit

$$I = \frac{V}{L\omega}$$

where $V$ and $I$ are rms values.

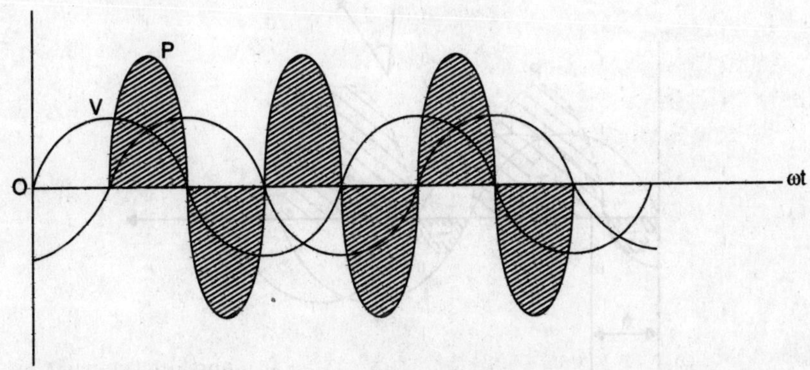


*FIGURE 6.8*

For the instaneous power

$$P = vi$$

$$= V_{max} \sin \omega t \cdot I_{max} \sin\left(\omega t - \frac{\pi}{2}\right)$$

$$= -V_{max} I_{max} \sin \omega t \cos \omega t$$

$$= -\frac{1}{2} V_{max} I_{max} \sin 2\omega t.$$

The average power is

$$P = -\frac{1}{2} V_{max} I_{max} \times (\text{average of } \sin 2\omega t)$$

$$= 0$$

The power is a pure sine wave of double frequency and maximum value

$$\frac{1}{2} E_{max} I_{max} = VI$$

The power is alternately positive and negative, the alternate lobes of the curve being of equal magnitude.

**Resistance and Inductance in Series**

Let $V$ = Total applied voltage

$V_1$ = Voltage drop along $R$

$V_2$ = Voltage drop along $L$

Then $V_1 = IR$ in phase with $I$

$V_2 = L\omega I$ leading $I$ by $\frac{\pi}{2}$

Current $$I = \frac{V}{\sqrt{(R^2 + (L\omega)^2)}}$$

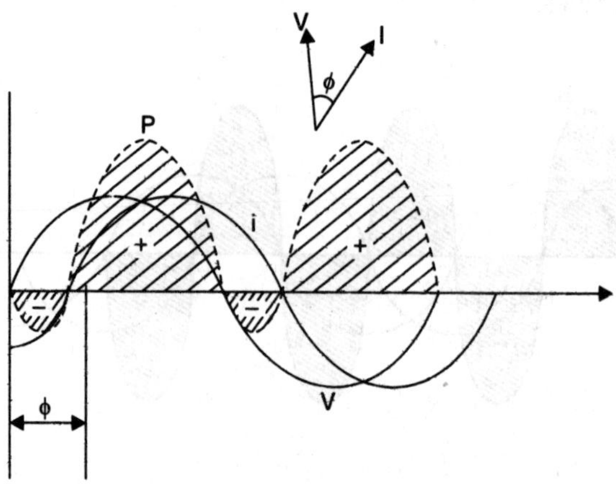


*FIGURE 6.9 (a)*

Current lags the voltage by an angle $\phi$

$$\phi = \tan^{-1}\frac{L\omega}{R}$$

Power
$$P = \frac{V_{max}}{\sqrt{2}}\frac{I_{max}}{\sqrt{2}}\cos\phi$$

$$= VI\cos\phi.$$

Thus in this case the average power, or the rate at which electrical energy is converted into heat, is $VI \cos\phi$ watts. The product $VI$ is called the apparent power. It is measured in volt-amperes (VA). The ratio of average power to apparent power is called the power factor.

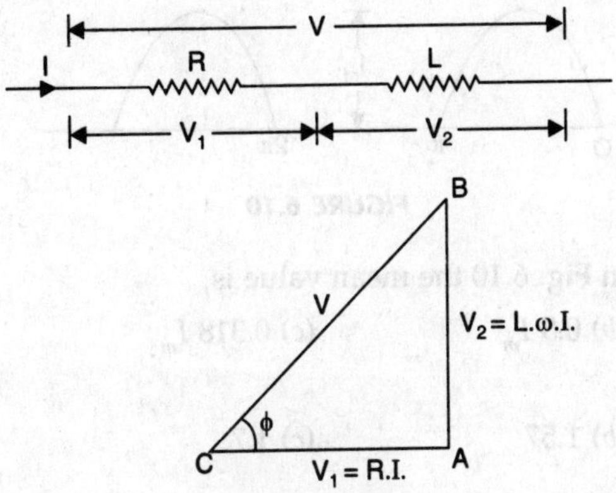


*FIGURE 6.9 (b)*

With sinusoidal voltage and current, the power factor is $\cos\phi$. Thus $VI$ must be multiplied by $\cos\phi$ in order to obtain true power. In *R-L* circuit the power curve is a periodic component of double frequency. However, the positive lobes of power are greater than the negative lobes. This is because positive power has always to be drawn from the supply on account of the $I^2R$ loss due to resistance.

The quantity $VI \sin\phi$ is called the reactive power. It is measured in reactive volt-amperes or vars (VAr).

## OBJECTIVE TYPE QUESTIONS

1. For a sine wave with peak value $I_m$ the rms value is

   (*a*) 0.818 $I_m$ (*b*) 0.707 $I_m$ (*c*) 0.5 $I_m$ (*d*) 1.414 $I_m$.

2. For a sine wave with peak value $V_{max}$ the mean value is

   (*a*) 0.707 $V_{max}$ (*b*) 0.636 $V_{max}$ (*c*) 0.500 $V_{max}$ (*d*) 0.414 $V_{max}$.

3. Form factor is the ratio

(*a*) $\frac{\text{RMS value}}{\text{Peak value}}$ (*b*) $\frac{\text{Mean value}}{\text{Peak value}}$ (*c*) $\frac{\text{RMS value}}{\text{Mean value}}$ (*d*) $\frac{\text{Mean value}}{\text{RMS value}}$.

4. Form factor for a sine wave is

(*a*) 0.667 (*b*) 0.707 (*c*) 1.11 (*d*) 1.414.

**Questions 5 to 8 Refer to Fig. 6.10**

5. For the half wave rectified sine wave shown in Fig. 6.10. RMS value is

(*a*) 0.707 $I_m$ (*b*) 0.5 $I_m$ (*c*) 0.3535 $I_m$ (*d*) 0.318 $I_m$.

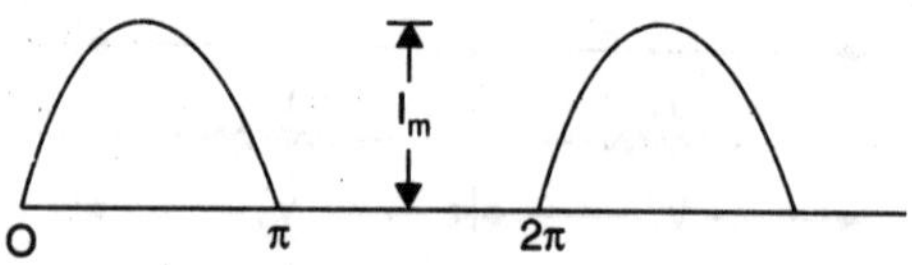


***FIGURE 6.10***

6. For the wave shown in Fig. 6.10 the mean value is

(*a*) 0.707 $I_m$ (*b*) 0.5 $I_m$ (*c*) 0.318 $I_m$ (*d*) 0.25 $I_m$.

7. The form factor is

(*a*) 1.11 (*b*) 1.57 (*c*) 1.7 (*d*) 2.0.

8. The peak factor is

(*a*) 1.41 (*b*) 2.0 (*c*) 2.82 (*d*) infinite.

**Questions 9 to 12 Refer to Fig. 6.11:**

9. For the full wave rectified sine wave shown in Fig. 6.11 RMS value is

(*a*) 0.707 $I_m$ (*b*) 0.636 $I_m$ (*c*) 0.5 $I_m$ (*d*) 0.318 $I_m$.

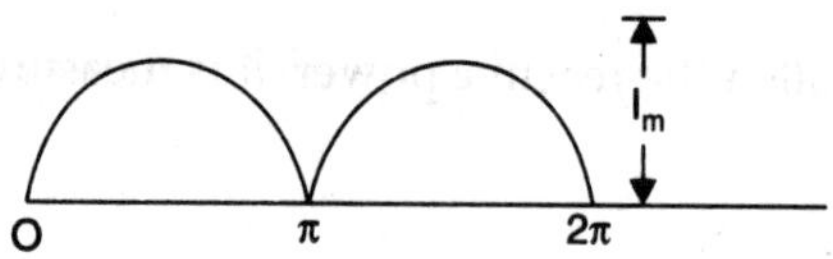


***FIGURE 6.11***

10. The mean value is

(*a*) 0.707 $I_m$ (*b*) 0.636 $I_m$ (*c*) 0.5 $I_m$ (*d*) 0.25 $I_m$.

11. The form factor is

(*a*) 1.0 (*b*) 1.11 (*c*) 2.0 (*d*) 2.22.

12. The peak factor is

(*a*) 1.11 (*b*) 1.25 (*c*) 1.41 (*d*) 2.82.

**Questions 13 to 16 Refer to Fig. 6.12:**

**13.** For the square wave shown in Fig. 6.12 RMS value is

(*a*) zero (*b*) $I_m$ (*c*) $\frac{I_m}{2}$ (*d*) infinite.

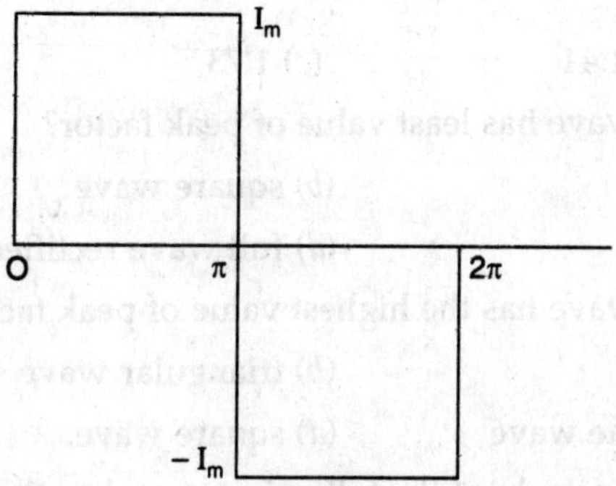


*FIGURE 6.12*

**14.** The mean value is

(*a*) zero (*b*) $I_m$ (*c*) $\frac{I_m}{2}$ (*d*) infinite.

**15.** The form factor is

(*a*) 2.0 (*b*) 1.0 (*c*) 0.5 (*d*) zero.

**16.** The peak factor is

(*a*) 2.0 (*b*) 1.0 (*c*) 0.5 (*d*) zero.

**Questions 17 to 20 refer to Fig. 6.13:**

**17.** For the triangular wave shown in Fig. 6.13 RMS value is

(*a*) $\sqrt{3}\, I_m$ (*b*) $\frac{I_m}{\sqrt{3}}$ (*c*) $0.578\, I_m$ (*d*) $0.289\, I_m$.

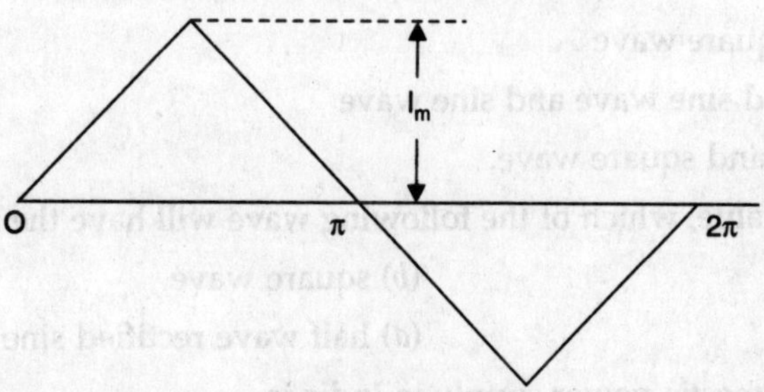


*FIGURE 6.13*

18. The mean value is

(*a*) 0.707 $I_m$ (*b*) 0.636 $I_m$ (*c*) 0.5 $I_m$ (*d*) 0.41 $I_m$.

19. Form factor is

(*a*) 1.0 (*b*) 1.16 (*c*) $\sqrt{3}$ (*d*) $\sqrt{2}$.

20. The peak factor is

(*a*) 1.0 (*b*) 1.41 (*c*) 1.73 (*d*) 2.

21. Which of the following wave has least value of peak factor?

(*a*) sine wave (*b*) square wave

(*c*) triangular wave (*d*) full wave rectified sine wave.

22. Which of the following wave has the highest value of peak factor?

(*a*) sine wave (*b*) triangular wave

(*c*) half wave rectified sine wave (*d*) square wave.

23. For the same peak value, which of the following wave has the least mean value?

(*a*) sine wave (*b*) square wave

(*c*) triangular wave (*d*) half wave rectified sine wave.

24. For the same peak value, which of the following wave has the highest mean value?

(*a*) sine wave (*b*) square wave

(*c*) half wave rectified sine wave (*d*) full wave rectified sine wave.

25. All of the following waves have form factor less than 1.5 EXCEPT

(*a*) sine wave (*b*) full wave rectified sine wave

(*c*) half wave rectified sine wave (*d*) square wave.

26. For the same peak value which of the following wave will have the least RMS value?

(*a*) sine wave (*b*) square wave

(*c*) triangular wave (*d*) half wave rectified sine wave.

27. For the same peak value, which of the following pairs of waves will have identical RMS value?

(*a*) half-wave rectified sine wave and full wave rectified sine wave

(*b*) sine wave and square wave

(*c*) full wave rectified sine wave and sine wave

(*d*) triangular wave and square wave.

28. For the same peak value, which of the following wave will have the highest RMS value?

(*a*) sine wave (*b*) square wave

(*c*) triangular wave (*d*) half wave rectified sine wave.

29. The frequency of domestic power supply in India is

(*a*) 25 Hz (*b*) 50 Hz (*c*) 100 Hz (*d*) 220 Hz.

30. The voltage of domestic supply is 220 V. This figure represents
(a) peak value (b) RMS value (c) average value (d) mean value.

31. The RMS value of half wave rectified sine wave is 100 volts. The RMS value of full wave rectified ac will be
(a) 100 V (b) 111 V (c) 141 V (d) 200 V.

32. For 100 volts RMS value triangular wave, the peak voltage will be
(a) 100 V (b) 111 V (c) 141 V (d) 173 V.

33. A sine wave of voltage varies from zero to a maximum of 100 V. How much is the voltage at the instant of 30° of the cycle?
(a) 25 V (b) 41.4 V (c) 50 V (d) 86.6 V.

34. In the above case the value at 270° will be
(a) 50 V (b) 100 V (d) – 86.6 V (d) – 100 V.

35. 1 Gigahertz is the same as
(a) $10^6$ Hz (b) 1000 MHz (c) 1000 kHz (d) $10^{-6}$ pico Hz.

36. The time period for the frequency of 1 MHz is
(a) 0.01 second (b) 1 milli second (c) 1 micro second (d) 1 pico second.

37. For a frequency of 400 Hz, the time period will be
(a) 0.25 second (b) 0.025 second (c) 0.0025 second (d) 0.00025 second.

38. The wavelength for a ratio wave of 30 GHz will be
(a) 1 mm (b) 1 cm (c) 10 cm (d) 1 metre.

39. For two waves with a frequency of 1 MHz, how much time is the phase angle of 36°?
(a) 10 micro seconds (b) 1 micro second (c) 0.1 micro second (d) 0.07 micro second.

40. What frequency will be the fourth harmonic of 12 MHz?
(a) 3 MHz (b) $\frac{3}{4}$ MHz (c) 48 MHz (d) 192 MHz.

41. What frequency is one octave above 220 Hz?
(a) 55 Hz (b) 110 Hz (c) 220 Hz (d) 440 Hz.

42. For two sinusoidal quantities to be in phase quadrature, the phase difference between the two must be
(a) zero (b) 90° (c) 120° (d) 180°.

43. The voltage in a circuit follows the law

$$v = 200 \sin \omega t.$$

If the frequency is 25 Hz, how long will it take for the voltage to rise to 100 volts?
(a) $\frac{1}{25}$ second (b) $\frac{1}{50}$ second (c) $\frac{1}{100}$ second (d) $\frac{1}{300}$ second.

44. The RMS value and mean value is the same in case of
(a) sine wave (b) square wave
(c) triangular wave (d) half wave rectified sine wave.

45. The current in a circuit is given by

$$i = 50 \sin 314\, t \text{ amperes}$$

The maximum value and frequency of current are

(*a*) 50 A; 314 Hz (*b*) 50 A; 50 Hz (*c*) $50\sqrt{2}$ A; 100 Hz (*d*) $50\sqrt{2}$ A; 314 Hz.

46. The time delay for a phase angle of 45° at the frequency of 500 Hz will be

(*a*) 0.25 ms (*b*) 2.5 ms (*c*) 25 ms (*d*) 250 ms.

47. Which of the frequency for the following time periods will be least

(*a*) 0.05 second (*b*) 5 milli seconds (*c*) 5 micro seconds (*d*) 5 nano seconds.

48. How much RMS current does a 300 W 120 V bulb take from the 120 V, 50 Hz power line?

(*a*) 4.1 A (*b*) 2.5 A (*c*) 1.6 A (*d*) 1.2 A.

49. Two waves of the same frequency have opposite phases when the phases angle between them is

(*a*) 0° (*b*) 90° (*c*) 180° (*d*) 360°.

**Questions 50 to 53 Refer to Data Given Below:**

A 60 Hz power line voltage of 120 V is applied across a resistance of 10 ohms.

50. The RMS current in the circuit is

(*a*) 168 A (*b*) 120 A (*c*) 16.8 A (*d*) 12 A.

51. What will be the frequency of current?

(*a*) 180 Hz (*b*) 120 Hz (*c*) 60 Hz (*d*) 50 Hz.

52. What will be the phase angle between the current and voltage?

(*a*) 180° (*b*) 120° (*c*) 90° (*d*) zero.

53. How much dc applied voltage would be necessary for the same heating effect in the resistance ?

(*a*) 168 V (*b*) 154 V (*c*) 120 V (*d*) 86.8 V.

54. Which of the following statement is false?

(*a*) when two sine waves are 90° out of phase, on has its peak value when the other is at zero

(*b*) a period of 2 micro seconds corresponds to a higher frequency than time period of one micro second

(*c*) when we compare the phase between two wave forms, they must have the same frequency

(*d*) a wave length of 2 metres corresponds to a lower frequency than a wave length of 1 metre.

55. The phase difference between voltage and current wave through a circuit element is given as 30°. The essential condition is that

(*a*) both waves must have zero at the same time

(*b*) both waves must have identical peak values

(*c*) both waves must have same frequency

(*d*) none of the above.

**56.** For 50 Hz power line the third harmonic is

(*a*) $16\frac{2}{3}$ Hz (*b*) $33\frac{1}{3}$ Hz (*c*) $83\frac{1}{2}$ Hz (*d*) 150 Hz.

**57.** The time delay for a phase angle of 45° at the frequency of 500 Hz will be

(*a*) 25 milli seconds (*b*) 2.5 milli seconds (*c*) 0.25 milli second (*d*) 0.025 milli second.

**58.** The frequency corresponding to a period of 5 nano seconds will be

(*a*) 0.2 kHz (*b*) 0.2 MHz (*c*) 0.2 GHz (*d*) none of the above.

**59.** An ac voltage of 50 Hz has a maximum value of 100 V. Its value $\frac{1}{600}$ second after the instant the current is zero, will be

(*a*) 10 V (*b*) 25 V (*c*) 50 V (*d*) 86.6 V.

**60.** The RMS value of 170 A peak will be

(*a*) 170 A (*b*) 120 A (*c*) 100 A (*d*) 50 A.

**61.** When two waves are in phase, they have peak values at an interval of

(*a*) 90° (*b*) 120° (*c*) 180° (*d*) none of the above.

**62.** When two waves are 90° out of phase

(*a*) both will have negative peak at the same time

(*b*) both will have positive peak at the same time

(*c*) one will have its peak value when the other will have its zero value

(*d*) one will have its peak value when the other will have its negative peak value.

**63.** If $V_1 = A \sin \omega t$ and

$V_2 = A \sin (\omega t - \phi°)$ then

(*a*) $V_1$ lags $V_2$ by $\phi°$ (*b*) $V_2$ leads $V_1$ by $\phi°$

(*c*) $V_1$ and $V_2$ are in phase (*d*) none of the above.

**64.** The power consumed in a circuit element will be least when the phase difference between the current and voltage is

(*a*) 0° (*b*) 30° (*c*) 90° (*d*) 180°.

**65.** Equation for voltage through a circuit is

$$v = 10 + 5 \cos (628\, t + 30°) \text{ volts}$$

The RMS value of the voltage is

(*a*) 5 V (*b*) 10 V (*c*) 10.6 V (*d*) 15 V.

**66.** The negative maximum of a cosine wave occurs at

(*a*) 45° (*b*) 90° (*c*) 180° (*d*) 270°.

**67.** A wave has RMS value of 10 V and frequency of 50 Hz. This can be represented as

(*a*) 10 sin 314*t* (*b*) 10 sin 100*t* (*c*) 14.1 sin 50*t* (*d*) 14.1 sin 314*t*.

**68.** The current $i = 180 \cos (\omega t - 161.5°)$ may be represented as sine function by the relation

(*a*) $i = 180 \sin (\omega t + 18.5°)$ (*b*) $i = 180 \sin (\omega t - 71.5°)$

(*c*) $i = 180 \sin (\omega t + 71.5°)$ (*d*) $i = 180 \sin (\omega t - 18.5°)$

**Questions 69 and 70 Refers to Data Given Below:**

The question of an emf is given by

$$e - I_m \sqrt{R^2 + 4\omega^2 R^2} \sin 2\omega t$$

**69.** In this case the amplitude is

(*a*) $I_m \sqrt{R^2 + 4\omega^2 R^2}$ (*b*) $I_m (R^2 + 4\omega^2 R^2)$

(*c*) $\sqrt{2}\, I_m (R^2 + 4\omega^2 R^2)$ (*d*) $I_m$.

**70.** In the above problem, the frequency in Hertz is

(*a*) $\omega$ (*b*) $2\omega$ (*c*) $\dfrac{\omega}{\pi}$ (*d*) $\dfrac{\omega}{2\pi}$.

**Questions 71 to 73 Refer to Data Given Below:**

The equation for alternating voltage is

$$v = 42.4 \sin 628\, t.$$

**71.** The average value of the voltage is

(*a*) 42.4 V (*b*) 30 V (*c*) 27 V (*d*) 21.2 V.

**72.** The frequency of the voltage is

(*a*) 628 Hz (*b*) 314 Hz (*c*) 50 Hz (*d*) 100 Hz.

**73.** The form factor of dc voltage is

(*a*) zero (*b*) infinite (*c*) unity (*d*) 0.5.

**74.** In a series circuit of $R = 20$ ohms and $L = 0.06$ henrys, the current lags the voltage by 80°. The value of $\omega$ will be

(*a*) 80 rad/s (*b*) 180 rad/s (*c*) 696 rad/s (*d*) 1893 rad/s.

**75.** In a series combination of $R$ and $L$

(*a*) current will lag the voltage by 90°

(*b*) current will lag the voltage by some angle between 0° and 90°

(*c*) current will lead the voltage by some angle between 0° and 90°

(*d*) current will lead the voltage by some angle between 90° and 180°.

**76.** In a series *R-L* circuit the instantaneous voltage is given by

$$v = \left(\sqrt{R^2 + (\omega L)^2}\right) I_m \sin\left(\omega t + \tan^{-1}\frac{\omega L}{R}\right)$$

The above equation indicates that

(*a*) current lags the voltage by 90° (*b*) current leads the voltage by 90°

(*c*) currrent leads the voltage by an angle $\tan^{-1}\dfrac{\omega L}{R}$

(*d*) current lags the voltage by an angle $\tan^{-1}\dfrac{\omega L}{R}$.

**77.** In the above case the magnitude of the impedance is

(*a*) $\dfrac{L}{R}$ (*b*) $LR$ (*c*) $\omega LR$ (*d*) $\sqrt{R^2+(\omega L)^2}$.

**Questions 78 and 79 Refer to Data Given Below:**

**78.** In a series circuit consisting of $R = 10\ \Omega$ and $L = 20$ mH, the current flowing is $i = 2 \sin 500\, t$ amperes.

In this circuit

(*a*) the current will lag the voltage by 45°

(*b*) the current will lead the voltage by 45°

(*c*) the current will be in phase with the voltage

(*d*) the current will lag the voltage by an angle between 45° and 90°.

**79.** The total applied voltage is given by

(*a*) $40 \sin 500\, t$ (*b*) $28.28 \sin 500\, t$

(*c*) $28.28 \sin (500\, t + 45°)$ (*d*) $28.28 \sin (500\, t + 60°)$

**Questions 80 and 81 Refer to Data Given Below:**

A series *RL* circuit has $L = 0.02$ henrys and an impedance of 17.85 ohms. With a sinusoidal voltage applied, the current lags the voltage by 63.4°.

**80.** The resistance in the circuit must have the value of

(*a*) 2 ohms (*b*) 4 ohms (*c*) 6 ohms (*d*) 8 ohms.

**81.** The $\omega$ must have the value of

(*a*) 800 radians/second (*b*) 400 radians/second

(*c*) 200 radians/second (*d*) 100 radians/second.

**82.** The RMS value of pure cosine function is

(*a*) zero (*b*) 0.707 of peak value

(*c*) $\dfrac{1}{0.707}$ of peak value (*d*) same as peak value.

83. The average power in a pure resistor of 10 ohms which carries a current $i = 14.14 \cos \omega t$ amperes will be

(*a*) zero (*b*) 500 W (*c*) 707 W (*d*) 1000 W.

**Questions 84 and 85 Refer to Fig. 6.14:**

For the wave form shown in Fig. 6.14 during first interval $y = 10\ e^{-200t}$

84. The average value of the wave form is

(*a*) 0.5 (*b*) 1.0 (*c*) 1.41 (*d*) 2.24.

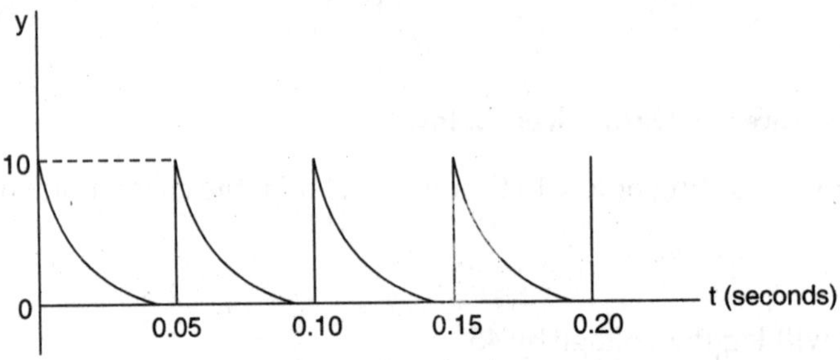


*FIGURE 6.14*

85. The RMS value of the wave form is

(*a*) 1.0 (*b*) 1.2 (*c*) 1.88 (*d*) 2.24.

**Questions 86 to 88 Refer to Triangular Wave shown in Fig. 6.15:**

86. The value of $Y_{rms}$ for the given wave is

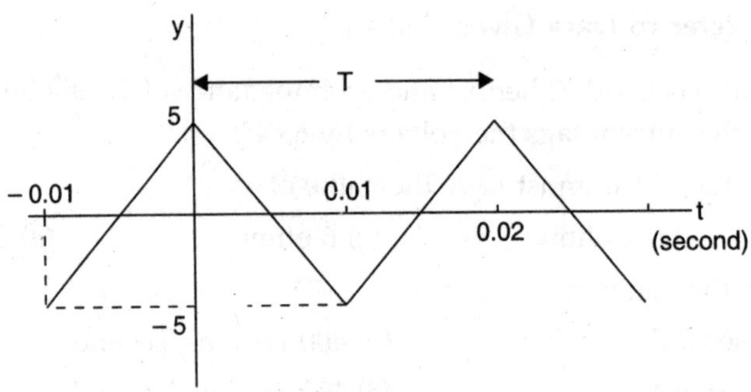


*FIGURE 6.15*

(*a*) 1.33 (*b*) 4.16 (*c*) 8.33 (*d*) 12.44.

87. The value of $Y_{average}$ for the given wave is

(*a*) zero (*b*) 2.5 (*c*) 5.0 (*d*) 10.0.

88. The form factor for the given wave is

(*a*) 1 (*b*) 1.10 (*c*) 1.16 (*d*) 1.76.

**Questions 89 to 91 Refer to Fig. 6.16:**

For the square wave shown in Fig. 6.16

**89.** The average value is

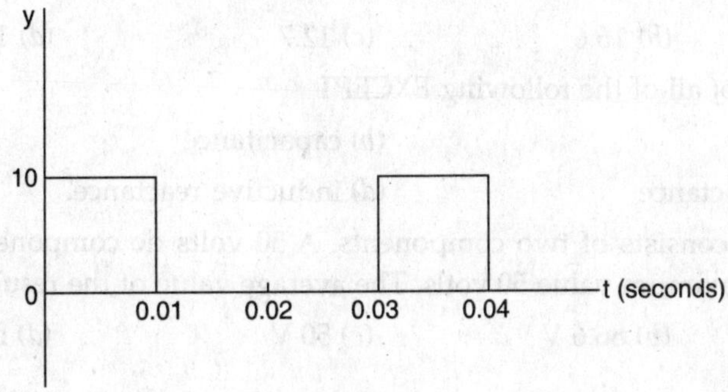


*FIGURE 6.16*

(*a*) 10 (*b*) 5 (*c*) 3.33 (*d*) 2.5.

**90.** For the square wave shown in Fig. 6.16 the rms value is

(*a*) 7.5 (*b*) 6.67 (*c*) 5.57 (*d*) 5.

**91.** The form factor is

(*a*) 1.1 (*b*) 1.3 (*c*) 1.67 (*d*) 1.78

**92.** The effective value of the function

$y = 50 + 30 \sin \omega t$ is

(*a*) 46.4 (*b*) 50 (*c*) 54.3 (*d*) 68.7.

**93.** The effective value of the function

$y = 50 + 141.4 \sin \omega t + 35.5 \sin 3\omega t$ volts will be

(*a*) 141.4 volts (*b*) 114.6 volts (*c*) 100 volts (*d*) 85.5 volts.

**94.** A 25 ohm resistor has an average power of 400 watts. The maximum value of the current function, if it is sinusoidal, will be

(*a*) 5.66 A (*b*) 6.66 A (*c*) 8.31 A (*d*) 9.99 A.

**95.** In the above case, in case the current function is triangular the maximum value will be

(*a*) 2.3 A (*b*) 4.68 A (*c*) 5.66 A (*d*) 6.93 A.

**96.** The effective value of the voltage given by

$v = 100 + 25 \sin 3\omega t + 10 \sin 5\omega t$ volts, will be

(*a*) 100 V (*b*) 101.8 V (*c*) 110 V (*d*) 135 V.

**97.** The effective value of voltage given by

$v = 100 + A \sin \omega t$ volts

is known to be 103.1 volts. The amplitude $A$ of the sine term will be

(*a*) 25 V (*b*) 35.5 V (*c*) 59.4 V (*d*) 66.5 V.

98. What averages power results in a 25 ohm resistor when it passes a current $i(t) = 2 + 3 \sin \omega t + 2 \sin 2\omega t + 1 \sin 3\omega t$ amperes?

(a) 116.4 W (b) 129.4 W (c) 250 W (d) 275 W.

99. A half wave rectified sine wave has an effective value of 20. Its average value will be

(a) 20 (b) 16.6 (c) 12.7 (d) 10.

100. Ohm is the unit of all of the following EXCEPT

(a) resistance (b) capacitance

(c) capacitive reactance (d) inductive reactance.

101. A voltage wave consists of two components. A 50 volts dc component and a sinusoidal component of maximum value 50 votls. The average value of the resultanet will be

(a) zero (b) 86.6 V (c) 50 V (d) none of the above.

**Questions 102 and 103 Refer to Fig. 6.17:**

102. The average value of the wave form shown in Fig. 6.17 is

(a) 10 (b) 15 (c) 25 (d) 55.

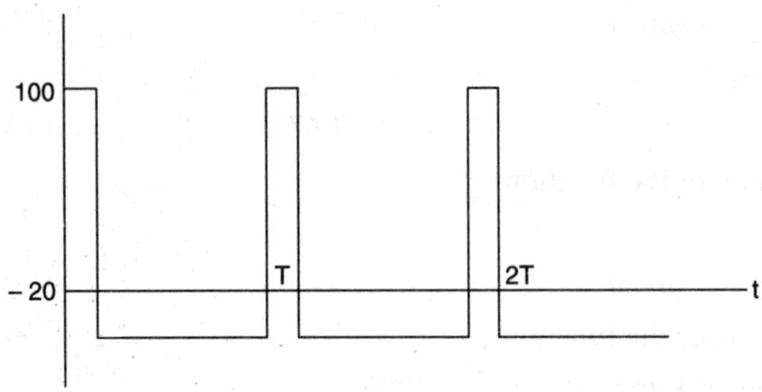


*FIGURE 6.17*

103. The RMS value of the wave form shown in Fig. 6.17 is

(a) 49.7 (b) 52.9 (c) 55.5 (d) 60.

**Questions 104 and 105 Refer to Fig. 6.18:**

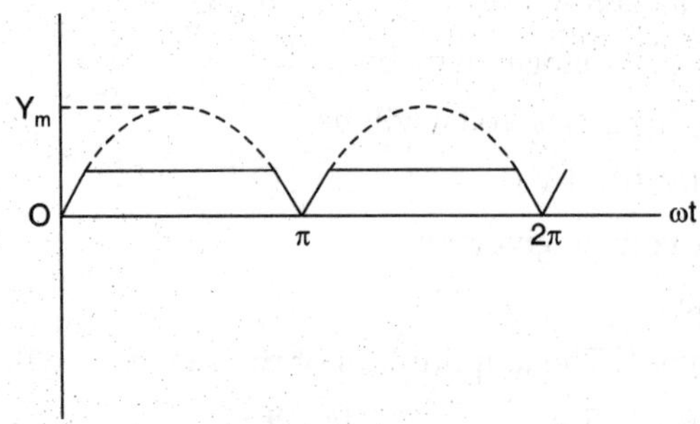


*FIGURE 6.18*

**104.** A full-wave rectified sine wave is clipped at one-half of its peak value. The RMS value of the rectified sine wave will be

(*a*) 0.5 $Y_m$ (*b*) 0.498 $Y_m$ (*c*) 0.478 $Y_m$ (*d*) 0.422 $Y_m$.

**105.** If the effective value of above rectified sine wave is 0.5 $Y_m$, the amplitude of the wave form at which it is clipped will be

(*a*) 0.581 $Y_m$ (*b*) 0.5 $Y_m$ (*c*) 0.432 $Y_m$ (*d*) none of the above.

**106.** In case the above wave form is clipped at 60°, the RMS value will be

(*a*) 0.75 $Y_m$ (*b*) 0.668 $Y_m$ (*c*) 0.558 $Y_m$ (*d*) 0.5 $Y_m$.

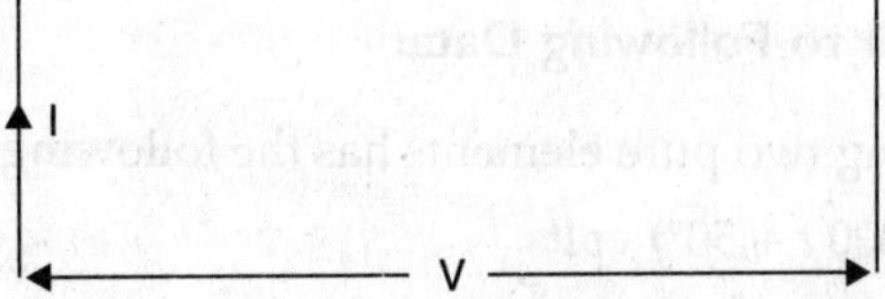


*FIGURE 6.19*

**107.** A current $i = 2 \sin 500\,t$ amperes is passed through the circuit shown in Fig. 6.19. The total applied voltage will be

(*a*) 14.14 sin 500*t* (*b*) 28.28 sin 500*t*

(*c*) 28.28 sin (500 *t* + 45°) (*d*) 28.28 cos (500 *t* + 45°)

**108.** Which of the following statement about the circuit is valid?

(*a*) $R = L$ (*b*) $R = \omega L$ (*c*) $RL = 1$ (*d*) $\omega RL = 1$.

**109.** The current

(*a*) lags the voltage by 45° (*b*) leads the voltage by 45°

(*c*) lags the voltage by 60° (*d*) leads the voltage by 60°.

**110.** A pure inductance of $L = 0.01$ H passes a current $i = 10 \cos 2000\,t$ amperes. The voltage across the element is

(*a*) 50 cos 2000 *t* (*b*) 50 cos (2000 *t* + 90°)

(*c*) 100 cos (2000 *t* + 90°) (*d*) 200 cos (2000 *t* + 90°).

**111.** When a voltage $v = 100 \sin 5000\,t$ volts is applied to a series circuit of $L = 0.05$ H and an unknown capacitance, the resulting current is $i = 2 \sin (5000\,t + 90°)$ amperes. The value of the capacitance is

(*a*) 66.7 pF (*b*) 6.67 pF (*c*) 0.667 μF (*d*) 6.67 μF.

**112.** When a current $i = 12 \sin 2000\,t$ amperes passes through a single element circuit the resulting voltage across the element is 200 sin (2000 *t* – 90°).

It can be concluded that

(*a*) the element is non-linear (*b*) the element has large leakage factor

(*c*) the element is a capacitor

(*d*) the element is a combination of resistance and inductance.

113. A series circuit consisting of two pure elements has the following applied voltage and current

$$v(t) = 150 \cos (200\, t - 30°) \text{ volts}$$

$$i(t) = 4.48 \cos (200\, t - 56.6°) \text{ amperes}$$

It can be concluded that

(*a*) one of the elements is a capacitor

(*b*) the circuit elements are a capacitor and a resistor

(*c*) the reactance of capacitor has numerical value half that of resistor

(*d*) the circuit elements are a resistor and an inductor.

## Questions 114 and 115 Refer to Following Data:

A series circuit containing two pure elements has the following current and applied voltage

$$v = 200 \sin (2000\, t + 50°) \text{ volts}$$

$$i = 4 \cos (2000\, t + 13.2°) \text{ amperes.}$$

114. From the above information it can be concluded that

(*a*) current leads the voltage by 36.8° (*b*) current lags the voltage by 36.8°

(*c*) current lags the voltage by 63.8° (*d*) current leads the voltage by 63.8°

115. The circuit elements must be

(*a*) resistance and capacitance (*b*) capacitance and inductance

(*c*) inductance and resistance.

116. The value of inductance must be

(*a*) 0.05 H (*b*) 10 H (*c*) 0.015 H (*d*) 30 H.

117. A parallel circuit with one branch of R = 5 ohms and a single unknown element in the other branch has the following applied voltage and total current

$$v = 10 \cos (50\, t + 60°) \text{ volts}$$

$$i = 5.38 \cos (50\, t - 8.23°) \text{ amperes.}$$

The unknown element could be

(*a*) inductor of 0.04 henry (*b*) inductor of 0.04 milli henry

(*c*) capacitor of 4 μF (*d*) capacitor of 4 *p*F.

118. A pure capacitor of C = 35 μF is in parallel with another single circuit element. The applied voltage and resulting current are as given below.

$$v = 150 \sin 3000\, t \text{ volts}$$

$$i = 16.5 \sin (3000\, t + 72.4°) \text{ amps.}$$

The other element is

(*a*) capacitor of 30 μF (*b*) inductor of 30 mH

(*c*) resistor of 30 ohms (*d*) none of the above.

**119.** The voltage $v$ and the current $i$ of a device are

$$v = 100 \sin 377\, t \text{ volts}$$

$$i = 10 \sin (377\, t + 30°) \text{ amp.}$$

The power $P$ as indicated by a wattmeter will be

(*a*) 1000 W (*b*) 774 W (*c*) 500 W (*d*) 433 W.

**120.** $(8 + j\,12)\,(j\,2) =$

(*a*) $4 + j\,6$ (*b*) $4 - j\,6$ (*c*) $6 + j\,4$ (*d*) $6 - j\,4$.

**121.** $(3 + j\,3)\,(2 + j\,2) =$

(*a*) 0.5 (*b*) 1.5 (*c*) 2.5 (*d*) 9.

**122.** $j\,5/(2 - j\,2) =$

(*a*) $-1.25 + j\,1.25$ (*b*) $1.25 + j\,1.25$ (*c*) $1.25 - j\,1.25$ (*d*) none of the above.

**123.** Which of the following is equal to $(-1 - j\,2)$?

(*a*) $(5 + j\,5)\,(1 - j\,1)$ (*b*) $(4 - j\,8)\,(2 + j\,2)$ (*c*) $(5 - j\,10)\,(3 + j\,4)$ (*d*) $j\,5\,(2 - j\,2)$.

**124.** The following sum will be equal to

(10 53.1° !) + $(4 + j\,2)$

(*a*) $1 + j\,10$ (*b*) $10 + j\,10$ (*c*) $10 + j\,10$ (*d*) $10 + j\,20$.

**125.** $(-5 + j\,5)$ – (7.07 135° !) =

(*a*) 0 (*b*) 1 (*c*) $1 + j\,1$ (*d*) $1 - j\,1$.

**126.** (2.83 45° !) – $(2 - j\,8)$ =

(*a*) 1 (*b*) 10 (*c*) $j\,10$ (*d*) $-j\,10$.

**127.** The complex number $20e^{j\pi/4}$ in polar form will be

(*a*) 20 20° ! (*b*) 20 45° ! (*c*) 10 45° ! (*d*) 10 22.5° !

**128.** Which complex number is equivalent to polar form 10 – 210°?

(*a*) $10\, e^{j\pi/6}$ (*b*) 8 120 !° (*c*) $10\, e^{-j\,7\pi/6}$ (*d*) $10e^{-j\,5\pi/6}$

**129.** The root of the complex number $\sqrt{5 + j\,8}$ will be

(*a*) 5 30° ! (*b*) 8 120°! (*c*) 3.07 29° ! (*d*) 30.7 119° !

**130.** Which of the following medium has the least value of permittivity?

(*a*) air (*b*) glass (*c*) mica (*d*) titanium dioxide.

**131.** Dielectric strength of air is nearly

(*a*) 0.3 kV/mm (*b*) 3 kV/mm (*c*) 30 kV/mm (*d*) 300 kV/mm.

**132.** Which of the following medium has dielectric strength less than that of air?

(*a*) paper (*b*) paraffin wax (*c*) glass (*d*) none of the above.

**133.** Which of the following medium has dielectric strength more than 20 kV/mm?

(*a*) paper (*b*) paraffin wax (*c*) mica (*d*) glass.

**134.** The energy stored in a capacitor, $E$ is given by all of the following relations EXCEPT

(*a*) $E = \frac{QV}{2}$ (*b*) $E = \frac{CV^2}{2}$ (*c*) $E = \frac{Q^2}{2C}$ (*d*) $E = \frac{QCV}{2}$.

**135.** A curve for the charging of a capacitor is shown in Fig. 6.20. In this graph the abscissa represents

(*a*) time (*b*) current (*c*) voltage (*d*) capacitance.

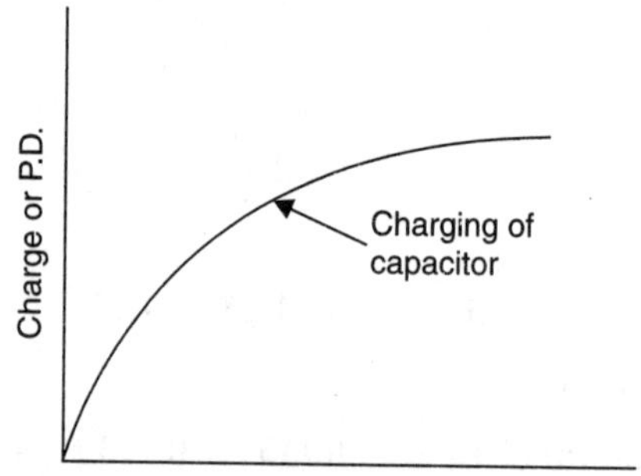


***FIGURE 6.20***

**136.** For a 16 SWG copper conductor the ratio $\frac{R_{AC}}{R_{DC}}$ will be maximum at which of the following frequency?

(*a*) 50 Hz (*b*) 1 kHz (*c*) 1 MHz (*d*) 10 MHz.

**137.** In the above case $R_{AC} > R_{DC}$ because

(*a*) in *AC*, current does not flow continuously

(*b*) high frequency *AC* has low amplitude

(*c*) conductors carrying *AC* have significant skin effect

(*d*) *AC* results in corona losses.

**138.** From Fig. 6.21 it can be concluded that

(*a*) $v$ lags $i$ by $\pi$ (*b*) $i$ lags $v$ by $\frac{\pi}{2}$ (*c*) $i$ leads $v$ by $\frac{\pi}{2}$ (*d*) $v$ leads $i$ by $\pi$.

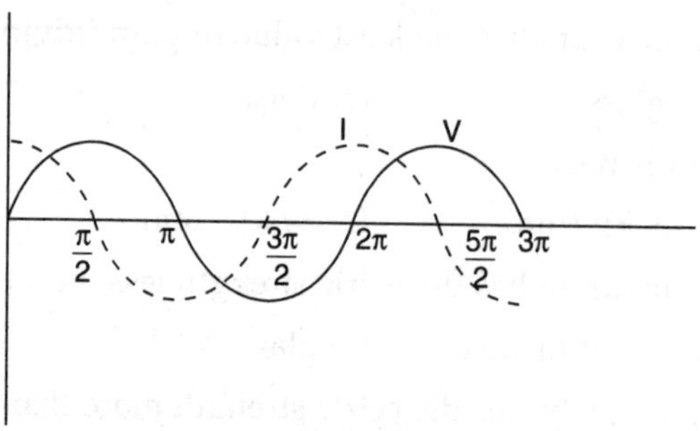


***FIGURE 6.21***

**139.** The current and voltage wave form of Fig. 6.21 is valid in case of
(*a*) pure resistors (*b*) pure inductors (*c*) pure capacitors (*d*) none of the above.

**Questions 140 to 143 Refer to Data Given Below:**

**140.** An alternating emf of 20 mV is applied to a series circuit in which $L = 400\ \mu H$, $C = 100\ \mu\mu F$ and $R = 10$ ohm. If the frequency of applied voltage is equal to the resonant frequency for the circuit, the current flowing through the circuit will be
(*a*) 20 μA (*b*) 2 mA (*c*) 20 mA (*d*) 2 A.

**141.** The voltage across capacitor will be
(*a*) 20 mV (*b*) 2 mV (*c*) 2 V (*d*) 4 V.

**142.** In the above circuit at resonance
(*a*) loud noise will be produced (*b*) the circuit will vibrate violently
(*c*) voltage across capacitor will be equal to voltage across inductor
(*d*) current through resistor will be zero.

**143.** The voltage magnification of the circuit will be nearly
(*a*) 10 (*b*) 20 (*c*) 100 (*d*) 200.

**144.** For the circuit shown in Fig. 6.22 the current through (O + $j$ 10) resistor will be
(*a*) $9 + j7$ (*b*) $9 - j\,7$ (*c*) $18 + j\,14$ (*d*) $18 - j\,14$.

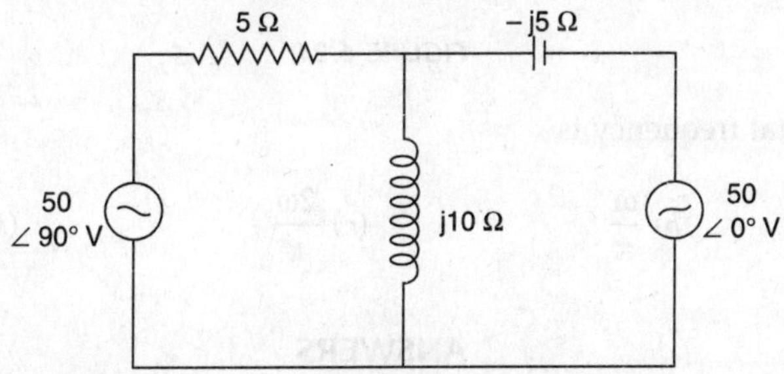


***FIGURE 6.22***

**145.** The Thevenin equivalent circuit shown in Fig. 6.23 will have $Z_{AB}$ =
(*a*) $1 + j\,2$ (*b*) $9 + j\,11$ (*c*) $9 - j\,11$ (*d*) $\frac{1}{2}\,(9 + j\,11)$.

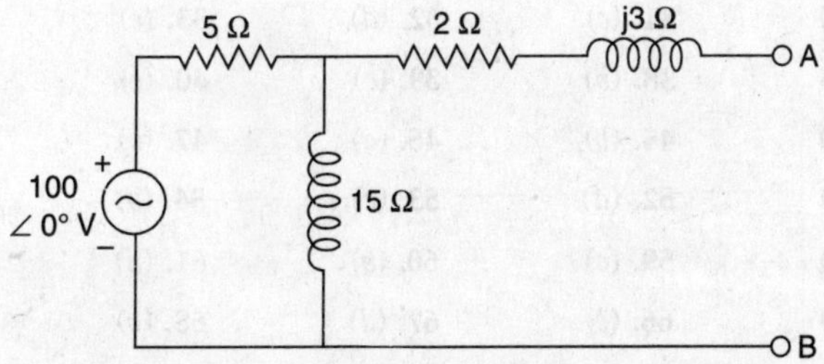


***FIGURE 6.23***

**146.** In the above case $V_{OC}$ =

(*a*) 50 (1 + *j*) (*b*) 50 (1 – *j*) (*c*) 25 (1 – *j*) (*d*) 25 (1 + *j*).

**147.** "The summation of instantaneous power in a network is zero". The above statement is associated with

(*a*) Super position theorem (*b*) Reciprocity theorem

(*c*) Maximum power theorem (*d*) Tellangen's theorem.

**148.** Tellangen's theorem is applicable to

(*a*) linear network only (*b*) linear and non-linear networks

(*c*) linear, non-linear, passive and active networks

(*d*) linear, non-linear, passive, active and time invariant or varying networks.

**149.** For the wave shown in Fig. 6.24 the magnitude of the dc component is

(*a*) 10 V (*b*) 5 √2 V (*c*) 5 V (*d*) 2.5 V.

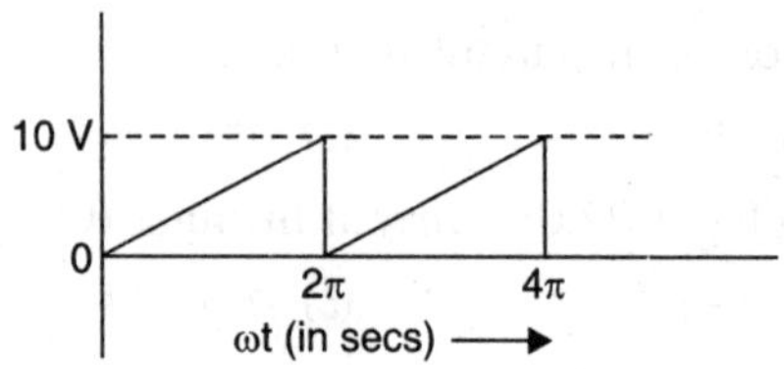


***FIGURE 6.24***

**150.** The fundamental frequency is

(*a*) ω (*b*) $\frac{\omega}{\pi}$ (*c*) $\frac{2\omega}{\pi}$ (*d*) $\frac{\omega}{2\pi}$.

## ANSWERS

| | | | | | | |
|---|---|---|---|---|---|---|
| **1.** (*b*) | **2.** (*b*) | **3.** (*c*) | **4.** (*c*) | **5.** (*b*) | **6.** (*c*) | **7.** (*b*) |
| **8.** (*b*) | **9.** (*b*) | **10.** (*b*) | **11.** (*b*) | **12.** (*c*) | **13.** (*b*) | **14.** (*b*) |
| **15.** (*b*) | **16.** (*b*) | **17.** (*c*) | **18.** (*c*) | **19.** (*b*) | **20.** (*c*) | **21.** (*b*) |
| **22.** (*c*) | **23.** (*d*) | **24.** (*b*) | **25.** (*c*) | **26.** (*d*) | **27.** (*c*) | **28.** (*b*) |
| **29.** (*b*) | **30.** (*b*) | **31.** (*c*) | **32.** (*d*) | **33.** (*c*) | **34.** (*d*) | **35.** (*b*) |
| **36.** (*c*) | **37.** (*c*) | **38.** (*b*) | **39.** (*c*) | **40.** (*c*) | **41.** (*d*) | **42.** (*a*) |
| **43.** (*d*) | **44.** (*b*) | **45.** (*b*) | **46.** (*a*) | **47.** (*a*) | **48.** (*b*) | **49.** (*c*) |
| **50.** (*d*) | **51.** (*c*) | **52.** (*d*) | **53.** (*c*) | **54.** (*b*) | **55.** (*c*) | **56.** (*d*) |
| **57.** (*c*) | **58.** (*c*) | **59.** (*c*) | **60.** (*b*) | **61.** (*d*) | **62.** (*c*) | **63.** (*b*) |
| **64.** (*c*) | **65.** (*c*) | **66.** (*c*) | **67.** (*d*) | **68.** (*b*) | **69.** (*a*) | **70.** (*c*) |
| **71.** (*c*) | **72.** (*d*) | **73.** (*c*) | **74.** (*d*) | **75.** (*b*) | **76.** (*d*) | **77.** (*d*) |

**78.** (*a*) **79.** (*c*) **80.** (*d*) **81.** (*a*) **82.** (*b*) **83.** (*c*) **84.** (*b*)
**85.** (*d*) **86.** (*c*) **87.** (*b*) **88.** (*c*) **89.** (*c*) **90.** (*c*) **91.** (*c*)
**92.** (*c*) **93.** (*b*) **94.** (*a*) **95.** (*d*) **96.** (*b*) **97.** (*b*) **98.** (*d*)
**99.** (*c*) **100.** (*b*) **101.** (*c*) **102.** (*a*) **103.** (*b*) **104.** (*d*) **105.** (*a*)
**106.** (*b*) **107.** (*c*) **108.** (*b*) **109.** (*a*) **110.** (*d*) **111.** (*c*) **112.** (*c*)
**113.** (*d*) **114.** (*d*) **115.** (*c*) **116.** (*c*) **117.** (*a*) **118.** (*c*) **119.** (*d*)
**120.** (*d*) **121.** (*b*) **122.** (*a*) **123.** (*c*) **124.** (*c*) **125.** (*a*) **126.** (*c*)
**127.** (*b*) **128.** (*c*) **129.** (*c*) **130.** (*a*) **131.** (*b*) **132.** (*d*) **133.** (*c*)
**134.** (*d*) **135.** (*a*) **136.** (*d*) **137.** (*c*) **138.** (*c*) **139.** (*c*) **140.** (*b*)
**141.** (*d*) **142.** (*c*) **143.** (*d*) **144.** (*c*) **145.** (*d*) **146.** (*a*) **147.** (*d*)
**148.** (*d*) **149.** (*d*) **150.** (*d*).

# Chapter 7 INDUCTANCE

Inductance is the ability of a conductor to produce induced voltage when the current varies. A long wire has more inductance than a short wire since more conductor length cut by magnetic flux produces more induced voltage. Similarly a coil has more inductance than the equivalent length of straight wire because the coil concentrates magnetic flux. Components manufactured to have a definite value of inductance are just coils of wire, therefore called inductors.

As the inductance is a measure of induced voltage, the amount of inductance has an important effect in any circuit where the current changes. The inductance is an additional characteristic of the circuit besides the resistance.

**Unit of Inductance.** The unit of inductance is henry. One henry is the amount of inductance that allows one volt to be induced when the current changes at the rate of one ampere per second. The formula is

$$L = \frac{V_L}{di/dt}$$

where $V_L$ is in volt and $\frac{di}{dt}$ is the current change in amperes per second.

**Inductance of coils.** In terms of physical construction, the inductance depends on how the coil is wound. Note the following factors:

1. A greater number $N$ of turns increases $L$ because more voltage can be induced.

   Actually $L$ increases in proportion to $N^2$. Thus double the number of turns in the same area and length increases the inductance four times.

2. More area $A$ for each turn increases $L$.

3. $L$ increases with the permeability of the core. For an air-core $\mu$ is 1. When a magnetic core, $L$ is increased by the $\mu$ factor as the magnetic flux is concentrated in the coil.

4. $L$ decreases with more length for the same turns, as the magnetic field then is less concentrated.

**Mutual Inductance.** When the current in an inductor changes, the varying flux can cut across any other inductor near by, producing induced voltage in both inductors. As shown in Fig. 7.1 the coil $L_1$ is connected to a generator that produces varying current in the turns. The winding $L_2$ is not connected to $L_1$ but the turns are linked by the magnetic field. A varying current in $L_1$, therefore, induces voltage across $L_1$ and across $L_2$. If all the flux of the current in $L_1$ links all the turns of the coil $L_2$ each turn in $L_2$ will have the same amount of induced voltage as each turn in $L_1$.

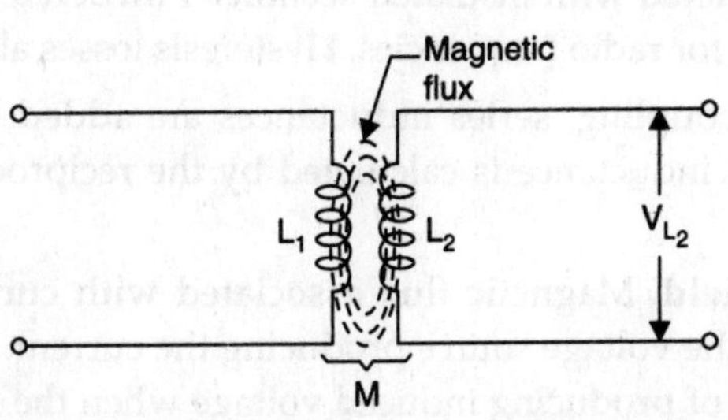


*FIGURE 7.1*

When the induced voltage produces current in $L_2$, its varying magnetic field induces voltage in $L_1$. The two coils $L_1$ and $L_2$ have mutual inductance, therefore, because current in one can induce voltage in other.

The unit of mutual inductance ($L_M$) is henry. Two coils have $L_M$ of one henry when a current change of one ampere per second in one coil induces one volt in the other coil.

**Co-efficient of coupling.** The fraction of total flux from one coil linking another coil is the co-efficient of coupling $K$ between the two coils. In case all the flux of $L_1$ links $L_2$, then $K$ equals 1. Thus

$$K = \frac{\text{Flux linkages between } L_1 \text{ and } L_2}{\text{Flux produced by } L_1}$$

$K$ is dimensionless number having value less than 1. $K$ is increased by placing the coils close together, possibly with one wound on the top of the other. A high value of $K$, called the tight coupling, allows the current in one coil to induce more voltage in the other coil. Loose coupling with low value of $K$, has the opposite effect. When $K = 0$ there will be no mutual inductance.

Air core coils wound on one form have value of $K$ equal to 0.05 to 0.3 approximately. Coils on a common iron core can be considered to have practically unity coupling with $K = 1$.

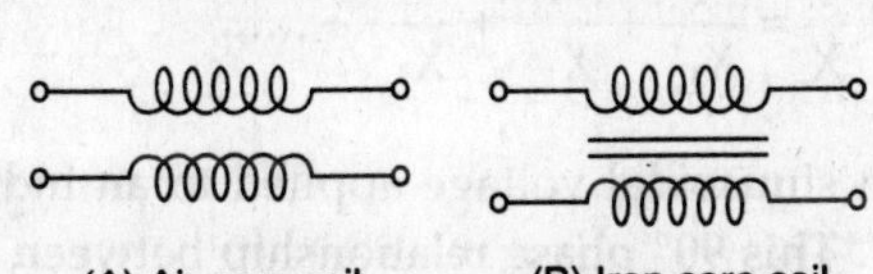


*FIGURE 7.2* Symbolic representation.

The mutual inductance $L_M$ is given by

$$L_M = K\sqrt{L_1 \times L_2}$$

**Core losses.** A magnetic core becomes warm or even hot, thereby confirming that the energy supplied to the coil is used up in heat. The two main effects are eddy current losses and hysteresis losses.

Eddy currents are induced in the iron core of an iductance causing wasted power that heats the core. Eddy current losses increase with higher frequency of alternating current. To reduce eddy-currents, the iron core is laminated with insulated sections. Powdered-iron and ferrite cores have minimum eddy-current losses for radio frequencies. Hysteresis losses also cause wastage of power.

Assuming no mutual coupling, series inductances are added like series resistances. For parallel inductances, the total inductance is calculated by the reciprocal formula, as for parallel resistances.

**Energy in magnetic field.** Magnetic flux associated with current in an inductance has electrical energy supplied by the voltage source producing the current. The energy is stored in the field, since it can do the work of producing induced voltage when the flux moves. The amount of electrical energy stored is given by

$$E = \frac{1}{2}LI^2$$

where $L$ is inductance in henries. $I$ is current in amperes. The factor of $\frac{1}{2}$ gives the average result of $I$ in producing energy.

**Inductive Reactance.** When alternating current flows in an inductance $L$, the amount of current is much less than the resistance alone would allow. The reason is that the current variations induce a voltage across $L$ that opposes the applied voltage. This additional opposition of an inductance to sine-wave alternating current is specified by the amount of its inductive reactance. The inductive reactance $X_L$ is given by

$$X_L = 2\pi fL$$

where $f$ is frequency, $X_L$ is expressed in ohms.

**Series and Parallel Reactances.** As reactance is an opposition in ohms, inductive reactances in series and parallel are combined the same way as ohms of resistance. Thus in case of series reactances

$$X_L = X_{L_1} + X_{L_2} + X_{L_3} + \ldots\ldots.$$

In parallel reactances

$$\frac{1}{X} = \frac{1}{X_{L_1}} + \frac{1}{X_{L_2}} + \frac{1}{X_{L_3}} + \ldots\ldots$$

**Inductive Circuits.** With sinusoidal voltage applied to an inductance, the current lags its induced voltage by exactly 90°. This 90° phase relationship between $i_L$ and $v_L$ is true in any sine wave ac circuit whether $L$ is in series or parallel, and whether $L$ is alone or combined with other components. This phase difference results because $v_L$ depends on the rate of change of $I_L$. Thus for a sine wave of $i_L$, the induced voltage is a cosine wave.

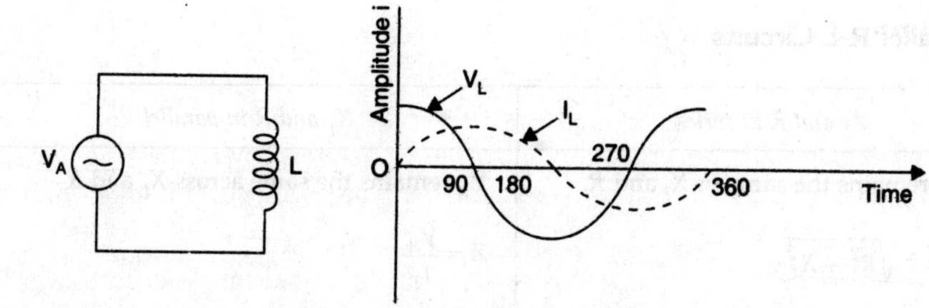


*FIGURE 7.3*

**Example 1.** *A pure inductance L = 0.02 henrys has an applied voltage*

$$v_{(t)} = 300 \sin 1000t \text{ volts}$$

*Determine the current i(t), the instantaneous power and the average power.*

**Sol. The current *i(t)* is given by**

$$i(t) = \frac{1}{L}\int v(t)\,dt$$

$$= \frac{1}{0.02}\int 300 \sin 1000t\,dt$$

$$= \frac{3000}{0.02}\left(\frac{-\cos 1000t}{1000}\right)$$

$$= -15 \cos 1000\,t \text{ amperes}$$

**The instantaneous power *p* is given by**

$$p = vi$$

$$= -300\,(7.5)\left(\tfrac{1}{2} \sin 2000\,t\right)$$

$$= -1125 \sin 2000\,t \text{ watts}$$

**The average power will be zero as shown in Fig. 7.4.**

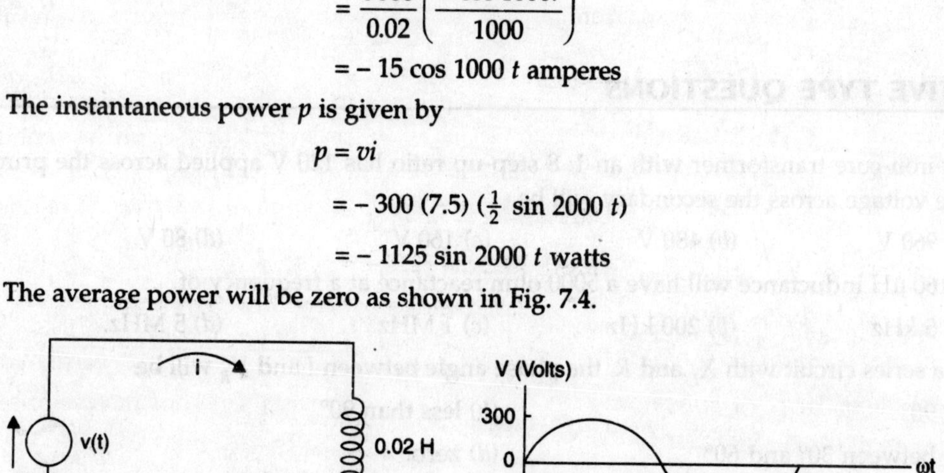


*FIGURE 7.4*

## Series and Parallel R-L Circuits

| $X_L$ and R in series | $X_L$ and R in parallel |
|---|---|
| 1. I remains the same in $X_L$ and R. | $V_A$ remains the same across $X_L$ and R |
| 2. $Z = \sqrt{R^2 + X_L^2}$ | $Z = \frac{V_A}{I_T}$ |
| 3. $V_L$ leads $V_R$ by 90° | $I_L$ lags $I_R$ by 90° |
| 4. $\tan\theta = \frac{X_L}{R}$ θ increases as more $X_L$ | $\tan\theta = -\frac{I_L}{-I_R}$ – θ decreases as more $X_L$ |
| makes circuit inductive | means $I_L$ |

**Skin Effect.** When current flows through the conductor cross-section, the centre of the conductor cross-section encounters slightly more inductance because of magnetic flux concentrated in the metal, compared with the edges where part of the flux is in air. This results in less current flowing through the conductor core. This effect is particularly significant at higher frequencies. For this reason, conductors for VHF currents are often made of hollow tubing. The skin effect increases the effective resistance as a smaller cross-sectional area is used for the current path in the conductor.

## OBJECTIVE TYPE QUESTIONS

**1.** An iron-core transformer with an 1: 8 step-up ratio has 120 V applied across the primary. The voltage across the secondary will be

(*a*) 960 V (*b*) 480 V (*c*) 160 V (*d*) 80 V.

**2.** A 160 μH inductance will have a 5000 ohm reactance at a frequency of

(*a*) 5 kHz (*b*) 200 kHz (*c*) 1 MHz (*d*) 5 MHz.

**3.** In a series circuit with $X_L$ and *R*, the phase angle between *I* and $V_R$ will be

(*a*) 90° (*b*) less than 90°

(*c*) between 30° and 60° (*d*) zero.

**4.** In a sine wave ac circuit with a 40 Ω *R* in series with a 30 Ω $X_L$ the total impedance will be

(*a*) 30 Ω (*b*) 40 Ω (*c*) 50 Ω (*d*) 55 Ω.

**5.** If an ohmmetre check on a 50 μH choke reads 3 ohms, the coil is probably

(*a*) open (*b*) partially open (*c*) normal (*d*) defective.

**6.** What will be the inductance of a coil that induces 1000 V when the current changes at the rate if 50 mA in 2μS?

(*a*) 0.04 H (*b*) 0.4 H (*c*) 0.04 mH (*d*) 0.4 mH.

7. The current through a 200 mH inductor changes from 0 to 100 mA in 2 μS. The voltage induced in the inductor will be
(a) 100 V (b) 1000 V (c) 10,000 V (d) 100,000 V.

8. Two 400 mH coils $L_1$ and $L_2$ have a co-efficient of coupling $K$ equal to 0.2. The value of $L_M$ will be
(a) 8 H (b) 80 H (c) 8 mH (d) 80 mH.

9. In case all the flux from the current in a coil $L_1$ links another coil $L_2$, the co-efficient of coupling will be
(a) zero (b) 2.0 (c) 0.5 (d) 1.

10. The energy in joules stored in the magnetic field of a 300 mH inductance with a 90 mA current will be
(a) 1.21 (b) $1.21 \times 10^{-3}$ (c) $1.21 \times 10^{-6}$ (d) $1.21 \times 10^{-9}$.

11. A coil with negligible resistance has 62.8 V across it with 0.01 A. The inductive reactance is
(a) 6.28 ohms (b) 62.80 ohms (c) 628.0 ohms (d) 6280 ohms.

12. In which case the inductive reactance is least?
(a) $L = 1$ H ; $f = 100$ Hz (b) $L = 0.5$ H ; $f = 100$ Hz
(c) $L = 1$ H ; $f = 159$ Hz (d) $L = 1$ H ; $f = 1000$ Hz.

13. Inductive reactance is measured in ohms because it
(a) reduces the amplitude of alternating current
(b) increases the amplitude of alternating current
(c) increases the amplitude of direct current
(d) has a back emf opposing a steady direct current.

14. Alternating current can induce voltage because alternating current has a
(a) high peak value (b) varying magnetic field
(c) stronger magnetic field than direct current
(d) constant magnetic field.

15. For the circuit shown in Fig. 7.5, the inductive reactance is
(a) 1000 Ω (b) 100 Ω (c) 10 Ω (d) zero.

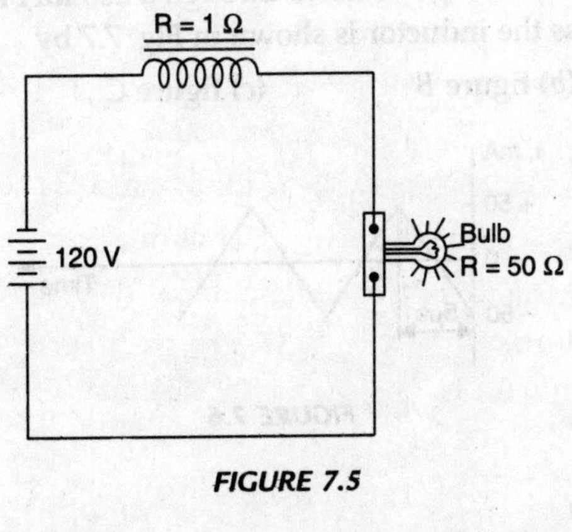


FIGURE 7.5

16. The inductive reactance of a 6 mH inductor at 41.67 kHz will be
(*a*) 15.70 ohms (*b*) 157.0 ohms (*c*) 1570 ohms (*d*) 15,700 ohms.

17. The inductive reactance of a 250 μH coil at 10 MHz will be
(*a*) 157 ohms (*b*) 15700 ohms (*c*) 1570000 ohms (*d*) none of the above.

**Questions 18 to 21 Refer to Data Given Below:**

For a 100 μH inductance $L_1$ and a 200 μH inductance $L_2$.

18. The total inductance of $L_1$ and $L_2$ in series without mutual coupling will be
(*a*) 100 μH (*b*) 150 μH (*c*) 200 μH (*d*) 300 μH.

19. The total inductance of $L_1$ and $L_2$ in parallel without mutual coupling will be
(*a*) 150 μH (*b*) 75 μH (*c*) 66.7 μH (*d*) 300 μH.

20. The combined inductance of $L_1$ and $L_2$ in series aiding with 10 μH mutual inductance will be
(*a*) 300 μH (*b*) 280 μH (*c*) 140 μH (*d*) 77 μH.

21. The value of coupling factor $K$ is
(*a*) 0.0707 (*b*) 0.0101 (*c*) 0.00707 (*d*) 0.7070.

22. In a sine-wave ac circuit with an inductive branch in parallel with a resistive branch
(*a*) the inductive and resistive braanch currents will be 180° out of phase
(*b*) the inductive and resistive branch current will be in phase
(*c*) the voltage across the inductance will lead the voltage across the resistance by 90°
(*d*) the resistive branch current will be 90° out if phase with the current through inductive branch.

23. A 100 ohm resistance is in series with a 100 ohm inductive reactance. The total impedance will be
(*a*) 200 ohms (*b*) 141.4 ohms (*c*) 100 ohms (*d*) 70.7 ohms.

24. Which of the following inductor will have the least eddy current losses?
(*a*) iron core (*b*) laminated iron core
(*c*) powdered iron core (*d*) air core.

25. The current wave shown in Fig. 7.6 flows through a 200 mH inductance. The wave form of induced voltage across the inductor is shown in Fig. 7.7 by
(*a*) figure *A* (*b*) figure *B* (*c*) figure *C* (*d*) figure *D*.

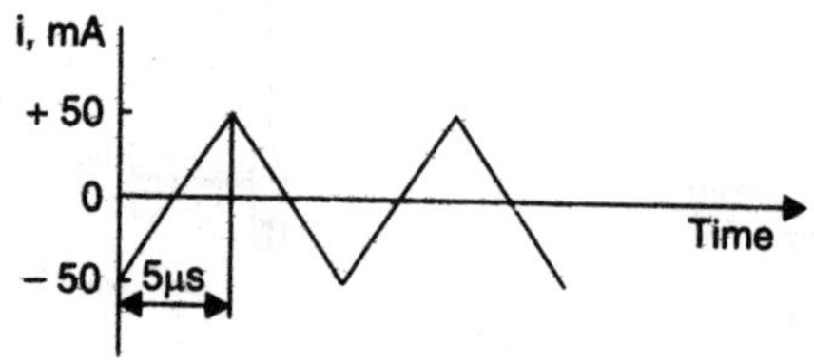


*FIGURE 7.6*

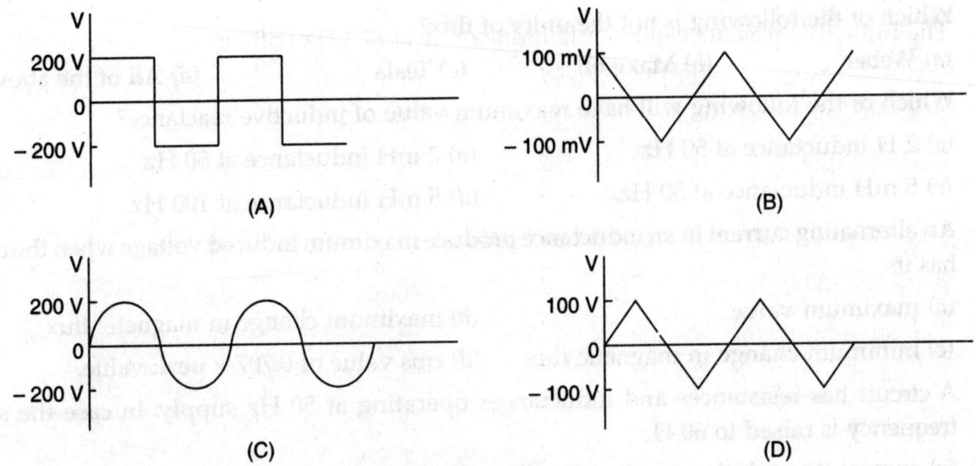


*FIGURE 7.7*

**Questions 26 to 28 Refer to Fig. 7.8:**

**26. Three inductances in parallel are shown in Fig. 7.8. If the value of equivalent inductance is 0.0755 H, the value of unknown inductance *L* will be**

(*a*) 0.1 μH (*b*) 0.1 mH (*c*) 0.1 H (*d*) 0.2 H.

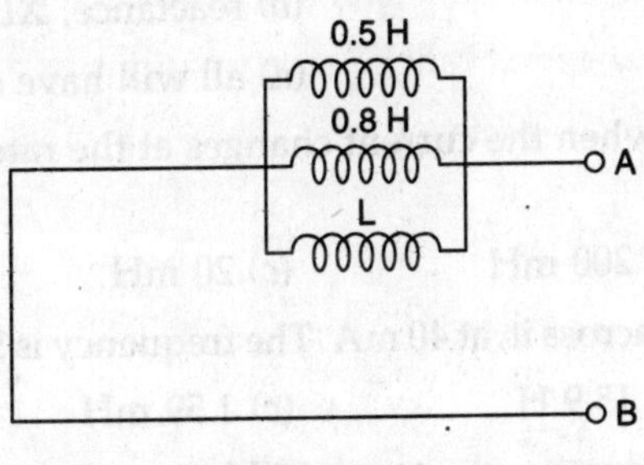


*FIGURE 7.8*

**27. For which value of *L* the value of equivalent inductance will be 0.5 H?**

(*a*) 0.15 H (*b*) 0.25 H (*c*) 0.3 H (*d*) none.

**28. What is the maximum value of equivalent inductance if inductance *L* is adjustable without limit?**

(*a*) 0.1 H (*b*) 0.2 H (*c*) 0.308 H (*d*) 0.411 H.

**29. Which of the following relation is incorrect?**

(*a*) reluctance × permeance = 1 (*b*) conductance × resistance = 1

(*c*) admittance × impedance = 1 (*d*) inductance × mutual inductance = 1.

**30. Which of the following circuit element stores energy in the electromagnetic field?**

(*a*) resistance (*b*) variable resistor (*c*) condenser (*d*) inductance.

**31. The power factor of a purely inductive circuit is**

(*a*) unity (*b*) leading (*c*) lagging (*d*) zero.

32. Which of the following is not the unity of flux?

(*a*) Weber (*b*) Maxwell (*c*) Tesla (*d*) All of the above.

33. Which of the following will have maximum value of inductive reactance?

(*a*) 2 H inductance at 50 Hz (*b*) 2 mH inductance at 50 Hz

(*c*) 5 mH inductance at 50 Hz (*d*) 5 mH inductance at 100 Hz.

34. An alternating current in an inductance produce maximum induced voltage when the current has its

(*a*) maximum value (*b*) maximum change in magnetic flux

(*c*) minimum change in magnetic flux (*d*) rms value of 0.707 × peak value.

35. A circuit has resistances and inductances operating at 50 Hz supply. In case the supply frequency is raised to 60 Hz

(*a*) current through the circuit will reduce

(*b*) current through the circuit will increase

(*c*) more noise will be produced

(*d*) power factor will become leading.

36. An inductive coil connected to a 200 V 50 Hz ac supply with 10 A current flows through it. The coil dissipates 500 watts. Which of the following will have least value in ohms?

(*a*) resistance *R* (*b*) reactance, *XL*

(*c*) impedance, *Z* (*d*) all will have equal value.

37. A coil induces 200 mV when the current changes at the rate of one ampere per second. The value of inductance is

(*a*) 2000 mH (*b*) 200 mH (*c*) 20 mH (*d*) 2 mH.

38. An inductance has 20 V across it, at 40 mA. The frequency is 5 kHz. The value of inductance is

(*a*) 1.59 H (*b*) 15.9 H (*c*) 1.59 mH (*d*) 15.9 mH.

39. In the above case the inductive reactance will be

(*a*) 5 Ω (*b*) 50 Ω (*c*) 500 Ω (*d*) 5000 Ω.

40. The inductance across the points *XY* shown in Fig. 7.9 will be

(*a*) 6 H (*b*) 12 mH (*c*) 2.4 mH (*d*) 4.2 mH.

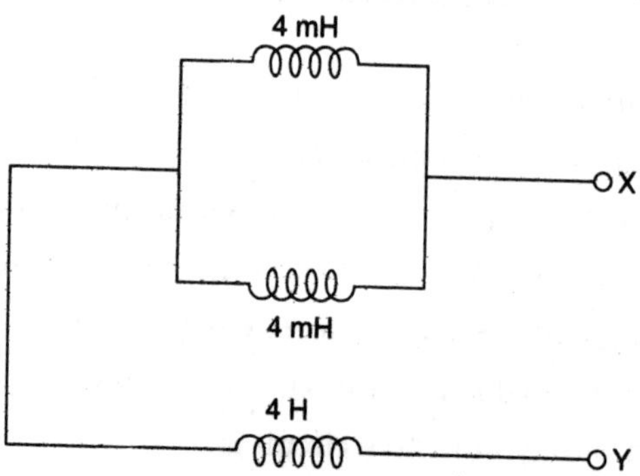


*FIGURE 7.9*

41. Inductances $L_1$ and $L_2$ shown in Fig. 7.10 are in the ratio of 2 to 1. If the equivalent inductance of all three is 0.7 H, what are the values of inductances $L_1$ and $L_2$?

(a) $L_1 = 0.1$ H ; $L_2 = 0.2$ H (b) $L_1 = 0.2$ H ; $L_2 = 0.3$ H

(c) $L_1 = 0.4$ H ; $L_2 = 0.6$ H (d) $L_1 = 0.6$ H ; $L_2 = 0.3$ H.

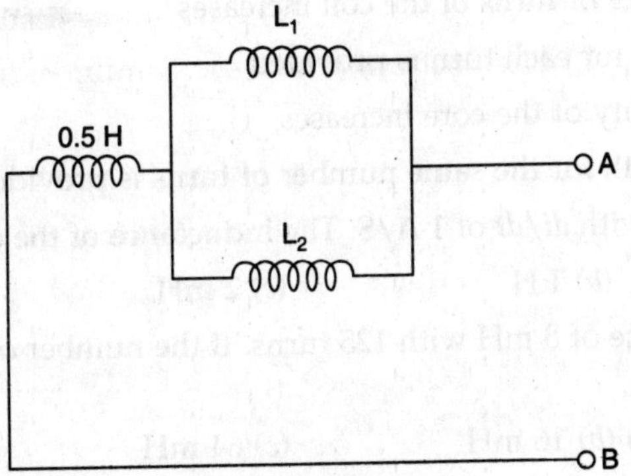


FIGURE 7.10

**Questions 42 to 44 Refer to Data Given Below:**

A pure inductor carries a current

$$i(t) = I_m \sin \omega t \text{ amperes.}$$

42. The average power will be

(a) infinity (b) $I_m^2 L$ (c) $\frac{1}{2} I_m^2 L$ (d) zero.

43. The stored energy will be maximum when $\omega t =$

(a) 0 (b) $\frac{\pi}{4}$ (c) $\frac{\pi}{2}$ (d) $\frac{2}{3}\pi$.

44. At $\omega t = \pi/2$ the energy stored will be

(a) $LI_m^2$ (b) $\frac{1}{2} LI_m^2$ (c) $\sqrt{2}\, LI_m^2$ (d) $\frac{1}{\sqrt{2}} LI_m^2$.

45. Three pure inductances are connected as shown in Fig. 7.11. The equivalent inductance to replace this circuit is

(a) 1.1 H (b) 0.5 H (c) 0.4 H (d) 0.29 H.

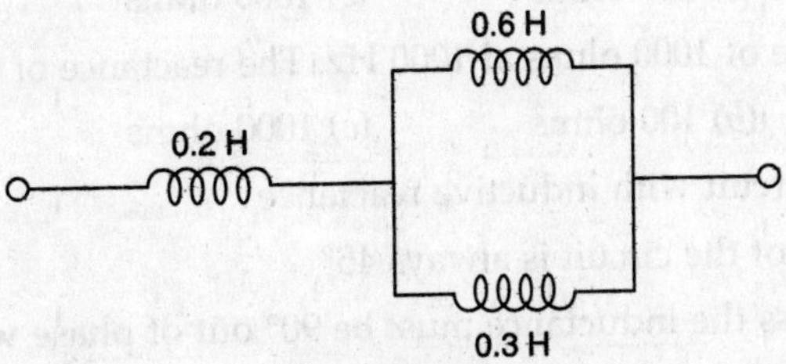


FIGURE 7.11

46. How much will be the inductance of a coil that induces 40 V when its current changes at the rate of 4 amperes per second?

(*a*) 0.1 mH (*b*) 1 mH (*c*) 0.1 H (*d*) 10 H.

47. The inductance of a coil will increase under all the following conditions EXCEPT

(*a*) when the number of turns of the coil increases

(*b*) when more area for each turn is provided

(*c*) when permeability of the core increases

(*d*) when more length for the same number of turns is provided.

48. A coil induces 2 V with $di/dt$ of 1 A/S. The inductance of the coil will be

(*a*) 2 H (*b*) 1 H (*c*) 2 mH (*d*) 1 mH.

49. A coil has inductance of 8 mH with 125 turns. If the number of turns is doubled, the inductance will be

(*a*) 4 mH (*b*) 16 mH (*c*) 64 mH (*d*) 128 mH.

50. Which of the following current change has least value in amperes per second?

(*a*) zero to 4000 mA in 2 seconds (*b*) zero to 50 mA in 5 micro seconds

(*c*) 100 to 150 mA in 5 micro seconds (*d*) 150 to 100 mA in 5 micro seconds.

51. Inductive reactance applies only to sine waves because it

(*a*) increases with lower frequencies (*b*) increases with lower inductance

(*c*) depends on the factor $2\pi$ (*d*) decreases with higher frequencies.

52. How many of the following inductors will require frequency of more than 1 MHz to give individual inductive reactance of 2000 ohms

(*i*) 250 mH (*ii*) 200 μH (*iii*) 800 μH (*iv*) 2 H

(*v*) 20 μH.

(*a*) two (*b*) three (*c*) four (*d*) none.

53. The time constant of $L$ = 240 mH in series with $R$ = 20 Ω will be

(*a*) 1.2 μS (*b*) 12 μS (*c*) 1.212 μS (*d*) 12 mS.

54. The dc resistance of 36 metre length of 30 gauge copper wire will be nearly

(*a*) 0.1 Ω (*b*) 1 Ω (*c*) 10 Ω (*d*) 100 Ω.

55. A 16 mH inductance has a reactance of 500 ohms. If two such inductors are connected in series without any mutual coupling, their total reactance will be

(*a*) 250 ohms (*b*) 125 ohms (*c*) 1000 ohms (*d*) 2000 ohms.

56. A coil has a rectance of 1000 ohms at 1000 Hz. The reactance of this coil at 10 kHz will be

(*a*) 10 ohms (*b*) 100 ohms (*c*) 1000 ohms (*d*) 10,000 ohms.

57. In a sine wave ac circuit with inductive reactance

(*a*) the phase angle of the circuit is always 45°

(*b*) the voltage across the inductance must be 90° out of phase with the applied voltage (*c*) the current through the inductance lags the induced voltage by 90°

(*d*) the current through the inductance and voltage across it are 180° out of phase.

**58.** Two 125 μH coils in series without mutual coupling have a total inductance of

(*a*) 125 μH (*b*) 250 μH (*c*) 62.5 μH (*d*) 500 μH.

**59.** Two 1000 ohm inductive reactances in parallel have an equivalent reactance of

(*a*) 250 ohms (*b*) 500 ohms (*c*) 1000 ohms (*d*) 2000 ohms.

**60.** How much will be the inductance of a coil with negligible resistance of the current is 0.1 A when connected across the 60 Hz 120 V power line

(*a*) 3.18 μH (*b*) 3.18 mH (*c*) 3.18 H (*d*) 318 H.

**61.** In a sine-wave ac circuit with $X_L$ and *R* in series

(*a*) the voltage across *R* will lead the voltage across $X_L$ by 90°

(*b*) the voltages across *R* and $X_L$ are in phase

(*c*) the voltages across *R* and $X_L$ are 180° out of phase.

(*d*) the voltage across *R* lags the voltage across $X_L$ by 90°.

**62.** A 1: 5 voltage step-up transformer has 120 V across the primary and a 600 ohm resistance across the secondary. Assuming 100 percent efficiency, the primary current will be

(*a*) 10 A (*b*) 5 A (*c*) 600 mA (*d*) $\frac{1}{5}$ A.

**63.** A coil has an inductive reactance of 100 ohms. If its inductance is doubled and the frequency is doubled, then the inductive reactance will be

(*a*) 100 ohms (*b*) 200 ohms (*c*) 400 ohms (*d*) 1600 ohms.

**64.** A 200 μH coil has a *Q* of 40 at 0.5 MHz. The value of $R_e$ will be

(*a*) 0.0157 ohm (*b*) 0.157 ohm (*c*) 1.57 ohm (*d*) 15.7 ohms.

**65.** A coil having 1000 Ω inductive reactance at 3 MHz and 10 Ω internal resistance has a *Q* of

(*a*) 3 (*b*) 10 (*c*) 100 (*d*) 1000.

**66.** A transformer with a 1: 6 voltage set up ratio has 720 V across 7200 Ω in the secondary. The secondary current will be

(*a*) 0.1 mA (*b*) 0.1 A (*c*) 0.6 A (*d*) 0.06 A.

**67.** Rest of the parameters remaining unchanged, which frequency will produce more hysteresis losses?

(*a*) 60 Hz (*b*) 60 MHz (*c*) 60 kHz (*d*) 600 kHz.

**68.** Two series coils each with an *L* of 250 μH have a total inductance of 550 μH when connected series aiding and 450 μH when series-opposing. The mutual inductance between the coils will be

(*a*) 5 μH (*b*) 10 μH (*c*) 15 μH (*d*) 25 μH.

**69.** In the above problem the coupling co-efficient will be

(*a*) 0.05 (*b*) 0.1 (*c*) 0.2 (*d*) 0.5.

70. An open coil has
(*a*) infinite resistance and zero inductance
(*b*) zero reistance and high inductance
(*c*) infinite resistance and normal inductance
(*d*) zero resistance and inductance.

71. An inductance has a reactance of 10,000 ohms at 10 kHz. At 20,000 Hz, its inductive reactance will be
(*a*) 500 ohms (*b*) 1000 ohms (*c*) 10,000 ohms (*d*) 20,000 ohms.

72. An $X_{L_1} = 1000\ \Omega$ and $X_{L_2} = 4000\ \Omega$ are connected in series across a 10 V, 60 Hz source. What will be current $X_{L_1}$?
(*a*) 1 μA (*b*) 2 μA (*c*) 1 mA (*d*) 2 mA.

73. In the above case the voltage across $X_{L_2}$ will be
(*a*) 2 V (*b*) 4 V (*c*) 8 V (*d*) 10 V.

74. Inductances $L_1$ and $L_2$ are
(*a*) 2.65 H and 10.6 H (*b*) 265 H and 106 H
(*c*) 10.6 mH and 2.65 mH (*d*) 10.6 μH and 2.65 μH.

75. A coil has inductance $L$ henries. With double the number of turns while length and area remains the same, the inductance will be
(*a*) same (*b*) double (*c*) quadruple (*d*) one-quarter.

76. The inductance required for a 2000 ohm reactance at 20 MHz will be
(*a*) 1.59 μH (*b*) 15.9 μH (*c*) 159 μH (*d*) 1590 μH.

**Questions 77 to 79 Refer to Data Given Below:**

A 30 ohm resistor and 40 ohm inductive reactance are in series with 100 V applied.

77. The current through the circuit will be
(*a*) 2 μA (*b*) 2 mA (*c*) 2 A (*d*) 20 A.

78. The voltage across resistance and inductive reactance respectively will be
(*a*) 100 V, 100 V (*b*) 60 V, 80 V (*c*) 80 V, 60 V (*d*) 50 V, 50 V.

79. $I$ lags $V_T$ by
(*a*) 0° (*b*) 47° (*c*) 53° (*d*) 90°.

80. When current in a conductor increases then according to Lenz's Law self-induced voltage will
(*a*) tend to decrease the amount of current
(*b*) aid the applied voltage
(*c*) produce current opposite to the increasing current
(*d*) aid the increasing current.

**81.** A coil with an inductive reactance of 2000 ohms is connected across a 10 V ac generator. The current will be

(*a*) 5 μA (*b*) 10 μA (*c*) 5 mA (*d*) 10 mA.

**82.** In the above porblem, the voltage drop across the coil will be

(*a*) 1 mA (*b*) 10 mA (*c*) 1 V (*d*) 10 V.

**83.** A 200 μH coil with an 8 Ω internal $R_e$ has an $X_L$ of 600 ohms. The $Q$ of the coil will be

(*a*) 5 ohm (*b*) 15 ohm (*c*) 25 ohm (*d*) 75 ohm.

**84.** A coil with a $Q$ of 50 has a 500 Ω $X_L$ at 4 MHz. Its internal $R_e$ will be

(*a*) 1 Ω (*b*) 10 Ω (*c*) 100 Ω (*d*) 1000 Ω.

**85.** A laminated iron core has reduced eddy-current losses because

(*a*) the laminations are stacked vertically

(*b*) the laminations are insulated from each other

(*c*) the magnetic flux is concentrated in the air gap of the core

(*d*) more wire can be used with less dc resistance in the coil.

**86.** With 10 V applied across an inductive reactance of 100 ohms, the current will be

(*a*) 100 μA (*b*) 1 mA (*c*) 10 mA (*d*) 100 mA.

**87.** An air-core coil has an inductive reactance of 700 ohms and resistance of 2 ohms. The $Q$ factor of the coil will be

(*a*) 50 (*b*) 100 (*c*) 250 (*d*) 350.

**88.** A 100 μH inductance is used as a choke at 10 MHz. At 15 MHz the choke

(*a*) will have negligible inductance

(*b*) will have less reactance

(*c*) will have more reactance

(*d*) none of the above.

**89.** A 100 ohm $R$ is in parallel with 100 Ω of $X_L$. The total impedance $Z$ will be

(*a*) 200 Ω (*b*) 141.4 Ω (*c*) 70.7 Ω (*d*) 50 Ω.

**90.** Current changing from 4 A to 6 A in one second induces 40 volts in a coil. The value of inductances is

(*a*) 20 H (*b*) 20 mH (*c*) 10 mH (*d*) 10 H.

**91.** A current of 100 mA through an inductive reactance of 100 ohm produces a voltage drop of

(*a*) 1 V (*b*) 6.28 V (*c*) 10 V (*d*) 100 V.

**92.** A 600 ohm resistance is in parallel with a inductive reactance of 300 ohms. The total impedance will not change with change in

(*a*) frequency (*b*) applied voltage (*c*) value of $R$ (*d*) value of inductance.

**93.** In a sine-wave ac circuit with a 90 Ω $R$ in series with a 90 Ω $X_L$, phase angle θ will be

(*a*) 0° (*b*) 30° (*c*) 45° (*d*) 60°.

**94.** If two waves have the frequency of 1000 Hz and one is at the maximum value when the other is at zero, the phase angle between them is

(*a*) 0° (*b*) 90° (*c*) 120° (*d*) 180°.

95. Air-core coils are paractically free from
    (*a*) eddy current losses (*b*) hysteresis losses
    (*c*) both (*a*) and (*b*) above (*d*) none of the above.

96. The value of co-efficient of coupling in air-core coils is usually
    (*a*) 0.05 to 0.3 (*b*) 0.3 to 0.55 (*c*) 0.55 to 0.75 (*d*) 0.80 to 0.95.

97. Carbon resistors are preferred to wire would resistors because these have
    (*a*) high strength (*b*) less noise
    (*c*) least stray inductance (*d*) low tolerance.

**Questions 98 to 99 Refer to Data Given Below:**

98. A solenoid 50 cm long and 10 cm in diameter is wound with 1500 turns. The inductance of the coil is
    (*a*) 0.45 μH (*b*) 0.45 mH (*c*) 4.5 mH (*d*) 4.5 H.

99. The energy stored in the magnetic field will be
    (*a*) $3.6 \times 10^{-3}$ joules (*b*) $3.6 \times 10^{-4}$ joules (*c*) $3.6 \times 10^{-6}$ (*d*) $3.6 \times 10^{-8}$ joules.

100. Two coils are wound close together on the same paxolin tube. Current is passed through the first coil and is varied at a uniform rate of 500 mA per second, inducing an emf of 0.1 V in the second coil. The second coil has 100 turns. The number of turns in the first coil if its inductance is 0.4 H, will be
    (*a*) 40 (*b*) 50 (*c*) 100 (*d*) 200.

## ANSWERS

| | | | | | | |
|---|---|---|---|---|---|---|
| **1.** (*a*) | **2.** (*d*) | **3.** (*b*) | **4.** (*c*) | **5.** (*c*) | **6.** (*a*) | **7.** (*c*) |
| **8.** (*d*) | **9.** (*d*) | **10.** (*b*) | **11.** (*d*) | **12.** (*b*) | **13.** (*a*) | **14.** (*b*) |
| **15.** (*d*) | **16.** (*c*) | **17.** (*b*) | **18.** (*d*) | **19.** (*c*) | **20.** (*b*) | **21.** (*a*) |
| **22.** (*d*) | **23.** (*b*) | **24.** (*d*) | **25.** (*a*) | **26.** (*c*) | **27.** (*d*) | **28.** (*c*) |
| **29.** (*d*) | **30.** (*d*) | **31.** (*d*) | **32.** (*c*) | **33.** (*d*) | **34.** (*b*) | **35.** (*a*) |
| **36.** (*b*) | **37.** (*b*) | **38.** (*b*) | **39.** (*c*) | **40.** (*a*) | **41.** (*d*) | **42.** (*d*) |
| **43.** (*c*) | **44.** (*b*) | **45.** (*c*) | **46.** (*d*) | **47.** (*d*) | **48.** (*a*) | **49.** (*c*) |
| **50.** (*a*) | **51.** (*c*) | **52.** (*a*) | **53.** (*d*) | **54.** (*c*) | **55.** (*c*) | **56.** (*d*) |
| **57.** (*c*) | **58.** (*b*) | **59.** (*b*) | **60.** (*c*) | **61.** (*d*) | **62.** (*b*) | **63.** (*c*) |
| **64.** (*d*) | **65.** (*c*) | **66.** (*b*) | **67.** (*b*) | **68.** (*d*) | **69.** (*b*) | **70.** (*a*) |
| **71.** (*d*) | **72.** (*d*) | **73.** (*c*) | **74.** (*a*) | **75.** (*c*) | **76.** (*b*) | **77.** (*c*) |
| **78.** (*b*) | **79.** (*c*) | **80.** (*c*) | **81.** (*c*) | **82.** (*d*) | **83.** (*d*) | **84.** (*b*) |
| **85.** (*b*) | **86.** (*d*) | **87.** (*d*) | **88.** (*c*) | **89.** (*c*) | **90.** (*a*) | **91.** (*c*) |
| **92.** (*b*) | **93.** (*c*) | **94.** (*b*) | **95.** (*c*) | **96.** (*a*) | **97.** (*c*) | **98.** (*b*) |
| **99.** (*a*) | **100.** (*d*) | | | | | |

# Chapter 8 CAPACITANCE

Capacitance is the capacity of a dielectric material to hold electric charge because free electrons cannot flow through an insulator. However, the charge must be applied by some source. The charging continues until the potential difference across the capacitor is equal to the applied voltage. The charging current is transient, as it flows only until the capacitor is charged to the applied voltage. Then there is no current in the circuit. The measure of how much charge can be stored is the capacitance C. More charge stored for a given amount of applied voltage means more capacitance. Physically a capacitance consists simply of two conductors separated by an insulator.

**Dielectric Constant.** This indicates the ability of an insulator to concentrate electric flux. Its numerical value is specified as the ratio of flux in the insulator compared with the flux in air or vacuum. The dielectric constant of air or vacuum is 1, since it is the reference. Mica has dielectric constant of 6 means it can provide a density of electric flux six time as great as that of air or vacuum for the same applied voltage and equal physical size.

**Dielectric Strength** is the ability of a dielectric to withstand a potential difference without arcing across the insulator. This voltage rating is important because rupture of the insulator provides a conducting path through the dielectric. Then it cannot store charge, as the capacitor has been short-circuited. The breakdown voltage increases with thickness of dielectric medium.

**Dielectric Materials**

| *Material* | *Dielectric constant K* | *Dielectric strength in kV/mm* |
|---|---|---|
| Air | 1 | 3.2 |
| Glass | 5 – 12 | 12 – 20 |
| Mica | 4 – 8 | 20 – 60 |
| Rubber | 2.5 | – |
| Micanite | 4.5 – 6 | 25 – 35 |
| Paper | 1.8 – 2.6 | – |
| Paraffin wax | 1.7 – 2.2 | 30 |
| Quartz | 4.6 | 8 |

**Unit of Capacitance.** The amount of charge $Q$ stored in the capacitance is proportional to the applied voltage. Thus

$$Q = CV$$

where $C$ is a physical constant, indicating the capacitance in terms of how much charge can be stored for a given amount of charging voltage when one coulomb is stored in the dielectric with a potential difference of one volt, the capacitance is one farad.

The common units are

1 microfarad = $10^{-6}$ Farad

1 micromicron farad = 1 picofarad

= $10^{-12}$ Farad.

**Commercial Capacitors.** Commercial capacitors are generally classified according to the dielectric. Most common are air, mica, paper, ceramic and electrolytic capacitor.

**Types of Capacitors**

| *Dielectric* | *Construction* | *Capacitance* | *Breakdown V* |
|---|---|---|---|
| Air | Meshed plate | 10 – 400 pF | 400 |
| Ceramic | Tubular | 0.5–1600 pF | 500 – 20,000 |
| . | Disk | 0.002 – 0.1 pF | |
| Electrolytic | Aluminium | 5 – 1000 μF | 10 – 450 |
| | Tantalum | 0.01 – 300 μF | |
| Mica | Stacked sheets | 10 – 500 pF | 500 – 20,000 |
| Paper | Rolled foil | 0.001 – 1 μF | 200 – 1600 |

**Capacitance Tolerance.** Ceramic disk capacitors for general applications usually have a tolerance of ± 20 percent. Paper capacitors usually have a tolerance of ± 10 percent. For closer tolerances, mica or ceramic tubular capacitors are used. These have a tolerance value of ± 2 to 20 percent. Silver plated mica capacitors are available with a tolerance of ± 1 percent.

**Voltage Rating.** This rating specifies the maximum potential difference that can be applied across the plates without puncturing the dielectric. Voltage ratings of general purpose paper, mica and ceramic capacitors are typically 200 to 500 volts. Ceramic capacitors with ratings of 1 kV to 5 kV are also available. Electrolytic capacitors are commonly used in 25, 150 and 450 V ratings 6 and 10 V rating capacitors are used in transistors.

The voltage ratings are for dc voltage applied. The breakdown rating is lower for ac voltage because of the internal heat produced by continuous charge and discharge.

**Charging of a Capacitor.** An arrangement for charging a capacitor is shown in Fig. 8.1. The capacitor is charged through a high resistance $R$ from a battery of $V$ volts. When switch $S$ is connected to '$p$' then $C$ is charged but when it is connected to '$q$', $C$ is short-circuited through $R$ thereby discharging $C$.

During charging the voltage across the capacitor, $v$, is given by

$$v = V\left(1 - e^{-\frac{t}{\lambda}}\right)$$

where $t$ is the time and $\lambda = CR$ = time constant. The variation of $v$ with time $t$ is shown in Fig. 8.1.

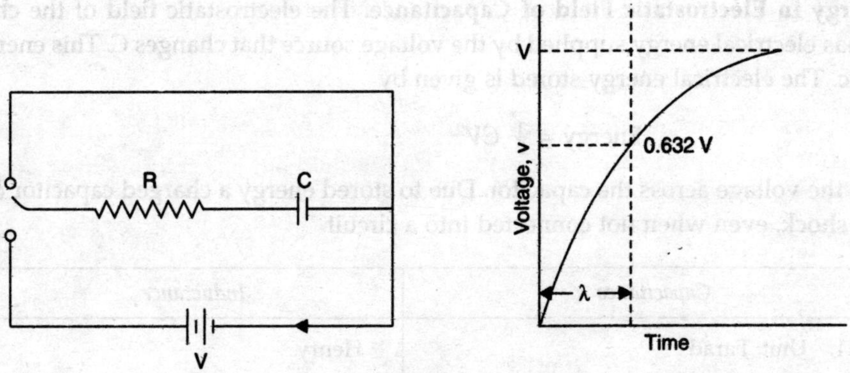


*FIGURE 8.1*

**Discharging of a Capacitor.** During discharging the voltage across capacitor is given by

$$v = V\,e^{-\frac{t}{\lambda}}$$

and current

$$i = I_m\,e^{-\frac{t}{\lambda}}$$

The fall in potential and discharging current are shown in Fig. 8.2.

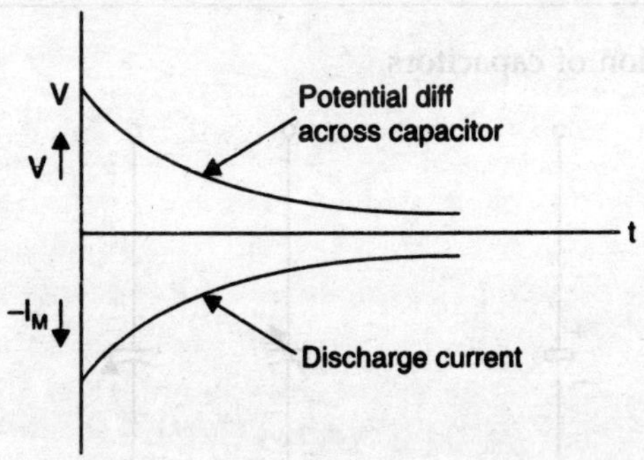


*FIGURE 8.2*

**Capacitors in Parallel.** When capacitors are connected in parallel, the total capacitance is given by

$$C_T = C_1 + C_2 + \ldots\ldots \text{ etc.}$$

**Capacitances in Series.** When capacitors are connected in series, the equivalent capacitance is given by

$$\frac{1}{C_T} = \frac{1}{C_1} + \frac{1}{C_2} + \frac{1}{C_3} + \ldots\ldots$$

Capacitors in series are used to provide a higher voltage break-down rating for the combination.

**Energy in Electrostatic Field of Capacitance.** The electrostatic field of the charge in the dielectric has electrical energy supplied by the voltage source that changes C. This energy is stored in dielectric. The electrical energy stored is given by

$$\text{Energy} = \frac{1}{2} CV^2$$

where $V$ is the voltage across the capacitor. Due to stored energy a charged capacitor can produce an electric shock, even when not connected into a circuit.

| *Capacitance* | *Inductance* |
|---|---|
| 1. Unit: Farad | Henry |
| 2. Stores energy $Q$ | Conducts current $I$ |
| 3. Needs dielectric as insulator | Needs wire conductor |
| 4. More plate area allows more $C$ | More turns allow more $L$ |
| 5. Dielectric with higher $K$ concentrates electric field for more $C$ | Core with higher $K$ concentrates magnetic field for more $L$ |
| 6. $\frac{1}{C_T} = \frac{1}{C_1} + \frac{1}{C_2} + \ldots\ldots$ in series | $L_T = L_1 + L_2$ in series |
| 7. $C_T = C_1 + C_2$ in parallel. | $\frac{1}{L_T} = \frac{1}{L_1} + \frac{1}{L_2}$ in parallel. |

Symbolic representation of capacitors.

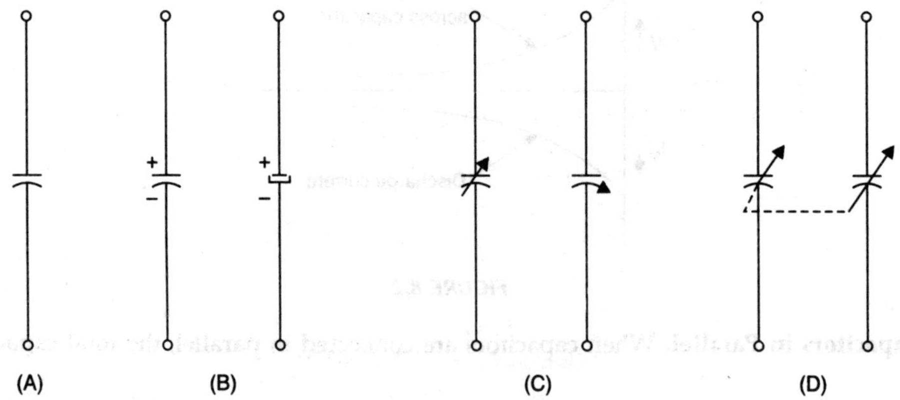


(A) Fixed value capacitor—air, paper, mica or ceramic
(B) Electrolytic, with polarity
(C) Variable
(D) Ganged on one shaft.

***FIGURE 8.3***

**Alternating Current in a Capacitive Circuit.** When alternating current flows through the circuit.

1. Smaller capacitance allows less current which means more capacitive reactance $X_C$ with more ohms of opposition.
2. Lower frequencies for the applied 8 voltage result in less current and more $X_C$. With a steady dc voltage source, which corresponds to a frequency of zero, the opposition of the capacitor is infinite and there is no current.

   The capacitive reactance $X_C$ is given by

$$X_C = \frac{1}{2\pi fC}$$

The following relations are valid for the circuit shown in Fig. 8.4

$$q_{(t)} < C\,v_{(t)}$$

$$i = \frac{dq}{dt} = C\frac{dv}{dt}$$

$$v_{(t)} = \frac{1}{e}\int i\,dt$$

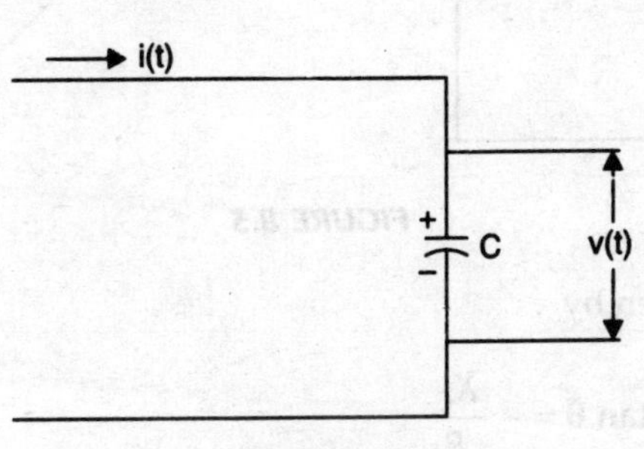


*FIGURE 8.4*

**Series and Parallel $X_C$.** Because capacitive reactance is an opposition in ohms, series or parallel reactances are combined in the same way as resistances. Thus, in series circuit.

$$X_{C_T} = X_{C_1} + X_{C_2} + X_{C_3} + \ldots\ldots$$

and

$$\frac{1}{X_{C_T}} = \frac{1}{X_{C_1}} + \frac{1}{X_{C_2}} + \ldots\ldots$$

for parallel combination.

**Comparison of $X_C$ and $X_L$**

| *Inductive reactance $X_L$* | *Capacitive reactance $X_C$* |
|---|---|
| Increases with more inductance | Decreases with more capacitance |
| Increases for higher frequencies | Decreases for higher frequencies |
| Allows more current for lower frequencies | Allows less current for lower frequencies |
| Passes direct current. | Blocks direct current. |

**Capacitive Circuits.** In a sine-wave ac circuit, the voltage across a capacitance lags its charge and discharge current by 90°. Therefore, capacitive reactance $X_C$ is a phasor quantity out of phase with its series resistance by – 90° because.

$$i_c = C\left(\frac{dv}{dt}\right)$$

**$X_C$ and R in Series.** With resistance in series with capacitive reactance, the voltage drop across the capacitive reactance lags the series current by 90°. The *IR* voltage drop has the same phase as *I*.

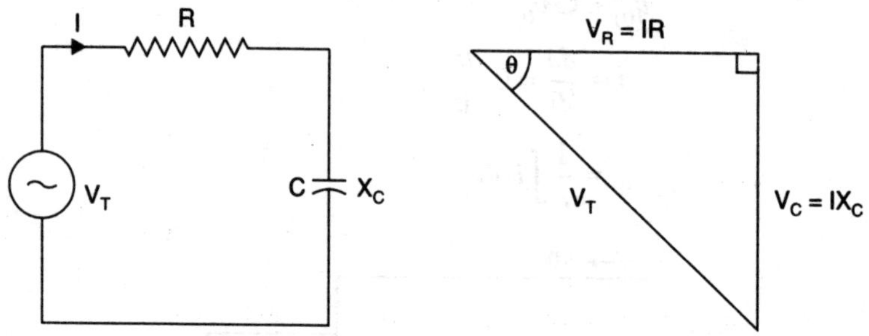


***FIGURE 8.5***

The phase angle is given by

$$\tan\theta = -\frac{X_C}{R}$$

The negative sign means the angle is clockwise from zero, to indicate that $V_T$ lags behind the leading *I*.

**Series and Parallel RC Circuits**

| *$X_C$ and R in Series* | *$X_C$ and R in Parallel* |
|---|---|
| *I* the same is $X_C$ and *R* | *V* the same across $X_C$ and *R* |
| $V_T = \sqrt{V_{R^2} + V_{C^2}}$ | $I_T = \sqrt{I_{R^2} + I_{C^2}}$ |
| $Z = \sqrt{R^2 + X_{C^2}}$ | $Z = \frac{V}{I_T}$ |
| $V_C$ lags $V_R$ by 90° | $I_C$ leads $I_R$ by 90° |
| $\tan\theta_Z = -\frac{X_C}{R}$ | $\tan\theta_I = \frac{I_C}{I_R}$ |
| $\theta_Z$ increases as more $X_C$ means more $V_C$. | $\theta_I$ decreases as more $X_C$ means less $I_C$. |

## OBJECTIVE TYPE QUESTIONS

1. In a capacitor the electric charge is stored in
   (*a*) dielectric (*b*) metal plates
   (*c*) both dielectric as well as metal plates (*d*) none of the above.
2. The unit of capacitance is
   (*a*) Farad (*b*) Farad/m (*c*) Farad/$m^2$ (*d*) Farad/$m^3$.
3. The value of dielectric constant for vacuum is taken as
   (*a*) zero (*b*) 1 (*c*) 100 (*d*) infinite.
4. Two plates each with an area of $2m^2$, are separated by 1 cm with a dielectric of air. The value of capacitance C will be
   (*a*) 1770 F (*b*) 17.70 F (*c*) 1770 μF (*d*) 1770 pF.
5. A capacitor charged to 100 V has 1000 micro coulombs of charge. The value of capacitance C will be
   (*a*) 10 pF (*b*) 100 pF (*c*) 10 μF (*d*) 10 F.
6. A constant current of 5 mA charges a 10 microfarad capacitor of one second. The voltage across the capacitor will be
   (*a*) 5 volts (*b*) 50 volts (*c*) 500 volts (*d*) 5000 volts.
7. Which of the following capacitor will have the least value of breakdown voltage?
   (*a*) electrolytic (*b*) ceramic (*c*) paper (*d*) mica.
8. The breakdown voltage for paper capacitors is usually
   (*a*) 10 to 50 volts (*b*) 200 to 1600 volts
   (*c*) 2000 to 4000 volts (*d*) more than 10,000 volts.
9. The capacitance value of paper capacitor is usually
   (*a*) 0.001 to 1 microfarad (*b*) 1 to 100 microfarad
   (*c*) 1000 to 1600 microfarad (*d*) 10 to 400 farad.
10. Air capacitors are generally available in the range
    (*a*) 1 to 10 pF (*b*) 10 to 400 pF (*c*) 10 to 400 μF (*d*) 1000 to 16000 μF
11. Dielectric constant for air is taken as
    (*a*) zero (*b*) 1 (*c*) 100 (*d*) infinite.
12. Which of the following material has least value of dielectric constant?
    (*a*) paper (*b*) oil (*c*) glass (*d*) ceramics.
13. Dielectric constant for mica is nearly
    (*a*) 1 (*b*) 3 to 8 (*d*) 30 to 80 (*d*) 100.
14. Which of the following material has the highest value of dielectric constant?
    (*a*) oil (*b*) glass (*c*) vacuum (*d*) ceramics.

15. The value of absolute permittivity (for electric flux in air or vacuum) in SI units is
(*a*) 8854 × $10^{-12}$ F/m (*b*) 885.4 × $10^{-12}$ F/m
(*c*) 88.54 × $10^{-12}$ F/m (*d*) 8.854 × $10^{-12}$ F/m.

16. How much charge is stored in a 2 μF capacitor with 50 volts across it?
(*a*) $10^{-6}$ coulomb (*b*) $10^{-4}$ coulomb (*c*) $10^{-2}$ coulomb (*d*) 1 coulomb.

17. If breakdown voltage of 10,000 volts is desired, which of the following capacitor will be the obvious choice?
(*a*) air (*b*) paper (*c*) electrolytic (*d*) ceramic.

18. Mica capacitors are generally available in the range
(*a*) 1 to 50 pF (*b*) 10 to 5000 pF (*c*) 10 to 5000 μF (*d*) 10 to 500 F.

19. Which of the following capacitor is marked for polarity?
(*a*) Mica (*b*) Paper (*c*) Electrolytic (*d*) Air.

20. A gang condenser is a
(*a*) ceramic capacitor (*b*) polarised capacitor
(*c*) variable capacitor (*d*) none of the above.

21. A constant current of 2 μA charges a capacitor for 20 seconds. The charge stored will be
(*a*) 4 micro coulombs (*b*) 40 micro coulombs
(*c*) 400 micro coulombs (*d*) 4000 micro coulombs.

22. In the above problem if the voltage across the charged capacitor is 20 V, the value of capacitance will be
(*a*) 2 F (*b*) 20 F (*c*) 2 millifarad (*d*) 2 microfarad.

23. A mica capacitor and a ceramic capacitor both have the same physical dimensions. Which will have more value of capacitance?
(*a*) mica capacitor (*b*) ceramic capacitor
(*c*) both will have identical value of capacitance
(*d*) it depends on applied voltage.

24. A paper capacitor is usually available in the form of
(*a*) meshed plates (*b*) disc (*c*) rolled foil (*d*) tubes.

25. Which of the following is incorrect?
(*a*) 1 microfarad × 1 microfarad = 1 picofarad
(*b*) 1 farad = $10^{12}$ × 1 picofarad
(*c*) 1 microfarad + 1 picofarad = $10^{-18}$ farad
(*d*) 1 micro microfarad = 1 picofarad.

26. Which of the following capacitor can be used for temperature compensation?
(*a*) ceramic capacitor (*b*) air capacitor
(*c*) paper capacitor (*d*) none of the above.

27. A capacitor in a circuit became hot and ultimately exploded due to wrong connections. Which type of capacitor it could be?

(*a*) electrolytic capacitor (*b*) paper capacitor
(*c*) ceramic capacitor (*d*) any of the above.

28. For closer tolerances which of the following capacitor is usually, preferred?

(*a*) ceramic disc capacitor (*b*) paper capacitor
(*c*) mica capacitor

29. Paper capacitors usually have a tolerance of

(*a*) ± 50% (*b*) ± 20% (*c*) ± 10% (*d*) ± 1%.

30. Which of the following capacitor will have the least variation?

(*a*) ceramic capacitor (*b*) paper capacitor
(*c*) ceramic capacitor (*d*) silver plated mica capacitor.

31. Which capacitor will be physically smaller for the same ratings?

(*a*) paper capacitor (*b*) ceramic capacitor (*c*) both will be of equal size.

32. What is the value of capacitance that must be connected in parallel with a 100 pico farad condenser to make an equivalent capacitance of 250 pF?

(*a*) 50 pF (*b*) 100 pF (*c*) 150 pF (*d*) 200 pF.

33. Which of the following statement is false?

(*a*) The stored energy in a capacitor increases with applied voltage.
(*b*) The leakage resistance of ceramic capacitors is generally high.
(*c*) A twirl cable has distributed capacitance between the conductors.
(*d*) The stored energy in a capacitor decreases with reduction in value of capacitance.

34. A capacitor consists of

(*a*) two insulators only (*b*) two conductors only
(*c*) two insulators separated by a conductor
(*d*) two conductors separated by an insulator.

35. If a 4 μF capacitor is charged to 100 V, the charge in coulombs will be

(*a*) 100 μC (*b*) 200 μC (*c*) 400 μC (*d*) 800 μC.

36. Capacitance increases with

(*a*) increase in plate area and the distance between the plates
(*b*) reduction in plate area and the distance between the plates
(*c*) increase in plate area and decrease in distance between the plates
(*d*) decrease in plate area and the value of applied voltage.

37. While testing a capacitor if an ohm metre reading immediately goes practically to zero and stays there, it can be concluded that the capacitor is

(*a*) alright (*b*) short-circuited (*c*) open-circuited (*d*) leaky.

38. The equivalent capacitance of the circuit shown in Fig. 8.6 will be

(*a*) 0.231 μF (*b*) 1/3 μF (*c*) 3 μF (*d*) 1 μF.

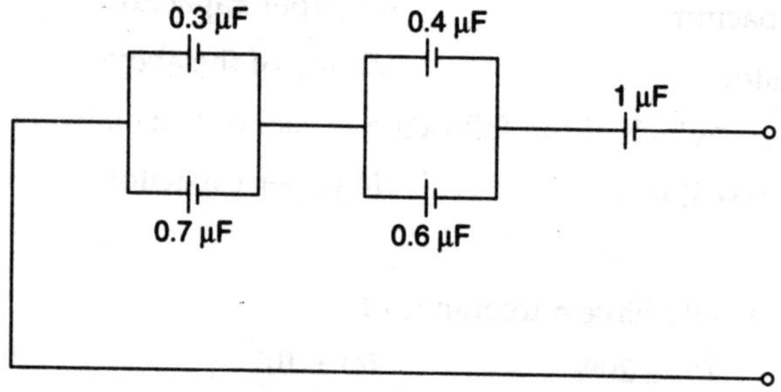


***FIGURE 8.6***

39. For a good 0.05 μF paper capacitor, the ohm metre observation should

(*a*) go quickly to about 100 ohms, approximately, and remain there

(*b*) show low resistance momentarily and back off to a very high resistance

(*c*) show high resistance momentarily and then a very low resistance

(*d*) not move at all.

40. Which of the following capacitor has relatively shorter shelf life?

(*a*) paper capacitor (*b*) mica capacitor

(*c*) ceramic capacitor (*d*) electrolytic capacitor.

41. While testing a capacitor with ohm metre, if the capacitor shows charging, but the final resistance reading is appreciably less than normal, it can be concluded that the capacitor is

(*a*) alright (*b*) leaky (*c*) open circuited (*d*) short-circuited.

42. Voltage applied across a ceramic dielectric produces an electrostatic field 100 times greater than in air, the dielectric constant *k* of the ceramic will be

(*a*) 10,000 (*b*) 100 (*c*) 50 (*d*) $33\frac{1}{3}$.

**Questions 43 to 45 Refer to Data Given Below:**

With 100 V across a capacitor, it stores 100 μC of charge. For the same capacitor when applied voltage is 200 V.

43. How much will be the voltage across the capacitor?

(*a*) 50 V (*b*) 100 V (*c*) 200 V (*d*) 250 V.

44. How much charge is stored in the capacitor?

(*a*) 50 micro coulombs (*b*) 100 micro coulombs

(*c*) 200 micro coulombs (*d*) 400 micro coulombs.

45. How much is the capacitance value?

(*a*) 1 μF (*b*) 0.50 μF (*c*) 1 pF (*d*) 0.01 pF.

46. A 10 μF capacitance is charged to 10 V. The energy stored will be
(*a*) 1 coulomb (*b*) 200 micro-coulombs
(*c*) 500 micro-coulombs (*d*) 10 micro-coulombs.

**Questions 47 and 48 Refer to Data Given Below:**

A 2 μF capacitor is charged by a constant 3 μA charging current for 4 seconds.

47. The charge stored in the capacitor will be
(*a*) 2 micro-coulombs (*b*) 12 micro-coulombs
(*c*) 30 micro-coulombs (*d*) 60 micro-coulombs.

48. The voltage across the capacitor will be
(*a*) 1 V (*b*) 2 V (*c*) 4 V (*d*) 6 V.

49. The equivalent capacitance of the circuit shown in Fig. 8.7 will be nearly
(*a*) 4 μF (*b*) 5 μF (*c*) 6 μF (*d*) 8 μF

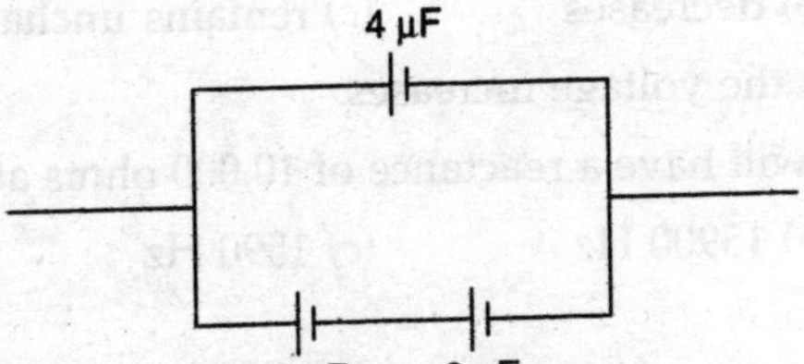


*FIGURE 8.7*

50. Which of the following statement is false?
(*a*) in a series circuit, the higher the value of $X_C$, the lesser is its voltage drop compared with the *IR* drop.
(*b*) in a parallel circuit, the voltage across $X_C$ lags 90° behind the capacitive branch current.
(*c*) when the frequency is doubled, the capacitive reactance reduces to half
(*d*) none of the above.

51. A 0.05 μF capacitor is charged to 264 V. It discharges through a 40 kΩ resistor. The time taken for $V_C$ to discharge down to 132 volts will be
(*a*) 1.4 S (*b*) 14 S (*c*) 1.4 mS (*d*) 1.4 μS.

52. Which of the following bridge is used to measure the value of capacitance of a condenser?
(*a*) Wheatstone bridge (*b*) Hay's bridge
(*c*) Schering bridge (*d*) Maxwell bridge.

53. The power dissipated in a pure capacitor is
(*a*) proportional to applied voltage (*b*) proportional to value of capacitance
(*c*) both (*a*) and (*b*) above (*d*) zero.

**54.** In an *RC* circuit consisting of 2 μF capacitor in series with a 1 MΩ resistor, a time period of 6 seconds will be equal to how many time constants?

(*a*) two (*b*) three (*c*) thirteen (*d*) thirty.

**55.** When $v_C$ across a 1 μF capacitor drops from 43 V to 42 V in 1 second, the discharge current $i_C$ will be

(*a*) 1 μA (*b*) 2 μA (*c*) 42 μA (*d*) 42 mA.

**56.** When *Q* is the charge and *C* the capacitance of a condenser, then the energy stored in the capacitor is given by

(*a*) $\frac{1}{2}\frac{C^2}{Q}$ (*b*) $\frac{1}{2}\frac{Q^2}{C}$ (*c*) $\frac{1}{2}\frac{C}{Q^2}$ (*d*) $\frac{1}{2}\frac{Q}{C^2}$.

**57.** The capacitor normally preferred for single phase motors is

(*a*) Paper capacitor (*b*) Mica capacitor (*c*) Ceramic capacitor (*d*) Electrolytic capacitor.

**58.** With higher frequencies, the amount of capacitive reactance

(*a*) increases (*b*) decreases (*c*) remains unchanged

(*d*) increase only when the voltage increases.

**59.** A capacitor of 0.01 μF will have a reactance of 10,000 ohms at a frequency of

(*a*) 159000 Hz (*b*) 15900 Hz (*c*) 1590 Hz (*d*) 159 Hz.

**60.** In a capacitive circuit

(*a*) an increase in applied voltage makes a capacitor charge

(*b*) an increase in applied voltage makes a capacitor discharge

(*c*) a decrease in applied voltage makes a capacitor charge

(*d*) a steady value of applied voltage causes discharge.

**61.** In a parallel plate capacitor, a dielectric slab is introduced, the potential difference between plates will be

(*a*) becomes zero (*b*) increase (*c*) decreas (*d*) remain unaltered.

**62.** Alternating current can flow in a capacitive circuit with ac voltage applied because

(*a*) of high peak value

(*b*) charging current flows when the voltage decreases

(*c*) discharge current flows when the voltage increases

(*d*) varying voltage produces charge and discharge current.

**Questions 63 and 65 Refer to Data Given Below:**

A capacitor with an $X_C$ of 2000 ohms is connected across a 10 V – 1000 Hz source.

**63.** The current in the circuit will be

(*a*) 1 mA (*b*) 5 mA (*c*) 50 mA (*d*) 100 mA.

**64.** What will be the frequency of the current?

(*a*) 100 Hz (*b*) 1000 Hz (*c*) 10000 Hz (*d*) 1 MHz.

65. What will be the $RC$ time constant for 1 MΩ resistance in series with a 2 μF capacitor for charge?

(a) 2 μS (b) 20 μS (c) 2 mS (d) 2S.

66. A combination of two 0.02 μF, 500 V capacitors in parallel will have the capacitance and breakdown rating of

(a) 0.04 μF, 1000 V (b) 0.04 μF, 500 V (c) 0.01 μF, 1000 V (d) 0.02 μF, 500 V.

67. For the circuit shown in Fig. 8.8 the equivalent capacitance across $AB$ will be

(a) 0.147 μF (b) 0.1 μF (c) 0.01 μF (d) 0.00343 μF.

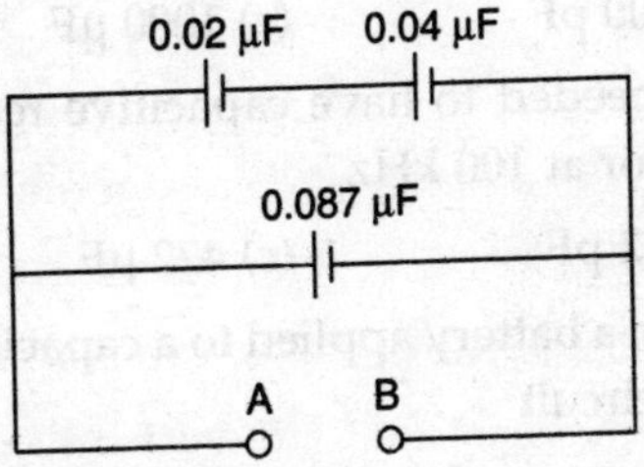


*FIGURE 8.8*

68. Three capacitors of values 2 μF, 3 μF and 5 μF are connected in parallel across an ac source. The maximum current will pass through

(a) 2 μF capacitor (b) 3 μF capacitor (c) 5 μF capacitor

(d) the same current will pass through all the capacitors.

69. Other factors remaining the same, which capacitor will have highest reactance?

(a) 0.1 pF (b) 0.5 pF (c) 0.1 μF (d) 0.5 μF.

70. Which of the following statement is false?

(a) a capacitor can store charge because it has a dielectric between two conductors

(b) the thinner the dielectric, the more the capacitance and the lower the voltage breakdown rating for a capacitor

(c) capacitors in series provide less capacitance but a higher voltage breakdown rating for the combination

(d) a six dot mica capacitor colour coded white, green, black, red and yellow has the capacitance value of 500 pF.

71. The capacity of capacitor bank used in power factor correction is expressed in terms of

(a) volts (b) kW (c) kVA (d) kVAR.

72. In radio, the gang condenser is

(a) ceramic capacitor (b) air capacitor

(c) paper capacitor (d) electrolytic capacitor.

73. The capacitor preferred for high frequency circuits is

(a) mica capacitor (b) air capacitor

(c) electrolytic capacitor (d) none of the above.

### Questions 74 to 77 Refer to Data Given Below:

Three equal capacitive reactances of 600 Ω each are in parallel.

**74.** The equivalent combined reactance will be

(*a*) 1800 Ω (*b*) 600 Ω (*c*) 200 Ω (*d*) 456.6 Ω.

**75** If the frequency of the applied voltage is 800 kHz, how much is the capacitance of each capacitor?

(*a*) 333 pF (*b*) 33.3 pF (*c*) 3.33 pF (*d*) 333 μF.

**76.** The equivalent combined capacitance of the t...ee in parallel will be

(*a*) 10 pF (*b*) 1000 pF (*c*) 1000 μF (*d*) 1333.3 pF.

**77.** What size of capacitor is needed to have capacitive reactance the same as the inductive reactance of a 6 mH inductor at 100 kHz

(*a*) 422 pF (*d*) 42.2 pF (*c*) 422 μF (*d*) 42.2 μF.

**78.** With steady dc voltage from a battery applied to a capacitance, after it charges to the battery voltage, the current in the circuit

(*a*) depends on the current rating of the battery

(*b*) is greater for larger values of capacitance

(*c*) is smaller for larger values of capacitance

(*d*) is zero for any capacitance value.

### Questions 79 to 81 Refer to Data Given Below:

Two capacitive reactances $X_{C_1} = 1000\ \Omega$ and $X_{C_2} = 4000\ \Omega$ are in series across a 10 V source.

**79.** The current in the circuit will be

(*a*) 2 μA (*b*) 20 μA (*c*) 2 mA (*d*) 20 mA

**80.** The voltage across $X_{C_1}$ will be

(*a*) 1 V (*b*) 2 V (*c*) 10 V (*d*) 20 V.

**81.** The voltage across $X_{C_2}$ will be

(*a*) 2 V (*b*) 4 V (*c*) 6 V (*d*) 8 V.

**82.** A capacitance of 0.02 μF is the same as

(*a*) $200 \times 10^{-12}$ F (*b*) $0.02 \times 10^{6}$ F (*c*) $0.02 \times 10^{-6}$ F (*d*) $0.02 \times 10^{-12}$ F.

**83.** A mica capacitor has $K = 8$, a thickness of 0.02 cm, plates of 6 cm$^2$ and five such sections in parallel. What will be the capacitance between the plates?

(*a*) 1062 pF (*b*) 10.62 pF (*c*) 1062 μF (*d*) 1062 F.

**84.** Three capacitors of following rating are connected in parallel:

5 microfarads 24 V

15 microfarads 56 V

25 microfarads 100 V.

The above bank will be suitable for operation on

(*a*) 24 V (*b*) 56 V (*c*) 100 V (*d*) 60 V.

**85.** If *R*, *L* and *C* denote resistance, inductance and capacitance respectively, then the quantity $CR^2$ has the dimensions of

(*a*) *R* (*b*) *L* (*c*) *C* (*d*) dimensionless.

**86.** A combination of two 0.02 μF, 500 V capacitors in series will have capacitance and breakdown rating of

(*a*) 0.01 μF, 500 V (*b*) 0.01 μF, 1000 V (*c*) 0.02 μF, 500 V (*d*) 0.04 μF, 500 V.

**87.** Which of the following capacitor will have the least amount of energy stored in it?

(*a*) 40 μF capacitor charged to 400 V (*b*) 1 μF capacitor charged to 5000 V

(*c*) 500 pF capacitor charged to 10 kV.

**88.** The sparking between two electrical contacts can be reduced by inserting a

(*a*) resistance in the line

(*b*) a capacitor in series with each contacts

(*c*) a capacitor in parallel with the contacts

(*d*) none of the above.

**89.** In the case of a lossy capacitor, its series equivalent resistance value will be

(*a*) very small (*b*) small (*c*) large.

**90.** The discharge curve of a capacitor is represented by

(*a*) figure *A* (*b*) figure *B* (*c*) figure *C* (*d*) figure *D*.

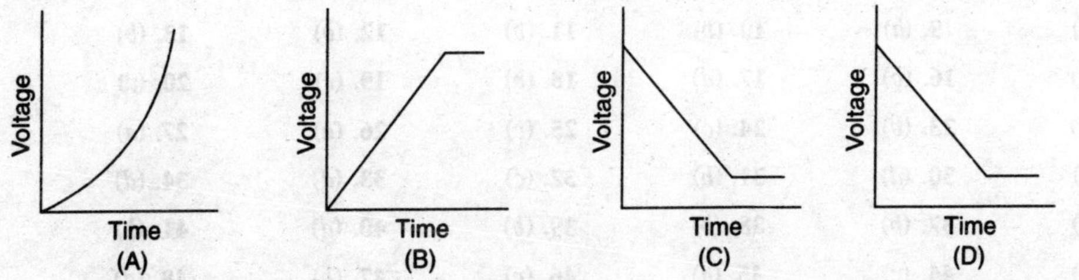


*FIGURE 8.9*

**91.** The capacitive reactance of a 100 pF capacitor at 10 MHz will be

(*a*) 15,900 ohms (*b*) 1590 ohms (*c*) 159 ohms (*d*) 15.9 ohms.

**92.** At which of the following frequency a 0.1 μF capacitor will have capacitive reactance of 1000 ohms?

(*a*) 1590 kHz (*b*) 159 kHz (*c*) 1590 Hz (*d*) 159 Hz.

**93.** The capacitive reactance of a capacitor at 5 MHz is 64 ohms. The capacitive reactance at 1.25 MHz will be

(*a*) 16 ohms (*b*) 32 ohms (*c*) 128 ohms (*d*) 256 ohms.

**94.** Two capacitive reactance of 100 Ω and 200 Ω are connected in series. The combined capacitive reactance of the circuit will be

(*a*) 300 Ω (*b*) 250 Ω (*c*) 66.7 Ω (*d*) 50 Ω.

**95.** The instantaneous value of charging current is produced by a 6 μF capacitance when its potential difference is increased by 50 volts in 1 second, will be

(*a*) 1200 μA (*b*) 600 μA (*c*) 300 μA (*d*) 150 μA.

**96.** For a 2.65 μF capacitor the capacitive reactance at 60 Hz will be

(*a*) 100 Ω (*b*) 1000 Ω (*c*) 1250 Ω (*d*) 2000 Ω.

**97.** The value of capacitance needed to have a 100 ohm capacitive reactance at 1 MHz will be

(*a*) 1590 pF (*b*) 159 pF (*c*) 15.9 pF (*d*) 15.9 μF.

**98.** The capacitive reactance for a capacitor at 3 MHz was 144 ohms. The capacitive reactance at 6 MHz will be

(*a*) 36 ohms (*b*) 72 ohms (*c*) 288 ohms (*d*) 576 ohms.

**99.** The value of capacitive reactance increases as the

(*a*) value of capacitance increases (*b*) frequency increases

(*c*) voltage increases (*d*) none of the above.

**100.** A 0.2 μF capacitor will have capacitive reactance of 796 ohms when the frequency is

(*a*) 10 MHz (*b*) 1 MHz (*c*) 10 kHz (*d*) 1 kHz.

## ANSWERS

| | | | | | | |
|---|---|---|---|---|---|---|
| **1.** (*a*) | **2.** (*a*) | **3.** (*b*) | **4.** (*d*) | **5.** (*c*) | **6.** (*c*) | **7.** (*a*) |
| **8.** (*b*) | **9.** (*a*) | **10.** (*b*) | **11.** (*b*) | **12.** (*b*) | **13.** (*b*) | **14.** (*d*) |
| **15.** (*d*) | **16.** (*b*) | **17.** (*d*) | **18.** (*b*) | **19.** (*c*) | **20.** (*c*) | **21.** (*b*) |
| **22.** (*d*) | **23.** (*b*) | **24.** (*c*) | **25.** (*c*) | **26.** (*a*) | **27.** (*a*) | **28.** (*c*) |
| **29.** (*c*) | **30.** (*d*) | **31.** (*b*) | **32.** (*c*) | **33.** (*d*) | **34.** (*d*) | **35.** (*c*) |
| **36.** (*c*) | **37.** (*b*) | **38.** (*b*) | **39.** (*b*) | **40.** (*d*) | **41.** (*b*) | **42.** (*b*) |
| **43.** (*c*) | **44.** (*c*) | **45.** (*a*) | **46.** (*c*) | **47.** (*b*) | **48.** (*d*) | **49.** (*c*) |
| **50.** (*a*) | **51.** (*c*) | **52.** (*c*) | **53.** (*d*) | **54.** (*b*) | **55.** (*a*) | **56.** (*b*) |
| **57.** (*d*) | **58.** (*b*) | **59.** (*c*) | **60.** (*a*) | **61.** (*c*) | **62.** (*d*) | **63.** (*b*) |
| **64.** (*b*) | **65.** (*d*) | **66.** (*b*) | **67.** (*b*) | **68.** (*c*) | **69.** (*a*) | **70.** (*d*) |
| **71.** (*d*) | **72.** (*b*) | **73.** (*a*) | **74.** (*c*) | **75.** (*a*) | **76.** (*b*) | **77.** (*a*) |
| **78.** (*d*) | **79.** (*c*) | **80.** (*b*) | **81.** (*d*) | **82.** (*c*) | **83.** (*a*) | **84.** (*a*) |
| **85.** (*b*) | **86.** (*b*) | **87.** (*c*) | **88.** (*c*) | **89.** (*c*) | **90.** (*d*) | **91.** (*c*) |
| **92.** (*c*) | **93.** (*c*) | **94.** (*a*) | **95.** (*c*) | **96.** (*b*) | **97.** (*a*) | **98.** (*b*) |
| **99.** (*d*) | **100.** (*d*). | | | | | |

# Chapter 9 AC CIRCUITS

**Series and Parallel Circuits.** A series circuit is shown in Fig. 9.1. When the source voltage is assumed to be constant, the phasor develops voltage across each of the impedances through which it passes. These voltages are the potential drops. Using Kirchhoff's law

$$\begin{aligned} V &= V_1 + V_2 + V_3 \\ &= IZ_1 + IZ_2 + IZ_3 \\ &= I\,(Z_1 + Z_2 + Z_3) \\ &= IZ_{eq}. \end{aligned}$$

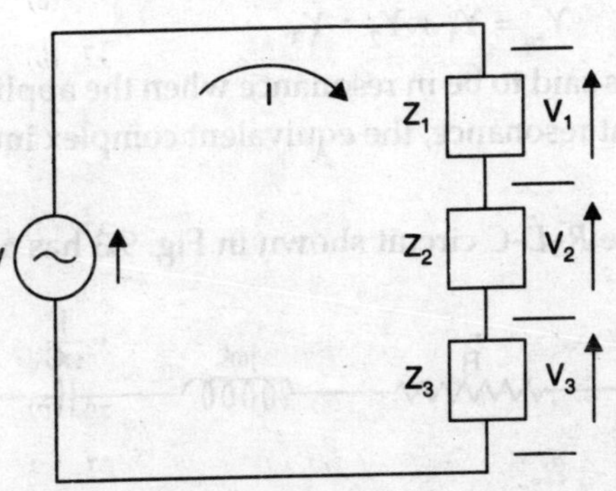


*FIGURE 9.1*

Hence, $$Z_{eq} = Z_1 + Z_2 + Z_3.$$

In case the three impendances are connected in parallel, the voltage across each impedance is *V* and according to Kirchhoff's law

$$I_T = I_1 + I_2 + I_3.$$

$$= \frac{V}{Z_1} + \frac{V}{Z_2} + \frac{V}{Z_3}$$

$$= V\left[\frac{1}{Z_1} + \frac{1}{Z_2} + \frac{1}{Z_3}\right]$$

and the equivalent impedance is given by

$$\frac{1}{Z_{eq}} = \frac{1}{Z_1} + \frac{1}{Z_2} + \frac{1}{Z_3} + .....$$

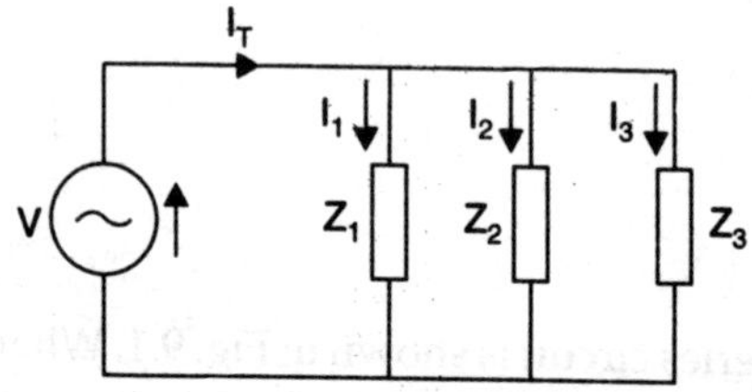


FIGURE 9.2

**Admittance.** The reciprocal of complex impedance $Z$ is complex admittance $Y$. Thus

$$Y = \frac{1}{Z} = \frac{I}{V}$$

$Y$ is expressed in siemens (S). Thus in case of parallel impedances

$$Y_{eq} = Y_1 + Y_2 + Y_3.$$

**Resonance.** A circuit is said to be in resonance when the applied voltage $V$ and the resulting current $I$ are in phase. Thus at resonance, the equivalent complex impedance of the circuit consists of only resistance $R$.

**Series Resonance.** The $R$–$L$–$C$ circuit shown in Fig. 9.3 has a complex impedance

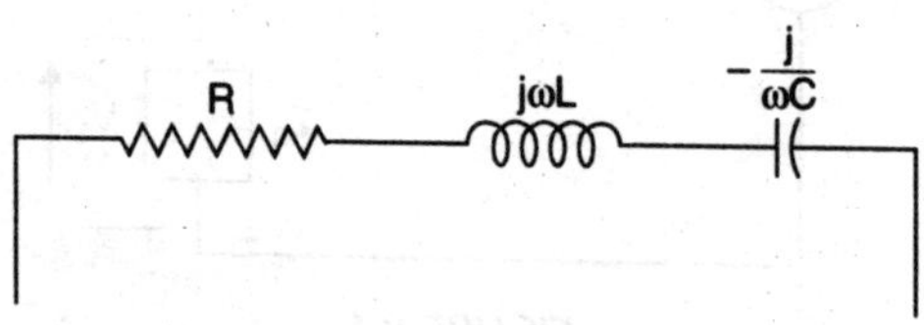


FIGURE 9.3

$$Z = R + j\left(\omega L - \frac{1}{\omega C}\right)$$

$$= R + jX.$$

The circuit is in resonance when $X = 0$

*i.e.,* $$\omega L = \frac{1}{\omega C}$$

or $$\omega = \frac{1}{\sqrt{LC}} = \omega_0.$$

The resonant frequency is given by

$$f_0 = \frac{1}{2\pi\sqrt{LC}} \text{ Hz}$$

In Fig. 9.4 the absolute value of $Z$ and its three components $R$, $X_L$ and $X_C$ are plotted as function of $\omega$. At $\omega = \omega_0$, the inductive and capacitive reactances are equal and

$$|Z| = \sqrt{R^2 + X^2}$$

$$Z = R$$

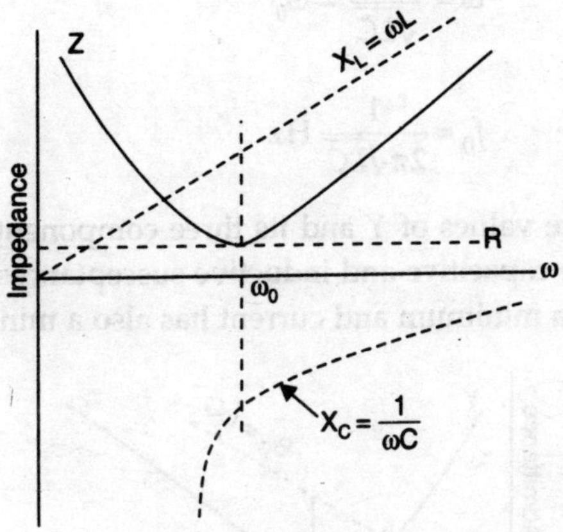


*FIGURE 9.4*

Thus at resonance the impedance $Z$ is a minimum. Also as $I = \frac{V}{Z}$, the current is maximum.

For frequencies below $\omega_0$ the capacitive reactance is greater than the inductive reactance, and the angle of the impedance is negative. At frequencies above $\omega_0$, the inductive reactance exceeds the capacitive reactance and the angle of $Z$ is positive and approaches $+ 90^{\circ}$ as $\omega >> \omega_0$.

**Parallel Resonance.** In Fig. 9.5 $R$, $L$ and $C$ are in parallel. In this case the admittance of the three elements is

$$Y = G + j\left(\omega C - \frac{1}{\omega L}\right)$$

$$= G + jB$$

where $$B = B_C - B_L$$

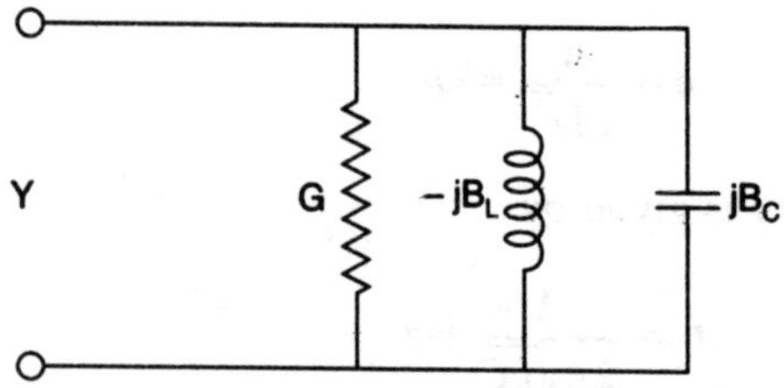


FIGURE 9.5

The circuit is in resonance, when $B = 0$.

*i.e.,* $$\omega C = \frac{1}{\omega L}$$

or $$\omega = \frac{1}{\sqrt{LC}} = \omega_0$$

and $$f_0 = \frac{1}{2\pi\sqrt{LC}} \text{ Hz.}$$

In Fig. 9.6 the absolute values of $Y$ and its three components $G_1$ $B_C$ and $B_L$ are plotted as functions of ω. At $\omega = \omega_0$ the capacitive and inductive susceptances are equal and $Y = G$. Thus at resonance the admittance is a minimum and current has also a minimum value.

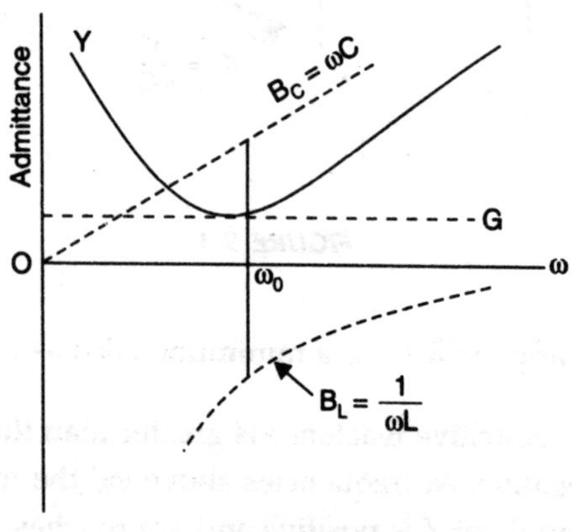


FIGURE 9.6

**Two Branch Circuit.** A two branch circuit is shown in Fig 9.7. The admittance $Y$ is given by

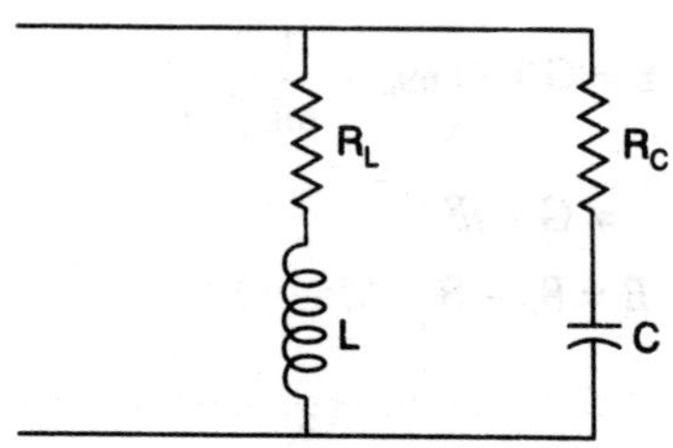


FIGURE 9.7

$$Y = Y_L + Y_C$$

$$= \frac{1}{R_L + jX_L} + \frac{1}{R_C - jX_C}$$

**The circuit will be resonant when the complex admittance is a real number. At resonance**

$$\omega_0 = \frac{1}{\sqrt{LC}} \sqrt{\frac{R_{L^2} - L/C}{R_{C^2} - L/C}}.$$

**Quality Factor.** **The quality factor is defined as**

$$Q = 2\pi \frac{\text{Maximum stored energy}}{\text{Energy dissipated per cycle}}$$

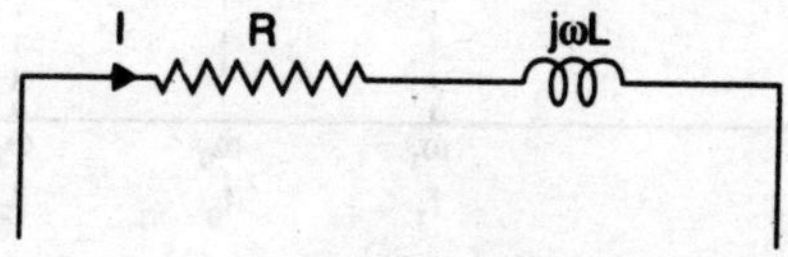

**FIGURE 9.8**

**For the $R - L$ circuit shown in Fig. 9.8 maximum stored energy = $\frac{1}{2}L(I_{max})^2$ energy dissipated per cycle**

$$= \left(\frac{I_{max}}{\sqrt{2}}\right)^2 R$$

**Hence**

$$Q = 2\pi \frac{\frac{1}{2}L(I_{max})^2}{\frac{(I_{max})^2}{2} R\left(\frac{1}{f}\right)}$$

$$= \frac{\omega L}{R}$$

**Likewisely in a $R$–C circuit shown in Fig. 9.9**

$$Q = \frac{1}{\omega RC}.$$

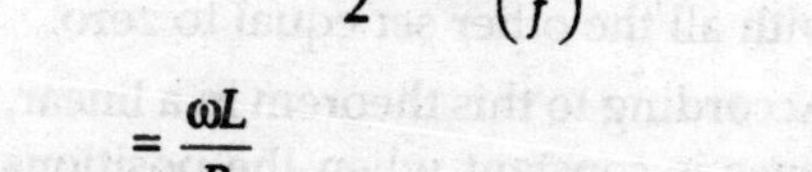

**FIGURE 9.9**

**In Fig. 9.10 the current of the *RLC* circuit is plotted as a function of ω or with an appropriate change of scale, as a function of *f*. At $\omega_0$ the current is maximum. The points, where the current is 0.707 of the maximum are indicated. The corresponding frequencies are $\omega_1$ and $\omega_2$.**

As the power delivered to the circuit is $I^2R$, at $I = 0.707\ I_0$, the power is one half of the maximum value which occurs at $\omega_0$. The points corresponding to $\omega_1$ and $\omega_2$ are called the half-power points. The distance between these points, measured in hertz, is known as band width, B.W.

Also
$$Q_0 = \frac{\omega_0}{\omega_2 - \omega_1} = \frac{f_0}{f_2 - f_1} = \frac{f_0}{\text{B.W.}}.$$

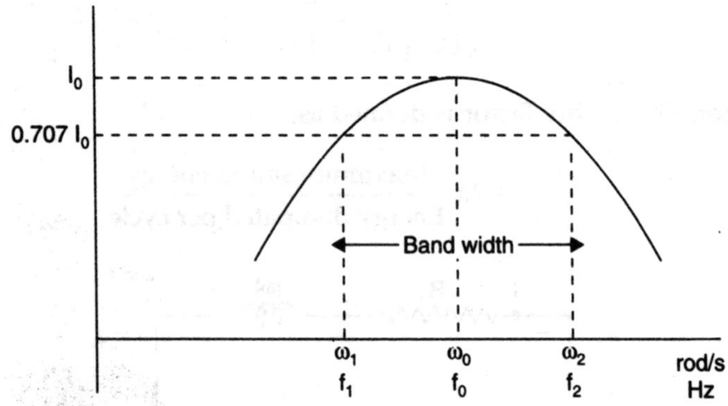


*FIGURE 9.10*

and the resonant frequency is the geometric mean of $\omega_1$ and $\omega_2$, so that

$$\omega_0 = \sqrt{\omega_1 \omega_2}$$

and
$$f_0 = \sqrt{f_1\, f_2}\,.$$

## NETWORK THEOREMS

**Superposition Theorem.** According to this theorem the response of any element of a linear bilateral network containing two or more sources is the sum of the responses obtained by each source acting separately and with all the other set equal to zero.

**Reciprocity Theorem.** According to this theorem in a linear, bilateral, single source network the ratio of excitation to response is constant when the positions of excitation and response are interchanged.

This theorem also applies to networks containing a single current source. Here the theorem states that the voltage which results at a pair of terminals mn due a current source acting at terminals *ab* is the same as the voltage at terminals *ab* when the current source is moved to terminals *mn*. It should be noted that the voltages at other points in the network will not remain the same.

**Compensation Theorem.** A network impedance Z which contains current *I* has a voltage drop given by *IZ*. Then according to compensation theorem this impedance may be replaced by a compensation emf, where the magnitude and phase of this source are equal to *IZ*.

Likewisely, if the voltage across an element or branch of a network containing an impedance Z is *V*, the element or branch may be replaced by a current source $I = \dfrac{V}{Z}$. Currents and voltages in all other parts of the network remain unchanged after the substitution of the compensation source. This theorem is also known as substitution theorem.

Fig. 9.11 shows a branch of a network containing impedances $Z_A$ and $Z_B$. If the current in this branch is $I_1$ the voltage drop across $Z_A$ is $I_1Z_A$ with polarity as shown in Fig. 9.11.

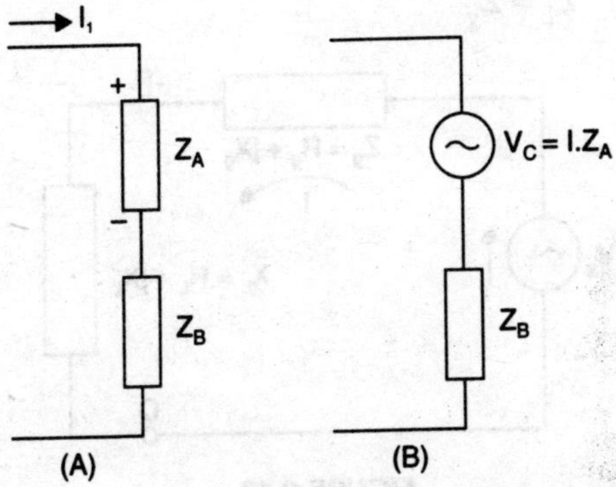


*FIGURE 9.11*

According to compensation theorem this source can be replaced by voltage source $V_C = I_1Z_A$ which replaces $Z_A.V_C$.

In case any change which affects $I_1$ occurs in the network, then the compensation source must be changed accordingly. For this reason the compensation source $V_C$ is called a dependent source.

This theorem is useful for networks where value of impedance is changed as in bridge and potentiometer circuits.

**Maximum Power Transfer Theorems.** These theorems determine the value of the load impedances which result in maximum power transfer across the terminals of an active network.

**Case I.** Load: Variable Resistance $R_L$. For maximum power transfer $R_L = |Z_g|$.

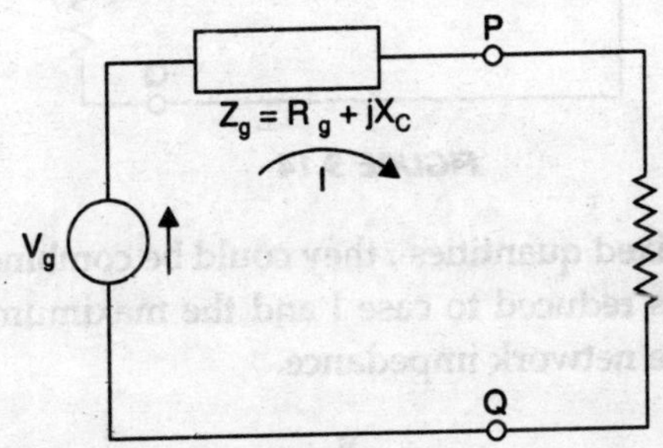


*FIGURE 9.12*

With a variable pure resistance load the maximum power is delivered across the terminals of the active network if the load resistance is made equal to the absolute value of the active network impedance.

In case the reactive component of the impedance in series with the source is zero, *i.e.*, $X_g = 0$, then the maximum power is transferred to the load when the load and source resistances are equal *i.e.*,

$$R_L = R_g.$$

**Case II.** Load: Impedance $Z_L$ with variable resistance and variable reactance.

In this case maximum power is transferred when

$$Z_L = Z_g$$

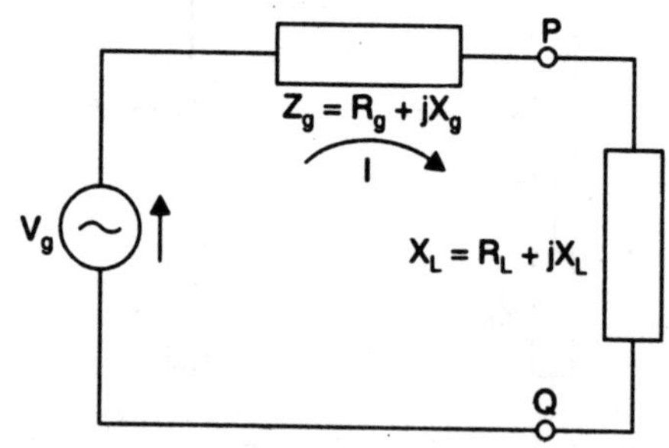


*FIGURE 9.13*

Load impedance = Complex conjugate of the network impedance $Z_g$.

**Case III.** Load: Impedance $Z_L$ with variable resistance and fixed reactance.

In this case for maximum power transfer

$$R_L = |Z_g + jX_L|$$

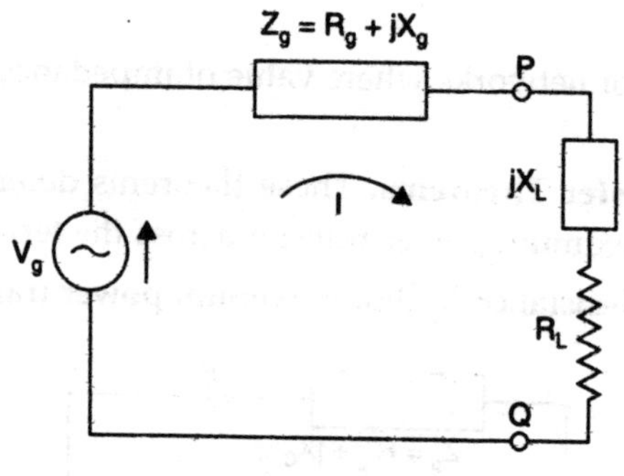


*FIGURE 9.14*

Since $Z_g$ and $X_L$ are both fixed quantities , they could be combined into a single impedance. Then with $R_L$ variable, case III is reduced to case I and the maximum power results when $R_L$ is equal to the absolute value of the network impedance.

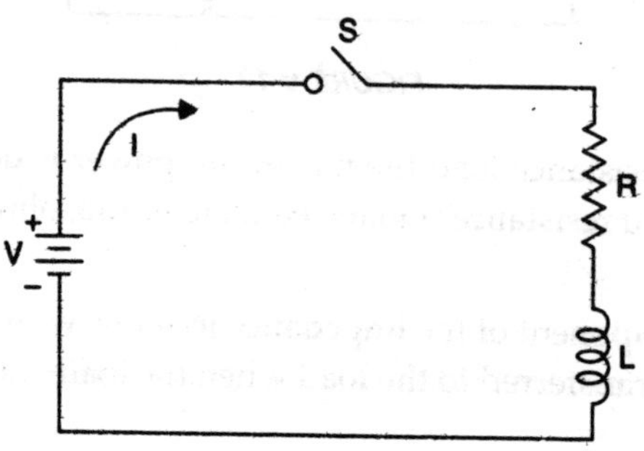


*FIGURE 9.15*

**Circuit Transients.** Whenever a circuit is switched from one condition to another either by a change in the applied voltage or change in a circuit parameter , there exists a transitional period during which the branch currents and voltage drops change from their former values to new ones. After this transition period, the circuit becomes 'steady'.

**DC Transients.** For the $R - L$ circuit shown in Fig. 9.16.

$$i = \frac{V}{R}\left[1 - e^{-\left(\frac{R}{L}\right)^t}\right]$$

This is an exponential rise equation for which plot of $i$ versus time is shown in Fig. 9.16.

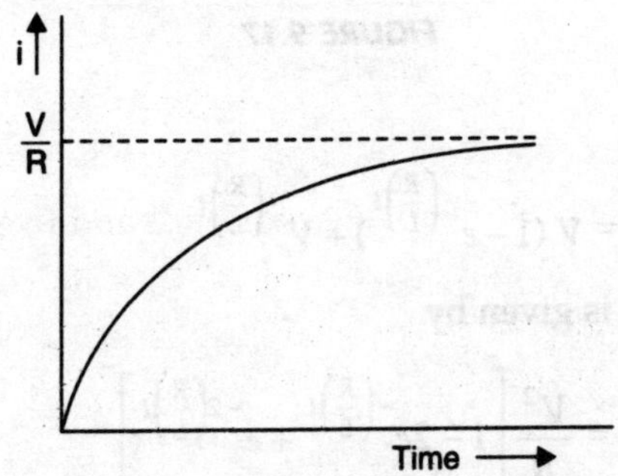


*FIGURE 9.16*

The time constant for the above function is the time at which the exponent of $e$ is unity. Thus in this case time constant is $\frac{L}{R}$.

At one time constant, the value of $i$ will be

$$i = (1 - e^{-1}) = 1 - 0.368$$

$$= 0.632$$

At this time the current will be 63.2% of its final value.

The voltage across the inductance will be

$$v_L = L\frac{di}{dt}$$

$$= V_e^{-\left(\frac{R}{L}\right)t}$$

and voltage across resistor

$$V_R = V^{(1 - e^{-\frac{R}{L}t})}$$

The resistor voltage has exponential rise while inductor voltage has exponential decay as shown in Fig. 9.17.

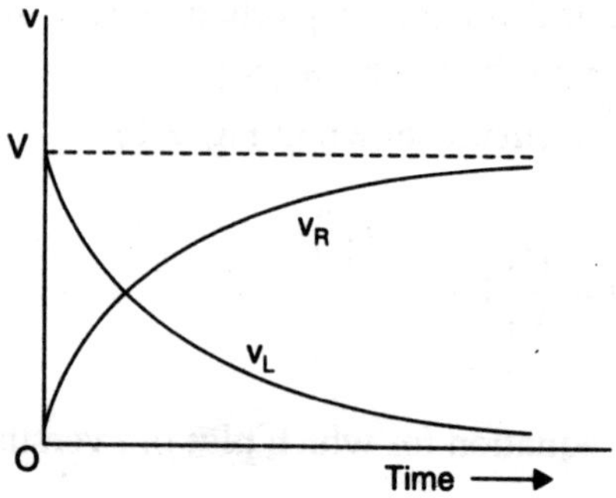


*FIGURE 9.17*

However,

$$v_R + v_L = V\left(1 - e^{-\left(\frac{R}{L}\right)t}\right) + V_e^{-\left(\frac{R}{L}\right)t}$$

Power in circuit elements is given by

$$P_R = \frac{V^2}{R}\left[1 - 2e^{-\left(\frac{R}{L}\right)t} + e^{-2\left(\frac{R}{L}\right)t}\right]$$

$$P_L = \frac{V^2}{R}\left[e^{-\left(\frac{R}{L}\right)t} - e^{-2\left(\frac{R}{L}\right)t}\right].$$

Total power

$$P_T = P_R + P_L$$

$$= \frac{V^2}{R}\left[1 - e^{-\left(\frac{R}{L}\right)t}\right].$$

**RG Transient.** In this case current $i$ is given by

$$i = \frac{V}{R}e^{-\frac{t}{RC}}$$

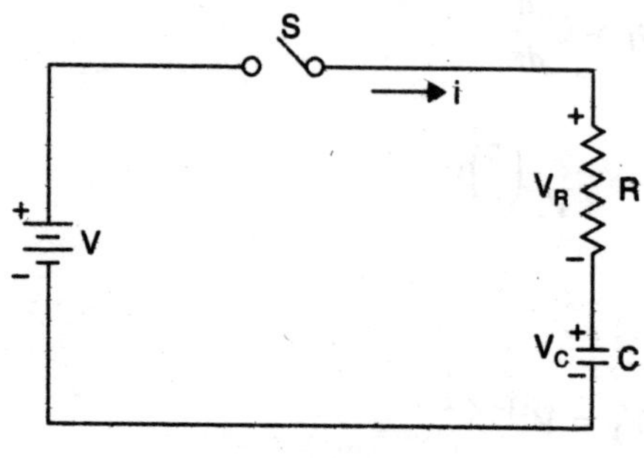


*FIGURE 9.18*

Transient voltages will be

$$v_R = Ve^{-\frac{t}{RC}}$$

$$v_C = V\left(1 - e^{-\frac{t}{RC}}\right)$$

Also instantaneous powers are given by

$$P_R = \frac{V^2}{R} e^{-\frac{2t}{RC}}$$

$$P_C = \frac{V^2}{R}\left(e^{-\frac{t}{RC}} - e^{-\frac{2t}{RC}}\right)$$

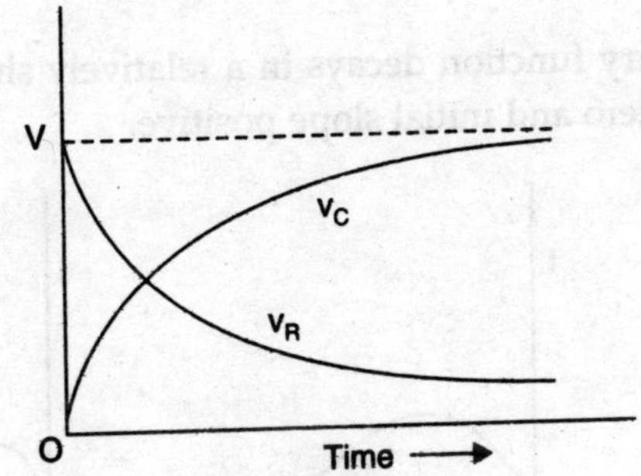


*FIGURE 9.19*

**RLC Transient.** For $R - L - C$ circuit shown in Fig. 9.20 the following integro-differential equation can be written

$$Ri + L\frac{di}{dt} + \frac{I}{C}\int i\,dt = V$$

While solving for $i$, three conditions are considered.

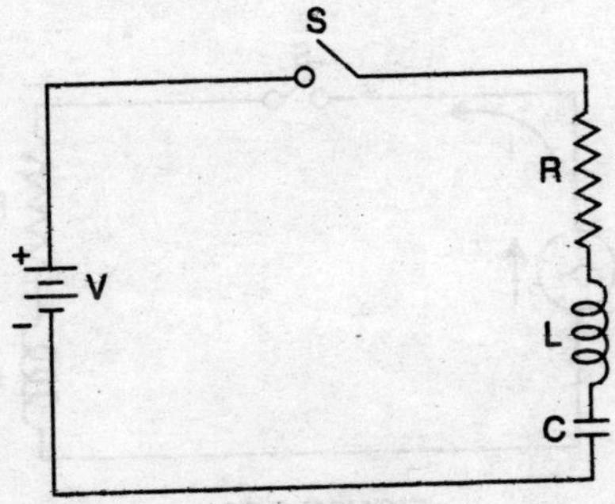


*FIGURE 9.20*

**Condition I.** $\left(\frac{R}{2L}\right)^2 > \frac{1}{LC}$

In this case the current *i* is given by

$$i = C_1 e^{Lt} C_1 (e^{\beta t} + C_2 e^{-\beta t}).$$

**Condition II.** $\left(\frac{R}{2L}\right)^2 = \frac{1}{LC}$

In this $\qquad i = e^{\alpha t} (C_1 + C_2 t).$

**Condition III.** $\left(\frac{R}{2L}\right)^2 < \frac{1}{LC}$

In this $\qquad i = e^{\alpha t} (C_1 \cos \beta t + C_2 \sin \beta t)$

The current in all cases contains the factor $e^{\alpha t}$ and since $\alpha = -\frac{R}{2L}$ the final value is zero, assuming that the complimentary function decays in a relatively short time. Fig. 9.21 shows the value of the *i* for initial values zero and initial slope positive.

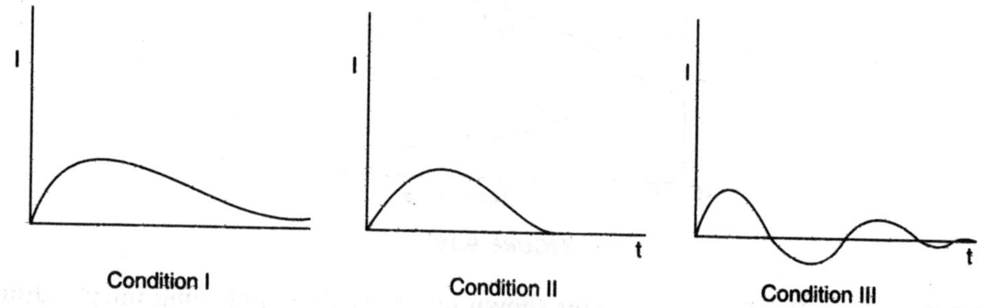


*FIGURE 9.21*

## AC Transients

1. **RL Sinusoidal Transient.** In this case the voltage function could be at any point in the period at the instant of closing the switch and therefore the phase angle $\phi$ can take any values from 0 to $2\pi$ rad/sec.

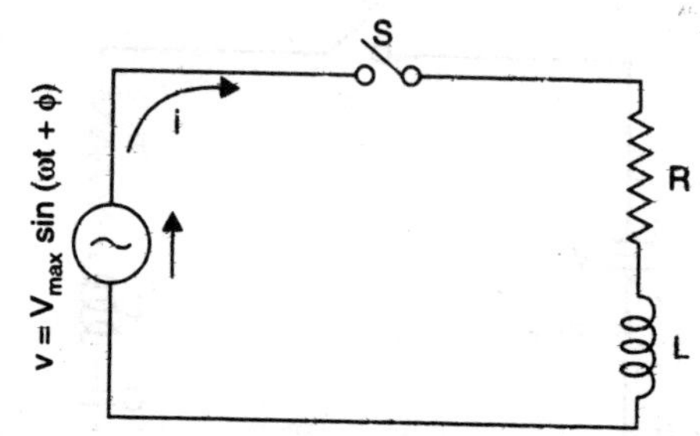


*FIGURE 9.22*

In this case

$$i = e^{-\left(\frac{R}{L}\right)t}\left[\frac{-V_{max}}{\sqrt{R^2+\omega^2L^2}}\sin\left(\phi - \tan^{-1}\frac{\omega L}{R}\right)\right]$$

$$+\frac{V_{max}}{\sqrt{R^2+\omega^2L^2}}\sin\left(\omega t + \phi - \tan^{-1}\frac{\omega L}{R}\right)$$

The first part of above equation contains the factor $e^{-\left(\frac{R}{L}\right)t}$ which has a value of zero in a relatively short time.

The second part of above equation is the steady state current which lags the applied voltage by $\tan^{-1}\frac{\omega L}{R}$.

**2. RG Sinusoidal Transient.** The basic equation for the circuit shown in Fig. 9.23 is

$$R_i + \frac{1}{C}\int i\,dt = V_{max}\sin(\omega t + \phi)$$

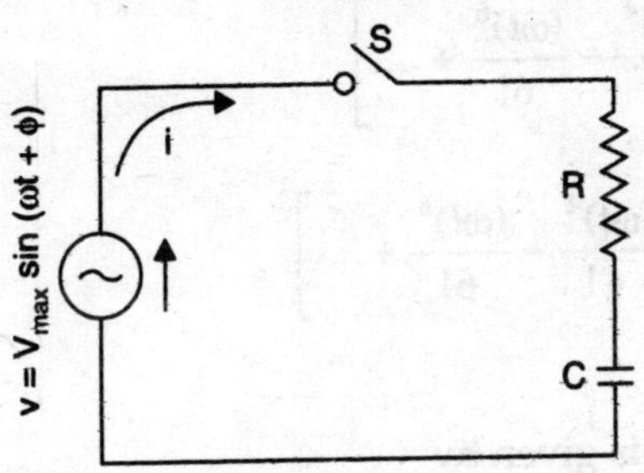


*FIGURE 9.23*

In this case the current is given by

$$i = e^{-\left(\frac{t}{RC}\right)}\left[\frac{V_{max}}{R}\sin\phi - \frac{V_{max}}{\sqrt{R^2+\left(\frac{1}{\omega C}\right)^2}}\sin\left(\phi + \tan^{-1}\frac{1}{\omega CR}\right)\right]$$

$$+\frac{V_{max}}{\sqrt{R^2+\left(\frac{1}{\omega C}\right)^2}}\sin\left(\omega t + \phi + \tan^{-1}\frac{1}{\omega CR}\right)$$

In this equation first term is the transient with the decay factor $e^{-\frac{t}{RC}}$. The second term is the steady state current which leads the applied voltage by $\tan^{-1}\frac{1}{\omega CR}$.

## OBJECTIVE TYPE QUESTIONS

1. A pure resistor of 5 ohms has an applied voltage $v(t)$ given by

$$v(t) = 10\left[1 - \frac{(\omega t)^2}{2!} + \frac{(\omega t)^4}{4!} - \frac{(\omega t)^6}{6!} + \ldots\right] \text{ volts}$$

The current in the circuit will be

(*a*) 10 A (*b*) 5 A (*c*) 2 cos $\omega t$ (*d*) 5 cos $\omega t$.

2. The voltage across a capacitor is given by

$$v = V_m\left[\omega t - \frac{(\omega t)^3}{3!} + \frac{(\omega t)^5}{5!} - \frac{(\omega t)^7}{7!} + \ldots\right]$$

The expression for current will be

(*a*) $i = V_m\left[1 - \frac{\omega t}{2!} + \frac{(\omega t)^2}{4!} + \frac{(\omega t)^3}{6!} + \ldots\right]$

(*b*) $i = V_m\left[1 - \frac{(\omega t)^2}{2!} + \frac{(\omega t)^4}{4!} - \frac{(\omega t)^6}{6!} + \ldots\right]$

(*c*) $i = \omega C V_m\left[1 - \frac{(\omega t)^2}{2!} + \frac{(\omega t)^4}{4!} - \frac{(\omega t)^6}{6!} + \ldots\right]$

(*d*) None of the above.

3. The current in a capacitor is given by

$$i = \frac{V_m}{R} e^{-\frac{t}{RC}}$$

The maximum stored energy in the electric field will be

(*a*) $C.V_m$ (*b*) $\frac{1}{2}$ $C.V_m$ (*c*) $\frac{1}{2}$ $C.V_m{}^2$ (*d*) $\frac{1}{2}$. $C.(V_m)^3$.

4. Three pure inductances are connected as shown in Fig. 9.24. What equivalent inductance will replace this circuit?

(*a*) 4 H (*b*) 4 mH (*c*) 2.2 mH (*d*) 1.2002 H.

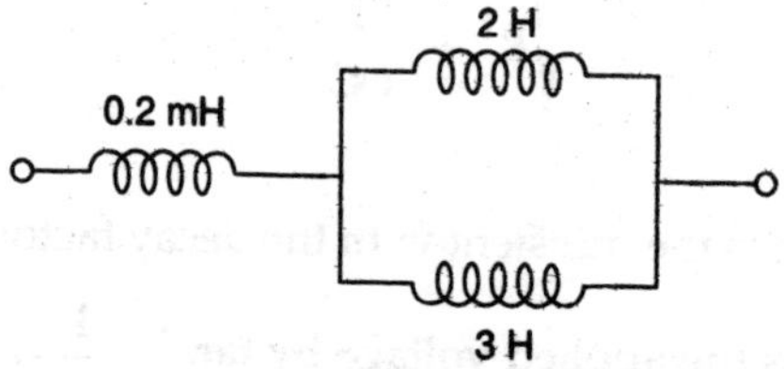


FIGURE 9.24

5. When a pure resistance $R$ has an applied voltage $v = V_m \sin \omega t$ the energy is given by

(a) $\frac{V_m}{2R}(1 - \sin \omega t)$ (b) $\frac{V_{m^2}}{2R}\left(1 - \frac{\sin \omega t}{\omega}\right)$

(c) $\frac{V_{m^2}}{2R}\left(t - \frac{\sin 2\omega t}{\omega}\right)$ (d) $\frac{V_{m^2}}{2R}\left(t - \frac{\sin 2\omega t}{2\omega}\right)$.

6. A single pure inductance of 6 millihenrys passes a current of waveform shown in Fig. 9.25 represents the instantaneous voltage $v(t)$?

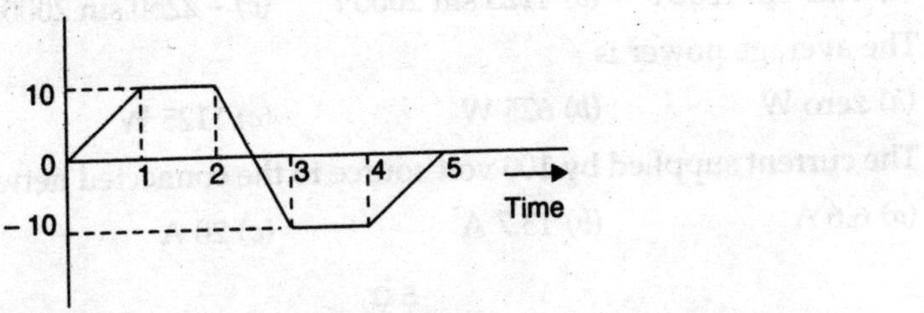


(a) figure $A$ (b) figure $B$ (c) figure $C$ (d) figure $D$.

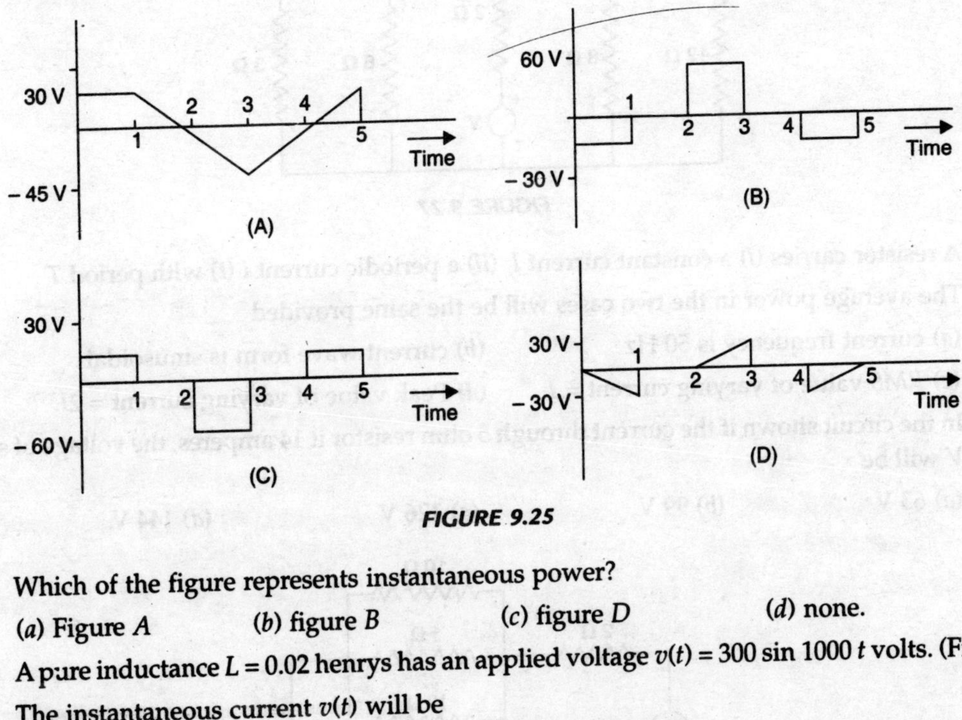


**FIGURE 9.25**

7. Which of the figure represents instantaneous power?

(a) Figure $A$ (b) figure $B$ (c) figure $D$ (d) none.

8. A pure inductance $L = 0.02$ henrys has an applied voltage $v(t) = 300 \sin 1000\,t$ volts. (Fig.9.26) The instantaneous current $v(t)$ will be

(a) $15 \cos 1000\,t$ (b) $15 \sin 1000\,t$ (c) $-15 \cos 1000\,t$ (d) $-15 \sin 1000\,t$.

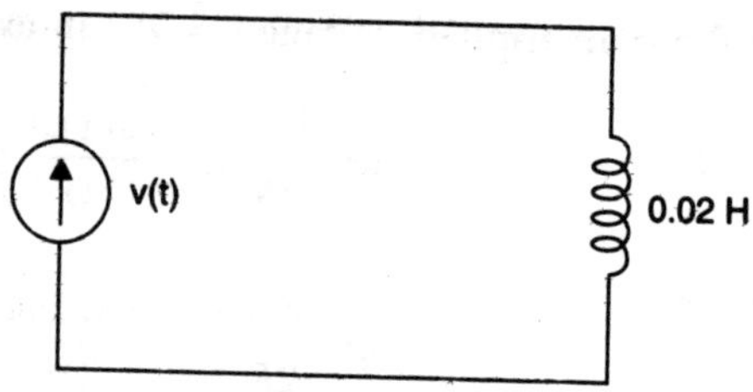


*FIGURE 9.26*

9. The instantaneous power is given by
   (*a*) 1125 sin 1000 *t* (*b*) 1125 sin 2000 *t* (*c*) – 2250 sin 2000 *t* (*d*) – 2250 sin 1000 *t*.
10. The average power is
   (*a*) zero W (*b*) 625 W (*c*) 1125 W (*d*) 2250 W.
11. The current supplied by 100 volt source to the connected network shown in Fig. 9.27 will be
   (*a*) 6.6 A (*b*) 13.7 A (*c*) 20 A (*d*) 27.4 A.

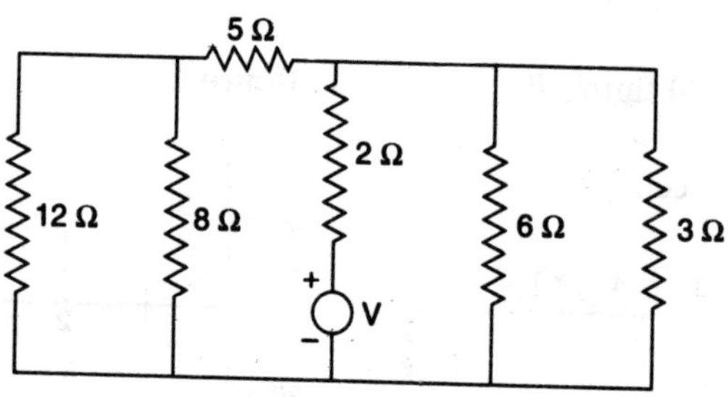


*FIGURE 9.27*

12. A resistor carries (*i*) a constant current *I*, (*ii*) a periodic current *i* (*t*) with period *T*
   The average power in the two cases will be the same provided
   (*a*) current frequency is 50 Hz (*b*) current wave form is sinusoidal
   (*c*) *RMS* value of varying current = *I* (*d*) Peak value of varying current = 2*I*.
13. In the circuit shown if the current through 5 ohm resistor it 14 amperes, the voltage of source *V* will be
   (*a*) 63 V (*b*) 99 V (*c*) 126 V (*d*) 144 V.

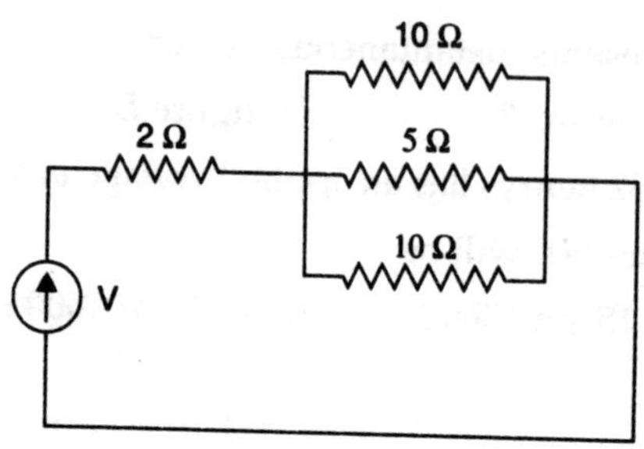


*FIGURE 9.28*

**14.** In the above case what will be the power dissipated in 2 ohm resistor?

(*a*) 392 watts (*b*) 784 watts (*c*) 980 watts (*d*) 1568 watts.

**15.** Two constant voltage sources act in the circuit shown in Fig. 9.29 The power supplied by 25 V source will be

(*a*) 15 watts (*b*) 30 watts (*c*) 45 watts (*d*) 75 watts.

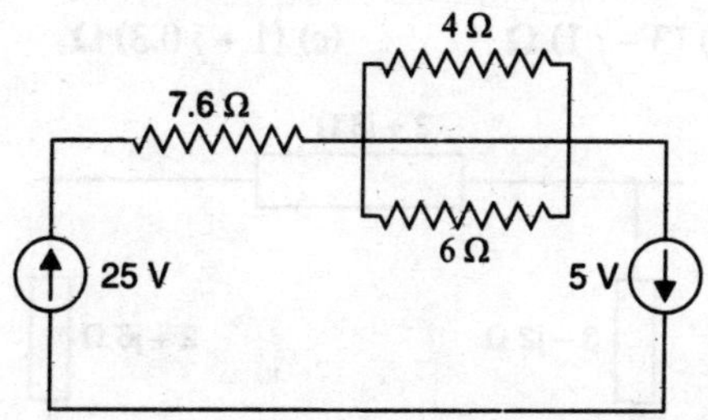


*FIGURE 9.29*

**16.** The power supplied by 5 V source will be

(*a*) 15 watts (*b*) 30 watts (*c*) 45 watts (*d*) 60 watts.

**17.** $V_A$ and $V_B$ are two constant voltage sources which act in the same circuit. What is the power delivered by each source Fig. 9.30?

(*a*) $V_A = V_B = 500$ W (*b*) $V_A = V_B = 250$ W

(*c*) $V_A = -200$ W, $V_B = 500$ W (*d*) $V_A = 500$ W, $V_B = -200$ W.

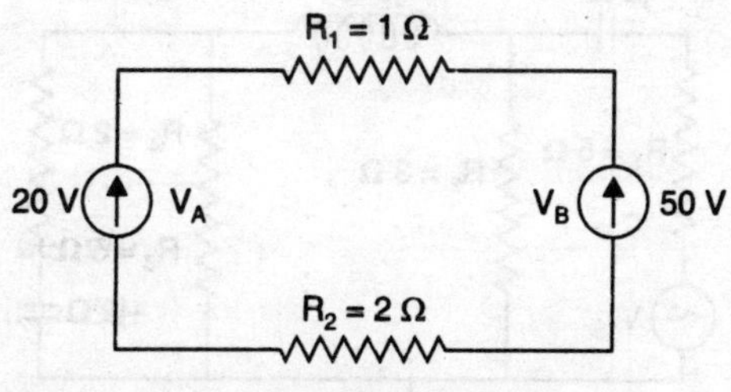


*FIGURE 9.30*

**18.** What is the value of resistor *R* in Fig. 9.31 when the power in the 5 ohm resistor is 20 watts ?

(*a*) 4 ohms (*b*) 8 ohms (*c*) 16 ohms (*d*) 32 ohms.

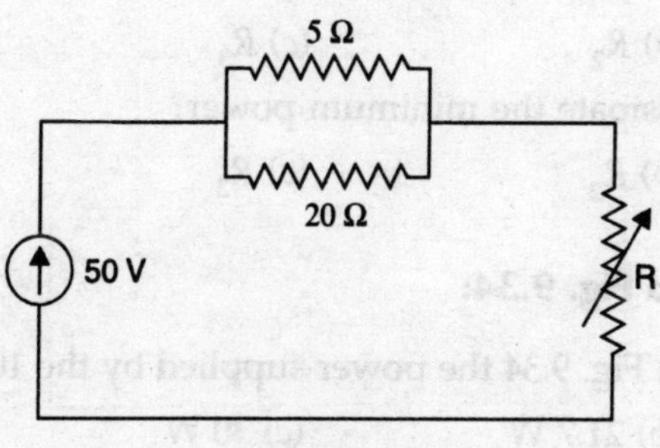


*FIGURE 9.31*

19. A 10 ohm resistor is in series with a parallel combination of two resistors of 15 Ω and 5 Ω. If the constant current in the 5 ohm resistor is 6 ampere, what total power is dissipated in the three resistors?

(*a*) 220 watts (*b*) 440 watts (*c*) 660 watts (*d*) 880 watts.

20. For the delta network shown in Fig. 9.32 which of the following will not be the impedance of equivalent star network?

(*a*) $(0.5 - j\,0.5)\ \Omega$ (*b*) $(3 - j\,1)\ \Omega$ (*c*) $(1 + j\,0.3)\ \Omega$ (*d*) $(1 + j\,3)\ \Omega$.

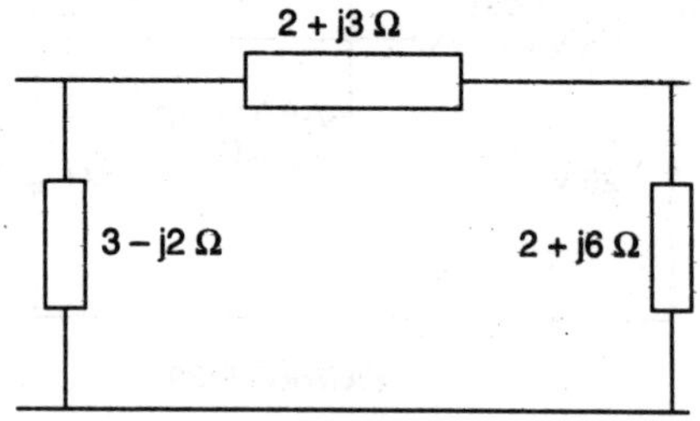


*FIGURE 9.32*

## Questions 21 to 23 Refer to Fig. 9.33:

21. In the circuit shown in Fig. 9.33 the power supplied by the source $V_1$ is

(*a*) 255 W (*b*) 295 W (*c*) 325 W (*d*) 354 W.

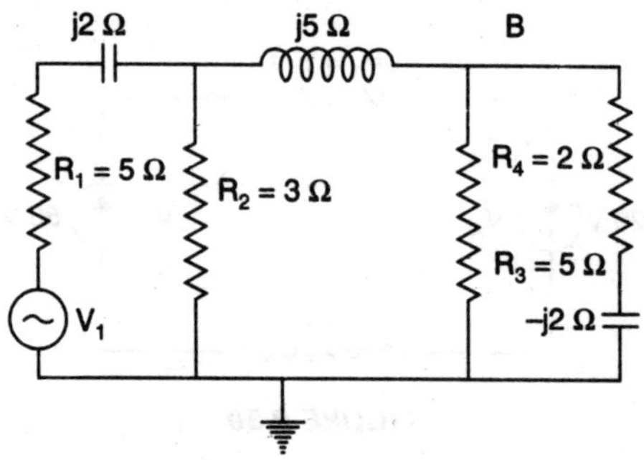


*FIGURE 9.33*

22. Which resistor will dissipate the maximum power?

(*a*) $R_1$ (*b*) $R_2$ (*c*) $R_3$ (*d*) $R_4$.

23. Which resistor will dissipate the minimum power?

(*a*) $R_1$ (*b*) $R_2$ (*c*) $R_3$ (*d*) $R_4$.

## Questions 24 to 27 Refer to Fig. 9.34:

24. In the circuit shown in Fig. 9.34 the power supplied by the 10 V source is

(*a*) 6.6 W (*b*) 21.7 W (*c*) 30 W (*d*) 36.7 W.

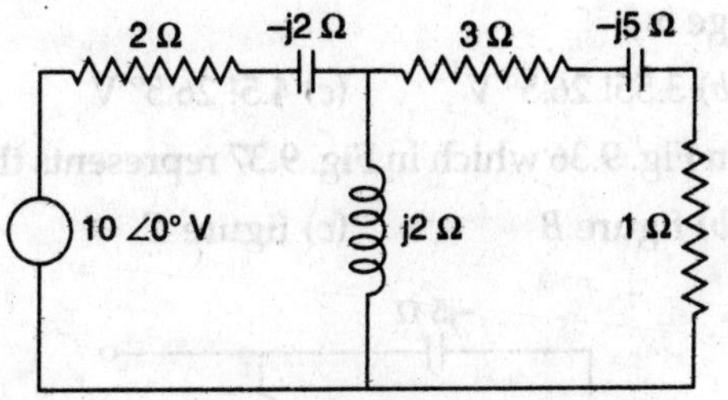


*FIGURE 9.34*

**25.** The power in 2 ohm resistor will be

(*a*) 36.7 W (*b*) 30.1 W (*c*) 27.8 W (*d*) 22.8 W.

**26.** The power in 1 ohm resistor will be

(*a*) zero W (*b*) 1.1 W (*c*) 2.22 W (*d*) 3.33 W.

**27.** The power in 3 ohm resistor will be

(*a*) 2.22 W (*b*) 4.44 W (*c*) 5.55 W (*d*) 6.66 W.

**28.** In a network a point common to two or more circuit elements is known as

(*a*) node (*b*) junction (*c*) inverter (*d*) bypass.

**29.** The number of elements that join at a node is

(*a*) 1 (*b*) 2 (*c*) 3 or more (*d*) infinite.

**30.** The rate of a current driving in at one node to the resulting voltage at another node, all other sources being set to zero, is known as

(*a*) admittance (*b*) driving point admittance

(*c*) transfer admittance (*d*) none of the above.

**Questions 31 and 32 Refer to Fig. 9.35:**

**31.** The Thevenin equivalent impedance of the circuit is

(*a*) $(1 + j\,5)\ \Omega$ (*b*) $(2.5 + j\,25)\ \Omega$

(*c*) $(6.25 + j\,6.25)\ \Omega$ (*d*) $(2.5 + j\,6.25)\ \Omega$.

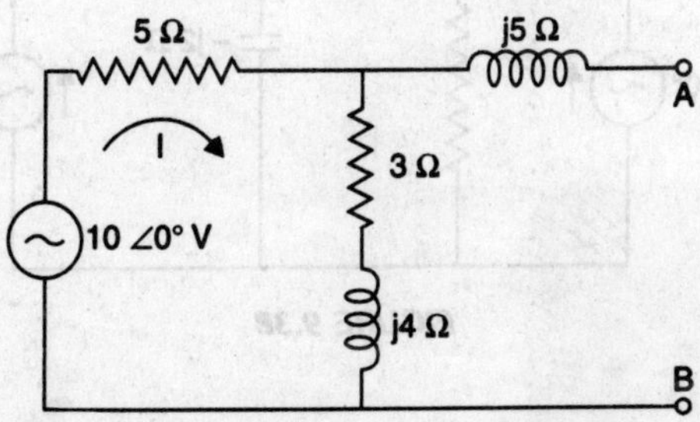


*FIGURE 9.35*

32. The open circuit voltage is

(*a*) 2.55! 26.5º V (*b*) 3.55! 26.5º V (*c*) 4.5! 26.5º V (*d*) 5.58! 26.5º V.

33. For the circuit shown in Fig. 9.36 which in Fig. 9.37 represents the Norton equivalent circuit ?

(*a*) figure *A* (*b*) figure *B* (*c*) figure *C* (*d*) figure *D*.

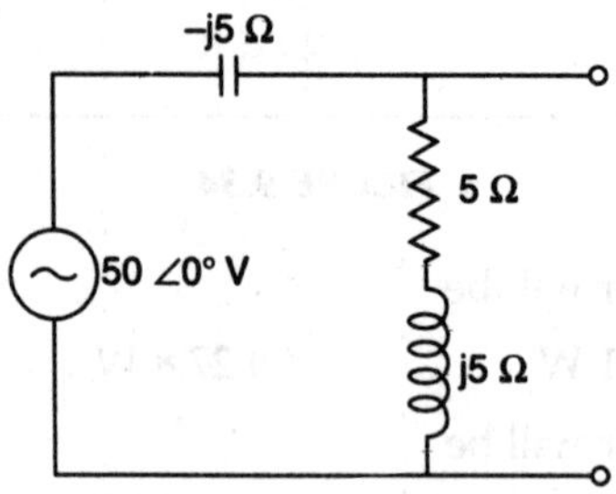


*FIGURE 9.36*

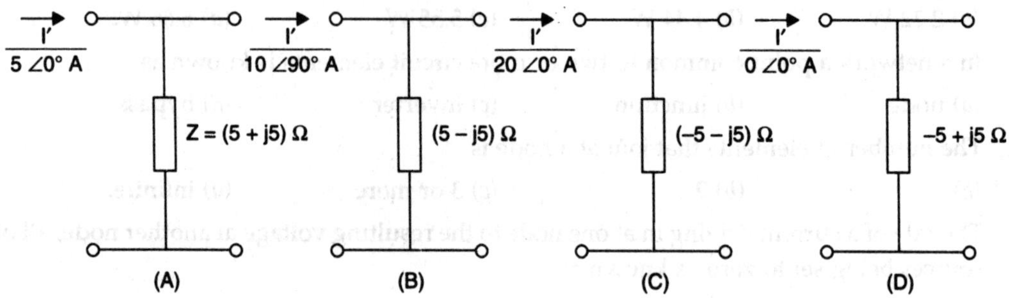


*FIGURE 9.37*

34. In the circuit shown in Fig. 9.38 what should be the value of $V_2$ such that its current will be zero

(*a*) 1! 180º V (*b*) 2! 180º V (*c*) 4! 180º V (*d*) 8! 180º V.

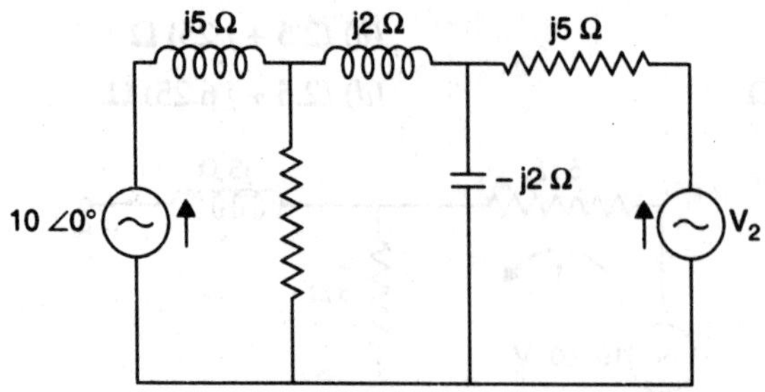


*FIGURE 9.38*

**Questions 35 to 37 Refer to Fig. 9.39:**

35. The voltage $V_{AB}$ will be

(*a*) 19.4! 0º V (*b*) 21.5! 10º V (*c*) 30.4! 25º V (*d*) 35.4! 45º V.

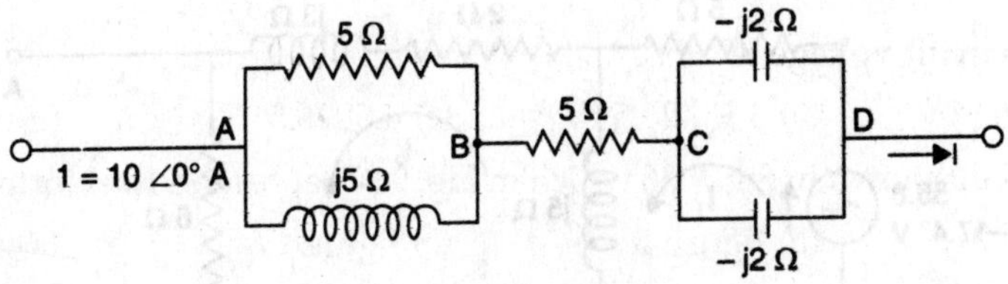


*FIGURE 9.39*

**36.** The voltage $V_{BC}$ will be

(*a*) 10! 0º V (*b*) 20! 0º V (*c*) 30! 0º V (*d*) 50! 0º V.

**37.** The voltage $V_{CD}$ will be

(*a*) 13.3! – 90º V (*b*) 26.6! – 45º V (*c*) 39.9! – 60º V (*d*) none of the above.

## Questions 38 to 40 Refer to Fig. 9.40:

**38.** In the circuit shown in Fig. 9.40 the value of *I* is

(*a*) 5! 0º (*b*) 5! 90º (*c*) 10! 0º (*d*) 10! 50º.

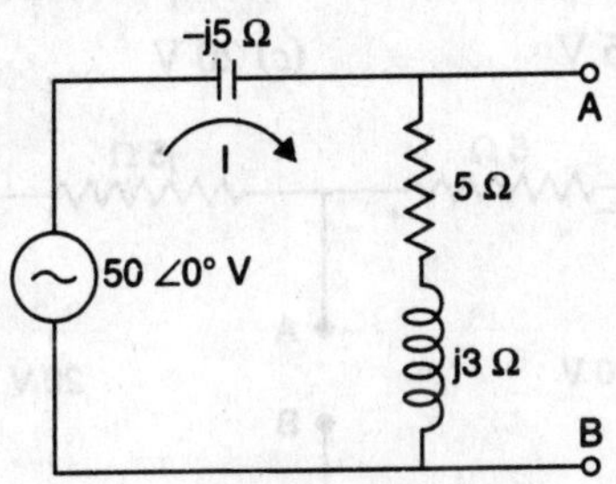


*FIGURE 9.40*

**39.** The voltage drop across (5 + *j* 5) Ω impedance will be

(*a*) 10.1! 45º V (*b*) 20.2! 45º V (*c*) 35.3! 45º V (*d*) 70.7! 45º V.

**40.** The driving point impedance at terminals *AB* is

(*a*) (5 + *j* 5) Ω (*b*) (5 – *j* 5) Ω (*c*) (– 5 + *j* 5) Ω (*d*) (– 5 + *j* 5) Ω.

## Questions 41 to 43 Refer to Fig. 9.41:

**41.** Which of the following statement is currect?

(*a*) the impedance of Thevenin as well as Norton equivalent circuit will be identical

(*b*) in the Thevenin equivalent circuit the voltage *V′* will be directed towards terminal *B*

(*c*) the Norton equivalent circuit will remain unaffected when 6 Ω resistor is not included in the given circuit

(*d*) none of the above.

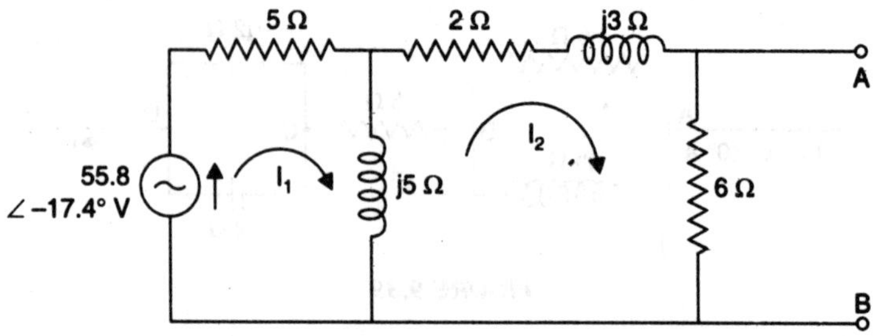


*FIGURE 9.41*

42. The impedance of the Thevenin equivalent circuit will be
(*a*) $(1 + j\,1)$ Ω (*b*) $(1.8 + j\,1.6)$ Ω (*c*) $(3.32 + j\,1.41)$ Ω (*d*) $(6.64 + j\,2.82)$ Ω.

43. The current source in Norton equivalent circuit will be
(*a*) 2.21! 19º A (*b*) 2.9! 9 – 33º A (*c*) 4.44! 23.4 A (*d*) 5.58! – 23.14 A.

**Questions 44 to 47 Refer to Fig. 9.42:**

44. For the circuit shown in Fig. 9.42 the Thevenin equivalent voltage of the source will be
(*a*) 6.25 V (*b*) 12.5 V (*c*) 20 V (*d*) 25 V.

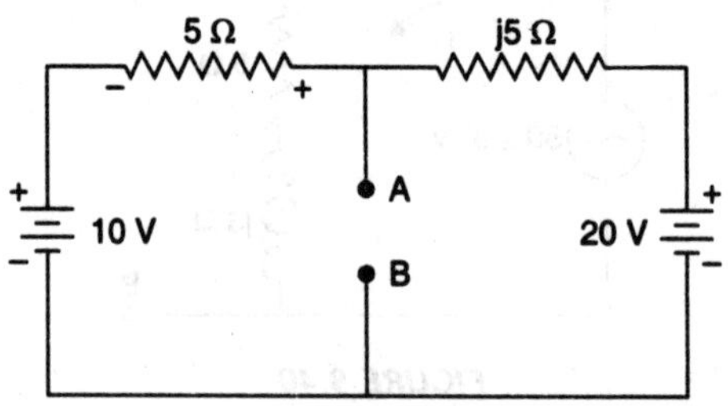


*FIGURE 9.42*

45. The impedance of the Thevenin's equivalent circuit will be
(*a*) 1.25 Ω (*b*) 2.0 Ω (*c*) 2.5 Ω (*d*) 3.75 Ω.

46. Three resistors $R_1$ = 1 Ω, $R_2$ = 5 Ω and $R_3$ = 10 Ω are connected in turn to terminals *AB*. In which case the power delivered tu the resistor will be least?
(*a*) $R_1$ (*b*) $R_2$
(*c*) $R_3$ (*d*) power delivered will be same in all cases.

47. In the above setting in how many resistors the power delivered will be more than 10 W?
(*a*) none (*b*) 1 (*c*) 2 (*d*) 3.

**Questions 48 to 50 Refer to Fig. 9.43:**

48. For the circuit shown in Fig. 9.43 the impedance in the Thevenin equivalent circuit will be
(*a*) 1.41 ohms (*b*) 3.3 ohms (*c*) 6 ohms (*d*) 9.43 ohms.

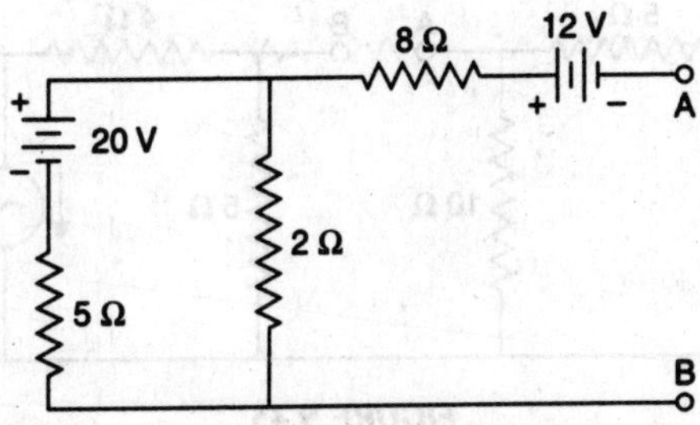


FIGURE 9.43

49. The voltage of the source will be

(a) 6.29 V (b) 8.88 V (c) 10.44 V (d) 12.45 V.

50. In the Thevenin equivalent circuit

(a) A will be positive (b) B will be positive

(c) either A or B will be positive (c) none of the above.

51. For the AC bridge shown in Fig. 9.44 the Thevenin equivalent circuit at terminals AB will have impedance of

(a) 11.1 $\angle 11.55°$ Ω (b) 28.8 $\angle -11.55°$ Ω (c) 88.7 $\angle 11.55°$ Ω (d) 166.6 $\angle -11.55°$ Ω.

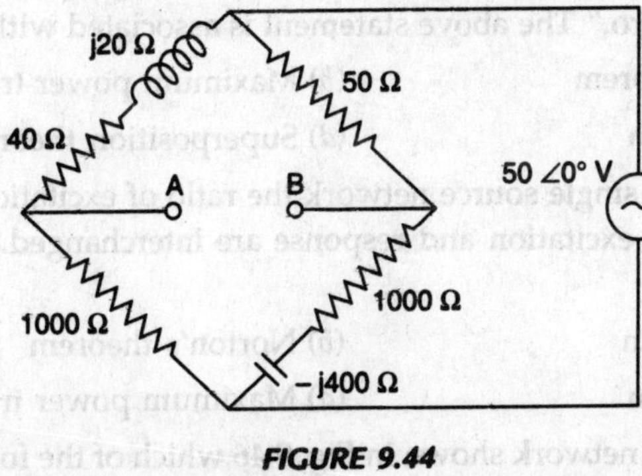


FIGURE 9.44

52. In the above case source voltage will be

(a) 0.192 $\angle 43.4°$ Ω (b) 0.192 $\angle -43.4°$ Ω (c) 0.384 $\angle -43.4°$ Ω (d) 0.384 $\angle 43.4°$ Ω.

**Questions 53 to 56 Refer to Fig. 9.45:**

53. For the circuit shown in Fig. 9.45 the impedance of Thevenin's equivalent circuit at AB will be

(a) 1.11 $\angle 0°$ Ω (b) 2.2 $\angle 0°$ Ω (c) 5.55 $\angle 0°$ Ω (d) 10.10 $\angle 0°$ Ω.

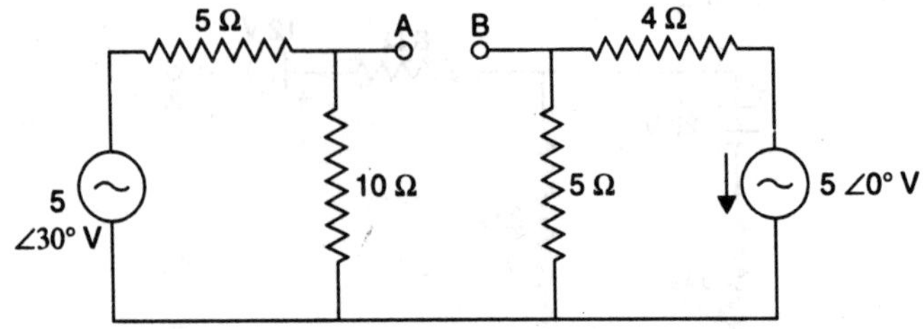


*FIGURE 9.45*

**54.** The source voltage for Thevenin equivalent circuit will be

(*a*) 4.9 $\angle 16°$ V (*b*) 5.9 $\angle -16.4°$ V (*c*) 5.9 $\angle 16.4°$ V (*d*) 11.1 $\angle 23°$ V.

**55.** The Norton equivalent circuit for the network at *AB* will have impedance

(*a*) 5.55 $\angle 0°$ Ω (*b*) 5.55 $\angle 90°$ Ω (*c*) 10.10 $\angle 0°$ Ω (*d*) 10.10 $\angle 90°$ Ω.

**56.** In the Norton equivalent circuit the current source *I′* will be

(*a*) 1.06 $\angle 16.4°$ A (*b*) 1.06 $\angle -16.4°$ A (*c*) 1.06 $\angle 32.4°$ A (*d*) 1.06 $\angle -32.4°$ A

**57.** " The response in any element of a linear bilateral network containing two or more sources is the sum of the responses obtained by each source acting separately and with all the other sources set equal to zero." The above statement is associated with

(*a*) Compensation theorem (*b*) Maximum power transfer theorem

(*c*) Thevenin's theorem (*d*) Superposition theorem.

**58.** " In a linear, bilateral, single source network the ratio of excitation to response is constant when the positions of excitation and response are interchanged." The above statement is associated with

(*a*) Reciprocity theorem (*b*) Norton's theorem

(*c*) Thevenin's theorem (*d*) Maximum power transfer theorem.

**59.** For the star connected network shown in Fig. 9.46 which of the following will not be one of the impedance in the equivalent delta network?

(*a*) $(15 + j\,5)$ Ω (*b*) $(5 + j\,15)$ Ω (*c*) $(10 + j\,30)$ Ω (*d*) $(10 - j\,30)$ Ω.

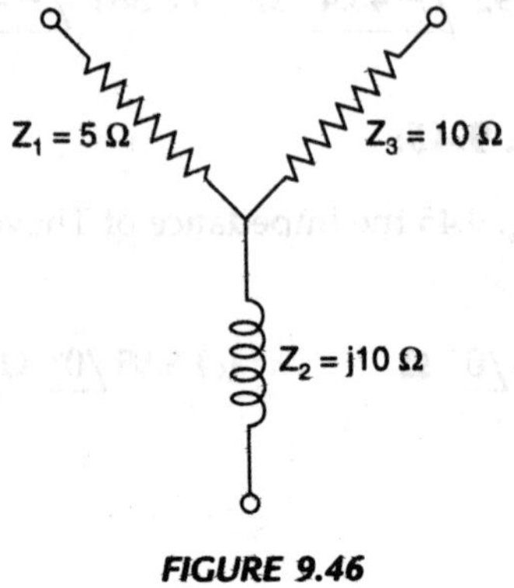


*FIGURE 9.46*

**60.** For the circuit shown in Fig. 9.47, $V_{AB}$ =

(*a*) 2 V (*b*) 4.4 V (*c*) 9.41 V (*d*) 11.22 V.

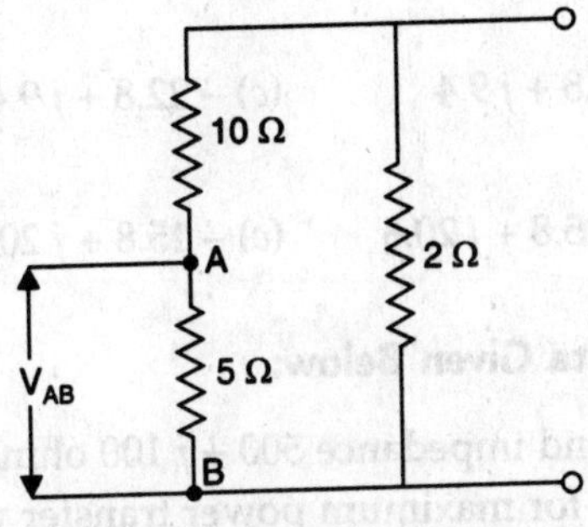


*FIGURE 9.47*

**61. For the network shown in Fig. 9.48 two voltage sources act on the load impedance connected to the terminals *AB*. If this load is variable in both reactance and resistance, what load $Z_L$ will receive maximum power.**

(*a*) 1.13 + *j* 1.5 (*b*) 1.13 – *j* 1.5 (*c*) 4.23 + *j* 1.15 (*d*) 14.23 – *j* 3.35.

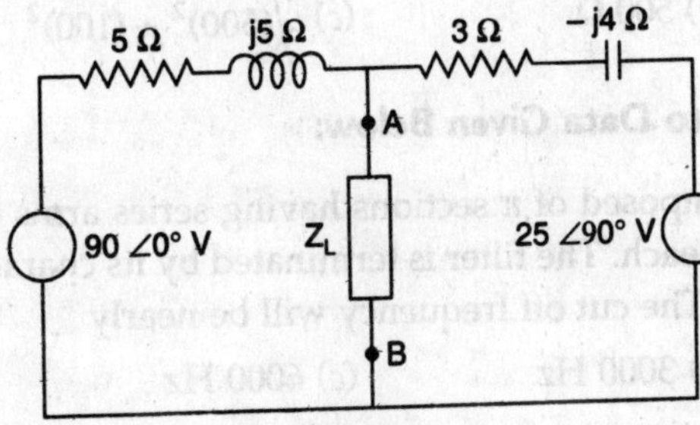


*FIGURE 9.48*

**62.** Under the above conditions the maximum power will be

(*a*) 1.11 W (*b*) 2.34 W (*c*) 3.68 W (*d*) 5.68 W.

## Questions 63 to 65 Refer to Fig. 9.49:

A closed circuit consisting of three impedances is shown in Fig. 9.49. One corner is connected to a source; currents are taken off at the other corners. That at *B* is 30 A at a power factor of 0.7 lagging that at C is 20 A at a power factor of 0.9 lagging. Let these currents be $I_1$ and $I_2$ respectively.

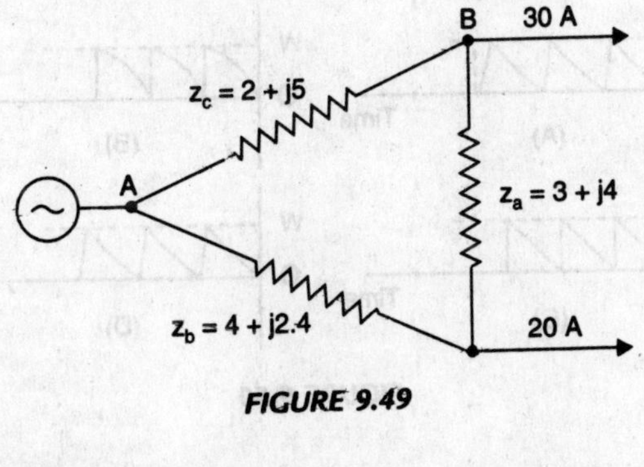


*FIGURE 9.49*

63. The current $I_A$ will be

(a) 5.3 + j 0.7 (b) – 5.3 + j 0.7 (c) 5.3 – j 0.7 (d) – 5.3 – j 0.7.

64. The current $I_B$ will be

(a) 22.8 – j 9.4 (b) 22.8 + j 9.4 (c) – 22.8 + j 9.4 (d) – 22.8 – j 9.4.

65. The current $I_C$ will be

(a) 15.8 + j 20.6 (b) – 15.8 + j 20.6 (c) – 15.8 + j 20.6 (d) 15.8 – j 20.6.

## Questions 66 to 68 Refer to Data Given Below:

A source has an emf 10 V and impedance 500 + j 100 ohms.

66. The impedance of the load for maximum power transfer will be

(a) – (500 + j 100) (b) – 500 + j 100 (c) 500 – j 100 (d) 500 + j 100.

67. The amount of power transferred in above case will be

(a) 0.5 mW (b) 0.05 mW (c) 0.05 W (d) 0.5 W.

68. In the above case the resistance of the purely resistive load for maximum power transfer will be

(a) 480 Ω (b) 500 Ω (c) $\sqrt{(500)^2 + (100)^2}$ (d) $\sqrt{(500)^2 - (100)^2}$.

## Questions 69 and 70 Refer to Data Given Below:

69. A low pass filter is composed of π sections having series arms of 63.6 mH inductance and shunt arms of 0.088 μF each. The filter is terminated by its characteristic impedance and the elements are loss free. The cut off frequency will be nearly

(a) 1500 Hz (b) 3000 Hz (c) 6000 Hz (d) 7500 Hz.

70. The attenuation per section at a frequency of 5000 Hz will be

(a) 19.1 dB (b) 29.1 dB (c) 39.1 dB (d) 49.1 dB.

71. The current function shown in Fig. 9.50 is a repeating saw tooth and exists in a pure resistor of 5 ohms. Which of the curve shown in Fig. 9.50 represents the instantaneous power $p(t)$?

(a) figure A (b) figure B (c) figure C (d) figure D.

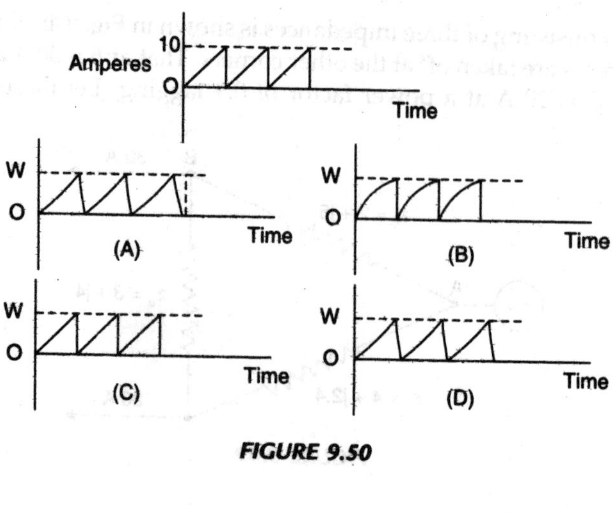


FIGURE 9.50

72. What will be the value of peak on power curve?

(*a*) 167 W (*b*) 225 W (*c*) 500 W (*d*) 625 W.

73. The average power will be

(*a*) 83.5 W (*b*) 167 W (*c*) 238 W (*d*) 315 W.

**Questions 74 to 76 Refer to Fig. 9.51:**

For the circuit shown in Fig. 9.51 the voltage $v$ is given by

$$v = 150 \sin \omega t.$$

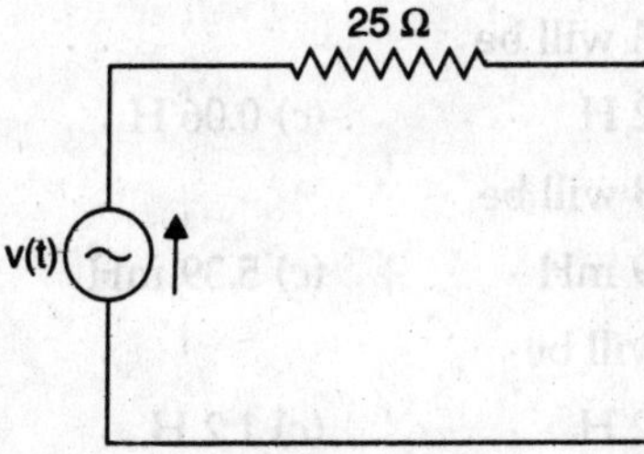


*FIGURE 9.51*

74. The current $i$ $(t)$ will be

(*a*) 150 cos $\omega t$ (*b*) 6 cos $\omega t$ (*c*) 6 sin $\omega t$ (*d*) 900 sin $\omega t$.

75. The instantaneous power $p$ $(t)$ is given by

(*a*) 900 sin $\omega t$ con $\omega t$ watts (*b*) 900 $\sin^2 \omega t$ watts

(*c*) 900 cos $\omega t$ (*d*) 900 sin $2\omega t$.

76. The average power $P$ is given by

(*a*) 150 W (*b*) 225 W (*c*) 300 W (*d*) 450 W.

**Questions 77 to 80 Refer to Fig. 9.52:**

In the circuit shown the current in the 5 ohm resistor is

$$i(t) = 6 \sin \omega t \text{ amperes.}$$

77. The current in 15 ohm resistor will be

(*a*) 80 sin $\omega t$ (*b*) 8 sin $\omega t$ (*c*) 3 sin $\omega t$ (*d*) 2 sin $\omega t$.

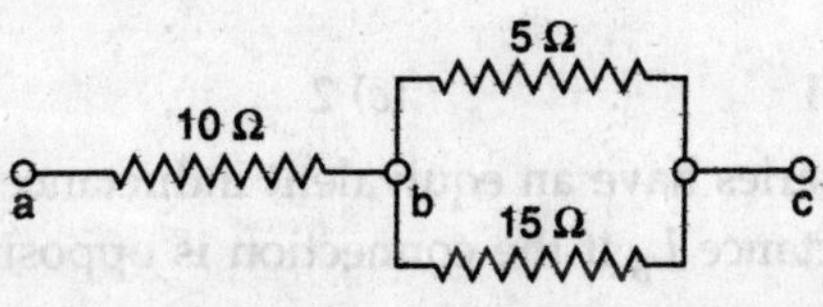


*FIGURE 9.52*

78. The current in the 10 ohm resistor will be

(*a*) 2 sin ωt (*b*) 3 sin ωt (*c*) 6 sin ωt (*d*) 8 sin ωt.

79. The instantaneous power *p* (*t*) is given by

(*a*) $p(t) = 180 \sin^2 \omega t$ (*b*) $p(t) = 180 \cos^2 \omega t$

(*c*) $p(t) = 90 \sin^2 \omega t$ (*d*) $p(t) = 90 \cos^2 \omega t$.

80. Average power in 5 ohm resistor will be

(*a*) 30 W (*b*) 45 W (*c*) 90 W (*d*) 320 W.

Coil *A* of a pair of coupled coils has a continuous current of 5 A and the corresponding fluxes $\phi_A$ and $\phi_B$ are 0.2 and 0.4 mWb respectively. The turns are $N_A = 500$ and $N_B = 1500$.

81. The self inductance of coil *A* will be

(*a*) 0.01 H (*b*) 0.02 H (*c*) 0.06 H (*d*) 0.09 H.

82. The self inductance of coil *B* will be

(*a*) 539 mH (*b*) 53.9 mH (*c*) 5.39 mH (*d*) 5.39 H.

83. The mutual inductance *M* will be

(*a*) 0.012 H (*b*) 0.12 H (*c*) 1.2 H (*d*) 12 H.

84. The value of coupling co-efficient is

(*a*) 0.112 (*b*) 0.225 (*c*) 0.333 (*d*) 0.667.

85. Two coupled coils with turns $N_1 = 100$ and $N_2 = 800$ have a coupling co-efficient of 0.85. With coil 1 open and a current of 5 amperes in coil 2, the flux $\phi_2$ is 0.35 mWb.

The self inductance of coil 1 is

(*a*) 288 mH (*b*) 28.8 mH (*c*) 8.75 mH (*d*) 0.875 mH.

86. The self inductance of coil 2 is

(*a*) 288 mH (*b*) 28.8 mH (*c*) 56 mH (*d*) 112 mH.

87. The mutual inductance *M* is

(*a*) 595 mH (*b*) 5.95 mH (*c*) 59.5 mH (*d*) 59.5 H.

**Questions 88 and 89:**

Two coupled coils of $L_1 = 0.8$ H and $L_2 = 0.2$ H have a coupling co-efficient $K = 0.9$.

88. The mutual inductance *M* will be

(*a*) 0.030 H (*b*) 0.72 H (*c*) 0.18 H (*d*) 0.36 H.

89. The turn ratio will be

(*a*) 0.5 (*b*) 1 (*c*) 2 (*d*) 20.

90. Two coils connected in series have an equivalent inductance $L_A$ if the connection is aiding and an equivalent inductance $L_B$ if the connection is opposing. The mutual inductance in terms of $L_A$ and $L_B$ will be

(*a*) $L_A - L_B$ (*b*) $\frac{1}{2}(L_A - L_B)$ (*c*) $\frac{1}{4}(L_A - L_B)$ (*d*) $\frac{1}{8}(L_A - L_B)$.

**91.** For the circuit shown in Fig. 9.53 for what value of $K$ the power output of 50 ! 0º voltage source will be 198 watts?

(*a*) 0.115 (*b*) 0.225 (*c*) 0.475 (*d*) 0.667.

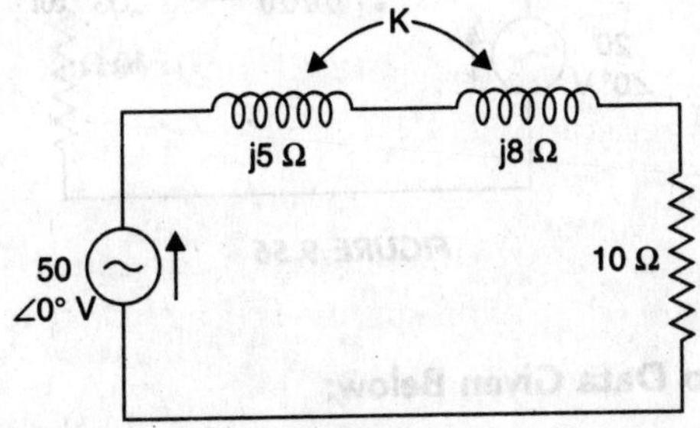


**FIGURE 9.53**

**92.** Two identical coils with $L = 0.02$ H have a coupling co-efficient $K = 0.8$. The value of mutual inductance will be

(*a*) 2 mH (*b*) 4 mH (*c*) 8 mH (*d*) 16 mH.

**93.** For the series circuit shown in Fig 9.54, find the value of $K$ so that the circuit is in series resonance

(*a*) $K = 0.112$ (*b*) $K = 0.224$ (*c*) $K = 0.448$ (*d*) $K = 0.999$.

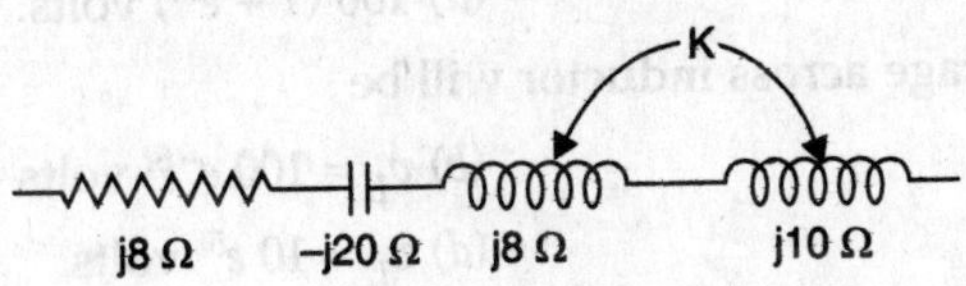


**FIGURE 9.54**

**94.** For the circuit shown in Fig. 9.55 if the power in the 5 ohm resistor is 45.2 watts, the mutual inductive reactance $j\omega M$ is

(*a*) 2 Ω (*b*) 4 Ω (*c*) $j4$ Ω (*d*) $j$ 8 Ω.

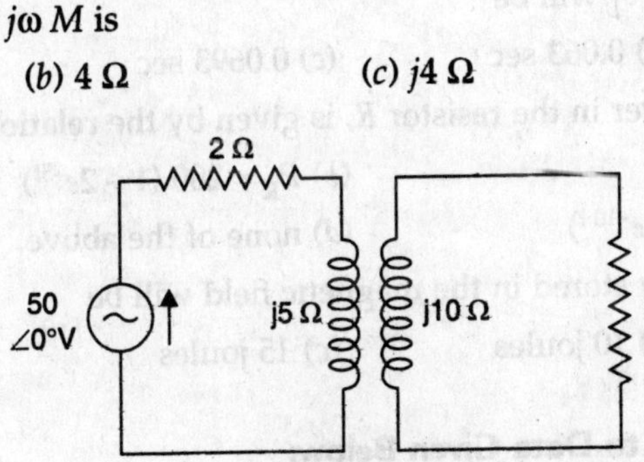


**FIGURE 9.55**

**95.** For the circuit shown in Fig. 9.56 if the power in the 10 ohm resistor is 32 watts what will be the value of $K$

(*a*) 0.119 (*b*) 0.227 (*c*) 0.338 (*d*) 0.791.

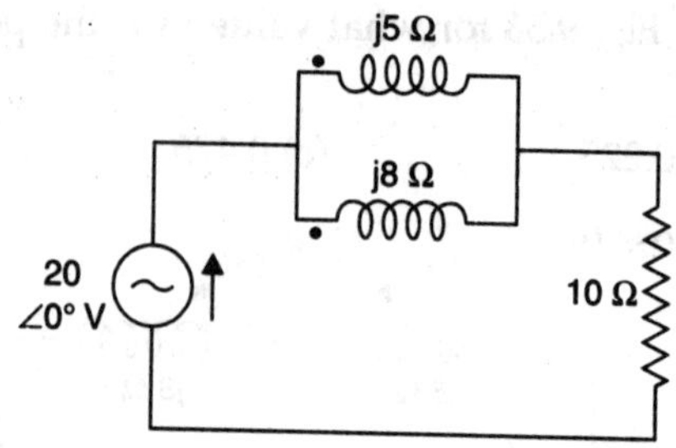


*FIGURE 9.56*

**Questions 96 to 98 Refer to Data Given Below:**

A series *RL* circuit has $R = 50$ ohms and $L = 10$ H. The voltage applied is 100 volts at $t = 0$ by the closing of a switch.

**96.** The equation for *i* will be

(*a*) $2(1 - e^{-5t})$ amperes (*b*) $2(1 + e^{-5t})$ amperes
(*c*) $2(1 - e^{+5t})$ amperes (*d*) $2(1 + e^{5t})$ amperes.

**97.** The equation for voltage across resistor *R* will be

(*a*) $v_R = 100e^{5t}$ volts (*b*) $v_R = 100\, e^{-5t}$ volts
(*c*) $100(1 - e^{-5t})$ volts (*d*) $100(1 + e^{5t})$ volts.

**98.** The equation for voltage across inductor will be

(*a*) $v_L = 100\, e^{5t}$ volts (*b*) $v_L = 100\, e^{-5t}$ volts
(*c*) $v_L = 100e^{-5t}$ volts (*d*) $v_L = 10\, e^{5t}$ volts.

**99.** The current at $t = 0.5$ seconds will be

(*a*) 0.918 A (*b*) 1.125 A (*c*) 1.836 A (*d*) 2.166 A.

**100.** The time at which $v_R = v_L$ will be

(*a*) 0.02 sec (*b*) 0.063 sec (*c*) 0.0693 sec (*d*) 0.1386 sec.

**101.** The instantaneous power in the resistor *R*, is given by the relation

(*a*) $P_R = 200(1 - e^{5t})$ (*b*) $P_R = 200(1 - 2e^{-5t})$
(*c*) $P_R = 200(1 - 2e^{-5t} + e^{-10t})$ (*d*) none of the above.

**102.** The steady state energy stored in the magnetic field will be

(*a*) 5 joules (*b*) 10 joules (*c*) 15 joules (*d*) 20 joules.

**Questions 103 to 105 Refer to Data Given Below:**

**103.** A series *RC* circuit with $R = 5000$ ohms and $C = 20$ μF has a constant voltage $V = 100$ V applied at $t = 0$ and the capacitor has no initial charge. The equation for instantaneous current *i* will be

(*a*) $i = 0.02\, e^{10t}$ amperes (*b*) $i = 0.02\, e^{-10t}$ amperes
(*c*) $i = 0.02\, e^{-5t}$ amperes (*d*) $i = 0.01\, e^{5t}$ amperes.

**104.** The transient voltage across the resistor is given by the relation

(a) $v_R = 100\, e^{-10t}$ volts (b) $v_R = 100 e^{10t}$ volts

(c) $v_R = 100\,(1 - e^{-10t})$ volts (d) $v_R = 100\,(1 + e^{-10t})$ volts.

**105.** The transient voltage across the capacitor is given by

(a) $v_C = 100\,(1 - e^{-10t})$ volts (b) $v_C = 100\,(1 + e^{-10t})$ volts

(c) $v_C = 100\,(1 - e^{10t})$ volts (d) $v_C = 100\,(1 + e^{10t})$ volts.

**106.** In the *RC* circuit shown in Fig. 9.57 the sinusoidal voltage applied is $v = 100 \sin(1000\,t + \phi)$ volts at a time when $\phi = 30^\circ$. What initial charge on the capacitor will cause the current to go directly into the steady state without a transient when the switch is closed?

(a) $6.68 \times 10^{-6}$ coulombs, + on top plate (b) $6.68 \times 10^{-6}$ coulombs, + on bottom plate

(c) $13.37 \times 10^{-6}$ coulombs, + on top plate (d) $13.37 \times 10^{-6}$ coulombs, + on bottom plate.

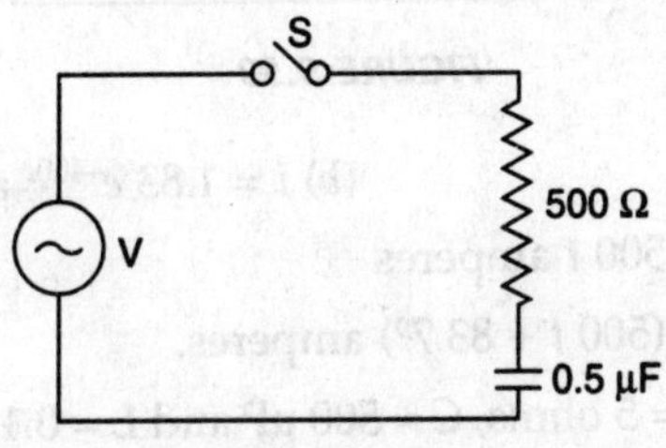


FIGURE 9.57

## Questions 107 and 108 Refer to Fig. 9.58:

**107.** In a two mesh network shown in Fig. 9.58 the switch is closed at $t = 0$. The resulting current *i*, will be

(a) $i_1 = 1.67\, e^{-6.67t}$ amperes (b) $i_1 = 1.67\, e^{6.67t}$ amperes

(c) $i_1 = 1.67\, e^{-6.67t} + 5$ amperes (d) $i_1 = 1.67\, e^{-6.67t} + 50$ amperes.

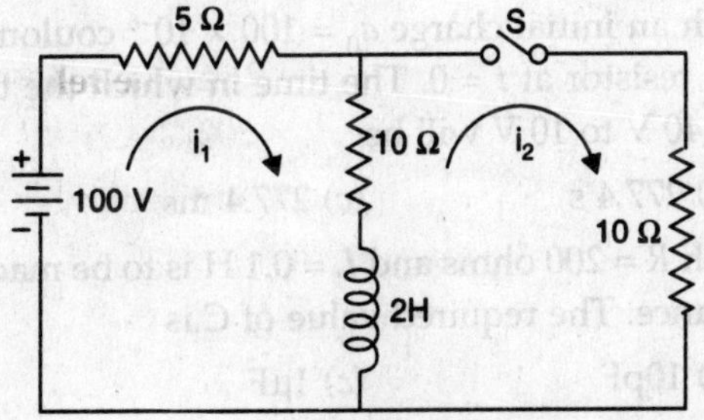


FIGURE 9.58

**108.** The current $i_2$ will be

(a) $0.555\, e^{6.67t}$ amperes (b) $0.555\, e^{-6.67t}$ amperes

(b) $0.555\, e^{-6.67t} + 5$ amperes (d) $-\,0.555\, e^{-6.66t} + 5$ amperes.

**109.** In a *RC* circuit shown in Fig. 9.59 the sinusoidal voltage source $v = 250 \sin (500t + \phi)$ volts is applied by closing the switch at a time when $\phi = 45°$. There is an initial charge $q = 5000 \times 10^{-6}$ coulombs on the capacitor with polarity shown in the diagram. The current in the circuit will be

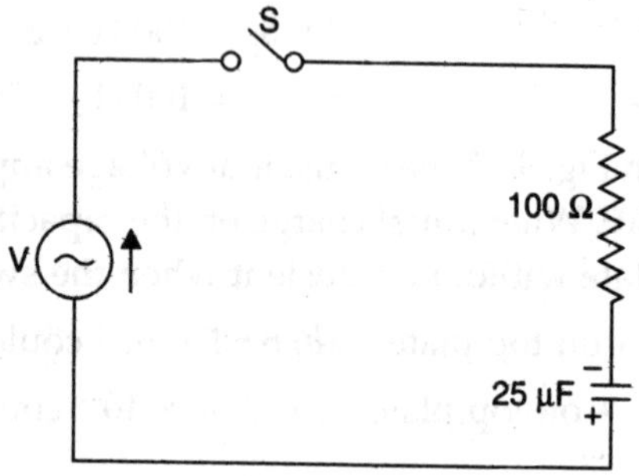


*FIGURE 9.59*

(*a*) $i = 1.83\, e^{400t}$ amperes (*b*) $i = 1.83\, e^{-400t}$ amperes

(*c*) $i = 1.83\, e^{-400t} + 1.955 \sin 500\, t$ amperes

(*d*) $i = 1.83\, e^{-400t} + 1.955 \sin (500\, t + 83.7°)$ amperes.

**110.** A series *RLC* circuit with $R = 5$ ohms, $C = 500\ \mu F$ and $L = 0.1$ H has a constant voltage $V = 10$ volts applied at $t = 0$. The resulting transient current will be

(*a*) $i = 0.72\, e^{25t}$ amperes (*b*) $i = 0.72\, e^{-25t}$ amperes

(*c*) $i = 0.72\, e^{-25t} \sin 139t$ amperes (*d*) $i = 0.72\, e^{-25t} \sin (139t + 113)$ amperes.

**111.** A series *RLC* circuit with $R = 50$ ohms, $L = 0.1$ H and $C = 50\ \mu F$ has a constant voltage $V = 100$ volts applied at $t = 0$. Assuming zero initial charge on the capacitor, the current transient will be

(*a*) $i = e^{-250t}\, 2.7 \sin (371t)$ amperes (*b*) $i = e^{250t}\, 2.7 \sin (371t)$ amperes

(*c*) $i = e^{250t}\, 2.7 \sin (371t + 111)$ amperes (*d*) none of the above.

**112.** A capacitor of 2 μF with an initial charge $q_0 = 100 \times 10^{-6}$ coulombs is connected across the terminals of a 100 ohm resistor at $t = 0$. The time in which the transient voltage across the resistor will drop from 40 V to 10 V will be

(*a*) 2.774 s (*b*) 277.4 s (*c*) 277.4 ms (*d*) 277.4 μs.

**113.** A series *RLC* circuit with $R = 200$ ohms and $L = 0.1$ H is to be made critically damped by the selection of the capacitance. The required value of C is

(*a*) 1 pF (*b*) 10pF (*c*) 1μF (*d*) 10 μF.

**114.** An *RLC* circuit has $R = 200\ \Omega$, $C = 5\ \mu F$ and $L = 0.1$ H. The natural frequency of the circuit will be

(*a*) 10 radians/second (*b*) 100 radians/second

(*c*) 1000 radians/second (*d*) 10000 radians/second.

**115.** A capacitor of 2 μF with an initial charge $q_0 = 100 \times 10^{-6}$ coulombs is connected across the terminals of a 100 ohm resistor at $t = 0$.

The time in which the transient voltage across the resistor will drop from 40 V to 10 V will be

(*a*) 2.774 μs (*b*) 27.74 μs (*c*) 277.4 μs (*d*) 2774 μs.

**116.** In the circuit shown in Fig. 9.60 the switch is closed on position 1 at $t = 0$, and then moved to position 2 after 1 millisecond. Find the time at which the current is zero and reversing in direction

(*a*) 0.66 ms (*b*) 1.261 ms (*c*) 1.261 s (*d*) 12.61 s.

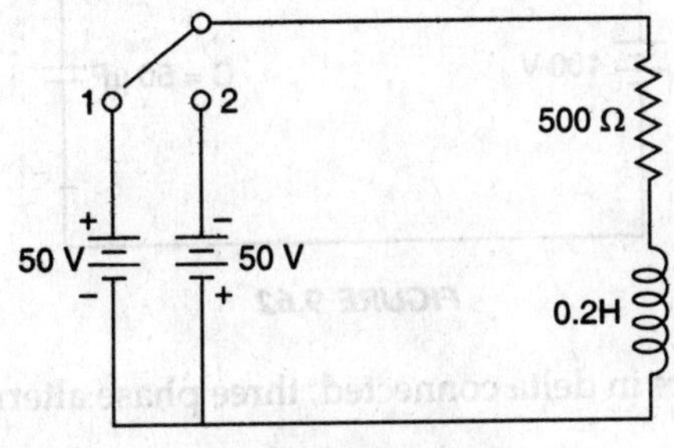


*FIGURE 9.60*

**117.** A constant voltage is applied to a series *RL* circuit by closing a switch. The voltage across *L* is 25 volts at $t = 0$ and drops to 5 volts at $t = 25$ ms. If $L = 2$H, what must be the value of *R* ?

(*a*) 33 Ω (*b*) 66 Ω (*c*) 128.8 Ω (*d*) 204.6 Ω.

## Questions 118 and 119 Refer to Data Given Below:

**118.** In a series *RL* circuit shown in Fig. 9.61 switch $S_1$ is closed at $t = 0$. After 4 milliseconds switch $S_2$ is opened. The current in the interval $0 < t < t'$ (where $t' = 4$ ms) will be

(*a*) $i = 2\,(1 - e^{-500t})$ (*b*) $i = 2\,(1 - e^{500t})$ (*c*) $i = 2\,(1 - e^{-250t})$ (*d*) $i = 2\,(1 - e^{250t})$.

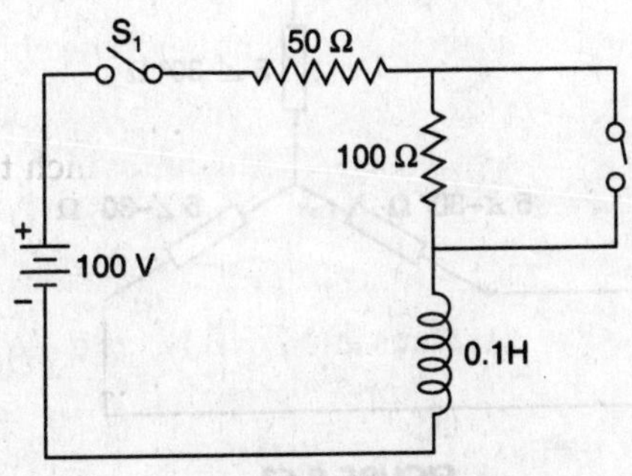


*FIGURE 9.61*

**119.** The current in the interval $t' < t$ (where $t' = 4$ ms) will be

(*a*) $i = 1.06\, e^{-1500t}$ amperes (*b*) $i = 1.06\, e^{-1500\,(t - t')}$ amperes

(*c*) $i = 1.06\, e^{-1500\,(t - t')} + 0.667$ amperes (*d*) none of the above.

**120.** The circuit shown in Fig. 9.62 has a constant voltage $V = 100$ V applied at $t = 0$. The current transient assuming zero initial charge on the capacitor will be

(*a*) $i = e^{-250t}$ amperes (*b*) $i = e^{-250t}$ (2.7 sin 371*t*) amperes

(*c*) $i = e^{250t}$ (2.7 sin 371*t*) amperes (*d*) $i = e^{250t}$ (2.7 cos 371*t*) amperes.

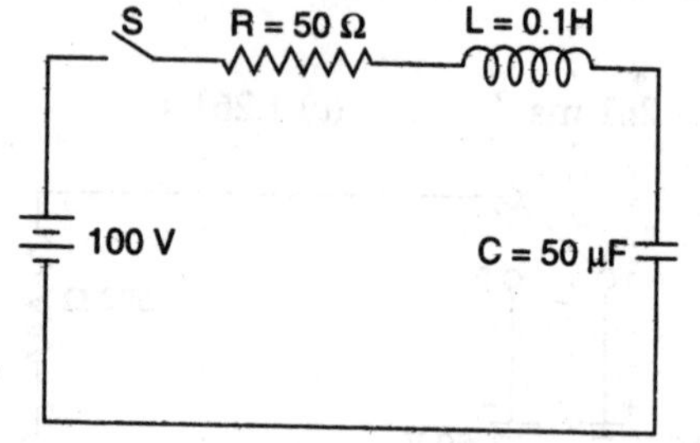


***FIGURE 9.62***

**121.** The full load coil currents in delta connected, three phase alternator with a rating of 25 kVA at 480 volts will be

(*a*) 17.35 A (*b*) 24.65 A (*c*) 30.1 A (*d*) 53.3 A.

**122.** The full load coil currents for star connected, three phase alternator with a rating of 25 kVA at 480 volts, will be

(*a*) 17.35 A (*b*) 24.65 A (*c*) 30.1 A (*d*) 53.3 A.

**123.** Three identical impedances of 5 !– 30º ohms are connected in star to a three phase three wire 150 volt *CBA* system. The line current $I_A$ will be

(*a*) 17.32 $\angle -60°$ A (*b*) 23.3 $\angle 60°$ A (*c*) 31.3 $\angle -180°$ A (*d*) 31.1 $\angle 180°$ A.

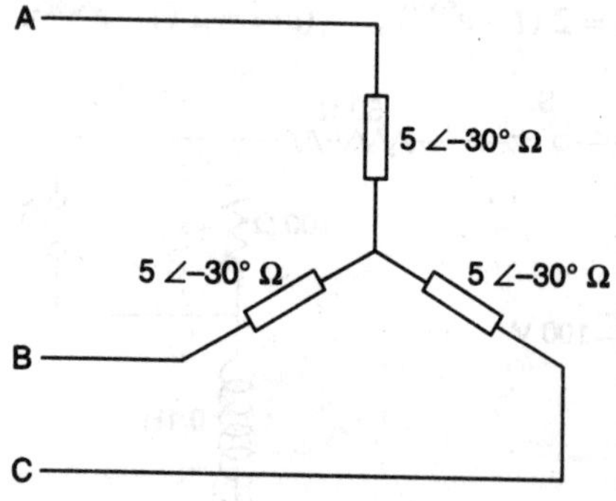


***FIGURE 9.63***

**124.** What will be the readings of two wattmeters in case of 2 wattmeter method

(*a*) 1850 watts each (*b*) 1800 W and 1900 W

(*c*) 1900 W and 2000 W (*d*) 2600 W and 1300 W.

**125.** Three identical impedances of 10! 30º ohms are in star connection. Another set of three identical impedances of 15 ! 0 ohms also are in star connection. Both star connections are on the same three phase, three wire 250 volt system.

The voltage required in the one line equivalent circuit will be

(*a*) 144.5 V (*b*) 176.1 V (*c*) 200 V (*d*) 220 V.

**126.** The line current will be

(*a*) 13 A (*b*) 17 A (*c*) 23.2 A (*d*) 31.7 A.

**127.** Total power of the circuit will be

(*a*) 3130 W (*b*) 6357 W (*c*) 8888 W (*d*) 9530 W.

**128.** A three phase four wire, 208 volt, star connected system has impedances as shown in Fig. 9.64 The line current $I_A$ will be

(*a*) $12 \angle 0°$ A (*b*) $12 \angle 90°$ A (*c*) $12 \angle 120°$ A (*d*) $12 \angle -60°$ A

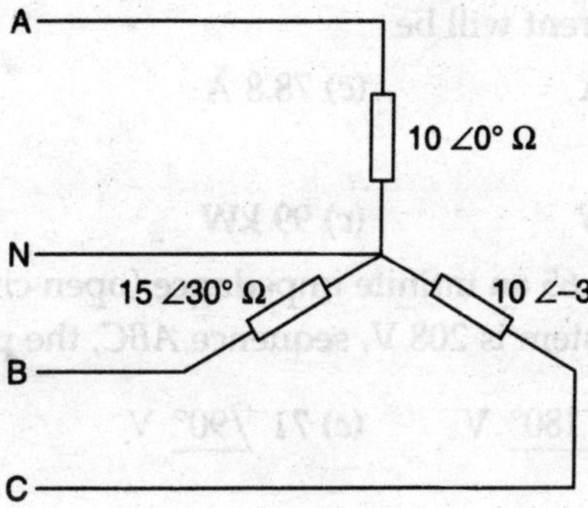


*FIGURE 9.64*

**129.** The line current $I_B$ will be

(*a*) $8 \angle 0°$ A (*b*) $8 \angle 60°$ A (*c*) $8 \angle -60°$ A (*d*) $8 \angle -120°$ A.

**130.** The magnitude of current $I_C$ will be

(*a*) 8 A (*b*) 10 A (*c*) 10.5 A (*d*) 12 A.

**131.** The current through the neutral will be

(*a*) $5.69 \angle 69.4°$ A (*b*) $5.69 \angle -69.4°$ A (*c*) $11.38 \angle 69.4°$ A (*d*) $11.38 \angle -69.4°$ A.

**132.** Total power will be

(*a*) 832 W (*b*) 1247 W (*c*) 1440 W (*d*) 3519 W.

**133.** A 50 HP induction motor with a full load efficiency of 85% and a power factor of 0.80 is connected to a three phase 480 volt system. The equivalent star connected impedance which can replace this motor will be

(*a*) $4.2 \angle 36.9°$ Ω (*b*) $4.2 \angle -36.9°$ Ω (*c*) $8.4 \angle -36.9°$ Ω (*d*) $8.4 \angle 36.9°$ Ω.

**134.** What will be the readings of two wattmeters used on a three wire, 240 volt system with a balanced delta-connected load of $20 \angle 80°$ ohms

(*a*) 1270 W, – 3840 W (*b*) –1270 W, 3840 W

(*b*) – 1710 W, 3210 W (*d*) 1710 W, – 3210 W.

**135.** A star connected, three phase, 400 volt alternator has a limit of 35 A per coil. What will be the kVA rating of the machine?

(*a*) 13.3 kVA (*b*) 26.6 kVA (*c*) 37.7 kVA (*d*) 41.4 kVA.

**136.** If the alternator supplies a line current of 20 A at a power factor of 0.65, what is the kVA per phase of the machine?

(*a*) 1.1 kVA (*b*) 2.2 kVA (*c*) 2.54 kVA (*d*) 5.08 kVA.

## Questions 137 and 138 Refer to Data Given Below:

Three identical impedances of $9 \angle -30°$ ohms in delta and three impedances of $5 \angle 45°$ ohms in star are both connected to the same three phase, three wire, 480 volt, *ABC* system.

**137.** The magnitude of the line current will be

(*a*) 44 A (*b*) 68.9 A (*c*) 78.8 A (*d*) 119.2 A.

**138.** Total power will be

(*a*) 33 kW (*b*) 66 kW (*c*) 99 kW (*d*) 132 kW.

**139.** For the circuit shown in Fig. 9.65 an infinite impedance (open circuit) in phase *B* of the star conected load exists. If the system is 208 V, sequence *ABC*, the phasor voltage $V_{OB}$ will be

(*a*) $284 \angle 150°$ V (*b*) $142 \angle 180°$ V (*c*) $71 \angle 90°$ V (*d*) $35.5 \angle 45°$ V.

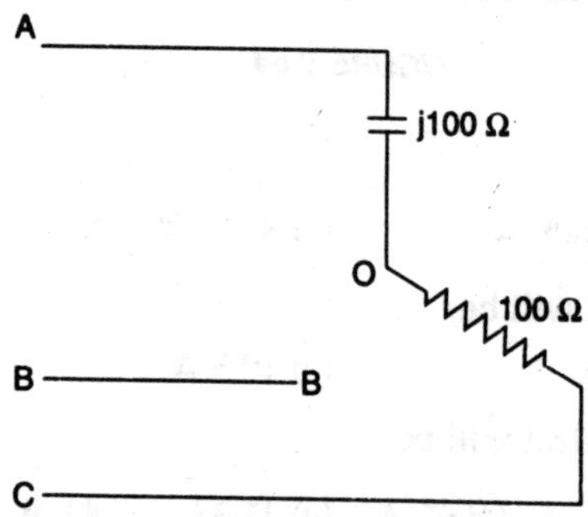


**FIGURE 9.65**

**140.** A star load with $Z_A = 12 \angle 45°$ ohms, $Z_B = 10 \angle 30°$ ohms and $Z_C = 8 \angle 0°$ ohms is connected to a four wire, 208 V system. The total power will be

(*a*) 1600 W (*b*) 2165 W (*c*) 3898 W (*d*) 6168 W.

**141.** The phasor diagram shown in Fig. 9.66 represents line currents and line-to-line voltages of a three-phase, three-wire, 346 V *ABC* system. If the line current magnitude is 10 A, the impedance of the star-connected load will be

(*a*) $5 \angle 90°$ Ω (*b*) $10 \angle 90°$ Ω (*c*) $15 \angle 90°$ Ω (*d*) $20 \angle 90°$ Ω.

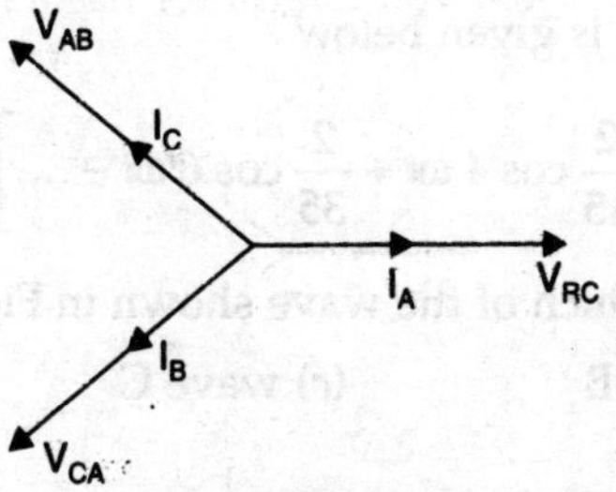


*FIGURE 9.66*

**142.** Two balanced delta loads with impedances of 20 $\angle -60°$ and 18 $\angle 45°$ ohms respectively are connected to a three phase 150 volt system.
The power in the low loads will be
(*a*) 1000 W each (*b*) 1690 W each
(*c*) 1690 W and 2650 W (*c*) 2650 W and 4300 W.

**143.** The two wattmeters in lines *A* and *B* of a 120 volt *CBA* system read 1500 and 500 watts respectively. The impedance of the balanced delta load will be
(*a*) 16.3 $\angle -41°$ Ω (*b*) 16.3 $\angle 41°$ Ω (*c*) 33 $\angle 42°$ Ω (*d*) 33 $\angle -42°$ Ω.

## Questions 144 and 145 Refer to Data Given Below

A three phase, three wire, 208 volt, *ABC* system supplies a delta connected load where

$$Z_{AB} = 5 \angle 0° \text{ ohms}$$
$$Z_{BC} = 4 \angle 30° \text{ ohms}$$
$$Z_{CA} = 6 \angle -15° \text{ ohms.}$$

**144.** Which of the following will not be the current flowing through any of the lines?
(*a*) 75 $\angle 90°$ A (*b*) 53.9 $\angle -68.2°$ A (*c*) 32 $\angle 231.3°$ A (*d*) 16 $\angle 112.2°$ A.

**145.** Total power will be
(*a*) 11 W (*b*) 26.6 W (*c*) 42.4 W (*d*) 56.4 W.

## Questions 146 to 148 Refer to Data Given Below

A voltage

$$v = 50 + 25 \sin 500t + 10 \sin 1500t + 5 \sin 2500t \text{ volts}$$

is applied to the terminals of a passive network and the resulting current is

$$i = 5 + 2.23 \sin (500t - 26.6°) + 0.566 \sin (1500t - 56.3°) + 0.186 \sin (2500t - 68.2°) \text{ amperes.}$$

**146.** The effective voltage will be
(*a*) 13.4 V (*b*) 22.2 V (*c*) 28.8 V (*d*) 53.6 V.

**147.** The effective current will be
(*a*) 1 A (*b*) 1.73 A (*c*) 2.25 A (*d*) 5.25 A.

**148.** The average power will be
(*a*) 101.5 W (*b*) 171.5 W (*c*) 276.5 W (*d*) 515.5 W.

**149.** A trigonometric Fourier series is given below

$$\frac{2V}{\pi}\left[1+\frac{2}{3}\cos 2\omega t-\frac{2}{15}\cos 4\,\omega t+\frac{2}{35}\cos 6\,\omega t-....\right]$$

The above series represents which of the wave shown in Fig. 9.67

(*a*) wave A (*b*) wave B (*c*) wave C (*d*) wave D.

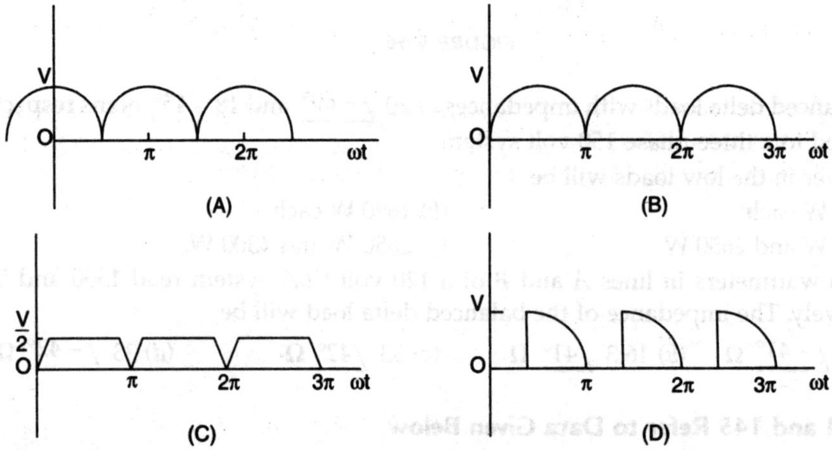


*FIGURE 9.67*

**150.** The following trigonometric Fourier series represents which of the wave shown in Fig. 9.67.

$$\frac{2V}{\pi}\left[1-\frac{2}{3}\cos 2\omega t-\frac{2}{15}\cos 4\,\omega t-\frac{2}{35}\cos 6\,\omega t-....\right]$$

(*a*) wave A (*b*) wave B (*c*) wave C (*d*) wave D.

**151.** A trigonometric Fourier series is given in Fig. 9.68. Which waveform does it represent?

(*a*) figure A (*b*) figure B

(*c*) figure C (*d*) none of the given figures.

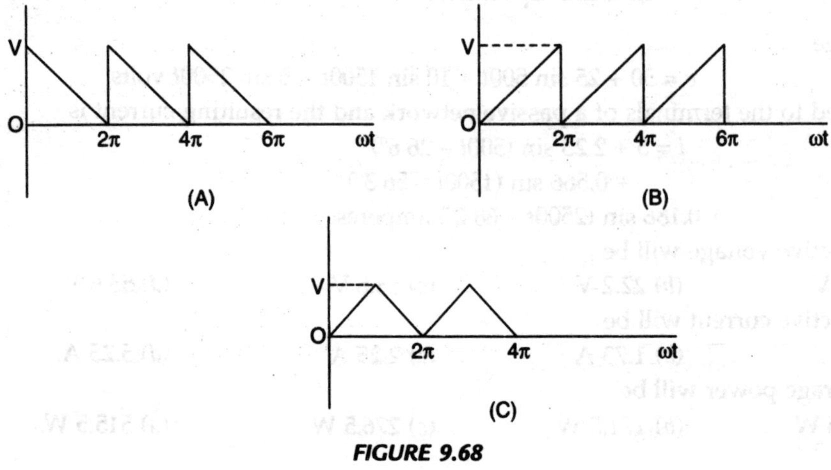


*FIGURE 9.68*

152. The trigonometric Fourier series for the waveform shown in Fig. 9.69 will have

(a) sine terms only (b) cosine terms only

(c) both sine and cosine terms (d) none of the above.

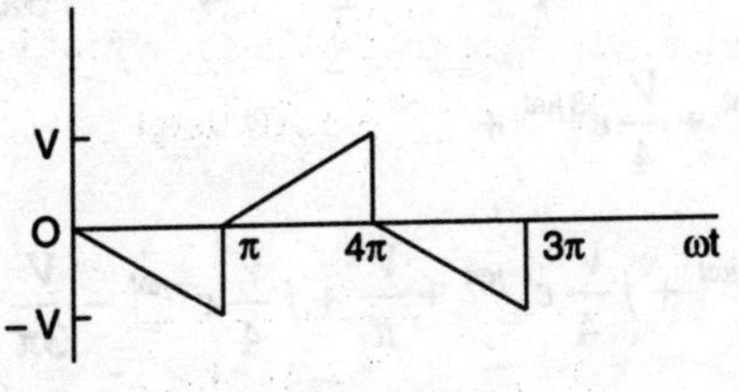


FIGURE 9.69

153. Which of the following series represents the waveform shown in above figure?

(a) $\frac{4V}{\pi^2}\left\{\cos\omega t+\frac{1}{9}\cos 3\omega t+\frac{1}{25}\cos 5\omega t+....\right\}-\frac{2V}{\pi}\left\{\sin\omega t+\frac{1}{3}\sin 3\omega t+\frac{1}{5}\sin 5\,\omega t+....\right\}$

(b) $\frac{4V}{\pi^2}\left\{\cos\omega t-\frac{1}{9}\cos 3\,\omega t+\frac{1}{25}\cos 5\omega t-....\right\}+\frac{2V}{\pi}\left\{\sin\omega t+\frac{1}{3}\sin 3\omega t+\frac{1}{5}\sin 5\,\omega t+....\right\}$

(c) $\frac{4V}{\pi^2}\left\{\cos\omega t+\frac{1}{9}\cos 3\,\omega t+\frac{1}{25}\cos 5\omega t+....\right\}-\frac{2V}{\pi}\left\{\sin\omega t-\frac{1}{3}\sin 3\,\omega t+\frac{1}{5}\sin 5\,\omega t\,....\right\}$

(d) none of the above.

154. Which of the following trigonometric Fourier series represents the half-wave rectified sine wave? (Fig. 9.70)

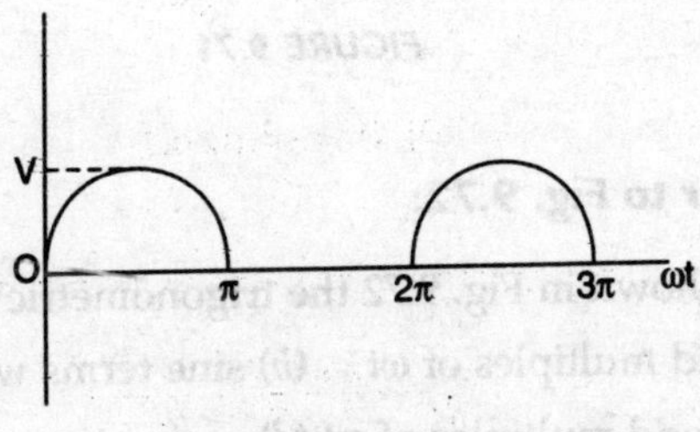


FIGURE 9.70

(a) $\frac{V}{\pi}\left[1+\frac{\pi}{2}\sin\omega t+\frac{2}{3}\cos 2\,\omega t-\frac{2}{15}\cos 4\,\omega t....\right]$

(b) $\frac{V}{\pi}\left[1+\frac{\pi}{2}\sin\omega t-\frac{2}{3}\cos 2\,\omega t-\frac{2}{15}\cos 4\,\omega t-....\right]$

(c) $\frac{V}{\pi}\left[1-\frac{\pi}{2}\sin\omega t+\frac{2}{3}\cos 2\,\omega t-\frac{2}{15}\cos 4\,\omega t+....\right]$

(d) none of the above.

**155.** Which of the following represents the exponential Fourier series for the half wave rectified sine wave shown in above figure?

(a) $-\frac{V}{15\pi}e^{j4\omega t}-\frac{V}{3\pi}e^{-j2\omega t}+j\frac{V}{4}e^{-j\omega t}+\frac{V}{\pi}-j\frac{V}{4}e^{j\omega t}-\frac{V}{3\pi}e^{j2\omega t}-\frac{V}{15\pi}e^{j4\omega t}-....$

(b) $\frac{V}{15\pi}e^{j\omega t}+\frac{V}{3\pi}e^{-2j\omega t}+\frac{V}{4}e^{3j\omega t}+....$

(c) $\frac{V}{15\pi}e^{j4\omega t}+\frac{V}{3\pi}e^{-2j\omega t}+j\frac{V}{4}e^{-j\omega t}+\frac{V}{\pi}+j\frac{V}{4}e^{-j\omega t}-\frac{V}{3\pi}e^{-j2\omega t}+\frac{V}{15\pi}e^{-j4\omega t}+....$

(d) none of the above.

**156.** The trigonometric Fourier series for the saw tooth wave shown in Fig. 9.71 is represented by which of the following relations?

(a) $\frac{2V}{\pi}[\sin\omega t+\frac{1}{2}\sin 2\omega t+\frac{1}{3}\sin 3\omega t....]$ (b) $\frac{2V}{\pi}[\sin\omega t-\frac{1}{2}\sin 2\omega t+\frac{1}{3}\sin 3\omega t....]$

(c) $\frac{2V}{\pi}[\cos\omega t+\frac{1}{2}\cos 2\omega t+\frac{1}{3}\cos 3\omega t+....]$ (d) $\frac{2V}{\pi}[\cos\omega t-\frac{1}{2}\cos 2\omega t+\frac{1}{3}\cos 3\omega t-....]$.

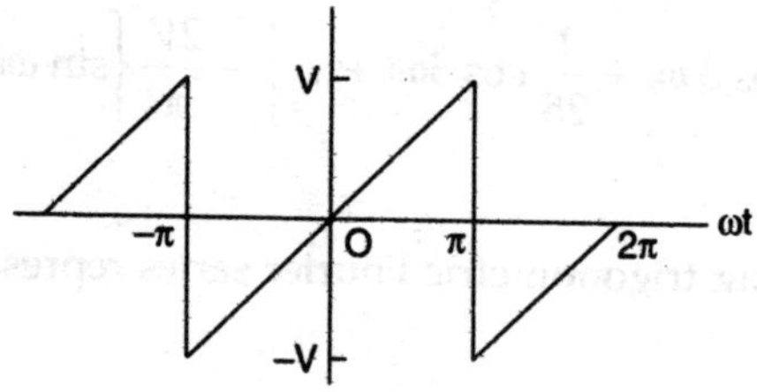


FIGURE 9.71

**Questions 157 to 159 Refer to Fig. 9.72:**

**157.** For the square wave shown in Fig. 9.72 the trigonometric Fourier series will contain.

(a) sine terms with odd multiples of $\omega t$ (b) sine terms with even multiples of $\omega t$

(c) cosine terms with odd multiples of $\omega t$ (d) cosine term with even multiples of $\omega t$.

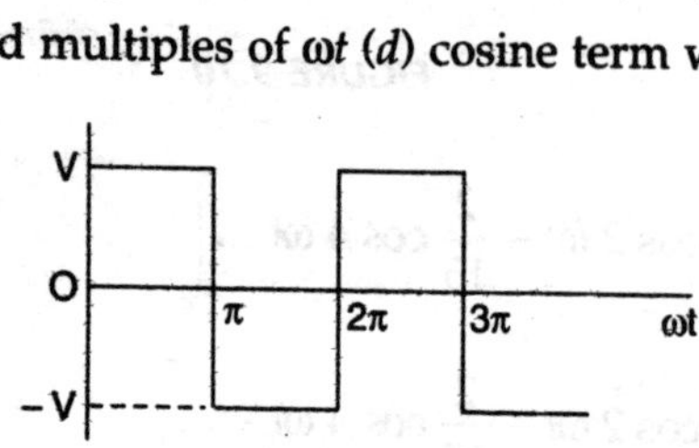


FIGURE 9.72

**158.** In the Fourier series for this wave, $a_n$ =

(a) 0 (b) 1 (c) $\pi$ (d) $n$.

159. In the Fourier series for this wave, $b_n$ =

(a) $\frac{V}{n}(1+\cos n\pi)$ (b) $\frac{V}{2n}(1+\cos n\pi)$ (c) $\frac{2V}{\pi n}(1+\cos n\pi)$ (d) $\frac{2V}{n\pi}(1-\cos n\pi)$.

**Questions 160 and 161 Refer to Fig. 9.73:**

160. For the circuit shown in Fig. 9.73, R = 5 ohms and L = 0.02 H. If the voltage applied is, $v = 100 + 50 \sin \omega t + 25 \sin 3\omega t$ volts. For $\omega$ = 500 radians/second, the effective value of the current will be

(a) 10.1 A (b) 20.25 A (c) 30.5 A (d) 40.5 A.

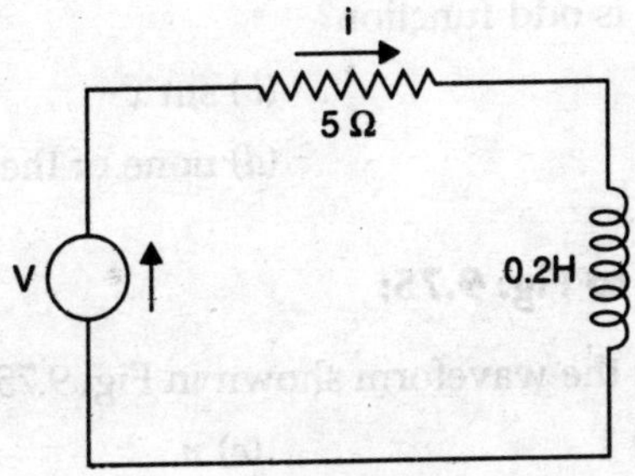


*FIGURE 9.73*

161. The power dissipated in 5 ohm resistor will be

(a) 1010 W (b) 1212 W (c) 2053 W (d) 2693 W.

162. A periodic function is said to have half wave symmetry when

(a) $f(x) = -f(x)$ (b) $f(x) = -f(-x)$

(c) $f(x) = -f(x + T/2)$ (d) $f(x) = f(xT)$.

(T = Period).

163. Which of the wave shown in Fig. 9.74 has half wave symmetry?

(a) figure A (b) figure B (c) figure C (d) none of the figures.

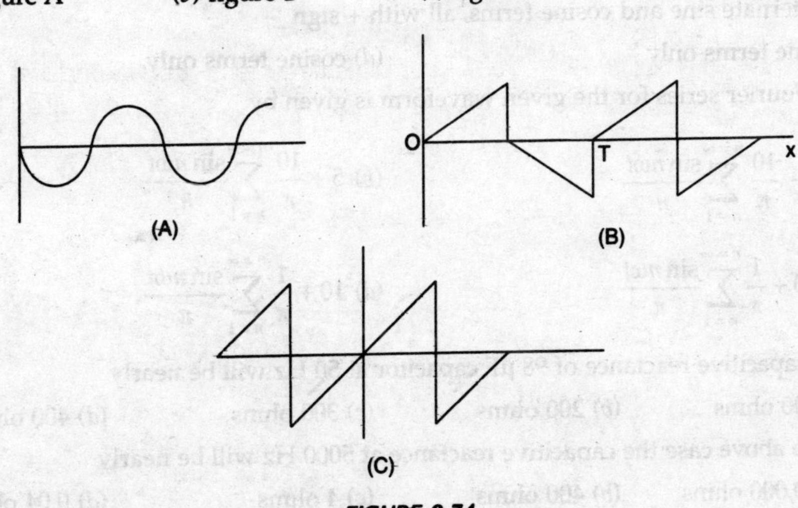


*FIGURE 9.74*

**164.** A function is said to even when

(*a*) $f(x) = \frac{1}{fx}$ (*b*) $f(x + 1) = \text{odd}$ (*c*) $fx = f(-x)$ (*d*) $fx = -f(x)$.

**165.** Which of the following function is even

(*a*) $2 + x^2 + x^4$ (*b*) $2 + 2x + 4x^3$ (*c*) $2 + 4x^3$ (*d*) $2 + x + x^3 + \frac{x^4}{4}$.

**166.** For a function of $f(x) = -f(-x)$ the function is said to be

(*a*) odd (*b*) symmetrical (*c*) periodic (*d*) harmonic.

**167.** Which of the following is odd function?

(*a*) cos *x* (*b*) sin *x*

(*c*) both sin *x* and cos *x* (*d*) none of the above.

**Questions 168 to 170 Refer to Fig. 9.75:**

**168.** In the Fourier series for the waveform shown in Fig. 9.75, $a_n$ =

(*a*) 0 (*b*) 1 (*c*) *n* (*d*) π.

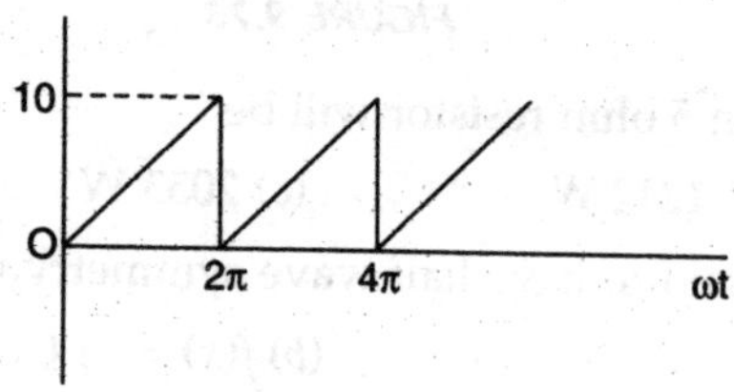


***FIGURE 9.75***

**169.** The Fourier series for this waveform will have

(*a*) alternate sine and cosine terms with alternate + and – signs

(*b*) alternate sine and cosine terms, all with + sign

(*c*) sine terms only (*d*) cosine terms only.

**170.** The Fourier series for the given waveform is given by

(*a*) $5 - \frac{10}{\pi}\sum_{n=1}^{n=\infty}\frac{\sin n\omega t}{n}$ (*b*) $5 + \frac{10}{\pi}\sum_{n=1}^{n=\infty}\frac{\sin n\omega t}{n}$

(*c*) $10 + \frac{1}{\pi}\sum_{n=1}^{n=\infty}\frac{\sin n\omega t}{\pi}$ (*d*) $10 + \frac{1}{\pi}\sum_{n=1}^{n=\infty}\frac{\sin n\omega t}{\pi}$.

**171.** The capacitive reactance of 98 μF capacitor at 50 Hz will be nearly

(*a*) 100 ohms (*b*) 200 ohms (*c*) 300 ohms (*d*) 400 ohms.

**172.** In the above case the capacitive reactance at 5000 Hz will be nearly

(*a*) 40,000 ohms (*b*) 400 ohms (*c*) 4 ohms (*d*) 0.04 ohms.

**173.** The inductive reactance of 20 μH inductor at 50 Hz will be nearly

(*a*) 0.00628 ohms (*b*) 0.0628 ohms (*c*) 0.6280 ohms (*d*) 6.280 ohms.

**174.** In the above case if frequency is raised to 5000 Hz, the inductive reactance will be nearly

(*a*) 0.00000628 ohms (*b*) 0.000628 ohms (*c*) 0.628 ohms (*d*) 628 ohms.

**175.** The inductive reactance in above problem will be nearly 1260 ohms at

(*a*) 5 kHz (*b*) 50 kHz (*c*) 1 MHz (*d*) 1 GHz.

**176.** The power consumed in a pure capacitive circuit is

(*a*) CV (*b*) $CV^2/2$ (*c*) $C^2V/2$ (*d*) zero.

### Questions 177 and 178 Refer to Data Given Below:

In the series circuit shown in Fig. 9.76 the impedance of the source is (5 + *j*3) Ω and the source frequency is 2000 Hz.

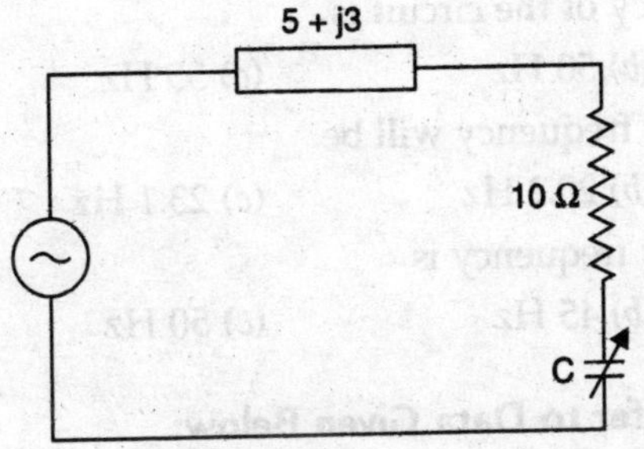


**FIGURE 9.76**

**177.** At what value of C will the power in the 10 Ω resistor be a maximum

(*a*) 13.3 μF (*b*) 26.6 μF (*c*) 41.4 μF (*d*) 54.5 μF.

**178.** In the above case what will be the power?

(*a*) 10 W (*b*) 11.1 W (*c*) 111 W (*d*) 181 W.

### Questions 179 and 180 Refer to Fig. 9.77:

**179.** For the circuit shown in Fig. 9.77 the resonance frequency is given by

(*a*) $\frac{1}{\sqrt{LC}}\sqrt{\frac{R+C}{L+C}}$ (*b*) $\frac{1}{\sqrt{LC}}\sqrt{\frac{RC+LC}{R+L+C}}$

(*c*) $\frac{1}{\sqrt{LC}}\sqrt{\frac{R_{L^2}-L/C}{R_{C^2}-L/C}}$ (*d*) $\frac{1}{\sqrt{LC}}\sqrt{\frac{R_{C^2}-L/C}{R_{L^2}-L/C}}$.

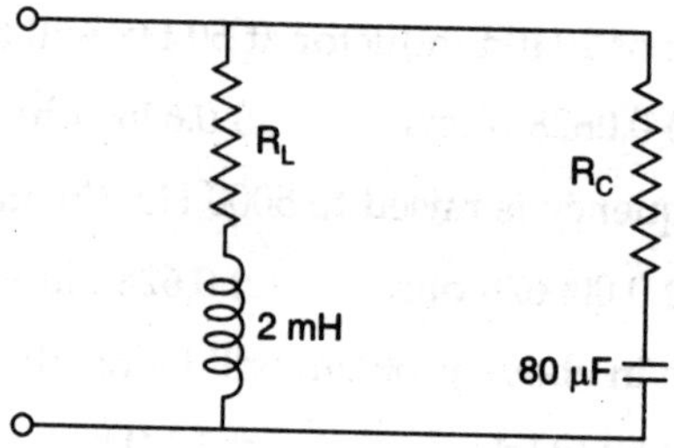


*FIGURE 9.77*

**180.** What must be the value of $R_L$ and $R_C$ for the circuit to be resonant?

(*a*) $R_L = 1\ \Omega$, $R_C = 10\ \Omega$ (*b*) $R_L = 2\ \Omega$, $R_C = 8\ \Omega$

(*c*) $R_L = R_C = 1\ \Omega$ (*d*) $R_L = R_C = 5\ \Omega$.

**Questions 181 to 183 Refer to Data Given Below:**

In a series $R - L - C$ circuit, $R = 100$ ohms, $L = 0.5$ H and $C = 40\ \mu$F

**181.** The resonant frequency of the circuit is

(*a*) 35.7 Hz (*b*) 50 Hz (*c*) 55 Hz (*d*) 67.7 Hz.

**182.** The lower half power frequency will be

(*a*) 17.3 Hz (*b*) 20.2 Hz (*c*) 23.1 Hz (*d*) 29.7 Hz.

**183.** The upper half power frequency is

(*a*) 41.5 Hz (*b*) 45 Hz (*c*) 50 Hz (*d*) 55 Hz.

**Questions 184 and 185 Refer to Data Given Below:**

A series circuit with $R = 5\ \Omega$, $C = 20\ \mu$F and a variable inductance $L$ has an applied voltage $V = 10\ \angle 0°$ volts with a frequency of 1000 radians per second. The value of $L$ is adjusted until the voltage across the resistor is a maximum.

**184.** The voltage across $L$ will be

(*a*) $10\ \angle 0°$ V (*b*) $10\ \angle 90°$ V (*c*) $100\ \angle 0°$ V (*d*) $100\ \angle 90°$ V.

**185.** The voltage across capacitor will be

(*a*) $10\ \angle 0°$ V (*b*) $10\ \angle -90°$ V (*c*) $100\ \angle -90°$ V (*d*) $100\ \angle 0°$ V.

**186.** A series circuit has $R = 5$ ohms, $L = 20$ mH and a variable capacitance C. If the frequency of applied voltage is 1000 Hz, what must be the value of C?

(*a*) 1.27 μH (*b*) 12.7 μF (*c*) 1.27 pF (*d*) 12.7 pF.

**Questions 187 to 190 Refer to Fig. 9.78:**

A three branch parallel circuit has a current locus as shown in Fig. 9.78.

**187.** The circuit constants for branch 1 will be

(*a*) $R = 8.05\ \Omega$, $L = 0.423$ mH (*b*) $R = 6.15\ \Omega$, $L = 0.626$ mH

(*c*) $R = 4.15\ \Omega$, $L = 0.555$ mH (*d*) $R = 2.15\ \Omega$, $L = 0.155$ mH.

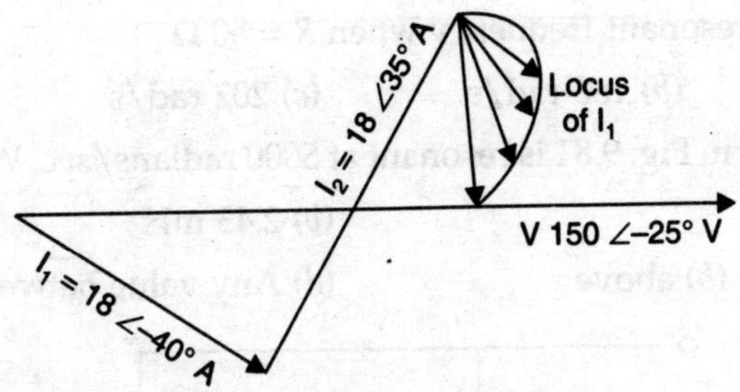


*FIGURE 9.78*

**188.** The circuit constants for branch 2 are

(*a*) $R = 2\ \Omega$, $C = 20\ \mu F$ (*b*) $R = 4.16\ \Omega$, $C = 27.7\ \mu F$

(*c*) $R = 6.64\ \Omega$, $C = 40\ \mu F$ (*d*) $R = 8\ \Omega$, $C = 50\ \mu F$.

**189.** Variable element in branch 3 is

(*a*) *L* (*b*) *C* (*c*) *R* (*d*) none.

**190.** In addition to variable element, branch 3 has

(*a*) $L = 2.74$ mH (*b*) $L = 4.55$ mH (*c*) $C = 2.74\ \mu F$ (*d*) $C = 4.55\ \mu F$.

**191.** In the circuit shown in Fig. 9.79, what will be the resonant frequency?

(*a*) 55 Hz (*b*) 68 Hz (*c*) 100 Hz (*d*) 159 Hz.

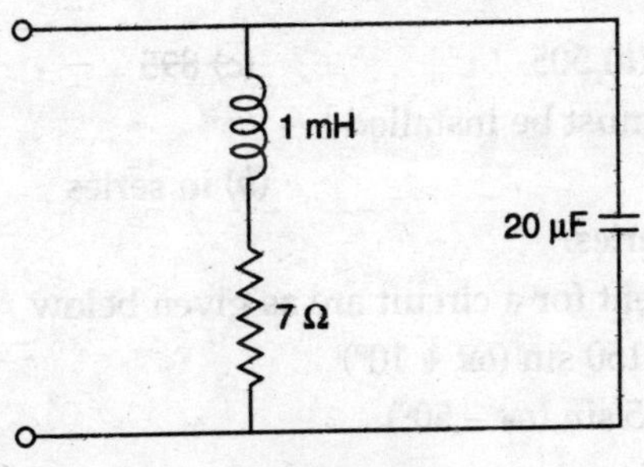


*FIGURE 9.79*

**192.** In the parallel circuit shown in Fig. 9.80 what will be the resonant frequency when $R_L = 0$

(*a*) 102 rad/s (*b*) 204 rad/s (*c*) 306 rad/s (*d*) 408 rad/s.

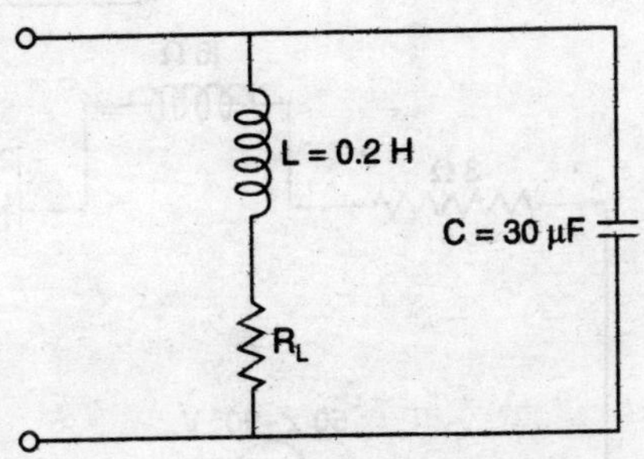


*FIGURE 9.80*

**193.** What will be the resonant frequency when $R = 50\ \Omega$

(*a*) 111 rad/s (*b*) 166 rad/s (*c*) 202 rad/s (*d*) 323 rad/s.

**194.** The circuit shown in Fig. 9.81 is resonant at 5000 radians/sec. What could be the value of *L* ?

(*a*) 0.066 mH (*b*) 2.43 mH

(*c*) Either of (*a*) or (*b*) above (*d*) Any value between (*a*) and (*b*).

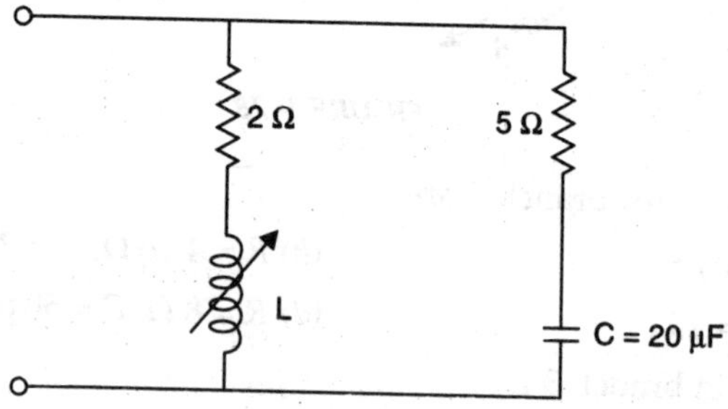


*FIGURE 9.81*

**195.** For a circuit having impedance $Z = (3 + j\,4)$ ohms the voltage applied is 100 ! 30º volts. The power factor will be

(*a*) 0.3 lagging (*b*) 0.6 lagging (*c*) 0.6 leading (*d*) 0.3 leading.

**196.** In the above circuit if power factor is to be improved to 0.9 lagging, the *VAR* of the capacitor must be

(*a*) 105 (*b*) 505 (*c*) 895 (*d*) 1015.

**197.** The above capacitor must be installed

(*a*) in parallel (*b*) in series

(*c*) either prallel or series.

**198.** The voltage and current for a circuit are as given below

$$v = 150 \sin (\omega t + 10^\circ)$$

$$i = 5 \sin (\omega t - 50^\circ)$$

The power angle is

(*a*) 30º (*b*) 45º (*c*) 50º (*d*) 60º.

**199.** For the circuit shown in Fig. 9.82 current *I* is given by

(*a*) $10\ \angle 143.1^\circ$ (*b*) $10\ \angle 143.1^\circ$ (*c*) $10\ \angle -143.1^\circ$ (*d*) $10\ \angle -43^\circ$.

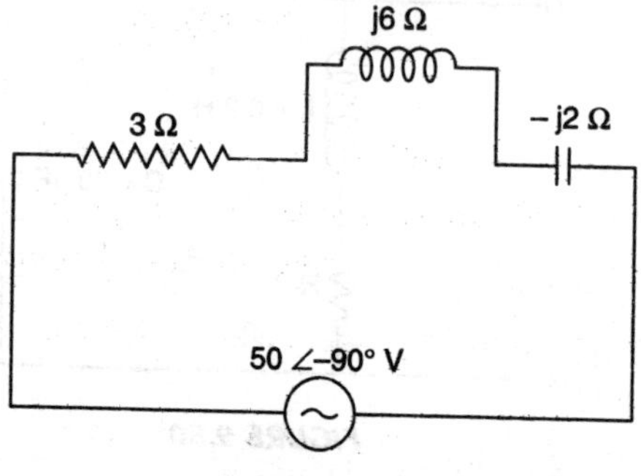


*FIGURE 9.82*

**200.** Power factor of the circuit is

(*a*) 0.4 (*b*) 0.5 (*c*) 0.6 (*d*) 0.75.

**201.** Power consumed by the circuit will be

(*a*) 200 W (*b*) 300 W (*c*) 400 W (*d*) 500 W.

### Questions 202 and 203 Refer to Data Given Below:

A 500 kVA transformer is at full load with an overall power factor of 0.6 lagging. The power factor is improved by adding capacitors until the overall power factor becomes 0.9 lagging.

**202.** The kVAR capacity of the capacitor required will be

(*a*) 400 (*b*) 254 (*c*) 140 (*d*) 124.

**203.** After correction of power factor, what percentage of the full load is the transformer carrying ?

(*a*) 100% (*b*) 86.6% (*c*) 75.5% (*d*) 66.7%.

### Questions 204 to 206:

When a voltage $v = 200 \sin (\omega t - 110^\circ)$ volts is applied to a circuit the resulting current is $i = 5 \sin (\omega t - 20^\circ)$.

**204.** The power consumed on the circuit will be

(*a*) 500 W (*b*) 400 W (*c*) 200 W (*d*) zero.

**205.** The power factor of the circuit is

(*a*) 0 (*b*) 0.5 lagging (*c*) 0.5 leading (*d*) 1.

**206.** VAR of the circuit is

(*a*) 1000 (*b*) 500 (*c*) 400 (*d*) 300.

**207.** Voltages and currents for circuits are given below. For which circuit the power factor will be lagging?

(*a*) $v = 340 \sin (\omega t - 60^\circ)$ volts

$i = 13.3 \sin (\omega t - 48.7^\circ)$ amperes

(*b*) $v = 200 \sin (\omega t - 110^\circ)$ volts

$i = 5 \sin (\omega t - 20^\circ)$ amperes

(*c*) $v = 14.14 \cos \omega t$ volts

$i = 17.1 \cos (\omega t - 14.05)$ milliamperes

(*d*) none of the above.

**208.** A series circuit has $R = 8\ \Omega$ and $X_C = 6\ \Omega$. The applied phasor voltage is $50 \angle -90^\circ$ volts. The power factor will be

(*a*) 0.6 leading (*b*) 0.8 leading (*c*) 0.8 lagging (*d*) 0.6 lagging.

**209.** A circuit takes 5040 volt amperes at a power factor of 0.894 leading when the applied phasor voltage is

$$V = 150 \angle 45^\circ \text{ volts.}$$

The impedance of the circuit is

(*a*) $(-4 - j\,2)\ \Omega$ (*b*) $(4 + j\,2)\ \Omega$ (*c*) $(4 - j\,2)\ \Omega$ (*d*) $(-4 + j\,2)\ \Omega$.

**Questions 210 and 211:**

A two element series circuit with an instantaneous current $i = 4.24 \sin (5000\,t + 45^{\circ})$ amperes has a power of 180 watts and a power factor of 0.8 lagging.

**210.** The resistance of the circuit must have the value

(*a*) 5 ohms (*b*) 10 ohms (*c*) 15 ohms (*d*) 20 ohms.

**211.** The inductance of the circuit must have the value

(*a*) 3 H (*b*) 0.3 H (*c*) 3 mH (*d*) 0.3 mH.

**212.** In the circuit shown in Fig. 9.83 if branch *A* takes 8 kVAR, the power of the circuit will be

(*a*) 2 kW (*b*) 4 kW (*c*) 6 kW (*d*) 8 kW.

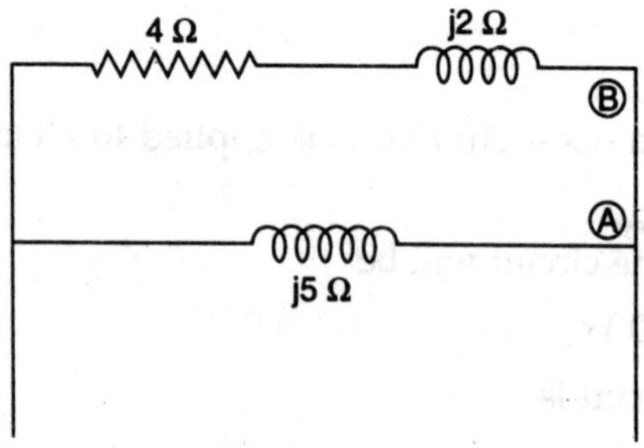


*FIGURE 9.83*

**213.** The power factor of the circuit will be

(*a*) 0.111 (*b*) 0.222 (*c*) 0.414 (*d*) 0.555.

**214.** A 250 kVA transformer is at full load with a power factor of 0.8 lagging. The power factor is to be corrected to 0.9 lagging by parallel capacitors.

What must be the kVAR capacity of the capacitors?

(*a*) 25 (*b*) 37.5 (*c*) 52.5 (*c*) 67.75.

**215.** What kW of new load at unity power factor may now be added without exceeding the rated transformer kVA?

(*a*) 10 kW (*b*) 20 kW (*c*) 30 kW (*d*) 40 kW.

**216.** A 100 kVA transformer is at 80% of full load with a power factor of 0.85 lagging. What kVA of load at 0.6 pf lagging may be added without exceeding the full load rating of the transformer?

(*a*) 2.13 kVA (*b*) 4.64 kVA (*c*) 18.5 kVA (*d*) 21.3 kVA.

**Questions 217 and 218 Refer to Data Given Below:**

**217.** An induction motor load of 1500 watts with a power factor of 0.75 lagging is combined with 500 volt amperes of synchronous motors with a power factor of 0.65 lead. What kVAR of capacitors is required to correct the overall power factor of the two groups of motors to 0.95 lagging?

(*a*) 117 VAR (*b*) 200 VAR (*c*) 314 VAR (*d*) 347 VAR.

**218.** What percent reduction in volt amperes will result?

(*a*) 6.3% (*b*) 12.6% (*c*) 18.9% (*d*) 24.5%.

**219.** A 65 kVA load with a lagging power factor is added to 25 kVA of synchronous motors with leading power factor of 0.6. If the overall power factor is 0.85 lagging, the power factor of the 65 kVA load must be

(*a*) 0.15 (*b*) 0.355 (*c*) 0.585 (*d*) 0.667.

### Questions 220 and 221:

A load of 300 kW with power factor of 0.65 lagging has the power factor improved to 0.9 lagging by the addition of parallel capacitors.

**220.** What kVAR of capacitors will be required?

(*a*) 51 (*b*) 102 (*c*) 153 (*d*) 204.

**221.** What percent reduction in kVA will result?

(*a*) 14% (*b*) 28% (*c*) 38% (*d*) 56%.

**222.** In the circuit shown in Fig. 9.84 the voltage across 30 ohm resistor is 45 volts. The reading of the ammeter *A* will be

(*a*) 10 A (*b*) 19.4 A (*c*) 22.4 A (*d*) 28 A.

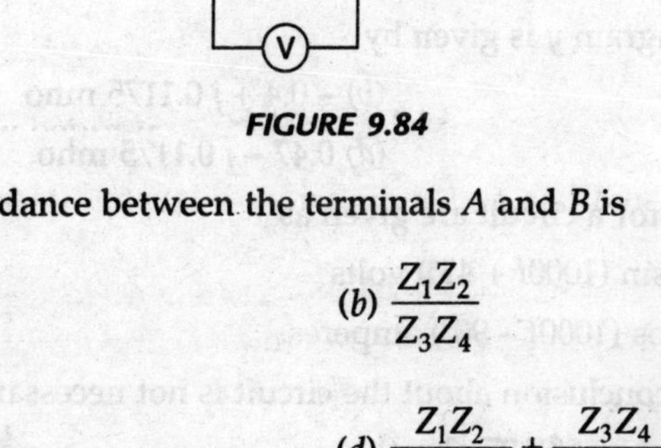


*FIGURE 9.84*

**223.** The equivalent impedance between the terminals *A* and *B* is

(*a*) $Z_1 + Z_2 + Z_3 + Z_4$ (*b*) $\frac{Z_1Z_2}{Z_3Z_4}$

(*c*) $\frac{Z_1Z_4}{Z_1 + Z_4} + \frac{Z_2Z_3}{Z_2 + Z_3}$ (*d*) $\frac{Z_1Z_2}{Z_1 + Z_2} + \frac{Z_3Z_4}{Z_3 + Z_4}$.

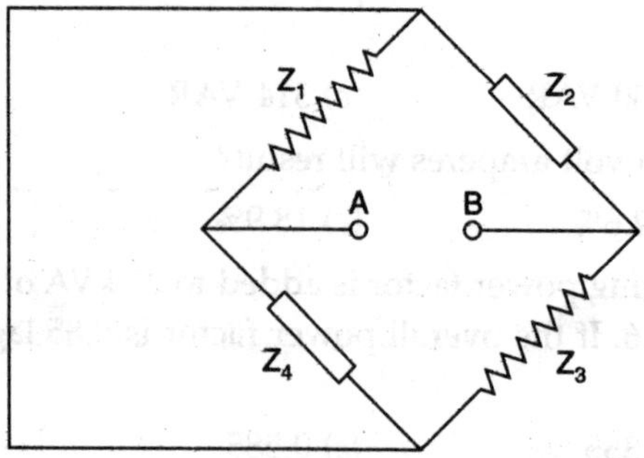


FIGURE 9.85

**224.** In the above circuit if $Z_1 = 500\ \Omega$, $Z_2 = 250\ \angle 30°\ \Omega$, $Z_3 = 1000\ \Omega$ and $Z_4 = 2000\ \angle -30°\ \Omega$ then equivalent impedance between terminals $A$ and $B$ will be

(*a*) $233\ \angle 24.05°\ \Omega$ (*b*) $465\ \angle 14.05°\ \Omega$ (*c*) $500\ \angle 50°\ \Omega$ (*d*) $596\ \angle 4.05°\ \Omega$.

**225.** For the phasor diagram shown in Fig. 9.86 the value of Z is

(*a*) $-2 - j\,0.5$ (*b*) $2 + j\,0.5$ (*c*) $2 - j\,0.5$ (*d*) $-2 + j\,0.5$.

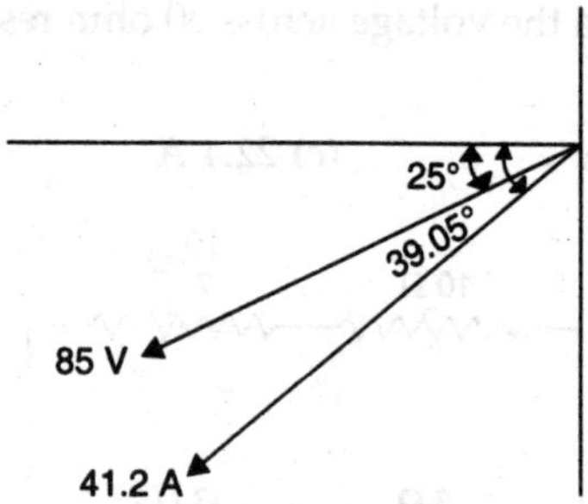


FIGURE 9.86

**226.** For the above phasor diagram $y$ is given by

(*a*) $0.47 + j\,0.1175$ mho (*b*) $-0.4 + j\,0.1175$ mho

(*c*) $-0.47 - j\,0.1175$ mho (*d*) $0.47 - j\,0.1175$ mho.

The voltage and current for a circuit are given as

$$v = 80 \sin (1000t + 45°) \text{ volts}$$

$$i = 8 \cos (1000t - 90°) \text{ amperes.}$$

**227.** Which of the following conclusion about the circuit is not necessarily valid?

(*a*) The circuit has resistance of 7.07 ohms

(*b*) The circuit has inductance of 7.07 mH

(*c*) The circuit has capacitance of 7.07 μF

(*d*) The current leads the voltage.

**228.** For a circuit if $v = 50 \sin (2000t - 25^\circ)$ volts and
$i = 8 \sin (2000t + 5^\circ)$ ampere, the value of capacitance must be
(*a*) 40 μF (*b*) 80 μF (*c*) 120 μF (*d*) 160 μF.

**229.** A series circuit has $R = 8$ ohms and $C = 30$ μF. At what frequency will the current lead the voltage by 30°?
(*a*) $f = 50$ Hz (*b*) $f = 500$ Hz (*c*) $f = 655$ Hz (*d*) $f = 1155$ Hz.

**230.** A 25 microfarad capacitor is in series with a resistor $R$ at a frequency of 60 Hz. The resulting current leads the voltage by 45°. Which of the following conclusion is not necessarily valid ?
(*a*) $X_C = 106$ ohms (*b*) $X_C = R$
(*c*) $R = 106$ ohms (*d*) capacitor is electrolytic.

**231.** If $i_1 = 50 \cos (\omega t + 75^\circ)$ amperes
$i_2 = 35.4 \cos (\omega t + 120^\circ)$ amperes.
The value of $(i_1 - i_2)$ is
(*a*) $25 \cos \omega t$ (*b*) $25 \cos (\omega t + 30^\circ)$
(*c*) $25 \sqrt{2} \cos (\omega t + 30^\circ)$ (*d*) $25 \sqrt{3} \cos (\omega t + 45^\circ)$.

**132.** The sum of three currents
$$i_1 = 32.6 \sin (\omega t - 145^\circ)$$
$$i_2 = 32.6 \sin (\omega t - 25^\circ)$$
$$i_3 = 32.6 \sin (\omega t + 95^\circ)$$
is equal to
(*a*) zero (*b*) $41.8 + j\,0.09$ (*c*) $41.8 - j\,0.09$ (*d*) $j\,0.09$ A.

**233.** For the circuit shown in Fig. 9.87 the ammeter will read
(*a*) 1.19 A (*b*) 5.53 A (*c*) 7.97 A (*d*) 11.9 A.

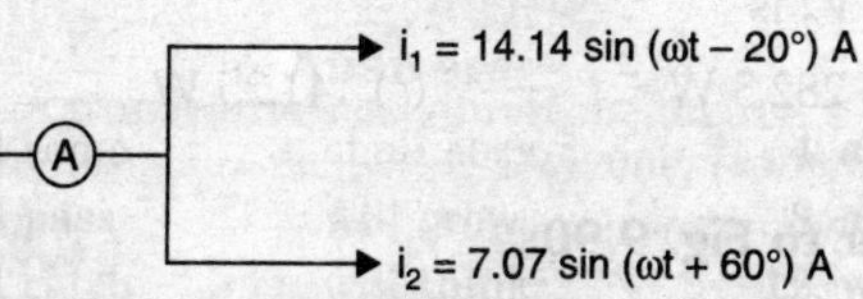


***FIGURE 9.87***

**234.** For the series circuit shown in Fig. 9.88 the effective value of the current is 5 amperes. What will be the reading of the voltmeter placed across the circuit?
(*a*) 10 V (*b*) 14.14 V (*c*) 20 V (*d*) 30 V.

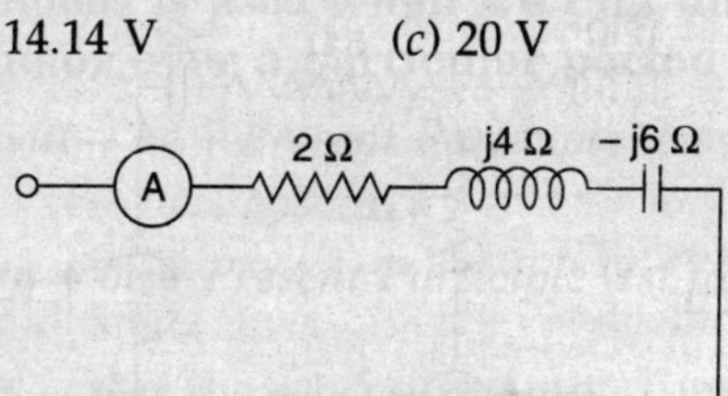


***FIGURE 9.88***

235. The voltage across the resistor will be

(a) 10 V (b) 15 V (c) 20 V (d) 30 V.

236. The voltage across the inductor will be

(a) 10 V (b) 14.14 V (c) 15 V (d) 20 V.

237. The voltage across the capacitor will be

(a) 10 V (b) 15 V (c) 20 V (d) 30 V.

**Questions 238 to 240 Refer to Fig. 9.89:**

238. Power dissipated in three resistors will be

(a) 471 W, 47.1 W, 4.71 W (b) 47.1 W, 471 W, 47.1 W

(c) 471 W, 47.1 W, 471 W (d) 471 W in each.

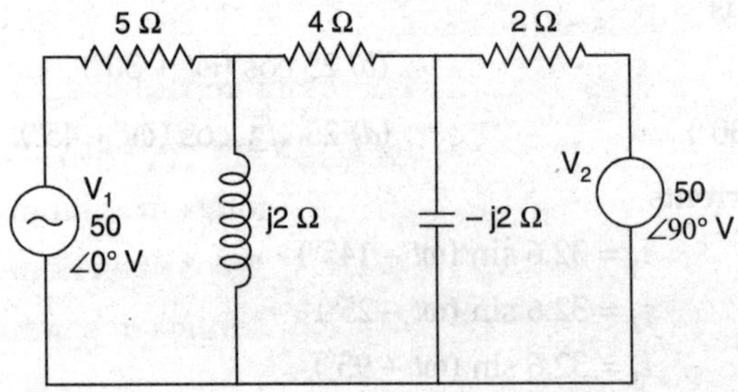


FIGURE 9.89

239. The power supplied by $V_1$ is

(a) 102 W (b) 211 W (c) 422 W (d) 844 W.

240. The power supplied by $V_2$ is

(a) 565 W (b) 282.5 W (c) 141.25 W (d) 70.75 W.

**Questions 241 and 242 Refer to Fig. 9.90:**

In the circuit shown in Fig. 9.90, $V_1$ and $V_2$ are two identical sources of $10 \angle 90°$.

241. The power supplied by $V_1$ is

(a) 6 W (b) 8.8 W (c) 11 W (d) 16 W.

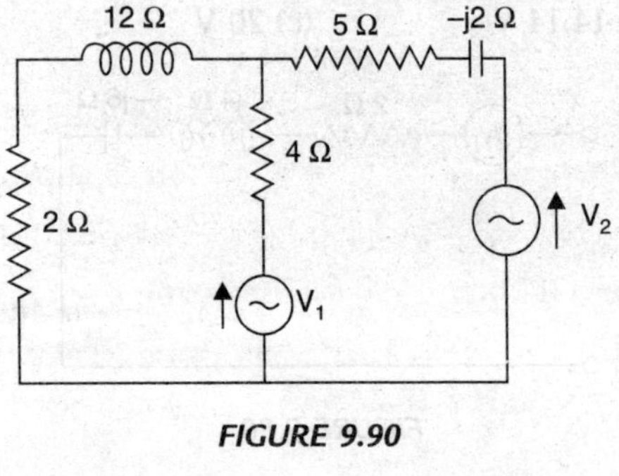


FIGURE 9.90

242. The power supplied by $V_2$ is

(a) 3.3 W (b) 6.6 W (c) 9.34 W (d) 11.11 W.

**Questions 243 to 245 Refer to Fig. 9.91:**

243. In the network shown in Fig. 9.91 the input impedance as seen by the 50 V source will be

(a) 20 Ω (b) 15 Ω (c) 10 Ω (d) 5 Ω.

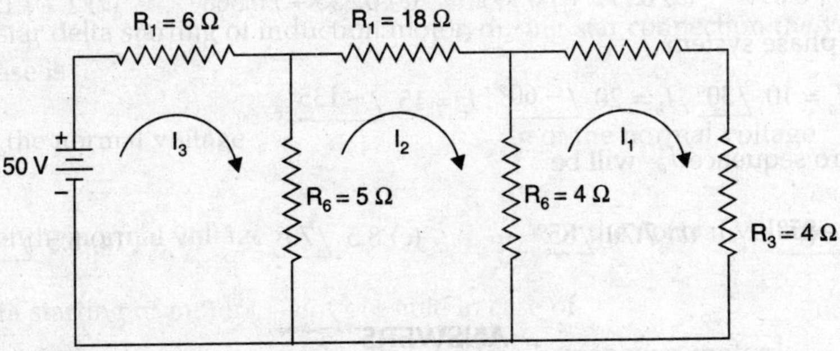


FIGURE 9.91

244. The mesh current $I_2$ will be

(a) 0.5 A (b) 1 A (c) 2 A (d) 3 A.

245. The mesh current $I_2$ will be

(a) 0.5 A (b) 0.75 A (c) 1 A (d) 1.25 A.

**Questions 246 to 248 Refer to Fig. 9.92:**

246. The power output of the voltage source shown in Fig. 9.92 will be

(a) 50 W (b) 100 W (c) 140 W (d) 280 W.

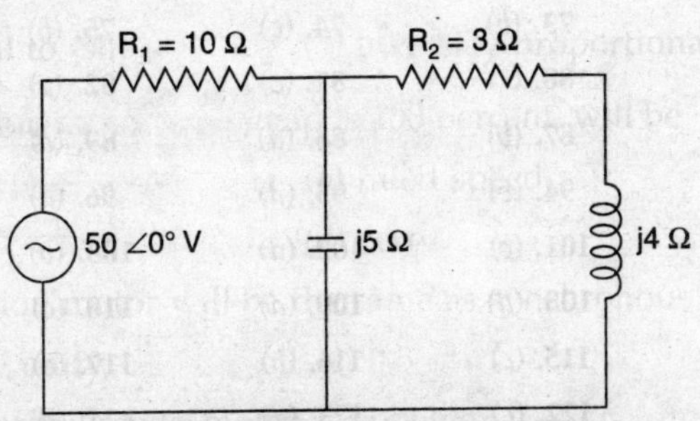


FIGURE 9.92

247. The power in 10 ohm resistor will be

(a) 50 W (b) 80 W (c) 100 W (d) 133 W.

248. The power in 3 ohm resistor will be

(a) 50 W (b) 60 W (c) 75 W (d) 100 W.

249. In a mesh connected system the impedances are

$$Z_{ab} = 1 + j\,2$$
$$Z_{bc} = 2 + j\,3$$
$$Z_{ca} = 3 + j\,4$$

Then for equivalent star connected system which of the following will not be one of the impedances?

(a) 0.514 + j 0.09 (b) 0.777 + j 0.991 (c) 0.333 + j 0.666 (d) 1 + j 1.33.

250. In a three phase system

$$I_a = 10 \angle 30°, I_b = 20 \angle -60°, I_c = 15 \angle -135°.$$

For the zero sequence $I_{a_0}$ will be

(a) 7.70 $\angle -85°$ (b) 7.70 $\angle 85°$ (c) 8.5 $\angle 70°$ (d) 8.5 $\angle -70°$.

## ANSWERS

| | | | | | | |
|---|---|---|---|---|---|---|
| 1. (c) | 2. (c) | 3. (c) | 4. (d) | 5. (d) | 6. (c) | 7. (d) |
| 8. (a) | 9. (c) | 10. (a) | 11. (d) | 12. (c) | 13. (c) | 14. (d) |
| 15. (d) | 16. (a) | 17. (c) | 18. (c) | 19. (d) | 20. (c) | 21. (d) |
| 22. (a) | 23. (c) | 24. (d) | 25. (c) | 26. (c) | 27. (d) | 28. (a) |
| 29. (c) | 30. (c) | 31. (d) | 32. (d) | 33. (b) | 34. (c) | 35. (d) |
| 36. (d) | 37. (a) | 38. (c) | 39. (d) | 40. (b) | 41. (a) | 42. (c) |
| 43. (c) | 44. (b) | 45. (d) | 46. (a) | 47. (b) | 48. (d) | 49. (a) |
| 50. (b) | 51. (c) | 52. (b) | 53. (c) | 54. (c) | 55. (a) | 56. (a) |
| 57. (d) | 58. (a) | 59. (d) | 60. (c) | 61. (c) | 62. (d) | 63. (b) |
| 64. (a) | 65. (d) | 66. (c) | 67. (c) | 68. (c) | 69. (b) | 70. (a) |
| 71. (d) | 72. (c) | 73. (b) | 74. (c) | 75. (b) | 76. (d) | 77. (d) |
| 78. (d) | 79. (a) | 80. (c) | 81. (c) | 82. (a) | 83. (b) | 84. (d) |
| 85. (d) | 86. (c) | 87. (b) | 88. (d) | 89. (c) | 90. (c) | 91. (c) |
| 92. (c) | 93. (a) | 94. (c) | 95. (d) | 96. (a) | 97. (c) | 98. (b) |
| 99. (c) | 100. (d) | 101. (c) | 102. (d) | 103. (b) | 104. (a) | 105. (a) |
| 106. (c) | 107. (c) | 108. (d) | 109. (d) | 110. (c) | 111. (a) | 112. (d) |
| 113. (d) | 114. (c) | 115. (c) | 116. (b) | 117. (c) | 118. (a) | 119. (c) |
| 120. (b) | 121. (a) | 122. (c) | 123. (a) | 124. (d) | 125. (a) | 126. (c) |
| 127. (d) | 128. (b) | 129. (c) | 130. (d) | 131. (a) | 132. (d) | 133. (a) |
| 134. (c) | 135. (b) | 136. (d) | 137. (d) | 138. (c) | 139. (a) | 140. (c) |
| 141. (d) | 142. (c) | 143. (a) | 144. (d) | 145. (c) | 146. (d) | 147. (d) |
| 148. (c) | 149. (a) | 150. (b) | 151. (a) | 152. (a) | 153. (a) | 154. (b) |
| 155. (a) | 156. (a) | 157. (a) | 158. (a) | 159. (d) | 160. (b) | 161. (c) |
| 162. (c) | 163. (b) | 164. (c) | 165. (a) | 166. (a) | 167. (b) | 168. (a) |

| | | | | | | |
|---|---|---|---|---|---|---|
| **169.** (*c*) | **170.** (*a*) | **171.** (*d*) | **172.** (*c*) | **173.** (*a*) | **174.** (*c*) | **175.** (*c*) |
| **176.** (*d*) | **177.** (*b*) | **178.** (*c*) | **179.** (*c*) | **180.** (*d*) | **181.** (*a*) | **182.** (*c*) |
| **183.** (*d*) | **184.** (*d*) | **185.** (*c*) | **186.** (*a*) | **187.** (*a*) | **188.** (*b*) | **189.** (*c*) |
| **190.** (*a*) | **191.** (*d*) | **192.** (*d*) | **193.** (*d*) | **194.** (*c*) | **195.** (*b*) | **196.** (*d*) |
| **197.** (*a*) | **198.** (*d*) | **199.** (*c*) | **200.** (*c*) | **201.** (*b*) | **202.** (*b*) | **203.** (*d*) |
| **204.** (*d*) | **205.** (*a*) | **206.** (*b*) | **207.** (*c*) | **208.** (*b*) | **209.** (*c*) | **210.** (*d*) |
| **211.** (*c*) | **212.** (*d*) | **213.** (*d*) | **214.** (*c*) | **215.** (*c*) | **216.** (*d*) | **217.** (*d*) |
| **218.** (*a*) | **219.** (*c*) | **220.** (*d*) | **221.** (*b*) | **222.** (*b*) | **223.** (*c*) | **224.** (*d*) |
| **225.** (*c*) | **226.** (*a*) | **227.** (*c*) | **228.** (*d*) | **229.** (*d*) | **230.** (*d*) | **231.** (*c*) |
| **232.** (*d*) | **233.** (*d*) | **234.** (*b*) | **235.** (*a*) | **236.** (*b*) | **237.** (*d*) | **238.** (*a*) |
| **239.** (*c*) | **240.** (*a*) | **241.** (*c*) | **242.** (*c*) | **243.** (*c*) | **244.** (*a*) | **245.** (*c*) |
| **246.** (*c*) | **247.** (*b*) | **248.** (*b*) | **249.** (*b*) | **250.** (*a*). | | |

# Chapter 10 ALTERNATORS

Alternators operate on the principle of electromagnetic induction. Thus alternators have armature winding and a magnetic field. In this case armature windings are mounted on stationary part called stator and field windings are provided on rotating element called rotor.

The stator is usually a cast-iron frame which supports the armature core having slots on its inner periphery which house the armature conductors. The poles for field are carried on rotor. These are alternate *N–S* poles. These poles are excited by direct current supplied through an exciter or an outside source. The current is supplied to rotating poles through two slip rings.

A prime mover supplies power for rotating the rotor. As a result emf is generated in the stator conductors. The frequency of emf generated depends in rotor speed.

Two types of rotors are used on alternators

(*i*) Salient pole type

(*ii*) Smooth cylindrical type.

Salient or projecting pole type rotors are used on low and medium speed alternators. In this large number of poles are bolted to the rotor. The poles and pole shoes are usually laminated to minimize heating due to eddy currents.

Smooth cylindrical type rotor is used with high speed prime movers such as steam or gas turbines. In this the rotor has smooth solid forged steel cylinder having a number of slots milled out at intervals for accommodating field coils.

**Demper Winding.** Generally alternators have slotted pole shoes for receiving copper bars of a grid or damper windings. These copper bars are short circuited at both ends by heavy copper rings. These dampers assist in preventing the hunting in generators.

**EMF Generated.** Alternators usually run at synchronous speeds. The frequency of emf generated is given by

$$f = \frac{PN}{120} \text{ Hz.}$$

where $N$ = Synchronous speed

$P$ = Number of magnetic poles.

| *No. of poles* | 2 | 4 | 6 | 12 | 24 | 36 |
|---|---|---|---|---|---|---|
| *Speed RPM* | 3000 | 1500 | 1000 | 500 | 250 | $166\frac{2}{3}$ |

Average emf per phase = $4 f \phi T$ volts

RMS value of emf/phase = $1.11 \times 4 \phi T$

$= 4.44 f \phi T$ volts

where $f$ is frequency of induced emf in Hz

$\phi$ flux per pole in webers

$$T = \frac{\text{No. of conductors in series per phase}}{2}$$

Actually available voltage/phase

$= 4.44\, k_c\, kd\, f \phi T$

where $kf$ = form factor = 1.11 for sinusoidal emf

$kd$ = distribution factor

$$= \frac{\sin \frac{m\beta}{2}}{m \sin \frac{\beta}{2}}$$

$k_c$ = pitch or coil span factor

$$= \cos \frac{\alpha}{2}.$$

**Pitch Factor.** When a coil spans over 180° (electrical), it is known as full pitched coil. Usually short pitched coils are used on alternators for saving in copper and improving the waveform of the generated emf. However, there is disadvantage of using short pitched coils as the total voltage around the coil is somewhat reduced. For this pitch factor is used which is defined as

$$\text{Pitch factor} = \frac{\text{Vector sum of the induced emf per coil}}{\text{Arithmetic sum of the induced emf per coil}}$$

Pitch factor is always less than unity.

**Distribution Factor** is defined as

$$\text{Distribution factor} = \frac{\text{emf with distributed winding}}{\text{emf with concentrated winding}}.$$

**Voltage Regulation.** With change in load there is a change in terminal voltage of alternator. The magnitude of this change depends not only on the load but also on the load power factor. The voltage regulation of an alternator is the rise in voltage when full load is removed divided by the rated terminal voltage.

$$\text{Thus percentage regulation} = \frac{E_0 - V}{V} \times 100.$$

In this $(E_0 - V)$ is only arithmetic difference and not the vectorial difference. In case of leading load power factor, terminal voltage will fall on removing the full load. In that case regulation is said to be negative. Further it is not necessary that the rise in voltage when load is thrown off, may be the same as fall in voltage when load is put on the alternator. Various methods for determining percent regulation are given below.

1. Synchronous impedance method
2. MMF method
3. Zero power factor or potier method.

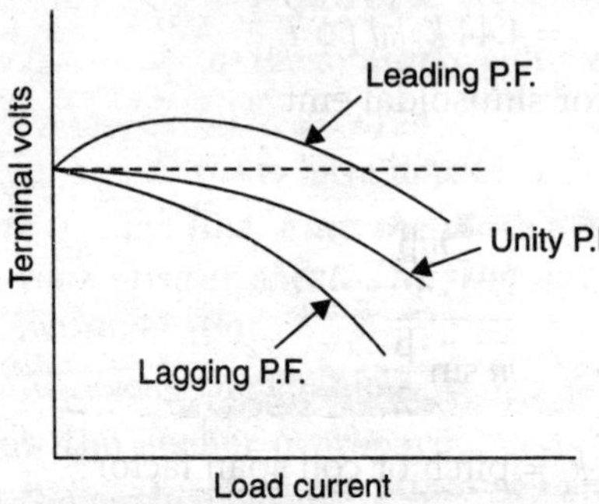


***FIGURE 10.1*** Voltage characteristics of an alternator.

All these methods require.
(*i*) Armature (or stator) resistance $R_a$.
(*ii*) Open circuit/No load characteristic.
(*iii*) Short circuit characteristic.

**Magnetisation Curve.** Graphical representation of exciting current and terminal voltage of the armature is known as magnetisation curve, shown in Fig. 10.2 while the alternator is kept in open circuit, by passing different exciting currents in the exciting windings, the corresponding armature voltages may be observed and graph plotted accordingly. For getting values at p.f. other than unity, necessary inductances are introduced.

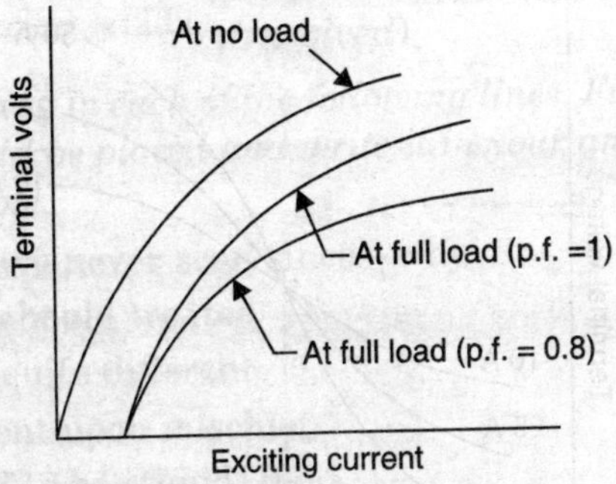


*FIGURE 10.2*

**Load Characteristics.** When exciting current and the speed remain constant, the terminal voltage changes with the load current in the armature and the relationship between the terminal voltage and load current of an laternator is known as its load characteristics. Fig. 10.3 shows the load characteristics of an alternator.

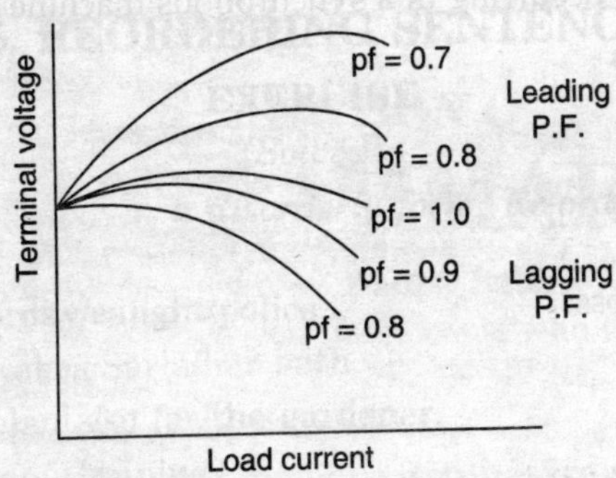


*FIGURE 10.3* Load characteristics.

It can be seen from Fig. 10.3 that when the armature current increases, the terminal voltage drops. This decrease in terminal voltage is due to many reasons, but is mainly due to

(*i*) armature reaction

(*ii*) resistance and reactance of armature winding.

In fact armature reaction weakens the magnetic field and resistance and reactance of the armature winding absorbs some of the generated voltage. The overall effect is decrease in generated voltage.

The magnitude of the effect of field weakening due to armature reaction depends primarily on the angle of lead or lag of the armature current. On leading power factor the magnetic field is strengthened instead of being weakened. The effect of power factor is shown in Fig. 10.4.

In case of three phase alternators the load should be balanced from the point of view of power factor as well as current.

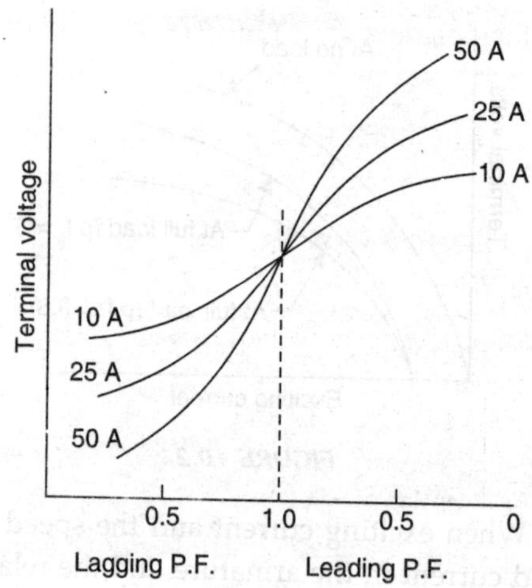


*FIGURE 10.4*

**Losses.** Various losses occurring in a synchronous machine are listed below:

**Exciting Current Loss**

(*i*) Field $I^2R$

(*ii*) Rheostat

(*iii*) Brushes

(*iv*) Exciter loss.

**Fixed Loss**

(*v*) Core loss

(*vi*) Bearing friction

(*vii*) Total windage

(*viii*) Brush friction.

**Direct Load Loss**

(*ix*) $I^2R$ loss in armature winding.

**Stray Load Loss**

(*x*) In iron

(*xi*) In conductors.

**Rating of Alternators.** When an alternator runs, losses occur and hence a rise in temperature takes place. This rise in temperature is a limiting factor. For a load drawing current $I$, the $I^2R$ loss in the armature limits the alternator output and depends upon the magnitude of the armature current $I$ and is independent of the power factor. For an alternator giving 10 kW output at unity power factor, the output will reduce to 8 kW when p.f. is 0.8 lagging. For this reason the alternators are rated in terms of kVA.

**Alternator Protection.** An alternator may require to be protected against the consequences of a variety of possible faults, including: Over-speed; Over-voltage; Over-current; unbalanced load;

insulation failure; bearing failure; prime mover failure; excessive coolant temperature, system surges and fire. The table below summarizes the protective means usually adopted:

**Methods of Alternator Protection**

| *Fault* | *Turbine and Engine driven alternators* | *Water wheel driven alternators* |
|---|---|---|
| Over-speed | Mechanical governor mounted on shaft | Mechanical governor mounted on shaft |
| Alternator over-voltage | — | Over-voltage relay |
| Over-current | Over-current relay or differential protection | Over current relay |
| Unbalanced load | Phase-unbalance network and relay, or alarm | Alarm |
| Stator insulation failure | Differential protection | Differential protection |
| Rotor insulation protection | Alarm from earth fault detector | Alarm from earth fault detector |
| Failure of prime mover | Reverse power relay | — |
| Surges | Supply network protection sufficient | Supply network protection sufficient |

## OBJECTIVE TYPE QUESTIONS

**1.** The alternating magnetic field can be split-up into two rotating fields of half its magnitude and travelling in opposite directions at synchronous speed. The principle associated with above statement is

(*a*) Feraris principle (*b*) Coulomb's law
(*c*) Faradays principle (*d*) Maxwells principle.

**2.** In three phase supply the phase angle between any two phases is

(*a*) 60° (*b*) 90° (*c*) 120° (*d*) 180°.

**3.** The speed of a 2 pole 50 Hz synchronous machine is

(*a*) 3000 rpm (*b*) 1500 rpm (*c*) 1000 rpm (*d*) 750 rpm.

**4.** The speed of a 4 pole 60 Hz synchronous machine will be

(*a*) 3600 rpm (*b*) 3000 rpm (*c*) 2400 rpm (*d*) 1800 rpm.

**5.** The speed of a *p*-pole synchronous machine in rpm is given by

(*a*) $120\,fp$ (*b*) $\sqrt{120\,fp}$ (*c*) $\dfrac{120p}{f}$ (*d*) $\dfrac{120f}{p}$.

**6.** An alternator is a

(*a*) homopolar machine (*b*) heteropolar machine
(*c*) either of (*a*) or (*b*) above (*d*) none of the above.

7. What is the largest size of alternator being manufactured in India?
(*a*) 110 MW (*b*) 210 MW (*c*) 250 MW (*d*) 500 MW.
8. Which of the following organisation is engaged in the manufacture of large size alternators for power plants, in India?
(*a*) Electricity Authority of India (*b*) National Thermal Power Corporation Ltd.
(*c*) Bharat Heavy Electricals Ltd. (*d*) Department of Science and Technology.
9. Which of the following is installed to excite the field system of large alternators in power plants?
(*a*) Batteries (*b*) Exciter (*c*) Slip rings (*d*) Governor.
10. An exciter is nothing but a
(*a*) dc shunt motor (*b*) dc series motor
(*c*) dc shunt generator (*d*) dc series generator.
11. The stator of an alternator consists of
(*a*) an iron core (*b*) stator winding
(*c*) both (*a*) and (*b*) above (*d*) none of the above.
12. Hydrogen is used in large alternators mainly to
(*a*) reduce eddy current losses (*b*) strengthen magnetic field
(*c*) cool the machine (*d*) reduce distortion of waveform.
13. For a 250 rpm 50 Hz alternator the number of poles will be
(*a*) 6 (*b*) 8 (*c*) 12 (*d*) 24.
14. An alternator coupled to which prime mover will usually have the highest rotating speed?
(*a*) Steam engine (*b*) Reciprocating diesel engine
(*c*) Francis turbine (*d*) Steam turbine.
15. In an alternator the voltage generated per phase is proportional to
(*a*) flux per pole (*b*) number of turns in the coil
(*c*) frequency of waveform (*d*) all of the above.
16. If $\phi$ be the flux, $N$ be the number of turns of the coil and $f$ be the frequency, then *RMS* value of the voltage generated per phase, in an alternator is given by
(*a*) $E_{ph} = 4 \times 1.11 \times \phi Nf$ (*b*) $E_{ph} = 4 \times 1.11 \times \phi N^2 f$
(*c*) $E_{ph} = 4 \times 1.11 \times \phi Nf^2$ (*d*) $E_{ph} = 4 \times 1.11 \times \phi N^2 f^2$.
17. A 6-pole alternator rotating at 1000 rpm has a single phase winding housed in three slots per pole, the slots in groups of three being 20° apart. If each slot contains 10 conductors and the flux per pole is $2 \times 10^{-2}$ Wb, the emf generated in case of sinusoidal waveform will be
(*a*) 191 V (*b*) 204 V (*c*) 382 V (*d*) 440 V.

**Questions 18 to 20 Refer to Fig. 10.5:**

18. In Fig. 10.5 which curve represents the open circuit voltage of an alternator?
(*a*) curve *A* (*b*) curve *B* (*c*) curve *C* (*d*) curve *D*.

19. In the above figure short circuit current is represented by

(a) curve A (b) curve B (c) curve C (d) curve D.

20. Curve A represents

(a) load current (b) short circuit voltage

(c) synchronous impedance (d) none of the above.

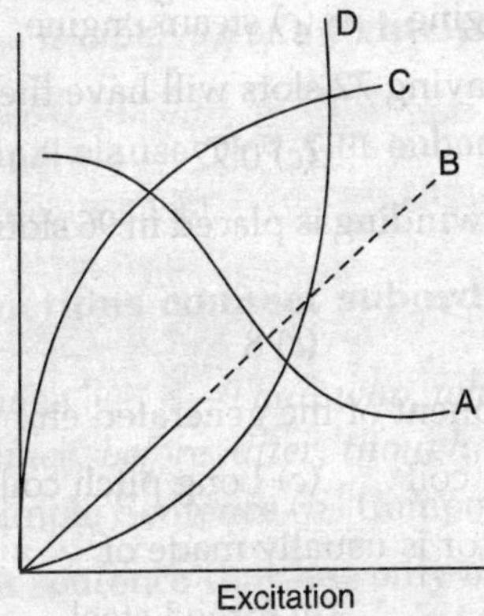


*FIGURE 10.5*

21. An alternator having 8 poles rotating at 250 rpm will have frequency of voltage generated as

(a) 60 Hz (b) 50 Hz (c) 25 Hz (d) $16\frac{2}{3}$ Hz.

22. Salient pole type alternators are generally used on

(a) high speed prime movers (b) low and medium speed prime movers

(c) hydrogen cooled prime movers (d) low voltage alternators.

23. The number of electrical degrees passed in one revolution of a six pole synchronous alternator is

(a) 120° (b) 360° (c) 720° (d) 1080°.

24. For the same kVA rating, a low voltage alternator will

(a) run at low speed (b) be large in size

(c) have better voltage regulation (d) be more efficient.

25. Turbo-alternators are generally designed to run at

(a) 15000 rpm (b) 5000 rpm (c) 3000 rpm (d) 1500 rpm.

26. The rotor preferred for alternators coupled to hydraulic turbines are

(a) salient pole type (b) cylindrical rotor type

(c) solid rotor type (d) any of the above.

27. The frequency of voltage generated in large alternators is

(a) in megacycles (b) in kilocycles (c) 60 Hz (d) 50 Hz.

28. The frequency of voltage generated in an alternator depends on
(*a*) number of poles (*b*) rotative speed
(*c*) both number of poles and rotative speed
(*d*) number of poles, rotative speed and kVA rating.

29. Which of the following prime mover is least efficient?
(*a*) gas turbine (*b*) petrol engine (*c*) steam engine (*d*) diesel engine.

30. A three phase, 8 pole alternator having 72 slots will have the distribution factor of
(*a*) 0.96 (*b*) 0.88 (*c*) 0.72 (*d*) 0.64.

31. For a 3 phase 4 pole alternator the winding is placed in 96 slots. The number of slots per pole per phase will be
(*a*) 2 (*b*) 4 (*c*) 8 (*d*) 16.

32. In which coil the harmonic component of the generated emf will be more?
(*a*) Short pitch coil (*b*) Full pitch coil (*c*) Long pitch coil (*d*) Same in all coils.

33. In case of turbo alternators the rotor is usually made of
(*a*) cast iron (*b*) forged steel
(*c*) laminated stainless steel (*d*) manganese steel.

34. Windage losses in an alternator
(*a*) vary directly with load current (*b*) vary directly with generated emf
(*c*) vary directly with (volt-amperes) (*d*) remain constant.

35. In case a three phase alternator supplies resistive load
(*a*) the armature reaction flux will be almost zero
(*b*) the armature reaction flux will be along the axis of the field
(*c*) the armature reaction flux will be at 90° with the field axis
(*d*) the armature reaction flux will be at 180° with the field axis.

36. In an alternator the eddy current and hysteresis losses occur in
(*a*) iron of field structure only (*b*) iron of armature structure only
(*c*) both (*a*) and (*b*) above (*d*) none of the above.

37. A 12 pole alternator will pass through how many electrical degrees in one complete revolution?
(*a*) 60° (*b*) 360° (*c*) 1080° (*d*) 2160°.

38. In which of the following case the waveform generated will be closest to sine waveform?
(*a*) distributed winding with short pitch coils
(*b*) distributed winding with full pitch coils
(*c*) concentrated winding with full pitch coils.

39. In an alternator for the same flux distribution and same number of turns, in which of the following coil the emf generated will be more thant the remaining?
(*a*) full pitch coils (*b*) short pitch coils (*c*) long pitch coils

**40.** The number of poles in turbo alternators is usually

(*a*) 2 (*b*) 4 (*c*) 12 (*d*) 50.

**41.** In salient pole alternators

(*a*) internal diameter of the stator is very large as compared to the axial length of the stator core

(*b*) internal diameter of the stator is equal to the axial length of the stator core

(*c*) internal diameter of the stator core is less than the axial length of the stator core.

**42.** A three phase alternator has 3 slots per pole per phase. Its breadth factor will be

(*a*) 0.760 (*b*) 0.800 (*c*) 0.850 (*d*) 0.960.

**43.** A winding having 24 slots, 4 poles with coil span of 1 to 6, will have pitch factor as

(*a*) 0.9659 (*b*) 0.8655 (*c*) 0.7355 (*d*) 0.6678.

**Questions 44 to 46 Refer to Data Given Below:**

A 3 phase, 50 Hz, 10 pole alternator has 90 slots. The flux per pole is 0.16 Wb. The winding is to be star connected to give a line voltage of 11000 volts.

**44.** The number of slots per pole-per phase is

(*a*) 2 (*b*) 3 (*c*) 4 (*d*) 5.

**45.** The breadth factor will be

(*a*) 0.9601 (*b*) 0.9061 (*c*) 0.9016 (*d*) 0.8861.

**46.** The number of armature conductors in series per phase required for the armature will be

(*a*) 2124 (*b*) 2672 (*c*) 3030 (*d*) 3726.

**47.** The rating of alternators is usually expressed in terms of

(*a*) full load current (*b*) rated voltage (*c*) kVA (*d*) kW.

**Questions 48 to 50 Refer to Data Given Below:**

A 3 phase 50 Hz 6600 V alternator is rated at 6600 kW at 0.8 power factor and full load efficiency of 90%.

**48.** The kVA rating of the alternator will be

(*a*) 750 kVA (*b*) 6600 kVA (*c*) 7200 kVA (*d*) 7500 kVA.

**49.** Current rating of the alternator is

(*a*) 4530 A (*b*) 4845 A (*c*) 5560 A (*d*) 6563 A.

**50.** The power input to alternator, in kW will be

(*a*) 6563 kW (*b*) 6667 kW (*c*) 6980 kW (*d*) 7777 kW.

**51.** Which harmonic will be totally eliminated in an alternator by using a fractional pitch of 4/5 ?

(*a*) ninth (*b*) seventh (*c*) fifth (*d*) third.

52. Dirt accumulation in generators can cause all of the following EXCEPT
(*a*) low power factor (*b*) poor voltage regulation
(*c*) overheating (*d*) flashovers.

53. Three phase alternators are usually star connected
(*a*) to obtain higher terminal voltage (*b*) to reduce windage losses
(*c*) to reduce conductor size (*d*) to save on copper.

54. The torque angle for an alternator is usually
(*a*) 0.1° to 0.3° (*b*) 1° to 3° (*c*) 10° to 30° (*d*) 30 ° to 60°.

55. If δ be the torque angle of an alternator, then for stable operation
(*a*) $\delta = 0$ (*b*) $\frac{d\delta}{dt} = 0$ (*c*) $\frac{d^2\delta}{dt^2} = 0$ (*d*) $\delta + \delta^2 + \delta^3 + ..... = 0$.

56. In an alternator is operating at leading power factor, then it can be concluded that
(*a*) the alternator is over-excited (*b*) the alternator is under-excited
(*c*) the torque angle of the alternator has negative value
(*d*) the residual magnetism of the poles is zero.

57. Which fractional pitch will eliminate the seventh harmonic from the voltage waveform of an alternator?
(*a*) 7/8 (*b*) 6/7 (*c*) 5/6 (*d*) none of the above.

58. Damper winding on alternators results in all of the following EXCEPT
(*a*) elimination of harmonic effects
(*b*) suppression of spontaneous hunting when supplying power to transmission line with high resistance to reactance ratio
(*c*) absorption of energy of oscillations when operating in parallel with other alternators
(*d*) increases instability of machine.

59. Short circuit ratio for turbo-alternators is usually
(*a*) 0.5 (*b*) between 0.5 and 0.7
(*c*) between 1.0 and 1.5 (*d*) between 2 and 3.

60. Low SCR of an alternator will result in all of the following EXCEPT
(*a*) poor regulation (*b*) lower stability limit
(*c*) easy parallel operation.

61. In a cylindrical rotor how much portion of the rotor is wound?
(*a*) Full (*b*) Two third (*c*) Half (*d*) One third.

62. On large alternators pilot exciter is used for
(*a*) battery charging (*b*) driving dc auxiliaries
(*c*) exciting field windings of main alternator
(*d*) exciting field windings of main excitor.

**63.** Breadth factor is given by

(*a*) $\dfrac{\sin m\frac{\psi}{2}}{m\sin\frac{\psi}{2}}$ (*b*) $\dfrac{\sin m\psi}{\sin\frac{m\psi}{2}}$ (*c*) $\dfrac{m\sin\frac{\psi}{2}}{\sin\frac{m\psi}{2}}$ (*d*) $\dfrac{m\sin\frac{m\psi}{2}}{\sin\frac{m\psi}{2}}$.

**64.** Dampers in large generators

(*a*) increase stability (*b*) reduce voltage fluctuations

(*c*) reduce frequency fluctuations (*d*) none of the above.

**65.** The permissible voltage variation for domestic power supply is

(*a*) 0.1% (*b*) 0.5% (*c*) 1% (*d*) 10%.

**66.** The speed of an alternator is changed from 3000 rpm to 1500 rpm. The generated emf/ phase will become

(*a*) double (*b*) unchanged (*c*) half (*d*) one fourth.

**67.** Zero power factor method for an alternator is generally used to determine

(*a*) efficiency of alternator (*b*) synchronous impedance of alternator

(*c*) voltage regulation of the alternator (*d*) all of the above.

**68.** When θ is the angle in electrical degrees between any point on the pole face and the centre of the pole, then the waveform generated will be more closer to sinusoidal wave when the air gap is proportional to

(*a*) $\sin\theta$ (*b*) $\cos\theta$ (*c*) $\dfrac{1}{\sin\theta}$ (*d*) $\dfrac{1}{\cos\theta}$.

**69.** In an alternator the armature reaction is mainly influenced by

(*a*) the power factor of the load (*b*) speed of the alternator

(*c*) SCR (*d*) total current drawn.

**70.** The armature reaction of an alternator will be cross-magnetizing when the power factor of the load is

(*a*) zero leading (*b*) zero lagging

(*c*) unity.

**71.** The effect of cross-magnetizing in an alternator can be reduced by

(*a*) using interpoles (*b*) shifting brush position

(*c*) reducing rotor speed (*d*) improved cooling.

**72.** When two alternators are to be operated in parallel, the blank lamp test is performed to ensure

(*a*) proper frequency matching (*b*) proper phase sequence

(*c*) proper voltage matching (*d*) all of the above.

**73.** In an alternator the effect of leakage flux is to make the winding

(*a*) inductive (*b*) resistive (*c*) capacitive (*d*) none of the above.

**74.** The Potier's triangle is used to separate

(*a*) iron losses and copper losses (*b*) copper losses and windage losses

(*c*) field mmf and armature mmf

(*d*) armature leakage reactance and armature reaction mmf.

**75.** For getting zero power factor characteristics for the Potier's diagram, the alternator is loaded by

(*a*) lamp load (*b*) brake load

(*c*) hydraulic dynamometer (*d*) synthronous motor.

**76.** The driving power from the primemover driving the alternator is lost but the alternator remains connected to the supply network and field supply is on. The alternator will

(*a*) get burnt (*b*) behave as an induction motor

(*c*) behave as a synchronous motor but will rotate in reverse direction to that corresponding to generator action.

(*d*) behave as a synchronous motor and will rotate in the same direction

(*e*) behave as an induction motor but will rotate in opposite direction.

**77.** If the input to the primemover of an alternator is constant but the excitation is changed, then the

(*a*) reactive component of the output is changed

(*b*) active component of the output is changed

(*c*) pf of the load remains constant.

**78.** Which of the following has significant influence on the power factor of an alternator?

(*a*) horsepower of primemover (*b*) speed of alternator

(*c*) connected load (*d*) excitation.

**79.** An alternator is supplying a load alone. The driving torque of the alternator is increased while excitation remains constant. It will result in

(*a*) leading power factor (*b*) unity power factor

(*c*) power factor less lagging (*d*) none of the above.

**80.** An alternator is connected to infinite bus bars. Its excitation is increased while its driving power remains constant. This is likely to result in

(*a*) more lagging power factor (*b*) less lagging power factor

(*c*) terminal voltage drop (*d*) none of the above.

**81.** In which case the terminal voltage rise of an alternator will be more?

(*a*) when unity power factor load is thrown off

(*b*) when leading power factor load is thrown off

(*c*) when lagging power factor load is thrown off.

**82.** An exciter for a large alternator is driven by

(*a*) same shaft as that of the alternator (*b*) belt drive from main shaft

(*c*) gear drive from main shaft (*d*) any of the above.

83. Which of the following load offers nearly unity power factor?
(*a*) tube light (*b*) induction-motor (*c*) arc furnace (*d*) GLS lamps.

84. The speed of primemover for an alternator is maintained constant by
(*a*) filters (*b*) coupling (*c*) brakes (*d*) governor.

85. Two alternators *A* and *B* are running at synchronous speed generating identical voltages. If alternator *A* falls in speed by a small amount δ. This causes resultant voltage $E_{syn}$ in the local circuit. The synchronizing current will be
(*a*) lagging $E_{syn}$ by nearly 90° (*b*) leading $E_{syn}$ by nearly 45°
(*c*) leading $E_{syn}$ by exactly 90° (*d*) in phase with $E_{syn}$.

86. If two alternators are running in proper synchronism and the voltage of one machine is suddenly increased
(*a*) the machine will burn (*b*) both machines will stop
(*c*) synchronizing torque will be produced to restore further synchronism.

87. Two alternators *A* and *B* are running in parallel. What will happen when the excitation of *A* is increased?
(*a*) alternator *B* will burn out (*b*) the wattless component will change
(*c*) power output will reduce (*d*) the machines will stop.

88. Which of the following is not an essential condition for parallel operation of two alternators ?
(*a*) identical terminal voltage for two machines
(*b*) same phase sequence for two machines
(*c*) same kVA rating for two machines
(*d*) same frequency of voltage generated.

89. Two alternators *A* and *B* have the same percentage impedance. They will share any additional load
(*a*) equally (*b*) in proportion to their power factor
(*c*) in proportion to their kVA ratings (*d*) in proportion to their SCR.

90. Two alternators *A* and *B* are running in parallel. Alternator *A* has its steam supply increased while its excitation remains unchanged.
This is likely to result in
(*a*) alternator *A* sharing more load (*b*) alternator *B* sharing more load
(*c*) power factor of *A* will become more lagging
(*d*) the speed of *A* will increase.

91. Two alternators *A* and *B* are sharing a resistive load equally. When the excitation of alternator *A* is increased
(*a*) power factor of *A* will become lagging and that of *B* will become leading
(*b*) power factor of *A* will become leading and that of *B* will become lagging
(*c*) power factor of both will remain unity
(*d*) power factor of both will remain identical and lagging.

92. Which of the following method is preferred for determination of voltage regulation of an alternator?

(*a*) MMF method (*b*) Synchronous impedance method
(*c*) EMF method (*d*) Potier triangle method.

93. Which of the following is the cheapest material?

(*a*) Cast irot (*b*) Aluminium (*c*) Copper (*d*) Mica.

94. Which of the following material has the highest scrap value as compared to its initial cost?

(*a*) silicon steel (*b*) copper (*c*) carbon (*d*) mica.

95. The armature reaction of an alternator influences

(*a*) windage losses (*b*) operating speed (*c*) waveform (*d*) emf/phase.

96. Superthermal power stations are being installed at all of the following places EXCEPT

(*a*) Singrauli (*b*) Farraka (*c*) Narora (*d*) Ramagundam.

97. Alternators used for electric power supply system on aircraft have frquency of

(*a*) $16\frac{2}{3}$ Hz (*b*) 25 Hz (*c*) 50 Hz (*d*) 400 Hz.

98. High frequency power lines on aircraft are used

(*a*) to reduce radio interference (*b*) to reduce bulk size alternator
(*c*) to minimise power consumption (*d*) to eliminate need for earthing.

99. Two alternators *A* and *B* are sharing an inductive load equally. What will happen when the excitation of *A* is increased?

(*a*) current delivered by *A* will increase while that by *B* will decreases
(*b*) current delivered by *A* will decrease while that by *A* will increase
(*c*) both will continue to share load equally.

100. Negative regulation can be expected in case of

(*a*) slow speed alternators (*b*) high speed alternators
(*c*) leading power factor load (*d*) lagging power factor load.

## ANSWERS

| | | | | | | |
|---|---|---|---|---|---|---|
| 1. (*a*) | 2. (*c*) | 3. (*a*) | 4. (*d*) | 5. (*d*) | 6. (*b*) | 7. (*d*) |
| 8. (*c*) | 9. (*b*) | 10. (*c*) | 11. (*c*) | 12. (*c*) | 13. (*d*) | 14. (*d*) |
| 15. (*d*) | 16. (*a*) | 17. (*c*) | 18. (*c*) | 19. (*b*) | 20. (*c*) | 21. (*d*) |
| 22. (*b*) | 23. (*d*) | 24. (*b*) | 25. (*c*) | 26. (*a*) | 27. (*d*) | 28. (*c*) |
| 29. (*c*) | 30. (*a*) | 31. (*c*) | 32. (*b*) | 33. (*b*) | 34. (*d*) | 35. (*c*) |
| 36. (*b*) | 37. (*d*) | 38. (*a*) | 39. (*a*) | 40. (*a*) | 41. (*a*) | 42. (*d*) |
| 43. (*a*) | 44. (*b*) | 45. (*a*) | 46. (*d*) | 47. (*c*) | 48. (*d*) | 49. (*d*) |
| 50. (*b*) | 51. (*c*) | 52. (*a*) | 53. (*d*) | 54. (*b*) | 55. (*b*) | 56. (*a*) |

**57.** (*b*) **58.** (*d*) **59.** (*b*) **60.** (*c*) **61.** (*b*) **62.** (*d*) **63.** (*a*)
**64.** (*a*) **65.** (*d*) **66.** (*c*) **67.** (*c*) **68.** (*d*) **69.** (*a*) **70.** (*c*)
**71.** (*b*) **72.** (*a*) **73.** (*a*) **74.** (*d*) **75.** (*d*) **76.** (*d*) **77.** (*a*)
**78.** (*c*) **79.** (*c*) **80.** (*a*) **81.** (*c*) **82.** (*d*) **83.** (*d*) **84.** (*d*)
**85.** (*a*) **86.** (*c*) **87.** (*b*) **88.** (*c*) **89.** (*c*) **90.** (*a*) **91.** (*a*)
**92.** (*d*) **93.** (*a*) **94.** (*b*) **95.** (*d*) **96.** (*c*) **97.** (*d*) **98.** (*b*)
**99.** (*a*) **100.** (*c*).

# Chapter 11 SYNCHRONOUS MOTORS

A synchronous motor runs at synchronous speed. These motors are inherently non-self starting. In these machines, stator, like that of the induction motor, produces a magnetic field rotating at synchronous speed. The poles on the (salient pole) rotor, excited by direct current in their field windings undergo magnetic attraction by the stator poles and are dragged round thereby. On no load the pole axes are practically coincident. When a retarding torque is applied to the shaft, the rotor tends to fall behind. In doing so the attraction of the stator on the rotor becomes tangential to an extent sufficient to develop a counter torque. The maximum load is that which retards the rotor so that the tangential attraction is a maximum. When the load is increased above this amount, the rotor is too much retarded and the attraction falls, the rotor then pulling out of step.

**Synchronous Speed.** For a multipolar machine the synchronous speed is given by

$$N = \frac{120f}{P}$$

where $f$ = frequency

$P$ = number of poles

$f$ = 50 Hz

| *No. of poles* | 2 | 4 | 6 | 8 | 12 | 24 |
|---|---|---|---|---|---|---|
| *Synchronous speed* | 3000 | 1500 | 1000 | 750 | 500 | 250 |

**Synchronous Reactance.** This is a concept applicable to synchronous motors. The resultant linkage of flux with any phase of the armature of a synchronous machine is due to the combined action of the field and armature currents. For a simple treatment, it is convenient to separate the resultant flux into components.

(*i*) the field flux due to field current alone, and

(*ii*) the armature flux due to armature current alone.

This separation does not affect qualitative matters but its quantitative validity rests on the assumption that the magnetic circuit has a constant permeability. In brief the simplifying assumptions are:

(*i*) All parts of the magnetic circuit have constant permeability so that the field and armature fluxes can be treated separately as proportional to their respective currents and their effects superposed.

(*ii*) The air gap is uniform, so that the armature flux is not affected by its position relative to the poles.

(*iii*) The field flux is sinusoidally distributed in the air gap.

(*iv*) The armature winding is uniformly distributed and carries balanced sinusoidal currents.

In synchronous motor, leakage reactance results from real self-flux linking the armature slot and overhang conductors, producing a corresponding volt-drop $I_x$. It operates during transients.

If stator resistance and leakage reactance are included, the synchronous machine may be depicted by the equivalent circuit of Fig. 11.1.

In this the resistance $r$, the true (leakage) reactance $x$ and the fictitious reactance $x_a$ for armature, reaction are separated from the 'ideal' machine in which the emf $E_t$ is induced by the flux produced field ampere-turns alone.

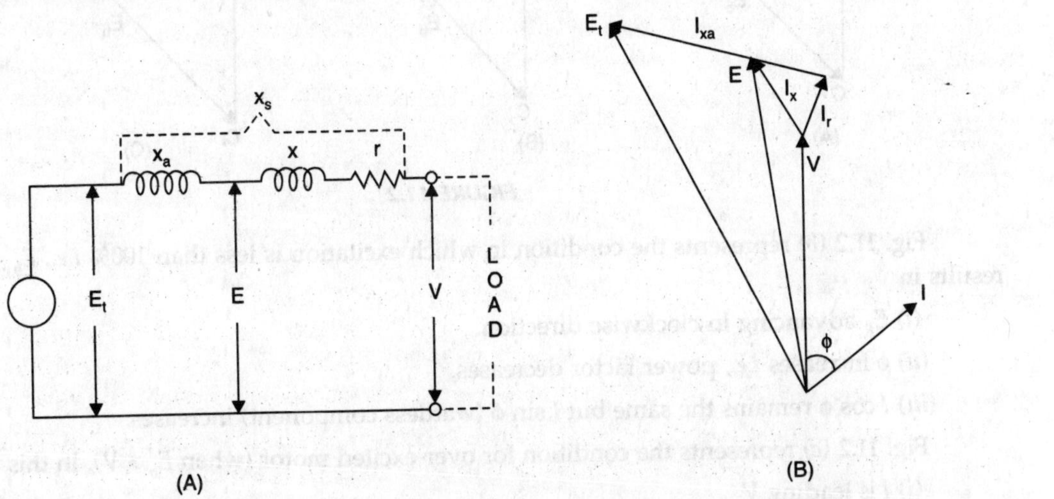


*FIGURE 11.1*

In $x_a$ a drop $Ix_a$ occurs to take account of armature reaction. The net emf is $E$. The terminal voltage is less than $t$ by the further complexor drops in resistance and leakage reactance, $I_r$ and $I_x$. Obviously $V$ can be obtained directly from $E_t$ and $I$ if $r$, $x$ and $x_a$ be combined to form an impedance $Z_s$, the components of which are $r$ and $(x + x_a) = x_s$. The reactance $x_s$ is termed the synchronous reactance and $z_s$ is the synchronous impedance.

In large machines, the leakage reactance is of the order 0.10 per unit and the resistance 0.01 per unit. The reactance $x_a$ is in general very much larger and the synchronous reactance $x_s$ may be as much as 1.75 per unit.

**Effect of excitation.** Whenever back emf $E_b$ of motor is equal in magnitude to the applied voltage $V$, the excitation of motor is said to be 100%. Fig. 11.2 (*a*) shows a synchronous motor with 100% excitation. In this

$$E_b = V$$

$\phi$ = the angle through which the armature current $I$ lags behind $V$

$$\theta = \tan^{-1} \frac{X_s}{R_a}.$$

The angle $\alpha$ is small in this case.

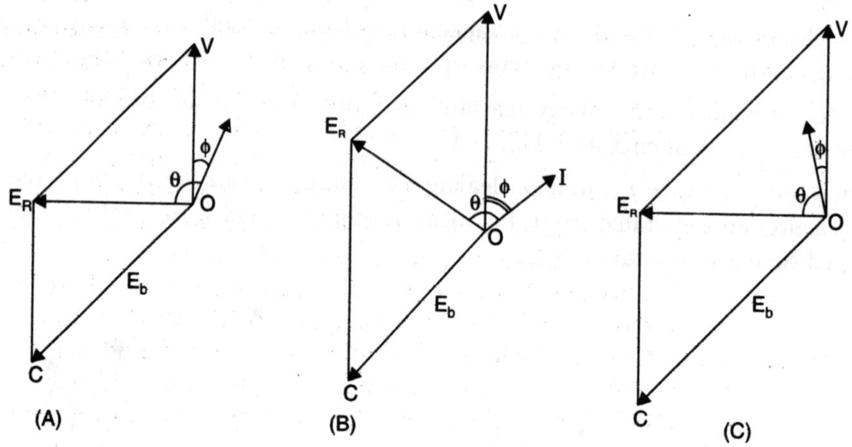


*FIGURE 11.2*

Fig. 11.2 (*b*) represents the condition in which excitation is less than 100% *i.e.,* $E_b < V$. This results in

(*i*) $E_R$ advancing in clockwise direction.

(*ii*) $\phi$ increases *i.e.,* power factor decreases.

(*iii*) $I \cos \phi$ remains the same but $I \sin \phi$ (wattless component) increases.

Fig. 11.2 (*c*) represents the condition for over-excited motor (when $E_b > V$). In this case

(*i*) $I$ is leading $V$

(*ii*) power factor is leading.

At unity power $I$ and $V$ will coincide and the current drawn by the motor will be least.

From the above now it can be concluded that

(*i*) the magnitude of the armature current varies with excitation.

(*ii*) for the same input, armature current varies between a wide-range and so causes the power factor also to vary accordingly.

(*iii*) when over excited the motor runs on leading power factor.

(*iv*) when under excited the motor runs on lagging power factor.

(*v*) the minimum armature current corresponds to unity power factor.

**V curves.** These curves represent variation of the armature current and field current for various loads on a synchronous motors. The input to a threephase synchronous motor is given by

$$P = 3\ VJ \cos \phi$$

where $V$ = applied voltage from the line to the neutral

$I$ = line current

$\cos \phi$ = the power factor

If $V$ and $P$ are kept constant, the power component ($I \cos \phi$) of the input current is also constant, although individually $I$ and $\cos \phi$ both vary with the excitation.

When excitation current is varied, the armature current also varies and the curves are obtained as shown in Fig. 11.3.

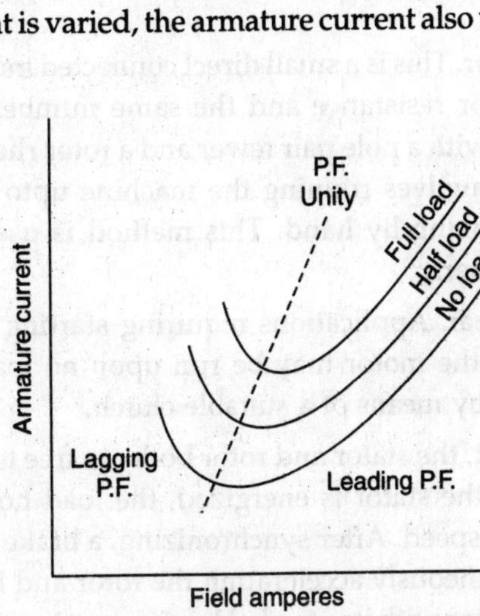


*FIGURE 11.3*

**Losses.** Various losses occurring in a synchronous motor are given below.

Power Input/Phase
$= VI_a \cos \phi$

Armature copper loss
$(I_a^2 R_4)$

Mech. power in armature
$E_b \cdot I_a \cos (\alpha - \phi)$

Iron and friction losses

Output (BHP)

**Hunting or Surging.** The phenomenon of hunting or surging occurs when a motor carries varying load or supply voltage frequency changes. When a synchronous motor runs normally under a load, its rotor falls back in phase by the coupling angle $\alpha$. As the load is further increased, this angle also increases so as to produce more torque for coping with the increased load. Now if the load suddenly drops, the motor is immediately pulled up or advanced to a new value of $\alpha$ corresponding to new load. But in this process the rotor over shoots and hence is again pulled back. In this way motor starts oscillating about its new position of equilibrium corresponding to new load. In case the time period of the oscillations happens to be equal to the natural time period of the machine, then mechanical resonance is set up and the motor starts hunting.

To stop the build up of these oscillations dampers or damping grids are used. These are short circuited copper bars embedded in the faces of the field poles of the motor. The oscillatory motion of the rotor sets up eddy currents in the dampers which flow in such a way as to suppress the oscillations.

**Starting.** The synchronous motor can develop synchronous torque only at synchronous speed. For starting it is therefore necessary to employ other methods of developing torque. Various methods are

1. **Pony induction motor.** This is a small direct connected induction motor, either a machine with a very low rotor resistance and the same number of poles as the synchronous machine, or a motor with a pole pair fewer and a rotor rheostat; or with a solid unwound rotor. The method involves running the machine upto a speed close to synchronous speed and synchronizing by hand. This method is used for large machines such as synchronous condensers.
2. **Clutch and brake gear.** Applications requiring starting torques greater than full-load torque are required, the motor may be run upon no load and synchronized, the load being then taken up by means of a suitable clutch.

   In a modified method, the stator and rotor both are free to rotate. The rotor is coupled to the load, and when the stator is energized, the load holds the rotor at rest while the stator runs up to full speed. After synchronizing, a brake is applied to the stator so as to slow it sown, simultaneously accelerating the rotor and load. When the stator has been brought to rest, the rotor with its coupled load is running normally, at synchronous speed. Considerable starting torque can be developed in this way.
3. **Starting as induction motor.** In this method the stator is supplied with three phase currents at normal or subnormal voltage. The induction torque developed in pole shoes and damper winding on the rotor is used for starting the motor. Thus the motor is designed as synchronous induction motor with a three-phase rotor winding on a non-salient pole core resembling that of a normal induction motor.

**Comparison of Induction and Synchronous Motors**

| | *Induction* | | *Salient pole synchronous motor* |
|---|---|---|---|
| | *Cage* | *Slip ring* | |
| 1 | 2 | 3 | 4 |
| Stator winding | Form wound insulated coils placed in slots | Same | Same, except that for equal rating there are usually fewer coils, thus providing air space between ends of coils, and improved insulation |

| | | | |
|---|---|---|---|
| Rotor winding | Cage winding carries starting and running current | Form wound insulated coils placed in slots, carry starting current at high voltage and running current at low voltage | (*i*) Cage winding carries starting current only<br>(*ii*) Definite pole, low voltage, dc exciting winding, easy to insulate, carries relatively small constant running current. |
| Slip rings | None | Three, that carry starting current at high voltage and running current at low voltage | Two, that carry direct current excitation at low voltage |
| Air gap | Small | Small | Three to five times as great as that in induction motors |
| Automatic control | Three switches starting, magnetizing, and running; an auto-transformer ; three control relays and two overload relays | Three or more switches : one primary and two or more to cutout secondary resistance, four or more operating relays and two overload relays : resistance grids | One or more switches for starting and running duty ; a field contactor ; two operating relays and two overload relays. But direct-on-line starting is becoming usual. |
| Starting torque | Less than running torque | Less than running torque | Zero |
| Speed | Less than synchronous speed | Less than synchronous speed | Synchronous speed |
| Power factor | Always lagging ; low values at light loads | Always lagging; low values at light loads | May be lagging or leading |
| Hunting | Not present | Not present | Generally tends to hunt |
| Cost | Cheapest | Costlier than cage induction motor | Costlier than induction motors |

## OBJECTIVE TYPE QUESTIONS

1. Synchronous motors are generally not self staring because
   (*a*) starters cannot be used on these machines
   (*b*) starting winding is not provided on these machines
   (*c*) the direction of rotation is not fixed
   (*d*) the direction of instantaneous torque reverses after half cycle.
2. Zero power factor test is generally conducted on synchronous machines of
   (*a*) miniature size (*b*) less than 10 kW (*c*) large size (*d*) multiple speed type.

3. Zero power factor test can be used to as certain
   (*a*) the temperature rise (*b*) the synchronous reactance
   (*c*) pull in torque (*d*) all of the above.
4. In case one phase of a three-phase synchronous motor is shortcircuited, the motor will
   (*a*) not start (*b*) run at $\frac{2}{3}$ of synchronous speed
   (*c*) run with excessive vibrations (*d*) take less than the rated load.
5. The load angle of a synchronous motor varies with
   (*a*) load only (*b*) load and speed
   (*c*) load, speed and power factor (*d*) none of the above.
6. Slip rings are not provided in case of
   (*a*) cage induction motor (*b*) slip ring induction motor
   (*c*) salient pole synchronous motor. (*d*) none of the above.
7. For the same rating which motor will have fewer coils?
   (*a*) cage induction motor (*b*) slip-ring induction motor
   (*c*) salient pole synchronous motor.
8. A pony motor is basically a
   (*a*) dc series motor (*b*) dc shunt motor
   (*c*) double winding ac/dc motor (*d*) small induction motor.
9. A synchronous motor can develop synchronous torque
   (*a*) only at synchronous speed (*b*) while over excited
   (*c*) when under loaded (*d*) below or above synchronous speed.
10. A synchronous motor can be started by
    (*a*) providing damper winding (*b*) pony motor
    (*c*) dc compound motor (*d*) any of the above.
11. A three phase synchronous motor will have
    (*a*) no slip rings (*b*) two slip rings (*c*) three slip rings (*d*) four slip rings.
12. As compared to an induction motor of same size the air gap in a synchronous motor is
    (*a*) less than half (*b*) half to three-fourth.
    (*c*) same (*d*) three to five times.
13. Cage winding in a synchronous motor carries
    (*a*) no current (*b*) starting current only
    (*c*) running current only (*d*) staring as well as running current.
14. Under which of the following machine hunting of synchronous motor is likely to occur?
    (*a*) small and constant load (*b*) over loading for long periods
    (*c*) over excitation (*d*) periodic variation of load.

**15.** The armature current for a synchronous motor has large values for

(*a*) low excitation only (*b*) high excitation only

(*c*) both high and low excitation.

**16.** When the excitation of an unloaded salient pole synchronous motor suddenly gets disconnected

(*a*) the motor stops (*b*) it runs as a reluctance motor at the same speed

(*c*) it runs as a reluctance motor at a lower speed.

**17.** A synchronous motor is switched on to supply with its field windings shorted on themselves. It will

(*a*) not start (*b*) start and continue to run as an induction motor

(*c*) start as induction motor and then run as a synchronous motor.

**18.** The synchronous watt is

(*a*) a unit to express the rating of synchronous motors

(*b*) kW as applicable to a synchronous motor

(*c*) the torque which under synchronous speed would develop a power of one watt

(*d*) rating of excitation power supplied to a synchronous motor.

**19.** When $V$ is the applied voltage, then breakdown torque of a synchronous motor varies as

(*a*) $V$ (*b*) $V^{3/2}$ (*c*) $V^2$ (*d*) $\frac{1}{V}$.

**20.** The power developed by a synchronous motor will be maximum when the load angle is

(*a*) zero (*b*) 45º (*c*) 90º (*d*) 120º.

**21.** In which motor stator and rotor fields rotate simultaneously?

(*a*) dc series motor (*b*) universal motor

(*c*) slip ring induction motor (*d*) synchronous motor.

**22.** A synchronous motor can be used as a synchronous capacitor when it is

(*a*) under loaded (*b*) over loaded (*c*) under excited (*d*) over excited.

**23.** For a fixed load the power drawn by a synchronous motor will be least when the power factor is

(*a*) zero (*b*) unity (*c*) leading (*d*) lagging.

**24.** If the power factor of a running synchronous motor is unity, it is likely to

(*a*) run on a speed less than the synchronous speed

(*b*) run on no load (*c*) run on full load

(*d*) vibrate violently.

**25.** A synchronous motor is running on a load with normal excitation. Now if the load on the motor is increased

(*a*) power factor as well as armature current will decrease

(*b*) power factor as well as armature current will increase (*c*) power factor will increase but armature current will decrease

(*d*) power factor will decrease and armature current will increase.

**26.** For *V*-curves for a synchronous motor the graph is drawn between

(*a*) armature current and power factor (*b*) power factor and field current

(*c*) terminal voltage and load factor (*d*) field current and armature current.

**27.** The back emf of a synchronous motor depends on

(*a*) speed (*b*) load (*c*) load angle (*d*) all of the above.

**28.** Torque angle for a synchronous motor is

(*a*) the angle through which the motor lags behind the synchronous speed

(*b*) the angle of lag from no load to full load conditions

(*c*) the angle between the rotating stator flux and rotor poles

(*d*) none of the above.

**29.** When a synchronous motor is run as strongly excited and on light load

(*a*) the active component of input current will be large and reactive component will be small

(*b*) the active component of input current will be small and reactive component will be large

(*c*) both active as well as reactive components of input current will be large

(*d*) both active as well as reactive components of input current will be small.

**30.** When *V* is the supply voltage and *R* is the rotor resistance per phase, the mechanical power developed by the synchronous motor will be proportional to

(*a*) $\frac{V}{R}$ (*b*) $\frac{V^2}{R}$ (*c*) $\frac{R}{V^2}$ (*d*) $\frac{V}{R^2}$.

**31.** In a synchronous motor which loss varies with load?

(*a*) core loss (*b*) copper loss

(*c*) bearing friction loss (*d*) windage loss.

**32.** The motor preferred as a drive for clock is

(*a*) dc series motor (*b*) dc shunt motor

(*c*) squirrel cage induction motor (*d*) synchronous motor.

**33.** Which of the following is *B* class insulation?

(*a*) wood (*b*) mica tape (*c*) treated paper (*d*) fabric of silk.

**34.** A synchronous motor can be made self-starting by providing

(*a*) damper winding on rotor poles (*b*) damper winding on stator

(*c*) damper winding on stator as well as rotor poles

(*d*) none of the above.

**35.** Alternators *A* and *B* are running in parallel. If the driving force of alternator *A* is cut off

(*a*) alternator *B* will run as motor (*b*) alternator *A* will run as motor

(*c*) alternator *A* will stop (*d*) both alternators will stop.

36. The running speed of a 24 pole 50 Hz synchronous motor will be
(*a*) 375 rpm (*b*) 300 rpm (*c*) 250 rpm (*d*) 200 rpm.

37. The size of a synchronous motor increases with increase in
(*a*) speed (*b*) flux density (*c*) frequency (*d*) none of the above.

**Questions 38 to 41 Refer to Fig. 11.4:**

The characteristics of a synchronous motor are shown in Fig. 11.4.

38. Power factor on full load is
(*a*) unity (*b*) 0.8 lagging (*c*) 0.8 leading (*d*) zero.

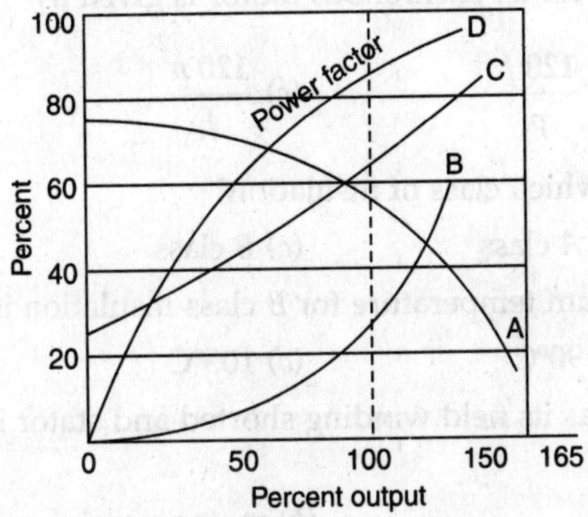


*FIGURE 11.4*

39. Curve *A* is likely to represent
(*a*) starting current (*b*) excitation voltage (*c*) reactive kVA (*d*) power factor.

40. Curve *B* is likely to represent
(*a*) excitation current (*b*) motor speed
(*c*) power angle (*d*) none of the above.

41. Curve C is likely to represent
(*a*) stator current (*b*) supply voltage (*c*) back emf (*d*) none of the above.

42. Which of the following always remains constant in case of synchronous motors?
(*a*) load, speed, excitation and torque (*b*) load, speed and excitation
(*c*) load and speed (*d*) speed only.

43. In a synchronous motor pull out torque represents
(*a*) the ability of the motor to maintain operation during the change over from induction to synchronous running
(*b*) the torque when motor is running on full load and then all of a sudden the load is switched off
(*c*) the torque on which the synchronous running of the motor depends
(*d*) none of the above.

44. In a induction motor which of the following loss remains nearly constant?
(a) windage loss (b) core loss
(c) brush friction loss (d) all of the above.

45. In a properly designed synchronous motor friction and windage loss is nearly
(a) 0.2% (b) 2.0% (c) 10% (d) 22%.

46. The oscillations in a synchronous motor can be damped out by
(a) maintaining constant excitation (b) running the motor on leading power factors
(c) providing damping bars in the rotor pole faces
(d) oscillations cannot be damped.

47. The synchronous speed for a synchronous motor is given by
(a) $\frac{200 f}{p}$ (b) $\frac{120 f}{p}$ (c) $\frac{120 p}{f}$ (d) $120 fp$.

48. Paper is considered as which class of insulation?
(a) Y class (b) A class (c) B class (d) F class.

49. The permissible maximum temperature for B class insulation is
(a) 75°C (b) 90°C (c) 105°C (d) 155°C.

50. A synchronous motor has its field winding shorted and stator is supplied variable voltage. This is likely to result in
(a) burning of rotor (b) motor running as a induction motor
(c) crawling of rotor (d) magnetic locking of rotor.

51. Which of the following class of insulation will have the highest permissible temperature?
(a) A (b) E (c) H (d) B.

52. The shaft of synchronous motor is made of
(a) mild steel (b) chrome steel (c) alnico (d) stainless steel.

53. For a synchronous motor the locus of armature current versus field current is known as
(a) open circuit characteristics (b) zero power factor characteristics
(c) O. curves (d) V. curves.

54. When the field of a synchronous motor is under excited, the power factor will be
(a) zero (b) unity (c) lagging (d) leading.

55. The speed regulation of a synchronous motor is always
(a) 1% (b) 0.5% (c) positive (d) zero.

56. The back emf set up in the stator of a three phase synchronous motor depends on
(a) rotor speed only (b) rotor excitation only
(c) rotor excitation as well as rotor speed (d) none of the above.

57. The ratio of no load speed to full load speed for a three phase 440 V, 200 kW motor will be
(a) 1.2 (b) 1.1 (c) 1.05 (d) 1.0.

**58.** The armature current in a synchronous motor will be least when power factor is

(*a*) zero (*b*) unity (*c*) leading (*d*) lagging.

**59.** The percentage slip in case of a synchronous motor is

(*a*) 1% (*b*) 100% (*c*) 0.5% (*d*) zero.

**60.** In a synchronous motor, the torque angle is defined as the angle between

(*a*) supply voltage and back emf

(*b*) magnetization current and excitation current

(*c*) rotating stator flux and rotor poles

(*d*) stator and rotor flux.

**61.** The maximum value of torque that a synchronous motor can develop without losing its synchronism, is called

(*a*) slip torque (*b*) pull out torque

(*c*) breaking torque (*d*) synchronising torque.

**62.** The back emf set up in the stator of synchronous motor depends on

(*a*) rotor speed (*b*) load (*c*) rotor excitation (*d*) coupling angle.

**63.** A synchronous motor working on leading power factor and not driving mechanical load is known as

(*a*) condenser (*b*) synchronous condenser

(*c*) invertor (*d*) convertor.

**64.** A synchronous motor is directly coupled to a mechanical load. It draws current at lagging power factor from a constant voltage supply. Now if the field excitation of the motor is increased, its power factor will

(*a*) increase (*b*) decrease (*c*) remain unaltered.

**65.** A synchronous motor is adjusted with such an excitation that it takes lagging power factor. Now if the applied voltage of the synchronous motor goes down, the power factor of the motor will

(*a*) become more lagging (*b*) become unity

(*c*) remain unaltered. (*d*) tend towards unity.

**66.** The factors associated with the operation of a synchronous motor are

I. Supply voltage

II. Rotor excitation

III. Coupling angle.

The maximum power developed in a synchronous motor will depend on

(*a*) I only (*b*) II only (*c*) III only (*d*) I, II and III.

**67.** The operating speed of a synchronous motor can be changed to new fixed value by

(*a*) changing the load (*b*) changing the supply voltage

(*c*) changing frequency (*d*) using brakes.

68. An unexcited single phase synchronous motor is
(*a*) ac series motor (*b*) reluctance motor (*c*) universal motor (*d*) repulsion motor.

69. A synchronous motor will always stop when
(*a*) supply voltage fluctuates (*b*) load in motor varies
(*c*) excitation winding gets disconnected (*d*) supply voltage frequency changes.

70. A synchronous motor while operating at rated speed but on no load will have
(*a*) negligible stator current (*b*) back emf nearly equal to supply voltage
(*c*) high stator current (*d*) none of the above.

71. Hunting in a synchronous motor takes place
(*a*) when supply voltage fluctuates (*b*) when load varies
(*c*) when power factor is unity (*d*) motor is under loaded.

72. A synchronous motor is provided with damper winding against mechanical load. The number of slip rings will be
(*a*) nil (*b*) 2 (*c*) 3 (*d*) 5.

73. A synchronous motor has better power factor as compared to that of an equivalent induction motor. This is mainly because
(*a*) synchronous motor has no slip
(*b*) stator supply is not required to produce magnetic field
(*c*) mechanical load on the rotor remains constant
(*d*) synchronous motor has large air gap.

74. A synchronous motor working at leading power factor can be used as
(*a*) noise generator (*b*) mechanical synchronizer
(*c*) phase advancer (*d*) voltage booster.

75. The accumulation of dust and dert on the slip rings of a motor is likely to result in
(*a*) increased mechanical wear (*b*) sparking
(*c*) resistance in path (*d*) none of the above.

76. The accumulation of dust and dert in motor bearings is likely to result in
(*a*) reduced friction (*b*) less noise
(*c*) increased wear and tear (*d*) none of the above.

77. Slip rings are usually made of
(*a*) copper or aluminium (*b*) silver or gold
(*c*) brass or steel (*d*) carbon or graphite.

78. When proper insulation and mechanical strength is the major consideration, it is more economical to locate ..... winding on rotor and ...... winding on stator
(*a*) low voltage ..... high voltage (*b*) high voltage ..... low voltage
(*c*) low voltage ..... low voltage (*d*) high voltage ..... high voltage.

**79.** Which of the following motor can be used as a phase advancer?

(*a*) slip ring induction motor (*b*) squirrel cage induction motor

(*c*) synchronous motor at leading power factor

(*d*) synchronous motor at lagging power factor.

**80.** An over excited synchronous motor is used for

(*a*) fluctuating loads (*b*) variable speed load

(*c*) low torque loads (*d*) power factor corrections.

**81.** The kilowatt rating of the exciter for a synchronous motor is about

(*a*) 3 to 5 percent of the kVA rating of the synchronous motor rating

(*b*) 10 to 12 percent of the kVA rating of the synchronous motor rating

(*c*) 15 to 18 percent of the kVA rating of the synchronous motor rating.

(*d*) 25 percent of the kVA rating of the synchronous motor rating.

**82.** When the voltage applied to a synchronous motor is increased, which of the following will reduce?

(*a*) Stator flux (*b*) Pull in torque

(*c*) Both (*a*) and (*b*) above (*d*) None of the above.

**83.** Two generators are running in parallel. One of the generators may run as a synchronous motor under which of the following condition?

(*a*) When the direction of rotation of generators is reversed

(*b*) When the speed of one of the generator is increased

(*c*) When the field of one of the generators is weakened

(*d*) When the phase sequence of one of the generator is reversed.

**84.** The speed of the synchronous motor

(*a*) increases with increase in load (*b*) decreases as load increases

(*c*) varies with power factor (*d*) always remains constant.

**85.** Dust and dirt accumulation on motor winding will result in

(*a*) shorting of coils (*b*) eddy current flow

(*c*) increased core losses (*d*) rise in winding temperature.

**86.** The efficiency of a properly designed synchronous motor will usually fall in the range

(*a*) 99% to 99.5% (*b*) 85% to 95% (*c*) 75% to 80% (*d*) 60% to 70%.

**Questions 87 to 89 Refer to Fig. 11.5:**

*V*-curves for a synchronous motor are shown in Fig. 11.5.

**87.** The power factor on the left of unity power factor line will be

(*a*) lagging (*b*) leading

(*c*) lagging or leading (*d*) depends on the load.

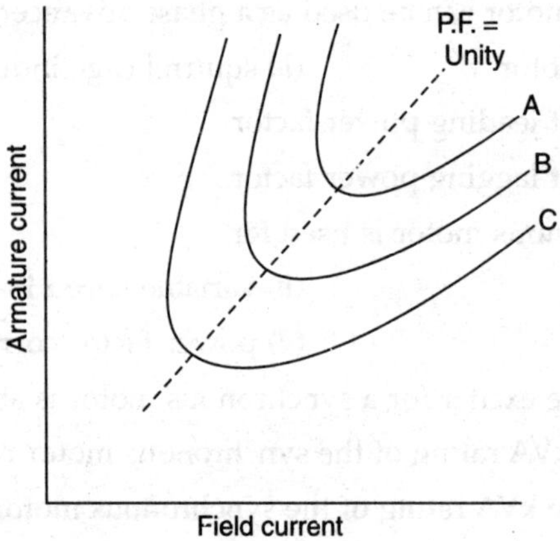


*FIGURE 11.5*

88. If curve represents 80% load on the machine, then curve for 50% load is likely to be
(*a*) *A* (*b*) *C*
(*c*) any of the two
(*d*) cannot be ascertained on the basis of information provided.

89. For the same power factor when the load on a synchronous motor decreases, which of the following increases?
(*a*) armature current (*b*) field current
(*c*) both (*a*) and (*b*) above (*d*) none of the above.

90. To safeguard the insulation from break down, an electrical machine should have proper
(*a*) voltage rating (*b*) current rating (*c*) power factor (*d*) speed.

91. To limit the operating temperature an electrical machine should have proper
(*a*) voltage rating (*b*) current rating (*c*) power factor (*d*) speed.

92. Amortisseur windings on synchronous motors are placed
(*a*) on the rotor pole faces (*b*) on the rotor slots
(*c*) on the stator (*d*) on all of the above.

93. Fig. 11.6 shows the performance curves for a synchronous motor. Which curve represents the variation of power factor?
(*a*) curve *A* (*b*) curve *B* (*c*) curve C (*d*) curve *D*.

94. Which curve represents the variation of reactive kVA?
(*a*) curve *A* (*b*) curve *B* (*c*) curve C (*d*) curve *D*.

95. Which of the following motor will have air gap more than that of the remaining motors?
(*a*) Cage induction motor (*b*) Slip ring induction motor
(*c*) Salient pole synchronous motor (*d*) All will have same air gap.

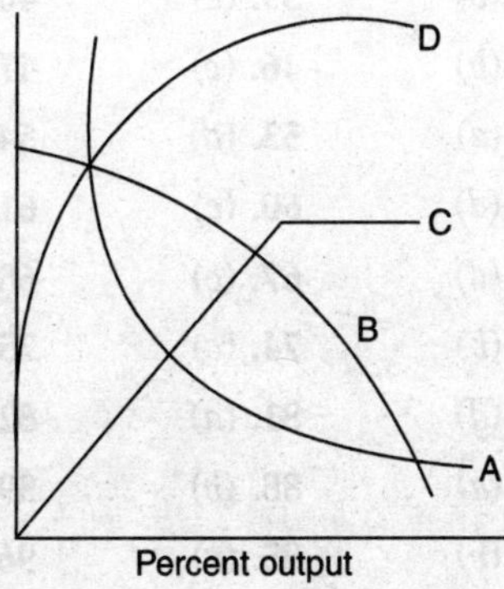


*FIGURE 11.6*

96. Slip rings in a synchronous motor carry
    (*a*) direct current (*b*) alternating current
    (*c*) no current.
97. The power rating of pony motor for starting an induction motor should be
    (*a*) more than that of synchronous motor (*b*) equal to that of synchronous motor
    (*c*) sufficient to over come bearing friction
    (*d*) sufficient to accelerate the mass of the machine.
98. Damper windings are generally provided on
    (*a*) synchronous motors (*b*) alternators
    (*c*) induction motor (*d*) three phase alternators.
99. Which of the following motor has tendency to hunt?
    (*a*) Synchronous motor (*b*) Slip ring induction motor
    (*c*) Cage induction motor (*d*) All of the above.
100. The clutch method of starting an induction motor is used
    (*a*) for high speed synchronous motors (*b*) for low voltage synchronous motors
    (*c*) when starting torque is high (*d*) to save energy in starting.

## ANSWERS

| | | | | | | |
|---|---|---|---|---|---|---|
| **1.** (*d*) | **2.** (*c*) | **3.** (*a*) | **4.** (*a*) | **5.** (*a*) | **6.** (*a*) | **7.** (*c*) |
| **8.** (*d*) | **9.** (*a*) | **10.** (*d*) | **11.** (*b*) | **12.** (*d*) | **13.** (*b*) | **14.** (*d*) |
| **15.** (*c*) | **16.** (*a*) | **17.** (*b*) | **18.** (*c*) | **19.** (*a*) | **20.** (*c*) | **21.** (*d*) |
| **22.** (*d*) | **23.** (*b*) | **24.** (*c*) | **25.** (*d*) | **26.** (*d*) | **27.** (*c*) | **28.** (*c*) |
| **29.** (*b*) | **30.** (*b*) | **31.** (*b*) | **32.** (*d*) | **33.** (*b*) | **34.** (*d*) | **35.** (*b*) |

| | | | | | | |
|---|---|---|---|---|---|---|
| **36.** (*c*) | **37.** (*b*) | **38.** (*b*) | **39.** (*c*) | **40.** (*d*) | **41.** (*a*) | **42.** (*d*) |
| **43.** (*a*) | **44.** (*d*) | **45.** (*b*) | **46.** (*c*) | **47.** (*b*) | **48.** (*a*) | **49.** (*c*) |
| **50.** (*b*) | **51.** (*c*) | **52.** (*a*) | **53.** (*d*) | **54.** (*c*) | **55.** (*d*) | **56.** (*b*) |
| **57.** (*d*) | **58.** (*b*) | **59.** (*d*) | **60.** (*c*) | **61.** (*d*) | **62.** (*c*) | **63.** (*b*) |
| **64.** (*a*) | **65.** (*d*) | **66.** (*d*) | **67.** (*c*) | **68.** (*b*) | **69.** (*c*) | **70.** (*a*) |
| **71.** (*b*) | **72.** (*d*) | **73.** (*b*) | **74.** (*c*) | **75.** (*a*) | **76.** (*c*) | **77.** (*c*) |
| **78.** (*b*) | **79.** (*c*) | **80.** (*d*) | **81.** (*a*) | **82.** (*d*) | **83.** (*c*) | **84.** (*d*) |
| **85.** (*d*) | **86.** (*b*) | **87.** (*a*) | **88.** (*b*) | **89.** (*d*) | **90.** (*a*) | **91.** (*b*) |
| **92.** (*a*) | **93.** (*d*) | **94.** (*b*) | **95.** (*c*) | **96.** (*a*) | **97.** (*d*) | **98.** (*a*) |
| **99.** (*a*) | **100.** (*c*). | | | | | |

Chapter 12

# INDUCTION MOTORS

The poly phase induction motor is used more extensively than any other form of ac motors. In this motor the current in the rotor is not drawn from the supply but is induced by the motion of the rotor conductors relative to the magnetic field produced by the stator currents. The stator carries the poly phase winding as in an alternator.

The stator winding is connected to the supply, and the poly phase currents circulating through it produce a magnetic field which rotates at synchronous speed. The lines of force of the stator field cut the rotor conductors and induce currents in them, hence by Lenz's law, the rotor follows after the stator field. The magnitude of the induced rotor current and hence the torque depends on the relative velocity of the rotor and the stator field. Thus even on no load the rotor speed will be less than synchronous speed, because some torque is required to overcome bearing friction and iron losses. As the load on the motor increases, the relative velocity of rotor to stator field must increase. Since the speed of stator field is fixed therefore rotor speed decreases.

**Slip**

Let $\omega_1$ = angular velocity of stator field, and therefore of rotor field

$\omega_2$ = actual angular velocity of rotor

then, absolute slip = $(\omega_1 - \omega_2)$

$$\text{Fractional slip} = \frac{\omega_1 - \omega_2}{\omega_1}$$

$$\text{Percentage slip} = \frac{\omega_1 - \omega_2}{\omega_1} \times 100.$$

The full load slip of modern motors varies from about 4 per cent for small motors to 1.5 or 2 percent for very large ones. The actual speed is thus very close to synchronous speed under normal working conditions.

### Cage Type and Phase Wound Rotors

Two types of rotor windings on 3 phase induction motors are commonly used:

(*i*) Squirrel cage (*ii*) Phase wound.

In squirrel cage rotors, a number of copper bars threaded through slots in a laminated core are used. The ends of the bars are connected to stout copper or alloy end rings. Thus these windings are permanently short-circuited on themselves and provision does not exist for incorporating any resistance or external circuit with rotor winding, for starting purposes.

In phase wound rotors, a distributed winding is provided. The separate phases are generally connected in star, and the three free ends connected each to one of three insulated slip rings mounted on the rotor shaft. Under normal running conditions these slip rings are short-circuited. However, for starting purposes they are connected to a three phase star connected resistor as shown in Fig. 12.1.

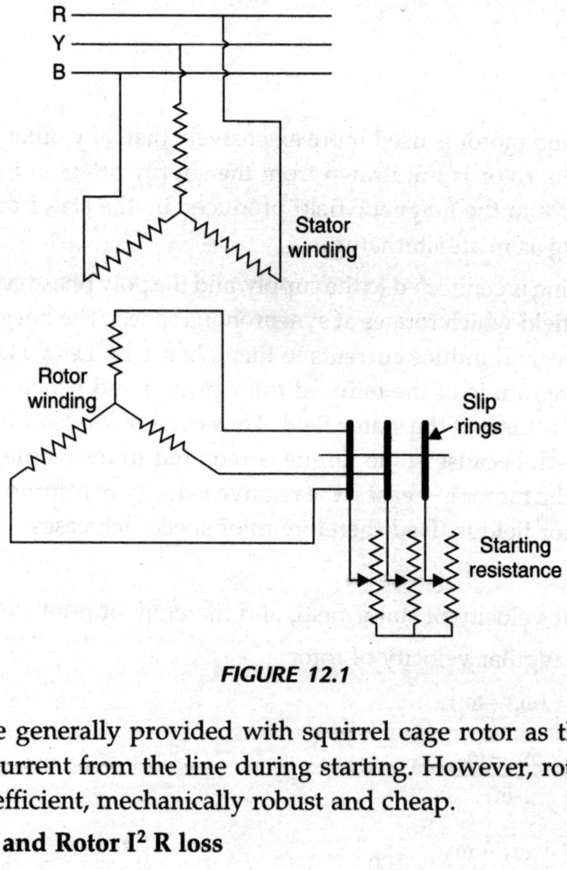


*FIGURE 12.1*

Small motors are generally provided with squirrel cage rotor as these motors have small torque and take large current from the line during starting. However, rotor resistance being low the motor is generally efficient, mechanically robust and cheap.

### Relation between Slip and Rotor $I^2$ R loss

$$\text{Fraction slip} = \frac{\omega_1 - \omega_2}{\omega_1} = \frac{\text{Rotor copper loss}}{\text{Input power}}$$

From the above relation it is evident that motors with a very low rotor resistance have a small slip, while those with high rotor resistance have a large slip.

**Torque.** The torque developed by the motor is given by

$$T = \frac{1}{2\pi n_s} \cdot \frac{m_2 k^2 s\, Ep_1 . R_2^2}{(R_2^2 + s^2 X_{2(0)}^2)} \text{ Nm.}$$

where $k$ is stand still voltage transformation ratio

$s$ is fractional slip

$X_{2(0)}$ is the rotor reactance per phase at stand still

$m_2$ is number of rotor phases

$n_s$ is synchronous speed.

For the above equation the following conclusions can be drawn

(*i*) Torque $\propto \dfrac{\text{Rotor copper loss}}{\text{Slip}}$

(*ii*) Fall in speed $(\omega_1 - \omega_2) \propto$ rotor intake $\propto$ output

(*iii*) Torque $\propto$ (applied voltage)$^2$

(*iv*) Starting torque $\propto \dfrac{R_2}{[R_2^2 + X_{2(0)}^2]}$

This is maximum when $R_2 = X_{2(0)}$.

Therefore, to obtain the maximum starting torque, sufficient resistance should be put in series with the rotor to make the total resistance per phase equal to the total reactance per phase Fig. 12.2 shows the variation of starting torque with rotor resistance. Squirrel cage motors and motors with permanently short circuited rotor winding have small starting torque whenever the rotor resistance is very small.

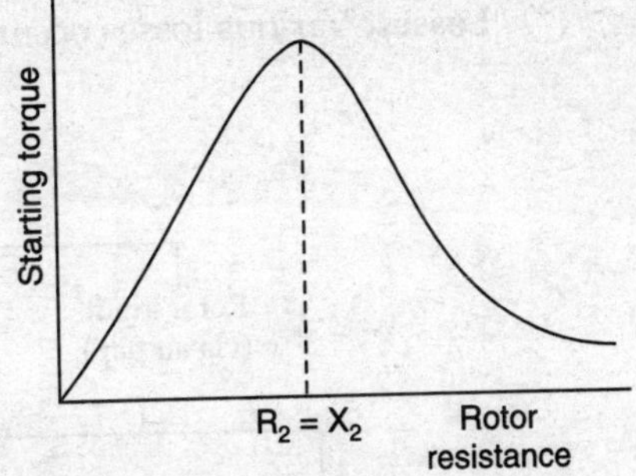

FIGURE 12.2

The condition for maximum torque under running conditions is

$$R_2 = sX_2$$

The relation between torque and slip is shown in Fig. 12.3.

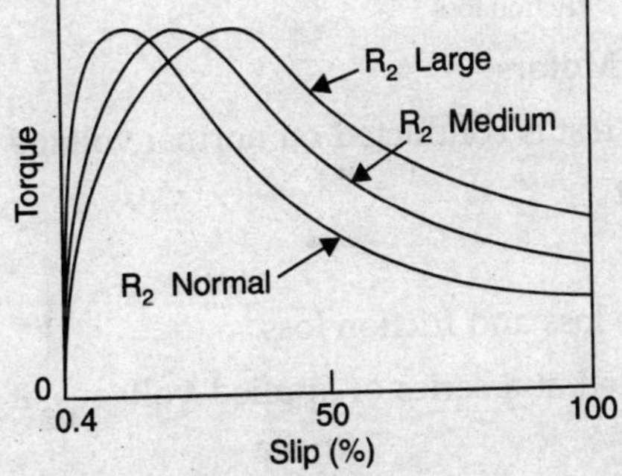

FIGURE 12.3

However, in case of cage motor, starting torque can be increased by introducing resistance in the rotor circuit.

**Equivalent Circuit.** The equivalent circuit of an induction motor is shown in Fig. 12.4. In this all quantities are referred to the primary (stator side) and the no load current is represented by the current taken by the shunted admittance $G_0$, $B_0$.

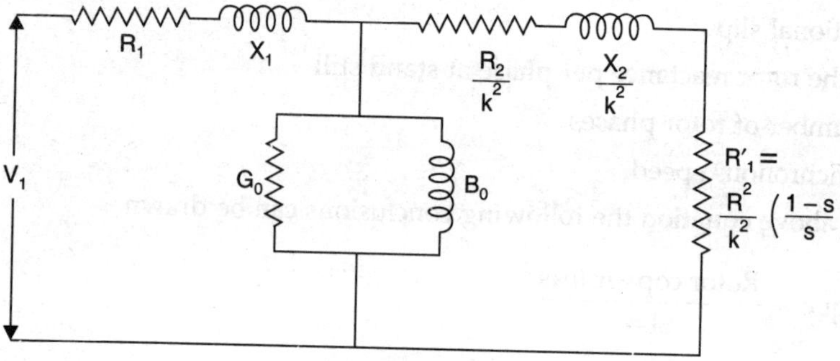


*FIGURE 12.4*

From the point of view of circuit relationships, it is correct to place this admittance between the primary and secondary constants, but for the purpose of analysis it is usual to place it across the supply terminals so that it does not enter into expressions for the torque or power developed. This diagram can be used to deduce circle diagram for the motor.

**Losses.** Various losses occurring in an induction motor are given below.

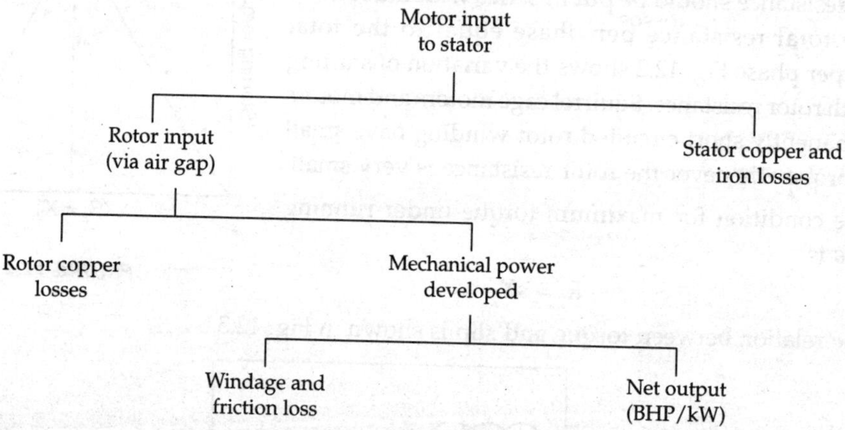


**Testing of Induction Motors**

1. **No load test.** This test is conducted on normal voltage and frequency. This test gives

   (*i*) no load current

   (*ii*) power factor

   (*iii*) sum of the core loss and friction loss.

If this load is conducted at a series of applied voltages, it is possible to separate iron and friction losses.

2. **Blocked rotor test.** This is also known as short-circuit test. The purpose of this test is to find

   (*a*) short-circuit current with normal voltage applied to the stator

   (*b*) power factor on short-circuit

   (*c*) total leakage reactance of the motor referred to stator

   (*d*) total resistance of the motor as referred to stator.

During this test the rotor is blocked (or allowed to rotate at slow speed) and the rotor windings are short-circuited at slip-rings if the motor has a wound rotor. A reduced voltage (10 to 20 percent of the rated voltage is applied to the stator terminals and is so adjusted that full load current flows in the stator. The values of current, voltage and power input on short circuit are measured.

From this test it can be seen that:

1. A straight line relation exists between voltage and short circuit current
2. Power factor on short-circuit is given by

$$Ws = \sqrt{3}\ V_{SL}\, I_{SL} \cos \phi_S$$

where $W_S$ = total power input on short-circuit

$V_{SL}$ = line voltage on short-circuit

$I_{SL}$ = line current on short-circuit.

3. Total power input during the test accounts for

   (*i*) Stator copper losses

   (*ii*) Rotor copper losses

   (*iii*) Core loss.

**Circle Diagram.** Circle diagram for an induction motor can be drawn using the data from

(*i*) no load test

(*ii*) blocked rotor test.

Various steps in construction of circle diagram are given below:

**Step 1.** Based on no load test $I_0$ and $\phi_0$ can be estimated. In Fig. 12.5, *OO′* represents $I_0$ and phase and $\phi_0$ is also shown.

**Step 2.** Based on blocked rotor test $I_{SN}$ corresponding to normal voltage of motor and $\phi_S$ can be determined. In Fig. 12.5 vector *OA* represents $I_{SN}$ in magnitude as well as phase.

Points *O′* and *A* lie on the circle. Centre *C* of the circle can be found by drawing perpendicular from the middle point of *O′A* which meets *O′D* at *C*.

*O′A* is the output line.

*O′P* represents fixed losses (core loss, friction and windage loss and a small fraction of stator $I^2R$ loss).

*AG* represents motor input on short-circuit.

Torque line $O'E$ separates the rotor copper loss and stator copper loss. Thus

$$\frac{AE}{EF} = \frac{\text{Rotor copper loss}}{\text{Stator copper loss}}$$

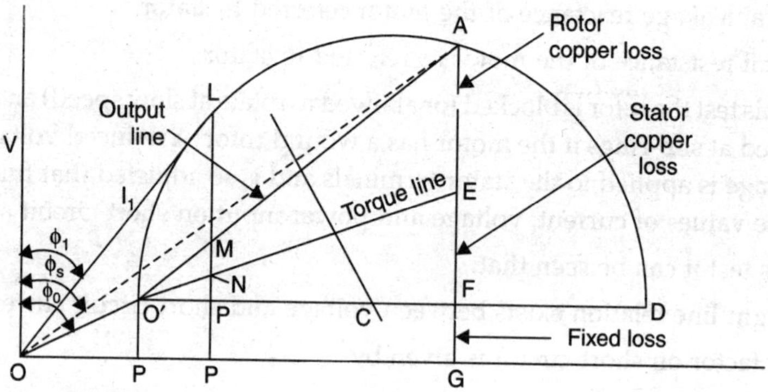


*FIGURE 12.5*

**Step 3.** Let the motor take current $OL$ during running. Then

Motor input = $\sqrt{3}.V_L.\ LK$

Mechanical output = $\sqrt{3}.V_L.\ ML$

Total losses = $\sqrt{3}.V_L.\ MR$

Fixed losses = $\sqrt{3}.V_L.\ PR$

Stator copper loss = $\sqrt{3}.V_L.\ PN$

Rotor copper loss = $\sqrt{3}.V_L.\ MN$

$$\text{Efficiency} = \frac{LM}{LR}$$

$$\text{Slip} = \frac{\text{Rotor copper loss}}{\text{Rotor input}} = \frac{MN}{NL}$$

$$\text{Power factor} = \frac{LR}{OL}.$$

**Crawling.** Sometimes squirrel-cage motor tends to run at a speed very much lower than synchronous speed (usually one-seventh). This is mainly due to fact that stator winding cannot be distributed in more than 3 or 4 slots per pole and phase. Due to this the wave of the stator flux instead of being pure sine wave is stepped. This wave in addition to fundamental wave carries odd harmonics—3rd, 5th, 7th and so on. It can be shown that 3rd harmonic wave produced by three phases gets neutralized. The field set up by the 5th harmonic rotates backwards while that due to 7th harmonic rotates forward.

The effect of these harmonics is shown in Fig. 12.6. The curve of total torque shows that the motor starts up from rest, the torque increases upto a maximum, then decreases, and finally reaches another maximum. Also the position of the first maximum is just below one-seventh synchronous speed. In case the frictional torque of the motor is such that it requires the motor to develop a torque approaching that in the neighbourhood of one seventh full speed, the motor will crawl at this speed and will be perfectly stable as long as working point is in neighbourhood of C.

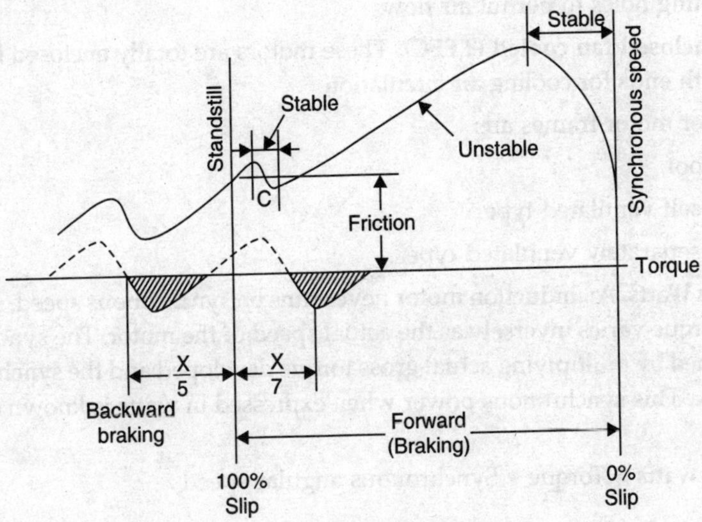


*FIGURE 12.6*

In case the field form of a motor is such that it results in production of large seventh harmonic, then the remedy lies in making the rotor resistance sufficiently high to enable the motor to develop a torque at one seventh speed greater than the frictional torque.

**Skew.** Slot of induction motor are intentionally given slight twist or skew, so that slots remain parallel to one another but not to the rotor axis. The skew reduces humming, thereby ensuring a quiet running of the motor. It also results in a smoother torque curve for different positions of the rotor. By this the effective transformation ratio between the stator and rotor is increased. The rotor resistance is also increased due to increased length of bars. It increases the motor impedance at a given slip and also incrases the slip for a given torque.

**Air gap.** Air gap between stator and rotor is the main source of the low power factor at which induction motor operates. Large air gap require large magnetising current necessary to set up air gap flux, which lags with respect to the applied voltage. Thus for reducing magnetising current small air gap is desired. However, there are some limitations for this. For this purpose the rotor and stator must be circular and concentric. Too much reduction in air gap may increase noise and may have effect on cooling also.

**Frame Structures of Motors.** Various frame structures commonly used for induction motors are given below:

1. **Open type.** In this the ends are provided with wide openings so that cooling is effective. However, such motors are suitable for dry moisture free climates.
2. **Drip proof.** In this the top half of the motor is covered by cast iron frame. However, lower half is wide open to permit circulation of air. The motor is protected against any drops of falling water or liquid.
3. **Totally enclosed.** This type of motor is enclosed but has sufficient openings in the form of ventilating holes to permit air flow.
4. **Totally enclosed fan cooled (TEFC).** These motors are totally enclosed but have fans at one or both ends for cooling air circulation.

   Some other motor frames are:
5. Splash proof.
6. Enclosed self ventilated type.
7. Enclosed separately ventilated type.

**Synchronous Watts.** An induction motor never runs on synchronous speed. At a given gross power output the torque varies inversely as the actual speed of the motor. The synchronous power is the wattage obtained by multiplying actual gross torque developed and the synchronous angular velocity of the motor. This synchronous power when expressed in watts is known as synchronous watts ($P_r$). Thus

Synchronous watts = Torque × Synchronous angular speed

$$P_r = \left(\frac{1}{1-s}\right) P_g$$

where $P_g$ is gross power output and $s$ is slip.

Also, Synchronous watts = Power transferred to the rotor

$$= \frac{3R_2 I_2^2}{s}$$

**Starting of Motors.** Various methods for starting induction motors are given below:

(*a*) **Direct-on-line Starting.** When a induction motor is directly put on line the starting current will be nearly equal to short circuit current. This will result in motor drawing as heavy as seven times the normal running current and the motor will develop a starting torque which may be nearly 2 times the full load value.

By drawing heavy current during starting, the other sections of the power distribution system are affected. For this reason, *DOL* starting is usually restricted to small horsepower motors only.

In general the following methods are used for starting squirrel cage induction motors:

(*a*) **Primary resistors or reactors.** These resistors or reactions are introduced in the circuit to reduce the voltage applied across the motor terminals so that initial current drawn by the motor is reduced.

This will also reduce the starting torque of the motor. As the motor picks up speed the resistance in the circuit is reduced. This method is used for smooth starting of small motors.

(*b*) **Autotransformer.** In this case autotransformer is used to vary the voltage during starting. Initially reduced voltage is applied to the motor, which is gradually increased as motor picks up speed.

(*c*) **Star-Delta Starter.** This method can be used only for motors designed to run with a delta connected stator winding. In this case the motor is connected in star while starting and is switched on to delta when it picks up speed.

A star-delta starter is equivalent to autotransformer of ratio $\frac{1}{\sqrt{3}}$ or 58% approximately.

This method is cheap and effective provided the starting torque is required not to be more than 1.5 times the full load torque.

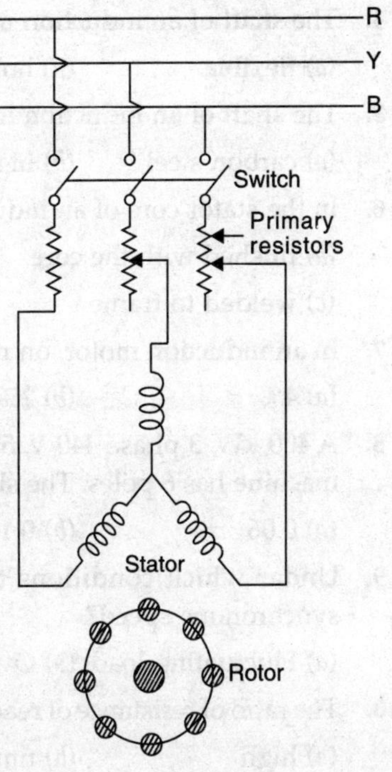


*FIGURE 12.7*

**Speed Control of Induction Motors.** A 3 phase induction motor virtually a constant speed motor. The speed regulation of motor may be within 5% (no load to full load). Several applications of motor need variable speed operation. For this purpose the following methods are used.

**Control from stator side**

(*a*) By changing the applied voltage

(*b*) By changing the applied frequency

(*c*) By changing the number of stator poles.

**Control from rotor side**

(*a*) Rotor rheostat control

(*b*) Cascade running of two motors

(*c*) Injecting emf in the rotor circuit.

## OBJECTIVE TYPE QUESTIONS

**1.** The frame of an induction motor is usually made of

(*a*) cast iron (*b*) silicon steel (*c*) aluminium (*d*) bronze.

**2.** Which of the following component of induction motor is usually fabricated out of silicon steel?

(*a*) Shaft (*b*) Bearings (*c*) Stator core (*d*) None of the above.

**3.** Which type of bearing is used to support rotor shaft of an induction motor?

(*a*) Bush bearing (*b*) Plummer block (*c*) Ball bearing (*d*) Magnetic bearing.

4. The shaft of an induction motor must be
(*a*) flexible (*b*) hollow (*c*) stiff (*d*) any of the above.

5. The shaft of an induction motor is made of
(*a*) carbon steel (*b*) high speed steel (*c*) stainless steel (*d*) cast iron.

6. In the stator core of an induction motor the winding is
(*a*) flushed with the core (*b*) housed in slots
(*c*) welded to frame (*d*) supported on bolts.

7. In an induction motor, on no load the slip is generally
(*a*) 4% (*b*) 2% (*c*) 1.5% (*d*) less than 1%.

8. A 400 kW, 3 phase 440 V, 50 Hz induction motor has a speed of 950 rpm on full load. The machine has 6 poles. The slip of the machine will be
(*a*) 0.06 (*b*) 0.10 (*c*) 0.04 (*d*) 0.05.

9. Under which conditions the operating speed of an induction motor will be closer to synchronous speed?
(*a*) Fluctuating load (*b*) Over load (*c*) Normal load (*d*) No load.

10. The ratio of resistance of reactance for an induction motor under starting conditions should be
(*a*) high (*b*) unity (*c*) less than unity (*d*) negligible.

11. In medium sized induction motors, the slip is generally around
(*a*) 0.04% (*b*) 0.4% (*c*) 4% (*d*) 14%.

12. A four pole, 50 Hz, 440 V, 3 phase induction motor with 3% slip will run on
(*a*) 1550 rpm (*b*) 1500 rpm (*c*) 1495 rpm (*d*) 1455 rpm.

13. In squirrel cage induction motors, the rotor slots are usually given slight skew in order to
(*a*) reduce magnetic hum (*b*) reduce windage losses
(*c*) reduce eddy currents (*d*) reduce accumulation of dirt and dust.

14. Under normal voltage an induction motor has a slip of 4%. If the motor develops the same torque when supply voltage is increased by 10%, the value of slip is likely to be
(*a*) 4.4% (*b*) 4.0% (*c*) 3.6% (*d*) 3.3%.

15. In case the air gap in an induction motor is increased
(*a*) the windage losses will increase
(*b*) the magnetising current of the rotor will decrease
(*c*) the power factor will decrease
(*d*) speed of motor will increase.

16. Slip rings are usually made of
(*a*) aluminium (*b*) copper (*c*) carbon (*d*) phosphor—bronze.

17. Which of the following is preferred rpm for four pole induction motors?
(*a*) 1485 rpm (*b*) 1460 rpm (*c*) 1440 rpm (*d*) 1420 rpm.

**18.** A 3-phase 440 V 50 Hz induction motor has 4% slip. The frequency of the rotor emf will be

(*a*) 0.2 Hz (*b*) 2 Hz (*c*) 50 Hz (*d*) 200 Hz.

**19.** The copper losses in the rotor of induction motor

(*a*) result in eddy currents (*b*) are lost as heat

(*c*) result in noise (*d*) are always negligible.

**20.** A four pole three phase 440 V induction motor has 4% slip at full load. If the motor runs at 720 rpm, the supply frequency must be

(*a*) 100 Hz (*b*) 50 Hz (*c*) 25 Hz (*d*) $16\frac{2}{3}$ Hz.

**21.** If $N_s$ is the synchronous speed and *s* the slip, then actual running speed of an induction motor will be

(*a*) *Ns* (*b*) *s.Ns* (*c*) (1 – *s*)*Ns* (*d*) (*Ns* – 1)*s*.

**22.** Which of the following application generally needs high speed motors?

(*a*) Lathe machines (*b*) Reciprocating air compressors

(*c*) Floor mills (*d*) Air blowers.

**23.** The efficiency of an induction motor can be expected to be nearly

(*a*) 99% (*b*) 95% to 98% (*c*) 80% to 90% (*d*) 60% to 90%.

**24.** Which of the following induction motor will need largest diameter shaft?

(*a*) 20 HP, 2880 rpm (*b*) 20 HP, 1440 rpm (*c*) 20 HP, 960 rpm (*d*) 20 HP, 730 rpm.

**25.** The number of slip rings on a squirrel cage induction motor is usually

(*a*) four (*b*) three (*c*) two (*d*) none.

**26.** The starting torque of a squirrel cage induction motor is

(*a*) negligible (*b*) low

(*c*) same as full load torque (*d*) slightly more than full load torque.

**27.** Fractional slip for induction motor is given by

(*a*) $\frac{\text{Rotor copper loss}}{\text{Input power}}$ (*b*) $\frac{\text{Rotor copper loss}}{\text{Core loss}}$

(*c*) $\frac{\text{Windage loss}}{\text{Core loss}}$ (*d*) $\frac{\text{Copper loss} - \text{Windage loss}}{\text{Core loss}}$.

**28.** The slip of an induction motor will be low when

(*a*) bearing friction is low (*b*) windage loss is least

(*c*) rotor resistance is low (*d*) rotor resistance is high.

**29.** Which figure represents the variation of starting torque with rotor resistance for an induction motor?

(*a*) figure *A* (*b*) figure *B* (*c*) figure C (*d*) figure D.

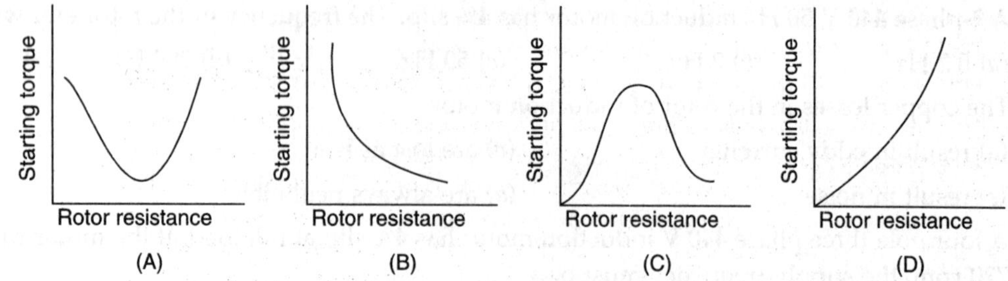


*FIGURE 12.8*

30. In a cage motor, if the material of greater resistivity than copper is used for end rings, this is likely to result in
(*a*) reduction in efficiency (*b*) increase in rotor temperature rise
(*c*) increase in slip (*d*) all of the above.

31. A double squirrel cage induction motor has
(*a*) two rotors moving in opposite directions
(*b*) two parallel windings in stator
(*c*) two parallel windings in rotor
(*d*) two series windings in stator.

**Questions 32 to 35 Refer to Data Given Below:**

A squirrel cage induction motor has a ratio of rotor standstill resistance to resistance per phase of 4 to 1, and the maximum torque is $2\frac{1}{4}$ times the normal full load torque.

32. The full load slip will be
(*a*) 1% (*b*) 3.7% (*c*) 5.7% (*d*) 12.7%.

33. The ratio of starting torque to normal torque will be highest when starting the motor by
(*a*) star delta starter (*b*) direct-on-line starter
(*c*) auto-transformer starter with 70% tapping
(*d*) auto transformer starter with 60% tapping.

34. The ratio of starting torque to normal torque will be least when starting the motor by
(*a*) star delta starter (*b*) direct-on-line starter
(*c*) autotransformer starter with 60% tapping
(*d*) autotransformer starter with 70% tapping.

35. The ratio of starting torque to normal torque in case of star-delta starter will be
(*a*) 0.35 (*b*) 0.67 (*c*) 1.07 (*d*) 1.35.

36. An induction motor having 8 poles runs on 727.5 rpm. If the supply frequency is 50 Hz, the emf in the rotor will have the frequency of
(*a*) 1.5 Hz (*b*) 48.5 Hz (*c*) 51.5 Hz (*d*) 75 Hz.

**37.** Which of the following method of starting an induction motor, does not result in reduced voltage applied to the motor while starting?

(*a*) Autotransformer starting (*b*) Star-delta starter

(*c*) DOL starter (*d*) All of the above.

**38.** During star-delta starting, the motor is initially run

(*a*) directly on line (*b*) in star

(*c*) in delta (*d*) any of the above.

**39.** During star delta starting of induction motor, during star connection the voltage applied to each phase is

(*a*) $\frac{1}{3}$ of the normal voltage (*b*) $\frac{1}{\sqrt{3}}$ of the normal voltage

(*c*) $\sqrt{3}$ of the normal voltage (*d*) $\frac{1}{3\sqrt{3}}$ of the normal voltage.

**40.** Star-delta starting of motors is not possible in case of

(*a*) high speed motors (*b*) low horsepower motors

(*c*) variable speed motors (*d*) single phase motors.

**41.** The term "clogging" is associated with

(*a*) induction motors (*b*) dc series motors

(*c*) compound generators (*d*) three phase transformers.

**42.** In which of the following motor speed can be controlled through rotor side?

(*a*) AC series motor (*b*) DC series motor

(*c*) Three phase slipring induction motor (*d*) Three phase squirrel cage induction motor.

**43.** In case of induction motors the torque is

(*a*) directly proportional to slip (*b*) inversely proportional to slip

(*c*) directly proportional to $(\text{slip})^2$ (*d*) inversely proportional to $(\sqrt{\text{slip}})$.

**44.** The speed of an induction motor when slip is 100 percent, will be

(*a*) synchronous speed (*b*) rated speed

(*c*) full load speed (*d*) zero.

**45.** The speed of air induction motor will be the same as synchronous speed when slip is

(*a*) infinite (*b*) unity (*c*) 0.5 (*d*) zero.

**46.** Stepless speed control of induction motor is possible by which of the following method?

(*a*) Changing the number of poles (*b*) Emf injection in rotor circuit

(*c*) Cascade operation (*d*) None of the above.

**Questions 47 to 51 Refer to Fig. 12.9:**

The torque slip characteristics of an induction motor are shown in Fig. 12.9.

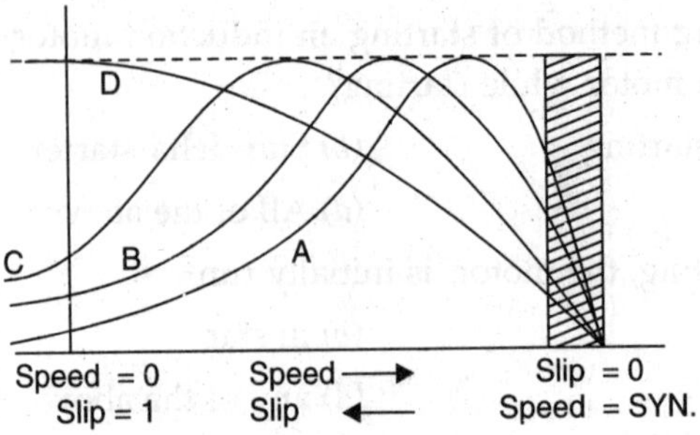


*FIGURE 12.9*

47. For the four curves shown, which of the following could be the variable parameter?
(*a*) speed (*b*) rotor resistance (*c*) load (*d*) none of the above.

48. Which curve represents the maximum torque at starting?
(*a*) curve *A* (*b*) curve *B* (*c*) curve *C* (*d*) curve *D*.

49. A motor with which characteristic will have the least rotor resistance?
(*a*) curve *A* (*b*) curve *B* (*c*) curve *C* (*d*) curve *D*.

50. For average induction motors the pull out torque is
(*a*) same as normal full load torque (*b*) one and a half times the full load torque
(*c*) two and a half times the full load torque
(*d*) five to six times the full load torque.

51. During which of the following test on induction motor, reduced voltage is applied?
(*a*) Load test (*b*) No load test
(*c*) Stator resistance test (*d*) Blocked rotor test.

## Questions 52 to 56 Refer to Data Given Below:

The results of no load and full load tests on a 3-phase mesh connected motor are as follows : The intake as measured by the two wattmeter method gave 3000 and – 550 W on no load, and 39000 and 23,500 W on full load. The motor has four poles, supply 50 Hz, full load speed 1460 RPM and the stator resistance per phase (hot) 0.13 ohm. No load current 25 A and full load current 92 A per line.

52. (Stray losses + No load stator $I^2R$ loss) will be
(*a*) 3550 W (*b*) 3000 W (*c*) 2450 W (*d*) 550 W.

53. No load stator $I^2R$ losses will be
(*a*) 82 W (*b*) 164 W (*c*) 400 W (*d*) 550 W.

54. Full load intake of the motor will be
(*a*) 35,000 W (*b*) 48,600 W (*c*) 62,500 W (*d*) 119,600 W.

55. Slip of the motor will be
(*a*) 1.3% (*b*) 2.6% (*c*) 4% (*d*) 4.4%.

56. The efficiency of the motor is nearly
(*a*) 90% (*b*) 95% (*c*) 92% (*d*) 88%.

57. Presence of fifth harmonic in induction motor is likely to cause
(*a*) reduction in slip (*b*) crawling
(*c*) cogging (*d*) reverse braking torque.

**Questions 58 to 60 Refer to Data Given Below:**

A 6-pole 3-phase induction motor has a stator winding housed in 3 slots per pole and phase and full pitch coils. Conductors in series per phase 486, stator resistance per phase 1.23 ohm, full load stator induced current per phase 11.2 A. Rotor with 57 round bars each $0.95 \times 10^{-2}$ m in diameter and 0.16 m long end rings $2 \times 10^{-2}$ m, square section, mean diameter 0.23 m $\rho_b$ for copper bars $1.7 \times 10^{-8}$ Ω m, $\rho_r$ for brass rings $6.8 \times 10^{-8}$ Ω m.

58. The current per bar at full load will be
(*a*) 56 A (*b*) 106 A (*c*) 292 A (*d*) 488 A.

59. The current per ring at full load will be
(*a*) 221.5 A (*b*) 443 A (*c*) 886 A (*d*) 1233 A.

60. If the total resistance per phase referred to stator is 2.13 ohms, the rotor copper loss will be
(*a*) 60 W (*b*) 125 W (*c*) 240 W (*d*) 340 W.

61. Rotor rheostat control method of speed control is used for
(*a*) slip ring induction motors only (*b*) squirrel cage induction motors only
(*c*) both (*a*) and (*b*) above (*d*) none of the above.

62. In the circle diagram for induction motor, the diameter of the circle represents
(*a*) line voltage (*b*) running torque (*c*) rotor current (*d*) slip.

63. For which motor the speed can be controlled from rotor side?
(*a*) slip ring induction motor (*b*) squirrel cage induction motor
(*c*) both (*a*) and (*b*) (*d*) none of the above.

64. If any two phases for an induction motor are interchanged
(*a*) the motor will burn (*b*) the motor will not run
(*c*) the motor will run at reduced speed (*d*) the motor will run in reverse direction.

65. "The rotor of a three phase induction motor rotates in the same direction as that of the revolving field". Which of the following law is associated with above phenomenon?
(*a*) Lenz's law (*b*) Faraday's law (*c*) Coulomb's law (*d*) Fleming's rule.

66. An induction motor is
(*a*) non-self starting (*b*) self starting with low torque
(*c*) self starting with high torque (*d*) self starting with zero torque.

67. The maximum torque in an induction motor depends on
(*a*) square of the supply voltage (*b*) frequency
(*b*) rotor inductive reactance (*d*) all of the above.

68. For a 250 HP induction motor, the rotor resistance per phase will be of the order of
(*a*) 0.016 ohm (*b*) 0.16 ohm (*c*) 1.6 ohm (*d*) 16 ohm.

69. In a three phase induction motor the frequency of the induced emf in the rotor circuit is
(*a*) maximum at synchronous speed (*b*) maximum at standstill
(*c*) minimum at standstill (*d*) zero at standstill.

70. In an induction motor the synchronous speed of revolving field depends on all of the following EXCEPT
(*a*) number of poles (*b*) supply frequency (*c*) flux density.

71. The tests on an induction motor are

I. No load test
II. Block rotor test
III. Stator resistance test.

For drawing complete circle diagram which tests are necessary?
(*a*) I and II (*b*) I and III
(*c*) II and III (*d*) I, II and III.

72. An induction motor has rotor resistance of 0.002 ohm per phase. If the resistance is increased to 0.004 ohm per phase the maximum torque will
(*a*) reduce to half (*b*) remain unaltered (*c*) increase by 100% (*d*) increase by 300%.

73. In three phase squirrel cage induction motors
(*a*) rotor conductor ends are short circuited through slip rings
(*b*) rotor conductors are short-circuited through end rings
(*c*) rotor conductors are kept open
(*d*) rotor conductors are connected to insulation.

74. Stable operation of an induction motor can be expected between
(*a*) 0% slip and 100% slip (*b*) 0% slip and 50% slip
(*c*) 0% slip and maximum torque slip (*d*) 0% slip and 1% slip.

75. In a three phase induction motor, the number of poles in the rotor winding is always
(*a*) zero (*b*) more than the number of poles in stator
(*c*) less than the number of poles in stator (*d*) equal to the number of poles in stator.

76. *DOL* starting of induction motors is usually restricted to
(*a*) high speed motors (*b*) high horsepower motors
(*c*) variable speed motors (*d*) low horsepower motors.

77. The terminals of a 3-phase slip ring induction motor are open circuited. In case the supply voltage is given to stator
(*a*) the motor will burn (*b*) the motor will not run
(*c*) the motor will run at reduced rpm (*d*) the motor will run at increased rpm.

**78.** 10 HP 3-phase induction motors with following rpm are available which motors will have higher cost as compared to the remaining?

(*a*) 2880 rpm (*b*) 1440 rpm (*c*) 960 rpm (*d*) 730 rpm.

**79.** No load current in an induction motor is nearly

(*a*) 0.1 percent of full load current (*b*) 1 to 5 percent of full load current

(*c*) 15 to 25 percent of full load current (*d*) 50 to 60 percent of full load current.

**Questions 80 to 83 Refer to Data Given Below:**

A 10 HP, 230 V, 3 phase, 50 Hz, 6 pole squirrel cage induction motor operates at a full load slip of 4 percent when rated frequency is applied

**80.** The full load speed, of the motor will be

(*a*) 1440 rpm (*b*) 1040 rpm (*c*) 960 rpm (*d*) 920 rpm.

**81.** The frequency of rotor current will be

(*a*) 2 Hz (*b*) 48 Hz (*c*) 50 Hz (*d*) 52 Hz.

**82.** Full load torque will be nearly

(*a*) 55 Nm (*b*) 73 Nm (*c*) 98 Nm (*d*) 122 Nm.

**83.** It is advisable to avoid line-starting of induction motor and use starter because

(*a*) starting torque is very high

(*b*) motor takes five to seven times its full load current

(*c*) it will pick up very high speed and may go out of step

(*d*) it will run in reverse direction.

**84.** The resistance of the rotor of a three phase induction motor is 0.003 ohms per phase. If the resistance is decreased to 0.001 ohms per phase, the maximum torque will

(*a*) increase by 9 times (*b*) increase by 3 times

(*c*) remain unaltered (*d*) reduce to one third.

**85.** When using star delta starter which of the following is reduced during starting

(*a*) current (*b*) power factor (*c*) volt-amperes (*d*) kW of motor.

**86.** The speed of a squirrel cage induction motor can be controlled by all of the following methods EXCEPT

(*a*) reducing supply voltage (*b*) changing rotor winding resistance

(*c*) changing number of poles (*d*) changing supply frequency.

**87.** When the input voltage and the frequency of a 3-phase induction motor is halved the following phenomenon is observed approximately

(*a*) the maximum torque is halved (*b*) the maximum torque remains the same

(*c*) the air gap flux is halved.

88. The crawling in an induction motor is caused by
(*a*) low voltage supply
(*b*) high loads
(*c*) improper design of machine
(*d*) harmonics developed in the motor.

89. A 400 kW, 3-phase, 440 V, 50 Hz induction motor has a speed of 950 r.p.m. on full load. The machine has 6 poles. The slip of the machine will be
(*a*) 0.06
(*b*) 0.10
(*c*) 0.04
(*d*) 0.05.

90. The power factor of an induction motor under no load conditions will be closer to
(*a*) 0.2 leading
(*b*) 0.2 lagging
(*c*) 0.5 leading
(*d*) unity.

91. A 3-phase, 400 V, 60 Hz, 960 rpm induction motor is run on 50 Hz supply. The motor will
(*a*) burn out
(*b*) not run
(*c*) run at more than 960 rpm
(*d*) run at less than 960 rpm.

92. The cogging of an induction motor can be avoided by
(*a*) using *DOL* starter
(*b*) autotransformer starter
(*c*) proper ventilation
(*d*) having number of rotor slots prime to the number of stator slots.

93. One phase of a three phase partly loaded induction motor power supply got opened up while the motor was in motion. This is likely to result in
(*a*) stalling of the motor
(*b*) burning of one phase winding of the motor
(*c*) burning of all three phases of winding of the motor
(*d*) undisturbed running of motor.

94. If an induction motor with certain ratio of rotor to stator slots, runs at $\frac{1}{7}$ of the normal speed, the phenomenon will be termed as
(*a*) hunting
(*b*) crawling
(*c*) cogging
(*d*) humming.

95. In case the line voltage of an induction motor is reduced by 10% of its rated value, the starting current will be
(*a*) reduced by 5%
(*b*) reduced by 10%
(*c*) increased by 5%
(*d*) increased by 10%.

96. Two of the power supply terminals to a 3-phase induction motor got interchanged during reconnection after maintenance of the motor. When put back into serivce, the motor will
(*a*) fail to rotate
(*b*) burn out
(*c*) rotate in the same direction as it was prior to maintenance
(*d*) rotate in the reverse direction to that prior to maintenance.

97. Single phasing preventers are used in
(*a*) step up transformers
(*b*) step down transformers
(*c*) induction motors
(*d*) all of the above.

98. Various methods for starting an induction motor are listed below. Which method requires six-stator terminals?
(*a*) rotor-rheostat
(*b*) autotransformer
(*c*) direct-on-line
(*d*) star-delta.

99. It is advisable to avoid line starting of induction motor and use starter because
(*a*) starting torque is very high
(*b*) motor takes five to seven times its full load current
(*c*) it will pick up very high speed and may go out of step
(*d*) it will run in reverse direction.

100. Somehow if an induction motor is made to run at synchronous speed, then
(*a*) slip will be infinite (*b*) rotor torque will be zero
(*c*) eddy currents will cause heating of rotor
(*d*) friction and windage losses will increase to disproportionately high values.

101. For a three phase induction motor having delta connected rotor, the number of slip rings will be
(*a*) zero (*b*) one (*c*) two (*d*) three.

102. The speed-load characteristics of an induction motor closely resembles the speed-load characteristics of which of the following machine?
(*a*) DC series motor (*b*) DC shunt motor (*c*) universal motor (*d*) none of the above.

103. In a slip-ring induction motor, resistance is connected in rotor phases
(*a*) to limit starting current (*b*) to increase starting torque
(*c*) to limit starting current and increase starting torque.

104. Which type of bearing is provided in small induction motors to support the rotor shaft?
(*a*) Bush bearings (*b*) Cast iron bearing (*c*) Ball bearings (*d*) None of the above.

105. In case of 100 kW induction motor, the slip is expected to be around
(*a*) 10% (*b*) 5% to 7% (*c*) 4% (*d*) 1.5% to 2%.

106. In an induction motor when the number of stator and rotor slots is equal and the motor refuses to start, the phenomenon is termed as
(*a*) crawling (*b*) cogging (*c*) short-circuiting (*d*) humming.

107. A pump induction motor is switched on to a supply 30% lower than its rated voltage. The pump runs. What will eventually happen? It will
(*a*) stall immediately (*b*) stall after some time
(*c*) continue to run at lower speed wihout damage
(*d*) get heated and consequently get damaged.

108. 5 Hp, 50 Hz, 3-phase, 440 V induction motors are available for following rpm. Which motor will be the costliest?
(*a*) 2880 rpm (*b*) 1440 rpm (*c*) 960 rpm (*d*) 730 rpm.

109. 5 HP, 50 Hz, 3-phase, 440 V, induction motors are available for the following rpm. Which motor will have highest efficiency?
(*a*) 2880 rpm (*b*) 1440 rpm (*c*) 730 rpm (*d*) 360 rpm.

**110.** 5 HP, 50 Hz, 3-phase, 440 V induction motors are available with following number of poles. Which motor will have highest value of power factor?

(*a*) 2 poles (*b*) 4 poles (*c*) 8 poles (*d*) 16 poles.

**111.** A 3-phase slip ring induction in motor has

(*a*) short-circuited rotor (*b*) wound rotor

(*c*) double cage rotor (*d*) any of the above.

**112.** The locking tendency in three phase induction motor is reduced by

(*a*) running the motor in reverse direction

(*b*) using auto-transformer for starting

(*c*) skewing in slots (*d*) making the rotor slots parallel to shaft.

**113.** In a squirrel cage induction motor starting current is usually

(*a*) less than the full load current (*b*) equal to the full load current

(*c*) slightly more than full load current (*d*) five to seven times the full load current.

**114.** A three phase squirrel-cage induction motor is useful when the motor has to start

(*a*) on no load (*b*) against light load (*c*) against full load (*d*) against over load.

**115.** The torque produced in an induction motor is nearly proportional to

(*a*) $\dfrac{1}{\text{applied voltage}}$ (*b*) applied voltage

(*c*) (applied voltage)$^2$ (*d*) $\dfrac{1}{(\text{applied voltage})^2}$.

**116.** In Fig. 12.10 which curve represents the torque-slip characteristics of a three phase induction motor?

(*a*) curve *A* (*b*) curve *B* (*c*) curve C (*d*) curve *D*.

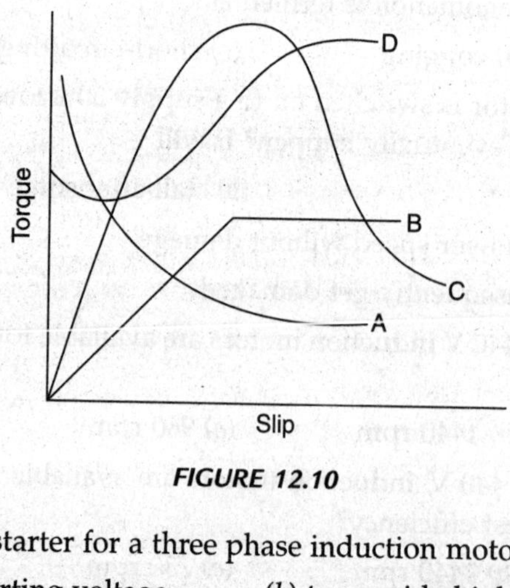


***FIGURE 12.10***

**117.** The main function of starter for a three phase induction motor is

(*a*) to provide high starting voltage (*b*) to provide high starting current

(*c*) to limit the starting current (*d*) to limit the starting voltage.

**118.** What will happen in case the starter is not used on an induction motor?

(*a*) The motor winding will burn (*b*) The motor will have tendency to crawl

(*c*) There will be heavy voltage drop along the line

(*d*) The motor will start with jerk, damaging the bearings.

**119.** Reduced voltage starters can be used with

(*a*) squirrel cage induction motors only (*b*) slip ring induction motors only

(*c*) both (*a*) and (*b*) above (*d*) none of the above.

**120.** A 3-phase, 50 Hz, 1440 rpm induction motor is specified to run at 380/220 volts. It can be concluded that

(*a*) motor can run on single phase supply at 220 V and three phase supply at 380 V

(*b*) motor simultaneously needs 380 V for stator and 220 V for rotor

(*c*) motor is double cage type capable of running at two speeds

(*d*) motor winding when connected in star will run on 380 volts and when connected in delta, will run on 220 volts.

**121.** A 3-phase, 5 HP, 400 volt motor on full load will draw a current of nearly

(*a*) 5 A (*b*) 7.5 A (*c*) 12.5 A (*d*) 20 A.

**122.** Which of the following is class C insulation?

(*a*) Teflon (*b*) Mica (*c*) Paper (*d*) Cloth.

**123.** Short circuit test on an induction motor cannot be used to determine

(*a*) copper losses (*b*) transformation ratio

(*c*) power scale of circle diagram

(*d*) windage losses.

**124.** In a three phase induction motor

(*a*) iron losses in stator will be negligible as compared to that in rotor

(*b*) iron losses in rotor will be negligible as compared to that in rotor

(*c*) iron losses in stator will be less than that in rotor

(*d*) iron losses in stator will be more than that in rotor.

**125.** Single phasing preventer is used on

(*a*) single phase motors (*b*) universal motors

(*c*) 3-phase induction motors with stator in star

(*d*) 3-phase induction motors with stator in delta.

**126.** Synchronous wattage of an induction motor implies

(*a*) stator input in watts (*b*) rotor input in watts

(*c*) rotor output in watts (*d*) total input in watts.

**127.** A motor rated to run at 440 V is running on normal voltage. Suddenly the voltage is dropped to 340 V. The motor will

(*a*) stop immediately (*b*) heat up excessively

(*c*) run in reverse direction (*d*) draw reduced current.

**Questions 128 to 130 Refer to Fig. 12.11:**

The speed-torque characteristics of an induction motor is shown in Fig. 12.11.

**128.** In this, point *A* represents

(*a*) frictional torque (*b*) rated torque (*c*) peak torque (*d*) starting torque.

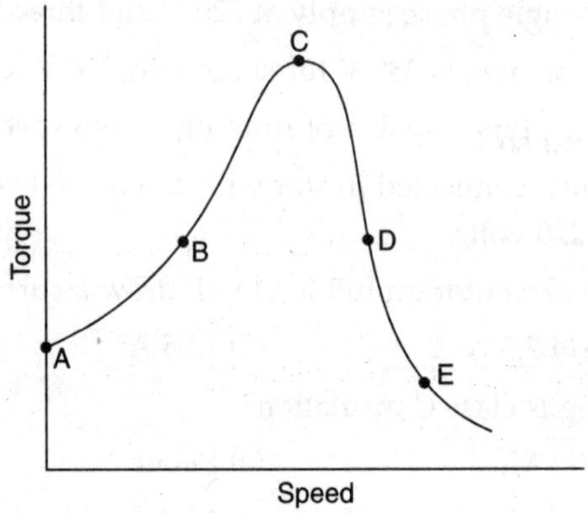


*FIGURE 12.11*

**129.** The maximum torque is represented by

(*a*) point *A* (*b*) point *B* (*c*) point *C* (*d*) point *D*.

**130.** The normal running condition of the motor is represented by

(*a*) point *A* (*b*) point *B* (*c*) point *C* (*d*) point *D*.

**131.** For a three phase induction motor to develop maximum torque, the phase difference between stator flux and the rotor current should be

(*a*) 0° electrical (*b*) 45° electrical (*c*) 90° electrical (*d*) 120° electrical.

**132.** Which of the following cannot be the cause of overheating of a three phase induction motor ?

(*a*) overloading (*b*) stalling of rotor

(*c*) too frequent starting and stopping (*d*) high voltage supply.

**133.** A three phase induction motor when connected in star takes 10 A. When connected in delta it will take

(*a*) 5 A (*b*) less than 10 A (*c*) 10 A (*d*) more than 10 A.

**134.** A 440 V, 50 Hz, 4 pole, three phase induction motor has 4% slip. The relative speed between rotor and stator magnetic field will be

(*a*) 1560 rpm (*b*) 1500 rpm (*c*) 1440 rpm (*d*) zero.

**135.** In case of 3 phase induction motors, plugging means
(*a*) putting the motor directly on line without a starter.
(*b*) locking of rotor due to harmonics
(*c*) starting the motor on a load which is more than the rated load
(*d*) interchanging two supply phases for quick stopping.

**136.** Which of the following is the best method for removing abrasive dust and particles from an electric motor?
(*a*) Rubbing with cloth (*b*) Using wire brush
(*c*) Passing compressed air (*d*) Vacuum suction.

**Questions 137 to 138 Refer to Data Given Below:**

The emf in the rotor of an 8 pole induction motor has a frequency of 2 Hz and that in the stator has a frequency of 50 Hz.

**137.** The motor will run at
(*a*) 700 rpm (*b*) 712.5 rpm (*c*) 720 rpm (*d*) 727.5 rpm.

**138.** The slip will be
(*a*) 2% (*b*) 3% (*c*) 3.5% (*d*) 4%.

**139.** When all the three phases of an induction motor burn out, the probable cause could be
(*a*) single phasing (*b*) unbalanced load
(*c*) unbalanced supply (*d*) overload.

**140.** In case of three phase induction motors, a fuse normally provides protection against
(*a*) high voltage (*b*) overload (*c*) short circuit (*d*) single phasing.

**141.** Which of the following test data is required to draw the circle diagram for an induction motor?
(*a*) Block rotor test only (*b*) No load test only
(*c*) Block rotor test and no-load test
(*d*) Block rotor test, no-load test and stator resistance test.

**142.** The slip of a 5 HP, 440 V, 50 Hz, 1440 rpm motor is $s = 1$. It can be concluded that
(*a*) motor is running at synchronous speed
(*b*) rotor rpm is 1440 (*c*) rotor rpm is $(100 - 1) \times 1440$
(*d*) rotor is not rotating.

**143.** Periodic low speeds—inching speeds can be obtained from
(*a*) slip ring induction motors only (*b*) squirrel cage induction motors only
(*c*) either of (*a*) and (*b*) above (*d*) none of the above.

**144.** The full load current for a 3-phase, 440 V, 20 HP induction motor will be closer to
(*a*) 5 A (*b*) 15 A (*c*) 25 A (*d*) 35 A.

**145.** The harmonic induced torques in induction motors can be reduced by
(*a*) using bush bearings (*b*) skewing
(*c*) large air gap (*d*) chording.

**146.** Which of the following is not a trade name of electric motors in India?
(*a*) Kirloskar (*b*) Tata (*c*) Jyoti (*d*) Crompton.

**147.** In three phase induction motors sometimes copper bars are placed deep in the rotor to
(*a*) improve power factor (*b*) improve efficiency
(*c*) reduce copper losses (*d*) improve starting torque.

**148.** In a three phase induction motor
(*a*) power factor at starting is the same as that while running
(*b*) power factor at starting is low as compared to that while running
(*c*) power factor at starting is high as compared to that while running.

**149.** In Fig. 12.12 which curve represents the variation of power factor with load?
(*a*) curve *A* (*b*) curve *B* (*c*) curve *C* (*d*) curve *D*.

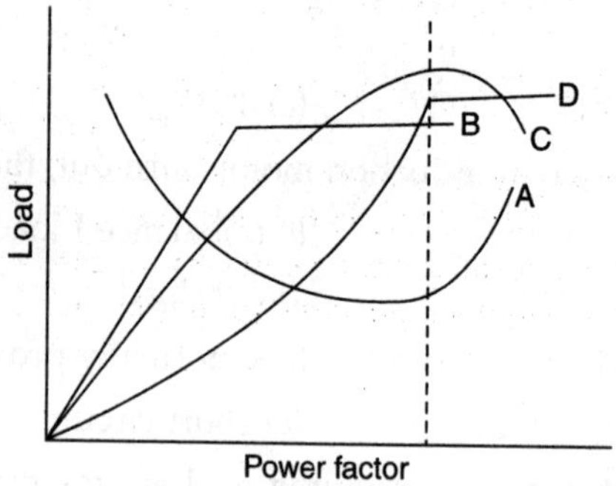


***FIGURE 12.12***

**150.** In an induction motor the magnetic field produced by the harmonics
(*a*) remains stationary with respect to the field produced by the fundamental harmonic
(*b*) rotates in forward direction at the rate of slip frequency
(*c*) rotates in the backward direction at the speed of rotor frequency
(*d*) none of the above.

**151.** The circle diagram for an induction motor cannot be used to determine
(*a*) frequency (*b*) output (*c*) power factor (*d*) efficiency.

**152.** If a particular application demands frequent starting, stopping and reversing, the motor to be recommended out of the following will be
(*a*) single phase capacitor start motor (*b*) three phase slip ring induction motor
(*c*) three phase squirrel cage induction motor
(*d*) DC series motor.

**153.** The current taken by a three phase induction motor during plugging will be

(*a*) normal full load current (*b*) more than blocked rotor current

(*c*) equal to blocked rotor current (*d*) less than blocked rotor current.

**154.** Blocked rotor test on induction motors is used to find out

(*a*) short-circuit current under rated voltage

(*b*) leakage reactance

(*c*) power factor on short-circuit (*d*) all of the above.

**155.** Lubricant used for ball bearings is usually

(*a*) graphite (*b*) grease (*c*) mineral oil (*d*) molasses.

**156.** An induction motor can run at synchronous speed when

(*a*) it is run on no load (*b*) it is run in reverse direction

(*c*) it is run on voltage higher than the rated voltage

(*d*) emf is injected in the rotor circuit.

**157.** For running an induction motor on synchronous speed, emf is injected in the rotor circuit. This injected emf must be

(*a*) dc (*b*) pulsating dc

(*c*) in phase with rotor emf (*d*) out of phase with rotor emf.

**158.** Imbalance in the shaft of induction motor may occur due to

(*a*) slip rings (*b*) poor quality of insulation

(*c*) non-uniform air gap (*d*) any of the above.

**159.** Which of the following can be reduced by skewing in induction motor?

(*a*) Noise (*b*) Vibrations

(*c*) Synchronous cusps (*d*) All of the above.

**160.** When a star-delta starter is used with three phase induction motor

(*a*) starting current reduces but starting torque increases

(*b*) starting current increases but starting torque reduces

(*c*) both starting current as well as starting torque reduces

(*d*) both starting current as well as starting torque increases.

**161.** Which motor is preferred for use in mines where explosive gases exist?

(*a*) Synchronous motor (*b*) DC shunt motor

(*c*) Induction motor (*d*) Air motor.

**162.** Ball bearings are usually made of

(*a*) cast iron (*b*) stainless steel (*c*) high speed steel (*d*) carbon-chrome steel.

**163.** The torque developed by a 3-phase induction motor least depends on

(*a*) shaft diameter (*b*) rotor emf

(*c*) rotor power factor (*d*) rotor current.

164. Under which of the following circumstances a wound rotor induction motor will be preferred over squirrel cage induction motor?

(a) When cost is the major consideration (b) When maintenance cost is to be kept low

(c) When wide range of speed control is desired

(d) None of the above.

165. In an induction motor if air gap is increased

(a) the power factor will be low (b) windage loss will be more

(c) bearing friction will reduce (d) copper loss will reduce.

166. The frame of an induction motor is generally made of

(a) aluminium (b) copper (c) cast iron (d) stainless steel.

167. In an induction motor

(a) Slip = $\frac{\text{Rotor speed–Synchronous speed}}{\text{Rotor speed}}$

(b) Slip = $\frac{\text{Synchronous speed–Rotor speed}}{\text{Synchronous speed}}$

(c) Slip = $\frac{\text{Synchronous speed–Rotor speed}}{\text{Rotor speed}}$

(d) Slip = $\frac{\text{Rotor speed–Synchronous speed}}{\text{Synchronous speed}}$.

168. The condition for maximum torque at standstill in case of induction motor is

(a) rotor resistance = 0 (b) rotor reactance at supply frequency = 0

(c) rotor resistance = rotor reactance at supply frequency

(d) rotor resistance × rotor reactance at supply frequency = 1.

169. When $R_2$ is the rotor resistance, $X_2$ the rotor reactance at supply frequency and $s$ the slip, then the condition for maximum torque under running conditions will be

(a) $sR_2X_2 = 1$ (b) $sR_2 = X_2$ (c) $R_2 = sX_2$ (d) $R_2 = s^2(X_2)$.

**Questions 170 and 171:**

170. If $T$ be the torque and $s$ be the slip if $s$ is very low, then

(a) $T = 0$ (b) $T \propto \frac{1}{s}$ (c) $T \propto s$ (d) $T \propto s^2$.

171. For high values of slip

(a) $T \propto s$ (b) $T \times s^2$ (c) $T \propto \frac{1}{s}$ (d) $T \propto \frac{1}{s^2}$.

172. The power factor of a three phase induction motor is usually
(*a*) unity (*b*) 0.8 leading (*c*) 0.8 lagging (*d*) zero.

173. How many different speeds can be obtained in cascade control of two induction motors?
(*a*) 2 (*b*) 4 (*c*) 8 (*d*) 16.

174. In which of the following case the speed of the motor will remain unchanged?
(*a*) frequency is doubled and number of poles reduced to half
(*b*) frequency is halved and number of poles doubled
(*c*) frequency is halved and number of poles is also halved
(*d*) frequency is doubled and number of poles remains unaltered.

175. In case of a double cage induction motor, the inner cage has
(*a*) high inductance and low resistance (*b*) low inductance and high resistance
(*c*) low inductance and low resistance (*d*) high inductance and high resistance.

176. Locked rotor test on induction motors is conducted to determine
(*a*) friction losses (*b*) windage losses (*c*) $I^2R$ losses (*d*) core losses.

177. Open circuit test is usually conducted on
(*a*) slip ring motors only (*b*) wound rotor motors only
(*c*) either of (*a*) and (*b*) above (*d*) none of the above.

178. When two motors of 12 and 8 poles respectively are connected to a 50 Hz supply, which of the following no load speed cannot be obtained?
(*a*) 1500 rpm (*b*) 750 rpm (*c*) 500 rpm (*d*) 450 rpm
(*e*) 300 rpm.

179. The low power factor of induction motor is due to
(*a*) the reactive lagging magnetizing current necessary to generate the magnetic flux
(*b*) stator reactance
(*c*) rotor leakage reactance (*d*) all of the above.

180. During back to back test
(*a*) both machines run as induction motors
(*b*) both machines run as induction generators
(*c*) one machine runs as induction motor and the other as induction generator
(*d*) one machine runs as induction motor and the other as synchronous generator.

**Questions 181 to 182 Refer to Fig. 12.13:**

181. In Fig. 12.13 no load test curves for an induction motor are drawn. Which curve is likely to represent the variation of power with voltage?
(*a*) curve *A* (*b*) curve *B* (*c*) curve *D* (*d*) curve *E*.

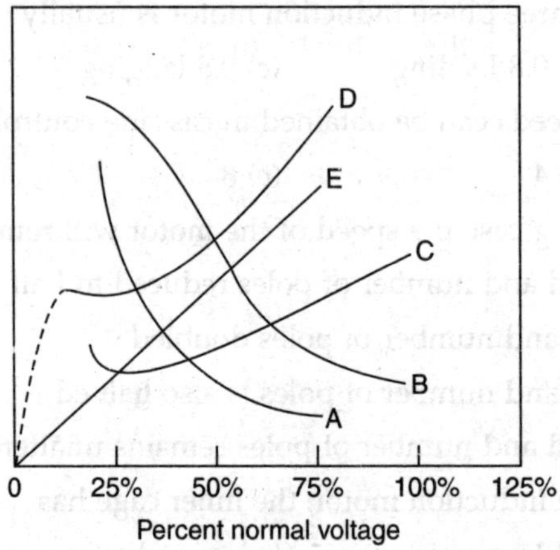


*FIGURE 12.13*

**182.** Which curve is likely to represent the variation of power factor?

(*a*) curve *B* (*b*) curve *C* (*c*) curve *E* (*d*) none of the curves.

**183.** Phase advancers are used with induction motors to

(*a*) improve power factor (*b*) reduce vibrations

(*c*) reduce copper losses (*d*) reduce noise.

**184.** During dynamic braking of induction motor

(*a*) fictitious load is applied on motor (*b*) stator winding is excited by direct current

(*c*) motor is used as induction generator (*d*) stator connections are reversed.

**185.** The operation of repeated starting and stopping of an induction motor for exact positioning of the driven equipment is called

(*a*) jogging (*b*) plugging (*c*) balancing (*d*) crawling.

**186.** A 400 V, 60 Hz, 3-phase induction motor is to be operated on 50 Hz supply. The desirable supply voltage is

(*a*) 330 V (*b*) 380 V (*c*) 400 V (*d*) 440 V.

**187.** No load current of an induction motor is nearly

(*a*) 1% to 2% (*b*) 2% to 5% (*c*) 10% to 20% (*d*) 50% to 55%.

**188.** The stator of a 4 pole 3 phase slip ring induction motor is fed from 50 Hz supply and the rotor is fed from 25 Hz supply. The speed of the rotor in rpm will be

(*a*) 2250 (*b*) 2125 (*c*) 1500 (*d*) 1200.

**189.** A stator rotor starter can be used in

(*a*) squirrel cage induction motors only (*b*) slip ring induction motors only

(*c*) both (*a*) and (*b*) above (*d*) none of the above.

**190.** In moist surroundings which frame of motor will be preferred?

(*a*) Drip proof (*b*) Splash proof (*c*) Totally enclosed fan cooled

(*d*) Open type.

**191.** The principle of operation of a three phase induction motor closely resembles to that of which of the following?

(*a*) DC series motor (*b*) Synchronous motor

(*c*) Transformer with open circuited secondary

(*d*) Transformer with short-circuited secondary.

**192.** Under which of the following conditions a squirrel cage induction motor would be preferred over a wound rotor induction motor?

(*a*) When cost is the major consideration (*b*) When higher starting torque is required

(*c*) When wide range of speed control is desired

(*d*) When external voltage is to be necessarily introduced into the rotor.

**193.** A three phase motor is designed to run at 1440 rpm. If the rotor of the motor is replaced by solid iron rotor, the rotor speed will be nerly

(*a*) 1000 rpm (*b*) 1400 rpm (*c*) 1140 rpm (*d*) zero.

**194.** In a three phase induction motor when rotor speed is more than the corresponding synchronous speed

(*a*) the motor runs as a generator (*b*) the motor acts as a synchronous generator

(*c*) the motor has negative slip

(*d*) the torque produced is more than the rated torque.

**195.** When a 3 phase induction motor is directly put on line, the starting current is usually

(*a*) $\frac{1}{5}$ to $\frac{1}{7}$ of the full load current (*b*) $\frac{1}{2}$ to $\frac{1}{3}$ of full load current

(*c*) same as full load current (*d*) 5 to 7 times the full load current.

**196.** For an induction motor the slip is *s* and the ratio of rotor resistance to rotor reactance is *k*.

The value of $\frac{\text{full load torque}}{\text{maximum torque}}$ for the motor will be

(*a*) $k^2s^2$ (*b*) $k^2 + s^2$ (*c*) $\frac{k}{\sqrt{k^2+s^2}}$ (*d*) $\frac{s}{\sqrt{k^2+s^2}}$.

**197.** A three phase induction motor takes full load current of *I* amperes when power factor is cos $\phi$. The component *I* sin $\phi$ will

(*a*) be equal to copper losses (*b*) over come iron losses

(*c*) produce magnetism (*d*) be dissipated as heat.

**198.** Which of the following statement is not valid when a squirrel cage induction motor operates under no load?

(*a*) power factor is low (*b*) slip is low

(*c*) rotor current is low (*d*) rotor induced emf is low.

199. A wound rotor induction motor can be distinguished from squirrel cage induction motor by the

(a) presence of cooling fan (b) size of rotor shaft

(c) presence of slip rings (d) type of starter.

200. An induction motor develops rated torque at 2% slip when rated voltage 400 V is supplied. If the supply voltage is increased to 440 volts, the slip to develop the rated torque will be

(a) – 2 p.c. (b) – 1.66 p.c. (c) 1.66 p.c. (d) 2 p.c.

## ANSWERS

| | | | | | | |
|---|---|---|---|---|---|---|
| 1. (a) | 2. (c) | 3. (c) | 4. (c) | 5. (a) | 6. (b) | 7. (d) |
| 8. (d) | 9. (d) | 10. (a) | 11. (c) | 12. (d) | 13. (a) | 14. (a) |
| 15. (b) | 16. (d) | 17. (c) | 18. (b) | 19. (b) | 20. (a) | 21. (c) |
| 22. (d) | 23. (c) | 24. (d) | 25. (d) | 26. (d) | 27. (a) | 28. (c) |
| 29. (c) | 30. (d) | 31. (c) | 32. (c) | 33. (b) | 34. (a) | 35. (a) |
| 36. (a) | 37. (c) | 38. (b) | 39. (b) | 40. (d) | 41. (a) | 42. (c) |
| 43. (a) | 44. (d) | 45. (d) | 46. (b) | 47. (b) | 48. (d) | 49. (a) |
| 50. (c) | 51. (d) | 52. (c) | 53. (a) | 54. (c) | 55. (b) | 56. (c) |
| 57. (d) | 58. (c) | 59. (c) | 60. (d) | 61. (a) | 62. (c) | 63. (a) |
| 64. (d) | 65. (a) | 66. (b) | 67. (d) | 68. (a) | 69. (b) | 70. (c) |
| 71. (d) | 72. (b) | 73. (b) | 74. (c) | 75. (d) | 76. (d) | 77. (b) |
| 78. (d) | 79. (c) | 80. (c) | 81. (a) | 82. (b) | 83. (b) | 84. (c) |
| 85. (a) | 86. (b) | 87. (a) | 88. (d) | 89. (d) | 90. (b) | 91. (d) |
| 92. (d) | 93. (d) | 94. (b) | 95. (b) | 96. (d) | 97. (c) | 98. (d) |
| 99. (b) | 100. (b) | 101. (d) | 102. (b) | 103. (b) | 104. (c) | 105. (d) |
| 106. (b) | 107. (d) | 108. (d) | 109. (a) | 110. (a) | 111. (b) | 112. (c) |
| 113. (d) | 114. (b) | 115. (c) | 116. (c) | 117. (c) | 118. (c) | 119. (c) |
| 120. (d) | 121. (b) | 122. (b) | 123. (d) | 124. (d) | 125. (d) | 126. (b) |
| 127. (b) | 128. (d) | 129. (c) | 130. (d) | 131. (b) | 132. (d) | 133. (d) |
| 134. (d) | 135. (d) | 136. (d) | 137. (c) | 138. (d) | 139. (d) | 140. (c) |
| 141. (d) | 142. (d) | 143. (a) | 144. (c) | 145. (d) | 146. (b) | 147. (d) |
| 148. (b) | 149. (c) | 150. (a) | 151. (d) | 152. (b) | 153. (b) | 154. (d) |
| 155. (b) | 156. (d) | 157. (c) | 158. (c) | 159. (d) | 160. (c) | 161. (d) |
| 162. (d) | 163. (a) | 164. (c) | 165. (a) | 166. (c) | 167. (b) | 168. (c) |
| 169. (c) | 170. (c) | 171. (c) | 172. (c) | 173. (b) | 174. (c) | 175. (a) |
| 176. (c) | 177. (a) | 178. (d) | 179. (d) | 180. (c) | 181. (c) | 182. (a) |
| 183. (a) | 184. (b) | 185. (a) | 186. (a) | 187. (c) | 188. (a) | 189. (b) |
| 190. (c) | 191. (d) | 192. (a) | 193. (b) | 194. (c) | 195. (d) | 196. (b) |
| 197. (c) | 198. (b) | 199. (c) | 200. (c). | | | |

# Chapter 13 SINGLE-PHASE MOTORS

Several appliances for domestic use have single phase motors which can operate on 220 V, 50 Hz supply. For example motors used in mixer-grinders, refrigerators, air conditioners, fans, etc. have single phase motors. These motors are generally of small size with their horsepower rating usually less than 1. Since the performance requirements vary considerably. Several types of motors have been developed. Based on their construction and method of starting, these motors may be classified as

1. Induction motors
   (*a*) split-phase
   (*b*) capacitor
   (*c*) shaded pole.
2. Repulsion motors.
3. AC series motors.
4. Unexcited synchronous motors.

**Split-Phase Motors.** Single phase motors are inherently non-self starting. To overcome this drawback and make the motor self starting, it is temporarily converted into a two phase motor during starting period. For this purpose, the stator is provided with two windings: one for starting and the other for running the motor. A phase difference between the two windings is created so that motor acts as a two phase motor and thus facilitate starting.

The starting and running windings are connected in parallel directly across the single phase line. In split phase motor, the starting winding is wound with small wire so that the winding has a high resistance and low inductance, whereas the running winding is wound with large diameter wire to have a low resistance and a high inductance. When motor is switched on, both windings get energized. However because of the higher inductance of the running winding with relation to that of the starting winding, the current flow in the running winding lags the current flow in the starting winding by approximately 30° electrical. Thus the phase is "split" to give the effect of two phases and a rotating field is set up in the stator which produces a starting torque and causes the rotor to rotate. When the motor has accelerated to approximately 70% of the running speed, a centrifugal mechanism, mounted on the rotor shaft opens a switch in the starting winding. With the starting winding disconnected, the motor continues to operate on the running winding alone.

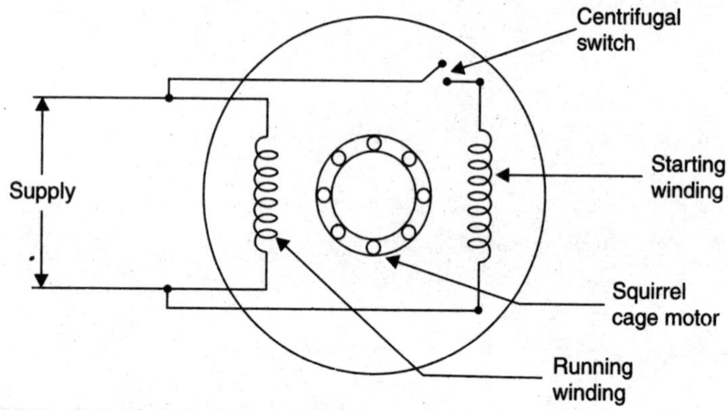


*FIGURE 13.1*

As the maximum split is restricted to about 30 electrical degrees, this motor has low starting torque. These motors are generally available in sizes ranging from $\frac{1}{20}$ to $\frac{1}{3}$ HP. These motors are primarily used to drive fans, blowers and pumps.

The torque speed characteristic of the motor is shown in Fig. 13.2.

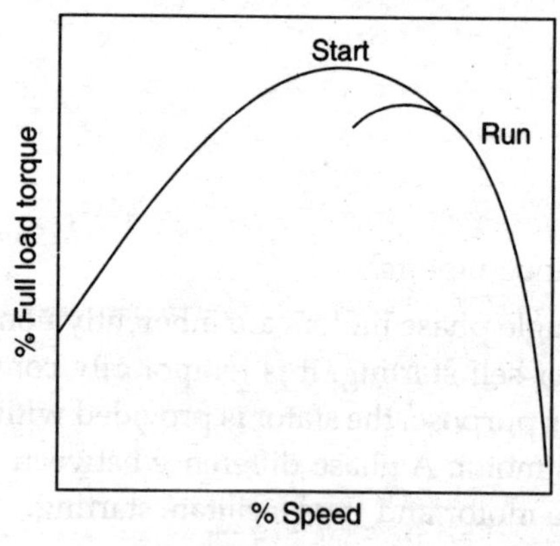


*FIGURE 13.2*

**Capacitor Start Motor.** In this case phase shift is produced by incorporating a capacitor in the starting winding. In this the current in the starting winding leads the voltage and in the running winding the current lags the voltage due to inductance of winding coils. By this arrangement phase difference of 90 electrical degrees can be achieved between the two windings.

These motors have high starting torque as compared to split phase motors. In this the capacitor is taken out of the circuit when motor attains about 70% of the normal speed.

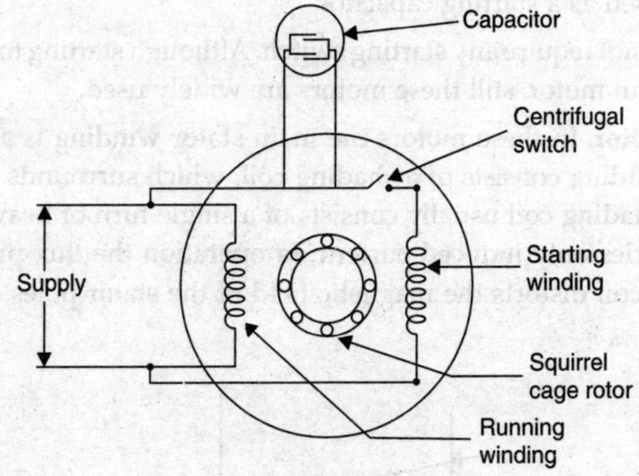


*FIGURE 13.3*

**Capacitor Start and Run Motor.** These motors are identical to capacitor start motors with the only difference that an additional capacitor is introduced in the running winding. The running capacitor is installed in series with the starting winding but in parallel with the starting switch. The operation of the capacitor start and run motor differs from that of the capacitor start-and-split phase motors in that the starting or auxiliary winding remains in the circuit at all times. At the instant of starting the starting-and-running capacitors are both in the circuit in series with the auxiliary winding so that the capacity of both capacitors is utilized during the starting period. At 70% of the rated speed when centrifugal mechanism disconnects the starting capacitor from the circuit, the motor continues to operate with both main and auxiliary windings in the circuit. The function of the running capacitor in the auxiliary winding is to correct power factor. Thus these motors have high starting torque and higher efficiency also.

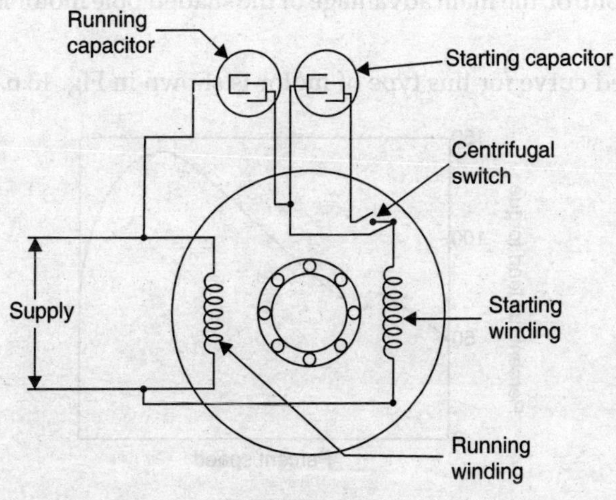


*FIGURE 13.4*

**Permanent Capacitor Motor.** This motor is similar to capacitor start and run motor except that starting capacitor and centrifugal switch is used. The capacitor is sized for power factor correction but is also used as a starting capacitor.

These motors do not require any starting switch. Although starting torque is low as compared to capacitor start and run motor, still these motors are widely used.

**Shaded Pole Motor.** In these motors the main stator winding is arranged to form salient poles. The auxiliary winding consists of a shading coil, which surrounds a portion of one side of each stator pole. The shading coil usually consists of a single turn of heavy copper wire which is short-circuited and carries only induced current. In operation the flux produced by the induced current in the shading coil distorts the magnetic field of the stator poles and thereby produces a small starting torque.

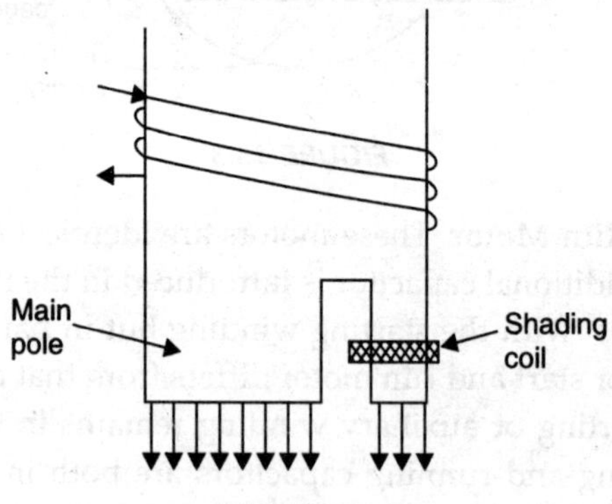


*FIGURE 13.5*

Shaded pole motors are widely used for fans which are mounted directly on the motor shaft. They are available in sizes ranging from $\frac{1}{125}$ HP through $\frac{1}{20}$ HP. In addition to its ready adaptability to speed control, the main advantage of the shaded pole motor is its simple construction and low cost.

The torque speed curve for this type of motor is shown in Fig. 13.6.

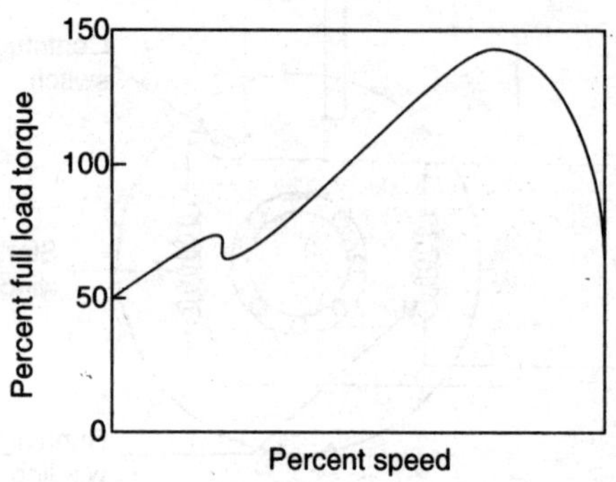


*FIGURE 13.6*

**Repulsion Induction Motor.** These motors are provided with two windings in the rotor.

1. A squirrel cage winding.
2. Commuted winding similar to that of a dc armature.

At start, the commuted winding supplies most of the torque, the squirrel cage winding being inactive because of its high reactance. When the rotor accelerates, the squirrel cage winding takes up a larger portion of the load.

This motor does not require short circuiting mechanism. The direction of rotation can be reversed by the usual brush shifting arrangement.

**AC Series Motor.** In case a dc series motor is connected to ac mains, it will rotate and exert unidirectional torque mainly due to the reason that the current flowing both in the armature and the field reverses at the same time. However, because of alternating current.

1. there will be excessive eddy current loss in the yoke and field cores, resulting is heating of the two.
2. vicious sparking will occur at brushes.
3. power factor will be low due to high inductances of armature and field circuits.

All above problems can be solved by appropriate modification. Eddy currents can be reduced by laminating the iron structure of the yoke and field core. Power factor can be improved by reducing the reactances of the windings. This is possible by reducing the number of turns.

These motors are provided with commutating poles for improving commutation. Fig. 13.7 shows a ac series motor with campole winding.

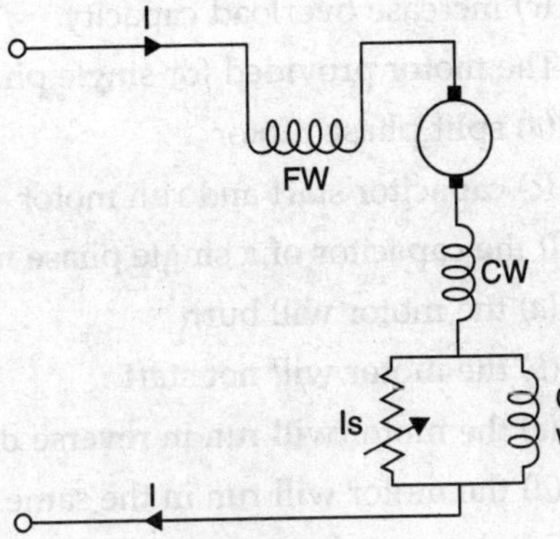


*FIGURE 13.7*

**Universal Motor.** This motor can be operated on dc as well as ac single phase supply. In fact this is a smaller version of ac series motor. As it is series wound therefore it has high starting torque and also runs at very high speeds on no loads. The speed-load characteristic of the motor is shown in Fig. 13.8.

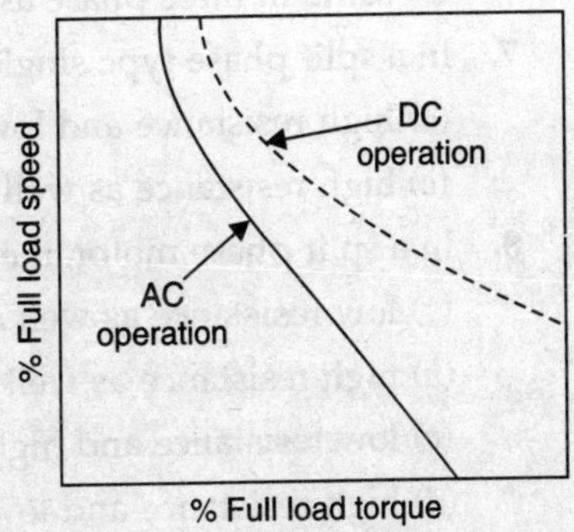


*FIGURE 13.8*

**Hysteresis Motor.** This motor has a smooth crome-steel cylinder having retentivity so that the hysteresis loss is high.

This motor does not have any winding. The rotor revolves synchronously because the rotor poles magnetically lock-up with the revolving stator poles of opposite polarity.

As this motor does not have teeth or winding therefore it is quiet calm in operation and free from vibrations. This motor is widely used for tape recorders, video, sound equipment etc.

## OBJECTIVE TYPE QUESTIONS

1. In capacitor start single phase motors
   (*a*) current in starting winding leads the voltage
   (*b*) current in starting winding lags the voltage
   (*c*) current in starting winding is in phase with voltage in running winding
   (*d*) none of the above.
2. In a capacitor start motor, the phase displacement between starting and running winding can be nearly
   (*a*) 10° (*b*) 30° (*c*) 60° (*d*) 90°
3. In a capacitor start and run motors the function of the running capacitor is series with the auxiliary winding is
   (*a*) to improve torque (*b*) reduce fluctuations in torque
   (*c*) increase overload capacity (*d*) improve power factor.
4. The motor provided for single phase motors is usually
   (*a*) split phase motor (*b*) capacitor start motor
   (*c*) capacitor start and run motor (*d*) universal motor.
5. If the capacitor of a single phase motor is short-circuited
   (*a*) the motor will burn
   (*b*) the motor will not start
   (*c*) the motor will run in reverse direction
   (*d*) the motor will run in the same direction at reduced rpm.
6. For the same horsepower, torque, rpm speed and operating voltage, the slip will be
   (*a*) more in single phase motor (*b*) more in three phase motor
   (*c*) same in three phase as well as single phase motor.
7. In a split phase type single phase motor the starting winding should have
   (*a*) high resistance and low inductance (*b*) low resistance and high inductance
   (*c*) high resistance as well as inductance (*d*) low resistance as well as inductance.
8. In a split phase motor, the running winding should have
   (*a*) low resistance as well as low inductance
   (*b*) high resistance as well as high inductance
   (*c*) low resistance and high inductance
   (*d*) high resistance and low inductance.
9. The maximum split that can be achieved with the split phase motor is nearly
   (*a*) 10° electrical (*b*) 30° electrical (*c*) 60° electrical (*d*) 90° electrical.
10. In a split phase motor
   (*a*) the starting winding is connected through a centrifugal switch
   (*b*) the running winding is connected through a centrifugal switch

(c) both starting as well as running windings are connected through a centrifugal switch
(d) centrifugal switch is used to control supply voltage.

**11.** The torque developed by a single phase motor in starting is
(a) more than the rated torque (b) rated torque
(c) less than the rated torque (d) zero.

**12.** Which of the following motor will give relatively high starting torque?
(a) Shaded pole motor (b) Capacitor run motor
(c) Split phase motor (d) Capacitor start motor.

**13.** A capacitor-start single phase induction motor is switched on to supply with its capacitor replaced by an inductor of equivalent reactance value. It will
(a) not start (b) start and run
(c) start and then stall (d) burn.

**14.** Which type of capacitor is preferred for capacitor start and run motor?
(a) Air capacitor (b) Paper capacitor (oil filled)
(c) Ceramic capacitor (d) Electrolytic capacitor.

**15.** Which of the following motor will have relatively higher power factor?
(a) Split phase motor (b) Shaded pole motor
(c) Capacitor start motor (d) Capacitor run motor.

**16.** In a shaded pole motor, the shading coil usually consists of
(a) a single turn of heavy copper wire which is in parallel with running winding
(b) a single turn of heavy copper wire which is short-circuited and carries only induced current
(c) a multilayer fine gauge copper wire in parallel with running winding
(d) none of the above.

**17.** Hermetically sealed motors are commonly used in
(a) ceiling fans (b) blowers (c) refrigerators (d) mixies.

**18.** A shaded pole single phase motor is generally not used due to all of the following reasons EXCEPT
(a) poor starting torque (b) low power factor
(c) low speed operation (d) irreversibility of direction of rotation.

**19.** In a shaded pole single phase motor, the revolving field is produced by the use of
(a) inductor (b) capacitor (c) resistor (d) shading coils.

**20.** A centrifugal switch is used to disconnect starting winding when motor has
(a) run for about 5 minutes (b) run for about 1 minute
(c) picked up about 50 to 70 percent of rated speed
(d) picked up about 10 to 25 percent of rated speed.

21. Three phase as well as single phase induction motors are available. For same voltage, rpm and hourse-power rating

(*a*) the two motors will have same p.f.

(*b*) single phase motor will have less p.f. as compared to three phase motor

(*c*) single phase motor will have higher p.f. as compared to three phase motor

(*d*) single phase motor will have leading p.f. while three phase motor will have lagging p.f.

22. If a particular application needs high speed and high starting torque, then which of the following motor will be preferred?

(*a*) Shaded pole type motor (*b*) Capacitor start motor

(*c*) Capacitor start and run motor (*d*) Universal motor.

23. In ceiling fan motors the rotor is usually supported in

(*a*) ball bearings (*b*) bush bearings (*c*) cast iron bearings

(*d*) no bearings are used in single phase motors.

24. A ball bearing is made of

(*a*) cast iron (*b*) mild steel (*c*) high speed steel (*d*) carbon chrome steel.

25. The balls for ball bearings are made by

(*a*) casting (*b*) rolling (*c*) turning (*d*) cold heading.

26. When $\alpha$ is the angle between the two winding currents of a capacitor start split phase induction motor, the starting torque of the motor will be proportional to

(*a*) $\sin \alpha$ (*b*) $\cos \alpha$ (*c*) $\operatorname{cosec} \alpha$ (*d*) $\tan \alpha$.

27. Which motor has the highest power to weight ratio?

(*a*) Universal motor (*b*) Capacitor start motor

(*c*) Capacitor run motor

28. A single phase motor is usually started by

(*a*) start delta starter (*b*) auto-transformer starter

(*c*) direct-on-line starter (*d*) none of the above.

29. Single phase capacitor start motors, single voltage, non-reversible with magnetic switch are commonly used in

(*a*) refrigerators (*b*) ceiling fans (*c*) toys (*d*) cranes.

30. In shaded pole motors, the necessary phase splitting is produced by

(*a*) induction (*b*) inductance (*c*) resistance (*d*) capacitance.

31. In a universal motor the width of the brush is nearly

(*a*) half the width of commutator segment

(*b*) equal to the width of the commutator segments

(*c*) equal to the width of two commutator segments

(*d*) equal to the width of four commutator segments.

**32.** Motorised handtools generally use

(*a*) capacitor start motors (*b*) reluctance motors
(*c*) shaded pole motor (*d*) universal motor.

**33.** If motor speed of 5000 rpm is desired which of the following motor will be the obvious choice?

(*a*) Reluctance motor (*b*) Hysteresis motor
(*c*) Universal motor (*d*) Capacitor start and run motor.

**34.** Shaded pole motors are built commercially in sizes

(*a*) $\frac{1}{250}$ HP to $\frac{1}{6}$ HP (*b*) $\frac{1}{4}$ HP to 1 HP (*c*) 1 HP to 2 HP (*d*) 2 HP to 5 HP.

**35.** The range of efficiency for shaded pole motors is

(*a*) 95% to 99% (*b*) 80% to 90% (*c*) 50% to 75% (*d*) 5% to 35%.

**36.** In a capacitor start single phase motor, when capacitor is replaced by a resistance

(*a*) the motor will consume less power (*b*) torque will increase
(*c*) motor will run in reverse direction
(*d*) motor will continue to run in same direction.

**37.** The power factor of a single phase induction motor is usually

(*a*) unity (*b*) unity to 0.8 leading
(*c*) always leading (*d*) lagging.

**38.** Which of the following motor does not have a centrifugal switch?

(*a*) Capacitor start motor (*b*) Capacitor run motor
(*c*) Split phase motor (*d*) None of the above.

**39.** The torque speed characteristics of a shaded pole motor is represented by

(*a*) figure *A* (*b*) figure *B* (*c*) figure C (*d*) figure *D*.

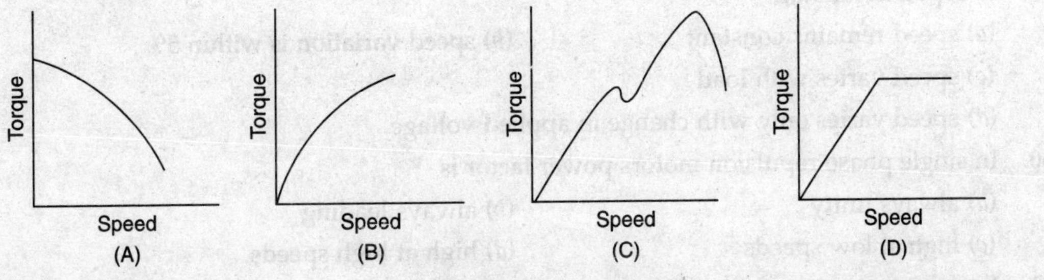


***FIGURE 13.9***

**40.** A shaded pole motor can be used for

(*a*) hair dryers (*b*) toys (*c*) circulators (*d*) any of the above.

**41.** What will happen when the capacitor of a capacitor start motor is replaced by an inductance of same reactance value?

(*a*) The motor will burn (*b*) The inductance will burn
(*c*) The motor will run in reverse direction (*d*) The motor will produce high torque.

42. A hysteresis motor works on the principle of
(*a*) hysteresis loss (*b*) magnetisation of rotor
(*c*) eddy current loss (*d*) electromagnetic induction.

43. Which of the following motor will give the highest starting torque?
(*a*) schrage motor (*b*) DC shunt motor
(*c*) universal motor (*d*) repulsion start and induction run motor.

44. Which of the following single phase motor has a commutator?
(*a*) Capacitor start motor (*b*) Capacitor run motor
(*c*) Split phase motor (*d*) Repulsion motor.

45. In repulsion motors by shifting the position of brushes
(*a*) only speed can be controlled
(*b*) speed and direction of rotation can be controlled
(*c*) speed, direction of rotation and torque can be controlled
(*d*) none of the above.

46. Which of the following single phase motor is reversible?
(*a*) reluctance motor (*b*) resistance start split phase motor
(*c*) capacitor start split-phase motor (*d*) none of the above.

47. For which of the following application a reluctance motor is preferred?
(*a*) Signalling and timing devices (*b*) Refrigerators
(*c*) Lifts and hoists (*d*) Electric shavers.

48. The rpm of a 2 pole hysteresis motor on 60 Hz supply will be
(*a*) 120 (*b*) 1500 (*c*) 1800 (*d*) 3600.

49. In repulsion motors
(*a*) speed remains constant (*b*) speed variation is within 5%
(*c*) speed varies with load
(*d*) speed varies only with change in applied voltage.

50. In single phase repulsion motors power factor is
(*a*) always unity (*b*) always leading
(*c*) high at low speeds (*d*) high at high speeds.

51. For a given output at rated rpm, a universal motor will require least voltage
(*a*) at 60 Hz (*b*) at 50 Hz (*c*) at 25 Hz (*d*) for dc supply.

52. The motor used on small lathes is usually
(*a*) synchronous motor 3 phase type (*b*) DC shunt motor
(*c*) single phase capacitor run motor (*d*) universal motor.

53. Which of the following is the unexcited single phase synchronous motor?
(*a*) Reluctance motor (*b*) Universal motor
(*c*) AC series motor (*d*) Repulsion motor.

**54.** Split phase motors are commonly available in sizes

(*a*) $\frac{1}{20}$ HP to $\frac{1}{3}$ HP (*b*) $\frac{1}{2}$ HP to 2 HP (*c*) 1 HP to 5 HP (*d*) above 5 HP.

**55.** The speed torque characteristic of a split phase motor is represented by

(*a*) figure *A* (*b*) figure *B* (*c*) figure *C* (*d*) figure *D*.

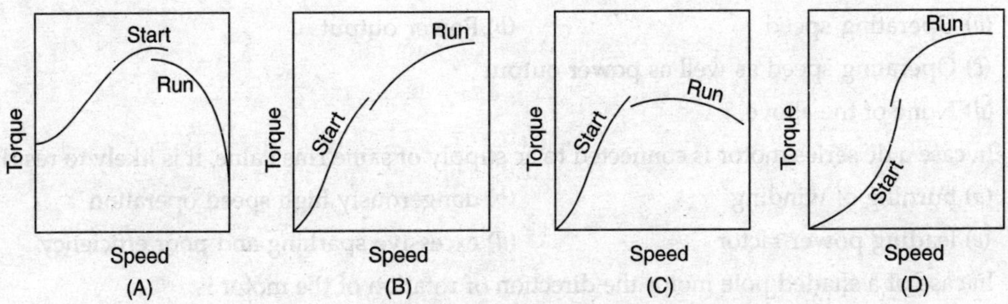


***FIGURE 13.10***

**56.** Which of the following motor is preferred for tape recorders?

(*a*) Shaded pole motor (*b*) Two value capacitor motor

(*c*) Hysteresis motor (*d*) Universal motor.

**57.** For which of the following application a two value capacitor motor is preferred?

(*a*) Toys (*b*) Refrigerators (*c*) Electric shavers (*d*) Record players.

**58.** A single phase induction motor is

(*a*) inherently self starting with high torque

(*b*) inherently self starting with low torque

(*c*) inherently non-self starting with low torque

(*d*) inherently non-self starting with high torque.

**59.** A schrage motor can run on

(*a*) zero slip (*b*) negative slip (*c*) positive slip (*d*) all of the above.

**60.** The speed regulation of standard split-phase motor is nearly same as that of

(*a*) universal motor (*b*) 3 phase induction motor

(*c*) synchronous motor (*d*) DC series motor.

**61.** In a capacitor start motor any reduction in the value of capacitance is likely to result in reduced

(*a*) speed (*b*) vibrations (*c*) rpm (*d*) starting torque.

**62.** Which of the following is not necessarily the characteristic of a shaded pole motor?

(*a*) Low starting torque (*b*) High power factor

(*c*) Large slip (*d*) Poor efficiency.

63. A universal motor can run on
(*a*) AC only (*b*) DC only (*c*) either AC or DC (*d*) none of the above.

64. Which of the following single phase motor is suitable for timing and control purposes?
(*a*) Universal motor (*b*) Series motor (*c*) Repulsion motor (*d*) Reluctance motor.

65. A universal motor can be operated on 220 V dc as well as 220 V single phase a.c. Which of the following will be the same in two cases?
(*a*) Operating speed
(*b*) Power output
(*c*) Operating speed as well as power output
(*d*) None of the above.

66. In case a dc series motor is connected to ac supply of same rms value, it is likely to result in
(*a*) burning of winding
(*b*) dangerously high speed operation
(*c*) leading power factor
(*d*) excessive sparking and poor efficiency.

67. In case of a shaded pole motor the direction of rotation of the motor is
(*a*) from shaded pole to main pole
(*b*) from main pole to shaded pole
(*c*) either of the above depending on voltage
(*d*) either of the above depending on power factor.

68. Which of the following is not valid in case of repulsion motors?
(*a*) High power factor
(*b*) Sparking at brushes
(*c*) High speed on no load
(*d*) Speed variation proportional to load.

69. The speed variation of split phase induction motor between no load and full load condition is
(*a*) 0.2 to 0.5 percent (*b*) 2 to 5 percent (*c*) 10 to 20 percent (*d*) 20 to 30 percent.

70. All of the following can cause overheating of a single phase motor EXCEPT
(*a*) overloads (*b*) wornout bearings (*c*) high voltage (*d*) shorted windings.

71. All of the following are not found in a shaded pole motor EXCEPT
(*a*) centrifugal switch
(*b*) rotor
(*c*) commutator
(*d*) capacitor.

72. In case of high speed universal motor which of the following needs more attention?
(*a*) Insulation in rotor
(*b*) Air gap
(*c*) Balancing of rotor
(*d*) End play.

73. The wattage rating for a ceiling fan motor will be in the range
(*a*) 10 to 20 watts (*b*) 50 to 150 watts (*c*) 250 to 500 watts (*d*) 500 to 750 watts.

74. The wattage of motor for driving domestic sewing machine will be around
(*a*) 5 to 10 watts (*b*) 10 to 30 watts (*c*) 40 to 75 watts (*d*) 100 to 150 watts.

75. The motor for which of the following application will have the least wattage
(*a*) Mixer grinder (*b*) Table fan (*c*) Electric shaver (*d*) Exhaust fan.

76. Which of the following single phase motor has relatively poor starting torque?
(*a*) repulsion motor (*b*) universal motor
(*c*) capacitor motor
(*d*) all single phase motors have zero starting torque.

77. The capacitor selected for capacitor run motor should be rated for
(*a*) peak value of voltage (*b*) rms value of voltage
(*c*) dc value of voltage (*d*) average value of voltage.

78. Which type of load is offered by cranes and hoists?
(*a*) Gradually varying load (*b*) non-reversing, no load start
(*c*) Reversing, light start (*d*) Reversing, heavy start.

79. The speed of a universal motor is generally reduced by using
(*a*) *V*-belts (*b*) gear trains
(*c*) brakes (*d*) chains.

80. *HRC* fuses in the line for single phase motors provide best protection against
(*a*) open-circuit (*b*) short-circuit (*c*) overload (*d*) voltage fluctuation.

81. Which of the following motor can be used for unity power factor?
(*a*) Shaded pole motor (*b*) capacitor run motor
(*c*) schrage motor (*d*) hysteresis motor.

82. The speed load characteristics of universal motors closely resembles the characteristics of which of the following motor?
(*a*) capacitor start motor (*b*) shaded pole motor
(*c*) 3 phase synchronous motor (*d*) none of the above.

83. When a dc series motor is connected to ac supply excessive eddy currents will be produced in
(*a*) yoke (*b*) yoke and field cores
(*c*) yoke, field cores and stator (*d*) yoke, field cores, stator and rotor shaft.

84. When a dc series motor is connected to ac supply, the power factor will be low because of
(*a*) high inductance of field and armature circuits
(*b*) induced current in rotor due to variation of flux
(*c*) fine copper wire winding (*d*) none of the above.

85. The direction of rotation of the universal motor can be reversed by reversing the flow of current through
(*a*) armature winding (*b*) field winding
(*c*) either armature winding or field winding
(*d*) none of the above.

86. Which universal motor is used for low horsepower rating?
(*a*) Concentrated pole, non-compensated type
(*b*) Distributed field compensated type (*c*) Either of (*a*) or (*b*) above.

87. In which single phase motor, the rotor has neither teeth nor winding?
(*a*) Reluctance motor (*b*) Split phase motor
(*c*) Hysteresis motor (*d*) Universal motor.

88. Which motor is normally free from mechanical and magnetic vibrations?
(*a*) Universal motor (*b*) Split phase motor
(*c*) Shaded pole motor (*d*) Hysteresis motor.

89. As hysteresis motors are free from mechanical and magnetic vibrations therefore these are considered as suitable for
(*a*) fans (*b*) blowers (*c*) mixer grinders (*d*) sound equipment.

90. For 50 Hz supply a hysteresis motor will run at
(*a*) 2880 rpm (*b*) 2950 rpm (*c*) 3000 rpm (*d*) 3120 rpm.

91. A reluctance motor
(*a*) is self starting (*b*) is constant speed motor
(*c*) needs no dc excitation (*d*) all of the above.

92. Reluctance motor can be used for
(*a*) recording instruments (*b*) timers
(*c*) signalling devices (*d*) any of the above.

93. In a hysteresis motor, the rotor must have high
(*a*) resistivity (*b*) retentivity (*c*) susceptibility (*d*) none of the above.

94. The rotor of a hysteresis motor is made of
(*a*) cast iron (*b*) aluminium (*c*) chrome-steel (*d*) copper.

95. Because of high retentivity of the rotor of hysteresis motor
(*a*) eddy current loss is high (*b*) eddy current loss is low.
(*c*) hysteresis loss is low (*d*) hysteresis loss is high.

96. In a universal motor, on no load the speed is limited by
(*a*) voltage (*b*) current flow
(*c*) friction and windage load (*d*) none of the above.

97. The electric motor used in portable drills is
(*a*) universal motor (*b*) capacitor run motor
(*c*) repulsion motor (*d*) hysteresis motor.

98. Which of the following applications always have some load whenever switched on?
(*a*) Fan motors (*b*) Vacuum cleaners (*c*) Pistol drills (*d*) All of the above.

99. In pistol drills gear train is used to
(*a*) vary the speed (*b*) reduce the speed
(*c*) increase the speed (*d*) change the direction or rotation.

**100.** The speed control of universal motor used for sewing machines is by

(*a*) varying the resistance (*b*) friction

(*c*) centrifugal mechanism (*d*) tapping the field.

## ANSWERS

| | | | | | | |
|---|---|---|---|---|---|---|
| **1.** (*a*) | **2.** (*d*) | **3.** (*d*) | **4.** (*c*) | **5.** (*b*) | **6.** (*a*) | **7.** (*a*) |
| **8.** (*c*) | **9.** (*d*) | **10.** (*a*) | **11.** (*a*) | **12.** (*d*) | **13.** (*a*) | **14.** (*b*) |
| **15.** (*d*) | **16.** (*b*) | **17.** (*c*) | **18.** (*c*) | **19.** (*d*) | **20.** (*c*) | **21.** (*b*) |
| **22.** (*d*) | **23.** (*a*) | **24.** (*d*) | **25.** (*d*) | **26.** (*a*) | **27.** (*a*) | **28.** (*c*) |
| **29.** (*c*) | **30.** (*a*) | **31.** (*c*) | **32.** (*d*) | **33.** (*c*) | **34.** (*a*) | **35.** (*d*) |
| **36.** (*d*) | **37.** (*d*) | **38.** (*b*) | **39.** (*c*) | **40.** (*d*) | **41.** (*c*) | **42.** (*a*) |
| **43.** (*a*) | **44.** (*d*) | **45.** (*b*) | **46.** (*c*) | **47.** (*a*) | **48.** (*d*) | **49.** (*c*) |
| **50.** (*d*) | **51.** (*d*) | **52.** (*c*) | **53.** (*a*) | **54.** (*a*) | **55.** (*a*) | **56.** (*c*) |
| **57.** (*b*) | **58.** (*c*) | **59.** (*d*) | **60.** (*b*) | **61.** (*d*) | **62.** (*b*) | **63.** (*c*) |
| **64.** (*d*) | **65.** (*c*) | **66.** (*d*) | **67.** (*b*) | **68.** (*a*) | **69.** (*b*) | **70.** (*c*) |
| **71.** (*b*) | **72.** (*c*) | **73.** (*b*) | **74.** (*d*) | **75.** (*c*) | **76.** (*c*) | **77.** (*a*) |
| **78.** (*d*) | **79.** (*b*) | **80.** (*b*) | **81.** (*c*) | **82.** (*d*) | **83.** (*b*) | **84.** (*a*) |
| **85.** (*c*) | **86.** (*a*) | **87.** (*c*) | **88.** (*d*) | **89.** (*d*) | **90.** (*c*) | **91.** (*d*) |
| **92.** (*d*) | **93.** (*b*) | **94.** (*c*) | **95.** (*d*) | **96.** (*c*) | **97.** (*a*) | **98.** (*d*) |
| **99.** (*b*) | **100.** (*a*). | | | | | |

Chapter **14** *TRANSFORMERS*

A transformer consists of a magnetic circuit linking with two distinct windings termed as the primary and secondary. When primary is connected to an alternating current, an alternating flux is set up and this flux linking with the secondary will induce an alternating emf in the secondary winding. This operation is known as mutual induction.

The transformers may be

(*i*) Core type with single magnetic circuit.

(*ii*) Shell type with double magnetic circuit.

(*iii*) Berry type with distributed magnetic circuit.

**EMF Equation.** In a transformer, the ratio

$$\frac{\text{Secondary induced emf, } E_2}{\text{Primary applied voltage, } E_1}$$

is known as transformation ratio.

Also, $$\frac{E_2}{E_1} = \frac{N_2}{N_1} = k$$

where $N_2$ = Number of turns in secondary

$N_1$ = Number of turns in primary.

In case $N_2 > N_1$, then $k > 1$ and the transformer is known as step-up transformer. If $N_2 < N_1$, then $k < 1$ and the transformer is known as step-down transformer.

**Transformer on no load.** In an ideal transformer there will be no losses and hence under no load conditions it should not drawn any current. However, in an actual transformer, on no load, iron loss occurs in the core and copper loss occurs in primary and secondary windings. Thus on no load, the primary carries a small current $I_0$ which has two components.

1. A magnetizing current, $I_\mu$ which lags so behind applied primary voltage, $V_1$. This is an idle component and its function is to produce the magnetic flux.
2. The working component $I_\omega$, which produces the necessary power to supply the hysteresis and eddy current losses in the iron. This component is in phase with $V_1$.

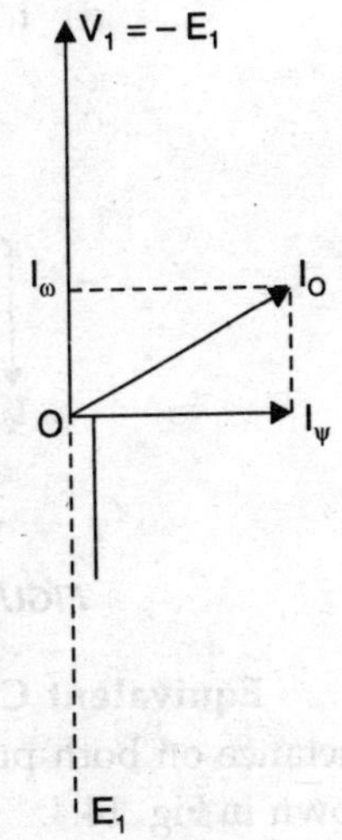

*FIGURE 14.1*

The no-load vector diagram is shown in Fig. 14.1. For high efficiency and better regulation $I_\mu$ and $I_\omega$ both must be small.

For low $I_\omega$

(*i*) low loss iron is used

(*ii*) high flux densities are not used.

For low $I_\mu$ a closed magnetic circuit without air gap is used.

**Transformer on Load.** Let us consider a non-inductive load on secondary side. In this case secondary current $I_2$ will be in phase with $E_2$.

The secondary mmf $N_2I_2$ will tend to produce a secondary flux, which if allowed to exist, would disturb the flux conditions existing at no load. This would result in primary mmf equal in magnitude but opposite in direction. This is provided by load current $I_1'$ which flows from the supply through the primary. Also

$$N_1I_1' = N_2I_2$$

$$\therefore \quad I_1' = \frac{N_2}{N} I_2$$

The total primary current is therefore the vector sum of the $I_1'$ and $I_0$ which can be taken as $I_1'$ for light loads. Thus approximately

$$I_1 = \frac{N_2}{N_1} I_2$$

or

$$\frac{I_2}{I_1} = \frac{N_1}{N_2}$$

Thus current transformation ratio is inverse of the voltage transformation ratio. The vector diagram for the transformer with non-inductive load is shown in Fig. 14.2.

In case the transformer secondary load has any phase angle $\phi$, the induced primary current $I_1'$ will be opposite in phase to the secondary current $I_2$, as shown in Fig. 14.3.

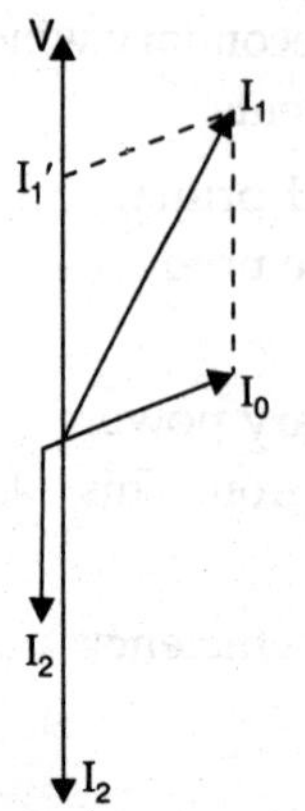


FIGURE 14.2

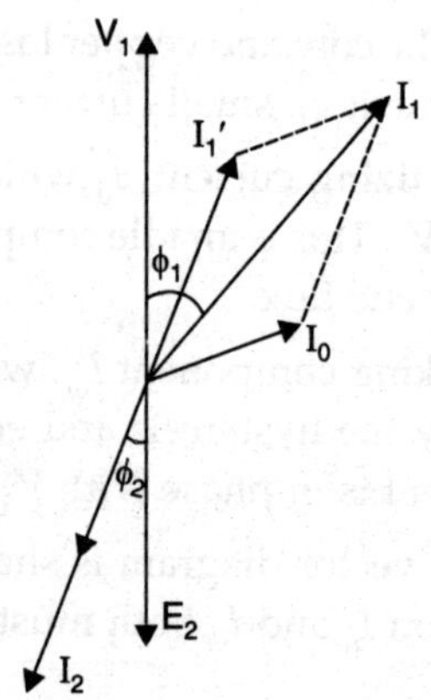


FIGURE 14.3

**Equivalent Circuit of a Transformer.** Transformer windings offer some resistance and reactance on both primary as well as secondary side. The equivalent circuit of a transformer is shown in Fig. 14.4.

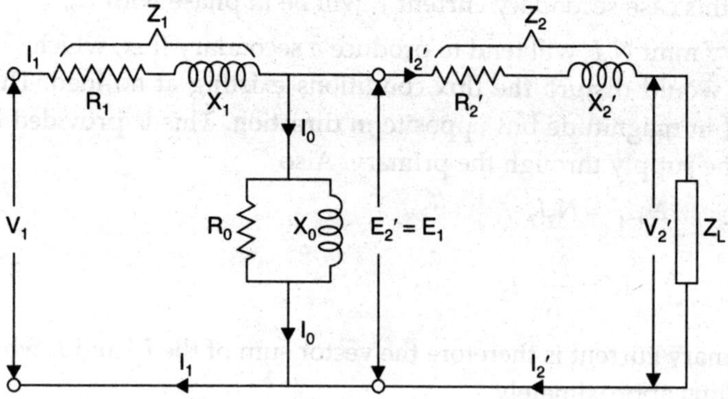


FIGURE 14.4

When $k$ is the transformation ratio then secondary resistance $R_2$ transferred to primary becomes

$$R_2' = \frac{R_2}{k^2}$$

Similarly,
$$X_2' = \frac{X_2}{k^2}$$

and
$$Z_2' = \frac{Z_2}{k^2}.$$

**Transformer Tests.** The performance of a transformer can be evaluated on the basis of circuit parameters. For this purpose two major tests are conducted on a transformer. These are:

1. **Open Circuit Test.** This test is used to determine no load loss or core loss and no load current $I_0$ which helps in finding $X_0$ and $R_0$. During this test usually high voltage winding is left open and the other side is connected to normal supply voltage and frequency. This test gives

(*i*) primary no load current.

(*ii*) core loss under no load condition.

Core losses of a transformer consist of

(*i*) hysteresis loss, $W_h = PB_{max}^{1.6} f$

(*ii*) eddy current loss, $W_e = QB_{max}^{2} f^2$.

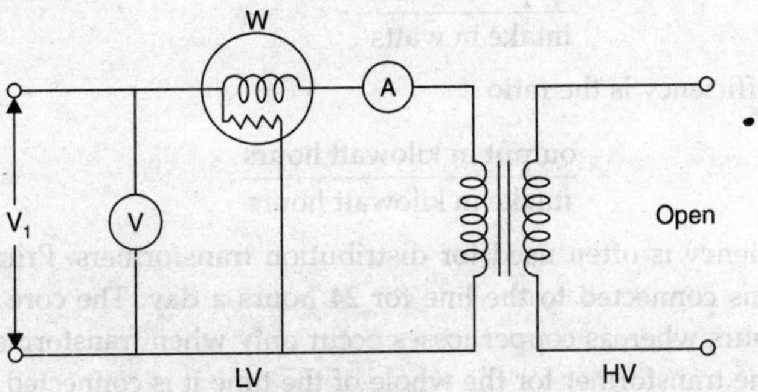


*FIGURE 14.5*

**Short Circuit Test.** This test is used to determine

(*i*) equivalent impedance, leakage reactance and total resistance of the transformer as referred to the winding in which the measuring instruments are placed.

(*ii*) copper loss at any load including full load.

During this test usually low voltage winding is short-circuited. A low voltage (5 to 10 percent of normal primary voltage) is applied to the primary side and is increased gradually till full load current flows through primary. As the voltage applied is small, therefore core losses will be small so that wattmeter reading can be considered to represent copper loss of the transformer.

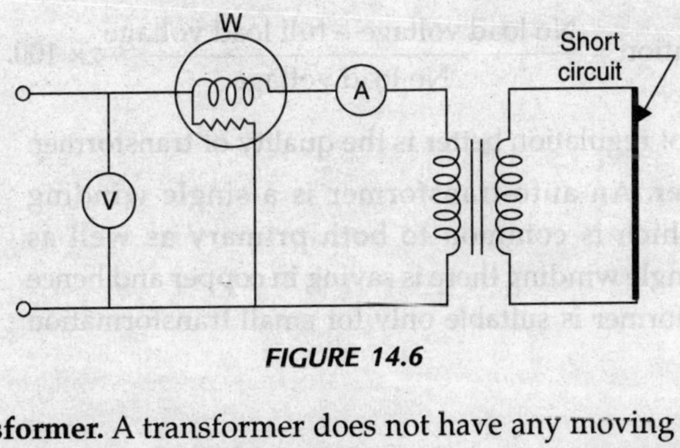


*FIGURE 14.6*

**Losses in a transformer.** A transformer does not have any moving parts, therefore it is free from friction and windage losses. However, following losses occur in a transformer:

**Iron loss.** This is also termed as core loss. This loss consists of

(*i*) hysteresis loss which is given by $W_h = \eta B_{max}^{1.6} fV$ watts

(*ii*) eddy current loss which is given by $W_e = PB_{max}^{2} f^2 t^2$ watts.

These losses can be reduced by using high silicon steel for the transformer core and by using very thin laminations.

**Copper loss.** This loss occurs due to ohmic resistance of the transformer winding. This loss occurs in primary as well as secondary winding. This loss is proportional to $(\text{current})^2$ or $(\text{kVA})^2$.

**All day efficiency.** The ordinary or commercial efficiency is defined as the ratio

$$\frac{\text{output in watts}}{\text{intake in watts}}$$

The all day efficiency is the ratio

$$\frac{\text{output in kilowatt hours}}{\text{intake in kilowatt hours}}.$$

All day efficiency is often used for distribution transformers. Primary winding of these transformers remains connected to the line for 24 hours a day. The core losses therefore occur during the all 24 hours whereas copper losses occur only when transformer is on load. Hence, if the load is not on the transformer for the whole of the time it is connected to the line, the all-day efficiency, for a given kVA loading, is less than the commercial efficiency.

**Maximum Efficiency.** The efficiency of a transformer is maximum when

$$\boxed{\text{copper losses} = \text{iron losses}}$$

If a transformer is intended to work most of the time on full load, it is designed to have the maximum efficiency at full load. If the load is variable, as in case of distribution transformers, it is usual to design it to have maximum efficiency at about three quarters of full load.

**Regulation.** Even when a transformer is supplied with rated primary voltage, the secondary voltage will change from no-load to full load conditions. This is mainly due to internal resistance and leakage reactance. The regulation of transformer is therefore defined as

$$\text{Percentage regulation} = \frac{\text{No load voltage} - \text{full load voltage}}{\text{No load voltage}} \times 100.$$

Lower the value of regulation better is the quality of transformer.

**Auto-transformer.** An auto-transformer is a single winding transformer part of which is common to both primary as well as secondary. Because of single winding there is saving in copper and hence it is cheaper. This transformer is suitable only for small transformation ratios.

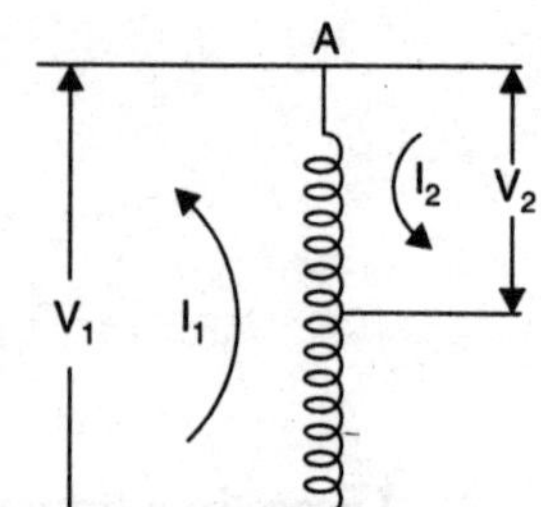


*FIGURE 14.7*

In this transformer when

$$\frac{N_1}{N_2} = \frac{I_2}{I_1} = m$$

then the ratio

$$\frac{\text{weight of copper on auto-transformer}}{\text{weight of copper on ordinary transformer}}$$

$$= \frac{m-1}{m}$$

Auto transformer can be used

1. as starter for induction motor
2. to boost voltage of a system.

**Current Transformer.** In case of dc, large currents are measured by using a shunt and measuring the potential difference across the shunt. This is not possible in case of ac. In case of ac it is usual to pass the current to be measured through the primary of the current transformer, the secondary being connected to the ammeter.

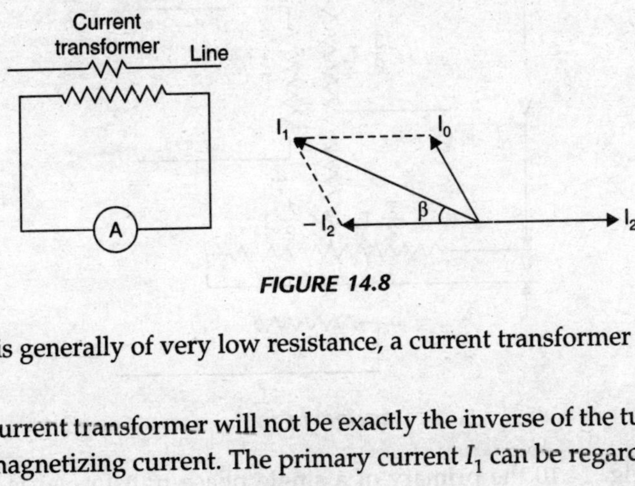


*FIGURE 14.8*

As ammeter is generally of very low resistance, a current transformer virtually works short circuited.

The ratio of current transformer will not be exactly the inverse of the turns ratio, mainly due to the effect of the magnetizing current. The primary current $I_1$ can be regarded as being made up of two components:

(*i*) a component exactly opposite to $I_2$

(*ii*) exciting current $I_0$.

$I_0$ causes slight error in turn ratio and phase shift α, known as phase angle of the transformer.

**Voltage Transformer.** When measuring very high voltages, the voltage is stepped down using a voltage transformer. The low voltage secondary of this transformer is connected to voltmeter.

When measuring power on a high voltage system both current and voltage transformers are used. A typical connection scheme is shown in Fig. 14.9.

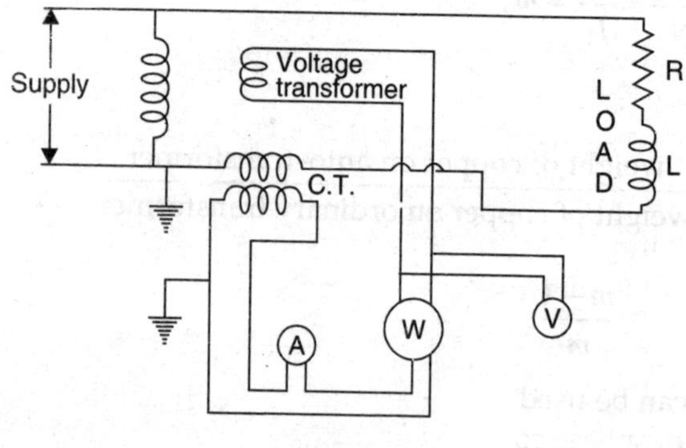


FIGURE 14.9

**Three phase to two phase transformation: Scott connections.** Sometimes it may be necessary to convert three phase supply into two phase supply. For this purpose Scott connections are used.

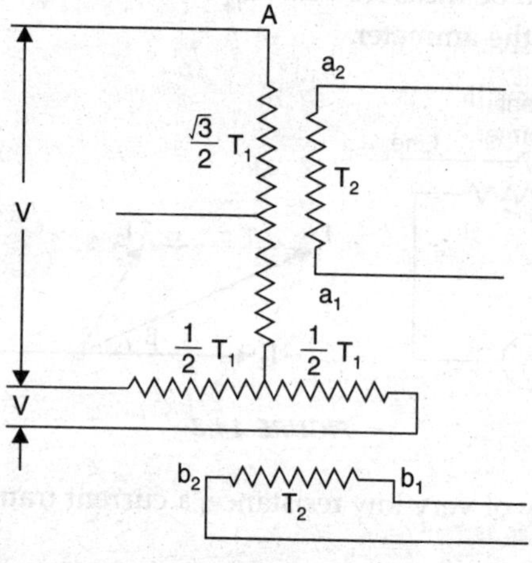


FIGURE 14.10

As shown in Fig. 14.10 the primary of a single phase transformer is connected between terminals *BC* of a three phase supply. The voltage of line being *V*, there will be a voltage *V* between *B* and *C* and between *A* and *B* or *C*. The voltage between *A* and the mid point of the phase at *S* will be a voltage $V\left(\frac{\sqrt{3}}{2}\right)$. If the primary of a second transformer having $\frac{\sqrt{3}}{2} = 0.87$ times the turns of the first, be connected across *AS*, the voltage of *S* will be unaltered and further across each turn of the second transformer there will be a voltage exactly equal to that per turn of the first. Thus if the two secondaries have equal numbers of turns, they will produce secondary terminal voltages equal in magnitude but in phase quadrature.

The neutral point of the three phase side can be located on the second or teaser transformer. The neutral must have a voltage $V/\sqrt{3}$ to $A$, and since the voltage $A$ to $S$ is $V(\sqrt{3})/2$, the neutral point will be $V\left[\frac{\sqrt{3}}{2}-\frac{1}{\sqrt{3}}\right]$ from $S$ or 0.288 $V$: that is, a number of turns above $S$ equivalent to 29% of the primary turns in the first or main transformer.

**Tertiary Winding.** Sometimes transformers are provided with a winding in addition to primary and secondary windings. This is known as tertiary winding. The purpose of providing this winding could be any of the following:

1. For an additional load which for some reason must be kept insulated from secondary winding.
2. To supply power for phase compensating devices such as condensers which are operated at a voltage which is neither equal to primary nor secondary voltage.
3. As a voltage coil in a testing transformer.
4. To load large split winding generators.
5. To inter connect three supply systems operating at different voltages.
6. In $Y/Y$ connected transformer, to allow sufficient earth fault current to flow.

**Parallel Operation.** For parallel operation of transformers the following conditions should be satisfied:

The transformer must have

1. the same voltage ratio
2. the same polarity
3. the same per unit impedance
4. the same phase sequence and zero relative phase displacement.

## OBJECTIVE TYPE QUESTIONS

**1.** In a transformer which of the following does not change?

(*a*) Voltage (*b*) Current (*c*) Frequency (*d*) All of the above.

**2.** In a transformer the energy is conveyed from primary to secondary

(*a*) through air (*b*) through cooling oil

(*c*) by the flux (*d*) none of the above.

**3.** The core losses of a transformer are ultimately dissipated in the form of

(*a*) light (*b*) electrical energy (*c*) magnetic hum (*d*) heat.

**4.** Core losses in a transformer include

(*a*) hysteresis and eddy current losses (*b*) hysteresis, eddy current and copper losses

(*c*) hysteresis, eddy current, copper and heat losses

(*d*) hysteresis, eddy current, copper, heat and radiation losses.

5. A transformer core is laminated to
   (*a*) reduce eddy current losses (*b*) reduce hysteresis losses
   (*c*) reduce copper losses (*d*) reduce all above losses.
6. Which loss is not common between a transformer and rotating machines?
   (*a*) Copper loss (*b*) Eddy current loss
   (*c*) Friction and windage loss (*d*) Hysteresis loss.
7. The sludge formation inside the oil cooled transformer is mainly due to
   (*a*) deterioration of insulation (*b*) variation of flux
   (*c*) fluctuation of load (*d*) heat and oxidation.
8. Which of the following impurity in transformer causes corrosion of metal parts?
   (*a*) Sulphur (*b*) Carbon (*c*) Abrasive particles (*d*) Dust and dirt.
9. Presence of moisture in transformer cooling oil
   (*a*) improves cooling rate (*b*) increases sludge formation
   (*c*) increases dielectric strength (*d*) reduces dielectric strength.
10. Fire point of an oil is
    (*a*) the temperature at which an oil will ignite and continue burning
    (*b*) always lower than flash point (*c*) limits the kVA rating of a transformer
    (*d*) directly depends on the viscosity of oil.
11. If flash point of transformer oil is 160°, the fire point could be
    (*a*) 100°C (*b*) 120°C (*c*) 161°C (*d*) 200°C.
12. The 'hum' in a transformer is developed in
    (*a*) core (*b*) winding (*c*) cooling oil (*d*) tank.
13. Magnetic hum is produced due to
    (*a*) eddy currents (*b*) magnetic forces (*c*) load variations (*d*) lightening etc.
14. Magneto striction is
    (*a*) very small extension, with corresponding reduction of cross-section, of sheet strips when magnetized
    (*b*) variations in flux causing magnetic fatigue of materials
    (*c*) the stresses developed due to difference in effect of variation of flux on core and insulating material
    (*d*) none of the above.
15. The degree of mechanical vibrations produced by the laminations of a transformer depends on
    (*a*) tightness of clamping (*b*) gauge of laminations
    (*c*) size of laminations (*d*) all of the above.
16. Star/star connections in case of transformers are used for
    (*a*) small, high voltage transformers (*b*) large, low voltage transformers
    (*c*) large, high voltage transformers (*d*) power supply transformers.

17. Delta/delta connections in case of transformers are used for
(*a*) small, high voltage transformers (*b*) large, low voltage transformers
(*c*) power supply transformers (*d*) six ring synchronous convertors.

18. Which type of transformer connection is preferred in case of power supply transformer?
(*a*) Star/star (*b*) Delta/delta (*c*) Star/delta (*d*) Zig zag/star.

19. The no load current drawn by a transformer is usually what percent of the full load current?
(*a*) 0.2 to 0.5 percent (*b*) 2 to 5 percent (*b*) 12 to 15 percent (*d*) 20 to 30 percent.

20. The voltage regulation of a transformer at full load and 0.8 power factor lagging is 2.5%. The voltage regulation at full load, 0.8 power factor leading could be
(*a*) 10% (*b*) 2.5% (*c*) zero (*d*) – 2.5%

21. In case there are butts on the edges of the laminations of the transformer, it is likely to result in
(*a*) noise (*b*) vibrations
(*c*) higher eddy current loss (*d*) higher hysteresis loss.

22. Which of the following represents zero phase displacement, three phase transformer?
(*a*) $Y_{v0}$ (*b*) $Y_{y6}$ (*c*) $D_{y1}$ (*d*) $D_{y11}$.

23. Star/double star connections in case of three phase transformers are generally used
(*a*) for power transformers (*b*) for reducing the effect of harmonics
(*c*) for supplying power to six ring synchronous converters
(*d*) for supplying power to unbalanced load.

24. In delta/delta connected transformers if one of the phase develops a fault,
(*a*) the entire unit has to be shut down (*b*) chances are that all the windings may burn
(*c*) the unit can be operated on 2/3 of its normal rating
(*d*) the unit can be operated on 58 percent of its rating.

25. The path of magnetic flux in a transformer should have
(*a*) high resistance (*b*) high reluctance (*c*) low resistance (*d*) low reluctance.

26. Which of the following is the most economical connection for small high voltage transformers ?
(*a*) Star-star connection (*b*) Delta-delta connection
(*c*) Zig-zag connection (*d*) Star-delta connection.

27. Which of the following statement is false?
(*a*) For the same kVA rating a auto-transformer has higher efficiency
(*b*) Oil in transformer serves the purpose of cooling as well as insulation
(*c*) A transformer is always free from windage losses
(*d*) none of the above.

28. An auto-transformer has
(*a*) windings common to primary and secondary
(*b*) main application in power transformers

(c) converts single phase supply into three phase supply
(d) automatically changes output voltage in proportion to load variations.

29. In which of the following applications, an auto-transformer can be used?
(a) Boosters (b) Static balancers
(c) Induction motor starters (d) All of the above.

30. Auto-transformers are generally not employed when
(a) input voltage varies considerably (b) load is not resistive
(c) voltage ratio exceeds 3: 1 (d) voltage form is not sinusoidal.

31. The advantage of using auto-transformer is
(a) saving in conductor material (b) reduced losses
(c) saving in core material (d) all of the above.

32. No load test on a transformer is carried out to determine
(a) copper loss (b) magnetising current
(c) magnetising current and core loss (d) efficiency of the transformer.

33. The dielectric strength of transformer oil is expected to be
(a) 1000 V (b) 33 kV (c) 100 kV (d) 330 kV.

34. Two transformers are connected in parallel. In case their per-unit impedance is not equal
(a) copper losses will reduced (b) one of the transformer will remain unloaded
(c) their p.f. will be different from that of the connected load
(d) dead short-circuit will occur.

35. Sumpener's test is conducted on transformers to determine
(a) temperature rise (b) stray losses (c) all day efficiency (d) none of the above.

**Questions 36 to 39 Refer to Data Given Below:**

36. A 100 kVA transformer has an iron loss of 1 kW and a copper loss on normal output current of 1.6 kW. The kVA loading at which the efficiency will be maximum, is
(a) 118.3 kVA (b) 100 kVA (c) 81.7 kVA (d) 66.6 kVA.

37. Total losses at maximum efficiency will be
(a) 0.5 kW (b) 1 kW (c) 2 kW (d) 4 kW.

38. The efficiency at unity power factor 81.7 kW output will be
(a) 97.61% (b) 96.11% (c) 94.41% (d) 91.47%

39. The efficiency under 81.7 kVA output under 0.8 lagging power factor will be
(a) 99.9% (b) 97.03% (c) 94.31% (d) 89.55%

40. Contamination of transformer oil is due to
(a) decomposition of oil (b) heating
(c) moisture (d) all of the above.

**41.** When the turn ratio $\frac{N_v}{N_s} = m$, the ratio of rating of an auto-transformer to that of equivalent two winding transformer is

(*a*) $\frac{m-1}{m}$ (*b*) $\frac{m}{m+1}$ (*c*) $\frac{m}{m-1}$ (*d*) $\frac{m-1}{m+1}$.

**42.** The permissible flux density in case of cold rolled grain oriented rolled steel is around

(*a*) 1.7 $Wb/m^2$ (*b*) 2.7 $Wb/m^2$ (*c*) 3.7 $Wb/m^2$ (*d*) 4.7 $Wb/m^2$.

**43.** During the short circuit test on a small transformer the frequency is increased from 50 Hz to 200 Hz. The copper losses will increase by a factor of

(*a*) 16 (*b*) 4 (*c*) 1 (*d*) $\frac{1}{4}$.

**44.** A 20 kW lighting transformer of ordinary efficiency 95% is on full load for 6 hours a day. The all day efficiency if the full load losses are equally divided between copper and iron losses, will be

(*a*) 98% (*b*) 94% (*c*) 89% (*d*) 82%.

**45.** The efficiency of a transformer will be maximum when

(*a*) copper losses = hysteresis losses (*b*) hysteresis losses = eddy current losses
(*c*) eddy current losses = copper losses (*d*) copper losses = iron losses.

**46.** For the same power rating which of the following arrangement will be costly?

(*a*) a three phase transformer (*b*) a bank of three single phase transformers.

**47.** No load current in a transformer

(*a*) lags behind the voltage by about 75° (*b*) leads the voltage by about 75°
(*c*) lags behind the voltage by about 15° (*d*) leads the voltage by about 15°.

**48.** Which of the following test can be conducted to ascertain the relative polarities of the terminals of a transformer?

(*a*) Short-circuit test (*b*) Back to back test (*c*) No load test (*d*) Phasing out test.

**49.** The purpose of providing an iron core in a transformer is to

(*a*) provide support to windings
(*b*) reduce hysteresis loss
(*c*) decrease the reluctance of the magnetic path
(*d*) reduce eddy current losses.

**50.** Tertiary winding is provided on transformers

(*a*) to load large split-winding generators (*b*) as a voltage coil in a testing transformer
(*c*) in star/star connected transformers, to allow sufficient earth fault current to flow for operation of protective gear.
(*d*) any of the above.

**51.** Harmonic voltages in a transformer may cause

(*a*) increased dielectric losses

(*b*) electrostatic interference with communication circuits

(*c*) resonance between the inductance of the transformer windings and the capacitance of the feeder to which they are connected

(*d*) all of the above.

52. Which of the following is not a part of transformer installation?

(*a*) Conservator (*b*) Breather (*c*) Buchholz relay (*d*) Exciter.

53. A SKAREL is

(*a*) treated insulating paper (*b*) gaseous insulating material

(*c*) synthetic liquid insulator (*d*) hygroscopic substance.

54. While conducting short-circuit test on a transformer the following side is short-circuited

(*a*) HV side (*b*) LV side (*c*) primary side (*d*) secondary side.

55. An auto-transformer having a transformation ratio of 0.7 supplies a 10 kW load. The power transferred inductively from primary to secondary is

(*a*) 10 kW (*b*) 7 kW (*c*) 3 kW (*d*) zero.

**Questions 56 and 57 refer to Data Given Below:**

A 100 kVA single-phase transformer, ratio 10,000/200, 50 Hz requires 300 V at the high voltage winding to circulate full load current with the low-voltage winding short-circuited. The intake is then 1000 W.

56. The percentage regulation on full load at 0.8 power factor lagging will be

(*a*) 10 p.c. (*b*) 5 p.c. (*c*) 2.5 p.c. (*d*) 1.25 p.c.

57. The secondary terminal voltage on full load 0.8 p.f. lagging will be

(*a*) 205 V (*b*) 203 V (*c*) 197 V (*d*) 195 V.

58. In a transformer the tappings are generally provided on

(*a*) primary side (*b*) secondary side (*c*) high voltage side (*d*) low voltage side.

59. Which of the following transformer will be smallest in size?

(*a*) 10 kVA, 400 Hz (*b*) 10 kVA, 200 Hz (*c*) 10 kVA, 50 Hz (*d*) 10 kVA, 25 Hz.

60. The use of higher flux density in the transformer design

(*a*) reduces weight per kVA (*b*) reduces iron losses

(*c*) reduces copper losses (*d*) increases part load efficiency.

61. If you impress a dc voltage of 230 volts to an unloaded 230 volts, 50 Hz transformer, do you expect the transformer to

(*a*) burn (*b*) give very high secondary voltage

(*c*) give secondary voltage according to the turn ratio

(*d*) none of the above.

62. For protection of transformer against overcurrents which protection is provided?

(*a*) Alarm (*b*) Buchholz relay (*c*) HV screening (*d*) protective gaps.

**63.** The chemical used in breather for transformer should have the quality of
(*a*) ionizing air (*b*) cleaning the transformer oil
(*c*) cooling the transformer oil (*d*) absorbing moisture.

**64.** The chemical used in breather is
(*a*) asbestos fibre (*b*) silica sand
(*c*) sodium hypochlorite (*d*) silica gel.

**65.** If a pump motor is run on two third of its supply voltage, it will
(*a*) continue to deliver same power (*b*) stall
(*c*) continue to run at lower speed (*d*) burn.

**66.** An ideal transformer has infinite values of primary and secondary inductances. The statement is
(*a*) true (*b*) false

**67.** Current transformer secondary is connected to an ammeter of 5 A range to measure line current of 1000 A. The ammeter is to be serviced. Can it be removed
(*a*) without interfering with supply (*b*) only after switching off the main current path.

**68.** The transformer ratings are usually expressed in terms of
(*a*) volts (*b*) amperes (*c*) kVA (*d*) kW.

**69.** In transformer the end coil is provided with extra insulation to
(*a*) prevent chemical action due to oil
(*b*) reduce leakage of current to body and earthing circuit
(*c*) reduce heating (*d*) take care of travelling wave surges.

**70.** The noise resulting from vibrations of laminations set by magnetic forces, is termed as
(*a*) magnetostriction (*b*) hum
(*c*) zoom (*d*) boo.

**71.** Hysteresis loss in a transformer varies as ($B_{max}$, maximum flux density).
(*a*) $B_{max}$ (*b*) $B_{max}^{1.6}$ (*c*) $B_{max}^{2}$ (*d*) $B_{max}^{2.4}$.

**72.** A transformer is working on 2000 V, 50 Hz, has 300 primary turns. The core has a mean magnetic path of 100 cm and cross-section 1000 cm$^2$, the iron having a relative permeability of 1800. The iron loss is 400 W. The primary no load current will be
(*a*) 0.11 A (*b*) 0.23 A (*c*) 0.37 A (*d*) 0.96 A.

## Questions 73 and 74 Refer to Data Given Below:

**73.** When a transformer is connected to a 1000 V 50 Hz supply the core loss is 1000 W, of which 700 W are hysteresis and 300 W are eddy-current loss. If the applied voltage is raised to 2000 V and the frequency to 100 Hz, the hysteresis loss will become
(*a*) 1400 W (*b*) 1166 W (*c*) 435 W (*d*) 275 W.

**74.** The eddy current loss will become
(*a*) 2400 W (*b*) 1800 W (*c*) 1400 W (*d*) 1200 W.

75. For a transformer having nominal voltage of 11 kV, the peak impulse voltage for testing will be
(*a*) 12 kV (*b*) 20 kV (*c*) 75 kV (*d*) 110 kV.

76. Material used for the construction of transformer core is usually
(*a*) wood (*b*) copper (*c*) aluminium (*d*) silicon steel.

77. The thickness of laminations used in a transformer is usually
(*a*) 0.4 mm to 0.5 mm (*b*) 4 mm to 5 mm
(*c*) 14 mm to 15 mm (*d*) 25 mm to 40 mm.

78. The function of conservator in a transformer is
(*a*) to protect against internal fault (*b*) to reduce copper as well as core losses
(*c*) to cool the transformer oil
(*d*) to take care of the expansion and contraction of transformer oil due to variation of temperature of surroundings.

79. The highest voltage for transmitting electrical power in India is
(*a*) 33 kV (*b*) 66 kV (*c*) 132 kV (*d*) 400 kV.

80. Scott connections are used to transform
(*a*) single phase supply into three phase supply
(*b*) single phase supply into two phase supply
(*c*) three phase supply into single phase supply
(*d*) star connected primary into delta connected secondary.

81. Zero phase sequence test on a transformer is a
(*a*) type test (*b*) routine test (*c*) supplementary test (*d*) high voltage test.

82. In a transformer the resistance between its primary and secondary is
(*a*) zero (*b*) 1 ohm (*c*) 1000 ohms (*d*) infinite.

83. Two transformers operating in parallel will share their load depending upon their
(*a*) ratings (*b*) leakage reactance (*c*) efficiency (*d*) p.u. impedance.

84. A transformer of highest rating is likely to find application in which segment of the power distribution system?
(*a*) Substations (*b*) Transmission (*c*) Generation (*d*) Distribution.

85. A transformer oil must be free from
(*a*) sludge (*b*) odour (*c*) gases (*d*) moisture.

86. During parallel operation of transformers, incorrect polarity is likely to result in
(*a*) open circuit (*b*) dead short-circuit
(*c*) reduced copper losses (*d*) none of the above.

87. Buchholz relay is used on
(*a*) mercury are rectifiers (*b*) transformers
(*c*) dc series motors (*d*) alternators.

**88.** A Buchholz relay can be installed on
(*a*) auto-transformers
(*b*) air cooled transformers
(*c*) welding transformers
(*d*) oil cooled transformers.

**89.** A Buchholz relay works on the principle of
(*a*) loss of dielectric strength of transformer oil due to presence of moisture
(*b*) magnetostriction in armature core causing violent vibrations
(*c*) violent gas generation before transformer breakdowns
(*d*) melting of lead wire at high currents.

**90.** The colour of the gas resulting from dissociation of transformer oil, as seen from window in the wall of the vessel is
(*a*) black or grey (*b*) white (*c*) yellow (*d*) pink.

**91.** Gas is usually not liberated due to dissociation of transformer oil unless the oil temperature exceeds
(*a*) 50°C (*b*) 80°C (*c*) 100°C (*d*) 150°C

**92.** The main reason for generation of harmonics in a transformer could be
(*a*) fluctuating load
(*b*) poor insulation
(*c*) saturation of core
(*d*) mechanical vibrations.

**93.** A transformer has maximum efficiency at full load when iron losses are 800 watts. The copper losses at half load will be
(*a*) 1600 W (*b*) 800 W (*c*) 400 W (*d*) 200 W.

**94.** Distribution transformers are generally designed for maximum efficiency around
(*a*) zero load (*b*) 25% load (*c*) 50% load (*d*) 90% load.

**95.** The full load copper loss of a single phase transformer is 400 W. The copper loss at full load and 0.8 power factor (lagging) will be
(*a*) 500 W (*b*) 400 W (*c*) 320 W (*d*) 256 W.

**96.** No load performance for a transformer is shown in Fig. 14.11. Which curve is likely to represent no load current?
(*a*) curve *A* (*b*) curve *B* (*c*) curve *C* (*d*) curve *D*.

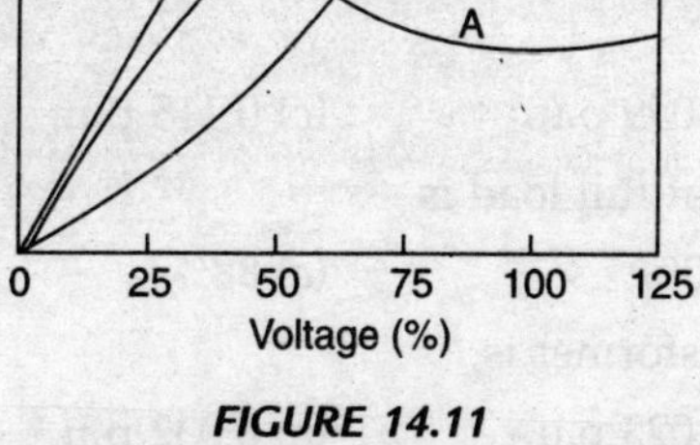


**FIGURE 14.11**

**97.** Usually in a transformer

(*a*) hysteresis losses are more than eddy-current losses

(*b*) hysteresis losses are less than eddy-current losses

(*c*) hysteresis losses are equal to eddy current losses

(*d*) hysteresis loss are equal to copper losses.

**98.** Which of the following is variable loss in a transformer?

(*a*) Eddy current loss (*b*) Copper loss

(*c*) Hysteresis loss (*d*) None of the above.

**99.** Which of the following property is not necessarily desirable in the material for transformer core?

(*a*) Low hysteresis loss (*b*) High permeability

(*c*) High thermal conductivity (*d*) Mechanical strength.

**100.** Transformers are provided with tap changer to adjust

(*a*) power factor (*b*) primary voltage

(*c*) secondary voltage (*d*) both primary as well as secondary voltage.

**101.** Helical coils can be used on

(*a*) high frequency transformers

(*b*) high voltage side of small capacity transformers

(*c*) high voltage side of high kVA rating transformers

(*d*) low voltage side of high kVA transformers.

**102.** High frequency transformers sometimes make use of ferrite cores because it has

(*a*) high resistance (*b*) high hysteresis (*c*) low permeability (*d*) high specific gravity.

**103.** For evaluating the iron losses of a transformer, the weight of which of the following is not necessarily required?

(*a*) Weight of core (*b*) Weight of yoke (*c*) Weight of winding.

**104.** A 300 kVA, 11000/44 V mesh/star three phase 50 Hz transformer gave the following test results

**Open circuit test.** On *LV* side at normal voltage and frequency: input 1.3 kW, 21.1 A.

**Short circuit test.** On *HV* side with voltage of 600 V: input 2.80 kW, 15 A. Full load power factor is 0.8 lagging

The per unit resistance is

(*a*) 0.0012 p.u. (*b*) 0.012 p.u. (*c*) 0.12 p.u. (*d*) 1.2 p.u.

**105.** The p.u. impedance is

(*a*) 0.058 p.u. (*b*) 0.099 p.u. (*c*) 0.115 p.u. (*d*) 0.155 p.u.

**106.** Efficiency of transformer at full load is

(*a*) 98% (*b*) 90% (*c*) 88% (*d*) 80%.

**107.** The regulation of the transformer is

(*a*) 0.012 p.u. (*b*) 0.023 p.u. (*c*) 0.032 p.u. (*d*) 0.043 p.u.

**108.** While conducting short-circuit test on a transformer the following side is short-circuited.
(*a*) *HV* side (*b*) *LV* side (*c*) Primary side (*d*) Secondary side.

**109.** If the frequency of supply voltage is doubled, which loss in a transformer will rise rapidly as compared to the remaining?
(*a*) Copper loss (*b*) Hysteresis loss (*c*) Eddy current loss.

**110.** Cross over windings are used on
(*a*) Current transformers (*b*) *HV* side of high kVA rating transformers
(*c*) *HV* side of low kVA rating transformer (*d*) *LV* side of high kVA rating transformers.

**111.** A substance is placed in strong magnetic field. Which of the following will have maximum influence on the capacity of the substance to get magnetized?
(*a*) Susceptibility (*b*) Permeability (*c*) Resistivity (*d*) Permittivity.

**112.** Harmonics in transformers result in
(*a*) increased $I^2R$ losses (*b*) increased core losses
(*c*) magnetic interference with communication circuits
(*d*) all of the above.

**113.** Transformers connected on one side of overhead transmission line may be subjected to overvoltages resulting from
(*a*) switching (*b*) faults
(*c*) lightening discharges (*d*) any of the above.

**114.** Voltage surges in transmission lines travel nearly at the speed of
(*a*) sound (*b*) light (*c*) 100 km/hr (*d*) none of the above.

**115.** Surge absorbers are basically
(*a*) resistors (*b*) series inductors (*c*) series capacitors (*d*) insulators.

**116.** An electric arc furnace requires
(*a*) high voltage; high current (*b*) low voltage; low current
(*c*) high voltage; low current (*d*) low voltage; high current.

**117.** An auto-transformer is set for 200: 100 volts. If the winding breaks at a place which is common to high voltage as well as low voltage side, then the output on low voltage side will be
(*a*) zero (*b*) 50 V (*c*) $100\sqrt{2}$ V (*d*) 200 V.

**118.** Tertiary winding is provided in transformers having
(*a*) mesh/mesh winding (*b*) mesh/star winding
(*c*) star/star winding (*d*) any of the above.

**119.** The core used in high frequency transformers is usually
(*a*) copper core (*b*) cast iron core (*c*) mild steel core (*d*) air core.

**120.** The full load copper loss of a transformer is 1600 watts. At half load, the copper loss will be
(*a*) 6400 W (*b*) 1600 W (*c*) 800 W (*d*) 400 W.

**121.** Which of the following is class *A* electromagnetic machine?

(*a*) DC machine (*b*) Transformer

(*c*) AC commutator machine (*d*) synchronous machine.

**122.** A transformer is classified as which class of machine?

(*a*) class *B* (*b*) class *C* (*c*) class *D* (*d*) class *E*.

**123.** In which class of machine the flux is pulsating?

(*a*) class *A* (*b*) class *B* (*c*) class *F* (*d*) all of the above.

**124.** In which class of machines the coils are fixed?

(*a*) class *F* (*b*) class *E* (*c*) class *B* (*d*) class *D*.

**125** In which of the following class of machines the connections are provided through tappings ?

(*a*) class *A* (*b*) class *C* (*c*) class *D* (*d*) class *E*.

**126.** The value of flux involved in the emf equation of a transformer is

(*a*) average value (*b*) rms value (*c*) instantaneous value (*d*) maximum value.

**127.** If the voltage regulation of a transformer is negative, it can be concluded that

(*a*) the load is resistive (*b*) the load is inductive

(*c*) the load is capacitive (*d*) the applied voltage is non-sinusoidal.

**128.** The direction of secondary current in a transformer must be such as to oppose any change in flux. The above statement is associated with

(*a*) Lenz's law (*b*) Faraday's law (*c*) Coulomb's law (*d*) Fleming's rule.

**129.** Silicon steel used in laminations mainly reduces

(*a*) hysteresis losses (*b*) eddy current losses

(*c*) copper losses (*d*) all of the above.

**Questions 130 to 132 Refer to Data Given Below:**

A 6600/400/110 V star/star/mesh, three phase transformer has a magnetizing current of 5.5 A and balanced three phase loads of 1000 kVA at 0.8 lagging p.f. on secondary and 200 kVA at 0.5 leading p.f. on the tertiary

**130.** The primary current will be

(*a*) 121.7 A (*b*) 89.7 A (*c*) 78.8 A (*d*) 42.9 A.

**131.** The primary load in kVA will be

(*a*) 6600 kVA (*b*) 1025 kVA (*c*) 950 kVA (*d*) 666 kVA.

**132.** Total primary load in kW will be

(*a*) 1200 kW (*b*) 1150 kW (*c*) 999 kW (*d*) 900 kW.

**133.** The secondary of a current transformer is always short-circuited under operating conditions because

(*a*) it protects the primary circuits (*b*) it is safe to human beings

(*c*) it avoids core saturation and high voltage induction.

**134.** For a properly designed transformer, the regulation should be closer to

(*a*) 1% (*b*) 10% (*c*) 100% (*d*) infinity.

**135.** Which winding of transformer has less cross-sectional area?

(*a*) Primary winding (*b*) Secondary winding

(*c*) Low voltage winding (*d*) High voltage winding.

**136.** Power transformers are generally designed to have maximum efficiency around

(*a*) no load (*b*) half load (*c*) near full load (*d*) 10% over load.

**137.** Some types of electromagnetic machines are given below.

I. Regulators

II. AC commutator machines

III. Transformers

IV. DC Machines.

Which of these machines are grouped under the same class?

(*a*) I and II only (*b*) I and III only (*c*) II and IV only (*d*) III and IV only.

**138.** Which of the following is the main advantage of an autotransformer over a two winding transformer?

(*a*) Hysteresis losses are reduced (*b*) Saving in winding material

(*c*) Eddy losses are totally eliminated (*d*) Copper losses are negligible.

**139.** During short-circuit test copper losses are negligible because

(*a*) the current on secondary side is negligible

(*b*) the voltage on secondary side does not vary

(*c*) the voltage applied on primary side is low

(*d*) full load current is not supplied to the transformer.

**140.** The normal colour of silica gel crystals is

(*a*) white (*b*) blue (*c*) pink (*d*) yellow.

**141.** When saturated with moisture the colour of silica gel crystals becomes

(*a*) grey (*b*) pink (*c*) yellow (*d*) green.

**142.** Tertiary winding is sometimes provided on transformers for

(*a*) any additional load that must be kept insulated from that of the secondary

(*b*) supplying power to phase compensating devices such as condensers, operated at some voltage not equal to that of primary or secondary

(*c*) interconnecting three supply systems operating at different voltages

(*d*) any of the above.

**143.** A 1200 kVA, 3.3 kV, star connected generator has a phase resistance of 0.18 ohm Full load IR drop in p.u. is

(*a*) 0.02 p.u. (*b*) 0.20 p.u. (*c*) 2.00 p.u. (*d*) 38 p.u.

**144.** A transformer is operating at unity load factor. The maximum efficiency of transformer can occur at

(*a*) zero power factor (*b*) unity power factor

(*c*) leading power factor (*d*) lagging power factor.

**145.** Two transformers are connected in parallel. These transformers do not have equal percentage impedance. This is likely to result in

(*a*) short-circuiting of one of the secondaries

(*b*) power factor of one of the transformer as leading while that of the other lagging

(*c*) transformer having higher copper losses will have negligible core losses

(*d*) loading of the transformers not in proportion to their kVA ratings.

**146.** Which transformer liquid when spilled on skin causes irritation?

(*a*) Askarel (*b*) Mineral oil

(*c*) Both (*a*) and (*b*) above (*d*) None of the above.

## Questions 147 to 149 Refer to Data Given Below:

A 500 kVA transformer (A) with 0.01 p.u. resistance and 0.05 p.u. reactance is connected in parallel with a 250 kVA transformer (B) with 0.015 p.u. resistance and 0.04 p.u. reactance. The secondary voltage of each transformer is 400 V on no load.

**147.** How will the transformers share the load of 750 kVA at 0.8 lagging power factor?

(*a*) (A) 470 kVA (B) 280 kVA

(*b*) (A) 370 kVA (B) 380 kVA

(*c*) (A) 375 kVA (B) 375 kVA

(*d*) (A) 280 kVA (B) 470 kVA.

**148.** The power factors of the two transformers will be

(*a*) (A) 0.8 lagging (B) 0.8 lagging

(*b*) (A) 0.762 lagging (B) 0.856 lagging

(*c*) (A) 0.688 lagging (B) 0.777 lagging

(*d*) (A) 0.555 lagging (B) 0.555 leading.

**149.** the kW load of the two transformers will be

(*a*) (A) 375 kW (B) 375 kW

(*b*) (A) 300 kW (B) 400 kW

(*c*) (A) 359 kW (B) 241 kW

(*d*) (A) 241 kW (B) 359 kW.

**150.** The oil preferred for use in transformers is

(*a*) mineral oil (*b*) cooking oil (*c*) rice -bran oil (*d*) petroleum jelly.

151. A transformer core is made of copper and the primary and secondary windings are made of mild steel. Such a transformer will have less

(*a*) eddy current losses (*b*) hysteresis losses

(*c*) copper losses (*d*) power factor.

152. The change in volume of transformer cooling oil due to variation of atmospheric temperature during day and night is taken care of by which part of the transformer?

(*a*) Breather (*b*) Conservator (*c*) Buchholz relay (*d*) Bushings.

153. The transformer laminations are insulated from each other by

(*a*) thin coat of varnish (*b*) mica strip

(*c*) paper (*d*) any of the above.

154. Which type of winding is used in 3 phase shell type transformer?

(*a*) Circular type (*b*) Sandwich type (*c*) Cylindrical type (*d*) Rectangular type.

155. The coil in battery ignition circuit of a car is a

(*a*) step up transformer (*b*) step down transformer

(*c*) current transformer (*d*) inverter.

**Questions 156 and 157 Refer to Data Given Below:**

The secondary of a 500 kVA, 4400/500 V transformer has 100 turns.

156. The emf per turn of primary will be

(*a*) 1 V (*b*) 2 V (*c*) 5 V (*d*) 12. 5 V.

157. The secondary current at unity power factor will be

(*a*) 100 A (*b*) 200 A (*c*) 500 A (*d*) 1000 A.

158. The phase difference between the primary voltage and secondary voltage of a transformer is usually

(*a*) zero degree (*b*) 90° (*c*) 120° (*d*) 180° .

159. The source of moisture for the contamination of transformer oil is

(*a*) atmospheric air (*b*) winding

(*c*) insulating material (*d*) core.

160. In an auto - transformer if a break occurs in the common winding, the voltage at the secondary terminals will be

(*a*) zero (*b*) half the primary voltage

(*c*) anywhere between half to full primary voltage

(*d*) same as primary voltage.

161. Electric current has tendency to constrain itself near the peripheral surface of the conductor. This phenomenon is known as

(*a*) magnetostriction (*b*) corona

(*c*) skin effect (*d*) pinning.

**162.** During open circuit test on a step up transformer
(*a*) primary is supplied full load current (*b*) primary is supplied rated voltage
(*c*) primary is supplied current at reduced voltage
(*d*) primary is supplied reacted kVA.

**163.** Open circuit test on transformers is conducted to determine
(*a*) copper losses (*b*) hysteresis losses
(*c*) eddy current losses (*d*) core losses.

**164.** Short-circuit test on transformers is conducted to determine
(*a*) copper losses (*b*) hysteresis losses
(*c*) eddy current losses (*d*) core losses.

**165.** Two three phase transformers are to be connected. Which of the following arrangement is impossible?
(*a*) Transformer A: primary Δ; secondary γ
Transformer B: primary Δ; secondary Δ
(*b*) Transformer A: primary γ; secondary γ
Transformer B: primary Δ; secondary Δ
(*c*) Transformer A: primary γ; secondary Δ
Transformer B: primary Δ; secondary γ .

**166.** For the parallel operation of two single phase transformers it is necessary that they should have
(*a*) same polarity (*b*) same efficiency
(*c*) same number of turns on secondary side
(*d*) same kVA rating.

**167.** While testing mineral oil as the transformer cooling medium, the maximum number of break downs permissible per cup is
(*a*) 1 (*b*) 2 (*c*) 5 (*d*) 10.

**168.** The transformer oil should have ........ volatility and........... viscosity?
(*a*) low..... low (*b*) high....... high (*c*) low....... high (*d*) high.......low.

**169.** In Scott connections the teaser transformer has tapping on
(*a*) 86% of primary (*b*) 58% of primary (*c*) 58% of secondary (*d*) 50% of secondary.

**Questions 170 to 173 Refer to Data Given Below:**

The no load current of a transformer is 5 A at 0.25 power factor, when supplied at 235 V, 50 Hz. The number of turns in the primary winding is 200.

**170.** The maximum value of flux in the core will be
(*a*) 5,293 lines (*b*) 52, 930 lines (*c*) 529, 300 lines (*d*) 5293,000 lines.

**171.** The core loss will be
(*a*) 14. 14 W (*b*) 29.38 W (*c*) 55. 66 W (*d*) 67.69 W.

**172.** The magnetising current will be
(*a*) 2.42 A (*b*) 4.84 A (*c*) 9.68 A (*d*) 12. 32 A.

**173.** The function of breather in a transformer is
(*a*) to cool the coils during reduced load (*b*) to provide oxygen inside the tank
(*c*) to filter the transformer cooling oil
(*d*) to arrest flow of moisture when outside air enters the transformer.

**174.** A 400 kVA 400 Hz transformer is to be operated at 50 Hz.
The kVA rating of the transformer
(*a*) could be 3200 kVA (*b*) could be $400\sqrt{8}$ kVA
(*c*) must be restricted to $100\sqrt{2}$ kVA (*d*) must be restricted to 50 kVA.

**175.** The secondary winding of which of the following transformer is always kept closed?
(*a*) Step up transformer (*b*) Step down transformer
(*c*) Current transformer (*d*) Potential transformer .

**176.** Clip on ammeter uses
(*a*) high voltage ratio transformer (*b*) step up transformer
(*c*) potential transformer (*d*) current transformer.

**177.** In a transformer if a wave more peaky than sine wave is applied
(*a*) voltage induced will be high (*b*) voltage induced will be low
(*c*) losses will reduce (*d*) current will increase.

**178.** The size of a transformer core will depend on
(*a*) frequency (*b*) area of the core
(*c*) flux density of the core material (*d*) (*a*) and (*b*) both.

**179.** Natural air cooling is generally restricted for transformers upto
(*a*) 1.5 MVA (*b*) 5 MVA (*c*) 15 MVA (*d*) 50 MVA.

**180.** The copper loss of a transformer actually occurs in
(*a*) winding (*b*) core
(*c*) insulating material (*d*) bushings.

**181.** For a properly designed transformer the efficiency can be expected in the range
(*a*) 95% to 99% (*b*) 85% to 90% (*c*) 80% to 84% (*d*) 70% to 75%.

**182.** Which of the following transformer will be smallest in size?
(*a*) 100 kVA, 440 V, 25 Hz (*b*) 100 kVA, 440 V, 50 Hz
(*c*) 100 kVA, 400 V, 100 Hz (*d*) 100 kVA, 400 V, 400 Hz.

**183.** A step down transformer 2000/200 V is inadvertantly connected to dc supply of 2000 V on HV side. This is likely to result in
(*a*) low voltage on secondary side (*b*) non - sinusoidal voltage on secondary side.
(*c*) burning of transformer winding (*d*) noise free operation of transformer.

**184.** A shell type transformer has
(*a*) high eddy current losses (*b*) reduced magnetic leakage
(*c*) negligible hysteresis losses (*d*) none of the above.

**185.** An induction motor and a transformer of same kVA rating are available. Which device will draw least current while operating on no load?
(*a*) induction motor (*b*) transformer
(*c*) both will draw identical current
(*d*) nothing can be concluded on the basis of given data.

**186.** If you impress a dc voltage of 230 volts on an unloaded 230 volts 50 Hz transformer, do you expect the transformer to
(*a*) burn (*b*) give very high secondary voltage
(*c*) give secondary voltage according to the turn ratio.

**187.** B class insulation can withstand a maximum temperature of
(*a*) 90°C (*b*) 105°C (*c*) 120°C (*d*) 130°C

**188.** Plain paper is categorized as which class of insulating material?
(*a*) class *Y* (*b*) class *A* (*c*) class *B* (*d*) class F.

**189.** Asbestos laminates form which class of insulation?
(*a*) class *H* (*b*) class *B* (*c*) class *A* (*d*) class *E*.

**190.** A transformer can have regulation closer to zero
(*a*) on full load (*b*) on over load
(*c*) on zero power factor (*d*) on leading power factor.

**191.** The magnetic coupling between primary and secondary sides of a transformer can be improved by
(*a*) using winding material of low resistivity
(*b*) using insulation of better quality between the windings
(*c*) using the magnetic core of low reluctance
(*d*) using oil of low viscosity.

**192.** When $V_0$ is the open circuit voltage and $V$ is the voltage on load in case of a transformer, the regulation of the transformer will be given by
(*a*) $\frac{V - V_0}{V} \times 100$ (*b*) $\frac{V - V_0}{V_0} \times 100$ (*c*) $\frac{V_0 - V}{V_0} \times 100$ (*d*) $\frac{V_0 - V}{V} \times 100$.

**193.** Which of the following is not an essential condition for running two transformers in parallel?
(*a*) Same polarity (*b*) Same kVA rating
(*c*) Same voltage ratio (*d*) Same percentage impedance drop.

**194.** A transformer transforms
(*a*) frequency (*b*) voltage (*c*) current (*d*) power.

**195.** The iron losses in a transformer occur in

(*a*) main body structure (*b*) winding

(*c*) core (*d*) cooling oil.

**196.** Which of the following is not the standard voltage for power supply in India?

(*a*) 11 kV (*b*) 33 kV (*c*) 66 kV (*d*) 122 kV.

**197.** Which of the following insulating material can withstand highest temperature?

(*a*) class *Y* (*b*) class *E* (*c*) class *H* (*d*) class *F*.

**198.** For a 10 kVA transformer, total losses occurring at full load will be around

(*a*) 5 W (*b*) 50 W (*c*) 500 W (*d*) 5000 W.

**199.** In a power transformer usually which winding is closer to the core?

(*a*) Primary winding (*b*) Secondary winding

(*c*) Low voltage winding (*d*) High voltage winding.

**200.** Which of the following statement about the transformer is correct?

(*a*) In a shell type transformer, the primary and secondary windings are wound on separate formers

(*b*) Laminated core is used in transformers to minimize copper losses

(*c*) Copper losses in a transformer can be determined by open circuit test

(*d*) Short circuit test on a transformer is conducted to determine variable losses.

## ANSWERS

| | | | | | | |
|---|---|---|---|---|---|---|
| **1.** (*c*) | **2.** (*c*) | **3.** (*d*) | **4.** (*a*) | **5.** (*a*) | **6.** (*c*) | **7.** (*d*) |
| **8.** (*a*) | **9.** (*d*) | **10.** (*a*) | **11.** (*d*) | **12.** (*a*) | **13.** (*b*) | **14.** (*a*) |
| **15.** (*d*) | **16.** (*a*) | **17.** (*b*) | **18.** (*c*) | **19.** (*b*) | **20.** (*d*) | **21.** (*c*) |
| **22.** (*a*) | **23.** (*c*) | **24.** (*d*) | **25.** (*d*) | **26.** (*a*) | **27.** (*d*) | **28.** (*a*) |
| **29.** (*d*) | **30.** (*c*) | **31.** (*d*) | **32.** (*c*) | **33.** (*b*) | **34.** (*c*) | **35.** (*a*) |
| **36.** (*c*) | **37.** (*c*) | **38.** (*a*) | **39.** (*b*) | **40.** (*d*) | **41.** (*a*) | **42.** (*a*) |
| **43.** (*c*) | **44.** (*c*) | **45.** (*d*) | **46.** (*b*) | **47.** (*a*) | **48.** (*d*) | **49.** (*c*) |
| **50.** (*d*) | **51.** (*d*) | **52.** (*d*) | **53.** (*c*) | **54.** (*b*) | **55.** (*c*) | **56.** (*c*) |
| **57.** (*d*) | **58.** (*d*) | **59.** (*a*) | **60.** (*a*) | **61.** (*a*) | **62.** (*a*) | **63.** (*d*) |
| **64.** (*d*) | **65.** (*b*) | **66.** (*b*) | **67.** (*b*) | **68.** (*c*) | **69.** (*d*) | **70.** (*b*) |
| **71.** (*b*) | **72.** (*c*) | **73.** (*a*) | **74.** (*d*) | **75.** (*c*) | **76.** (*d*) | **77.** (*a*) |
| **78.** (*d*) | **79.** (*d*) | **80.** (*c*) | **81.** (*c*) | **82.** (*d*) | **83.** (*d*) | **84.** (*b*) |
| **85.** (*d*) | **86.** (*b*) | **87.** (*b*) | **88.** (*d*) | **89.** (*c*) | **90.** (*a*) | **91.** (*d*) |
| **92.** (*c*) | **93.** (*d*) | **94.** (*c*) | **95.** (*b*) | **96.** (*b*) | **97.** (*a*) | **98.** (*b*) |
| **99.** (*c*) | **100.** (*c*) | **101.** (*d*) | **102.** (*a*) | **103.** (*c*) | **104.** (*b*) | **105.** (*a*) |

| | | | | | | |
|---|---|---|---|---|---|---|
| 106. (*a*) | 107. (*d*) | 108. (*b*) | 109. (*c*) | 110. (*c*) | 111. (*a*) | 112. (*d*) |
| 113. (*d*) | 114. (*b*) | 115. (*b*) | 116. (*d*) | 117. (*d*) | 118. (*c*) | 119. (*d*) |
| 120. (*d*) | 121. (*a*) | 122. (*c*) | 123. (*c*) | 124. (*d*) | 125. (*c*) | 126. (*d*) |
| 127. (*c*) | 128. (*a*) | 129. (*a*) | 130. (*b*) | 131. (*b*) | 132. (*d*) | 133. (*c*) |
| 134. (*a*) | 135. (*d*) | 136. (*c*) | 137. (*b*) | 138. (*b*) | 139. (*c*) | 140. (*b*) |
| 141. (*b*) | 142. (*d*) | 143. (*a*) | 144. (*b*) | 145. (*d*) | 146. (*a*) | 147. (*a*) |
| 148. (*b*) | 149. (*c*) | 150. (*a*) | 151. (*d*) | 152. (*b*) | 153. (*d*) | 154. (*b*) |
| 155. (*a*) | 156. (*c*) | 157. (*d*) | 158. (*d*) | 159. (*a*) | 160. (*d*) | 161. (*c*) |
| 162. (*b*) | 163. (*d*) | 164. (*a*) | 165. (*a*) | 166. (*a*) | 167. (*c*) | 168. (*a*) |
| 169. (*a*) | 170. (*b*) | 171. (*b*) | 172. (*b*) | 173. (*d*) | 174. (*d*) | 175. (*c*) |
| 176. (*d*) | 177. (*a*) | 178. (*d*) | 179. (*a*) | 180. (*a*) | 181. (*a*) | 182. (*a*) |
| 183. (*c*) | 184. (*b*) | 185. (*b*) | 186. (*a*) | 187. (*d*) | 188. (*a*) | 189. (*b*) |
| 190. (*d*) | 191. (*c*) | 192. (*c*) | 193. (*b*) | 194. (*d*) | 195. (*c*) | 196. (*d*) |
| 197. (*c*) | 198. (*c*) | 199. (*c*) | 200. (*d*). | | | |

# Chapter 15 CONTROL SYSTEMS

## Definitions

**System.** A system is an arrangement, set or collection of things connected or related in such a manner as to form an entirety or whole. A system is also defined as an arrangement of physical components connected or related in such a manner as to form and/or act as an entire unit.

**Control System.** The word control is usually taken to mean **regulate, direct** or command. Accordingly a control system is an arrangement of physical components connected or related in such a manner as to command, direct or regulate itself or another system.

**Input.** The input is the stimulus or excitation applied to a control system from an external energy source, usually in order to produce a specified response from the control system.

**Output.** The output is the actual response obtained from a control system. It may or may not be equal to the specified response implied by the input.

**Open Loop Control System.** An open loop control system is one in which the control action is independent of the output.

**Closed Loop Control System.** A closed loop control system is one in which the control action is somehow dependent on the output. These are also known as feedback control systems.

**Feedback.** Feedback is that property of a closed loop system which permits the output or some other controlled variable of the system, to be compared with the input to the system, so that the appropriate control action may be formed as some function of the output and input. A feedback is said to exist in a system when a closed sequence of cause and effect relations exists between system variables.

### Characteristics of Feedback

1. Increased accuracy.
2. Reduced sensitivity of the ratio of output to input to variations in system characteristics.
3. Reduced effects of non-linearities and distortion.
4. Increased bandwidth.
5. Tendency towards oscillation or instability.

**Representation through Model.** For solving a system problem, the specifications or description of the system configuration and its components must be put into a form amenable to analysis, design and evaluation. Three basic models for systems are:

1. Differential equations and other mathematical relations.
2. Block diagrams.
3. Signal flow graphs.

**Analogous Systems.** For mathematical relations analogies are drawn between features of a system and features of some known elements or properties. Some analogous systems are given below:

**Table 1. Translational and Rotational Mechanical Systems**

| *Translational* | *Rotational* |
|---|---|
| Force, $f$ | Torque, $T$ |
| Acceleration, $a$ | Angular acceleration, $\alpha$ |
| Velocity, $u$ | Angular velocity, $\omega$ |
| Displacement, $x$ | Angular displacement, $x$ |
| Mass, $M$ | Moment of inertia, $I\theta$ |
| Damping co-efficient, $D$ | Rotational damping co-efficient, $D\theta$ |
| Compliance, $K$ $\left(\dfrac{1}{\text{stiffness}}\right)$. | Torsional compliance, $K\theta$ |

**Table 2. Force-Voltage Analogy**

| *Mechanical System* | *Electrical System* |
|---|---|
| Force, $f$ | Voltage, $v$ |
| Velocity, $u$ | Current, $i$ |
| Displacement, $x$ | Charge, $q$ |
| Mass, $M$ | Inductance, $L$ |
| Damping co-efficient, $D$ | Resistance, $R$ |
| Compliance, $K$ | Capacitance, $C$ |

**Table 3. Force-Current Analogy**

| *Mechanical System* | *Electrical System* |
|---|---|
| Force, $f$ | Current, $i$ |
| Velocity, $u$ | Voltage, $v$ |
| Displacement, $x$ | Flux linkage, $\phi$ |
| Mass, $M$ | Capacitance, C |
| Damping co-efficient, $D$ | Conductance, G |
| Compliance, $K$ | Inductance, $L$ |

**Table 4**

| *Electrical Systems* | *Thermal Systems* | *Liquid-level Systems* | *Pneumatic Systems* |
|---|---|---|---|
| Charge, $q$ | Heat flow, J | Liquid flow, $m^3$ | Air flow, $m^3$ |
| Current, $A$ | Heat flow rate, J/s | Liquid flow rate, $m^3/s$ | Air flow rate, $m^3/s$ |
| Voltage, $V$ | Temperature, ºC | Head, m | Pressure, $N/m^3$ |
| Resistance, Ω | Resistance, $°CsJ^{-1}$ | Resistance, $m^{-2}s$ | Resistance, $N\text{-}ms^{-1}$ |
| Capacitance, $F$ | Capacitance | Capacitance | Capacitance |

**Example 1.** *In the system shown in Fig. 15.1 a sinusoidally varying torque $T_1 = T_0 \sin \omega t$ is applied to wheel 1, which is engaged with wheel 2. Assume that the shafts of both the wheels are inertia less and supported on frictionless bearings. Steady state angular velocity of wheel 1 is to be determined.*

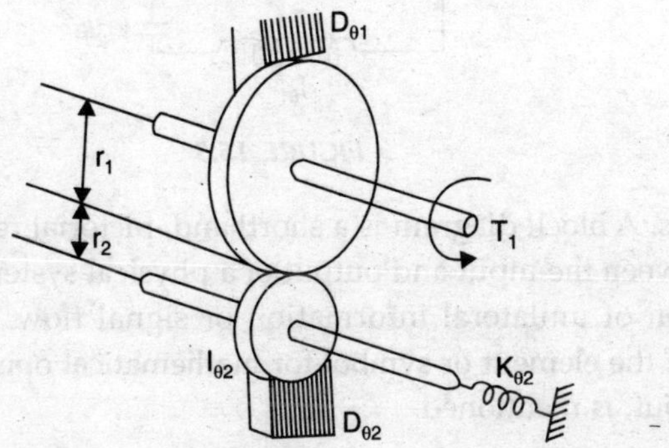


*FIGURE 15.1*

**Sol.** In this case let us use torque-voltage analogy. The analogous electrical circuit is shown in Fig. 15.2. In the electrical quantities have been written in terms of their analogous mechanical quantities. In this circuit ideal transformer is incorporated.

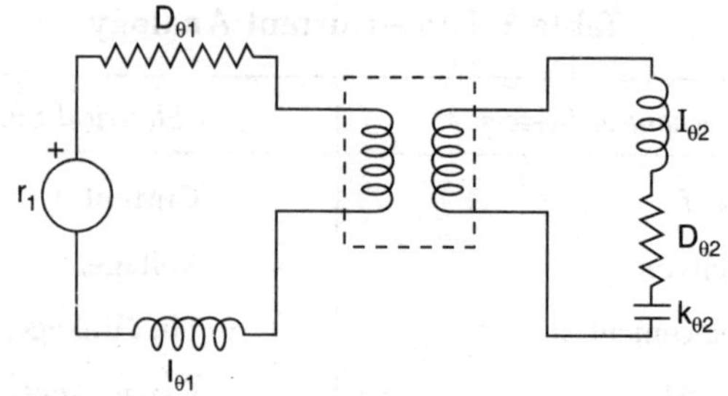


*FIGURE 15.2*

Above circuit with ideal transformer removed is shown in Fig. 15.3. In this all the elements in the secondary circuit have been transformed to the primary side. Steady state angular velocity $\omega_1$ is given by

$$\omega_1 = \frac{T_0}{\sqrt{(D_\theta + a^2 D_{\theta 2})^2 + [\omega(I_{\theta 1} + a^2 I_{\theta 2}) - (a^2/\omega k_{\theta 2})]^2}}$$

$$= \sin(\omega t - \psi)$$

where $$\Psi = \tan^{-1}\left[\frac{\omega(I_{\theta 1} + a^2 I_{\theta 2}) - (a^2/\omega k_{\theta 2})}{D_{\theta 1} + a^2 D_{\theta 2}}\right]$$

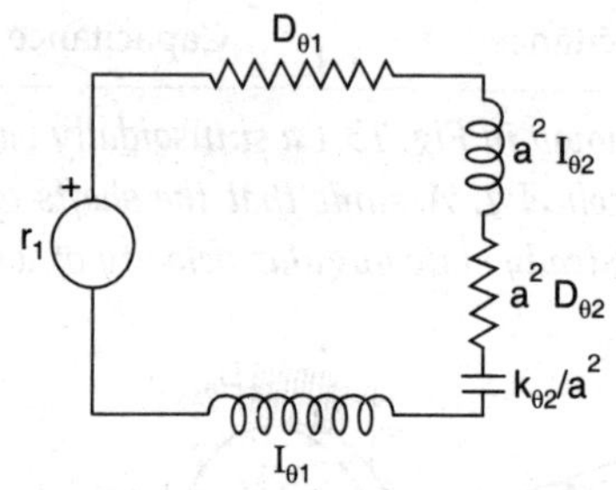


*FIGURE 15.3*

**Block Diagrams.** A block diagram is a shorthand, pictorial representation of the cause and effect relationship between the input and output of a physical system. In a block diagram, arrows represent the direction of unilateral information or signal flow. Inside the rectangle usually description or name of the element or symbol for mathematical operation to be performed on the input to yield the output, is mentioned.

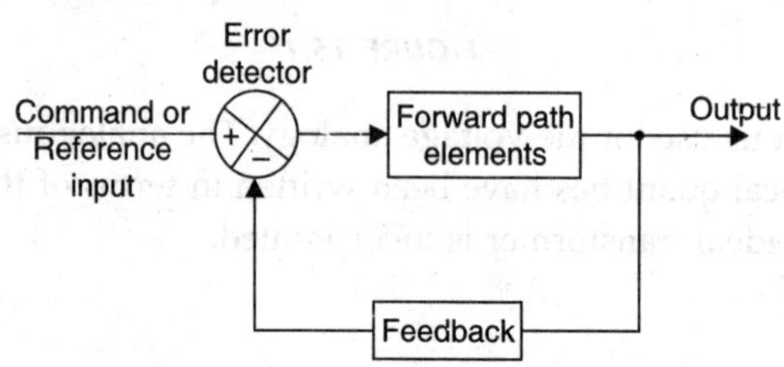


*FIGURE 15.4* Block diagram of servomechanism.

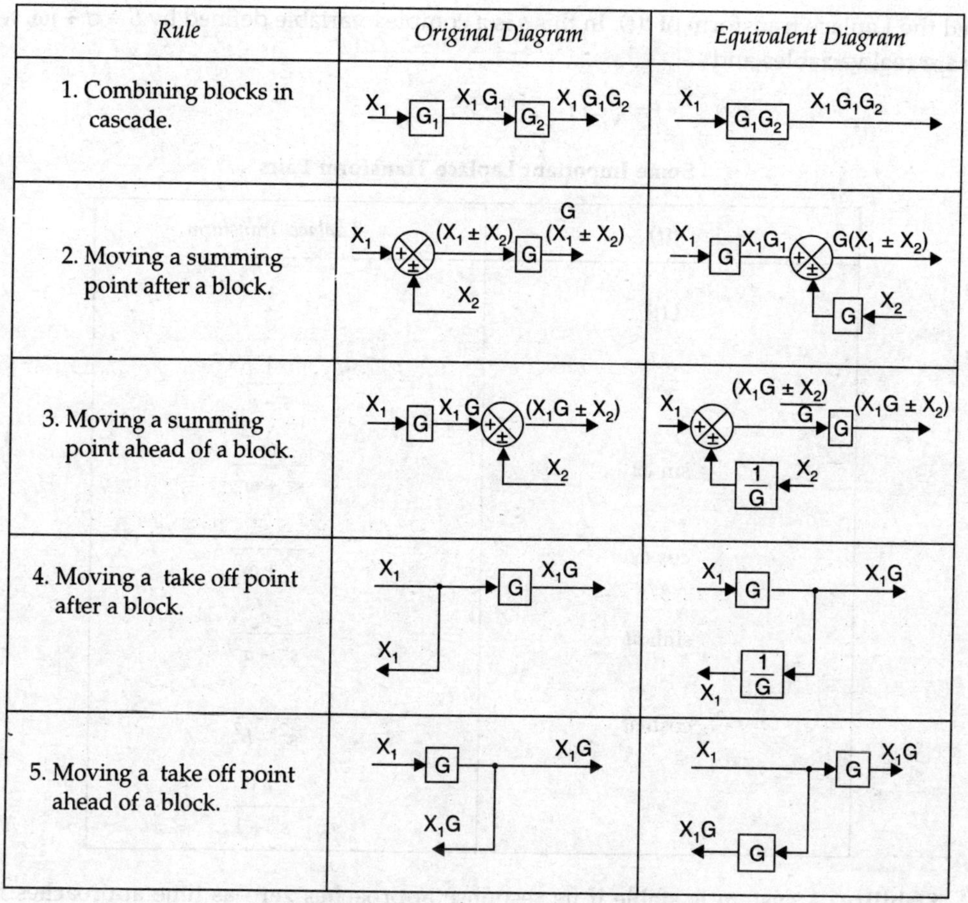


***FIGURE 15.5*** Rules for block diagram algebra.

**Servomechanism.** A servomechanism is a power-amplifying feedback control system in which the controlled variable $C$ is mechanical position or a time derivative of position such as velocity or acceleration.

**Regulator.** A regulator or regulating system is a feedback control system in which the reference input or command is constant for long periods of time, often for the entire interval during which the system is operational.

**Laplace Transform.** Let $f(t)$ be a real function of a real variable $t$ defined for $t > 0$, then

$$\mathscr{L}[f(t)] \equiv F(s) \equiv \lim_{\substack{T \to \infty \\ \varepsilon \to 0}} \int_{\varepsilon}^{T} f(t)e^{-st}\, dt$$

$$= \int_{0+}^{\infty} f(t)e^{-st}\, dt$$

$$0 < \varepsilon < T$$

is called the Laplace transform of *f*(*t*). In this *s* is a complex variable defined by $S \equiv \sigma + j\omega$, where $\sigma$ and $\omega$ are real variables and

$$j = \sqrt{-1}.$$

**Some Important Laplace Transform Pairs**

| *f*(*t*) | *Laplace Transform* |
|---|---|
| $L(t)$ | $\frac{1}{s}$ |
| $e^{at}$ | $\frac{1}{s-a}$ |
| $\sin \omega t$ | $\frac{\omega}{s^2+\omega^2}$ |
| $\cos \omega t$ | $\frac{s}{s^2+\omega^2}$ |
| $\sinh at$ | $\frac{a}{s^2-a^2}$ |
| $\cosh at$ | $\frac{s}{s^2-b^2}$ |
| $t^n$ | $\frac{n!}{s^n+1}$ |

**Stability.** A system is stable if its response approaches zero as time approaches infinity. Alternatively a system is stable if every bounded input produces a bounded output.

**Characteristic Root Locations.** If a system has some roots with real parts equal to zero, but none with positive real parts, the system is said to be **marginally stable.** However, in case the real parts of the roots of the characteristic equation have negative real parts, the system is said to be stable.

**Routh Stability Criterion.** This method is used for the determination of stability of systems with *n*th order characteristic equation of the form

$$a_n s^n + a_{n-1} s^{n-1} + \ldots\ldots + a_1 s + a_0 = 0$$

A Routh table is prepared as given below:

| | | | | |
|---|---|---|---|---|
| $s^n$ | $a_n$ | $a_{n-2}$ | $a_{n-4}$ | ... |
| $s^{n-1}$ | $a_{n-1}$ | $a_{n-3}$ | $a_{n-5}$ | ... |
| ... | $b_1$ | $b_2$ | $b_3$ | |
| ... | $c_1$ | $c_2$ | $c_3$ | |
| ... | — | — | — | ... |

where $a_n$, $a_{n-1}$, ....., $a_0$ are the co-efficients of the characteristic equation, and

$$b_1 = \frac{a_{n-1}a_{n-2} - a_n a_{n-3}}{a_{n-1}},\ b_2 = \frac{a_{n-1}a_{n-4} - a_n a_{n-5}}{a_{n-1}} \text{ and so on}$$

$$c_1 = \frac{b_1 a_{n-3} - a_{n-1}b_2}{b_1},\ c_2 = \frac{b_1 a_{n-5} - a_{n-1}b_3}{b_1} \text{ and so on.}$$

**The table is continued horizontally and vertically until only zeros are obtained. Any row can be multiplied by a constant before the next row is computed without disturbing the properties of the table. All the roots of this characteristic equation have negative real parts if and only if the elements of the first column of the Routh table have the same sign. Otherwise, the number of roots with positive real parts is equal to the number of changes of sign.**

**Example.** *Determine whether the system represented by the following equation is stable or not?*
*$2s^4 + 8s^3 + 10s^2 + 10s + 20 = 0$.*

**Sol.** The Routh table is given below

| | | | |
|---|---|---|---|
| $s^4$ | 2 | 10 | 20 |
| $s^3$ | 8 | 10 | 0 |
| $s^2$ | $\frac{15}{2}$ | 20 | 0 |
| $s$ | $-\frac{34}{3}$ | 20 | 0 |

$$b_1 = \frac{a_{n-1}a_{n-2} - a_n a_{n-3}}{a_{n-1}} = \frac{8\times 10 - 2\times 10}{8} = \frac{60}{8} = \frac{15}{2}$$

$$b_2 = \frac{a_{n-1}a_{n-4} - a_n a_{n-5}}{a_{n-1}} = \frac{8\times 20 - 2\times 0}{8} = 20$$

$$b_3 = \frac{a_{n-1}a_{n-6} - a_n \cdot a_{n-7}}{a_{n-1}} = \frac{8\times 0 - 2\times 0}{8} = 0$$

$$c_1 = \frac{b_1 a_{n-3} - a_{n-1}b_2}{b_1} = \frac{\frac{15}{2}\times 10 - 8\times 20}{\frac{15}{2}}$$

$$= -\frac{34}{3}$$

$$c_2 = \frac{b_1 a_{n-5} - a_{n-1}b_3}{b_1} = \frac{\frac{15}{2}\times 20 - 8\times 0}{\frac{15}{2}}$$

$$= 20$$

$$c_3 = 0.$$

As there is change of sign in the elements of first column (s) therefore the system is unstable.

**Properties of Transfer Functions**

1. The transfer function of a system is the Laplace transform of its impulse response. Accordingly if the input to a system with transfer function $P(s)$ is an impulse and all initial values are zero, the transform of the output is $P(s)$.
2. The system transfer function can be determined from the system differential equation by taking the Laplace transform and ignoring all terms arising from initial values.
3. The system differential equation can be obtained from the transfer function by replacing the $s$ variable with the differential operator $D$ defined by

$$D = \frac{d}{dt}.$$

4. In case of time-invariant linear system the stability can be determined from the characteristic equation. The denominator of the system transfer function set equal to zero is the characteristic equation. Consequently if all the roots of the denominator have negative real parts, the system is stable.
5. The roots of the numerator are the system zeros and the roots of the denominator are the system poles.

The system transfer function can then be specified to within a constant by specifying the system poles and zeros. This constant, denoted by $k$ is the system gain factor.

**Signal Flow Graphs.** A signal flow graph is a pictorial representation of the simultaneous equations describing a system. It graphically displays the transmission of signals through the system, as does the block diagram. Following points may be noted.

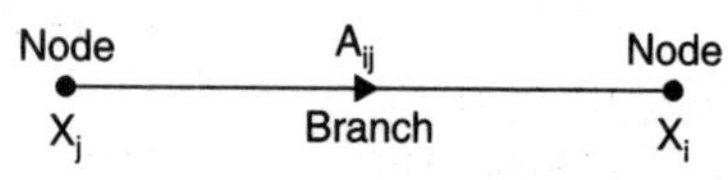


*FIGURE 15.6* Signal flow graph.

1. In signal flow graphs $A_{ij}$ is a mathematical operator mapping $X_j$ and $X_i$, and is called the transmission function.
2. The variables $X_i$ and $X_j$ are represented by a small dot called a node and the transmission function $A_{ij}$ is represented by a line with an arrow called a branch.
3. Branches are always unidirectional.
4. The arrow denotes the direction of signal flow.
5. The value of the variable designated by a node is equal to the sum of all signals entering the node. Thus

$$X_i = \sum_{j=1}^{n} A_{ij} X_j .$$

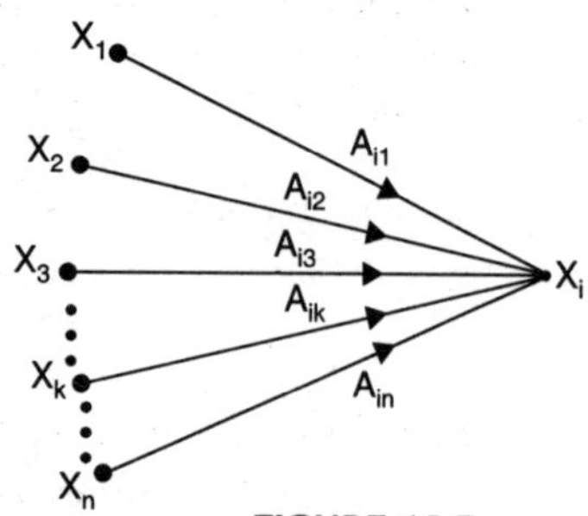


*FIGURE 15.7*

6. The value of the variable designated by a node is transmitted on every branch leaving that node. Thus $X_i = A_{ik} X_k$

$$i = 1, 2, \ldots\ldots, n, k \text{ fixed}.$$

7. A cascaded (series) connection of $n - 1$ branches with transmission functions $A_{21}, A_{32}, A_{43}, \ldots, A_{n(n-1)}$ can be replaced by a single branch with a new transmission function equal to the product of the old ones. That is

$$X_n = A_{21} \cdot A_{32} \cdot A_{43} \cdots\cdots A_{n(n-1)} \cdot X_1$$

The signal flow equivalence is represented as shown in Fig. 15.9,

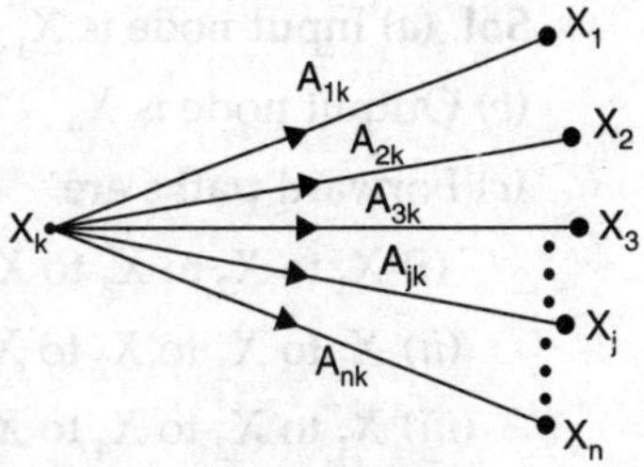


*FIGURE 15.8*

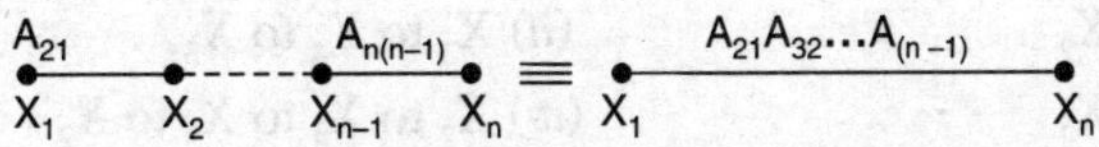


*FIGURE 15.9*

8. In SFG a path is a continuous, unidirectional succession of branches along which no node is passed more than once.
9. An **input node** or **source** is a node with only outgoing branches.
10. An **output node** or **sink** is a node with only incoming branches.
11. The path from input node to the output node is known as **forward path.**
12. A path that originates and terminates on the same node is known as **feedback loop.**
13. The **gain** of a branch is the transmission function of that branch when the transmission function is a multiplicative operator.
14. **Path gain** is the product of the branch gains encountered in traversing a path.
15. **Loop gain** is the product of the branch gains of the loop.

**Example.** *For the signal flow graph shown in Fig. 15.10, identify the following:*

(*a*) *Input node* (*b*) *Output node*

(*c*) *Forward paths* (*d*) *Feedback paths*

(*e*) *Self loop*

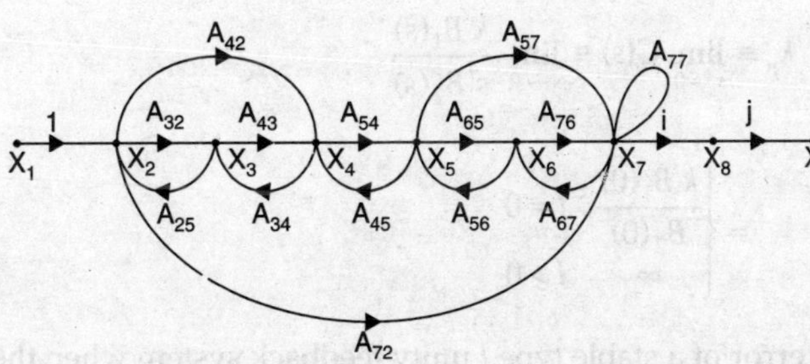


*FIGURE 15.10*

*Also determine.*

(*f*) *Loop gains of the feedback loops*

(*g*) *Path gains of the forward paths.*

**Sol.** (*a*) Input node is $X_1$

(*b*) Output node is $X_9$

(*c*) Forward paths are

(*i*) $X_1$ to $X_2$ to $X_3$ to $X_4$ to $X_5$ to $X_6$ to $X_7$ to $X_8$ to $X_9$

(*ii*) $X_1$ to $X_2$ to $X_7$ to $X_8$ to $X_9$

(*iii*) $X_1$ to $X_2$ to $X_4$ to $X_5$ to $X_6$ to $X_7$ to $X_8$ to $X_9$.

(*d*) Feedback paths are

(*i*) $X_2$ to $X_3$ to $X_2$ (*ii*) $X_3$ to $X_4$ to $X_3$

(*iii*) $X_4$ to $X_5$ to $X_4$ (*iv*) $X_2$ to $X_4$ to $X_3$ to $X_2$

(*v*) $X_2$ to $X_7$ to $X_5$ to $X_4$ to $X_3$ to $X_2$

(*vi*) $X_5$ to $X_6$ to $X_5$ (*vii*) $X_6$ to $X_7$ to $X_6$

(*viii*) $X_5$ to $X_6$ to $X_7$ to $X_5$ (*ix*) $X_7$ to $X_7$

(*x*) $X_2$ to $X_7$ to $X_6$ to $X_5$ to $X_4$ to $X_3$ to $X_2$.

(*e*) Self loop $X_7$ to $X_7$.

(*f*) (*i*) $A_{32}\,A_{23}$ (*ii*) $A_{43}\,A_{34}$

(*iii*) $A_{54}\,A_{45}$ (*iv*) $A_{65}\,A_{56}$

(*v*) $A_{76}\,A_{67}$ (*vi*) $A_{65}\,A_{76}\,A_{57}$

(*vii*) $A_{77}$ (*viii*) $A_{42}\,A_{34}\,A_{23}$

(*ix*) $A_{72}\,A_{57}\,A_{45}\,A_{34}\,A_{23}$ (*x*) $A_{72}\,A_{67}\,A_{56}\,A_{45}\,A_{34}\,A_{23}$.

(*g*) (*i*) $A_{32}\,A_{43}\,A_{54}\,A_{65}\,A_{76}$ (*ii*) $A_{72}$

(*iii*) $A_{42}\,A_{54}\,A_{65}\,A_{76}$.

## System Classification by Type

**Position Error Constant, $k_p$** of a type $l$ unity feedback system is defined as

$$k_p \equiv \lim_{s\to 0} G(s) \equiv \lim_{s\to 0} \frac{k\,B_1(s)}{s^l B_2(s)}$$

$$= \begin{cases} \dfrac{k\,B_1(0)}{B_2(0)} & l = 0 \\ \infty & l > 0 \end{cases}$$

The steady-state error of a stable type $l$ unity feedback system when the input is a unit-step function

$$(e(\infty)) = 1 - c(\infty)$$

is related to the position error constant by

$$e(\infty) = \lim_{t\to\infty} e(t) = \frac{1}{1+k_p}$$

**Velocity Error Constant,** $k_v$ **of a stable type** $l$ **unity feedback system is defined as**

$$k_v \equiv \lim_{s \to 0} s\,G(s) = \lim_{s \to 0} \frac{kB_1(s)}{s^{l-1}B_2(s)}$$

$$= \begin{cases} 0 & l = 0 \\ \dfrac{kB_1(0)}{B_2(0)} & l = 1 \\ \infty & l > 1 \end{cases}$$

**The steady-state error of a stable type** $l$ **unity—feedback system when the input is a unit ramp function is related to the velocity error constant by**

$$e(\infty) = \lim_{t \to \infty} e(t) = \frac{1}{k_v}$$

**Acceleration Error Constant,** $k_a$ **of a stable type** $l$ **unity-feedback system is defined as**

$$k_a \equiv \lim_{s \to 0} s^2 G(s) \equiv \lim_{s \to 0} \frac{kB_1(s)}{s^{l-2}B_2(s)}$$

$$= \begin{cases} 0 & l = 0, 1 \\ \dfrac{kB_1(0)}{B_2(0)} & l = 2 \\ \infty & l > 2 \end{cases}$$

**The steady-state error constant of a stable type** $l$ **unity feedback system when the input is a unit parabolic function is related to the acceleration error constant by**

$$e(\infty) = \lim_{t \to \infty} e(t) = \frac{1}{k_a}.$$

**System Classification**

| *Input* / *System Type* | *Unit Step* $k_p$ | *Unit Step* *Steady State Error* | *Unit Ramp* $k_v$ | *Unit Ramp* *Steady State Error* | *Unit Parabola* $k_a$ | *Unit Parabola* *Steady State Error* |
|---|---|---|---|---|---|---|
| Type 0 | $\frac{kB_1(0)}{B_2(0)}$ | $\frac{1}{1+k_p}$ | 0 | $\infty$ | 0 | $\infty$ |
| Type 1 | $\infty$ | 0 | $\frac{kB_1(0)}{B_2(0)}$ | $\frac{1}{k_v}$ | 0 | $\infty$ |
| Type 2 | $\infty$ | 0 | $\infty$ | 0 | $\frac{kB_1(0)}{B_2(0)}$ | $\frac{1}{k_a}$ |

**Nyquist Analysis.** Nyquist analysis method is a frequency response method. It is basically a graphical procedure for determining absolute and relative stability of a closed loop control systems. In this the system is put into canonical form and the formation is available directly from the graph of the sinusoidal openloop transfer function *GH* ($j\omega$). This technique is also useful for obtaining information about transfer functions of components or systems from experimental frequency data.

Any complex variable $s = \sigma + j\omega$ is dependent on two independent quantities, the real and imaginary parts of *s*. Due to this *s* cannot be represented by a line. Therefore for plotting complex function *P*(*s*) with $s = \sigma + j\omega$, two 2-dimensional graphs are used

(*i*) graph of $j\omega$ versus $\sigma$ called the *s* plane

(*ii*) graph of imaginary parts of *P*(*s*) versus real parts of *P*(*s*) called *P*(*s*) plane.

**Mapping** is the correspondence between points in the two planes. Thus points in the *s*-plane can be mapped into points of the *P*(*s*) plane by the function *P*. In general, only a very specific locus of points in the *s*-plane is mapped into the *P*(*s*) plane. For the Nyquist stability plots this locus is called the Nyquist path.

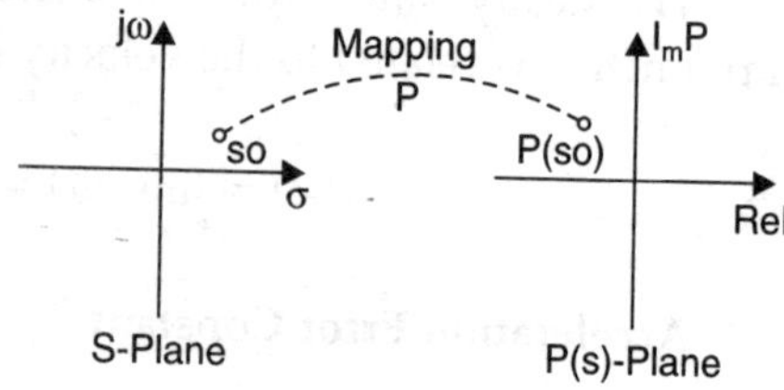


*FIGURE 15.11*

## Some Properties of the Mapping

1. *P*(*s*) is a single valued function. Thus every point in the *s*-plane maps into one and only one point in the *P*(*s*) plane.
2. *s*-plane contours avoid singular points of *P*(*s*).
3. *P*(*s*) is analytic except possibly at a finite number of points (singularities) in the *s*-plane.
4. Every closed contour in the *s*-plane maps into a closed contour in the *P*(*s*) plane.
5. *P*(*s*) is a conformal mapping. That is, the angle between and the direction of any two intersecting curves in the *s*-plane are preserved by the mapping of these curves into the *P*(*s*) plane.

**Nyquist Path** is a closed contour in the *s*-plane which completely encloses the entire right half of the *s*-plane.

In order that the Nyquist path should not pass through any poles of *P*(*s*), small semi-circles along the imaginary axis or at the origin of *P*(*s*) are required in the path if *P*(*s*) has poles in the $j\omega$ axis or at the origin. The radii of these small circles are interpreted as approaching zero in the limit. In order to enclose the right half of the *s*-plane at infinity, a large semi-circular path is drawn in the *RHP* and the radius *R* of this semi-circle is interpreted as being infinite in the limit.

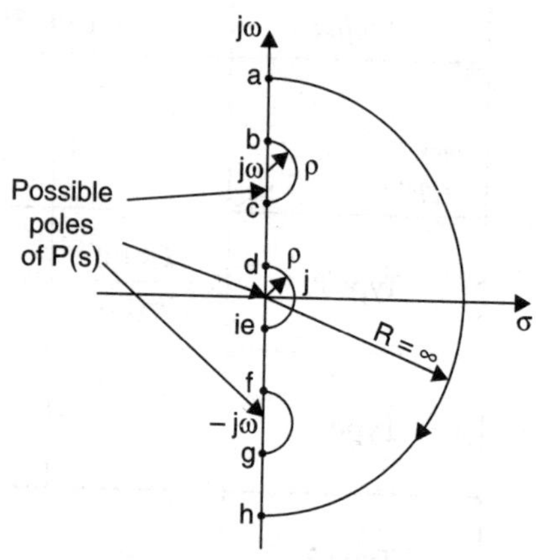


*FIGURE 15.12*

**Nyquist Stability Criterion.** A closed-loop control system is considered to be absolutely stable if the roots of the characteristic equation have negative real parts. Equivalently, the poles of the closed-loop transfer function, or the zeros of the denominator of the closed-loop transfer function must lie in the left $1 + GH(s)$ half plane (*LHP*). The Nyquist stability criterion establishes the number of zeros of $1 + GH(s)$ in the *RHP* directly from the Nyquist Stability plot of $GH(s)$. Thus when

$P_0$ = number of poles of $GH$ in the $RHP \geq 0$

$N$ = total number of *CW* encirclements of the $(-1, 0)$ point (*i.e.*, $GH(s) = -1$) in the $GH(s)$ plane.

when $N > 0$, the number of zeros $Z_0$ of $1 + GH$ in the *RHP* is determined by

$$Z_0 = N + P_0$$

when $N \leq 0$, the $(-1, 0)$ point is not enclosed by the Nyquist stability plot. Hence $N < 0$ if the region to the right of the contour in the prescribed direction does not include the $(-1, 0)$ point.

Whenever $N \leq 0$ and $P_0 = 0$, the system is absolutely stable.

**Root Locus Analysis.** The root-locus techniques permit accurate computation of the time-domain response in addition to yielding readily available frequency response information.

This method requires that only the location of the poles and zeros of $GH$ be known and does not require factorization of the characteristic polynomial.

Let us consider a canonical feedback control system, shown in Fig. 15.13 for which the closed loop transfer function is given by

$$\frac{C}{R} = \frac{G}{1+GH}$$

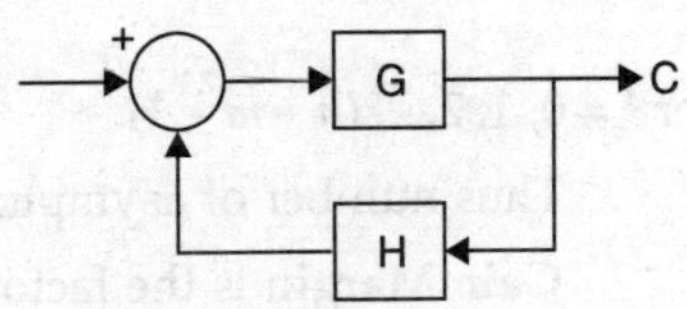


*FIGURE 15.13*

Let the open loop transfer function $GH$ be represented by

$$GH = \frac{KN(s)}{D(s)}$$

$$= \frac{K(s^m + a_{m-1}s^{m-1} + \ldots + a_0)}{(s^n + b_{n-1}s^{n-1} + \ldots + b_0)}$$

where $N(s)$ and $D(s)$ are the finite polynomials in the complex variable $s$, $m \leq n$ and $K$ is the open-loop gain factor. The closed loop transfer function then becomes

$$\frac{C}{R} = \frac{G}{1+\dfrac{KN}{D}} = \frac{GD}{D+KN}$$

The closed loop poles are roots of the characteristic equation

$$D(s) + KN(s) = 0$$

A locus of these roots plotted in the $s$-plane as a function of $K$ is called a root-locus.

**Number of Loci.** The number of loci, *i.e.*, the number of branches of the root-locus, is equal to the number of poles of the open-loop transfer function $GH$.

**Real Axis Loci.** Those sections of the root-locus on the real axis in the $s$-plane are determined by counting the total number of finite poles and zeros of $GH$ to the right of the points in question.

**When K > 0.** Points of the root-locus on the real axis lie to the left of an odd number of finite poles and zercs.

**When K < 0.** Points of the root-locus on the real axis lie to the left of an even number of finite poles and zeros.

**Asymptotes.** Asymptotes emanate from a point in the *s*-plane on the real axis called the centre of asymptotes $\sigma_c$, given by

$$\sigma_c = -\frac{\sum_{i=1}^{n} p_i - \sum_{i=1}^{m} z_i}{n-m}$$

where $-p_i$ are the poles

$-z_i$ are the zeros

*n* is the number of poles of *GH*

*m* is the number of zeros of *GH*.

The angles between the asymptotes and the real axis are given by

$$\beta = \begin{cases} \dfrac{(2l+1)\,180}{n-m} & \text{degrees for } K > 0 \\[2ex] \dfrac{(2l)\,180}{n-m} & \text{degrees for } K < 0 \end{cases}$$

for $l = 0, 1, 2,..., (n - m - 1)$.

Thus number of asymptotes = $(n - m)$.

**Gain Margin** is the factor by which the design value of the gain factor *k* can be multiplied before the closed-loop system becomes unstable. Thus

$$\text{Gain Margin} = \frac{\text{Value of } K \text{ at imaginary axis cross - over}}{\text{Design value of } K}$$

In case root locus does not cross the $j\omega$ axis, the gain margin is infinite.

**Phase Margin** for a feedback system is computed from the relation

$$\phi_{Din} = 180^{\circ} + \arg GH\,(j\omega_1).$$

**Bode Analysis.** Like Nyquist analysis, Bode analysis also uses graphical representations of *GH* ($j\omega$), however, in this case two graphs are plotted—one for the magnitude of *GH* ($j\omega$) and the other for phase angle of *GH* ($j\omega$), both plotted as functions of frequency ω. In this usually logarithmic scales are used for frequency axes and for *GH* ($j\omega$). In this decibel (*db*) magnitude of frequency is used. Thus *db* magnitude varsus log ω plot is called the Bode magnitude plot, and the phase angle versus log ω plot is the Bode phase angle plot. The Bode magnitude plot is also known as Log-modulus plot.

## OBJECTIVE TYPE QUESTIONS

1. In a refrigerator, the input for thermostat is
   (*a*) reference voltage (*b*) reference current
   (*c*) reference temperature (*d*) reference heat.
2. A part of the human temperature control system is
   (*a*) leg movement (*b*) ear
   (*c*) perspiration system (*d*) digestive system.
3. A system in which the control action is independent of the output is known as
   (*a*) open loop system (*b*) closed-loop system
   (*c*) dependent system (*d*) responding system.
4. In a closed loop control system
   (*a*) control action is independent of the output
   (*b*) output is independent of input (*c*) there is no feedback
   (*d*) control action is dependent of the output.
5. An automatic toasier is a
   (*a*) open loop control system (*b*) closed loop control system
   (*c*) partially closed loop control system (*d*) not a system.
6. 'A man driving an automobile' forms a
   (*a*) man made control system (*b*) biological control system
   (*c*) control system whose components are both man made and natural
   (*d*) none of the above.
7. 'Pointing of an object with a finger' is a
   (*a*) man made control system (*b*) biological control system
   (*c*) man made and natural control system
   (*d*) none of the above.
8. Which of the following distinguishes a closed loop system from open-loop system?
   (*a*) Output pattern (*b*) Input pattern (*c*) Servomechanism (*d*) Feedback.
9. For a dc generator input could be in form of
   (*a*) rotational speed (*b*) dc voltage induced
   (*c*) dc current (*d*) any of the above.
10. Which of the following is a closed loop system?
    (*a*) electric switch (*b*) dc generator
    (*c*) car starter (*d*) auto pilot for an aircraft
11. While a man walks along a path, the control action is determined by
    (*a*) legs (*b*) hands (*c*) eyes (*d*) brain.

12. In the above case of eyes are closed
(a) input will reduce to zero (b) output will reduce to zero
(c) feedback loop will be broken (d) none of the above will happen.

13. Any externally introduced input signal affecting the controlled output is called a
(a) signal (b) feedback (c) stimulus (d) gain control.

14. A power amplifying feedback control system is which the controlled variable is mechanical position or a time derivative of position such as velocity or acceleration, is known as
(a) servomechanism (b) regulator (c) timer (d) automation.

15. Which of the following represents the dimensional units for thermistor?
(a) Input: temperature; output: temperature
(b) Input temperature; output: electrical quantity like ohms, volts or amperes
(c) Input: volts; output: temperature
(d) Input: amperes; output: volts.

16. Which of the following represents the dimensional units for a thermocouple?
(a) Input: volts; output: amperes
(b) Input: absolute temperature; output: temperature ºC
(c) Input: temperature difference; output: volts
(d) Input: volts; output: volts.

17. Which of the following differential equation represents ohms law?
(a) $v = I\dfrac{dR}{dt}$ (b) $I = R\dfrac{dv}{dt}$ (c) $R = \dfrac{dI}{dt} \times \dfrac{dv}{dt}$ (d) $v = R\dfrac{dq}{dt}$.

18. Which of the following is true for the diffusion equation

$$\frac{\partial T}{\partial x} = K\frac{\partial T}{\partial t}$$

The above equation is
(a) linear differential equation (b) partial differential equation
(c) time dependent differential equation (d) all of the above.

19. The function shown in Fig. 15.14 is
(a) unit step function
(b) unit ramp function
(c) unit impulse function
(d) none of the above.

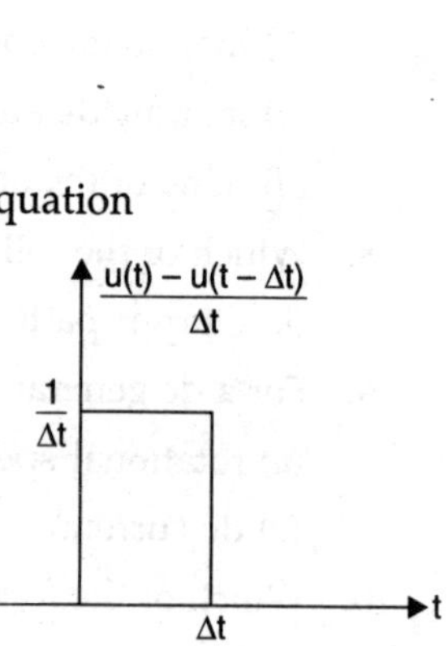

FIGURE 15.14

20. Which of the following term is first degree in the dependent variable $y = y(t)$?
(a) $\cos t$ (b) $\tan y$ (c) $e^{-y}$ (d) $t^2 y$.

21. The forced response of

$$\frac{d^2y}{dt^2} + 3\frac{dy}{dt} + 2y = x$$

where $x = \sin t$ will be

(a) $y_b(t) = \frac{1}{10}\left[e^{-t} + e^{-2t} + \sin t\right]$
(b) $y_b(t) = \frac{1}{10}\left[5e^{-t} + 2e^{-2t}\right]$

(c) $y_b(t) = \frac{1}{10}\left[5e^{-t} + 2e^{-2t} + \sin t - 3\cos t\right]$
(d) $y_b(t) = \frac{1}{10}\left[5e^{t} + 2e^{2t} - \sin t - \cos t\right]$.

22. A servomechanism usually consists of
(a) power amplifier (b) error actuated signal
(c) mechanical output (d) all of the above.

23. Under force-voltage analogy, mass is analogous to
(a) resistance (b) charge (c) inductance (d) current.

24. Under force-voltage analogy, resistance is analogous to
(a) velocity (b) momentum
(c) displacement (d) viscous friction co-efficient.

25. Under force-voltage analogy, velocity is analogous to
(a) current (b) charge (c) capacitance (d) time constant.

26. Under force-voltage analogy, spring constant is analogous to
(a) capacitance (b) reciprocal of capacitance
(c) reciprocal of inductance (d) reciprocal of resistance.

27. Under force-current analogy, velocity is analogous to
(a) current (b) voltage (c) magnetic flux (d) charge.

28. Under force-current analogy, displacement is analogous to
(a) capacitance (b) inductance
(c) resistance (d) magnetic flux linkage.

29. Under force-current analogy, reciprocal of inductance is analogous to
(a) mass (b) viscous friction co-efficient
(c) spring constant (d) velocity.

30. Under force-current analogy, viscous-friction co-efficient is analogous to
(a) resistance (b) conductance (c) inductance (d) none of the above.

31. Under force-current analogy, mass is analogous to
(a) current (b) capacitance
(c) reciprocal of capacitance (d) reciprocal of conductance.

**32.** Which of the following is incorrect pair under force-voltage analogy?

(*a*) Spring constant-capacitance (*b*) Mass-inductance

(*c*) Velocity-current (*d*) Displacement-charge.

**33.** Under force-current analogy which of the following is incorrect pair?

(*a*) Displacement—capacitance (*b*) Velocity—voltage

(*c*) Spring constant—reciprocal of inductance

(*d*) Viscous—friction co-efficient—reciprocal of resistance.

**34.** Under torque voltage analogy, moment of inertia is analogous to

(*a*) voltage (*b*) resistance (*c*) inductance (*d*) charge.

**35.** Under torque — voltage analogy, angular velocity is analogous to

(*a*) charge (*b*) current

(*c*) resistance (*d*) reciprocal of capacitance.

**36.** Under torque — voltage analogy, charge is analogous to

(*a*) viscous friction co-efficient (*b*) coulomb's friction co-efficient

(*c*) angular velocity (*d*) angular displacement.

**37.** Under torque — voltage analogy, viscous friction co-efficient is analogous to

(*a*) resistance (*b*) reciprocal of resistance

(*c*) capacitance (*d*) reciprocal of capacitance.

**38.** Under torque — voltage analogy, torsional spring stiffness is analogous to

(*a*) capacitance (*b*) reciprocal of capacitance

(*c*) inductance (*d*) reciprocal of inductance.

**39.** Under torque — current analogy, moment of inertia is analogous to

(*a*) capacitance (*b*) reciprocal of capacitance

(*c*) inductance (*d*) reciprocal of inductance.

**40.** Under torque — current analogy, viscous friction co-efficient is analogous to

(*a*) resistance (*b*) reciprocal of resistance

(*c*) capacitance (*d*) reciprocal of inductance.

**41.** Under torque — current analogy, reciprocal of inductance is analogous to

(*a*) torsional spring stiffness (*b*) torque

(*c*) angular velocity (*d*) co-efficient of viscosity.

**42.** Under torque — current analogy, angular displacement is analogous to

(*a*) capacitance (*b*) resistance

(*c*) magnetic flux linkage (*d*) voltage.

**43.** Under torque — current analogy, voltage is analogous to

(*a*) angular velocity (*b*) angular acceleration

(*c*) angular displacement (*d*) none of the above.

**44.** If viscous friction co-efficient and reciprocal of resistance are considered analogous, the analogous system being considered is

(*a*) force—voltage analogy (*b*) torque—voltage analogy

(*c*) force—current analogy (*d*) torque—current analogy

**45.** If displacement and charge are considered to be analogous quantities, the analogy being considered is

(*a*) force—voltage (*b*) torque—voltage (*c*) force—current (*d*) torque—current.

**46.** If torsional spring stiffness and reciprocal of capacitance are considered as analogous quantities, the system being considered is

(*a*) torque—current (*b*) torque—voltage

(*c*) force—current (*d*) force—voltage.

**47.** For system shown in Fig. 15.15 (*a*) the analogous system is represented in Fig. 15.15 (*b*) by

(*a*) figure *A*

(*b*) figure *B*

(*c*) figure *C*

(*d*) figure *D*.

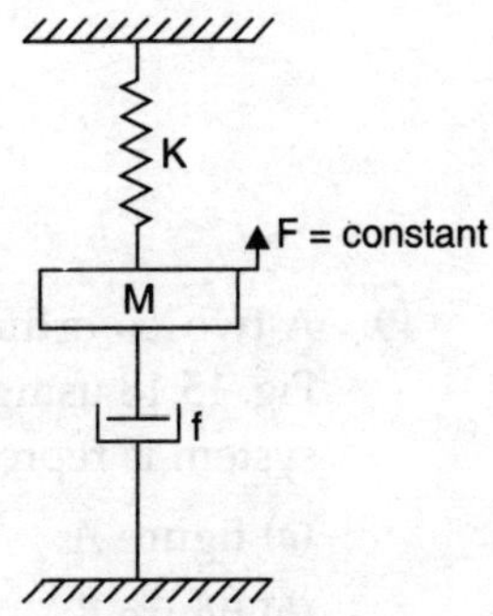


FIGURE 15.15 (*a*)

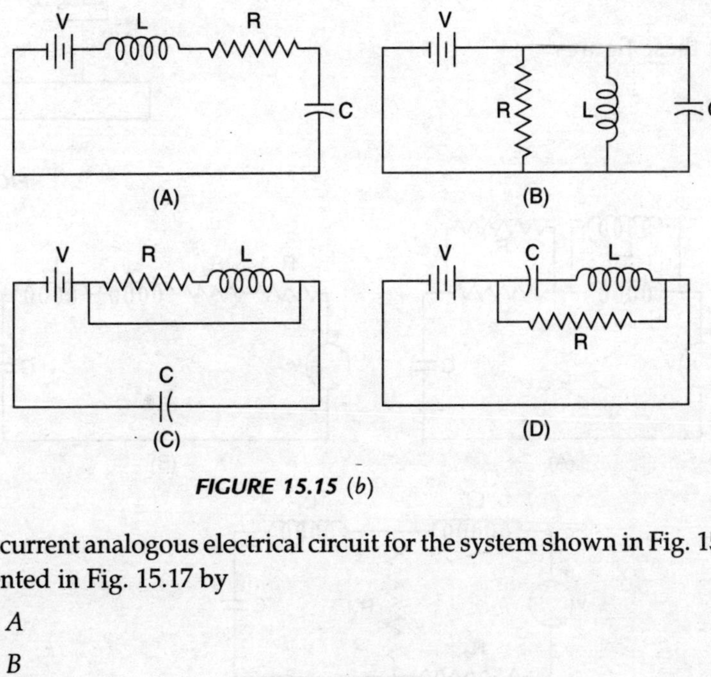


FIGURE 15.15 (*b*)

**48.** The force current analogous electrical circuit for the system shown in Fig. 15.16 is represented in Fig. 15.17 by

(*a*) figure *A*

(*b*) figure *B*

(*c*) figure *C*

(*d*) none of the above.

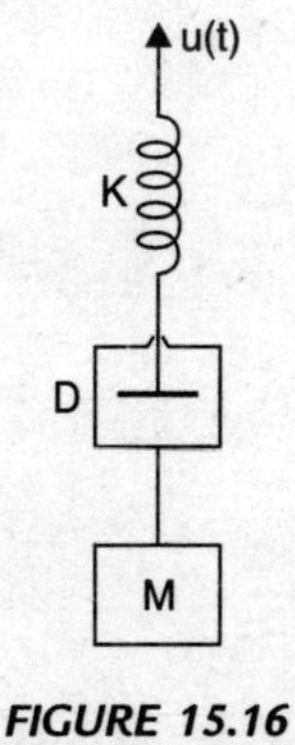


FIGURE 15.16

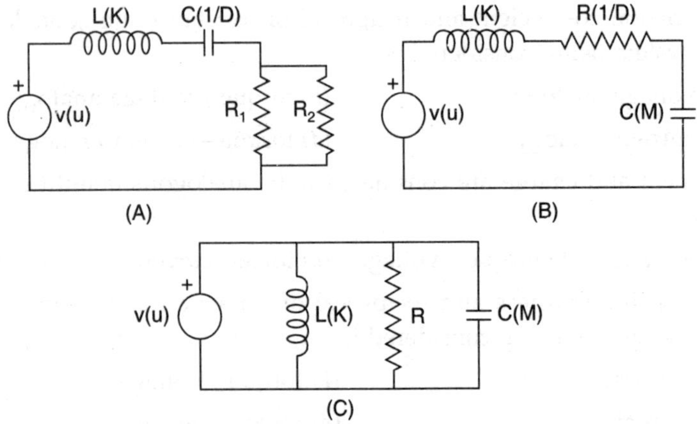


*FIGURE 15.17*

**49.** A two co-ordinate mechanical system is shown in Fig. 15.18 using force-voltage analogy, the analogous system is represented in Fig. 15.19 by

(*a*) figure *A*

(*b*) figure *B*

(*c*) figure *C*

(*d*) none of these figures.

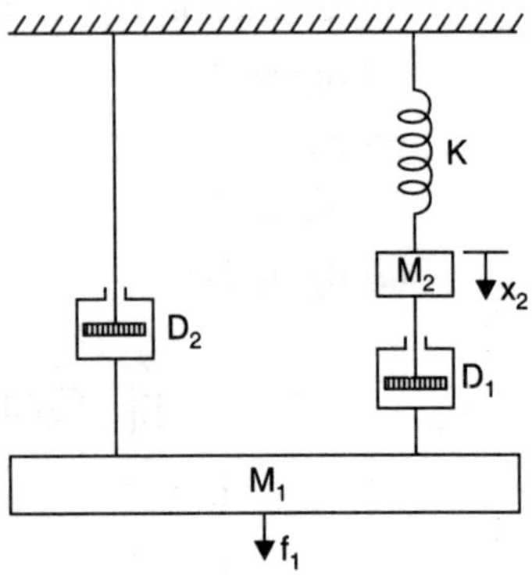


*FIGURE 15.18*

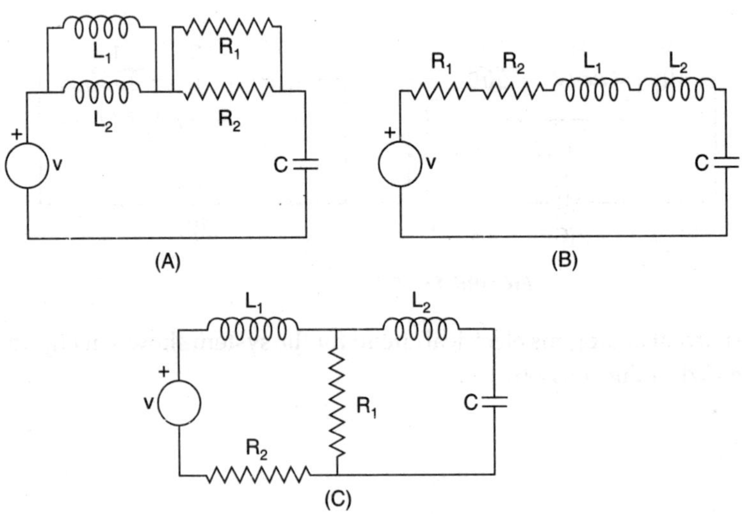


*FIGURE 15.19*

**50.** For the above system, using force-current analogy the analogous circuit is represented by

(*a*) figure *A* (*b*) figure *B* (*c*) figure *C* (*d*) none of these figures.

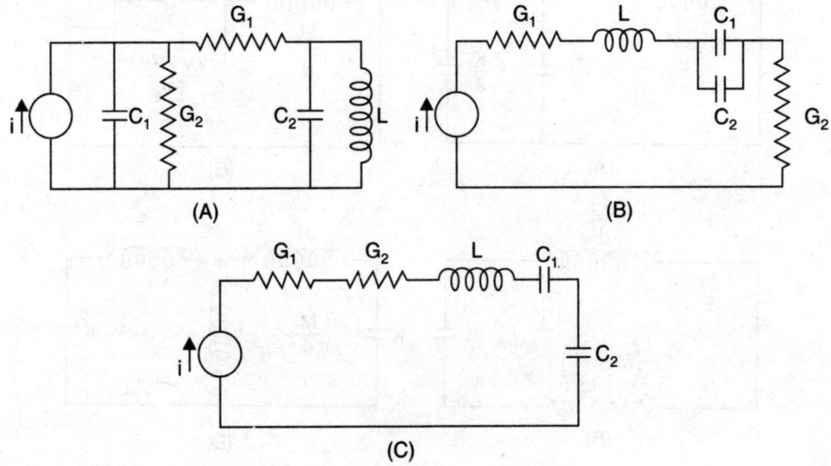


*FIGURE 15.20*

**51.** Which of the following forces resist motion?

(*a*) Inertia force (*b*) Spring force (*c*) Damping force (*d*) All of the above.

**52.** In linear systems damping force is considered to be proportional to

(*a*) displacement (*b*) velocity (*c*) acceleration (*d*) any of the above.

**53.** The restoring force of a spring is considered proportional to

(*a*) displacement (*b*) velocity (*c*) acceleration (*d*) coulomb friction.

**54.** The mass $M$ in mechanical translation system is analogous to which of the following quantity in rotational system?

(*a*) Torque (*b*) Angular velocity

(*c*) Moment of inertia (*d*) Torsional compliance.

**55.** Fig. 15.21 shows a rotary machine with mass $M$ and moment of inertia $I_\theta$ about its shaft in the centre. It is supported on two springs with identical compliance (1/stiffness) $k$ and translational motion is possible in the vertical direction only.

Which of the following figure represents the electrical analogous circuit for this system?

(*a*) figure *A* (*b*) figure *B*

(*c*) figure *C* (*d*) figure *D*.

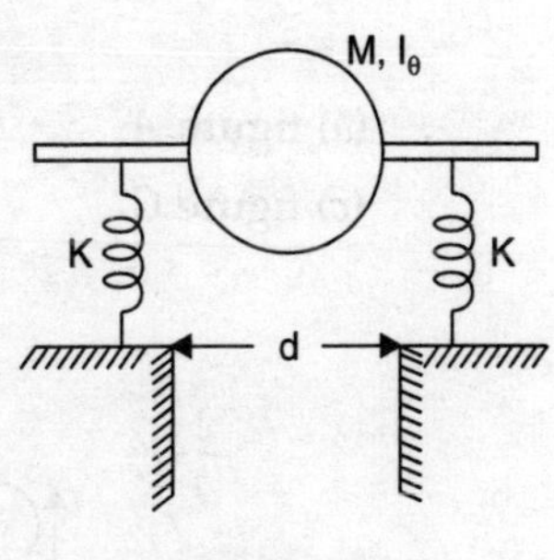


*FIGURE 15.21*

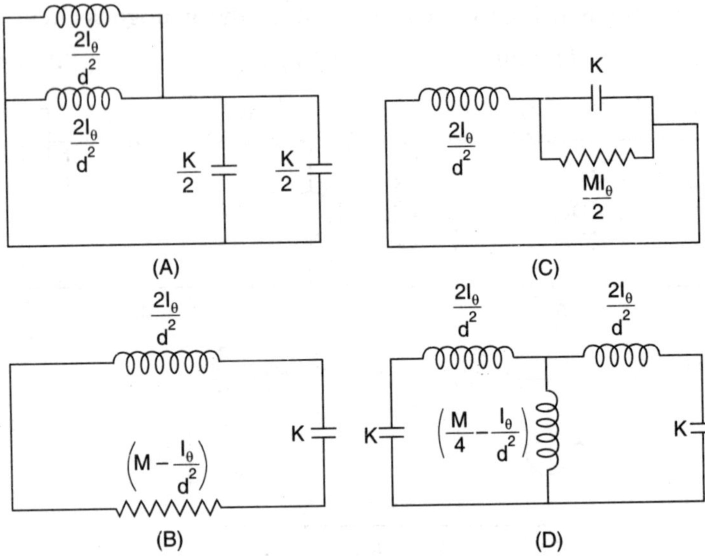


*FIGURE 15.22*

**56.** A mechanical system is shown in Fig. 15.23. The force-current analogous circuit is represented by (in Fig. 15.24).

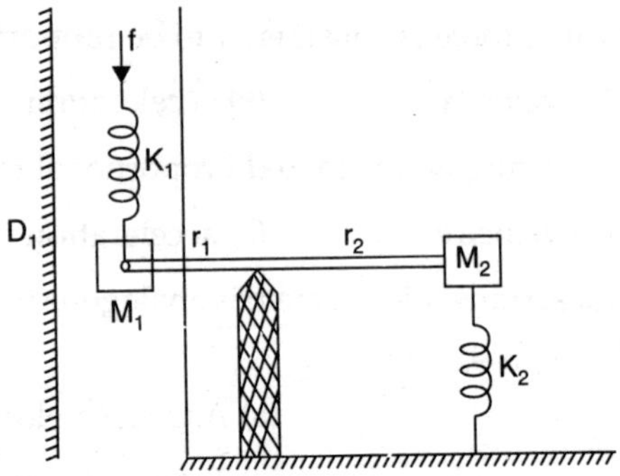


*FIGURE 15.23*

(*a*) figure *A* (*b*) figure *B*

(*c*) figure C (*d*) none of the above figures.

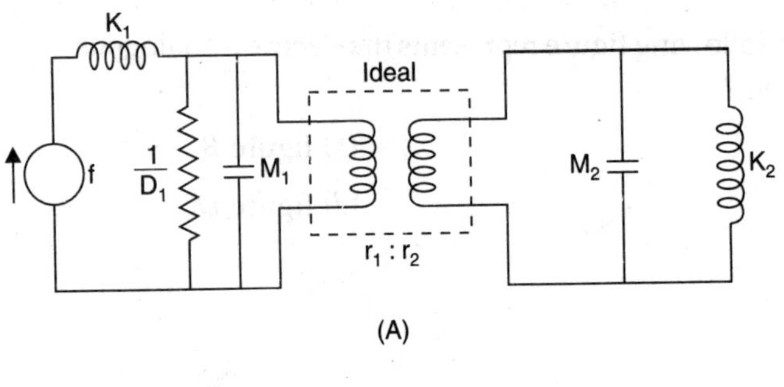


(A)

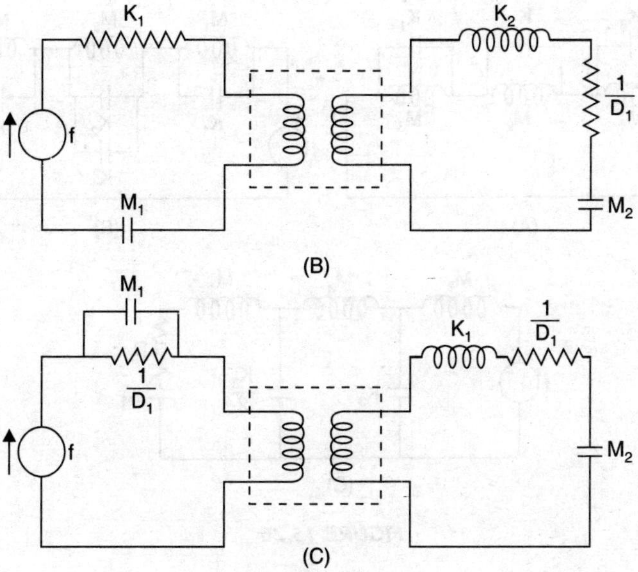


*FIGURE 15.24*

**57.** For the mechanical system shown in Fig. 15.25 the force-voltage analogous circuit is represented by (in Fig. 15.26).

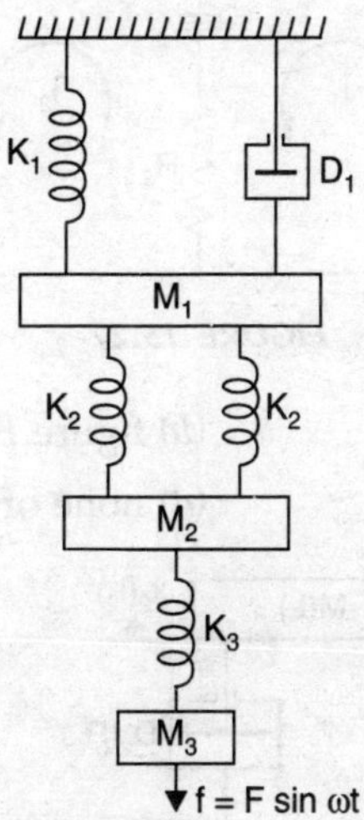


*FIGURE 15.25*

(*a*) figure *A* (*b*) figure *B*

(*c*) figure *C* (*d*) none of the above figures.

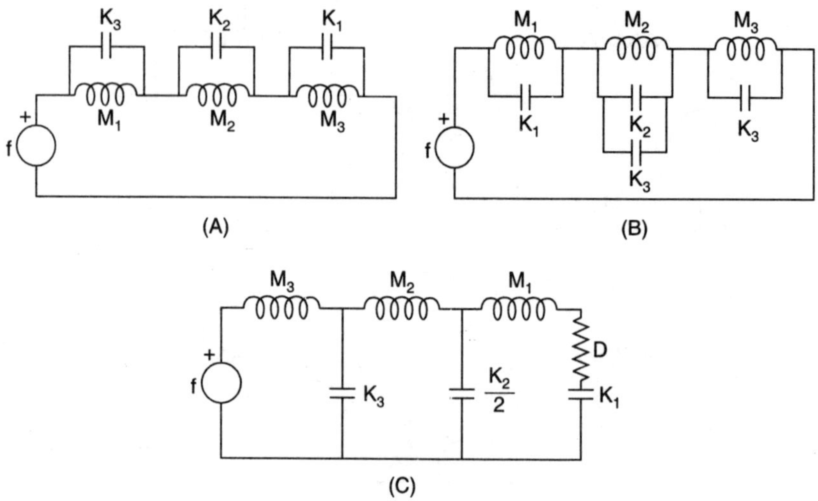


*FIGURE 15.26*

**58.** The force voltage analogous system for the electrical circuit shoes in Fig. 15.27 is given by (in Fig. 15.28).

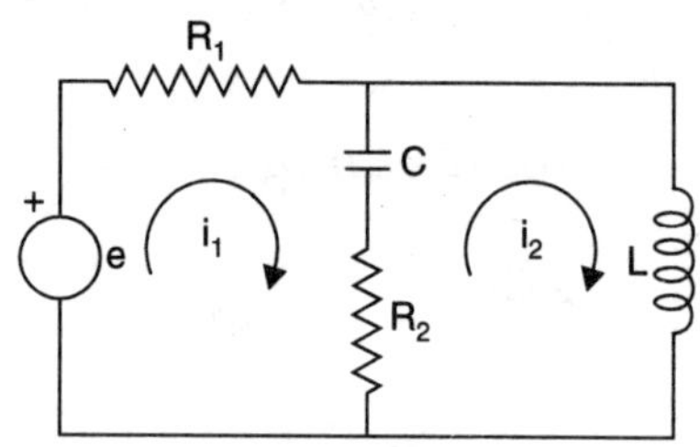


*FIGURE 15.27*

(*a*) figure *A* (*b*) figure *B*

(*c*) figure *C* (*d*) none of the above figure.

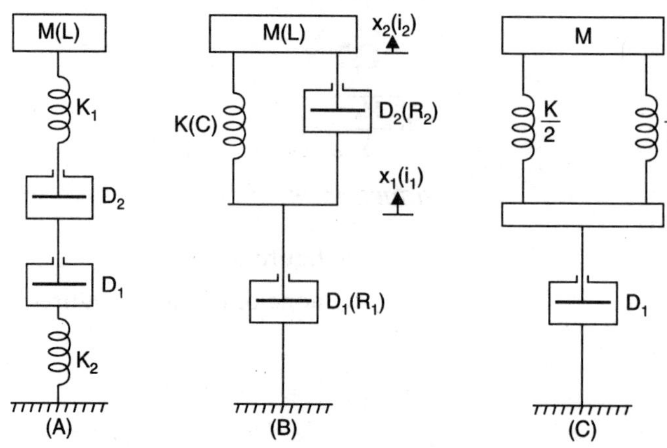


*FIGURE 15.28*

59. For the system shown in Fig. 15.29 (*a*) electric analog circuit is represented in Fig. 15.29 (*b*) by

(*a*) figure *A*

(*b*) figure *B*

(*c*) figure *C*

(*d*) none of the above figures.

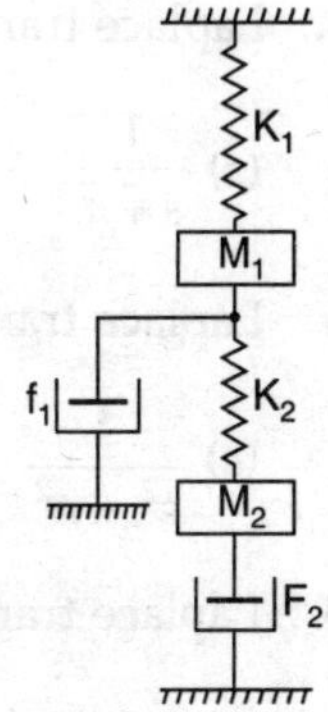


**FIGURE 15.29 (*a*)**

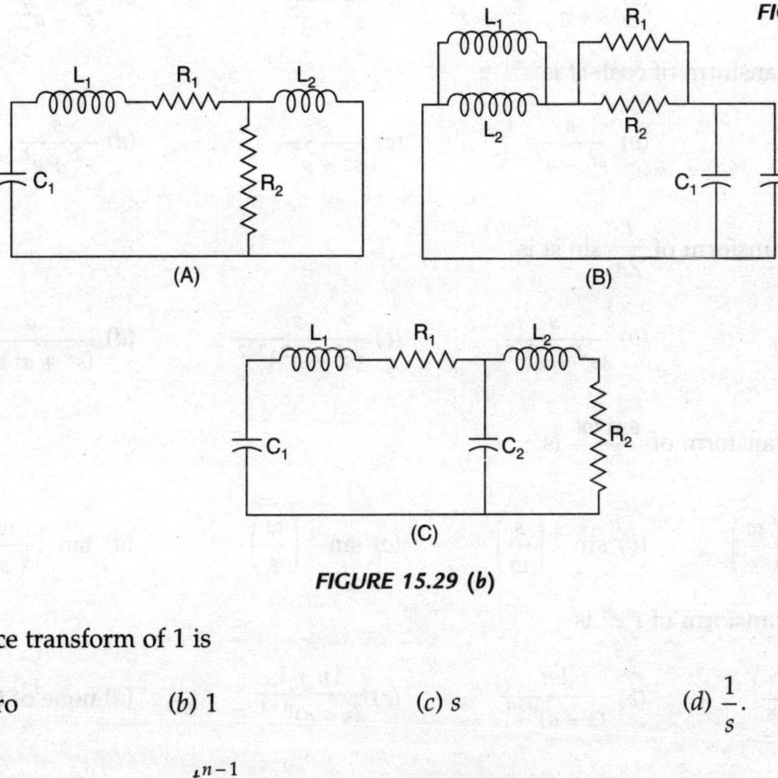


**FIGURE 15.29 (*b*)**

60. Laplace transform of 1 is

(*a*) zero (*b*) 1 (*c*) *s* (*d*) $\frac{1}{s}$.

61. Laplace transform of $\frac{t^{n-1}}{(n-1)!}$ is (*n* is positive integer)

(*a*) 1 (*b*) $s^n$ (*c*) $\frac{1}{s^{n+1}}$ (*d*) $\frac{1}{s^n}$.

62. Laplace transform of $e^{at}$ is

(*a*) $\frac{1}{s+a}$ (*b*) $\frac{1}{s-a}$ (*c*) $\frac{1}{s^a}$ (*d*) $\frac{1}{a^s}$.

**63.** Laplace transform of sin *at* is

(*a*) $\dfrac{1}{s+a}$ (*b*) $\dfrac{1}{s^2+a^2}$ (*c*) $\dfrac{a}{s^2+a^2}$ (*d*) $\dfrac{a}{s^2-a^2}$.

**64.** Laplace transform of cos *at* is

(*a*) $\dfrac{1}{s^2+a^2}$ (*b*) $\dfrac{s}{s^2+a^2}$ (*c*) $\dfrac{s}{s^2-a^2}$ (*d*) $\dfrac{1}{s^2-a^2}$.

**65.** Laplace transform of sinh *at* is

(*a*) $\dfrac{a}{s-a}$ (*b*) $\dfrac{a}{s+a}$ (*c*) $\dfrac{a}{s^2+a^2}$ (*d*) $\dfrac{a}{s^2-a^2}$.

**66.** Laplace transform of cosh *at* is

(*a*) $\dfrac{a}{s^2+a^2}$ (*b*) $\dfrac{a}{s^2-a^2}$ (*c*) $\dfrac{s}{s^2-a^2}$ (*d*) $\dfrac{s}{s^2+a^2}$.

**67.** Laplace transform of $\dfrac{t}{2a}$ sin *at* is

(*a*) $\dfrac{s}{s^2+a^2}$ (*b*) $\dfrac{s}{(s^2+a^2)^2}$ (*c*) $\dfrac{s}{(s^2+a^2)^{3/2}}$ (*d*) $\dfrac{a}{(s^2+a^2)^{3/2}}$.

**68.** Laplace transform of $\dfrac{\sin \omega t}{t}$ is

(*a*) $\tan^{-1}\left(\dfrac{\omega}{s}\right)$ (*b*) $\sin^{-1}\left(\dfrac{s}{\omega}\right)$ (*c*) $\sin^{-1}\left(\dfrac{\omega}{s}\right)$ (*d*) $\tan^{-1}\left(\dfrac{\omega}{s}\right)$.

**69.** Laplace transform of $t^n e^{at}$ is

(*a*) $\dfrac{!n}{(s-a)^n}$ (*b*) $\dfrac{!n}{(s-a)^{n+1}}$ (*c*) $\dfrac{!n+1}{(s-a)^{n+1}}$ (*d*) none of the above.

**70.** Laplace transform of $\dfrac{e^{-at}-e^{-bt}}{b-a}$ is

(*a*) $\dfrac{1}{(s+a)(s-b)}$ (*b*) $\dfrac{1}{(s-a)(s+b)}$ (*c*) $\dfrac{1}{(s-a)(s-b)}$ (*d*) $\dfrac{1}{(s+a)(s+b)}$.

**71.** Laplace transform of $(1-\cos \omega_n t)$ is

(*a*) $\dfrac{\omega_n^2}{s(s^2+\omega_n^2)}$ (*b*) $\dfrac{s}{\omega_n(s^2+\omega_n^2)}$ (*c*) $\dfrac{s^2}{\omega n^2(s^2+\omega_n^2)}$ (*d*) $\dfrac{\omega_n}{s^2(s^2+\omega_n^2)}$.

72. Laplace transform of $1-e^{-\frac{t}{T}}$ is

(a) $\frac{1}{(1+Ts)}$ (b) $\frac{1}{s(1+Ts)}$ (c) $\frac{1}{s(1-Ts)}$ (d) $\frac{1}{(1-Ts)}$.

73. $\frac{1}{s^2}$ is the Laplace transform of

(a) $t^{-1}$ (b) $t^{-2}$ (c) $t$ (d) $t^2$.

74. Laplace transform of $t^n$ is

(a) $s^{n+1}$ (b) $\frac{s^{n+1}}{n+1}$ (c) $\frac{n}{s^{n+1}}$ (d) $\frac{!n}{s^{n+1}}$.

75. $\frac{1}{s^2-a^2}$ is the Laplace transform of

(a) cos $at$ (b) $\frac{1}{a}$ cos $at$ (c) $\frac{1}{a}$ sinh $at$ (d) $\frac{1}{a}$ cosh $at$.

76. A unit step function on integration results in a

(a) unit step function (b) unit ramp function
(c) unit parabolic function (d) unit doublet.

77. A unit ramp function on integration results in

(a) unit doublet (b) unit parabolic function
(c) unit step function (d) unit impulse function.

78. A unit impulse function on differentiation results in

(a) unit doublet (b) unit triplet
(c) unit parabolic function (d) unit ramp function.

79. The transient function of the system and response in terms of Laplace transform are same for which of the following input function?

(a) Ramp function (b) Parabolic function
(c) Unit step function (d) Unit impulse function.

80. In Laplace transform and Fourier integral

(a) frequency domain and time domain are related
(b) only time domain is related (c) only frequency domain is related
(d) none of the above is related.

**81.** Laplace transform of a unit doublet is

(a) $s$ (b) $\frac{1}{s}$ (c) $\frac{1}{s^2}$ (d) $\frac{1}{s^3}$.

**82.** Laplace transform of a function shifted by a time $T$ is ($F's$ is the Laplace transform of the function occurring at $t = 0$)

(a) $F(s) = e^{T^s} F'(s)$ (b) $F(s) = e^{-T^3} F'(s)$ (c) $F(s) = Te^{-Ts} F'(s)$ (d) $F(s) = Ts\, e^{T^s} F'(s)$.

**83.** If $F(s) = \frac{2}{(s+1)(s+3)^2}$, then $f(t)$ will be

(a) $\frac{e^{-t}}{2} - e^{-3t} + t$ (b) $\frac{e^{-t}}{2} - \frac{e^{-3t}}{2} - t$ (c) $\frac{e^{-t}}{2} - \frac{e^{-3t}}{2} - t\,e^{-3t}$ (d) $\frac{e^{-t}}{2} - \frac{e^{-3t}}{2} + te^{3t}$

**84.** Laplace transform of $\left[\frac{\sin \omega t}{t}\right]$ is

(a) $\sin^{-1}\left(\frac{\omega}{s}\right)$ (b) $\cos^{-1}\left(\frac{\omega}{s}\right)$ (c) $\tan^{-1}\left(\frac{\omega}{s}\right)$ (d) $\cos^{-1}\left(\frac{s}{\omega}\right)$.

**85.** Laplace transform of $(t^3 - 2t + 1)$ is

(a) $\frac{(6 + 2s^2 + s^3)}{s}$ (b) $\frac{6 + 2s^2 + s^3}{s^2}$ (c) $\frac{6 - 2s^2 + s^3}{s^3}$ (d) $\frac{6 - 2s^2 + s^3}{s^4}$.

**86.** The function shown in Fig. 15.30 is

(a) an arbitrary periodic function (b) an even function

(c) an odd function (d) none of the above.

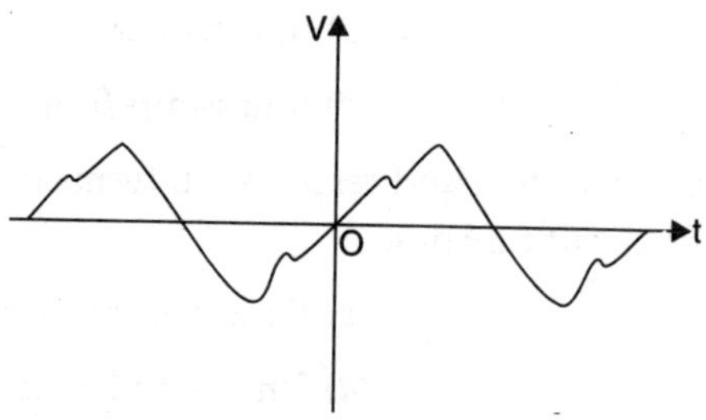


*FIGURE 15.30*

**87.** For a function $f_c(\theta)$, the symmetry conditions are $f_c(\theta) = f_c(-\theta)$

The function will have

(a) sine terms only (b) cosine terms only

(c) both sine and cosine terms (d) neither sine nor cosine terms.

88. The function shown in Fig. 15.31
(a) linear function
(b) odd function
(c) even function
(d) non-periodic function.

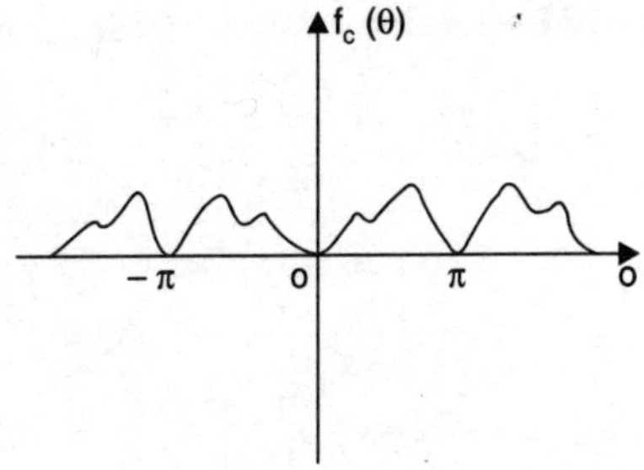


FIGURE 15.31

89. The expression for the saw-tooth voltage shown in Fig. 15.32 can be written as

(a) $f(t) = \frac{E}{T} tU(t)$

(b) $f(t) = -EU(t-T)$

(c) $f(t) = \frac{E}{T}(t-T)\,U(t-T)$

(d) $f(t) = \frac{E}{T} tU(t) - EU(t-T) - \frac{E}{T}(t-U)\,U(t-T)$.

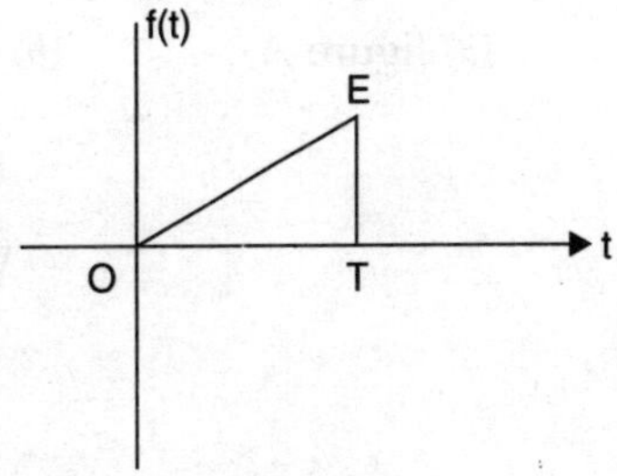


FIGURE 15.32

90. The Laplace transform of the saw tooth voltage shown in Fig. 15.32 will be

(a) $\frac{E}{T}[1+(Ts+1)\in^{-T^s}]$ (b) $\frac{E}{T}[1-(Ts+1)\in^{-T^s}]$

(c) $\frac{E}{Ts}[1+(Ts+1)\in^{-T^s}]$ (d) $\frac{E}{Ts^2}[1-(Ts+1)\in^{-T^s}]$.

91. For the R.C circuit shown in Fig. 15.33, where $v_i$ and $v_o$ are the input and output variables respectively, the block diagram is represented by (in Fig. 15.34).
(a) figure A (b) figure B
(c) figure C
(d) none of the above figures.

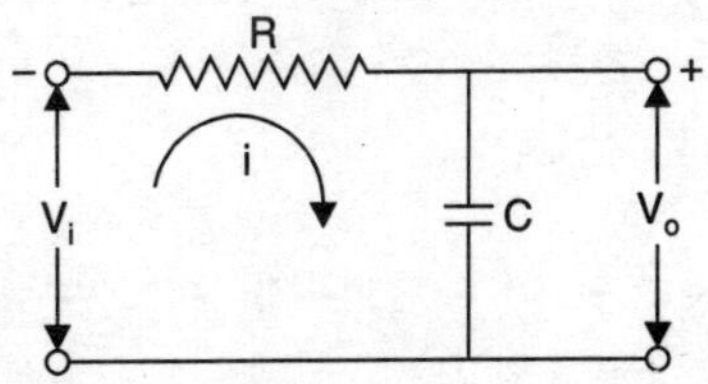


FIGURE 15.33

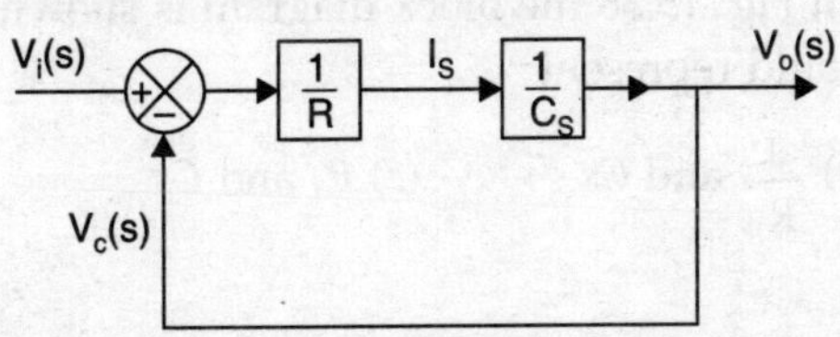


(A)

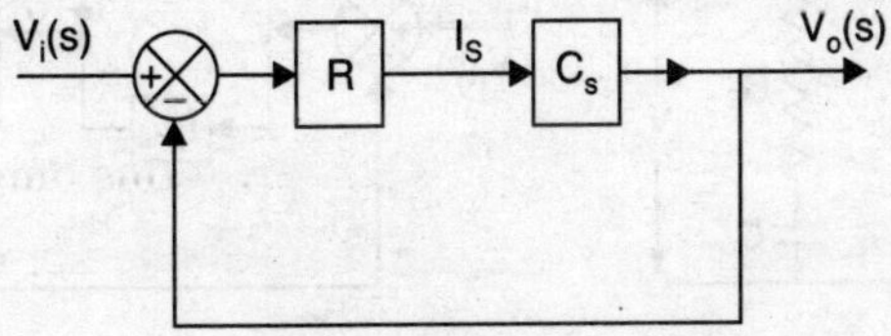


(B)

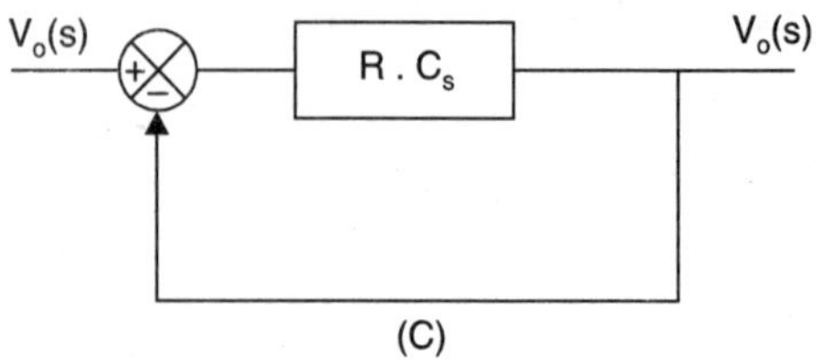


FIGURE 15.34

**92.** The signal flow graph for this circuit is represented by

(a) figure A (b) figure B (c) figure C (d) none of these figures.

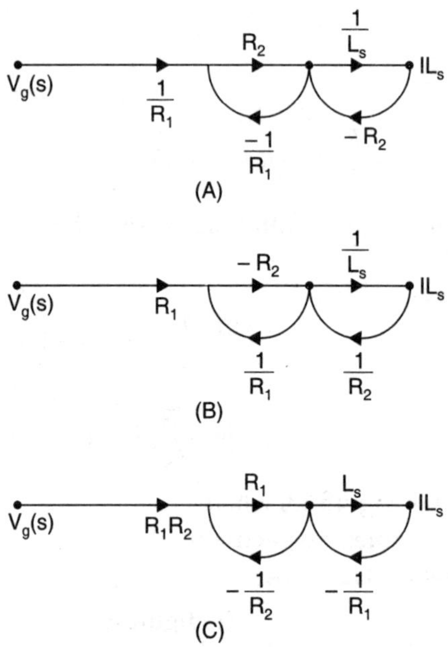


FIGURE 15.35

**93.** For the circuit shown in Fig. 15.36 the block diagram is shown with two blocks marked $X$ and $Y$. These blocks should represent

(a) $\frac{1}{R_1}$ and $\frac{1}{Cs}$ (b) $\frac{1}{R_1}$ and $Cs$ (c) $R_1$ and $Cs$ (d) $R_1$ and $\frac{1}{Cs}$.

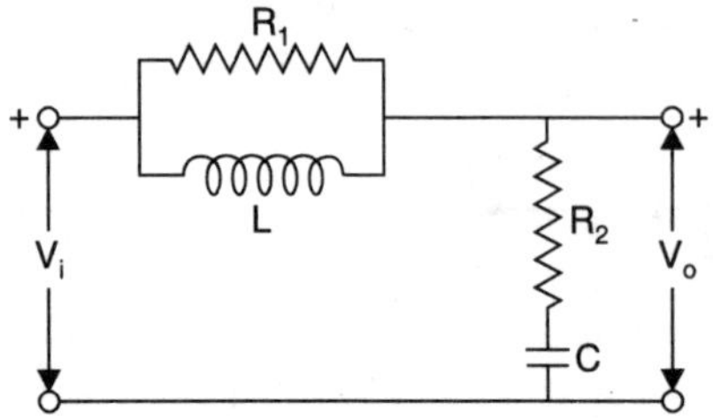


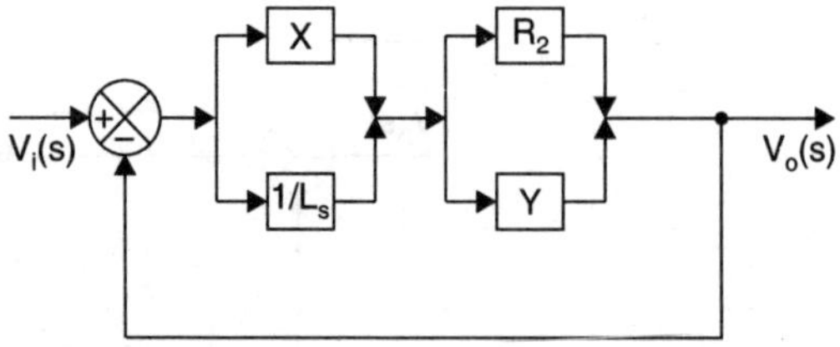


FIGURE 15.36

94. For the circuit shown in Fig. 15.37 $v_g$ and $i_L$ are the input and output variables respectively. The block diagram for this circuit is also shown. In this block diagram, three blocks have been marked X, Y and Z. These should read respectively

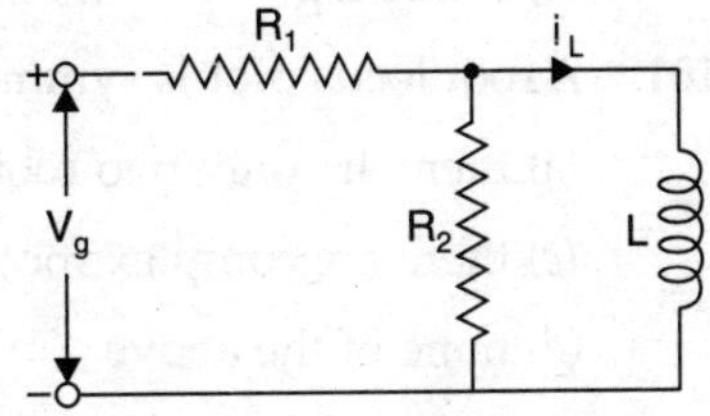


(a) $\frac{1}{R_1}, \frac{1}{R_2}$ and $\frac{1}{Ls}$  (b) $R_1, R_2$ and $\frac{1}{Ls}$

(c) $\frac{1}{R_1}, R_2$ and $\frac{1}{Ls}$  (d) $\frac{1}{R_1}, \frac{1}{R_2}$, and $Ls$.

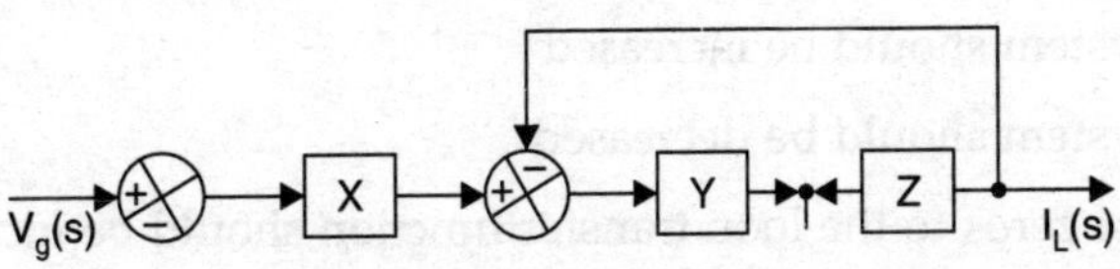


*FIGURE 15.37*

95. Which of the following system is unstable?

(a) $G(s)\,H(s) = \dfrac{k}{(T_1s+1)(T_2s+1)}$  (b) $G(s)\,H(s) = \dfrac{k(s+1)}{s^3(s+4)(s+5)}\ k > 99$

(c) $G(s)\,H(s) = \dfrac{k(s+2)}{(s+1)(s-3)}\ k > 2$  (d) $G(s)\,H(s) = \dfrac{k}{(Ts+1)^3} - 1 < 1 < 8.$

96. Which of the following error-detector characteristic should be considered while selecting error detectors?

(a) Resolution  (b) Rotational limits

(c) Moment of inertia  (d) All of the above.

97. The device commonly used as error detectors in instruments is

(a) microsyns  (b) vernistats  (c) resolvers  (d) any of the above.

98. The algebraic sum of all angles of vectors coming from poles and zeros to any point on root locus will be

(a) 180º  (b) odd multiple of 180º

(c) even multiple of 180º  (d) non of the above.

99. In root locus plot the angle of asymptote is given by

(a) $\dfrac{360^\circ}{\text{number of poles}}$  (b) $\dfrac{360^\circ}{\text{number of zeros}}$

(c) $\dfrac{360^\circ}{\text{number of poles + number of zeros}}$  (d) $\dfrac{360^\circ}{\text{number of poles - number of zeros}}$.

**100.** Which of the following compensator will be used to increase the band width of a control system?

(*a*) Phase lag (*b*) Phase lead (*c*) Phase lag lead (*d*) None of the above.

**101.** A root locus plot is symmetrical about the real axis because

(*a*) there are only two roots (*b*) there are several imaginary roots

(*c*) there are complex roots occurring in conjugate pairs

(*d*) none of the above.

**102.** For making an unstable system stable

(*a*) gain of the system should be increased

(*b*) gain of the system should be decreased

(*c*) the number of zeros to the loop transfer function should be increased

(*d*) the number of poles to the loop transfer function should be increased.

**103.** When the poles of a system lie on the $j\omega$ axis, it can be concluded that the system is

(*a*) stable (*b*) marginally stable

(*c*) conditionally stable (*d*) unstable.

**104.** For system gain to be zero, the roots will move towards

(*a*) poles (*b*) zeros (*c*) origin (*d*) none of the above.

**105.** Which of the following technique gives quick transient and stability response?

(*a*) Root locus (*b*) Bode (*c*) Nyquist (*d*) Nicholas.

**106.** In root locus plot different roots have the same

(*a*) gain (*b*) phase

(*c*) both gain and phase (*d*) gain margin and phase margin.

**107.** The starting points for root loci are

(*a*) closed loop poles (*b*) closed loop zeros (*c*) open loop poles (*d*) open loop zeros.

**108.** A root loci will end at

(*a*) closed loop poles (*b*) closed loop zeros

(*c*) open loop poles (*d*) open loop zeros.

**109.** The number of root loci branches which do not terminate at zeros is given by

(*a*) the number of zeros (*b*) the number of poles

(*c*) the number of zeros + the numbers of poles

(*d*) the number of zeros – the number of poles.

**110.** The effect of adding poles and zeros can be determined quickly by

(*a*) Nicholar chart (*b*) Nyquist plot (*c*) Bode plot (*d*) Root locus.

**111.** The Bode plot is used to analyse

(*a*) all phase network (*b*) lag lead network

(*c*) minimum phase network (*d*) maximum phase network.

**112.** For a transfer function $G(s) = s$, the Bode plot will be

(*a*) zero magnitude and zero phase shift

(*b*) 20 $dB$/decade and phase shift of $\pi/2$

(*c*) constant magnitude and constant phase shift angle

(*d*) none of the above.

**113.** The unity circle of Nyquist plot corresponds to the 0 $dB$ line for the Bode plot for

(*a*) high frequencies (*b*) low frequencies

(*c*) medium frequencies (*d*) all frequencies.

**114.** The slope in Bode plot is expressed as (per unit change in log ω)

(*a*) 0 *db*/octave (*b*) 6 *db*/octave (*c*) – 6 *db*/octave (*d*) – 10 *db*/octave.

**115.** The effect of addition of pole and zero on phase margin and gain margin can be most conveniently seen in

(*a*) Nyquist plot (*b*) Bode plot (*c*) Root locus (*d*) Routh Hartwiz.

**116.** Which of the following will increase the steady state accuracy?

(*a*) Phase lead compensator (*b*) Phase lag compensator

(*c*) Differentiator (*d*) Integrator.

**117.** The negative real axis of Nyquist plot is plotted on Bode plot as

(*a*) a negative 180º phase (*b*) a positive 180º phase

(*c*) 10 *db* line (*d*) 0 *db* line.

**118.** In Nyquist plot the magnitude at phase cross-over is 0.4. The gain margin of the system will be

(*a*) 0.4 (*b*) 4 percent (*c*) – 0.4 (*d*) 1/0.4.

**119.** The advantage of Nyquist stability test is

(*a*) closed loop stability can be predicted from open loop results

(*b*) it indicates how to stablise an unstable system

(*c*) is applicable to experimental results or frequency response of open loop system

(*d*) all of the above.

**120.** Nyquist criterion is used to find

(*a*) relative stability only (*b*) absolute stability

(*c*) both (*a*) and (*b*) above (*d*) none of the above.

**121.** Nichol's chart is a plot between

(*a*) frequency and gain (*b*) amplitude and phase

(*c*) phase and frequency (*d*) amplitude and gain.

**122.** Nichol's charts are used for determining

(*a*) closed loop frequency response (*b*) open loop frequency response

(*c*) both (*a*) and (*b*) above (*d*) none of the above.

**123.** In a Routh Hurwitz array if all the terms in a row become zero, the characteristic will have a pair of roots

(*a*) at orgin (*b*) on negative axis

(*c*) on positive axis (*d*) which are negative of each other.

**124.** Frequency domain analysis is preferred when dealing with systems having input as

(*a*) ramp and parabolic (*b*) non sinusoidal

(*c*) sinusoidal with fixed frequency

(*d*) sinusoidal with variable frequency and amplitude.

**125.** The disadvantage of the Routh's criteria is

(*a*) it assumes that the characteristic equation is available in the polynomial form

(*b*) it provides information about absolute stability only

(*c*) (*a*) and (*b*) above (*d*) none of the above.

**126.** Apply Routh Hurwitz criteria and determine which of the following does not have two roots with positive real parts?

(*a*) $s^4 + 5s^3 + 2s + 10 = 0$ (*b*) $2s^5 + s^4 + 6s^3 + 3s^2 + s + 1 = 0$

(*c*) $s^5 - s^4 - 2s^3 + 3s^2 - 8s + 8 = 0$ (*d*) $s^4 + 2s^3 + 7s^2 + 10s + 10 = 0$.

**127.** In the above question how many equations have two roots with zero and real parts?

(*a*) none (*b*) one (*c*) two (*d*) three.

**128.** In the above question how many equations have two or more than two roots with negative real parts?

(*a*) none (*b*) one (*c*) two (*d*) three.

**129.** Which of the following is used for Nyquist plot?

(*a*) Pole-zero plot (*b*) Closed loop function

(*c*) Open loop function (*d*) Characteristic equation.

**130.** By increasing the band width of a system

(*a*) response will become slower (*b*) response will become faster

(*c*) response will remain unaffected.

**131.** Viscous friction can be added to a servomechanism by

(*a*) increasing shaft rpm (*b*) reducing shaft rpm

(*c*) coupling a disc to shaft which is immersed in a viscous medium

(*d*) increasing the temperature of the oil.

132. The overall transfer function of any number of non-interacting components in series is
(*a*) the sum of the individual transfer functions
(*b*) the sum of the reciprocals of individual transfer functions
(*c*) the product of the individual transfer functions
(*d*) none of the above.

133. According to Nyquist criterion, a system is absolutely stable if all the roots of its characteristic equation are
(*a*) in the left hand half of the *s*-plane (*b*) in the right hand half of the *s*-plane
(*c*) are positive (*d*) complex with positive real parts.

134. The choice of error detector in a control application depends on
(*a*) the nature of the controlled variable (*b*) the type of control system
(*c*) the accuracy desired (*d*) all of the above.

**Questions 135 and 136 Refer to Data Given Below:**

A system is described by the differential equation

$$(D^4 + 5D^3 + 2D^2 + 6D + 1)\,\theta_0 = 4\theta_i$$

135. The system is
(*a*) stable (*b*) unstable (*c*) marginally stable (*d*) none of these.

136. How many roots with positive real parts are present?
(*a*) none (*b*) one (*c*) two (*d*) three.

137. The limit of the gain *k* for which the forward path

$$\frac{\theta_0(D)}{\epsilon(D)} = \frac{k(1+0.01D)}{D(1+D)(.1+0.2D)}$$

will result in a stable system is
(*a*) 1.6 (*b*) 3.2 (*c*) 6.4 (*d*) 12.8.

138. The equation of a servomechanism damped solely by pure viscous friction, is given by

$$(10D^2 + 20D + 1000)\theta_0 = 1000\,\theta_i$$

where $\theta_0$ = load shaft displacement, redians
$\theta_i$ = input shaft displacement, redians
The value of the damping ratio will be
(*a*) 0.1 (*b*) 0.16 (*c*) 0.21 (*d*) 0.25.

139. What fraction of derivative of error will be needed in order to produce a damping ratio of 0.5?
(*a*) 0.008 (*b*) 0.08 (*c*) 0.04 (*d*) 0.02.

**140.** The transfer function for the circuit shown in Fig. 15.38, when there is no external load, will be

$$\left(a = \frac{R_2}{R_1 + R_2}\right)$$

(*a*) $\frac{\theta_0}{\theta_i} = a\frac{1 + R_1CD}{1 + aR_1CD}$ (*b*) $\frac{\theta_0}{\theta_i} = a\frac{1 - R_1CD}{1 - aR_1CD}$

(*c*) $\frac{\theta_0}{\theta_i} = a\frac{1 - R_1CD}{1 + aR_1CD}$ (*d*) $\frac{\theta_0}{\theta_i} = a\frac{1 + R_1CD}{1 - aR_1CD}$.

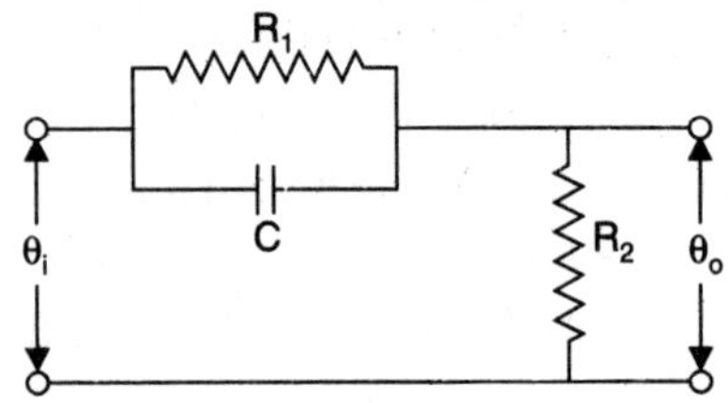


*FIGURE 15.38*

**141.** The transfer function for the circuit shown in Fig. 15.39 when there is no external load, will be

$$\left(\frac{L}{R} = T_1 \text{ and } RC = T_2\right)$$

(*a*) $\frac{\theta_0}{\theta_i} = \frac{1}{T_1T_2s^2 + 1}$ (*b*) $\frac{\theta_0}{\theta_i} = \frac{1}{T_1T_2s^2 + T_1s}$

(*c*) $\frac{\theta_0}{\theta_i} = \frac{1}{T_1T_2s^2 + T_1s + 1}$ (*d*) $\frac{\theta_0}{\theta_i} = \frac{1}{T_1T_2s^2 + T_2s + T_1}$.

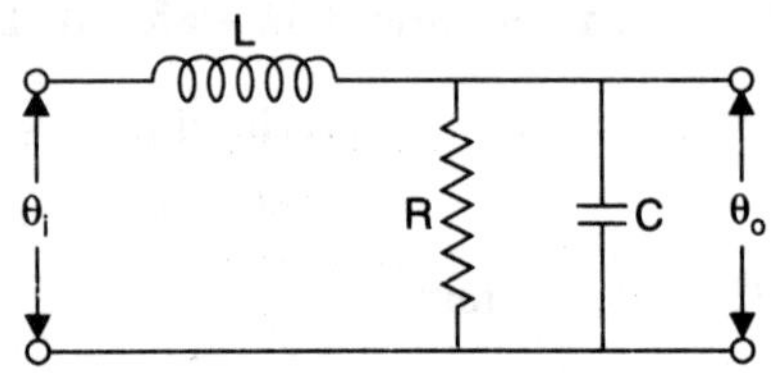


*FIGURE 15.39*

**Question Refers to Fig. 15.40:**

**142.** The transfer operator of the passive network shown in Fig. 15.40 will be

(*a*) $\frac{1}{C(R_1 + R_2)}$ (*b*) $\frac{CR_2}{C(R_1 + R_2)D}$

(*c*) $\frac{CR_2D + 1}{C(R_1 + R_2)D + 1}$ (*d*) $\frac{CR_1D + 1}{CR_1R_2D + 1}$.

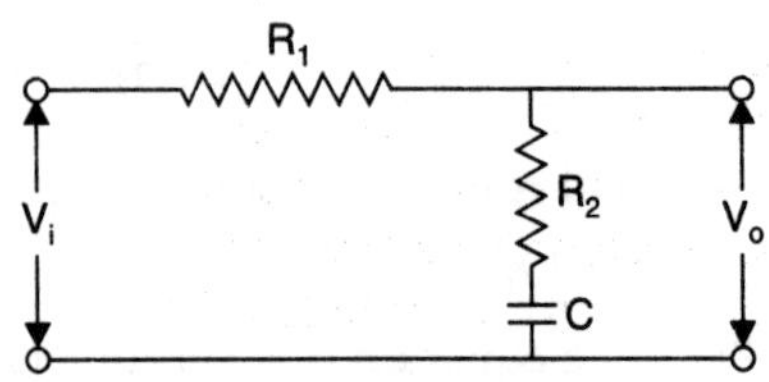


*FIGURE 15.40*

**Questions 143 to 147 Refer to Data Given Below:**

The equation of a servo is given by

$$(10D^2 + 42D + 90)\ \theta_0 = 90\ \theta_i$$

where $\theta_i$ = displacement of input shaft, radians

$\theta_0$ = displacement of load shaft, radians.

**143.** The undamped natural frequency in cycles/second will be

(*a*) 0.0477 (*b*) 0.477 (*c*) 4.77 (*d*) 47.7.

**144.** The damping ratio will be

(*a*) 0.1 (*b*) 0.3 (*c*) 0.5 (*d*) 0.7.

**145.** The damping factor is

(*a*) 0.5 (*b*) 1.1 (*c*) 1.6 (*d*) 2.1.

**146.** The damped natural frequency in cycles/second will be

(*a*) 0.77 (*b*) 0.66 (*c*) 0.34 (*d*) 0.17.

**147.** The logarithmic decrement is

(*a*) 0.00306 (*b*) 0.0306 (*c*) 0.306 (*d*) 0.612.

**148.** By adjusting the second derivative input signal, it will adjust

(*a*) time constant and suppress oscillations

(*b*) gain of the system

(*c*) damping of the system (*d*) all of the above.

**149.** A servomechanism with step displacement input will form

(*a*) type 1 system (*b*) type 2 system (*c*) type 3 system (*d*) type 0 system.

**150.** In type 1 system a constant output velocity at steady state will be possible when

(*a*) there is no error (*b*) there is constant steady state error

(*c*) there is variable steady state error (*d*) there is fluctuating error.

**151.** AC servomotors resemble which of the following

(*a*) universal motor (*b*) three phase synchronous motor

(*c*) two phase induction motor (*d*) DC series motor.

**152.** The phase difference between the two windings of ac servo-motors is

(*a*) 45º (*b*) 90º (*c*) 120º (*d*) 180º.

**153.** The torque speed characteristic of two phase induction motor is largely affected by

(*a*) voltage (*b*) speed

(*c*) X/R ratio (*d*) supply voltage frequency.

**154.** An amplidyne is virtually a

(*a*) dc generator (*b*) two phase transformer

(*c*) rectifier (*d*) rotating amplifier.

**155.** Which of the following is used to avoid the loading effect on the potentiometer?

(*a*) Trimmer (*b*) Battery

(*c*) Alternator (*d*) Any source of voltage.

**156.** A synchros is

(*a*) an electronic rectifier (*b*) an electromagnetic transducer

(*c*) a frequency transformer (*d*) none of the above.

**157.** A synchro is used to

(*a*) accelerate a rotating shaft

(*b*) convert an angular position of a shaft into an electrical signal

(*c*) convert linear motion into angular position

(*d*) amplify low frequency signals.

**158.** The construction of a synchro resembles which of the following?

(*a*) DC series motor (*b*) Universal motor

(*c*) Split phase motor (*d*) Three phase alternator.

**159.** The stator of a synchro is made of

(*a*) cast iron (*b*) copper

(*c*) laminated silicon steel (*d*) cadmium alloy.

**160.** The rotor of a synchro is supplied

(*a*) AC voltage (*b*) DC voltage

(*c*) no voltage (*d*) AC super imposed on DC voltage.

**161.** A stepper motor is a

(*a*) pneumatic device (*b*) hydraulic device

(*c*) electro mechanical device (*d*) none of the above.

**162.** Input to a stepper motor is in the form of

(*a*) mechanical inertia (*b*) electric pulses

(*c*) frictional force (*d*) viscous damping force.

**163.** The output of a stepper motor is in the form of

(*a*) angular movements (*b*) linear movements

(*c*) either of (*a*) or (*b*) above (*d*) none of the above.

**164.** A stepper motor can be used in control system for

(*a*) computers (*b*) tape drives (*c*) capstan drives (*d*) none of the above.

**165.** The main advantage of hydraulic system over electrical system is

(*a*) smaller size (*b*) no danger of electric shock

(*c*) easy balancing (*d*) no cooling problems.

**166.** The disadvantage of hydraulic system as compared to electrical system is

(*a*) low efficiency (*b*) bulkiness (*c*) fluid leakage (*d*) all of the above.

**167.** Which of the following is used to prevent leakage from hydraulic systems?

(*a*) Ball bearing (*b*) Plug (*c*) Seal (*d*) Socket.

**168.** Which of the following uses hydraulic control device?

(*a*) Automobiles (*b*) Machine tools (*c*) Ships (*d*) All of the above.

**169.** In pluton turbine hydraulic device is used for

(*a*) power generation (*b*) power transmission

(*c*) speed governing (*d*) creation of head.

**170.** A *LVDT* is a (an)

(*a*) electrostatic device (*b*) electrodynamic device

(*c*) electromechanical device (*d*) electromagnetic device.

**171.** 'Selsyn' is the same as

(*a*) stepper motor (*b*) synchros (*c*) hydraulic motor (*d*) transformer.

**172.** 'Synchros' are commonly used as transmitters of

(*a*) digital data (*b*) angular data (*c*) computed data (*d*) graphical data.

**173.** Which of the following can be used for high power amplification?

(*a*) SCR (*b*) Push pull amplifier

(*c*) Amplidyne (*d*) Any of the above.

**174.** A reduction gear is used to

(*a*) increase the viscous friction (*b*) increase the moment of inertia

(*c*) increase the torque developed (*d*) increase the driven shaft speed.

**175.** A control system having unity damping factor, will give

(*a*) critically damped response (*b*) oscillatory response

(*c*) undamped response (*d*) no response.

**176.** Which of the following will not decrease as a result of introduction of negative feedback?

(*a*) Instability (*b*) Band width (*c*) Overall gain (*d*) Distortion.

**177.** The resolution of a potentiometer should be

(*a*) zero (*b*) very high (*c*) medium (*d*) infinite.

**178.** Magnetic amplifiers are used for

(*a*) voltage amplification (*b*) current amplification

(*c*) power amplification (*d*) none of the above.

**179.** Compensating winding is provided on amplidyne to

(*a*) increase the control voltage (*b*) increase the control current

(*c*) compensate the effect of magnetic field set up by the load current

(*d*) none of the above.

**180.** Which of the following motor is preferred for self-balancing instruments?

(*a*) dc series motor (*b*) dc servo motor (*c*) ac servo motor (*d*) ac induction motor.

**181.** Which of the following device is used for conversion of co-ordinates?

(*a*) Microsyn (*b*) Selsyn

(*c*) Synchro resolver (*d*) Any of the above.

**182.** In a system, non-linearity may be introduced by

(*a*) back lash in gears (*b*) end positions in detectors

(*c*) saturation effects in amplifiers (*d*) any of the above.

**183.** Non-linearity may be introduced in a servomotor due to

(*a*) air resistance (*b*) back lash (*c*) bearing friction (*d*) saturation.

**184.** the control system used in radars is

(*a*) discrete data control system (*b*) continuous control system

(*c*) relay control system (*d*) none of the above.

**185.** Regenerative feedback implies

(*a*) feedback with negative sign (*b*) feedback with positive sign

(*c*) feedback with step input (*d*) feedback with oscillations.

**186.** The system damping factor can be reduced by using

(*a*) positive feedback (*b*) negative feedback

(*c*) rate feedback (*d*) integral feedback.

**187.** Dynamic character of a system is due to

(*a*) friction (*b*) coupling

(*c*) storage of energy (*d*) all of the above.

**188.** The time required for the response to reach half the final value for first time is

(*a*) delay time (*b*) pick-up time (*c*) rise time (*d*) decay time.

**189.** The complex frequencies where the response becomes infinite are known as

(*a*) poles (*b*) conjugate roots

(*c*) ordinates (*d*) canonical frequencies.

**190.** Magnetic amplifiers are used to amplify

(*a*) voltage (*b*) current (*c*) power (*d*) all of the above.

**191.** The power amplification of a magnetic amplified is increased by using

(*a*) dc current in control winding (*b*) resistive load circuit

(*c*) negative feedback (*d*) all of the above.

**192.** The output of a feedback control system must be a function of

(*a*) reference and input (*b*) reference and output

(*c*) output and feedback signal (*d*) input and feedback signal.

**193.** The resolution of a potentiometer depends on

(*a*) size of wire (*b*) material of wire (*c*) type of contact (*d*) all of the above.

**194.** A 1000 turns potentiometer is excited by 10 V source. The percentage resolution will be

(*a*) 10% (*b*) 1% (*c*) 0.1% (*d*) 0.01%.

**195.** The system response can be tested better with

(*a*) sinusoidal input signal (*b*) unit impulse input signal

(*c*) ramp input signal (*d*) exponentially decaying signal.

196. The system steady state error can be minimized by
(a) increasing damped frequency (b) decreasing natural frequency
(c) decreasing damped frequency (d) increasing system gain constant.

197. A potentiometer can be used as a
(a) measuring device (b) controlling device
(c) regulating device (d) computational device.

198. With feedback, the transient response in a system
(a) will decay slowly (b) rise slowly
(c) rise quickly (d) decay quickly.

199. For a second order differential equation in case the damping ratio is 1,
(a) poles will be complex conjugate (b) poles will be unequal
(c) poles will be positive (d) poles will be equal, negative and real.

200 A system has only two poles each having a negative real part. The damping ratio of the system will be
(a) zero (b) infinitely high (c) non-zero (d) any of the above.

201. In the type 1 system the velocity lag error is
(a) inversely proportional to bandwidth of the system
(b) directly proportional to gain constant
(c) inversely proportional to gain constant
(d) independent of gain constant.

202. Which of the following is not valid in case of signal flow graph?
(a) in signal flow graphs signals travel along branches only in the marked direction
(b) nodes are arranged from right to left in a sequence
(c) signal flow graph is applicable to linear systems only
(d) for signal flow graph the algebraic equations must be in the form of cause and effect relationship.

203. As compared to a standard induction motor, ac servomotor has
(a) high inertia and low starting torque (b) low inertia and high starting torque
(c) high inertia and high starting torque (d) low inertia and low starting torque.

204. A 'drag cup' type motor has
(a) low power rating (b) low starting torque
(c) low damping (d) low inertia.

205. Underlying principle of a gyroscope is
(a) first law of thermodynamics (b) Newton's third law
(c) law of conservation of momentum (d) law of conservation of energy.

206. The transfer function of a first order system is $\frac{a}{s+a}$. Its setting time will be nearly

(a) $a/2$ (b) $a/4$ (c) $2/a$ (d) $4/a$.

207. The steady state error on a control system can be reduced by increasing its

(a) gain constant (b) time constant

(c) both gain constant and time constant

(d) gain constant, time constant and damping frequency.

208. In case the time constant of a system is $T$, the corner frequency will be

(a) zero (b) 1 (c) $\frac{1}{T}$ (d) $\log_e T$.

209. In the above problem, at corner frequency the phase angle of the function will be

(a) zero (b) multiple of $\frac{\pi}{3}$ (c) multiple of $\frac{\pi}{4}$ (d) multiple of $\frac{\pi}{2}$.

210. The eigen values of a linear system are the locations of

(a) zeros of the system (b) poles of the system

(c) finite poles (d) none of the above.

211. A regulator is a

(a) 0 type system (b) 1 type system (c) 2 type system (d) 3 type system.

212. The phase margin is the amount of angle to make a system

(a) stable (b) damped (c) exponential (d) unstable.

213. The gain margin is the factor by which the gain of the system can be increased to make it

(a) damped (b) oscillatory (c) unstable (d) stable.

214. In case the gain margin is positive and the phase margin is negative, the system can be termed as

(a) stable (b) unstable (c) probalistic (d) none of the above.

215. Which of the following is an open loop control system?

(a) Stroboscope (b) Field controlled dc motor

(c) Ward Leonard control (d) Metadyne.

**Questions 216 to 219 Refer to Fig. 15.41:**

Pole zero locations are shown in Fig. 15.41.

216. The unstable system is represented by

(a) figure $A$ (b) figure $B$ (c) figure $C$ (d) figure $D$.

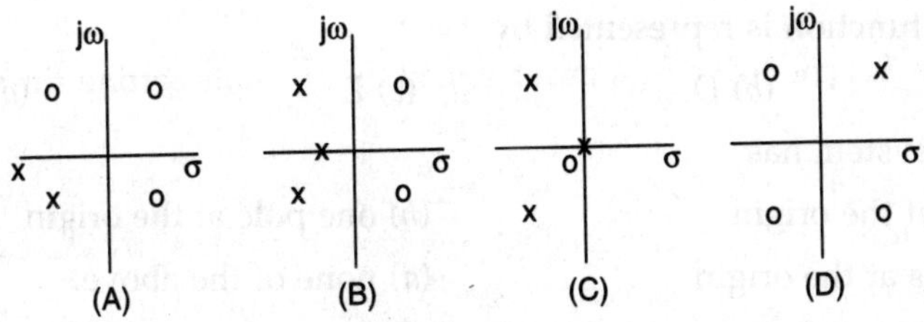


*FIGURE 15.41*

**217.** Figure *A* represents

(*a*) stable system (*b*) unstable system

(*c*) asymptotic system (*d*) non-asymptotic system.

**218.** Figure C represents

(*a*) asymptotic system (*b*) non-asymptotic system

(*c*) stable system (*d*) unstable system.

**219.** A stable system is represented by

(*a*) figure *A* (*b*) figure *B* (*c*) figure *C* (*d*) figure *D*.

**Questions 220 to 228 Refer to Fig. 15.42:**

**220.** Figure *A* represents

(*a*) unit impulse function (*b*) unit step function

(*c*) parabolic function (*d*) ramp function.

**221.** A unit step function is represented by

(*a*) curve *A* (*b*) curve *B* (*c*) curve C (*d*) curve *E*.

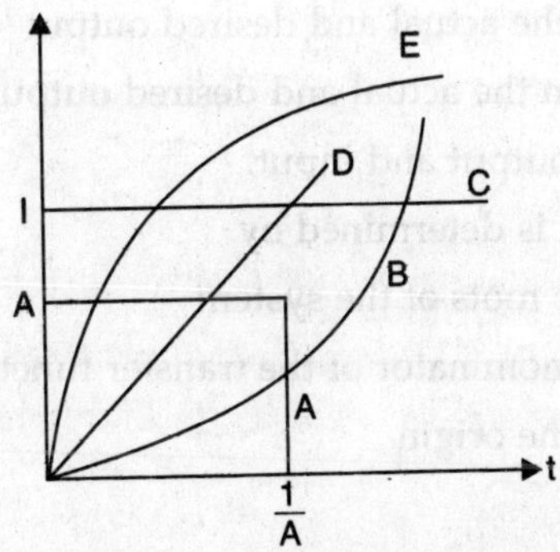


*FIGURE 15.42*

**222.** In the given figure *D* represents

(*a*) unit impulse function (*b*) parabolic function

(*c*) unit impulse function (*d*) ramp function.

**223.** A parabolic function is represented by

(*a*) *A* (*b*) *D* (*c*) *E* (*d*) *B*.

**224.** The type *O* system has

(*a*) no pole at the origin
(*b*) one pole at the origin
(*c*) two poles at the origin
(*d*) none of the above.

**225.** The type 1 system has

(*a*) no pole at origin
(*b*) simple pole at origin
(*c*) two poles of the origin
(*d*) none of the above.

**226.** The type 2 system has

(*a*) constant position error and zero velocity error
(*b*) zero position error and constant velocity error
(*c*) constant position error as well as velocity error
(*d*) zero position error as well as velocity error.

**227.** The transfer function of a system can be used to evaluate

(*a*) time constant
(*b*) feedback
(*c*) type of system
(*d*) output for a given input.

**228.** The effect of error rate damping is

(*a*) to reduce steady state error
(*b*) faster response
(*c*) faster response and reduction in steady state error
(*d*) faster response and larger settling time.

**229.** The overshoot is an indication of

(*a*) no error between the actual and desired output
(*b*) least error between the actual and desired output
(*c*) largest error between the actual and desired output
(*d*) resonance between output and input.

**230.** The order of the system is determined by

(*a*) the number of stable roots of the system
(*b*) power of *s* in the denominator of the transfer function
(*c*) number of poles at the origin
(*d*) none of the above.

**231.** A second order system has transfer function of $1/s^2$. The phase shift will be

(*a*) $0°$ (*b*) $90°$ (*c*) $180°$ (*d*) $-180°$.

**232.** In a stable control system, backlash can cause

(*a*) under damping
(*b*) over damping
(*c*) low-level oscillations
(*d*) all of the above.

233. A non-return valve is analogous to
(a) pneumatic *NOR* gate
(b) pneumatic *AND* gate
(c) pneumatic *RL* circuit
(d) pneumatic rectifier.

234. A control system with excessive noise, is likely to suffer from
(a) oscillations
(b) vibrations
(c) loss of gain
(d) saturation in amplifying stages.

235. In a control system, noise can be reduced by
(a) increasing bandwidth
(b) reducing bandwidth
(c) reducing bandwidth and attenuating frequencies at which external signal gets coupled into the system
(d) none of the above.

236. A two pressure valve is analogous to
(a) pneumatic rectifier
(b) pneumatic *OR* gate
(c) pneumatic *NOR* gate
(d) pneumatic *NAND* gate.

237. Which of the following can measure pressure error?
(a) Strain gauge
(b) Selsyn
(c) Potentiometer
(d) All of the above.

238. The frequency range is specified by
(a) resonance
(b) modulation index
(c) bandwidth
(d) peak value.

239. For a system, zero initial condition means
(a) zero stored energy
(b) no initial movement of moving parts
(c) system is at rest and no energy is stored in any of its components
(d) input reference signal is zero.

240. The term conformity is closely associated with
(a) single phase transformers
(b) potentiometers
(c) metadyne
(d) synchros.

241. 'Drag type' motors have generally
(a) high inertia
(b) low inertia
(c) low damping
(d) low starting torque.

242. A servo system will become absolutely unstable if
(a) there is zero feedback
(b) there is no feed back path
(c) there is no stable component
(d) feedback path becomes zero.

243. The setting time in the second order linear system is
(a) $\frac{1}{4}$ of time constant
(b) $\frac{1}{2}$ of time constant
(c) 2 times the time constant
(d) 4 times the time constant.

**244.** A system has forward transfer function G.

If now negative unity feedback is inserted, the forward transfer function will be modified as

(a) $\frac{G+1}{G-1}$ (b) $\frac{G}{1-G}$ (c) $\frac{G}{1+G}$ (d) $\frac{G-1}{G+1}$.

**245.** Two blocks have transfer functions $G_1$ and $G_2$. If the blocks are connected in series, their resultant will be

(a) $G_1 + G_2$ (b) $G_1G_2$ (c) $\frac{G_1G_2}{G_1+G_2}$ (d) $\frac{1}{G_1}+\frac{1}{G_2}$.

**246.** Steady state error is the difference between input and output at

(a) $t = \frac{T}{2}$ (b) $t = T$ (c) $t = 2T$ (d) $t = \infty$.

**247.** As the type of a system increases

(a) steady state error for a particular function increases

(b) steady state error for a particular function decreases

(c) stability problem decreases (d) none of the above.

**248.** A controller is basically a

(a) comparator (b) clipper (c) amplifier (d) sensor.

**249.** The input to a controller is always a

(a) servo signal (b) sensed signal (c) error signal (d) any of the above.

**250.** In a control system the output of controller is given to

(a) amplifier (b) sensor

(c) final control element (d) none of the above.

**251.** Which of the following cannot be the final control element?

(a) servomotor (b) potentiometer

(c) electro-pneumatic converter (d) all of the above.

**Questions 252 to 254 Refer to Fig. 15.43:**

**252.** Which of the figure shown represents an unstable system?

(a) figure A (b) figure B (c) figure C (d) figure D.

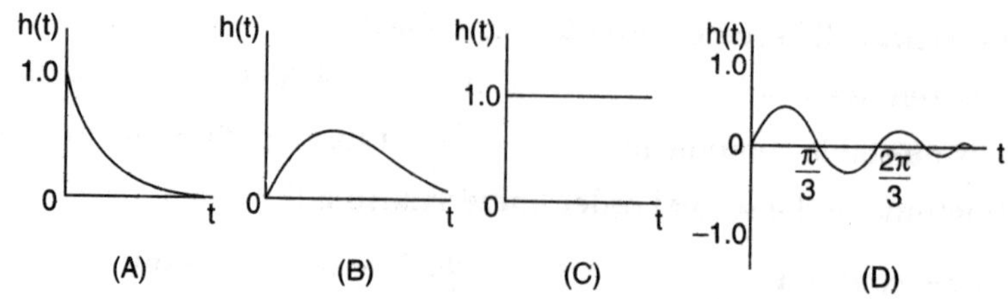


*FIGURE 15.43*

**253.** Which of the figure can have the relation $h(t) = t^{-t} \sin 3t$?

(*a*) figure *A* (*b*) figure *B* (*c*) figure *C* (*d*) none of the figures.

**254.** Which of the figure can have the relation $h(t) = te^{-t}$?

(*a*) figure *A* (*b*) figure *B* (*c*) figure *C* (*d*) none of the figures.

**255.** If a step function is applied to the input of a system and the output is of the form $y = t$, the system is

(*a*) stable (*b*) conditionally stable

(*c*) unstable

(*d*) cannot be ascertained on the basis of information provided.

**256.** The roots of the characteristic equations for some systems are given below. Which system is unstable?

(*a*) $-1, -2$ (*b*) $-1 + j, -1 - j$ (*c*) $-1, +1$ (*d*) $-2 + 3j, -2 - 3j, -2$.

**257.** The roots of the characteristic equations for some systems are given below. Which system is marginally stable?

(*a*) $2, -1, -3$ (*b*) $-3, -2, 0$ (*c*) $-6, -4, 7$ (*d*) $-j, j, -1, 1$.

**258.** The roots of the characteristic equations for some systems are given below. Identify the system which is marginally stable?

(*a*) $-2 + j, -2 - j, 2j, -2j$ (*b*) $2, -1, -3$

(*c*) $-j, j, -1, 1$ (*d*) $-1, +1$.

**259.** The characteristic equation of a system is given as

$$s^4 + 6s^3 + 11s^2 + 6s + k = 0$$

For which of the following restriction on the parameter $k$, the system will become stable?

(*a*) $k = 0$ (*b*) $k > 0, k < 10$ (*c*) $k > 5, k < -5$ (*d*) $k = 1$.

**260.** The characteristic equation for a system is

$$s^2 + ks + 2k - 1 = 0.$$

The system will be stable when

(*a*) $k = 0$ (*b*) $k > \frac{1}{2}$ (*c*) $k < \frac{1}{2}$ (*d*) $k < -\frac{1}{2}$.

**261.** Which of the following transfer function represents stable system?

(*a*) $P(s) = \dfrac{(s+2)(s-2)}{(s+1)(s-1)(s+4)}$ (*b*) $P(s) = \dfrac{(s-1)}{(s+2)(s+4)}$

(*c*) $P(s) = \dfrac{5(s+10)}{(s+5)(s^2 - s + 10)}$ (*d*) $P(s) = \dfrac{(s-1)}{(s+2)(s^2+4)}$.

**262.** Which of the signal shown in Fig. 15.44 is different from the remaining?

(*a*) figure *A* (*b*) figure *B* (*c*) figure *C* (*d*) figure *D*.

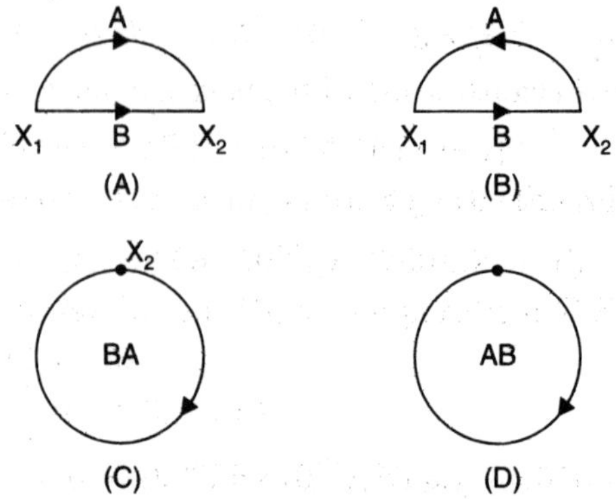


*FIGURE 15.44*

**263.** Which of the following is not type 1 system?

(*a*) $G = \frac{1}{s}, H = 1$ (*b*) $G = \frac{5}{s(s+3)}, H = \frac{s+1}{s+2}$

(*c*) $G = \frac{2}{s^2 + 2s + 5}, H = s + 5$ (*d*) $G = \frac{24}{(2s+1)(4s+1)}, H = \frac{4}{4s\,(3s+1)}$.

**264.** The system shown in Fig. 15.45 is a

(*a*) type 0 system (*b*) type 1 system (*c*) type 2 system (*d*) type 3 system.

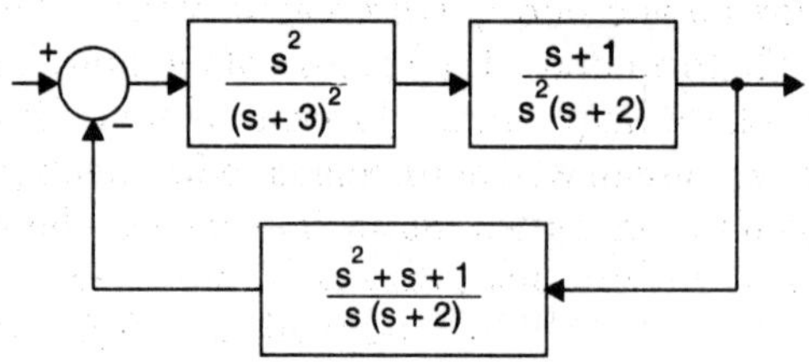


*FIGURE 15.45*

**265.** The system having

$$G = \frac{4}{s(s+3)}, H = \frac{1}{s} \text{ is}$$

(*a*) type 0 (*b*) type 1 (*c*) type 2 (*d*) type 3.

### Questions 266 to 271 Refer to Fig. 15.46:

A stable system is shown in Fig. 15.46

**266.** The position error vector is

(*a*) 0 (*b*) 1 (*c*) 100 (*d*) ∞.

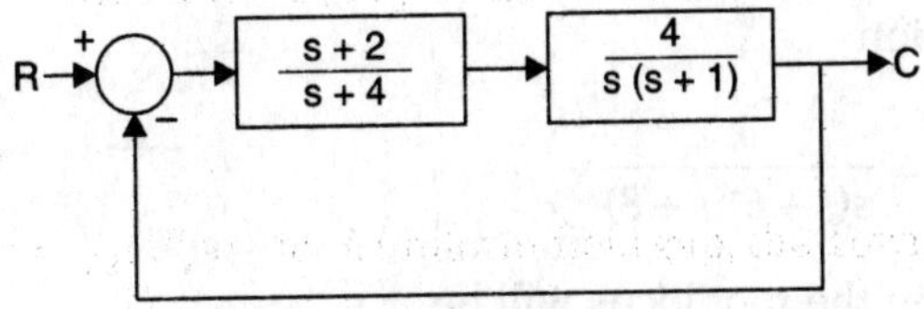


FIGURE 15.46

**267.** The velocity error constant $k_v$ is

(a) 0 (b) 1 (c) 2 (d) ∞.

**268.** The acceleration error constant is

(a) 0 (b) 1 (c) 2 (d) ∞.

**269.** For a unit-step input, the steady state error will be

(a) 0 (b) 1 (c) $\frac{1}{2}$ (d) ∞.

**270.** For a unit ramp input, the steady state error will be

(a) 0 (b) $\frac{1}{2}$ (c) 1 (d) 2.

**271.** For a unit parabolic input, the steady state error will be

(a) 0 (b) $\frac{1}{2}$ (c) 2 (d) ∞.

**272.** A system has open loop transfer function as

$$G = \frac{k(s+2)}{(s+1)(s+4)}$$

The closed-loop transfer function of the unity negative feedback control system will be

(a) $\frac{1}{(s+1)(s+4)}$ (b) $\frac{(s+2)}{(s+1)(s+4)}$

(c) $\frac{k(s+2)}{(s+1)(s+4)}$ (d) $\frac{k(s+2)}{(s+1)(s+4)+k(s+2)}$.

**273.** In the above case the characteristic equation will be

(a) $k + 1 = 0$ (b) $(s + 1)(s + 4) = 0$

(c) $(s + 1)(s + 4) + k = 0$ (d) $(s + 1)(s + 4) + k(s + 2) = 0$.

**274.** For the transfer function

$$GH = \frac{k(s+1)}{s^2(s+9)}$$

the breakaway point in the root locus will be

(a) $\sigma b = 0, 1$ (b) $\sigma b = 1, -1$ (c) $\sigma b = -1, -2$ (d) $\sigma b = 0, -3$,

**275.** For the transfer function

$$GH = \frac{k}{s(s+6)(s+8)}$$

the breakaway point in the root locus will be

(*a*) $\sigma b = -2.25, -7.07$ (*b*) $\sigma b = 2.25, -7.07$ (*c*) $\sigma b = -2.25, 7.07$ (*d*) $\sigma b = 2.25, 7.07$.

**276.** For the transfer function

$$GH = \frac{k(s+5)}{(s+2)(s+4)}$$

the breakaway point in the root locus will be

(*a*) – 3.27, 6.73 (*b*) – 3.27, – 6.73 (*c*) 3.27, 6.73 (*d*) 3.27, – 6.73.

**277.** A negative feedback control system is characterised by

(*a*) reduced bandwidth (*b*) increased accuracy

(*c*) reduced distortion (*d*) increased non-linearity.

**278.** If *R* and *f* are input and the feedback signal in a negative feedback control system then error signal *E* is given by

(*a*) $E = R + F$ (*b*) $E = R - F$ (*c*) $E = R \pm F$ (*d*) none of these.

**279.** If the system in the last question is positive feedback one them the error signal *E* is given by

(*a*) $E = R - F$ (*b*) $E = R \pm F$ (*c*) $E = \frac{R.F}{R+F}$ (*d*) $E = R + F$.

**280.** A control system is said to be linear provided

(*a*) it is a negative feedback system

(*b*) it is a positive feedback system

(*c*) it input *X* and output *Y* relationship is expressed as $\sum_{i=0}^{m} a_i(t)\frac{d^i y}{dt^i} = \sum_{i=0}^{n} b_i(t)\frac{d^i x}{dt^i}$

(*d*) none of these.

**281.** Free response of control system is one which is characterised by

(*a*) input approaching infinite with time (*b*) input being a finite value

(*c*) zero input (*d*) all initial conditions being zero.

**282.** "Free response of control system explicitly depends upon initial condition of the system". This statement

(*a*) is absolutely true (*b*) is absolutely false

(*c*) can be both true and false (*d*) none of these.

**283.** Free response in mathematically given by

(a) $\sum_{i=0}^{m} a_i(t)\dfrac{d^iy}{dt^i} = 0$ (b) $\sum_{i=0}^{m} a_i(t)\dfrac{d^iy}{dt^i} = k$

(c) $\sum_{i=0}^{m} a_i(t)\dfrac{d^iy}{dt^i} = \infty$ (d) none of these.

**284.** Forced response of a feedback control system is one characterised by

(a) all initial conditions being zero (b) input being zero

(c) output depending upon input and initial conditions

(d) none of these.

**285.** "Forced response of a control system explicitly depend upon the input of the system and all initial conditions are being assumed to be zero". This statement is

(a) true (b) false (c) partially true (d) none of these.

**286.** Transient response of a control system is the one

(a) which approaches zero when time approaches infinity

(b) which does not depend on time

(c) which is time dependent

(d) none of these.

**287.** Steady state response of a control system is the one

(a) which is time dependent (b) which is time independent

(c) which does not approach zero when time approaches infinity

(d) none of these.

**288.** Mark the correct statement

(a) free response + forced response = total response

(b) total response = steady state response

(c) total response = free response = stead state response

(d) none of these.

**289.** A causal system is one in which the output depends upon

(a) input as well as output

(b) present input and past input

(c) present as well as future input

(d) present and past input and past output.

**290.** A control system response is represented by

$Y(t) = X(t + T)$ with $T > 0$. Is it causal?

(a) Yes (b) No (c) Not necessarily (d) None of these.

**291.** A control system response is represented by a convolution integral

$$Y(t) = \int_{-\infty}^{\infty} w(t, T).X(T)dT$$

Is it a linear system?

(*a*) Yes it is a linear system (*b*) No it is not a linear system

(*c*) The above equation does not represent a control system

(*d*) None of these.

**292.** Fig. 15.47 shows a control system the output $Y(t)$ is given by

(*a*) $Y(t) = 5 (\cos t + 2 \sin 2t)$

(*b*) $Y(t) = 5 (\sin t - 2 \cos 2t)$

(*c*) $Y(t) = (5 \cos t - 10 \sin 2t)$

(*d*) none of these.

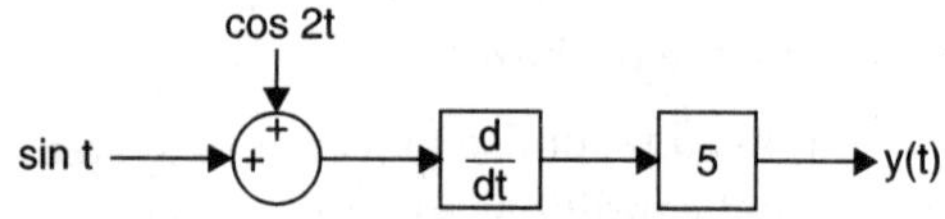


*FIGURE 15.47*

**293.** If some one says the transient response of a certain control system is given by $Y(t) = e^{-2t} + e^{-t} + t$. Will you believe it?

(*a*) Yes (*b*) No (*c*) Both are possible (*d*) Both are impossible.

**294.** One of the following expressions represents a unit step. What is that?

(*a*) $u(t - t_0) = \begin{bmatrix} 1, & t > t_0 \\ 0, & t \le t_0 \end{bmatrix}$ (*b*) $\int_{-\infty}^{t} u(\tau - t_0)\, d\tau = \begin{bmatrix} t - t_0 & t > t_0 \\ 0 & t \le t_0 \end{bmatrix}$

(*c*) $u(t - t_0) = 0$ for $t > t_0$ (*d*) none of these.

**295.** One of the following function represents a unit impulse. It is,

(*a*) $\int_{-\infty}^{t} u(t - t_0)\, d\tau = \begin{bmatrix} t - t_0, & t > t_0 \\ 0, & t \le t_0 \end{bmatrix}$ (*b*) $\delta(t) = \lim_{\Delta t \to 0} \left[\frac{u(t) - u(t - \Delta t)}{\Delta t}\right]$

(*c*) $\delta(t) = \lim_{\Delta t \to \infty} \left[\frac{u(t) - u(t - \Delta t)}{\Delta t}\right]$ (*d*) $\delta(t) = \lim_{\Delta t \to 0} \left[\frac{u(t - \Delta t) - u(t)}{\Delta t}\right]$.

**296.** Area under a unit impulse function is

(*a*) $\infty$ (*b*) 0 (*c*) unity (*d*) none of these.

**297.** Below are given some expressions related to impulse functions.

Mark the correct expression.

(*a*) $\int_{-\infty}^{\infty} \delta(t)\, dt = 1$ (*b*) $\int_{0}^{\infty} \delta(t)\, dt = 1$

(*c*) $\int_{-\infty}^{\infty} \delta(t - t_0)\delta(t) = \delta(t)$ (*d*) $\int_{-\infty}^{\infty} \delta(t - t_0)\delta(t) = \delta(t_0)$.

**298.** Unit impulse response of a system is the one when

(*a*) the input to the δ/*s* is a unit impulse

(*b*) the input to the δ/*s** is a unit impulse and all initial conditions are zero

(*c*) the output is a unit impulse for all inputs

(*d*) none of these

[* δ/*s* = system]

**299.** Unit impulse response of a system is no thing but its

(*a*) Forced responsed (*b*) Free response

(*c*) Steady state response (*d*) None of these.

**300.** Unit impulse response is also

(*a*) Free response (*b*) Forced response

(*c*) Transient response (*d*) Total response.

**301.** Unit ramp response is also

(*a*) Free response (*b*) Transient response

(*c*) Forced response (*d*) Steady state response.

**302.** A control system is represented by the second order differential equation

$$\frac{d^2y}{dt^2} + 2\xi w_n \left(\frac{dy}{dt}\right) + w_n^2\, y = w_n^2\, X$$

The damping coefficient (α) of such a system is

(*a*) $\alpha = \xi w_n$ (*b*) $\alpha = (1/\xi w_n)$ (*c*) $(\xi/w_n)$ (*d*) none of these.

**303.** Time constant (τ) of such a system is given by

(*a*) $\tau = (\xi w_n)$ (*b*) $\tau = (1/\xi w_n)$ (*c*) $\tau = (1/2\xi w_n)$ (*d*) $\tau = (\xi_n/\xi)$.

**304.** Damped natural frequency of such a system is given by

(*a*) $w_d = \xi w_n$ (*b*) $w_d = w_n(1 - \xi^2)$ (*c*) $w_d = w_n\sqrt{1-\xi^2}$ (*d*) $w_d = \sqrt{w_n(1-\xi^2)}$.

**305.** A control system represented by the second order differential equation

$$\frac{d^2y}{dt^2} + 2\xi w_n \left(\frac{dy}{dt}\right) + w_n^2 y = w_n^2 X$$

Such a system will be critically damped when

(*a*) $\xi = 1$ (*b*) $\xi = 0$ (*c*) $\xi = 0.5$ (*d*) $\xi = \infty$.

**306.** The condition $\xi < 1$ for such a system, makes it

(*a*) Critically damped (*b*) Under damped

(*c*) Over damped (*d*) None of these.

**307.** Laplace transform *F*(*s*) of any time function δ(*t*) is given by

(*a*) $F(s) = \int_{-\infty}^{\infty} \delta(t) e^{-st}\, dt$ (*b*) $F(s) = \int_0^{\infty} \delta(t) e^{-st}\, dt$

(*c*) $F(s) = \int_0^{\infty} \delta(t) e^{st}\, dt$ (*d*) None of these.

**308.** Inverse Laplace of $F(s)$ is given by

(a) $£^{-1} F(s) = \frac{1}{2\pi i}\int_{C-j\infty}^{C+j\infty} \delta(s)\, e^{st}\, ds$ (b) $£^{-1} F(s) = \int_{-\infty}^{\infty} F(s)\, e^{st}\, ds$

(c) $£\, F(s) = \frac{1}{2\pi i}\int_{-\infty}^{\infty} F(s)\, e^{st}\, ds$ (d) None of these.

**309.** Laplace transform of damped cosine wave expressed by $e^{-at} \cos \omega t$ is given by

(a) $\frac{S+a}{(S+a)^2 - w^2}$ (b) $\frac{S-a}{(S+a)^2 + w^2}$ (c) $\frac{S+a}{(S+a)^2 + w^2}$ (d) None of these.

**310.** If $F_1(s)$ and $F_2(s)$ are Laplace transform of $F_1(t)$ and $F_2(t)$ resp, then Laplace transform of $[AF_1(t) + BF_2(t)]$ will be

(a) $AF_1(s) + BF_2(s)$ (b) $\frac{1}{A}F_1(s) + \frac{1}{B}F_2(s)$ (c) $BF_1(s) + AF_2(s)$ (d) None of these.

**311.** Transfer function of a simple $RC$ integrator circuit shown in Fig. 15.48 is given by

(a) $\frac{1}{S+a}$ (b) $\frac{1}{S-a}$

(c) $\frac{a}{S+a}$ (d) $\frac{a}{S-a}$.

where $\boxed{a = \frac{1}{RC}}$

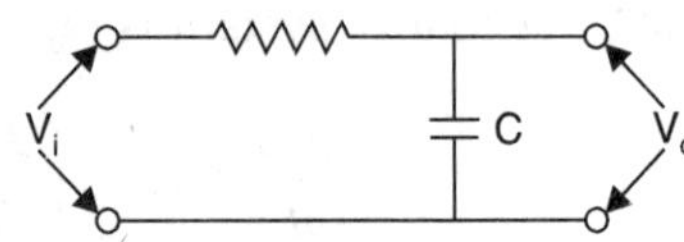


FIGURE 15.48

**312.** Transfer function of simple differentiator circuit show in Fig. 15.49 given by

(a) $\frac{S}{S+a}$ (b) $\frac{a}{S+a}$

(c) $\frac{1}{S+a}$ (d) None of these.

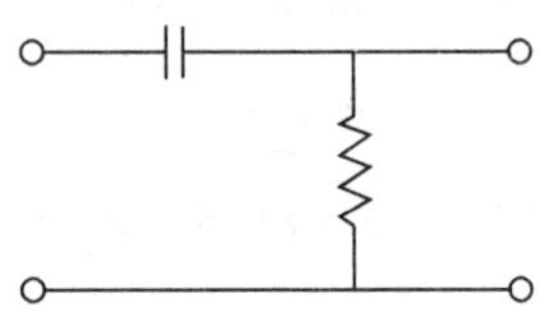

FIGURE 15.49

**313.** Transfer function of a lead compensator is given by

(a) $\frac{S+a}{S+b}$ with $a > b$ (b) $\frac{S+a}{S+b}$ with $a < b$ (c) $\frac{S-a}{S+b}$ with $a < b$ (d) none of these.

**314.** For a control system given by

$$P(s) = \frac{G(S+1)(S+3)}{S^2 (S+2)(S+4)(S+5)} \text{ it is a}$$

(a) type 5 system (b) type 2 system (c) type 3 system (d) none of these.

**315.** A control system represented in a block systematic as shown in Fig. 15.50. What is the type of the system?

(a) type 1 (b) type 2

(c) type 0 (d) none of these.

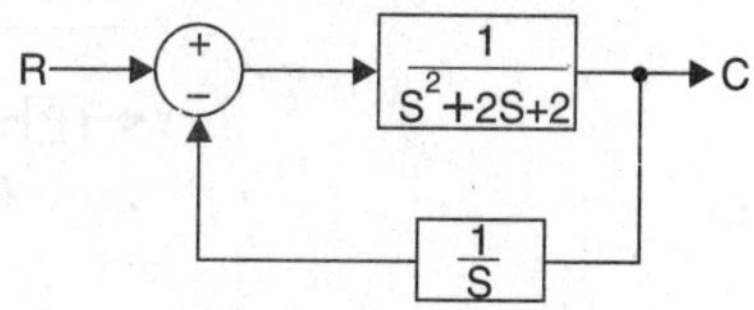


*FIGURE 15.50*

**316.** The positional error constant of a type 1 canonical feedback system is given by

(a) $\lim_{S\to 0} G(s)$ (b) $\lim_{S\to 0} \delta G(s)$ (c) $\lim_{S\to \infty} G(s)$ (d) None of these.

**317.** Position error constant of a type 2 canonical feedback system is

(a) two (b) zero (c) infinity (d) none of these.

**318.** A stable canonical feedback system has two poles at the origin. The steady-state error when the input to the system is unit step, is

(a) zero (b) infinity

(c) in determinate from given data (d) none of these.

**319.** A stable unity feedback system has one pole at the origin. An input of a unit parabolic function will give a steady state error of

(a) zero (b) infinity (c) in determinate (d) none of these.

**320.** A control system has a transfer function given by

$$T = \frac{A_1 + kA_2}{A_3 + kA_4}, (A_1, A_2, A_3, A_4) \text{ are}$$

Polynomial in 'δ'.

What is the sensitivity of 'T' w.r.t. parameter 'k'.

(a) $\frac{k(A_2A_3 - A_1A_4)}{(A_3 + kA_4)(A_1 + kA_2)}$ (b) $\frac{k(A_3 + kA_4)(A_1 + kA_2)}{(A_2A_3 - A_1A_4)}$

(c) $\frac{(A_2A_3 - A_1A_4)}{(A_3 + kA_4)(A_1 + kA_2)}$ (d) $\frac{kA_1A_2}{A_3A_4}$.

**321.** The transfer function of the system represented in Fig. 15.51 is

(a) $\frac{P_1}{1 - P_1P_2}$ (b) $\frac{P_1}{1 + P_1P_2}$

(b) $1 + P_1P_2$ (d) none of these.

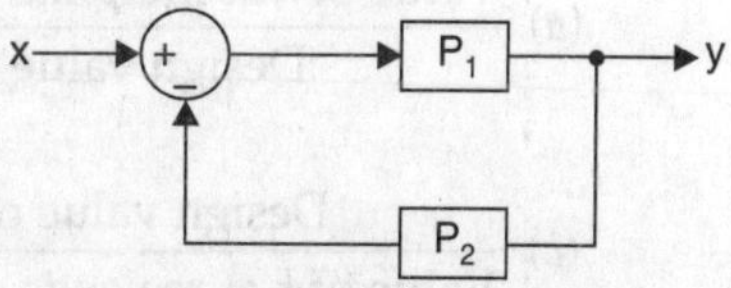


*FIGURE 15.51*

**322.** Fig. 15.52 (a) and (b) shows different arrangements

(a) Both arrangement are equivalent (b) Both are different

(c) Neither of the two

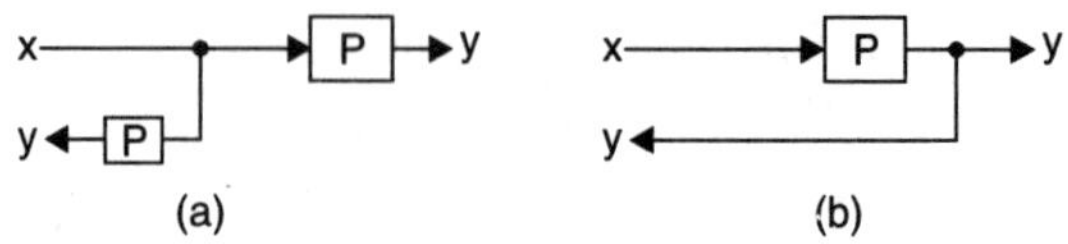


*FIGURE 15.52*

**323.** Gain margin of a control $s/s$ is expressed mathematically by one of the following expressions

(*a*) $\text{G.M.} = -\dfrac{1}{|\text{G.H.}(j\omega\pi)|}$ (*b*) $\text{G.M.} = |\text{G.H.}(j\omega\pi)|$

(*c*) $\text{G.M.} = -\dfrac{1}{|\text{G.H.}(j\omega\pi)|}$ (*d*) None of these.

**324.** Phase margin can be expressed in terms of gain cross-over frequency by

(*a*) P.M. = [180 + Arg. G.H. ($j\omega_1$)] degree (*b*) P.M. = [180 – Arg. G.H. ($j\omega_1$)] degree

(*c*) P.M. = [Arg. G.H. ($j\omega_1$)] degree (*d*) None of these.

where G.H. $\omega_1$ is the cross-over frequency.

**325.** Fig. 15.53 given below shows a Nyquist plot.

Find gain margin?

(*a*) 0.1 (*b*) 10 (*c*) 100 (*d*) none of these.

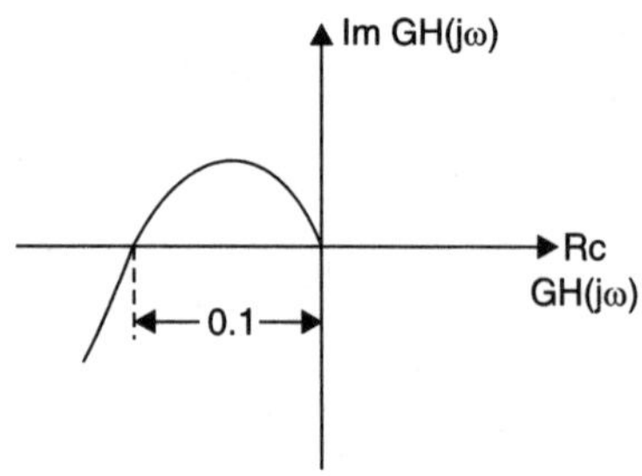


*FIGURE 15.53*

**326.** Gain margin can be determined from the root-locus plot by

(*a*) $\dfrac{\text{Value of } k \text{ at imaginary cross-over}}{\text{Design value of } k}$ (*b*) $\dfrac{\text{Design value of } k}{\text{Value of } k \text{ at imaginary cross-over}}$

(*c*) $\dfrac{\text{Design value of } k}{\text{Value of } k \text{ at real axis cross-over}}$ (*d*) $\dfrac{\text{Value of } k \text{ at real axis cross-over}}{\text{Design value of } k}$.

**327.** A higher value of gain factor '*k*' at the imaginary cross-over of the root-locus plot of control system

(*a*) brings it closer to instability (*b*) takes it farther from instability

(*c*) is always undesirable (*d*) none of these.

**328.** In Fig. 15.54 shown below the real axis portion of the root locus of a certain system. The portion shown tells about the gain factor '*k*' of the system

(*a*) $k = 0$ (*b*) $k = 1$ (*c*) $k > 0$ (*d*) $k < 0$.

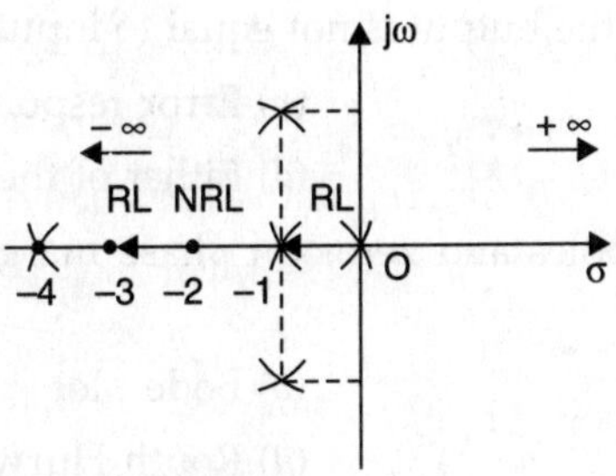


***FIGURE 15.54***

**329.** The angles made by asymptotes in a root locus plot are given by (assume $k > 0$)

(*a*) $\dfrac{(2l+1)\,180}{(n-m)}$ degree (*b*) $\dfrac{(2l)\,180}{(n-m)}$ degree

(*c*) $(2l + 1)$ 180 degree (*d*) $(2l)$ 180 degree.

**330.** The low frequency behaviour of a control system can be modified by using a cascade compensator which

(*a*) adds a dipole (pole–zero pair) at a low frequency

(*b*) cancels low frequency poles (*c*) cancles high frequency zeros

(*d*) none of these.

**331.** Is it feasible to use cancellation technique with compensator in the feedback path

(*a*) No (*b*) Yes (*c*) Sometimes (*d*) None of these.

**332.** Closed loop transfer function of a system is given by

$$\frac{C(S)}{R(S)} = \frac{(s+1)}{(s^2+s+1)}$$

What are the time constant and damped natural frequency respectively?

(*a*) 2, 0.86 (*b*) 0.5, 1 (*c*) 0.86, 2 (*d*) None of these.

**333.** Which of the following will increase the steady state accuracy?

(*a*) Phase lead compensator (*b*) Phase lag compensator

(*c*) Differentiator (*d*) Integrator.

**334.** In case of magnitude of the polar plot at phase cross-over is 'A' then the gain margin is

(*a*) zero (*b*) A (*c*) – A (*d*) 1/A.

**335.** The root loci of a system has three asymptoters the system can have

(*a*) Five poles, two zeros (*b*) Four poles, one zero

(*c*) Three poles (*d*) All of the above.

**336.** The Bode plot is applicable to

(*a*) minimum phase network (*b*) all phase network

(*c*) maximum phase network (*d*) none of these.

**337.** The initial response when the output is not equal to input is called

(*a*) Transient response (*b*) Error response

(*c*) Dynamic response (*d*) Either of the above.

**338.** The effect of addition of poles and zeros on phase margin and gain margin can be most conveniently seen in

(*a*) Nyquist plot (*b*) Bode plot

(*c*) Root locus (*d*) Routh Hurwitz criteria.

**339.** In root locus plot the angle of asymptotes is 360° divided by

(*a*) the number of poles (*b*) the no. of zeros

(*c*) number of poles – number of zeros (*d*) number of poles + number of zeros.

**340.** The gain margin of the given transfer function

$$G(S) = \frac{0.75}{(s+1)(s+2)} \text{ will be}$$

(*a*) 4 dB (*b*) 8 dB (*c*) 12 dB (*d*) 16 dB.

**341.** Which of the following is the most important technique used for stability and transient response of the system?

(*a*) Bode plot (*b*) Root locus

(*c*) Routh Hurwitz criteria (*d*) Nyquist plot.

**342.** In root locus plot different roots have the same

(*a*) phase (*b*) gain

(*c*) both (*a*) and (*b*) (*d*) phase and gain margin.

**343.** If the bandwidth of the system is increased then the response will

(*a*) become faster (*b*) become slower

(*c*) Remain unaffected (*d*) can't say.

**344.** If the characteristic equation of a closed-loop system is

$$1+\frac{k}{\delta(s+1)(s+2)}=0,$$

the centroid at the asymptotes in root – locus plot will be

(*a*) zero (*b*) 2 (*c*) – 3 (*d*) – 2.

**345.** A control system working under unknown random actions is called

(*a*) computer control system (*b*) digital control system

(*c*) stochastic control system (*d*) adaptic control system.

**346.** The on-off controller is a ..... system.

(*a*) digital (*b*) linear (*c*) non-linear (*d*) discontinuous.

**347.** ...... is the serious disadvantage of electronic control system

(*a*) temperature sensitiveness (*b*) low reliability

(*c*) operational difficulty (*d*) none of these.

**348.** Under which of the following conditions, the transfer function technique is considered as inadequate?

(*a*) system having stability problems (*b*) system having multiple input disturbances

(*c*) system having complexities and non-linearities

(*d*) all of the above.

**349.** With feedback system

(*a*) the transient response decays at constant rate

(*b*) the transient response gets magnified

(*c*) the transient response decays slowly

(*d*) the transient response decays more slowly.

**350.** A servomechanism with step displacement input will form ..... system

(*a*) type 3 (*b*) type 2 (*c*) type 1 (*d*) type 0.

**351.** The phase difference between the two windings of A.C. servometer is

(*a*) 30° (*b*) 60° (*c*) 90° (*d*) 120°.

**352.** Integral error compensation in a control system

(*a*) minimizes steady state error (*b*) increases steady state error

(*c*) doesn't have any effect on steady state error

(*d*) is equal to steady state error

**353.** The effect of error rate damping

(*a*) to reduce steady state error (*b*) to delay the response

(*c*) to provide large settling time (*d*) none of these.

**354.** Synchros' are popularly used as transmitter of

(*a*) Digital data (*b*) Mathematical data

(*c*) Angular data (*d*) All of the above

**355.** For type 0 system, the steady state acceleration error is

(*a*) zero (*b*) unity

(*c*) between zero and unity (*d*) infinity.

**356.** Which of the following reduces with feedback

(*a*) system stability (*b*) system gain (*c*) either of above (*d*) none of these.

**357.** Type 0 system has ... at the origin

(*a*) no pole (*b*) three poles (*c*) two poles (*d*) one pole

**358.** The damping factor of the system is unity. The system is

(*a*) over damped (*b*) critically damped (*c*) under damped (*d*) unstable.

**359.** A bimetallic thermostat is a ..... controller

(*a*) on-off (*b*) zero team (*c*) one team (*d*) two team.

**360.** Polar plots for negative and positive frequencies

(*a*) can never be symmetrical (*b*) may or may not be symmetrical

(*c*) are always symmetrical (*d*) none of these.

**361.** The characterstic equation of an armature controlled DC motor is of .... order equation.

(*a*) zero (*b*) first (*c*) second (*d*) third.

**362.** For frequency domain a analysis sinusoidal signals considered have ..... frequency.

(*a*) logarithmic (*b*) some (*c*) differential (*d*) any of these.

**363.** A signal flow graph is a

(*a*) topological representation of set of differential equation

(*b*) polar graph (*c*) log graph

(*d*) special type of graph to analyse modern control system.

**364.** In an absolutely stable system oscillations

(*a*) are always present (*b*) can't be set-up all

(*c*) can be set-up and may be controlled

(*d*) set-up automatically and get damped out eventually.

**365.** The negative real axis of Nyquist plot corresponds to the

(*a*) negative 180° phase line of Bode plot

(*b*) positive 180° phase line of Bode plot

(*c*) 0 dB line of Bode plot (*d*) none of these.

**366.** Which of the following is the main cause of absolute instability?

(*a*) Error detector where the two signals are compared

(*b*) Parameters of controlled system

(*c*) Parameters of controlling system

(*d*) Parameters of feedback system.

## ANSWERS

| | | | | | | |
|---|---|---|---|---|---|---|
| **1.** (*c*) | **2.** (*c*) | **3.** (*a*) | **4.** (*d*) | **5.** (*a*) | **6.** (*c*) | **7.** (*b*) |
| **8.** (*d*) | **9.** (*a*) | **10.** (*d*) | **11.** (*c*) | **12.** (*c*) | **13.** (*c*) | **14.** (*a*) |
| **15.** (*b*) | **16.** (*c*) | **17.** (*d*) | **18.** (*d*) | **19.** (*c*) | **20.** (*d*) | **21.** (*c*) |
| **22.** (*d*) | **23.** (*c*) | **24.** (*d*) | **25.** (*a*) | **26.** (*b*) | **27.** (*b*) | **28.** (*d*) |
| **29.** (*c*) | **30.** (*b*) | **31.** (*b*) | **32.** (*a*) | **33.** (*a*) | **34.** (*c*) | **35.** (*b*) |
| **36.** (*d*) | **37.** (*a*) | **38.** (*b*) | **39.** (*a*) | **40.** (*b*) | **41.** (*a*) | **42.** (*c*) |

| | | | | | | |
|---|---|---|---|---|---|---|
| **43.** (*c*) | **44.** (*d*) | **45.** (*a*) | **46.** (*b*) | **47.** (*a*) | **48.** (*b*) | **49.** (*c*) |
| **50.** (*a*) | **51.** (*d*) | **52.** (*b*) | **53.** (*a*) | **54.** (*c*) | **55.** (*d*) | **56.** (*a*) |
| **57.** (*c*) | **58.** (*b*) | **59.** (*c*) | **60.** (*d*) | **61.** (*d*) | **62.** (*b*) | **63.** (*c*) |
| **64.** (*b*) | **65.** (*d*) | **66.** (*c*) | **67.** (*b*) | **68.** (*d*) | **69.** (*b*) | **70.** (*d*) |
| **71.** (*a*) | **72.** (*b*) | **73.** (*c*) | **74.** (*d*) | **75.** (*c*) | **76.** (*b*) | **77.** (*b*) |
| **78.** (*a*) | **79.** (*d*) | **80.** (*a*) | **81.** (*a*) | **82.** (*b*) | **83.** (*c*) | **84.** (*c*) |
| **85.** (*d*) | **86.** (*c*) | **87.** (*a*) | **88.** (*c*) | **89.** (*d*) | **90.** (*d*) | **91.** (*a*) |
| **92.** (*a*) | **93.** (*a*) | **94.** (*c*) | **95.** (*b*) | **96.** (*d*) | **97.** (*d*) | **98.** (*b*) |
| **99.** (*d*) | **100.** (*c*) | **101.** (*c*) | **102.** (*a*) | **103.** (*b*) | **104.** (*a*) | **105.** (*a*) |
| **106.** (*a*) | **107.** (*c*) | **108.** (*d*) | **109.** (*d*) | **110.** (*c*) | **111.** (*c*) | **112.** (*b*) |
| **113.** (*d*) | **114.** (*c*) | **115.** (*b*) | **116.** (*d*) | **117.** (*a*) | **118.** (*d*) | **119.** (*d*) |
| **120.** (*c*) | **121.** (*a*) | **122.** (*c*) | **123.** (*d*) | **124.** (*d*) | **125.** (*c*) | **126.** (*d*) |
| **127.** (*c*) | **128.** (*d*) | **129.** (*c*) | **130.** (*b*) | **131.** (*c*) | **132.** (*c*) | **133.** (*a*) |
| **134.** (*d*) | **135.** (*b*) | **136.** (*b*) | **137.** (*c*) | **138.** (*a*) | **139.** (*b*) | **140.** (*a*) |
| **141.** (*c*) | **142.** (*c*) | **143.** (*b*) | **144.** (*d*) | **145.** (*d*) | **146.** (*c*) | **147.** (*c*) |
| **148.** (*a*) | **149.** (*a*) | **150.** (*b*) | **151.** (*c*) | **152.** (*b*) | **153.** (*c*) | **154.** (*d*) |
| **155.** (*a*) | **156.** (*b*) | **157.** (*b*) | **158.** (*d*) | **159.** (*c*) | **160.** (*a*) | **161.** (*c*) |
| **162.** (*b*) | **163.** (*c*) | **164.** (*d*) | **165.** (*a*) | **166.** (*c*) | **167.** (*c*) | **168.** (*d*) |
| **169.** (*c*) | **170.** (*c*) | **171.** (*b*) | **172.** (*b*) | **173.** (*c*) | **174.** (*c*) | **175.** (*a*) |
| **176.** (*b*) | **177.** (*d*) | **178.** (*c*) | **179.** (*c*) | **180.** (*c*) | **181.** (*c*) | **182.** (*d*) |
| **183.** (*d*) | **184.** (*a*) | **185.** (*b*) | **186.** (*c*) | **187.** (*c*) | **188.** (*a*) | **189.** (*a*) |
| **190.** (*c*) | **191.** (*c*) | **192.** (*b*) | **193.** (*a*) | **194.** (*c*) | **195.** (*b*) | **196.** (*d*) |
| **197.** (*d*) | **198.** (*d*) | **199.** (*d*) | **200.** (*c*) | **201.** (*c*) | **202.** (*b*) | **203.** (*b*) |
| **204.** (*d*) | **205.** (*c*) | **206.** (*d*) | **207.** (*a*) | **208.** (*c*) | **209.** (*c*) | **210.** (*d*) |
| **211.** (*a*) | **212.** (*d*) | **213.** (*c*) | **214.** (*b*) | **215.** (*b*) | **216.** (*d*) | **217.** (*c*) |
| **218.** (*b*) | **219.** (*b*) | **220.** (*a*) | **221.** (*c*) | **222.** (*d*) | **223.** (*d*) | **224.** (*a*) |
| **225.** (*b*) | **226.** (*d*) | **227.** (*d*) | **228.** (*c*) | **229.** (*c*) | **230.** (*c*) | **231.** (*d*) |
| **232.** (*c*) | **233.** (*d*) | **234.** (*d*) | **235.** (*c*) | **236.** (*b*) | **237.** (*a*) | **238.** (*c*) |
| **239.** (*c*) | **240.** (*b*) | **241.** (*b*) | **242.** (*d*) | **243.** (*d*) | **244.** (*b*) | **245.** (*b*) |
| **246.** (*d*) | **247.** (*b*) | **248.** (*a*) | **249.** (*c*) | **250.** (*c*) | **251.** (*b*). | **252.** (*c*) |
| **253.** (*d*) | **254.** (*b*) | **255.** (*c*) | **256.** (*c*) | **257.** (*b*) | **258.** (*a*) | **259.** (*b*) |
| **260.** (*b*) | **261.** (*b*) | **262.** (*a*) | **263.** (*c*) | **264.** (*c*) | **265.** (*c*) | **266.** (*d*) |
| **267.** (*c*) | **268.** (*a*) | **269.** (*a*) | **270.** (*b*) | **271.** (*d*) | **272.** (*d*) | **273.** (*d*) |
| **274.** (*d*) | **275.** (*a*) | **276.** (*b*) | **277.** (*d*) | **278.** (*b*) | **279.** (*c*) | **280.** (*c*) |
| **281.** (*c*) | **282.** (*a*) | **283.** (*a*) | **284.** (*a*) | **285.** (*a*) | **286.** (*a*) | **287.** (*c*) |
| **288.** (*a*) | **289.** (*b*) | **290.** (*b*) | **291.** (*a*) | **292.** (*c*) | **293.** (*b*) | **294.** (*a*) |
| **295.** (*b*) | **296.** (*c*) | **297.** (*a*) | **298.** (*b*) | **299.** (*a*) | **300.** (*b*) | **301.** (*d*) |
| **302.** (*a*) | **303.** (*b*) | **304.** (*c*) | **305.** (*a*) | **306.** (*b*) | **307.** (*b*) | **308.** (*a*) |
| **309.** (*c*) | **310.** (*a*) | **311.** (*c*) | **312.** (*a*) | **313.** (*b*) | **314.** (*b*) | **315.** (*b*) |

| | | | | | | |
|---|---|---|---|---|---|---|
| **316.** (*a*) | **317.** (*a*) | **318.** (*c*) | **319.** (*a*) | **320.** (*b*) | **321.** (*a*) | **322.** (*b*) |
| **323.** (*a*) | **324.** (*a*) | **325.** (*a*) | **326.** (*b*) | **327.** (*a*) | **328.** (*b*) | **329.** (*c*) |
| **330.** (*a*) | **331.** (*a*) | **332.** (*a*) | **333.** (*a*) | **334.** (*b*) | **335.** (*d*) | **336.** (*d*) |
| **337.** (*a*) | **338.** (*a*) | **339.** (*b*) | **340.** (*c*) | **341.** (*c*) | **342.** (*c*) | **343.** (*c*) |
| **344.** (*b*) | **345.** (*c*) | **346.** (*a*) | **347.** (*c*) | **348.** (*d*) | **349.** (*d*) | **350.** (*c*) |
| **351.** (*c*) | **352.** (*a*) | **353.** (*a*) | **354.** (*c*) | **355.** (*d*) | **356.** (*a*) | **357.** (*a*) |
| **358.** (*c*) | **359.** (*a*) | **360.** (*c*) | **361.** (*d*) | **362.** (*c*) | **363.** (*a*) | **364.** (*d*) |
| **365.** (*a*) | **366.** (*a*). | | | | | |

# Chapter 16 POWER PLANT ENGINEERING

Energy exists in various forms like mechanical energy, electrical energy, thermal energy and so on. One form of energy can be converted into another form by suitable arrangements. Out of these forms, electrical energy is preferred due to the following reasons.

(*i*) It can be easily transported from one form to another.

(*ii*) Losses in energy transportation are minimum.

(*iii*) It can be easily controlled and regulated to suit requirements.

(*iv*) It can be easily converted into other forms of energy particularly mechanical and thermal.

(*v*) It can be easily subdivided.

**Sources of Energy.** Some of the sources of energy are

1. **Fuels**

   **(*a*) Solid fuels**—wood, charcoal, peat, coke, briquetted coal, pulverised coal.

   **(*b*) Liquid fuels**—petroleum based like petrol, diesel, etc. Non-petroleum fuels like ethyl alcohol, methyl alcohol, benzene, etc.

   **(*c*) Gaseous fuels**—producer gas, blast furnace gas, coke oven gas, water gas.

2. **Energy of water**—as used in hydroelectric plants where potential energy is converted into electrical energy of water.

3. Nuclear energy.
4. Solar energy.
5. Wind power.
6. Wave power.
7. Tidal power.
8. Geothermal energy.
9. Fuel cells.
10. Thermo electric power.
11. Thermionic convertor.
12. Magneto hydrodynamic (*MHD*) generator.

**Thermal Power Plants.** These plants may be classified as

1. **Vapour power cycles.** These cycles may operate on

(*i*) Rankine cycle (*ii*) Reheat cycle

(*iii*) Regenerative cycle (*iv*) Reheat–regenerative cycle

(*v*) Supercritical pressure cycle

(*vi*) Binary vapour cycle.

2. **Gas power cycles.** These cycles may be classified as

(*i*) Spark ignition cycles (*ii*) Compression ignition cycles

(*iii*) Gas turbine cycles

(*a*) closed cycle (*b*) open cycle.

Factors to be considered while selecting the site for a Thermal Power plant are:

1. **Availability of raw material.** A 100 MW plant will consume nearly 20,000 to 25,000 tonnes of coal per month. At an average a plant burns about 0.4 to 0.5 kg of coal per kWh generated. Coal can be transported by train, trucks, etc.

2. **Availability of water.** Water is required for two purposes:

(*a*) Make up water for main plant cycle.

(*b*) Cooling water for condenser. A 200 MW plant may need as much as 50 million litres of water per hour and another 0.8 to 1 million litres per day as make up water. Whenever water from free running stream is not available, condenser water can be recirculated after cooling in cooling towers.

(*c*) **Load centre.** Transmission cost is reduced when plant is located near the load centre. However, with present grid system this factor has lost significance.

Other factors are

(*d*) Availability of land (*e*) Topography of land

(*f*) Subsoil conditions (*g*) Pollution problems

(*h*) Ash disposal problems.

**Coal.** Coal contains fixed carbon, volatile matter, sulphur ash and moisture. Ash content may vary from 1% to 20%.

Proximate analysis of coal gives percentage of moisture, ash, volatile matter and fixed carbon. Ultimate analysis gives percentage of constituents like carbon, sulphur, oxygen, etc.

Coal may be classified according to:

(*a*) **Rank**—degree of metamorphism or progressive alteration in the natural series from lignite to anthracites. Various ranks of coal are

I. Anthracite II. Bituminous

III. Sub-bituminous IV. Lignite.

(*b*) **Grade**—quality determined by designation, calorific value, ash, ash softening temperature and sulphur various grades are designated as $A_4$, $A_6$, $A_{10}$, $S_{1.0}$, $S_{2.0}$, etc.

(*c*) **Type or variety**—determined by the nature of the original plant material and subsequent alteration thereof.

**Transfer of coal.** Within a power plant, coal may be transferred using any of the following equipment:

1. Scraper or flight conveyor
2. Vee bucket elevator and conveyor
3. Pivoted bucket conveyor
4. Belt conveyor
5. Skip hoist
6. Screw conveyor
7. Grab bucket conveyor
8. Mass flow conveyor
9. Chutes.

**Coal burning equipment.** Fuels burn in a confined space called furnace. The principal requisites for proper combustion of fuel in a furnace are:

1. Supply of requisite quantity of air to the furnace.
2. Efficient mixing of air and fuel.
3. High flame temperature to maintain ignition.
4. Enough time to complete the burning process within the furnace enclosure.

Various method for burning coal are:

(*i*) **Hand firing**—as in locomotive boiler.

(*ii*) **Stoker firing**

(*a*) **Travelling grate stoker.** This consists of an endless belt which runs on two sets of sprocket wheels. The upper level of the grate carries coal from front to the rear side of the furnace. Air is supplied from underneath the grate so as to facilitate combustion. Secondary air is supplied through nozzles on the furnace wall.

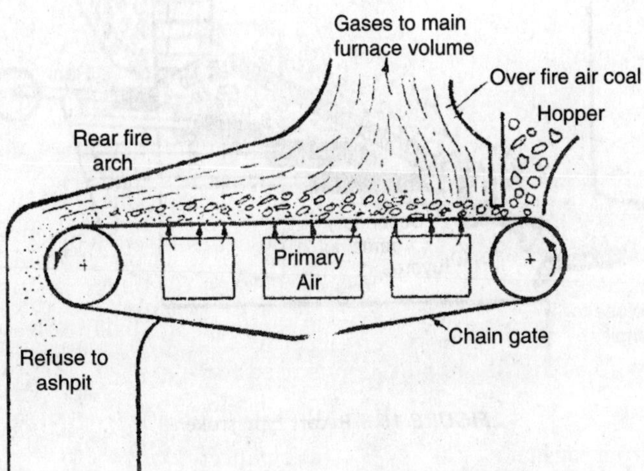


*FIGURE 16.1* Chain grate stoker.

(*b*) **Spreader stoker.** In these stokers grate is stationary and the coal is handled by the revolving blades of the feeder units of drum type and flung out into the furnace on the top of the fire. The grate is usually of perforated iron bars. Air supply is through furnace bottom so that air gets preheated and thus facilitates combustion.

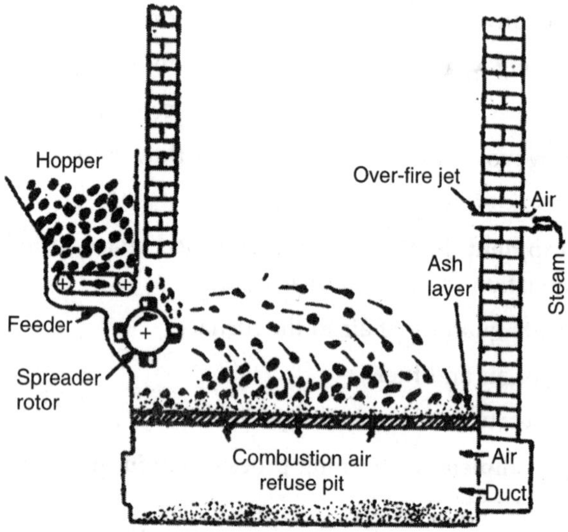


*FIGURE 16.2* Spreader stoker.

(*c*) **Retort type stoker.** This is a underfeed type stoker. Coal is supplied through ram and ash is collected at the opposite end of the furnace.

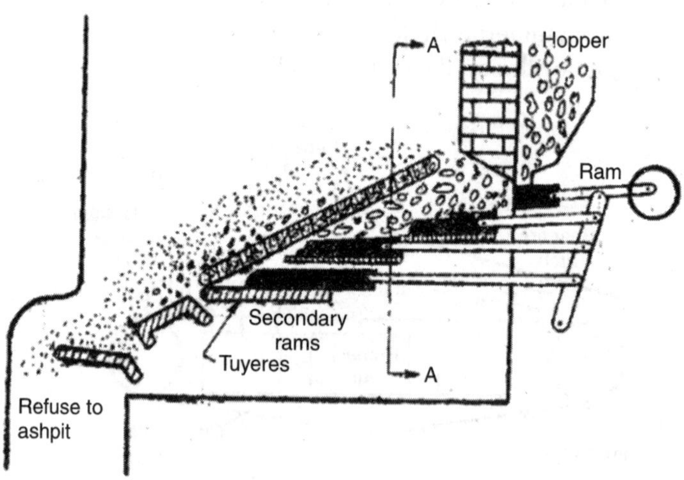


*FIGURE 16.3* Retort type stoker.

**Pulverised Fuel.** Low grade fuels are burned by pulverising the coal into fine particles. Four types of burners used for pulverised fuel firing are shown in Fig. 16.4.

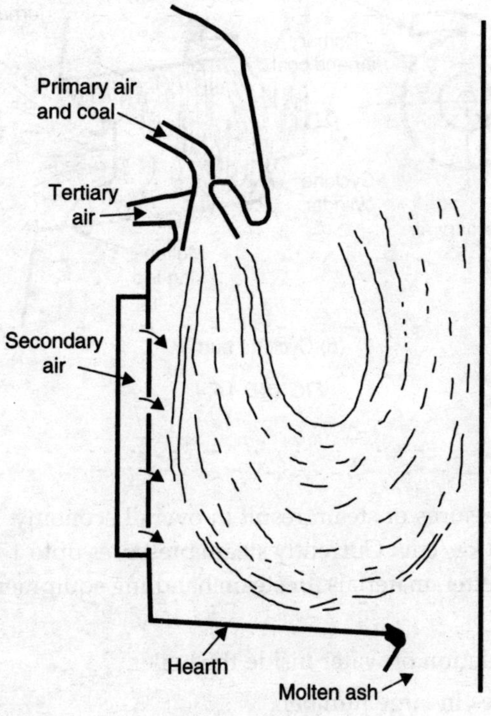


(a) Long flame burner

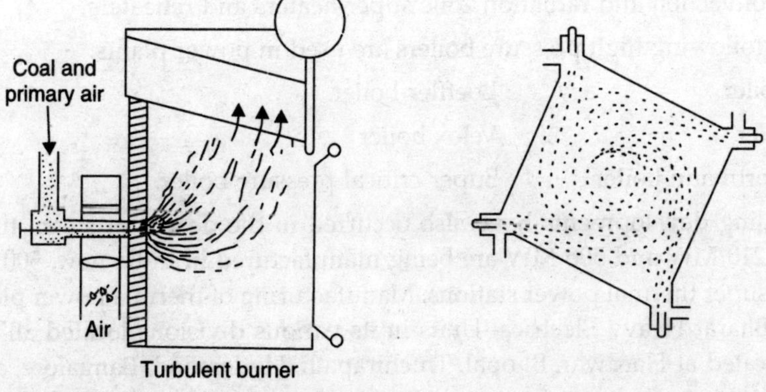


(b) Turbulent burner

(c) Tangential burner

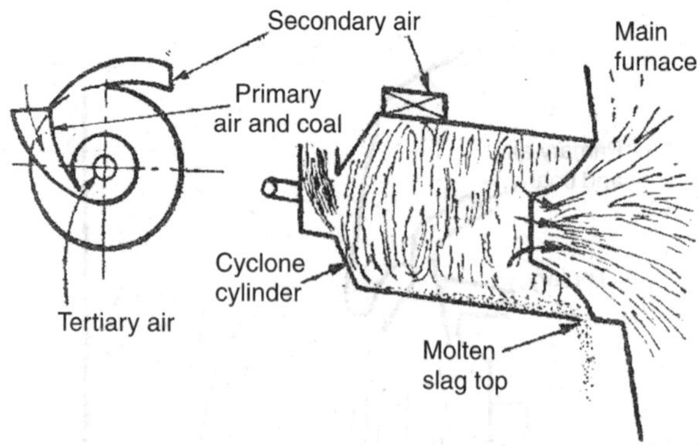


(d) Cyclone burner

*FIGURE 16.4*

## Trends in Modern Boilers

**Pressures.** Higher pressures of steam result in overall economy. Modern power plants use steam at pressures above 100 kg/cm$^2$. Currently steam pressures upto 170 kg/cm$^2$ have been used. This necessitates the use of better materials for steam handling equipment. In general use of higher steam pressures means.

(*i*) use of forced circulation of water inside the boiler.

(*ii*) use of smaller tubes in large number.

(*iii*) use of multiple tube circuits.

(*iv*) pressurised combustion.

(*v*) use of water walls.

(*vi*) use of convection and radiation zone super heaters and reheaters.

In general, following high pressure boilers are used in power plants.

| | |
|---|---|
| La Mant boiler | Loeffler boiler |
| Benson boiler | Velox boiler |
| Schmidt Hartmann boiler | Super critical pressure boiler. |

Corresponding developments have also occurred in the design of steam turbines. Single units of 110 MW, 210 MW and 500 MW are being manufactured in India now. 500 MW units are being installed in super thermal power stations. Manufacturing of thermal power plants in India is confined to M/s Bharat Heavy Electrical Units in its various divisions located all over India. Its major units are located at Hardwar, Bhopal, Truchirapalli, Hyderabad, Bangalore, and Jhansi.

**Hydro Power Plants.** These plants utilise the potential energy of water by converting it into mechanical energy and thereby driving the generators. For continuous flow of water to operate the hydraulic turbines, sufficient quantity of water must be available, throughout the year. For this purpose generally dams are constructed to store water and regulate its flow. The plants in which water is available for power generation throughout the year are known as run-of-river plants without pondage. Such plants generally have high flow rates during rainy season and low flow rates during off season, particularly in summer months. To overcome this difficulty run-of-river plants with pondage are used where water is stored during rainy season and utilised during off season.

Depending upon the head available, the plants may be classified as low head, medium head and high head plants.

Geographical, meteorological and geological considerations play an important role in the selection of the site for a hydroelectric power plant.

**Hydrograph.** A hydrograph is a plot of discharge through a river versus time. A hydrograph may be plotted for several weeks or even months. It indicates the variation in flow and also the water available from the stream at different times of the day, week or year. Extreme conditions of flow can be studied from hydrograph. A hydrograph also helps in the studies of the effect of storage on flow. Behaviour of flash stream on a hydrograph is indicated by the steep rise and fall of the curve. Flashy behaviour of a stream may be due to impervious soil surfaces, impervious subsurface strata, irregular distribution and slope. On the other hand, soils and rock formation conducive to seepage, gentle slopes, etc. show relatively steady flow and minimum variation in the shape of the curve.

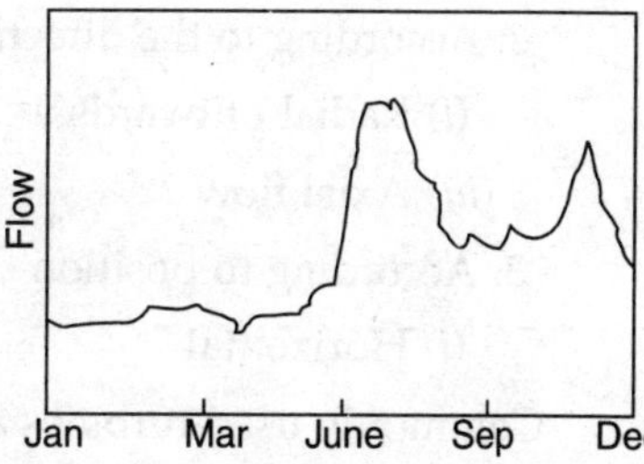


*FIGURE 16.5*

**Flow Duration Curve.** A flow duration curve is a plot of flow available during a period versus the fraction of time. The flow may be expressed in cubic metres per second, per week or in any other convenient unit of time, knowing the head, total energy of flow can be computed. A flow duration curve may be used to determine the minimum and the maximum conditions of flow. The flow duration can be improved by storing water on the upstream side which may be used during off periods.

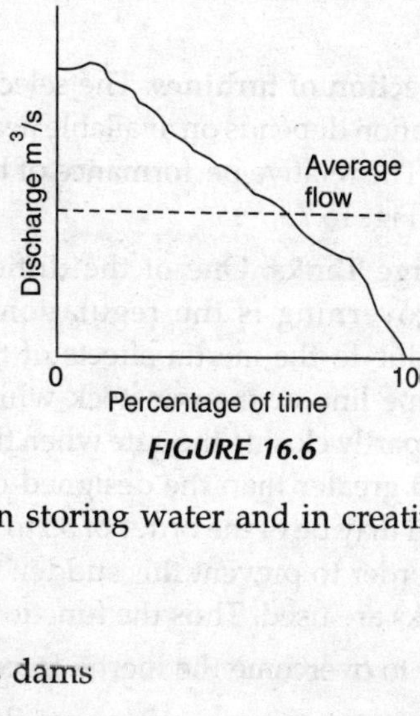


*FIGURE 16.6*

**Dams.** The function of a dam is two fold—it helps in storing water and in creating head. The dams are classified as:

1. **Masonry dams.** These may be

   (*a*) Solid gravity dams (*b*) Buttress dams

   (*c*) Arched dams.

2. Rock filled dams

3. Earth dams.

**Turbines.** Turbines used in hydro-power stations may be classified as follows.

1. According to the action of water as

   (*i*) ***Impulse turbines***

   (*a*) Pelton wheel (*b*) Girard turbine

   (*c*) Jonvel turbine (*d*) Banki turbine

   (*e*) Turgo turbine.

***(ii) Reaction turbines***

(*a*) Fournyron turbine (*b*) Francis turbine

(*c*) Thomson turbine (*d*) Kaplan turbine

(*e*) Mixed flow turbine.

2. According to the direction of flow as

(*i*) Radial outwards (*ii*) Radial inwards

(*iii*) Axial flow (*iv*) Mixed flow.

3. According to position of shaft as

(*i*) Horizontal (*ii*) Vertical.

Commonly used turbines are:

(*i*) Pelton wheel for high heads

(*ii*) Francis turbine for medium heads.

(*iii*) Kaplan turbines for low heads.

Now bulb turbines are being considered for developing power available from small streams like canals.

**Selection of turbines.** The selection of a turbine for a hydro station depends on available head, quantity of water and load. The relative performance of hydraulic turbines is shown in Fig. 16.7.

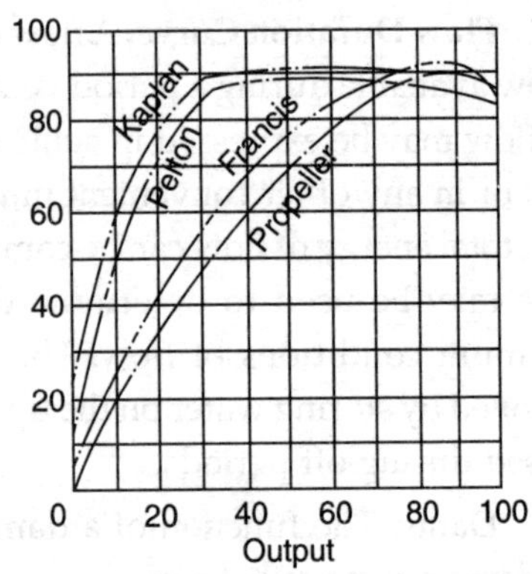


*FIGURE 16.7*

**Surge Tanks.** One of the difficulties in hydraulic turbine governing is the regulation of the increase of pressure due to the inertia effects of the water column in supply pipe line or the penstock which is caused by the governor partly closing the gate when the turbine is running at a speed greater than the designed one. The pressure so developed may be of the order of 35 to 70 kg/sq cm or even more. In order to prevent this sudden increase of pressure, surge tanks are used. Thus the functions of surge tanks are

(*i*) to overcome the inertia forces in the supply pipe line during the fall of load.

(*ii*) to act as reservoir or auxiliary storage during increase of load.

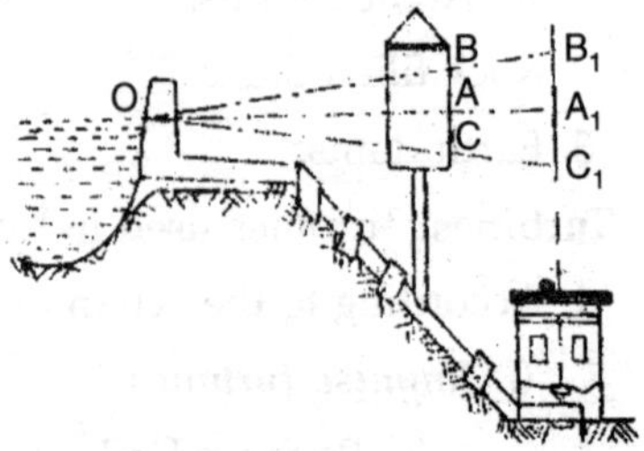


*FIGURE 16.8* Surge tanks.

**Pumped storage.** Providing daily or yearly peak load is one of the most important problems of any power system. The daily peak load requires generating plant to be kept standby for operation during a very short period each day. Obviously to meet this maximum demand provision of generating capacity has to be made and the revenue earnings for this are relatively small.

One of the methods for meeting this peak load demand is to instal pumped storage plants. In these plants power is generated during peak load period. The water flowing through turbines during this period is stored in the tail water pond.

During off peak periods this water is pumped back to the head water pond. During off peak period spare power available in the main plant is used to pump water upwards. Reversible turbine pump units have been developed so that the same unit can be used as a turbine for power generation and also as pump for pumping water to high head pond.

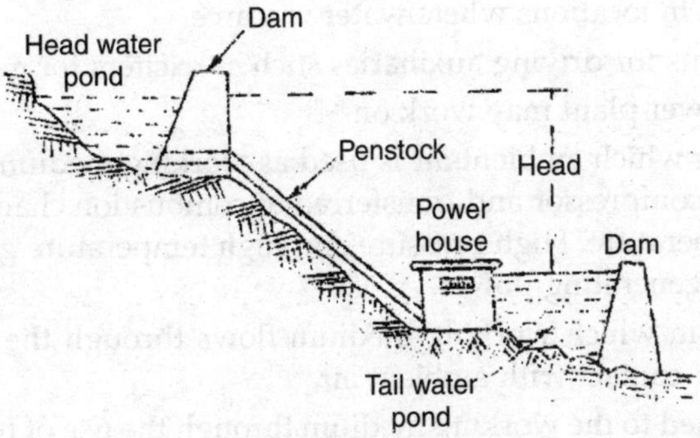


*FIGURE 16.9*

One pumped storage plant has been installed at Kadamparai in (Tamilnadu) where 4 units each of 100 MW capacity have been installed. Net head available is more than 300 metres.

**Internal Combustion Engine Plants.** Depending upon fuel used, the internal combustion engines may be designated as gas engines, petrol engines or diesel engines. In general, diesel engines are often used for generation of power because of higher thermal efficiency and low cost of diesel oil as compared to that of petrol. The advantages offered by a diesel engine power plant are:

1. It can be easily started from cold conditions.
2. It does not require any initial warning time and the engines can be directly loaded.
3. Overall efficiency is better than that of steam power plant.
4. Limited quantity of make up water is required for engine cooling water system.
5. As the calorific value of diesel is 50% higher than that of coal, therefore, less quantity of fuel has got to be transported for same power generation.
6. A diesel engine power plant needs limited space and can be easily installed closer to load centre.
7. Less labour is required for operating the power plant.
8. Installation and commissioning of such plants does not require much time.
9. Total cost of the plant per kilowatt of installed capacity is almost 20% to 30% less than that of a steam plant of equivalent size.
10. Investment in spare parts for plant is less.

However, single unit capacity of a diesel engine unit is restricted to 10 MW. The cost of diesel is more than the cost of coal even on the basis of calorific values.

Diesel engines of large size generally used for power generation are generally started by compressed air. These engines are invariably super charged for higher specific power production and better thermal efficiency.

Large size engines used for power generation need water cooling of the cylinder. This water is recirculated after cooling in a spray pond or cooling tower. Air and oil filters are provided on engines which need periodic checking. The alternator is generally directly coupled to the engine shaft.

**Gas Turbine Power Plants.** Gas turbines in power systems, in general, find the following applications:

1. Fully automatic booster stations at the end of transmission lines.
2. Standby and peak load plants.
3. Plants for use in locations where water is scarce.
4. In power plants for driving auxiliaries such as exciters for generator, pumps, etc.

A gas turbine power plant may work on

1. **Open cycle** in which ambient air is used as working medium. The air is compressed in an axial flow compressor and transferred to combustion chamber where fuel is burnt to raise the temperature. High pressure and high temperature gases expand in the turbine unit, thereby generating power.
2. **Closed cycle** in which working medium flows through the closed path and does not come in direct contact with ambient air.

The heat is supplied to the working medium through the use of heat exchangers. As a result fuels with low calorific values like wood, lignite, peat, etc. can be used. The working medium may be any fluid with better thermodynamic properties.

Thermal efficiency of a gas turbine power plant can be improved by

1. reducing the power in compression of air in compressors by resorting to multi-stage compression with inter cooling. Axial flow compressors are invariably used for better efficiency and lower power consumption.
2. use of air preheaters utilising the heat of exhaust gases from the turbine. This results in lesser quantity of fuel required in combustion for heating.
3. multistage expansion of gases by using two or more stages of expansion. Reheaters are used to heat gases coming out of one of the stage of turbine.

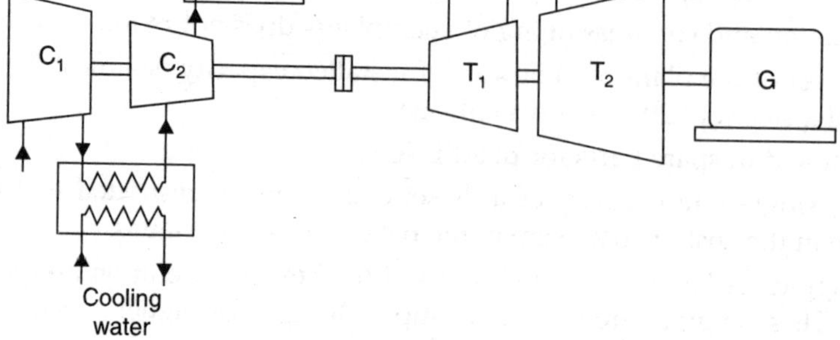


*FIGURE 16.10* Gas turbine plant with inter coolers, air preheaters and multi-stage expansion.

**Combination gas turbine cycles.** The exhaust gases leaving the gas turbine are usually at temperatures of the order of 520°C to 450°C. This heat of gases can be utilised in combination steam power plants for preheating water or generation of low pressure steam. Also boiler for steam power plant can be used for heating gases for turbines. Thus a combination of gasturbine power plant and steam power plant can be used for the reasons of overall economy.

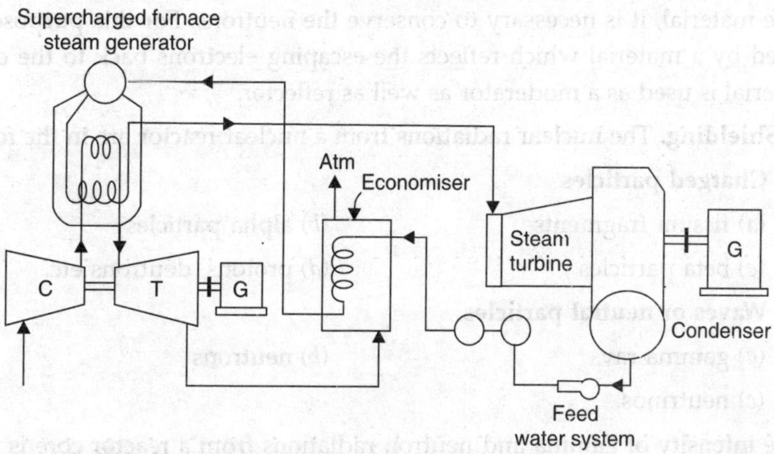


***FIGURE 16.11*** Combination gasturbine-steam power plant.

**Nuclear Power Plants.** These plants use heavy elements like Uranium ($U^{235}$) or Thorium $Th^{232}$ which are subjected to the process of fission (breaking up of nuclei of heavy atoms into nearly equal parts with release of huge amount of energy) in a reactor. The heat energy released in the reactor is used in raising steam which is used as a working medium for steam turbines. Following are the major parts of a nuclear reactor:

**1. Fuels.** Uranium ($U^{235}$) is a naturally occurring fuel (less than 1% in the uranium ore). $U^{233}$ and $Pu^{239}$ are produced in the nuclear reactor as a result of fission process. Fuel is placed in a nuclear reactor in specific shapes taking into account the heat transfer, corrosion, structural strength, etc. As the heat produced from fuel is transferred to coolant which surrounds the fuel elements, a protective coating or cladding must separate the fuel from the coolant stream. Fuel element cladding must satisfy the following conditions:

1. It should have high thermal conductivity.
2. It must not be affected by the radiation to which it is subjected.
3. It should not have a tendency to absorb neutrons.
4. It should have enough strength to withstand a high temperature within the reactor.
5. It should be able to resist corrosion effectively.

**2. Moderator.** The neutrons emitted as a result of fission emerge at a very high speed and have high kinetic energy. These neutrons are called fast neutrons. The function of a moderator is to slow down these neutrons. However, a moderator should not absorb these neutrons.

A moderator should have low parasitic capture, non-corrosiveness, high melting point (if used as solid), high thermal conductivity and chemical and radiation stability.

$H_2O$, $D_2O$, He (gas), Be and C (graphite) are used as moderators.

**3. Coolant.** The function of coolant is to transfer heat produced inside the reactor to a heat exchanger for utilisation in the generation of power. A coolant should have low parasitic capture, non-corrosiveness, low melting point, high boiling point, high specific heat, high specific gravity, low viscosity and non-toxicity.

$$H_2O, D_2O, Na, Bi, He, CO_2, Hg$$

are some of the coolants used in nuclear reactors.

**4. Reflector.** In order to keep the size of the reactor small and hence the amount of the fissionable material, it is necessary to conserve the neutrons. For this purpose the reactor core is surrounded by a material which reflects the escaping electrons back to the core. Sometimes the same material is used as a moderator as well as reflector.

**5. Shielding.** The nuclear radiations from a nuclear reactor are in the form of:

1. **Charged particles**
   (*a*) fission fragments
   (*b*) alpha particles
   (*c*) beta particles
   (*d*) protons, deutrons etc.
2. **Waves or neutral particles**
   (*a*) gamma rays
   (*b*) neutrons
   (*c*) neutrinos.

The intensity of gamma and neutron radiations from a reactor core is far greater than the human body can tolerate. For this purpose shielding is provided on nuclear reactors to reduce the radiation intensity to acceptable limits.

**Classification of Nuclear Reactors.** Nuclear reactors may be classified on the basis of neutron energies as:

(*i*) **Fast reactors.** In these reactors fission occurs with high energy neutrons, in the absence of a moderator.

(*ii*) **Intermediate reactors.** With the use of some moderator the reactors are known as intermediate reactors.

(*iii*) **Thermal reactors.** When the energy of neutrons is reduced to low electron voltages *i.e.* thermal range, the reactors are known as thermal reactors.

On the basis of fuel reproduction characteristics, the nuclear reactors may be classified as

1. **Non-regenerative type.** These reactors do not create an appreciable amount of replacement fuel as the fuel is burned. Reactors using highly enriched fuel containing 90% or more of $U^{235}$ in the fuel are of non-regenerative type.
2. **Regenerative type.** In these reactors the fuel is slightly enriched. These reactors do not replace the used fuel fully.
3. **Breeder type.** In this type of reactor, the fissionable type of fuel produced is more than the fuel consumed.

On the basis of fuel-moderator arrangement used the reactors may be classified as:

1. **Homogeneous reactors.** In these reactors the fuel is finely divided and homogeneously mixed with moderator.
2. **Heterogeneous reactors.** In these reactors fuel is in the form of rods, plates or any other shape and is placed in the matrix of moderator.

On the basis of their functions. The reactors may be classified as

1. **Research reactors**

   (*a*) Isotope production (*b*) Material testing

   (*c*) As neutron source.

2. **Plutonium production**

3. **Power reactors**

   (*a*) *For stationary power plants*

   (*i*) Central station (*ii*) Package reactors.

   (*b*) *Mobile reactors*

   (*i*) Naval reactors (*ii*) Aircraft reactors

   (*iii*) Merchant ship reactors.

Nuclear reactors can also be classified on the basis of moderator, coolant and fuel used.

## OBJECTIVE TYPE QUESTIONS

1. Which of the following is unconventional source of electrical power?
   (*a*) Coal (*b*) Diesel oil (*c*) Geothermal energy (*d*) Nuclear power.
2. Which of the following fuel is generally used in liquid form?
   (*a*) Peat (*b*) Charcoal (*c*) Coke (*d*) Pulverised coal.
3. Which of the following fuel is produced as a result of destructive distillation of some soft coals in absence of air?
   (*a*) Pulverized coal (*b*) Coke (*c*) Peat (*d*) Charcoal.
4. Which of the following variety of coal is usually burned in the form of small particles?
   (*a*) Anthracite coal (*b*) Coke (*c*) Pulverised coal (*d*) Cooking coal.
5. Specific gravity of petroleum fuels is generally
   (*a*) 2.2 to 2.7 (*b*) 1.9 to 2.2 (*c*) 1 to 1.8 (*d*) 0.77 to 0.98.
6. As the temperature of a petroleum fuel increases, its
   (*a*) specific gravity decreases (*b*) viscosity increases
   (*c*) viscosity decreases (*d*) both viscosity and specific gravity increase.
7. Which of the following fuel is generally used in gaseous form?
   (*a*) Petrol (*b*) Neptha (*c*) Light diesel oil (*d*) Methane.
8. Producer gas is produced from
   (*a*) calcium carbide (*b*) blast furnace exhaust
   (*c*) incomplete combustion of coke charcoal in air
   (*d*) mixture of methane and water gas.

9. Two major constituents of producer gas are
   (*a*) nitrogen and oxygen
   (*b*) nitrogen and carbon monoxide
   (*c*) nitrogen and carbon dioxide
   (*d*) nitrogen and hydrogen.
10. Water gas is a mixture of
   (*a*) carbon dioxide and oxygen
   (*b*) oxygen and hydrogen
   (*c*) hydrogen, nitrogen and oxygen
   (*d*) carbon monoxide, carbon dioxide and hydrogen.
11. Which cycle would you consider for a thermal power plant?
   (*a*) Otto cycle (*b*) Carnot cycle (*c*) Dual cycle (*d*) Rankine cycle.
12. Which part of a thermal power plant will be at the lowest pressure of the cycle?
   (*a*) Boiler (*b*) Super heater (*c*) Turbine (*d*) Condenser.
13. Which part of a boiler will have highest pressure?
   (*a*) Economiser (*b*) Super heater (*c*) Steam drum (*d*) Feed water.
14. In which part of the cycle, steam is converted into water?
   (*a*) Boiler (*b*) Super heater (*c*) Economiser (*d*) Condenser.
15. Water to be converted into steam is supplied to the boiler at
   (*a*) atmospheric pressure
   (*b*) less than atmospheric pressure
   (*c*) slightly above atmospheric pressure
   (*d*) more than steam pressure within the boiler.
16. The pressure of steam in condenser is
   (*a*) atmospheric pressure
   (*b*) more than atmospheric pressure
   (*c*) slightly less than atmospheric pressure
   (*d*) much less than atmospheric pressure.
17. An ideal fluid for Rankine cycle should have
   (*a*) low specific gravity
   (*b*) low liquid specific heat
   (*c*) high liquid specific heat
   (*d*) high boiling point.
18. For steam, as pressure increases
   (*a*) critical temperature decreases
   (*b*) critical volume increases
   (*c*) latent heat of vapourisation increases
   (*d*) latent heat of vapourisation decreases.
19. The critical temperature of water is
   (*a*) 0°C (*b*) 100°C (*c*) 212°C (*d*) 374.6°C.
20. Latent heat of vapourisation for steam at critical pressure is
   (*a*) 1000 kJ/kg
   (*b*) 100 kJ/kg
   (*c*) 1 kJ/kg
   (*d*) zero.

21. Which part of steam power plant must be airtight?
(a) Coal handling plant
(b) Water treatment plant
(c) Condenser
(d) Economiser.

22. In reheat cycle for thermal power plants
(a) water is alternately heated and cooled in boiler
(b) steam is heated twice before being supplied to turbine
(c) second heating of steam occurs within the turbine due to blade friction, etc.
(d) steam after expansion in turbine is reheated in boiler for further expansion.

23. As steam expands in turbine
(a) its pressure increases
(b) its specific volume increases
(c) its boiling point increases
(d) its temperature increases.

24. Which of the following is not necessarily the advantage of a reheat cycle in thermal power plants?
(a) Higher thermal efficiency
(b) Smaller condenser
(c) Low steam pressure
(d) Small boiler.

25. Bleeding of steam is done in
(a) regenerative cycle
(b) binary vapour cycle
(c) non-condensing power plants
(d) all of the above.

26. In regenerative cycle, bled steam is
(a) discharged to atmosphere
(b) condensed in steam condenser
(c) used to heat feed water for boiler
(d) is mixed with steam supplied to turbine.

27. Which of the following method can be used to increase the thermal efficiency of a steam power cycle?
(a) Higher steam pressure
(b) Reheating of steam
(c) Regeneration
(d) All of the above.

28. Critical pressure for steam is
(a) 100 kg/cm$^2$
(b) between 100 kg/cm$^2$ and 150 kg/cm$^2$
(c) between 150 kg/cm$^2$ and 200 kg/cm$^2$
(d) between 200 kg/cm$^2$ and 250 kg/cm$^2$.

29. In Binary vapour cycle
(a) steam at two different pressures is used
(b) steam is expanded in two turbines
(c) two fluids are used
(d) vapour and liquid are used as working substances.

30. Mercury is used in binary vapour cycles because it has
(a) high boiling point
(b) high specific gravity
(c) high liquid specific heat
(d) all of the above.

**31.** Specific gravity of mercury is

(*a*) 1 (*b*) 1.36 (*c*) 7.36 (*d*) 13.6.

**32.** The boiling point of mercury is

(*a*) 100°C (*b*) 254.6°C (*c*) 355°C (*d*) 1355°C.

**33.** In a binary vapour cycle using mercury and water

(*a*) mercury is used for high pressure and steam for low pressure

(*b*) mercury is used for low pressure and steam for high pressure

(*c*) both mercury and steam are used for same pressure range

(*d*) both mercury and steam are intimately mixed together before expansion.

**34.** In binary vapour cycle, heat rejected by mercury vapour is

(*a*) rejected to atmosphere (*b*) used to heat air

(*c*) used to heat water (*d*) wasted in mercury condenser.

**35.** In binary vapour cycle, condenser is not provided for

(*a*) mercury vapour cycle (*b*) steam cycle

(*c*) both (*a*) and (*b*) above (*d*) none of the above.

**36.** Which of the following is the objection to the use of mercury as a working fluid in power plants?

(*a*) high initial cost (*b*) toxic vapours

(*c*) 'none-wetting' of heating surfaces (*d*) all of the above.

**37.** Which of the following state in India is rich in coal reserves?

(*a*) Rajasthan (*b*) U.P. (*c*) Bihar (*d*) Tamil Nadu.

**38.** The largest size of generator installed in India has capacity of

(*a*) 1000 MW (*b*) 780 MW

(*c*) 500 MW (*d*) 220 MW.

**39.** Which of the following is not an important part of steam power plant?

(*a*) Ball inill (*b*) Exhaust silencer

(*c*) Cooling pond (*d*) Water treatment plant.

**40.** Which of the following property of coal has major influence on the design and operation of a steam power plant?

(*a*) Density of coal (*b*) Carbon content of coal

(*c*) Ash content of coal (*d*) Shape and size of coal.

**41.** Ash content of Indian coals is around

(*a*) 0.1% to 1% (*b*) 2% to 8%

(*c*) 10% to 25% (*d*) 25% to 40%.

**42.** Heating value of coal is around

(*a*) 1000 kcal/kg (*b*) 2000 kcal/kg (*c*) 6000 kcal/kg (*d*) 10000 kcal/kg.

43. Sulphur content in coal all over the world may vary from
(a) 0.01% to 0.1% (b) 0.1% to 1% (c) traces to 8% (d) 10% to 25%.

44. In which of the following plant, hand firing of coal is invariably used?
(a) Super thermal power plants (b) High pressure power plants
(c) Steam locomotives (d) None of the above.

45. Stoker firing is used for
(a) untreated coal (b) pulverized coal
(c) lignite (d) peat.

46. A variety of coal having ash content 25%, moisture 30%, calorific value 3500 kcal/kg, will be termed as
(a) Grade 1 fuel (b) Superior quality fuel
(c) Medium quality fuel (d) Low grade fuel.

47. A cyclone burner is used for
(a) lumped coal (b) pulverised coal
(c) gaseous fuels (d) all of the above.

48. When pH value of water is 7 it is supposed to be
(a) acidic (b) alkaline
(c) neutral (d) free from dissolved hydrogen.

49. The range of pH scale is
(a) 0–14 (b) 1–10 (c) 7–14 (d) 1–100.

50. Which of the following pH value of water represents most alkaline water?
(a) 1 (b) 7 (c) 10 (d) 13.

51. Presence of oxygen in boiler water is likely to result in
(a) corrosion (b) scale formation (c) embrittlement (d) foaming.

52. The presence of which of the following impurity in water is likely to result in foaming?
(a) Dissolved carbon dioxide (b) Oil
(c) Calcium carbonate (d) Magnesium carbonate.

53. Turbidity in boiler feed water is likely to result in
(a) sludge formation (b) corrosion (c) scale formation (d) foaming.

54. Which of the treatment can be used to remove oil from boiler feed water?
(a) Coagulation (b) Filteration (c) Surface blow off (d) Any of the above.

55. Which of the following treatment can be used to remove turbidity from boiler feedwater?
(a) Coagulation (b) Filteration (c) Sedimentation (d) Any of the above.

56. Carbon dioxide can be removed from boiler feed water by
(a) deaeration (b) deionization (c) acid neutralizer (d) sedimentation.

57. Particles less than 0.1 mm in size, in exhaust gases, are termed as
(*a*) fly ash (*b*) cinders (*c*) dust (*d*) soot.

58. In order to minimise atmospheric pollution, which of the following plant is installed in thermal power plants?
(*a*) Water treatment plant (*b*) Cooling towers
(*c*) Electrostatic precipitators (*d*) Soot blowers.

59. In a thermal power plant cooling towers are used to
(*a*) condense low pressure steam
(*b*) cool condensed steam
(*c*) cool water used in condenser for condensing steam
(*d*) cool feed water for boiler.

60. Which of the following is a high pressure boiler?
(*a*) Lancashire boiler (*b*) La-Mount boiler (*c*) Cornish boiler (*d*) Nestler boiler.

61. Steam engine used in locomotive is
(*a*) single acting, condensing type (*b*) single acting, non-condensing type
(*c*) double acting, non-condensing type (*d*) double acting, condensing type.

62. Steam pressure in locomotive boiler is
(*a*) 5 to 6 kg/cm$^2$ (*b*) 10 to 15 kg/cm$^2$ (*c*) 20 to 35 kg/cm$^2$ (*d*) 50 to 100 kg/cm$^2$.

63. In a steam locomotive electricity for lighting is supplied by
(*a*) battery (*b*) steam engine
(*c*) small turbo generator
(*d*) there is no provision for light in locomotive.

64. The overall efficiency of a locomotive power plant is
(*a*) 5 to 15 percent (*b*) 20 to 30 percent
(*c*) 30 to 50 percent (*d*) around 65 percent.

65. Which of the following variety of coal has lowest percentage of carbon?
(*a*) Anthracite coal (*b*) Bituminous coal
(*c*) Lignite
(*d*) All have nearly same percentage of carbon.

66. The calorific value of lignite can be expected to be around
(*a*) 1000 kcal/kg (*b*) 2000 kcal/kg
(*c*) 5000 kcal/kg (*d*) 7500 kcal/kg.

67. The calorific value of diesel oil can be expected to be around
(*a*) 1000 kcal/kg (*b*) 5000 kcal/kg
(*c*) 7500 kcal/kg (*d*) 10,000 kcal/kg.

68. Percentage of carbon in diesel oil can be expected to be around
(*a*) 50% (*b*) 66% (*c*) 86% (*d*) 96%.

69. The calorific value of natural gas can be expected to be around
(*a*) 500 kcal/$m^3$ (*b*) 1500 kcal/$m^3$ (*c*) 3500 kcal/$m^3$ (*d*) 5500 kcal/$m^3$.

70. The major constituent of methane gas is
(*a*) carbon monoxide (*b*) methane (*c*) oxygen (*d*) hydrogen.

71. Site for which of the following plant is decided on the basis of the contours of the land only ?
(*a*) Steam power plant (*b*) Diesel power plant
(*c*) Hydro-electric power plant (*d*) Nuclear power plant.

72. Which of the following plant has the highest cost per MW installed?
(*a*) Steam power plant (*b*) Hydro-electric power plant
(*c*) Diesel power plant (*d*) Nuclear power plant.

73. In an analysis report of a sample of coal, the percentage of one of the constituents is specified as 2.5%. This constituent could be
(*a*) carbon (*b*) ash (*c*) moisture (*d*) sulphur.

74. Rank of coal is based on
(*a*) fixed carbon and heating value (*b*) ash content
(*c*) size (*d*) ash and moisture content.

75. LPG stands for
(*a*) Low pressure gas (*b*) Light phosgene gas
(*c*) Liquified petroleum gas (*d*) Lubricated petroleum gas.

76. The blades of a steam turbine are made of
(*a*) cast iron (*b*) stainless steel (*c*) high speed steel (*d*) alloy steel.

77. Bagasse is
(*a*) a variety of coal (*b*) a fuel consisting of wood, etc.
(*c*) fibrous portion of sugar cane left after extracting the juice
(*d*) a kind of rice straw.

78. Boilers using bagasse as fuel use
(*a*) spreader stokers (*b*) underfeed stokers
(*c*) chain grate stokers (*d*) pulverised fuel burners.

79. Temporary hardness of water is caused by
(*a*) bicarbonates of calcium and magnesium
(*b*) sulphates of calcium and magnesium
(*c*) chlorides of calcium and magnesium
(*d*) nitrates of calcium and magnesium.

**80.** Safety valve generally used on high pressure boilers is

(*a*) dead weight type (*b*) lever type (*c*) spring type (*d*) any of the above.

**81.** The analysis of coal which gives the composition in respect of moisture, volatile matter, ash and fixed carbon is known as

(*a*) ultimate analysis (*b*) proximate analysis

(*c*) orsat analysis (*d*) chemical analysis.

**82.** Analysis of flue gases can be obtained by

(*a*) gas calorimeter (*b*) bomb calorimeter (*c*) orsat apparatus (*d*) none of the above.

**83.** In a coal fired boiler approximately what percentage of heat is lost in flue gases?

(*a*) 15% (*b*) 35% (*c*) 55% (*d*) 65%.

**84.** Low grade fuels have

(*a*) low moisture content (*b*) low ash content

(*c*) low calorific value (*d*) all of the above.

**85.** Which of the following plant will take least time to start from cold conditions?

(*a*) Steam power plant (*b*) Nuclear power plant

(*c*) Hydro-electric power plant (*d*) Diesel engine power plant.

**86.** The overall efficiency of a diesel power plant can be expected around

(*a*) 15% (*b*) 35% (*c*) 55% (*d*) 75%.

**87.** The cost of transmission and distribution will be least for which type of power plant?

(*a*) Diesel engine power plant (*b*) Steam power plant

(*c*) Nuclear power plant (*d*) Hydro-electric power plant.

**88.** Which of the following power plant can be installed closest to load centre?

(*a*) Hydro-electric power plant (*b*) Steam power plant

(*c*) Diesel power plant (*d*) Nuclear power plant.

**89.** Deaeration of boiler feed water is done to reduce

(*a*) weight of water to be handled (*b*) priming difficulties in feed pumps

(*c*) corrosion

(*d*) amount of heat required to convert water into steam.

**90.** In oil burners air is supplied

(*a*) to assist in flow of oil (*b*) to prevent clogging of burner

(*c*) to improve combustion efficiency (*d*) to uniformly spray oil in combustion chamber.

**91.** 'Coking' is

(*a*) formation of lumps or masses of coke in boiler furnaces at high temperatures

(*b*) heating of coal in absence of air, driving out carbon dioxide and leaving behind the residue of carbon

(*c*) burning of coal in furnace

(*d*) producing lumps of coal from fine powders with the help of a binder.

**92.** 'Caking' is

(*a*) making cakes of coal out of fine powders

(*b*) a variety of free burning coals

(*c*) coal produced from burning of wood in inert atmosphere

(*d*) in boiler furnace some coals become plastic and form lumps or masses of coal.

**93.** Which of the following power plant offers the least running cost?

(*a*) Steam power plant (*b*) Diesel power plant

(*c*) Hydro-electric power plant (*d*) Nuclear power plant.

**94.** Which of the following power plant has least expenditure on transportation of fuel?

(*a*) Steam power plant (*b*) Diesel power plant

(*c*) Hydro-electric power plant (*d*) Nuclear power plant.

**95.** Which of the following power plant has least overall efficiency?

(*a*) Steam power plant (*b*) Diesel power plant

(*c*) Nuclear power plant (*d*) Hydro-electric power plant.

**96.** The overall efficiency of a steam power plant can be expected to be around

(*a*) 10% (*b*) 25% (*c*) 50% (*d*) 65%.

**97.** The boilers using lignite as fuel do not use

(*a*) underfeed stokers (*b*) travelling grate stokers

(*c*) spreader stokers (*d*) U flame pulverised burner.

**98.** The calorific value of methane gas can be determined by

(*a*) Bomb calorimeter (*b*) Orsat's apparatus

(*c*) Bunsens calorimeter (*d*) Exhaust gas calorimeter.

**99.** In the Bomb calorimeter the fuel to be tested is burned

(*a*) at constant volume (*b*) at constant pressure

(*c*) at constant temperature (*d*) in open atmosphere.

**100.** In closed cooling system

(*a*) hot water is recirculated after cooling

(*b*) water does not flow

(*c*) air and water cooling is simultaneously used

(*d*) constant supply of fresh water for cooling is required.

## Hydro-Electric Power Plants

**101.** The advantage of a hydro-electric power station is

(*a*) low operating cost (*b*) free from pollution problems

(*c*) no fuel transportation problems (*d*) all of the above.

**102.** Which of the following location is not associated with hydro-electric stations

(*a*) Koyna (*b*) Bhakhra (*c*) Tarapur (*d*) Nagarjun sagar.

**103.** Low head hydro-electric plants invariably make use of

(*a*) Francis turbines (*b*) Kaplan turbines (*c*) Pelton wheels (*d*) Any of the above.

**104.** Pelton wheels are installed on

(*a*) run-of river plants with pondage (*b*) run-of-river plants without pondage

(*c*) base load plant (*d*) high head plants.

**105.** A hydrograph is a plot of discharge through a river versus

(*a*) time (*b*) velocity of water (*c*) rate of flow of water(*d*) none of the above.

**106.** In pumped storage systems, water is pumped

(*a*) upstream during peak periods (*b*) upstream during off peak periods

(*c*) upstream during off season (*d*) none of the above.

**107.** For which of the following data would you select a Kaplan turbine?

| | *Capacity (MW)* | *Head (m)* | *Speed (PM)* |
|---|---|---|---|
| (*a*) | 20 | 20 | 72 |
| (*b*) | 20 | 150 | 360 |
| (*c*) | 40 | 180 | 360 |
| (*d*) | 100 | 550 | 360 |

**108.** A Francis turbine is

(*a*) inward flow reaction turbine (*b*) inward flow impulse turbine

(*c*) outward flow reaction turbine (*d*) outward flow impulse turbine.

**109.** A Kaplan turbine is suitable for

(*a*) high head, low discharge (*b*) high head, high discharge

(*c*) low head, high discharge (*d*) low head, low discharge.

**110.** Cavitation is caused be

(*a*) high fluid velocities (*b*) low pressures

(*c*) high pressures (*d*) low fluid velocities.

**111.** Which of the following turbine is different from the remaining?

(*a*) Propeller turbine (*b*) Francis turbine (*c*) Pelton wheel (*d*) Kaplan turbine.

**112.** In axial flow turbines

(*a*) only part of the available head is converted into velocity before the water enters the wheel

(*b*) water is admitted over part of the circumference

(*c*) it is possible that the wheel may run full

(*d*) it is possible to regulate the flow.

**113.** Which turbine is different from the remaining?

(*a*) Pelton wheel (*b*) Banki turbine (*c*) Girand turbine (*d*) Kaplan turbine.

**114.** If the specific speed of a turbine is 6, then which type of turbine should it be?

(*a*) Francis turbine (*b*) Kaplan turbine (*c*) Pelton wheel (*d*) Thomson turbine.

**115.** If the specific speed of a turbine is 800, the turbine should be

(*a*) Francis turbine (*b*) Pelton wheel (*c*) Kaplan turbine (*d*) Girand turbine.

**116.** An impulse turbine

(*a*) always operates submerged (*b*) makes use of a draft tube

(*c*) operates by initial conversion of head into kinetic energy

(*d*) all of the above.

**117.** A Pelton wheel is

(*a*) inward flow impulse turbine (*b*) outward flow impulse turbine

(*c*) inward flow reaction turbine (*d*) axial flow impulse turbine.

**118.** The running away speed of a Pelton wheel is

(*a*) actual operating speed (*b*) on load speed

(*c*) full load speed (*d*) on load speed when governor mechanism fails.

**119.** In a reaction turbine

(*a*) flow can be regulated without loss

(*b*) the work done is purely by the change in kinetic energy

(*c*) only part of the head is converted into velocity head before the water enters the wheel

(*d*) none of the above.

**120.** A Fourneyron turbine is

(*a*) outward flow reaction turbine (*b*) inward flow reaction turbine

(*c*) outward flow impulse turbine (*d*) inward flow impulse turbine.

**121.** A surge tank is provided in turbine installation

(*a*) to smoothen the flow of water (*b*) to minimise friction losses

(*c*) to prevent occurrence of hydraulic jump

(*d*) to relieve pressure due to water hammer.

**122.** Hammer blow in pipes occurs when

(*a*) pipe is hit with a sledge hammer (*b*) pipe bursts under high fluid pressure

(*c*) excessive leakage occurs in pipe

(*d*) flow of fluid through pipe is suddenly brought to rest by the closing of valve

**123.** Hydraulic jump occurs when

(*a*) flow is super critical and adequate down stream depth is available

(*b*) flow is subcritical and adequate down stream depth is available

(*c*) flow is turbulent (*d*) flow is laminar.

124. The lowest portion of storage basin from which from where the water is not drawn usually, is

(*a*) bottom storage (*b*) dead storage (*c*) spinning reserve (*d*) sub-soil storage.

125. The maximum continuous power available from a hydro-electric plant under the most adverse hydraulic conditions, is known as

(*a*) base power (*b*) firm power (*c*) primary power (*d*) secondary power.

126. Koyna dam is located in

(*a*) Gujarat (*b*) Karnataka (*c*) Maharashtra (*d*) Madhya Pradesh.

127. The dam constructed on river Chambal is

(*a*) Jawahar sagar dam (*b*) Rana Pratap sagar dam

(*c*) Gandhi sagar dam (*d*) All of the above.

128. Bhakhra dam has been constructed on river

(*a*) Sutlej (*b*) Yamuna (*c*) Ravi (*d*) Chenab.

129. In a Buttress dam, the inclination of the deck with horizontal is kept between

(*a*) 10° to 15° (*b*) 15° to 20° (*c*) 25° to 30° (*d*) 35° to 45°.

130. The arch dam is not suitable for

(*a*) narrow sites (*b*) wide sites (*c*) small discharge (*d*) low heads.

131. The magnitude of earthquake pressure on a dam varies as a function of

(*a*) Frequency of vibrations (*b*) Amplitude

(*c*) Both (*a*) and (*b*) above (*d*) Richards scale number.

132. In order to keep out debris and other materials carried by water which are likely to damage turbine, provision is usually made for

(*a*) trash racks (*b*) radial gates (*c*) trunnious (*d*) ensign valves.

133. The portion of the power canal that extends from the intake works to the power plant, is known as

(*a*) head race (*b*) main storage (*c*) tail race (*d*) diversion canal.

134. A portion of the power canal which extends from the power house to the receipent water course, is known as

(*a*) tailrace (*b*) exhaust (*c*) discharge channel (*d*) tail surge channel.

135. In a gravity dam, for safety against overturning, the resultant of all the forces acting on a dam must pass

(*a*) outside the base (*b*) through first third of the base

(*c*) through middle third of the base (*d*) outside the middle third of the base.

136. In case the resultant of all the forces acting on a dam passes outside the base, the dam would overturn unless

(*a*) it can resist tensile stress (*b*) anchor blocks are provided

(*c*) friction pads are provided (*d*) weight of the dam is increased.

**137.** Which of the following is a reaction turbine?

(*a*) Banki turbine (*b*) Jonval turbine

(*c*) Girard turbine (*d*) None of the above.

**Questions 138 and 139 refer to Fig. 16.12:**

The performance curves for four turbines are shown in Fig. 16.12. These are Kaplan turbine, Pelton wheel, Francis turbine and Propeller turbine.

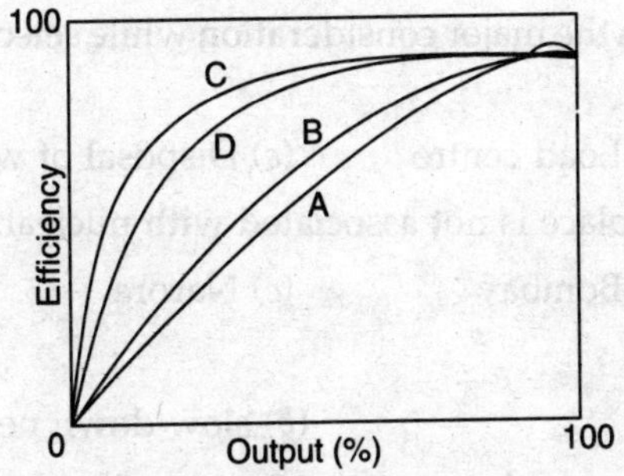


***FIGURE 16.12***

**138.** Which curve is likely to represent the performance of Kaplan turbine?

(*a*) curve *A* (*b*) curve *B* (*c*) curve *C* (*d*) curve *D*.

**139.** Which curve is likely to represent the performance of Propeller turbine?

(*a*) curve *A* (*b*) curve *B* (*c*) curve *C* (*d*) curve *D*.

**140.** Reversible turbines are used in

(*a*) low head plants (*b*) run-off-river plants

(*c*) pumped storage plants (*d*) none of the above.

**141.** The state in which Hirakund dam is established is

(*a*) Bihar (*b*) West Bengal (*c*) Orissa (*d*) Madhya Pradesh.

**142.** Nagarjun Sagar dam has been constructed on river

(*a*) Periyar (*b*) Cauvery (*c*) Krishna (*d*) Mahanadi.

**143.** A draft tube is used for discharging water from a

(*a*) Pelton wheel (*b*) Francis turbine

(*c*) All impulse turbines (*d*) None of the above.

**144.** The graph shown in Fig. 16.13 is a

(*a*) hydrograph

(*b*) unit hydrograph

(*c*) flow duration curve

(*d*) load curve.

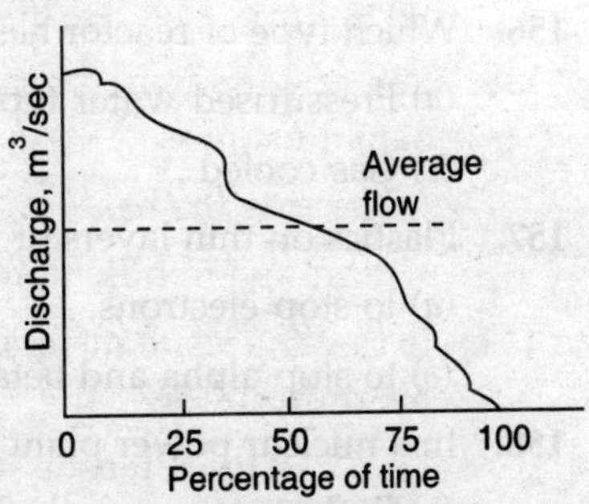


***FIGURE 16.13***

**145.** A pumped storage plant in India exists in
(*a*) Rajasthan (*b*) Tamil Nadu (*c*) Assam (*d*) Jammu and Kashmir.

**146.** Jet dispersers are used in
(*a*) high head plants (*b*) low head plants
(*c*) reaction turbine installations (*d*) all of the above.

## Nuclear Power Plants

**147.** Which of the following is the major consideration while selecting the site for a nuclear power plant?
(*a*) Safety (*b*) Load centre (*c*) Disposal of waste (*d*) Availability of water.

**148.** Which of the following place is not associated with nuclear power plants?
(*a*) Kota (*b*) Bombay (*c*) Narora (*d*) Kodambakkam.

**149.** A moderator is used to
(*a*) slow down electrons (*b*) slow down neutrons
(*c*) accelerate neutrons (*d*) direct flow of electrons.

**150.** A moderator should have
(*a*) low parasitic capture (*b*) high thermal conductivity
(*c*) radiation stability (*d*) all of the above.

**151.** Which of the following material can be used as a moderator in a nuclear power plant?
(*a*) Graphite (*b*) Helium (*c*) Heavy water (*d*) Any of the above.

**152.** Chemical representation for heavy water is
(*a*) $H_2O$ (*b*) $H_2O_2$ (*c*) $H_3O$ (*d*) $D_2O$.

**153.** The reactor in which fissionable fuel produced is more than the fuel consumed, is known as
(*a*) Fast reactor (*b*) Breeder reactor (*c*) Thermal reactor (*d*) None of the above.

**154.** Which of the following material is not used as moderator?
(*a*) Graphite (*b*) Beryllium (*c*) Aluminium (*d*) Light water.

**155.** Isotopes of uranium
(*a*) $U_{235}$ (*b*) $U_{234}$ (*c*) $U_{238}$ (*d*) all of the above.

**156.** Which type of reactor has been installed in Tarapur Atomic Power plant?
(*a*) Pressurised water type (*b*) Boiling water type
(*c*) Gas cooled (*d*) None of the above.

**157.** Plastics on thin layers of metals are used in nuclear plants
(*a*) to stop electrons (*b*) to slow down neutrons
(*c*) to stop alpha and beta particles (*d*) to arrest stray neutrons.

**158.** In a nuclear power plant which of the following is used to slow down the fast neutrons?
(*a*) Reflector (*b*) Shield (*c*) Control rod (*d*) Moderator.

**159.** Which of the following is produced by neutron irradiation of $U_{235}$?

(*a*) Plutonium (*b*) Uranium 233 (*c*) Uranium 238 (*d*) Uranium 239.

**160.** Which of the following is a primary fuel?

(*a*) $U_{233}$ (*b*) $U_{235}$ (*c*) $Pu_{239}$ (*d*) none of the above.

**161.** Control rods for nuclear reactors are made of

(*a*) graphite (*b*) concrete (*c*) cadmium (*d*) lead.

**162.** A nuclear reactor is said to be critical when

(*a*) neutrons attain maximum velocity

(*b*) power produced is maximum

(*c*) production of neutron is exactly balanced by loss of neutrons through leakage

(*d*) the reactor attains maximum energy and can explode at any moment.

**163.** Which of the following material is used for reactor shielding?

(*a*) Graphite (*b*) Carbon (*c*) Copper (*d*) Concrete.

**164.** Which moderator is used in fast breeder reactors?

(*a*) Water (*b*) Carbon dioxide (*c*) Graphite (*d*) No moderator is used.

**165.** Coolant used in fast breeder reactors is

(*a*) water (*b*) liquid metal (*c*) carbon dioxide (*d*) liquid oxygen.

**166.** Alpha particles

(*a*) carry positive charge (*b*) carry negative charge

(*c*) are neutral (*d*) none of the above.

**167.** Gamma rays are

(*a*) electro magnetic waves (*b*) positively charged ions

(*c*) negatively charged ions (*d*) none of the above.

**168.** ${}_{2}^{4}He$ represents helium which contains

(*a*) 2 protons and 4 neutrons (*b*) 4 neutrons and 2 protons

(*c*) 2 electrons and 2 protons (*d*) 2 protons and 2 neutrons.

**169.** 1 atomic mass unit is equal to

(*a*) $1.66 \times 10^{-31}$ kg (*b*) $1.66 \times 10^{-27}$ kg (*c*) $1.66 \times 10^{-25}$ kg (*d*) $1.66 \times 10^{-23}$ kg.

**170.** Which of the following is not a naturally occurring isotope of Uranium?

(*a*) $U_{238}$ (*b*) $U_{235}$ (*c*) $U_{234}$ (*d*) $U_{233}$.

**171.** A CANDU reactor is

(*a*) heavy water cooled and moderated reactor

(*b*) fast breeder reactor

(*c*) pressurised water reactor

(*d*) none of the above.

## Gas Turbine Power Plants

172. Brayton cycle cannot be used in reciprocating engines even for same compression ratio and work output because
(*a*) Otto and diesel cycles are more efficient
(*b*) Brayton cycle is less efficient
(*c*) Brayton cycle needs large air-fuel ratio
(*d*) Large volumes at low pressures cannot be efficiently handled in reciprocating engines.

173. A gas turbine power plant suits as
(*a*) peak load plant (*b*) base load plant (*c*) casual run plant (*d*) any of the above.

174. In aircraft using gas turbine the cycle used is
(*a*) simple (*b*) regeneration
(*c*) reheating (*d*) reheating with regeneration.

175. A jet aircraft has
(*a*) no propeller (*b*) one propeller
(*c*) two propellers (*d*) one propeller for air and one for exhaust.

176. Compression ratio for gas turbines is in the range
(*a*) 5 to 10 (*b*) 11 to 15 (*c*) 16 to 20 (*d*) 20 to 30.

177. A gas turbine plant operates between 1 ata and 9 ata with minimum and maximum cycle temperatures of 25°C and 1000°C. A two stage compressor is provided with perfect intercooling and work is equally divided between the two stages. The intermediate pressure will be
(*a*) 3 ata (*b*) 4.5 ata (*c*) 5 ata (*d*) 6 ata.

178. The air standard efficiency of a gas turbine cycle is 55%. Which one of the following may be carnot efficiency between the temperature limits of the cycle?
(*a*) 27.5% (*b*) 40% (*c*) 65% (*d*) 100%.

179. The blades of a gas turbine are made of
(*a*) cast iron (*b*) stainless steel (*c*) high speed steel (*d*) alloy steel.

180. Samples of exhaust gases from following engines have been collected. The sample shows presence of 10% oxygen by volume. To which engine this sample refers to?
(*a*) Petrol engine (*b*) Diesel engine (*c*) Gas engine (*d*) Jet propulsion unit.

181. The efficiency of a gas turbine increases as
(*a*) gas inlet temperature increases (*b*) speed increases
(*c*) exhaust temperature increases (*d*) compression ratio decreases.

182. In which plant any carbonaceous material having some heat value can be used as fuel?
(*a*) Gas engine (*b*) Closed cycle gas turbine
(*c*) Open cycle gas turbine with regeneration
(*d*) Open cycle gas turbine with reheating.

**183.** Regeneration in gas turbine practice implies

(*a*) discharging gases to atmosphere at low pressure

(*b*) recycling combustion gases after reheating

(*c*) utilising heat of exhaust gases through heat exchanger

(*d*) none of the above.

**184.** Free piston engines generally find applications in

(*a*) gas turbines (*b*) supercharging of diesel engines

(*c*) mining installations (*d*) none of the above.

**185.** The advantage of gas turbine over reciprocating engines is

(*a*) perfect balancing of rotor (*b*) continuous and uniform power

(*c*) usually small working pressures (*d*) all of the above.

**186.** The type of air compressor generally used in gas turbines is

(*a*) axial flow type (*b*) lobe type (*c*) sliding vane type (*d*) reciprocating type.

**187.** Which of the following has higher air-fuel ratio as compared to the remaining?

(*a*) Gas engine (*b*) Petrol engine (*c*) Diesel engine (*d*) Gas turbine.

**188.** If the air standard efficiency of a thermodynamic cycle is given as

$$\eta = 1 + \frac{k\,(r-1)}{(r^k - 1)}$$

where $r$ = compression ratio and $k = \frac{c_p}{c_v}$, the cycle is

(*a*) Lenoir cycle (*b*) Brayton cycle (*c*) Atkinson cycle (*d*) None of the above.

**189.** Brayton cycle processes are

(*a*) two isentropic and two constant volume

(*b*) two isentropic and two constant pressure

(*c*) constant pressure, constant volume and two adiabatics

(*d*) constant pressure, constant volume and two isothermals.

**190.** Erricson cycle processes are

(*a*) two isothermal and two constant pressure

(*b*) two isothermal and two constant volume

(*c*) two isothermal and two isentropic

(*d*) two adiabatic, a constant volume and a constant pressure.

**191.** The cycle generally used for gas turbines is

(*a*) Otto cycle (*b*) Brayton cycle (*c*) Carnot cycle (*d*) Dual cycle.

**192.** The air fuel ratio for jet engines is of the order of

(*a*) 10 : 1 (*b*) 15 : 1 (*c*) 20 : 1 (*d*) 60 : 1.

193. Intake air filters are provided on air compressors
(a) to remove dust and dirt from suction air
(b) to reduce the temperature of the suction air
(c) to reduce suction pressure
(d) all of the above.

194. Free air is air at
(a) atmospheric conditions at any specific location
(b) 1 kg/cm$^2$ and atmospheric temperature
(c) 1 kg/cm$^2$ and 0°C (d) 1 kg/cm$^2$ and 20°C.

195. Which of the following is a non petroleum fuel?
(a) Benzol (b) Methyl alcohol (c) Bunker oil (d) All of the above.

196. In atmospheric air
(a) weight of oxygen = 0.23 × weight of air
(b) weight of oxygen = 0.32 × weight of air
(c) weight of air = 0.23 × weight of oxygen
(d) weight of air = 2.33 × weight of oxygen.

## Diesel Engine Power Plants

197. A turbocharger is used
(a) in gas turbines for compression of air
(b) for supercharging of reciprocating diesel engines
(c) for starting a jet propulsion unit
(d) in rockets for producing air-fuel mixtures.

198. Oil rings for pistons are slotted to
(a) reduce friction (b) reduce the bulk
(c) provide the escape for the oil that the slot edges cut from the cylinder walls
(d) remove oil from cylinder walls.

199. In isochronous governors, the speed drop is
(a) zero (b) 1% (c) 10% (d) 25%.

200. The minimum power of self starter for a diesel engine must be
(a) 1% of rated horsepower of engine
(b) 10% of rated horsepower of engine
(c) more than friction horsepower
(d) equal to IHP of the engine.

201. Crank case explosion normally occurs in
(a) four stroke CI engines (b) four stroke SI engines
(c) two stroke CI engines (d) two stroke SI engines.

**202.** In compression ignition engines, swirl is
(*a*) circular motion imparted to exhaust gases
(*b*) directional spray of fuel
(*c*) circular motion imparted to suction air
(*d*) radial motion imparted to fuel air mixture.

**203.** Which of the following factors will not assist in smooth running of a compression ignition engine?
(*a*) moderate rate of pressure rise (*b*) short delay period
(*c*) minimum of after burning (*d*) high ignition temp.

**204.** Morse test is conducted on
(*a*) horizontal engines (*b*) vertical engines
(*c*) single cylinder engines (*d*) multi-cylinder engines.

**205.** For the same heat input and the same maximum pressure, the most efficient cycle is
(*a*) Brayton cycle (*b*) Otto cycle (*c*) Diesel cycle (*d*) Dual cycle.

**206.** For the same heat input and within the same temperature limits, the most efficient cycle is
(*a*) Carnot cycle (*b*) Otto cycle (*c*) Diesel cycle (*d*) Brayton cycle.

**207.** Cetane number for high speed diesel available in India is
(*a*) 10 to 15 (*b*) 15 to 25 (*c*) 25 to 35 (*d*) 40 to 55.

**208.** The charge of a two stroke diesel engine consists of
(*a*) air + diesel fuel + lubricating oil (*b*) air + diesel fuel
(*c*) air + lubricating oil (*d*) air only.

**209.** If a engine has a spark plug, it can be concluded that the engine is a
(*a*) petrol engine (*b*) diesel engine (*c*) carnot engine (*d*) joule engine.

**210.** In a two stroke engine there is one power stroke in
(*a*) 90° of crank rotation (*b*) 180° of crank rotation
(*c*) 360° of crank rotation (*d*) 720° of crank rotation.

**211.** If in a petro engine by mistake diesel fuel is supplied
(*a*) the engine will run smoothly
(*b*) the engine will run but at reduced speed
(*c*) the engine will run with reduced efficiency
(*d*) the engine will not run.

**212.** The scavanging efficiency of a four stroke diesel engine is usually in the range
(*a*) 50% to 60% (*b*) 60% to 75% (*c*) 75% to 90% (*d*) 95% to 100%.

**213.** The volumetric efficiency of a well designed diesel engine is usually in the range
(*a*) 75% to 90% (*b*) 60% to 75% (*c*) 50% to 65% (*d*) 35% to 50%.

**214.** Thermal efficiency of a diesel engine may be expected to be in the range

(a) 30% to 40% (b) 40% to 55% (c) 55% to 80% (d) 80% to 95%.

**215.** In a diesel engine, during idling air-fuel may be of the order of

(a) 5 (b) 10 (c) 15 (d) 30.

**216.** Which of the following factor will not assist in getting high power output from a diesel engine ?

(a) high charge density (b) large quantity of excess air

(c) fine atomisation of fuel (d) high compression ratio.

**217.** If a diesel engine shows excessive fuel consumption, the probable cause could be

(a) low compression ratio (b) leaks in fuel lines

(c) injection pressure low (d) any of the above.

**218.** If a diesel engine shows loss of power, the probable cause could be

(a) poor compression (b) exhaust restricted

(c) clogged air cleaner (d) any of the above.

**219.** In diesel engines, fuel ignition accelerators are added to

(a) reduce combustion knock (b) accelerate combustion knock

(d) increase delay period (d) reduce exhaust temperature.

**220.** In diesel engines, the injection pressure is of the order of

(a) 2 kg/cm$^2$ (b) 20 kg/cm$^2$ (c) 100 kg/cm$^2$ (d) 400 kg/cm$^2$.

**221.** If in a diesel engine the noise level is to be reduced, the design features which will not be affected are

(a) Combustion system (b) Exhaust muffling

(c) Intake muffling (d) Fuel injection system.

**222.** Pour point is the temperature at which oil

(a) will ignite (b) will give off vapours enough for ignition

(c) will barely flow (d) will start cracking.

**223.** The compression ratio for diesel engines is generally in the range

(a) 5 to 10 (b) 10 to 14 (c) 14 to 22 (d) 22 to 30.

**224.** In a diesel engine the consumption of diesel per BHP per hour is nearly

(a) 100 grams (b) 200 grams (c) 300 grams (d) 450 grams.

**225.** Piston rings are made by

(a) copper (b) cast iron (c) aluminium (d) carbon steel.

**226.** A mixture containing 60% cetane and 40% Iso-octane will have

(a) Cetane number 60 (b) Octance number 60

(c) Octance number 40 (d) None of the above.

227. For same output and same compression ratio, four stroke spark ignition engines as compared to two stroke engines have
(a) higher fuel consumption (b) lower thermal efficiency
(c) higher thermal efficiency (d) higher exhaust temperatures.

228. Which of the following acts as antiknock for compression ignition engines?
(a) Amyl nitrate (b) Tetra ethyl load (c) Hexadecane (d) Trimethyl pentane.

229. Total sulphur content in diesel oil must not exceed
(a) 10% (b) 1% (c) 0.5% (d) 0.05%.

230. Which of the following acts as ignition accelerator for diesel engine fuels?
(a) Acetone peroxide (b) Hydrogen peroxide
(c) *n*-heptane (d) None of the above.

231. If a diesel engine has excess oil consumption, the probable reason could be
(a) wornout piston rings (b) cylinder bore out of round
(c) lubricating oil level too high (d) any of the above.

232. Exhaust gases from a engine possess
(a) potential energy (b) kinetic energy (c) chemical energy (d) solar energy.

233. In case of a four stroke four cylinder diesel engine the heat carried away by exhaust gases will be around
(a) 5 to 7 percent (b) 7 to 10 percent (c) 10 to 20 percent (d) 20 to 35 percent.

234. Calorific value of diesel oil is in the range
(a) 1000 to 1200 kcal/kg (b) 5000 to 6500 kcal/kg
(c) 10,000 to 12000 kcal/kg (d) 100,000 to 120,000 kcal/kg.

235. Flash point of normal diesel oil is in the range
(a) 32°F to 100°F (b) 100°F to 240°F (c) 250°F to 400°F (d) 400°F to 600°F.

236. Cetane is
(a) $C_{10}H_{20}$ (b) $C_{12}H_{24}$ (c) $C_{14}H_{30}$ (d) $C_{16}H_{34}$.

237. Alpha methyl naphthalene is
(a) $C_{11}H_{10}$ (b) $C_{12}H_{11}$ (c) $C_{14}H_{13}$ (d) $C_{24}H_{21}$.

238. Specific gravity of diesel oil is
(a) 2.27 (b) 1.55 (c) 1.00 (d) 0.86.

239. Theoretical quantity of air required for combustion of 1 kg of diesel oil is nearly
(a) 1 kg (b) 1.45 kg (c) 5 kg (d) 14.5 kg.

240. The self ignition temperature of diesel is around
(a) 40°C (b) 100°C (c) 150°C (d) 350°C.

241. In atmospheric air, weight of nitrogen is
(a) 2.33 × weight of oxygen (b) 1.33 × weight of oxygen
(c) 0.33 × weight of oxygen (d) 0.769 × weight of air.

242. As a result of burning of 1 kg of carbon with minimum quantity of air required the percentage of carbon dioxide in the exhaust gases will be

(a) 10% (b) 14.5% (c) 29% (d) 58%.

243. The ratio of air required for complete combustion of carbon to carbon dioxide and that to carbon monoxide is

(a) 0.5 (b) 1.0 (c) 2.0 (d) 2.33.

244. The percentage of oxygen in air by weight is around

(a) 11% (b) 23% (c) 32% (d) 77%.

245. A 1000 HP diesel can be started by

(a) electric motor (b) battery system (c) cranking by hand (d) compressed air.

246. In gas diesel engines

(a) air and gas are supplied during suction

(b) only air is supplied during suction

(c) only gas is supplied during suction

(d) air and part of gas is supplied during suction.

247. Thermionic converters are based on the principle of

(a) Seebeck effect (b) Microwave theory

(c) Transfer of electrons (d) Faradays law.

248. Fuel cells convert

(a) electrical energy into mechanical energy

(b) chemical energy into electrical energy

(c) mechanical energy into electrical energy

(d) none of the above.

249. When two dissimilar materials are joined to form a loop and the two junctions maintained at different temperatures, an emf will set up in the loop. This phenomenon is known as

(a) Seebeck effect (b) Peltier effect (c) Thermionic effect (d) Voltaic effect.

250. In magnetohydrodynamic plants

(a) rotating conductors are used

(b) gaseous or liquid conductor is used

(c) electrolytic solution is used

(d) kinetic energy of high velocity gases is used for conversion into electrical energy.

251. The major advantage of MHD plant is

(a) large number of moving parts (b) it can easily respond to load variations

(c) capital cost is low (d) no fuel is required.

252. Crankcase dilution is

(a) dilution of fuel in crankcase (b) dilution of lubricating oil in crankcase

(c) addition of water in crankcase (d) dilution of mixture of fuel and air.

**253.** Air consumption of a four stroke diesel engine per rated BHP/hour will be

(*a*) 1 kg (*b*) 5 to 7 kg (*c*) 10 to 14 kg (*d*) 15 to 20 kg.

**254.** The volumetric analysis of a sample of exhaust gases is as follows:

| | |
|---|---|
| Carbon dioxide | 15% |
| Carbon monoxide | 2.2% |
| Oxygen | 1.6% |
| Nitrogen | 81.2% |

If this analysis is converted into analysis by weight the percentage of carbon dioxide will

(*a*) remain unaltered (*b*) reduce

(*c*) increase (*d*) depend in other factors.

**255.** In a four stroke engine if a valve opens 250° before BDC and closes 10°after T-D-C, the valve should be

(*a*) Inlet valve (*b*) Exhaust valve (*c*) Puppet valve (*d*) Spring valve.

**256.** Diesel engines are supercharged

(*a*) to increase air-fuel ratio (*b*) to reduce combustion temperature

(*c*) to improve heat transfer (*d*) to increase specific power output.

**257.** For the same heat input and same compression ratio

(*a*) Otto cycle and diesel cycle are equally efficient

(*b*) Otto cycle is less efficient as compared to diesel cycle

(*c*) Otto cycle is more efficient as compared to diesel cycle

(*d*) Efficiency of a cycle depends on air-fuel ratio only.

**258.** The air standard efficiency of a diesel engine depends on

(*a*) compression ratio (*b*) speed

(*c*) torque (*d*) all of the above.

**259.** In case engines of different horsepower, rpm, and cylinder dimensions are to be compared for their performance, the parameter of comparison could be

(*a*) fuel consumption per hour (*b*) mean effective pressure

(*c*) exhaust temperature (*d*) specific speed.

**260.** Piston rings are plated with chromium, cadmium or phosphate to

(*a*) prevent clogging (*b*) improve heat transfer

(*c*) reduce wear and eliminate scuffing (*d*) reduce clearance between ring and cylinder.

**261.** The function of piston rings in a diesel engine is

(*a*) to prevent lubricating oil from entering the combustion space

(*b*) to prevent leakage of combustion chamber products past piston

(*c*) to transfer heat from piston to cylinder walls

(*d*) all of the above.

**262.** The variation of mechanical efficiency with load for a diesel engine, is represented in Fig. 16.14 by

(*a*) curve *A*

(*b*) curve *B*

(*c*) curve *C*

(*d*) curve *D*.

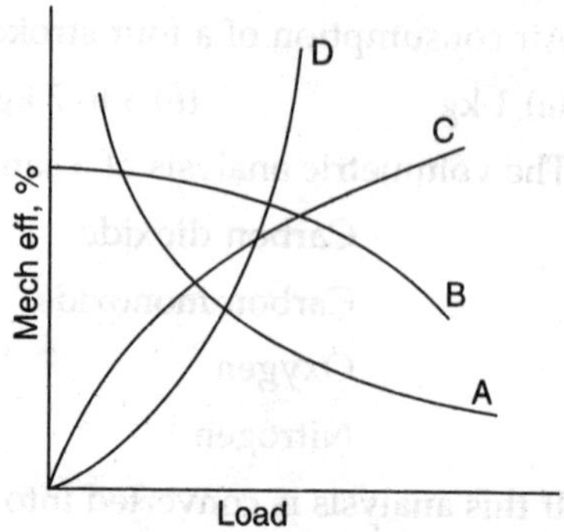


*FIGURE 16.14*

**263.** For air filtration which of the following medium is not used?
(*a*) Oil (*b*) Cloth (*c*) Felt (*d*) Paper.

**264.** A two stroke engine can be identified by
(*a*) piston size (*b*) absence of valves (*c*) cooling system (*d*) lubrication system.

**265.** In a piston the maximum temperature occurs at
(*a*) ring section (*b*) gudgeon pin (*c*) top centre (*d*) bottom.

**266.** Tidal power plant is being installed in India in
(*a*) Kerala (*b*) Karnataka (*c*) Gujarat (*d*) West Bengal.

**267.** Which of the following is the major constituent of solar cells?
(*a*) Sulphur (*b*) Ferrite (*c*) Silicon (*d*) Carbon.

**268.** Which of the following state has the highest per capita energy consumption?
(*a*) Maharashtra (*b*) Bihar (*c*) Uttar Pradesh (*d*) J and K.

**269.** The source of power for satellite is
(*a*) thermionic convertor (*b*) microwave energy reflector
(*c*) solar cells (*d*) none of the above.

**270.** Which public sector unit is engaged in the manufacture of thermal power plants in India?
(*a*) Heavy Engineering Corporation Ltd. (*b*) National Thermal Power Corporation Ltd.
(*c*) Department of Science and Technology
(*d*) Bharat Heavy Electricals Ltd.

**271.** Nuclear power plants in India are manufactured under the supervision of
(*a*) HAL (*b*) BARC (*c*) NTPC (*d*) BHEL.

**272.** Generators coupled to steam turbines usually run at
(*a*) 15000 rpm (*b*) 5000 rpm (*c*) 3000 rpm (*d*) 1500 rpm.

**273.** A generator coupled to which prime mover will usually have low rmp?
(*a*) Steam turbine (*b*) Gas turbine (*c*) Hydraulic turbine (*d*) Diesel engine.

**274.** The function of exciter for a generator is
(*a*) to energise the magnetic pole (*b*) to switch on the flow of current
(*c*) to amplify the voltage induced in the coils
(*d*) none of the above.

**275.** An exciter is basically a

(*a*) dc series motor (*b*) induction motor (*c*) dc shunt generator (*d*) none of the above.

## ANSWERS

| | | | | | | |
|---|---|---|---|---|---|---|
| **1.** (*c*) | **2.** (*a*) | **3.** (*b*) | **4.** (*c*) | **5.** (*d*) | **6.** (*c*) | **7.** (*d*) |
| **8.** (*c*) | **9.** (*b*) | **10.** (*d*) | **11.** (*d*) | **12.** (*d*) | **13.** (*b*) | **14.** (*d*) |
| **15.** (*d*) | **16.** (*d*) | **17.** (*c*) | **18.** (*d*) | **19.** (*d*) | **20.** (*d*) | **21.** (*c*) |
| **22.** (*d*) | **23.** (*b*) | **24.** (*c*) | **25.** (*a*) | **26.** (*c*) | **27.** (*d*) | **28.** (*d*) |
| **29.** (*c*) | **30.** (*d*) | **31.** (*d*) | **32.** (*c*) | **33.** (*a*) | **34.** (*c*) | **35.** (*a*) |
| **36.** (*d*) | **37.** (*c*) | **38.** (*c*) | **39.** (*b*) | **40.** (*c*) | **41.** (*c*) | **42.** (*c*) |
| **43.** (*c*) | **44.** (*c*) | **45.** (*a*) | **46.** (*d*) | **47.** (*b*) | **48.** (*c*) | **49.** (*a*) |
| **50.** (*d*) | **51.** (*a*) | **52.** (*b*) | **53.** (*a*) | **54.** (*d*) | **55.** (*d*) | **56.** (*a*) |
| **57.** (*a*) | **58.** (*c*) | **59.** (*c*) | **60.** (*b*) | **61.** (*c*) | **62.** (*b*) | **63.** (*c*) |
| **64.** (*a*) | **65.** (*c*) | **66.** (*c*) | **67.** (*d*) | **68.** (*c*) | **69.** (*a*) | **70.** (*b*) |
| **71.** (*c*) | **72.** (*d*) | **73.** (*d*) | **74.** (*a*) | **75.** (*c*) | **76.** (*d*) | **77.** (*c*) |
| **78.** (*c*) | **79.** (*a*) | **80.** (*c*) | **81.** (*b*) | **82.** (*c*) | **83.** (*a*) | **84.** (*c*) |
| **85.** (*d*) | **86.** (*b*) | **87.** (*a*) | **88.** (*c*) | **89.** (*c*) | **90.** (*d*) | **91.** (*b*) |
| **92.** (*d*) | **93.** (*c*) | **94.** (*c*) | **95.** (*a*) | **96.** (*b*) | **97.** (*a*) | **98.** (*c*) |
| **99.** (*a*) | **100.** (*a*) | **101.** (*d*) | **102.** (*c*) | **103.** (*b*) | **104.** (*d*) | **105.** (*a*) |
| **106.** (*b*) | **107.** (*a*) | **108.** (*a*) | **109.** (*c*) | **110.** (*b*) | **111.** (*c*) | **112.** (*c*) |
| **113.** (*d*) | **114.** (*c*) | **115.** (*c*) | **116.** (*c*) | **117.** (*d*) | **118.** (*d*) | **119.** (*c*) |
| **120.** (*a*) | **121.** (*d*) | **122.** (*d*) | **123.** (*a*) | **124.** (*b*) | **125.** (*b*) | **126.** (*c*) |
| **127.** (*d*) | **128.** (*a*) | **129.** (*d*) | **130.** (*b*) | **131.** (*c*) | **132.** (*a*) | **133.** (*a*) |
| **134.** (*a*) | **135.** (*c*) | **136.** (*a*) | **137.** (*d*) | **138.** (*c*) | **139.** (*a*) | **140.** (*c*) |
| **141.** (*c*) | **142.** (*c*) | **143.** (*b*) | **144.** (*c*) | **145.** (*b*) | **146.** (*a*) | **147.** (*c*) |
| **148.** (*d*) | **149.** (*b*) | **150.** (*d*) | **151.** (*d*) | **152.** (*d*) | **153.** (*b*) | **154.** (*c*) |
| **155.** (*d*) | **156.** (*b*) | **157.** (*c*) | **158.** (*d*) | **159.** (*a*) | **160.** (*b*) | **161.** (*c*) |
| **162.** (*c*) | **163.** (*d*) | **164.** (*d*) | **165.** (*b*) | **166.** (*a*) | **167.** (*a*) | **168.** (*d*) |
| **169.** (*b*) | **170.** (*d*) | **171.** (*a*) | **172.** (*d*) | **173.** (*a*) | **174.** (*a*) | **175.** (*a*) |
| **176.** (*a*) | **177.** (*a*) | **178.** (*c*) | **179.** (*d*) | **180.** (*d*) | **181.** (*a*) | **182.** (*b*) |
| **183.** (*c*) | **184.** (*d*) | **185.** (*d*) | **186.** (*a*) | **187.** (*d*) | **188.** (*a*) | **189.** (*b*) |
| **190.** (*a*) | **191.** (*b*) | **192.** (*d*) | **193.** (*a*) | **194.** (*a*) | **195.** (*d*) | **196.** (*a*) |
| **197.** (*b*) | **198.** (*c*) | **199.** (*a*) | **200.** (*c*) | **201.** (*d*) | **202.** (*c*) | **203.** (*d*) |
| **204.** (*d*) | **205.** (*a*) | **206.** (*a*) | **207.** (*d*) | **208.** (*d*) | **209.** (*a*) | **210.** (*c*) |
| **211.** (*d*) | **212.** (*d*) | **213.** (*a*) | **214.** (*a*) | **215.** (*d*) | **216.** (*b*) | **217.** (*d*) |

| | | | | | | |
|---|---|---|---|---|---|---|
| 218. (*d*) | 219. (*a*) | 220. (*c*) | 221. (*d*) | 222. (*c*) | 223. (*c*) | 224. (*b*) |
| 225. (*b*) | 226. (*d*) | 227. (*c*) | 228. (*c*) | 229. (*c*) | 230. (*a*) | 231. (*d*) |
| 232. (*b*) | 233. (*d*) | 234. (*c*) | 235. (*c*) | 236. (*d*) | 237. (*a*) | 238. (*d*) |
| 239. (*d*) | 240. (*d*) | 241. (*b*) | 242. (*c*) | 243. (*c*) | 244. (*b*) | 245. (*d*) |
| 246. (*b*) | 247. (*c*) | 248. (*b*) | 249. (*a*) | 250. (*b*) | 251. (*b*). | 252. (*b*) |
| 253. (*b*) | 254. (*b*) | 255. (*b*) | 256. (*d*) | 257. (*c*) | 258. (*a*) | 259. (*b*) |
| 260. (*c*) | 261. (*d*) | 262. (*c*) | 263. (*d*) | 264. (*b*) | 265. (*c*) | 266. (*c*) |
| 267. (*c*) | 268. (*a*) | 269. (*c*) | 270. (*b*) | 271. (*b*) | 272. (*c*) | 273. (*c*) |
| 274. (*a*) | 275. (*c*). | | | | | |

Chapter **17**

# Economics of Power Generation

The purpose of a power station is to deliver power at the lowest possible cost per unit. The total cost is made up of fixed charges consisting of interest on capital investment, taxes depreciation, insurance, management, etc.., and the operating expenses such as cost of fuel, water, oil repair, maintenance etc. The cost of power generation can be minimised by

1. Choosing equipment that will be available for operation during the largest possible percentage of the time in a year.
2. Reducing the amount investment in a plant.
3. So selecting the station as to reduce the cost of fuel, labour, etc.
4. Simplifying operation by having uniformity of design throughout the station.
5. Fixing the responsibility of operation in the hands of fewer men.

**Prediction of Load.** Some of the terms often used are given below.

**Connected Load.** It is the sum of the continuous ratings of all the equipments connected or installed at the premises of the consumer.

**Maximum Demand.** It is the greatest demand of load on the power station during a given period.

**Demand Factor.** It is the ratio of maximum demand to the connected load of a consumer.

**Average Load.** The average load on the power station in a given period (day, month or year) is known as the average load or average demand.

**Load Factor.** It is the ratio of average load to the maximum demand during a given period. Thus

$$\text{Load Factor} = \frac{\text{Average load}}{\text{Maximum demand}}$$

**Diversity Factor.** It is the ratio of the sum of individual maximum demands to the combined maximum demand on the power station. Thus

$$\text{Diversity factor} = \frac{\text{Sum of individual maximum demands}}{\text{Maximum demand on power station}}$$

**Plant Use Factor.** It is the ratio of kWh generated to the maximum kWh that the plant could produce utilising its full generation capacity.

$$\text{Plant use factor} = \frac{\text{Station output in kWh}}{\text{Plant capacity} \times \text{Hours of use}}$$

**Load Curve.** It is a plot of load versus time. The time may be a day, a week, a month or a year. The area under load curve for a specified period gives the energy consumed.

**Load Duration Curve.** When the load elements of a load curve are arranged in the order of descending magnitudes, the curve thus obtained is called a load duration curve.

The area under load duration curve is equal to the area under load curve.

**Variable Loads.** The potential consumers for a power station are domestic consumers, industrial consumers and business houses and offices. Peak demand of domestic consumers usually occurs during evening and night hour. This demand is usually more in summer months as compared to winter months. The peak demand of industrial consumers occurs during day time, however, it may continue in some cases during evening and night hours. Offices etc. also contribute to load during day time. During evening and night hours street lighting load in urban areas is also substantial.

The overall effect of all the loads connected to a power station is to provide a variable load on the plant. To meet with variable demand, a station is usually provided with a number of generators so that some of the units can be put off during off peak hours. Additional units can be put into operation whenever demand rises. However, total generating capacity of a plant must be more than the predicted peak demand.

Following methods can be used to meet with peak load demand:

1. **Peak load plants.** These plants need not be very efficient, however, these should be capable of being started within short time. Diesel engine plants, gas turbine plants, pumped storage plants can be used to meet with peak load demand.
2. **Purchasing power.** A power station may have arrangement with neighbouring stations so that power can be purchased during peak load periods.
3. **Use of accumulator.** Steam can be collected in steam accumulators during off-peak periods and the same can be used during peak demand for power generation. However, cost of steam accumulators and insulation is usually very high.
4. **Load shedding.** When no other alternative is available, load shedding may be resorted to. This may be done by

(i) *Reducing system voltage.* 1% drop in system voltage results in almost similar kW load drop on the plant.

(ii) *Reducing frequency.* By reducing system frequency, system load can be reduced.

(iii) *Shutting off loads.* In this some of the loads are switched off for sometime to reduce the peak demand.

**Selection of Type of Generation.** The following points are considered in selecting the type of generation for a power station.

1. Capacity of the power plant.
2. Probable load factor.
3. Cost of fuel and transportation facilities.
4. Space requirement.
5. Interest and depreciation.
6. Availability of water.
7. Degree of reliability.

**Cost of Electrical Energy.** The capital cost of a power plant consists of

1. Cost of land.
2. Cost of buildings.
3. Cost of generating equipment and accessories.
4. Cost of transmission and distribution network.
5. Cost of designing and planning the power station.

The total cost of power generation can be split into two parts:

**1. Fixed Cost.** This consists of

(*a*) Interest on capital (*b*) taxes

(*c*) Depreciation (*d*) Insurance

(*e*) Management—salary of managerial staff.

**2. Variable Cost.** This is also known as operating cost. This consists of

(*a*) cost of fuel

(*b*) cost of maintenance, repairs, etc.

(*c*) cost of oil, grease, water and consumable stores

(*d*) wages of employees.

The depreciation on the plant is charged by any of the following methods.

1. Straight line method
2. Sinking fund method
3. Diminishing value method.

**Base Load Supply.** The load duration curve can be conveniently split into two parts—base load and peak load.

A plant for base load should have

(*i*) high efficiency (*ii*) high availability.

A plant for peak load should have

(*i*) ability to start and take peak load during short time.

(*ii*) low capital cost in view of the small annual output with the efficiency only a secondary consideration.

(*iii*) a site near the load centre of accessible transmission facilities to avoid additional transmission costs.

In general following plants are preferred for base load operations:

1. Nuclear power plant.
2. Hydro-electric power plant.
3. Steam power plant.

Following power plants are preferred for peak load operations:

1. Diesel engine power plant.
2. Gas turbine power plant.
3. Pumped storage plant.

**Tariff.** A tariff is the rate of charge per kilowatt hour of the electrical energy supplied to a consumer. As has been mentioned earlier the total cost of production of electrical energy consists of some fixed charges and some variable charges. So a tariff should be adjusted in such a way so that the total receipts must balance the total expenditure incurred in generating the power. Various types or tariffs have been suggested as given below:

1. **Uniform Rate Tariff.** In this a fixed rate per kWh is charged from consumers. The consumption of energy at the consumer's premises is recorded by means of a kWh meter also known as energy meter. In this case, besides operating costs, an equitable contribution towards fixed charges of the generating station is merged with each unit of energy consumed.
2. **Two Part Tariff.** This tariff consists of two parts—one fixed charges, usually on the basis of maximum demand of consumer in kW and variable charges on the basis units or kWh of electrical energy consumed. This method suffers from the disadvantage that an additional meter has to be installed to register the peak load. Sometimes this is dispensed with by recording the maximum kW capacity of the connected loads at the premises of the consumer.
3. **Block Rate Tariff.** In this case fixed charges are merged into the unit charges for one or two blocks of consumption; all units in excess being charged for at low or higher rates. In case the power plant has surplus generating capacity incentive may be offered to consumers by charging at reduced rate for higher consumption blocks. Thus charges may be stipulated as Rs. $X$ per kWh for first 100 kWh and Rs. $Y$ per kWh for kWh beyond 100 kWh, $Y$ being less than $X$. However, if the generating capacity of the plant is in inadequate to meet with the consumer demand, penalty may be imposed on consumers of larger power by making $Y$ greater than $X$ in the tariff.
4. **Three Part Tariff.** It is an extension of the two part tariff in that it adds to the consumer bill some fixed charges irrespective of the energy consumed or maximum demand. In this case, even when the power consumption of the consumer is zero, some charges are payable merely because of the connection provided for power consumption.
5. **Power Factor Tariff.** In ac power systems the size of the power plant is decided by the kVA ratings. In case the power factor of the equipment installed at the consumer premises is low, the energy bill based on kW consumption will be low, whereas the power station

will have the equipment of larger kVA rating. In order to discourage such consumers, power factor tariff is introduced. This can be on any of the following patterns:

(*i*) **Maximum kVA demand Tariff.** In this case instead of kW, the kVA consumption is recorded and the charges are estimated partly or fully on this demand.

(*ii*) **Sliding Scale.** In this case the average power factor is fixed say at 0.8 lagging. Now if the power factor of a consumer falls below this by say 0.01 or multiples thereoff, additional charges are imposed. A discount may be offered.

## OBJECTIVE TYPE QUESTIONS

1. Connected load is
(*a*) the rating in kW of the installed electrical load of the consumer
(*b*) the maximum load that a consumer puts on at any time
(*c*) part of the load which always remains on at the consumer ends
(*d*) average load of the consumer during a specified period.

2. Demand factor is the ratio of
(*a*) $\frac{\text{maximum demand}}{\text{average demand}}$ (*b*) $\frac{\text{maximum demand}}{\text{connected load}}$
(*c*) $\frac{\text{connected load}}{\text{maximum demand}}$ (*d*) $\frac{\text{average demand}}{\text{maximum demand}}$.

3. The area under a load curve gives
(*a*) maximum demand (*b*) average demand
(*c*) energy consumed (*d*) none of the above.

4. The wattage of immersion heater is usually in the range
(*a*) 50 W to 150 W (*b*) 150 W to 500 W (*c*) 750 W to 2000 W (*d*) 5 kW to 15 kW.

5. The wattage of a hair dryer is usually around
(*a*) 5 W (*b*) 50 W (*c*) 500 W (*d*) 5000 W.

6. Which of the following domestic appliance usually has least electric load?
(*a*) Hot plate (*b*) Refrigerator (*c*) Electric iron (*d*) Electric shaver.

7. Which of the following domestic appliance usually offers the highest electrical load?
(*a*) Washing machine (*b*) Vacuum cleaner
(*c*) Toaster (*d*) Television set.

8. Diversity factor is the ratio
(*a*) $\frac{\text{maximum demand of consumers}}{\text{average demand}}$
(*b*) $\frac{\text{sum of individual maximum demands of consumers}}{\text{maximum demand of consumers at the particular time}}$

(*c*) $\dfrac{\text{maximum demand of consumers}}{\text{average demand of consumers}}$

(*d*) none of the above.

**9.** The capacity factor of a power station is

(*a*) $\dfrac{\text{maximum demand}}{\text{average demand}}$ (*b*) $\dfrac{\text{average demand on station}}{\text{maximum demand on station}}$

(*c*) $\dfrac{\text{maximum installed capacity of station}}{\text{average demand of the station}}$

(*d*) $\dfrac{\text{average demand of the station}}{\text{maximum installed capacity of station}}$.

**10.** Which load shown in Fig. 17.1 has the load factor different from that of the remaining?

(*a*) load *A* (*b*) load *B* (*c*) load *C* (*d*) load *D*.

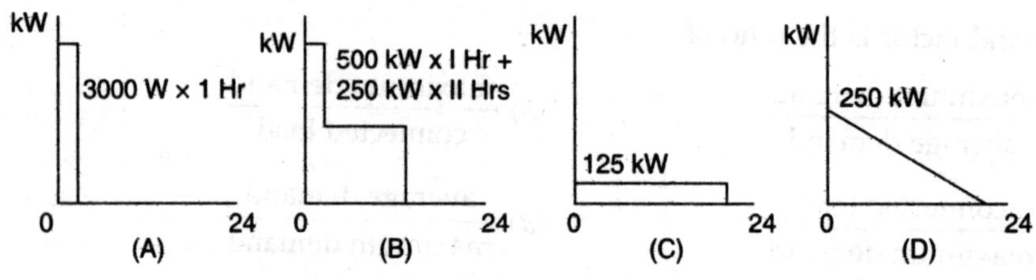


***FIGURE 17.1***

**11.** Spinning reserve is the

(*a*) generating capacity connected to the bus bars and ready to take load when switched on

(*b*) additional power that can be generated at short notice

(*c*) capacity of standby unit

(*d*) none of the above.

**12.** The reserved generating capacity available for service under emergency conditions, which is not kept in operation but in working order, is known as

(*a*) firm power (*b*) cold reserve (*c*) spinning reserve (*d*) hot reserve.

**13.** Which industry consumes maximum electric power per tonne of product?

(*a*) Steel industry (*b*) Cement industry

(*c*) Aluminium industry (*d*) Copper industry.

**14.** Which of the following industry has least electric power consumption per tonne of product ?

(*a*) Vegetable oil (*b*) Sugar (*c*) Caustic soda (*d*) Soap.

**15.** A bulk supply is taken by a large industrial consumer at the rate of Rs. 10 per kW plus 2.5 paise per kWh. The overall charge per kWh for a load factor of 80% will be

(*a*) 4.79 paise (*b*) 3.64 paise (*c*) 3.41 paise (*d*) 3.01 paise.

### Questions 16 to 19 Refer to Data Given Below

The annual peak load on a 60 MW power station is 50 MW. The power station supplies four loads having maximum demand of 20 MW, 17 MW, 10 MW and 9 MW. The annual load factor is 60%

16. The average load on the plant is
(*a*) 1500 kW (*b*) 2500 kW (*c*) 3000 kW (*d*) 6000 kW.

17. Total energy supplied per year is
(*a*) 13140,000 kWh (*b*) 26280,000 kWh (*c*) 39420,000 kWh (*d*) 52580,000 kWh.

18. Diversity factor is
(*a*) 0.8 (*b*) 1.12 (*c*) 1.25 (*d*) 1.55.

19. The capacity factor of the station is
(*a*) 0.03 (*b*) 0.06 (*c*) 0.12 (*d*) 0.15.

### Questions 20 to 23 Refer to Data Given Below

A generating station has a connected load of 40 MW and a maximum demand of 20 MW. Total units generated are $4.36 \times 10^7$ per annum.

20. The demand factor is
(*a*) 2.0 (*b*) $\frac{2}{3}$ (*c*) $\frac{1}{2}$ (*d*) $\frac{1}{3}$.

21. Average demand is
(*a*) 6600 kW (*b*) 5000 kW (*c*) 4000 kW (*d*) 2500 kW.

22. Load factor is
(*a*) 0.75 (*b*) 0.50 (*c*) 0.25 (*d*) 0.20.

23. The highest point on a load curve represents
(*a*) average demand (*b*) diversified demand
(*c*) peak demand (*d*) none of the above.

24. Which of the following is not a fixed cost on a power station?
(*a*) Insurance charges (*b*) Interest on capital
(*c*) Depreciation (*d*) Cost of fuel.

25. Which support for overhead transmission line has the least life?
(*a*) Fabricated steel structures (*b*) Wooden poles
(*c*) RCC poles (*d*) Steel poles.

26. The useful life for a diesel generating station is expected to be around
(*a*) 15 to 20 years (*b*) 25 to 50 years (*c*) 50 to 75 years (*d*) 75 to 100 years.

27. The expected useful life for a hydroelectric station is
(*a*) 10 years (*b*) 12 years (*c*) 25 years (*d*) 100 years.

28. A 100 MW power station delivers 100 MW for 2 hours, 50 MW for 5 hours, 10 MW for 5 hours and is shut down for the rest of the period each day. It is also shut down for maintenance for 45 days each year. The annual load factor will be nearly

(*a*) 10% (*b*) 16% (*c*) 21% (*d*) 27%.

29. A power station has annual load factor of 50% and capacity factor of 40%. If the maximum demand is 15 MW. The reserve capacity of the plant will be

(*a*) 1250 kW (*b*) 2500 kW (*c*) 3750 kW (*d*) 4750 kW.

**Questions 30 and 31 Refer to Fig. 17.2**

30. For the load curves shown in Fig. 17.2 which load has the least value of load factor?

(*a*) Load *A* (*b*) Load *B* (*c*) Load *C* (*d*) Load *D*.

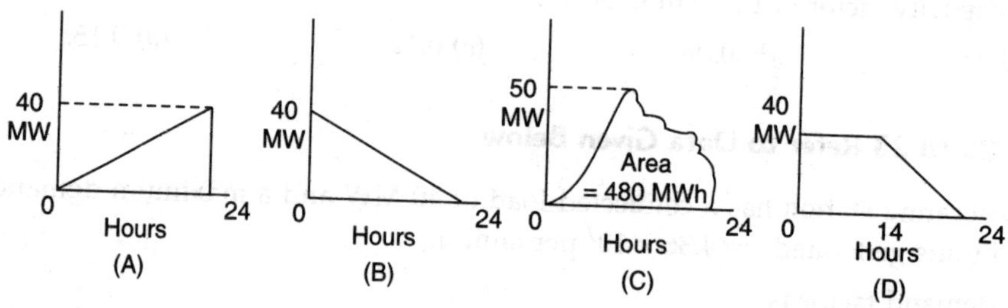


*FIGURE 17.2*

31. Which load has the highest value of load factor?

(*a*) Load *A* (*b*) Load *B* (*c*) Load *C* (*d*) Load *D*.

**Questions 32 to 35 Refer to Data Given Below**

A power station supplies loads of 5 MW, 6 MW, 8 MW 12 MW, and 15 MW. The maximum demand on the station is 20 MW. The annual load factor of the station is 50%.

32. Total kWh supplied annually will be

(*a*) 876000,000 (*b*) 87600,000 (*c*) 8760,000 (*d*) 876,000.

33. The diversity factor is

(*a*) 0.44 (*b*) 0.88 (*c*) 1.23 (*d*) 2.3.

34. The station demand factor is

(*a*) 0.44 (*b*) 0.55 (*c*) 0.66 (*d*) 0.88.

35. Which of the following is not a method for estimating depreciation charges?

(*a*) Straight line method (*b*) Sinking fund method

(*c*) Diminishing value method (*d*) Halsey's 50-50 formula.

36. A consumer has a load of 500 kW at power of 0.8. The tariff applied for the calculation of the consumer's bill is (A) kVA + BkW + C kWh, where A = Rs. 100 per kVA , B = 0 and C = 30 paise per kWh. If the consumer load factor is 90%, the annual bill of the consumer will be nearly

(*a*) Rs. $12.475 \times 10^3$ (*b*) Rs. $12.475 \times 10^4$ (*c*) Rs. $12.475 \times 10^5$ (*d*) Rs. $12.475 \times 10^6$.

**37.** A power station has a maximum demand of 30 MW and a connected load of 50 MW. Total kWh produced annually is $25 \times 10^7$. The load factor will be nearly

(*a*) 0.95 (*b*) 0.85 (*c*) 0.65 (*d*) 0.05.

**38.** In the above case the demand factor will be

(*a*) 0.95 (*b*) 0.85 (*c*) 0.75 (*d*) 0.60.

**39.** The expected useful life for underground cables is usually

(*a*) 1 year (*b*) 5 years (*c*) 10 years (*d*) 40 years

**40.** A bulk supply is taken by a industrial consumer at Rs. 100 per kW plus 15 paise per kWh. The variation of cost per kWh with load factor is represented by

(*a*) figure *A* (*b*) figure *B* (*c*) figure C (*d*) figure *D*.

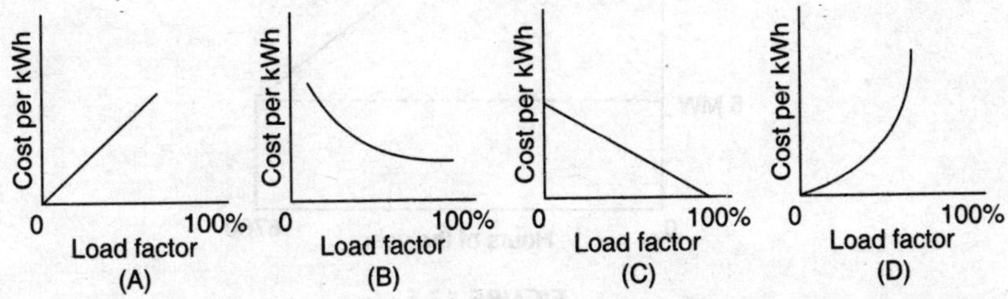


***FIGURE 17.3***

## Questions 41 to 44 Refer to Fig. 17.4

**41.** Various curves for depreciation and wear and tear of machine are shown in Fig. 17.4.

Which curve represents diminishing value method?

(*a*) curve *A*

(*b*) curve B

(*c*) curve C

(*d*) curve *D*.

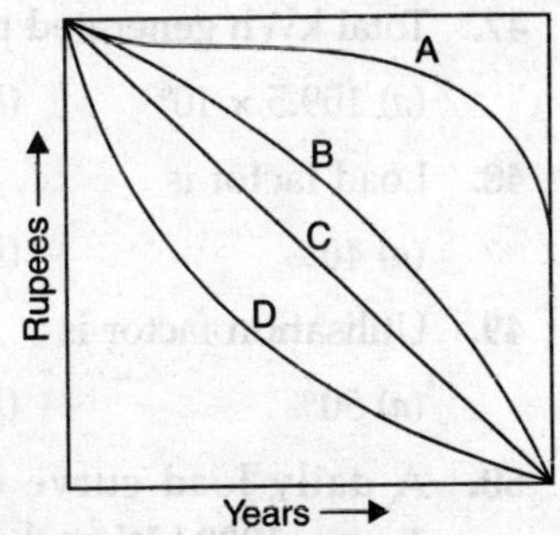


***FIGURE 17.4***

**42.** Actual wear and tear of machinery is approximately represented by

(*a*) curve *A* (*b*) curve *B* (*c*) curve C (*d*) curve *D*.

**43.** Sinking fund method is represented by

(*a*) curve *A* (*b*) curve *B* (*c*) curve C (*d*) curve *D*.

**44.** Which method of depreciation provides for heavier instalments during initial period and lighter instalments during subsequent period?

(*a*) Sinking fund method (*b*) Straight line method

(*c*) Diminishing value method (*d*) None of the above.

**Questions 45 to 49 Refer to Fig. 17.5**

A plant having installed capacity of 25 MW has annual load duration curve shown in Fig. 17.5.

**45.** The average demand is

(*a*) 10 MW (*b*) 11.5 MW (*c*) 12 MW (*d*) 12.5 MW

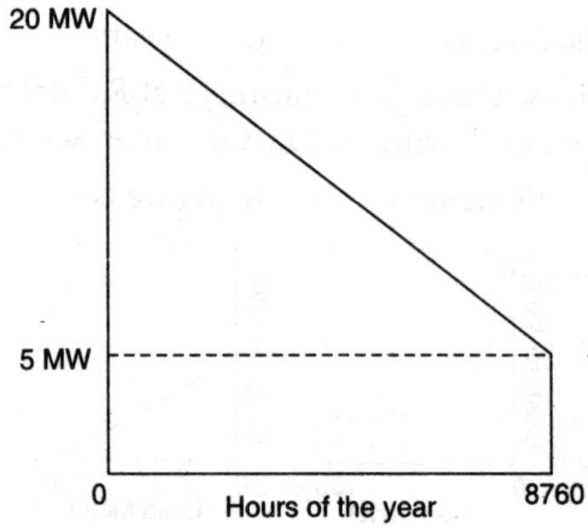


*FIGURE 17.5*

**46.** The plant factor is

(*a*) 25% (*b*) 50% (*c*) 75% (*d*) 100%.

**47.** Total kWh generated per day are

(*a*) $109.5 \times 10^6$ (*b*) $109.5 \times 10^7$ (*c*) $109.5 \times 10^8$ (*d*) $109.5 \times 10^9$.

**48.** Load factor is

(*a*) 40% (*b*) 50% (*c*) 62.5% (*d*) 75%.

**49.** Utilisation factor is

(*a*) 50% (*b*) 60% (*c*) 75% (*d*) 80%.

**50.** A daily load curve exhibited a 15 minute peak of 5000 kW is drawn to a scale of 1 cm = 1000 kW and 1 cm = 2 hours. The total area under the load curve is 21 square centimetres.

The average demand is

(*a*) 1750 kW (*b*) 2250 kW (*c*) 2750 kW (*d*) 3750 kW.

**51.** The load factor is

(*a*) 15% (*b*) 25% (*c*) 35% (*d*) 55%.

**52.** A plant mainly supplying power to domestic consumers will have peak load during

(*a*) morning (*b*) around noon (*c*) 6 PM to 10 PM (*d*) 10 AM to 6 PM.

**53.** Which of the following plant is invariable used as a base load plant?

(*a*) Gas turbine plant (*b*) Pumped storage plant

(*c*) Nuclear power plant (*d*) Diesel engine plant.

**54.** The average consumption of power per tonne of production of aluminium is nearly

(*a*) 100 kWh (*b*) 1000 kWh (*c*) 5000 kWh (*d*) 20,000 kWh.

## Questions 55 to 58 Refer to Data Given Below

A power station supplies the following loads:

| | |
|---|---|
| Industrial consumer | 1500 kW |
| Commercial establishment | 750 kW |
| Domestic power | 100 kW |
| Domestic light | 450 kW |

The maximum demand on the station is 2500 kW. Total kWh generated per annum is $8.76 \times 10^6$.

**55.** The diversity factor is

(*a*) 0.889 (*b*) unity (*c*) 1.12 (*d*) 1.22.

**56.** Average demand is

(*a*) 500 kW (*b*) 1000 kW (*c*) 1500 kW (*d*) 2500 kW.

**57.** Load factor is

(*a*) 0.80 (*b*) 0.60 (*c*) 0.50 (*d*) 0.40.

**58.** If the maximum demand on the plant is equal to the plant capacity, then

(*a*) load factor is zero (*b*) capacity factor will be zero

(*c*) plant will have 50% reserve capacity

(*d*) load factor and the capacity factor will have the same value.

**59.** Transportation of fuel is a major problem in case of

(*a*) hydro-electric power plants (*b*) thermal power plants

(*c*) nuclear power plants (*d*) diesel power plants.

**60.** Most of the power generation in India is through

(*a*) thermal power plants (*b*) nuclear power plants

(*c*) diesel power plants (*d*) gas turbine power plants.

**61.** The size of generating sets in super thermal power plants is

(*a*) 50 MW (*b*) 100 MW (*c*) 210 MW (*d*) 500 MW.

**62.** A super thermal power plant is being installed as

(*a*) Durgapur (*b*) Delhi (*c*) Singarauli (*d*) Narora.

**63.** If the maximum demand on a plant is equal to the plant capacity, then

(*a*) load factor will be unity (*b*) load factor will be nearly 50%

(*c*) diversity factor will be unity (*d*) plant reserve capacity will be zero.

**64.** For a load curve for 24 hours having shape as shown in Fig. 17.6 the shape of load duration curve will be as shown in (*a*) figure *A* (*b*) figure *C* (*c*) figure *B* (*d*) figure *D*.

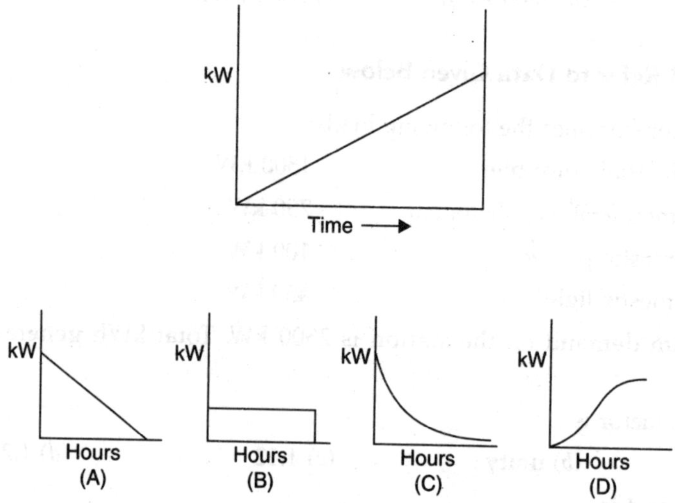


*FIGURE 17.6*

**Questions 65 to 69 Refer to Fig. 17.7**

Load curve for a plant having two generators of 50 MW each, is shown in Fig. 17.7.

**65.** The maximum demand on the plant is

(*a*) 100 MW (*b*) 80 MW (*c*) 50 MW (*d*) 40 MW.

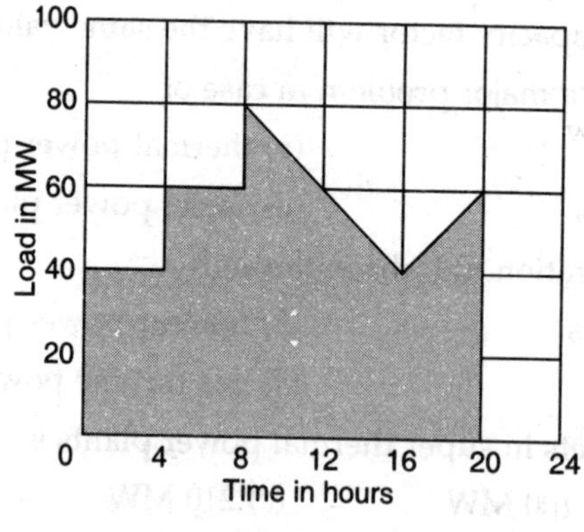


*FIGURE 17.7*

**66.** Total units generated in a day are

(*a*) 5800,000 kWh (*b*) 11600,000 kWh (*c*) 580,000 kWh (*d*) 1160,000 kWh.

**67.** Average load on the plant is

(*a*) 80 MW (*b*) 60 MW (*c*) 48.3 MW (*d*) 40.3 MW.

**68.** Load factor of the plant is nearly

(*a*) 80% (*b*) 72% (*c*) 60% (*d*) 36%.

**69.** Plant use factor is nearly

(*a*) 48% (*b*) 58% (*c*) 68% (*d*) 78%.

**70.** Highly skilled engineers are essentially required for running

(*a*) gas turbine power plants (*b*) solar power plants

(*c*) hydro-electric power plants (*d*) nuclear power plants.

**71.** Which plant needs least period for installation?

(*a*) Nuclear power plant (*b*) Thermal power plant

(*c*) Hydro-electric power plant (*d*) Diesel engine power plant.

**72.** Maintenance cost is usually high in case of

(*a*) thermal power plants (*b*) hydro-electric power plants

(*c*) diesel engine power plants (*d*) nuclear power plants.

**73.** Most efficient plants are normally used as

(*a*) base load plants (*b*) peak load plants (*c*) either of (*a*) or (*b*).

**74.** The knowledge of diversity factor helps in determining

(*a*) average load (*b*) kWh generated (*c*) number of units (*d*) plant capacity.

**75.** The cost per unit generated will be least when

(*a*) diversity factor is zero (*b*) total losses = 50% of power generated

(*c*) load factor is unity (*d*) none of the above.

**Questions 76 to 78 Refer to Fig. 17.8:**

Load duration curve for a power station is shown in Fig. 17.8. There are two sources of power having annual costs as given below.

Source A = Rs. (75,000 + 80 × kW + 0.02 kWh)

Source B = Rs. (50,000 + 50 × kW + 0.03 kWh)

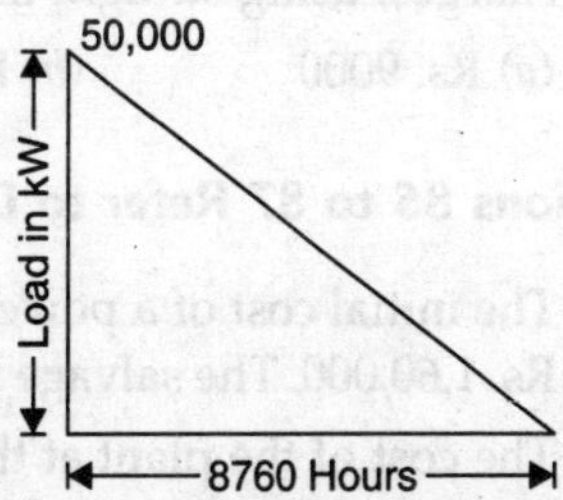


***FIGURE 17.8***

**76.** Which source should supply the base load?

(*a*) Source *A* (*b*) Source *B*

(*c*) Either of the two

(*d*) Cannot be ascertained on the basis of information given.

**77.** The minimum capacity of source *B* should be

(*a*) 10 MW (*b*) 17.242 MW (*c*) 25.0 MW (*d*) 32.758 MW.

**78.** Source *B* should operate for minimum of

(*a*) 1000 hours per annum (*b*) 2020 hours per annum

(*c*) 3020 hours per annum (*d*) 1980 hours per annum.

79. The quantity of coal burnt per kWh of electrical power generated in a thermal power plant is

(*a*) about 10 kg (*b*) about 5 kg (*c*) more than 2 kg (*d*) less than 1 kg.

**Questions 80 and 81 Refer to Data Given Below:**

The relation between water evaporated ($E$ kg) and coal consumed ($K$ kg) and kWh generated per 8 hours shift for a steam generating station is as follows:

$$E = 13500 + 7.5 \text{ kWh}$$

$$K = 5000 + 2.9 \text{ kWh.}$$

80. To what limiting value does the water evaporated per kg of coal consumed approaches as the station output increases?

(*a*) 1.3 kg (*b*) 2.6 kg (*c*) 2.9 kg (*d*) 7.5 kg.

81. How much coal per hour would be required to keep the station running at no load?

(*a*) zero (*b*) 125 kg (*c*) 625 kg (*d*) 1250 kg.

82. Which of the following plant is invariably used for peak load?

(*a*) Steam turbine plant (*b*) Nuclear power plant

(*c*) Pumped storage plant (*d*) None of the above.

83. Which of the following is not a operating cost?

(*a*) Cost of fuel (*b*) Cost of maintenance

(*c*) Salaries of operating staff (*d*) Salaries of high officials.

84. A transformer costing Rs. 200,000 has a useful life of 20 years. The annual depreciation charges, using straight line method, when the scrap value of Rs. 20,000 will be

(*a*) Rs. 9000 (*b*) Rs. 9,900 (*c*) Rs. 10,000 (*d*) Rs. 11,000.

**Questions 85 to 87 Refer to Data Given Below:**

The initial cost of a power plant is Rs. 14,00,000. The cost of additional capital equipment is Rs. 1,60,000. The salvage value of the plant at the end of 25 years is expected to be Rs. 60,000.

85. The cost of the plant at the end of 10 years by straight line method will be

(*a*) Rs. 960,000 (*b*) Rs. 1060,000 (*c*) Rs. 860,000 (*d*) 745,000.

86. The cost of the plant at the end of 20 years by diminishing value of method will be

(*a*) Rs. 100,000 (*b*) Rs. 1,15,615 (*c*) Rs. 11,65,150 (*d*) Rs. 7,94,650.

87. If the rate of compound interest for sinking fund method is 5%, the depreciated cost of the plant will be least when estimated by

(*a*) straight line method (*b*) sinking fund method

(*c*) diminishing value method.

88. Load shedding is done

(*a*) to reduce peak demand (*b*) to improve power factor

(*c*) to run the equipment efficiently (*d*) to repair the machine.

**89.** When a plant resorts of load shedding, it can be concluded that
(*a*) daily load factor is unity (*b*) diversity factor is zero
(*c*) peak demand is more than the installed capacity
(*d*) plant is under repairs.

**90.** In India domestic power is supplied at a frequency of
(*a*) 25 Hz (*b*) 50 Hz (*c*) 60 Hz (*d*) 220 Hz.

**Questions 91 to 93 Refer to Data Given Below:**

A central station has annual factors as given below:

Load factor = 58.5%

Capacity factor = 40.9%

Use factor = 45.2%.

The reserve carried over and above the peak load is 8900 kW.

**91.** The installed capacity of the plant must be
(*a*) 14,633 kW (*b*) 17,979 kW (*c*) 29,582 kW (*d*) 39,688 kW.

**92.** The actual number of hours of operation per annum is
(*a*) 1566.4 hrs (*b*) 3333.3 hrs (*c*) 4848.5 hrs (*d*) 7904.5 hrs.

**93.** Total annual energy production will be
(*a*) $106 \times 10^4$ kWh (*b*) $106 \times 10^6$ kWh (*c*) $106 \times 10^8$ kWh (*d*) $106 \times 10^{10}$ kWh.

**94.** Plant and machinery for which type of plant has relatively least useful life?
(*a*) Hydro-electric plant (*b*) Nuclear power plant
(*c*) Thermal power plant (*d*) Diesel engine plant.

**95.** For which of the following plant a single unit of 100 MW is generally not available?
(*a*) Steam power plant (*b*) Nuclear power plant
(*c*) Diesel engine plant (*d*) All of the above.

**96.** Production and distribution of electricity in states, is controlled by
(*a*) Central electricity authority (*b*) Department of science and technology
(*c*) State government (*d*) State electricity boards.

**97.** The essential requirement of peak load plant is that
(*a*) it should be capable of starting quickly
(*b*) it should run at high speed
(*c*) it should produce high voltage
(*d*) it should be small in size.

**98.** In India production and transmission of electrical energy is confined to
(*a*) government departments (*b*) public sector
(*c*) joint sector (*d*) private sector.

**99.** Which public sector undertaking is associated with erection and sometimes running of thermal power plants?

(*a*) BHEL (*b*) BEL (*c*) NTPC (*d*) SAIL.

**100.** A plant mainly supplying power to domestic consumers will have high load during

(*a*) winter (*b*) summer

(*c*) autumn (*d*) depends on other consideration.

**101.** Which of the following is not necessarily an advantage of interconnecting various power stations?

(*a*) Increased reliability (*b*) Reduction in total installed capacity

(*c*) Economy in operation of plants (*d*) Improved frequency of power supplied.

**102.** The necessary condition for parallel operation of alternators is

(*a*) their terminal voltages must be equal to the busbar voltage

(*b*) the phase sequence of voltages of the incoming machine should be the same as that of the busbar

(*c*) the frequency of the machine should be equal to that of busbars

(*d*) all of the above.

**103.** In a distribution system, in order to improve power factor, the synchronous capacitors are installed

(*a*) at the sending end (*b*) at the receiving end

(*c*) in between the sending and receiving end

(*d*) synchronous capacitors are never installed for power factor improvement.

**104.** A pilot exciter is provided on generators

(*a*) to provide requisite starting torque to generator

(*b*) to provide requisite starting torque to main exciter

(*c*) to excite the poles of main exciter

(*d*) none of the above.

**105.** Large capacity generators are invariably

(*a*) natural air cooled (*b*) forced air cooled

(*c*) water cooled (*d*) hydrogen cooled.

**106.** The size of a power transformer is usually expressed in

(*a*) volts (*b*) kW (*c*) kVA (*d*) kVAr.

**107.** For economical load sharing between two generators

(*a*) their voltages must be identical (*b*) their kVA ratings must be identical

(*c*) their kW ratings must be identical (*d*) their incremental rates must be identical.

**108.** Power factor can be improved by using

(*a*) static capacitors (*b*) synchronous compensators

(*c*) phase advancers (*d*) any of the above.

**109.** Rating of static capacitors is expressed in terms of

(*a*) volts (*b*) kW (*c*) kVAr (*d*) kWh.

**110.** All of the following plants are used for peak load EXCEPT

(*a*) gas turbine plant (*b*) diesel engine plant

(*c*) steam power plant (*d*) pumped storage plant.

**111.** Load factor is usually

(*a*) high for base load plant and low for peak load plant

(*b*) low for base load plant and high for peak load plant

(*c*) low for both base load as well as peak load plant

(*d*) high for both base load as well as peak load plant.

**112.** Which of the following is the disadvantage due to low power factor?

(*a*) High cost of equipment for a given load

(*b*) Poor voltage regulation

(*c*) Increased transmission losses

(*d*) All of the above.

**113.** Which of the following equipment offers nearly unity power factor?

(*a*) Arc lamp (*b*) Induction motor (*c*) Tube light (*d*) GLS lamps.

**114.** Power factor of induction motor is usually around

(*a*) 0.95 (*b*) 0.8 (*c*) 0.55 (*d*) 0.35.

**115.** An induction motor has relatively high power factor at

(*a*) no load (*b*) 25% load (*c*) rated rpm (*d*) near full load.

## Questions 116 to 119 Refer to Fig. 17.9:

A power station has a maximum demand of 20 MW and annual load duration curve as shown in Fig. 17.9. The load is supplied by two generators, one of 12.5 MW and the other of 10 MW.

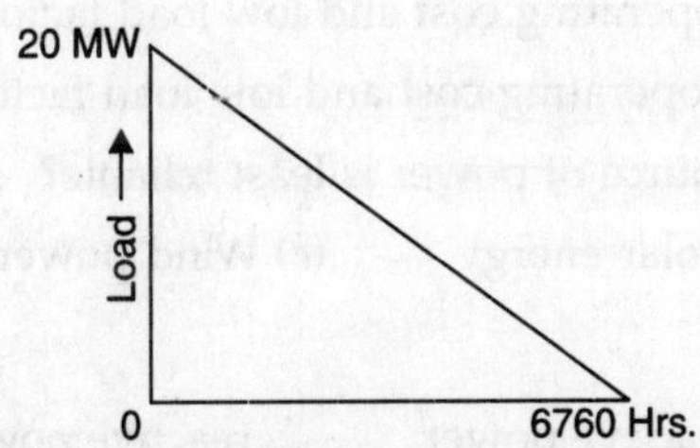


*FIGURE 17.9*

**116.** Which generator should be used for base load?

(*a*) 10 MW generator (*b*) 12.5 MW generator

(*c*) Any of the two generators

(*d*) Cannot be ascertained on the basis of information provided.

**117.** Load factor for the supply system is

(*a*) zero (*b*) 0.5 (*c*) 0.667 (*d*) unity.

**118.** In case 10 MW generator is used to supply peak load, its load factor on the basis of time for which it worked, will be

(*a*) 18.75% (*b*) 37.5% (*c*) 50% (*d*) 67.7%.

**119.** For the base load generator we have

I. Load factor II. Capacity factor

III. Plant use factor

which factor have identical values?

(*a*) I and II (*b*) II and III (*c*) I and III (*d*) I, II and III.

## Questions 120 to 123 Refer to Data Given Below:

A generating station has a maximum demand of 20 MW, a load factor of 60%, plant capacity factor of 40% and a plant use factor of 80%.

**120.** The average demand is

(*a*) 20 MW (*b*) 16 MW (*c*) 12 MW (*d*) 8 MW.

**121.** Installed capacity is

(*a*) 30 MW (*b*) 25 MW (*c*) 20 MW (*d*) 12 MW.

**122.** Reserve capacity of the plant is

(*a*) 8 MW (*b*) 10 MW (*c*) 18 MW (*d*) 12 MW.

**123.** Daily kWh produced are

(*a*) 14,400 (*b*) 28,800 (*c*) 144,000 (*d*) 288,000.

**124.** Base load plants usually have

(*a*) high capital cost, high operating cost and high load factor

(*b*) high capital cost, low operating cost and high load factor

(*c*) low capital cost, low operating cost and low load factor

(*d*) low capital cost, high operating cost and low load factor.

**125.** Which of the following source of power is least reliable?

(*a*) MHD (*b*) Solar energy (*c*) Wind power (*d*) Geothermal power.

**126.** Power factor is given by

(*a*) $\frac{\text{kVA}}{\text{kW}}$ (*b*) $\frac{\text{active power}}{\text{reactive power}}$ (*c*) $\frac{\text{reactive power}}{\text{active power}}$ (*d*) $\frac{\text{kW}}{\text{kVA}}$.

**127.** Major reason for low power factor in supply system is due to installation of

(*a*) dc motors (*b*) synchronous motors

(*c*) single phase motors (*d*) induction motors.

**128.** An over excited synchronous motor on no load known as

(*a*) alternator (*b*) generator

(*c*) induction motor (*d*) synchronous condenser.

**129.** The most economical power factor for a consumer is generally

(*a*) 0.95 lagging (*b*) 0.95 leading (*c*) 0.5 lagging (*d*) 0.5 leading.

**130.** The normal life of a capacitor is expected to be

(*a*) 6 months (*b*) one year (*c*) 8 to 10 years (*d*) 50 to 100 years.

**131.** A synchronous condenser is virtually a

(*a*) dc generator (*b*) induction motor

(*c*) over excited synchronous motor (*d*) under excited synchronous condenser is

**132.** The disadvantage of synchronous condenser is

(*a*) continuous losses in motor (*b*) high maintenance cost

(*c*) noisy (*d*) all of the above.

**133.** If a consumer is charged at Rs. $x$ per kVA of maximum demand and the expenditure incurred on power factor correction equipment is Rs. $y$ per kVAR per annum, then most economical power factor is given by cos $\phi$ =

(*a*) $1-\frac{y}{x}$ (*b*) $1-\frac{x}{y}$ (*c*) $\sqrt{1-\left(\frac{y}{x}\right)^2}$ (*d*) $\sqrt{1-\left(\frac{x}{y}\right)^2}$.

**134.** Which of the following relation is valid?

(*a*) $(kW)^2 = (kVA)^2 + (kVAR)^2$ (*b*) $(kVA)^2 = (kW)^2 + (kVAR)^2$

(*c*) $(kVAR)^2 = (kW)^2 + (kVA)^2$ (*d*) kVAR = kW + kVA.

**135.** No load power factor of an induction motion is around

(*a*) zero (*b*) 0.2 to 0.3 lagging (*c*) 0.2 to 0.3 leading (*d*) 0.8 to 0.85 lagging.

**136.** The advantage of static capacitor for power factor improvement is

(*a*) low losses (*b*) little maintenance cost

(*c*) case in installation (*d*) all of the above.

**137.** The disadvantage of static capacitor for power factor improvement is

(*a*) short service life (*b*) easily damaged by high voltage

(*c*) cannot be repaired (*d*) all of the above.

**138.** Maximum demand tariff is generally not applied to domestic consumers because

(*a*) they cannot afford to pay (*b*) they consume less power

(*c*) their load factor is low (*d*) their maximum demand is low.

**139.** The connected load of a domestic consumer will be closer to

(*a*) 5 kW (*b*) 50 kW (*c*) 100 kW (*d*) 200 kW.

**140.** The ideal tariff for any type of consumer is

(*a*) block rate tariff (*b*) two part tariff (*c*) three part tariff (*d*) any of the above.

**141.** In diminishing value method, the depreciation charges are
(*a*) heavy in early years (*b*) heavy in later years
(*c*) light in early years (*d*) same in all years.

**142.** Which of the following constitutes running cost for a power plant?
(*a*) Cost of fuel (*b*) Cost of lubricants (*c*) Cost of wages (*d*) All of the above.

**143.** For the same maximum demand, if load factor is decreases, the cost of generation will
(*a*) increase (*b*) decrease (*c*) remain unaltered.

**144.** The knowledge of diversity factor helps in determining
(*a*) average load (*b*) peak load (*c*) kWh generated (*d*) plant capacity.

**145.** The efficiency of a plant is of least concern when it is selected as
(*a*) base load plant (*b*) peak load plant
(*c*) casual run plant (*d*) either of (*b*) or (*c*) above.

**146.** In an interconnected system, the diversity factor of the whole system
(*a*) increases (*b*) decreases
(*c*) remains unaltered
(*d*) cannot be predicted on the basis of information provided.

**147.** Generators for base load plants are usually designed for maximum efficiency around
(*a*) 20% over load (*b*) full load (*c*) 75% full load (*d*) No load.

**148.** Cost of generation of power reduces as
(*a*) diversity factor reduces and load factor increases
(*b*) diversity factor increases and load factor decreases
(*c*) both diversity factor as well as load factor increase
(*d*) both diversity factor as well as load factor decrease.

**149.** The base load on a plant occurs for
(*a*) 2 to 4 hours a day (*b*) 4 to 8 hours a day (*c*) 12 hours a day (*d*) 24 hours a day.

**150.** The usual value of
(*a*) diversity factor is less than 1 and demand factor is more than one
(*b*) diversity factor is more than 1 and demand factor is less than 1
(*c*) diversity factor as well as demand factor is less than unity
(*d*) diversity factor as well as demand factor is more than unity.

**151.** A generating station has four generators of capacities 20 MW, 50 MW, 50 MW and 100 MW. A spare generating set is to be installed. According to Indian Electricity Supply Act (1948) the capacity of the spare generating set must be
(*a*) 1 MW (*b*) 20 MW (*c*) 50 MW (*d*) 100 MW.

**152.** The area under daily load curve divided by 24 hours gives
(*a*) peak demand (*b*) average load (*c*) least load (*d*) total kWh generated.

**Questions 153 to 155 Refer to Data Given Below:**

A 60 MW plant has annual load factor of 50%. The capital cost of the building and equipment is Rs. 5 million. The annual cost of fuel, oil, taxation and wages of operating staff is Rs. 0.9 million. The annual cost of organisation and interest on cost of site etc. is Rs. 500,000. The interest and depreciation on building and equipment are 10% per annum. If the annual working cast of the power station is to be represented by the formula

Rs. (*a* + *b* kW + *c* kWh)

**153.** The value of *a* will be

(*a*) Rs. 10,000 (*b*) Rs. 50,000 (*c*) Rs. 500,000 (*d*) Rs. 1000,000.

**154.** The value of *b* will be

(*a*) Rs. 4.17 (*b*) Rs. 8.34 (*c*) Rs. 16.68 (*d*) Rs. 33.34.

**155.** The value of *c* will be

(*a*) Re. 0.00034 (*b*) Re. 0.0034 (*c*) Re. 0.034 (*d*) Re. 0.34.

**156.** A consumer takes a steady load of 200 kW at a power factor of 0.85 lagging for 8 hours power per day and 315 days per annum. His annual payment at Rs. 80 per annum per kVA plus Rs. 8 per kW plus 2 paise per reactive kVAh will be

(*a*) Rs. 4,030 (*b*) Rs. 8,060 (*c*) Rs. 16,120 (*d*) Rs. 32,240.

**157.** The electric meter installed at the premises of domestic consumers reads

(*a*) voltage (*b*) kW (*c*) kWh (*d*) kVArh.

**158.** The energy meter installed for power consumption measurement work on the principle of

(*a*) electromagnetic induction (*b*) electrostatic induction

(*c*) mechanical torque (*d*) none of the above.

**159.** The energy meter installed at the premises of electric power consumers gives

(*a*) instantaneous reading (*b*) average reading

(*c*) rms reading (*d*) cumulative reading.

**160.** Generators for peak load plants are usually designed for maximum efficiency at

(*a*) 20% overload (*b*) full load

(*c*) 50 to 75% full load (*d*) 25 to 50% full load.

**Questions 161 to 164 Refer to Data Given Below:**

Incremental costs of fuel in rupees per MWh for two generators in a power station are as given below

$$C_1 = 200 + W_1$$
$$C_2 = 160 + 1.2\,W_2$$

**161.** If both the generators are operating and minimum load on a generator is to be 20 MW, a load of 50 MW will be divided between the two generators as

(*a*) $W_1$ = 20 MW; $W_2$ = 30 MW (*b*) $W_1$ = 25 MW; $W_2$ = 25 MW

(*c*) $W_1$ = 30 MW; $W_2$ = 20 MW (*d*) $W_1$ = 40 MW; $W_2$ = 10 MW.

162. At 50 MW load incremental fuel cost for generator $W_2$ will be

(a) Rs. 160/MWh (b) Rs. 196/MWh (c) Rs. 200/MWh (d) Rs. 236/MhW.

163. If the maximum load on a generator is restricted as 150 MW, in the incremental rate for generator $W_1$ at this load will be

(a) Rs. 300/MWh (b) Rs. 325/MWh (c) Rs. 350/MWh (d) Rs. 390/MWh.

164. The difference in incremental rate for the two generators at 300 MW load will be

(a) Rs. 1/MWh (b) Rs. 10/MWh (c) Rs. 50/MWh (d) Rs. 100/MWh.

**Questions 165 to 167 Refer to Fig. 17.10:**

165. The input output curves for two generators are shown in Fig. 17.10.

From the curves it can be concluded that

(a) both generators are equally efficient

(b) generator *A* is more efficient than generator *B* at a load of 50 MW

(c) both generators are run at same speed

(d) for a load of more the 50 MW generator *B* is more efficient than generator *A*.

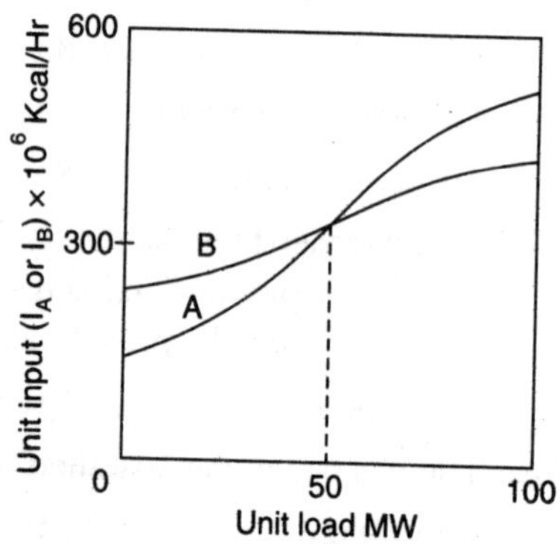


*FIGURE 17.10*

166. When total load on the station is 100 MW, the load will be shared as

(a) $L_A$ = 20 MW, $L_B$ = 80 MW (b) $L_A$ = 40 MW, $L_B$ = 60 MW

(c) $L_A$ = 50 MW, $L_B$ = 50 MW (d) $L_A$ = 60 MW, $L_B$ = 40 MW.

167. At no load

(a) generator *A* will consume less energy as compared to generator *B*

(b) generator *B* will consume less energy as compared to generator *A*

(c) both generators will consume same energy

(d) none of the above is true.

168. If the tariff for electrical energy charges provides incentive by way of reduced charges for higher consumption, then it can be concluded that

(a) power is generated through hydroelectric plant

(b) load factor is unity

(c) station has more than 2 generators (d) plant has sufficient reserve capacity.

169. By reducing system voltage by 5%, the load on the power station will

(a) increase by 5% (b) remain unaltered (c) reduce by 5% (d) reduce by 10%.

170. According to Indian Electricity Act the voltage at consumer terminals is to be maintained within

(a) ± 1% of declared value (b) ± 5% of declared value

(c) ± 10% of declared value (d) ± 20% of declared value.

**171.** Which of the following appliance will be least affected as a result of change in supply voltage frequency?

(*a*) Room heater (*b*) Mixer grinder (*c*) Celling fan (*d*) Electric clock.

**Questions 172 to 175 Refer to Data Given Below:**

Two tariffs are offered.

(*a*) Rs. 200 plus 10 paise per unit (*b*) a flat rate of 50 paise per unit.

**172.** At what consumption is tariff (*a*) preferable

(*a*) 100 kWh (*b*) 200 kWh (*c*) below 500 kWh (*d*) above 500 kWh.

**173.** At what consumption the charges by tariff (*a*) will become double of charges by tariff (*b*)

(*a*) 100 kWh (*b*) 222 kWh (*c*) 333 kWh (*d*) 444 kWh.

**174.** At what consumption the charges by tariff (*b*) will become double of the charges by tariff (*a*) ?

(*a*) 500 kWh (*b*) 1000 kWh (*c*) 1333 kWh (*d*) 1777 kWh.

**175.** For consumption of 2000 kWh the difference in charges by two tariffs will be

(*a*) Rs. 200 (*b*) Rs. 400 (*c*) Rs. 600 (*d*) Rs. 800.

## ANSWERS

| | | | | | | |
|---|---|---|---|---|---|---|
| **1.** (*a*) | **2.** (*b*) | **3.** (*c*) | **4.** (*c*) | **5.** (*b*) | **6.** (*d*) | **7.** (*c*) |
| **8.** (*b*) | **9.** (*d*) | **10.** (*b*) | **11.** (*a*) | **12.** (*b*) | **13.** (*c*) | **14.** (*a*) |
| **15.** (*b*) | **16.** (*c*) | **17.** (*b*) | **18.** (*b*) | **19.** (*b*) | **20.** (*c*) | **21.** (*b*) |
| **22.** (*c*) | **23.** (*c*) | **24.** (*d*) | **25.** (*b*) | **26.** (*a*) | **27.** (*d*) | **28.** (*c*) |
| **29.** (*c*) | **30.** (*c*) | **31.** (*d*) | **32.** (*b*) | **33.** (*d*) | **34.** (*a*) | **35.** (*d*) |
| **36.** (*c*) | **37.** (*a*) | **38.** (*d*) | **39.** (*d*) | **40.** (*b*) | **41.** (*d*) | **42.** (*a*) |
| **43.** (*b*) | **44.** (*c*) | **45.** (*d*) | **46.** (*b*) | **47.** (*a*) | **48.** (*c*) | **49.** (*d*) |
| **50.** (*a*) | **51.** (*c*) | **52.** (*c*) | **53.** (*c*) | **54.** (*d*) | **55.** (*c*) | **56.** (*b*) |
| **57.** (*d*) | **58.** (*d*) | **59.** (*b*) | **60.** (*a*) | **61.** (*d*) | **62.** (*c*) | **63.** (*d*) |
| **64.** (*a*) | **65.** (*b*) | **66.** (*d*) | **67.** (*c*) | **68.** (*c*) | **69.** (*a*) | **70.** (*d*) |
| **71.** (*d*) | **72.** (*a*) | **73.** (*a*) | **74.** (*d*) | **75.** (*c*) | **76.** (*a*) | **77.** (*b*) |
| **78.** (*c*) | **79.** (*d*) | **80.** (*b*) | **81.** (*c*) | **82.** (*c*) | **83.** (*d*) | **84.** (*a*) |
| **85.** (*a*) | **86.** (*b*) | **87.** (*c*) | **88.** (*a*) | **89.** (*c*) | **90.** (*b*) | **91.** (*c*) |
| **92.** (*d*) | **93.** (*b*) | **94.** (*d*) | **95.** (*c*) | **96.** (*d*) | **97.** (*a*) | **98.** (*b*) |
| **99.** (*c*) | **100.** (*b*) | **101.** (*d*) | **102.** (*d*) | **103.** (*b*) | **104.** (*c*) | **105.** (*d*) |
| **106.** (*c*) | **107.** (*d*) | **108.** (*c*) | **109.** (*c*) | **110.** (*c*) | **111.** (*a*) | **112.** (*d*) |
| **113.** (*d*) | **114.** (*b*) | **115.** (*d*) | **116.** (*d*) | **117.** (*b*) | **118.** (*c*) | **119.** (*d*) |
| **120.** (*c*) | **121.** (*a*) | **122.** (*b*) | **123.** (*d*) | **124.** (*b*) | **125.** (*c*) | **126.** (*d*) |

| | | | | | | |
|---|---|---|---|---|---|---|
| **127.** (*d*) | **128.** (*d*) | **129.** (*a*) | **130.** (*c*) | **131.** (*c*) | **132.** (*d*) | **133.** (*c*) |
| **134.** (*b*) | **135.** (*b*) | **136.** (*d*) | **137.** (*d*) | **138.** (*d*) | **139.** (*a*) | **140.** (*c*) |
| **141.** (*a*) | **142.** (*d*) | **143.** (*a*) | **144.** (*d*) | **145.** (*d*) | **146.** (*a*) | **147.** (*b*) |
| **148.** (*c*) | **149.** (*d*) | **150.** (*b*) | **151.** (*d*) | **152.** (*b*) | **153.** (*c*) | **154.** (*b*) |
| **155.** (*b*) | **156.** (*d*) | **157.** (*c*) | **158.** (*a*) | **159.** (*d*) | **160.** (*c*) | **161.** (*a*) |
| **162.** (*b*) | **163.** (*c*) | **164.** (*b*) | **165.** (*d*) | **166.** (*c*) | **167.** (*a*) | **168.** (*d*) |
| **169.** (*c*) | **170.** (*c*) | **171.** (*a*) | **172.** (*d*) | **173.** (*b*) | **174.** (*c*) | **175.** (*c*). |

Chapter **18**

# TRANSMISSION AND DISTRIBUTION

Energy generated at the power station has to be carried to the consumer's premises through a network of transmission lines. Transmission network consists of conductors, supporting poles or towers, voltage compensation devices like booster's transformers etc. and associated equipment for safety, etc. Further as the losses incurred in transmitting power reduce considerably as the voltage of power transmission increases, therefore all transmitting networks use step up and step down transformers. With grid connection system trend in India is now for bulk generation of power. Naturally this has necessitated use of higher voltages for transmitting power. 132 kV, 220 kV and 400 kV are the common voltages for transmitting bulk power. The general arrangement for transmitting power is shown in Fig. 18.1.

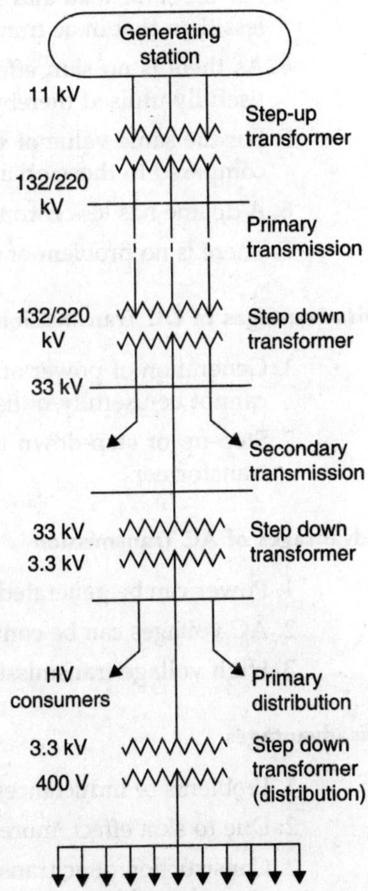


*FIGURE 18.1* Transmission and distribution of electrical power.

**Generating Stations.** The electrical power is generated in generating stations in three phase alternators. Generation voltages are 6.6 kV and 11 kV in most cases. This voltage is stepped up and supplied to primary distribution network.

**Primary Transmission.** High voltages of the order of 66 kV, 132 kV, 220 kV and 400 kV are used for transmitting power by 3 phase 3 wire overhead system. This is supplied to substations usually at the outskirts of a major distribution centre or city.

**Secondary Transmission.** The primary transmission voltage is reduced to low values of the order of 3.3 kV, 11 kV or 33 kV for secondary transmission.

**Primary Distribution.** Some consumers use high voltage equipment. Such consumers are supplied power directly from 11 kV lines. Bulk consumers also take power from HV and step down by installing transformer at their premises.

**Secondary Distribution.** Domestic and majority of other consumers take electrical power either at 220 V single phase or 400 V 3 phase. Suitable distribution transformers are installed closer to consumer centres and power is distributed by providing connections from low tension transmission lines.

## Comparison of AC and DC Transmission

### Advantages of DC Transmission

1. It requires only two conductors.
2. There is no problem of inductance, capacitance and phase displacement which is common in ac transmission.
3. For the same load and sending end voltage, the voltage drop in dc transmission lines is less than that in ac transmission.
4. As there is no skin effect on conductors, therefore entire cross-section of conductor is usefully utilised thereby affecting saving in material.
5. For the same value of voltage insulating materials on dc lines experience less stress as compared to those on ac transmission lines.
6. A dc line has less corona loss and reduced interference with communication circuits.
7. There is no problem of system instability so common in ac transmission.

### Disadvantages of DC Transmission

1. Generation of power at high dc voltages is difficult due to commutation problems and cannot be usefully utilised at consumer ends.
2. Step-up or step-down transformation of dc voltages is not possible in equipment like transformer.

### Advantages of AC Transmission

1. Power can be generated at high voltages as there is no commutation problem.
2. AC voltages can be conveniently stepped up or stepped down.
3. High voltage transmission of ac power reduces losses.

### Disadvantages

1. Problems of inductances and capacitance exist in transmission lines.
2. Due to skin effect, more copper is required.
3. Construction of ac transmission lines is more complicated as well as costly.
4. The effective resistance of ac transmission lines is increased due to skin effect.

**Conductors for Various Systems of Transmission.** When the size of conductor-for two wire dc transmission system is taken as unity, the size of conductors for various power transmission systems dc as well as ac are given in table below. These sizes are given for two different conditions.

1. Same maximum voltage with respect to earth—a conditon valid for overhead conductors.
2. Same maximum voltage between conductors—a condition valid for cables.

| System | Same maximum voltage to earth | Same maximum voltage between conductors |
|---|---|---|
| **1. DC systems** | | |
| (a) Two wire system | 1 | 1 |
| (b) Two wire system with mid-point earthed | 0.25 | 1 |
| (c) Three wire system | 0.3125 | 1.25 |
| **2. AC systems** | | |
| (a) Single phase systems | | |
| (i) 2 wire system | $\frac{2}{\cos^2 \phi}$ | $\frac{2}{\cos^2 \phi}$ |
| (ii) 2 wire with mid-point earthed | $\frac{0.5}{\cos^2 \phi}$ | $\frac{2}{\cos^2 \phi}$ |
| (iii) 3 wire | $\frac{0.625}{\cos^2 \phi}$ | $\frac{2.5}{\cos^2 \phi}$ |
| (b) Two phase systems | | |
| (i) 2 phase, 4 wire | $\frac{0.5}{\cos^2 \phi}$ | $\frac{2}{\cos^2 \phi}$ |
| (ii) 2 phase, 3 wire | $\frac{1.457}{\cos^2 \phi}$ | $\frac{2.194}{\cos^2 \phi}$ |
| (c) Three phase systems | | |
| (i) 3 phase, 3 wire | $\frac{0.5}{\cos^2 \phi}$ | $\frac{1.5}{\cos^2 \phi}$ |
| (ii) 3 phase, 4 wire | $\frac{0.583}{\cos^2 \phi}$ | $\frac{1.75}{\cos^2 \phi}$ |

**Economic Size of conductor: Kelvin's Law.** The most economical size of conductor is that for which the variable part of the annual charges is equal to the cost of energy losses per year.

## Limitations of Kelvin's Law

1. The law assumes a linear relation between the cost on account of interest and depreciation on the capital outlay which is not necessarily always valid. Moreover, it is difficult to calculate these values.
2. Actual energy loss on a transmission line cannot be estimated without actual loas curves. Load curves are not available at the planning stages.
3. The conductor size estimated according to this law may not be the optimum as various aspects of safety etc. have not been taken into account.

4. The law does not take into account some of the aspects like safe current density, mechanical strength, corona loss, etc.

**Conductor Materials.** A material for conducting electrical power should have the following properties.

1. High electrical conductivity.
2. Low cost.
3. Low specific gravity.
4. High tensile strength.

Commonly used materials for conductors are:

1. Copper. 2. Aluminium.
3. Aluminium Conductor Steel Reinforced (ACSR).
4. Galvanised steel. 5. Cadmium copper.

Transmission lines are carried on supports which may be

1. **Wooden Poles** for short spans of 50 to 75 metres. These have short life of 20-25 years and are generally not used for voltages more than 11 kV.
2. **Steel Poles.** These have longer life and more strength as compared to wooden poles.
3. **RCC Poles.** These poles have longer life and can be used for longer spans. However, these are bulky.
4. **Steel Towers.** These are structures fabricated at the site usually used for high voltage lines (33 kV and above).

**Insulators.** A insulator should have high mechanical and electrical strength and should be non-porous free from impurities and cracks and should have high ratio of puncture strength to flash over.

Pin type insulators are used upto 33 kV. Suspension type insulators are used for voltages higher than 33 kV. Strain insulators are installed at the dead end of the line or whenever there is corner or sharp curve.

**String Efficiency.** A string of suspension insulators consists of a number of porcelain discs connected in series through metallic links. Fig. 18.2 shows 3 strings of suspension insulators. The porcelain portion of each disc is in between two metal links. Shunt capacitance exists between porcelain discs as shown. The string efficiency is defined as the ratio of voltage across the whole strong to the product of number of discs and the voltage across the disc nearest to the conductor. Thus

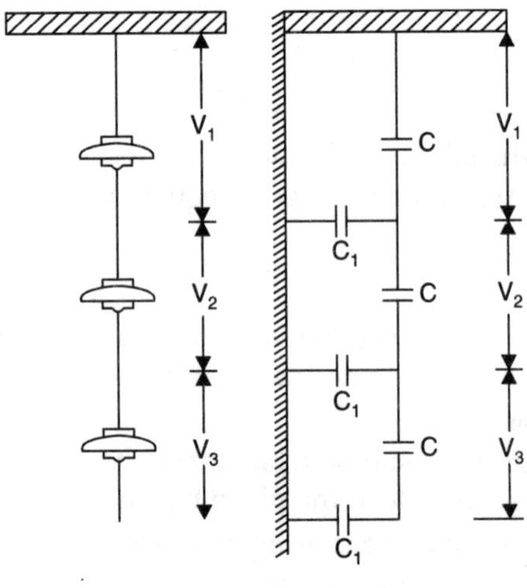


*FIGURE 18.2*

$$\text{String efficiency} = \frac{\text{Voltage across the string}}{n \times \text{voltage across disc nearest to conductor}}$$

where $n$ is the number of discs in the string.

String efficiency can be improved by

(*i*) using longer cross-arms

(*ii*) grading the insulators

(*iii*) using a guard ring.

**Corona.** The phenomenon of violet glow, hissing noise and production of ozone gas in an overhead transmission line is known as corona. The factors affecting corona are:

1. *Atmosphere.* There is more tendency during stormy weather.
2. *Conductor size.* Rough and irregular surfaces give rise to more corona because unevenness of the surface decreases the value of breakdown voltage stranded conductors having rough surface have more tendency for corona.
3. *Spacing between conductors.* Shorter distance between conductors increases tendency for corona.
4. *Line voltage.* Higher voltages have increased chances of corona.

**Skin Effect.** The tendency of alternating current to concentrate near the surface of the conductor is known as skin-effect.

## Classification of Overhead Transmission Lines

1. **Short Transmission Lines.** When the length of an overhead transmission line is upto 50 km and the line voltage less than 20 kV, it is known as short transmission line. Due to smaller length and lower voltage, the capacitance effects are small and hence can be neglected. Thus resistance and inductance are the major parameters considered for these lines.
2. **Medium Transmission Lines.** These lines are 50 km to 150 km in length and the range of voltage is 20 kV to 100 kV. Due to sufficient length and voltage of the line, the capacitance effects cannot be neglected.
3. **Long Transmission Lines.** These are more than 150 km in length and carry voltages higher than 100 kV.

**Voltage Regulation.** The difference in voltage at the receiving end of a transmission line between conditions of no load and full load is called voltage regulation. Thus

$$\%\text{ voltage regulation} = \frac{\text{Sending end voltage} - \text{Receiving end voltage}}{\text{Receiving end voltage}}$$

**Generalized constants.** The sending end voltage and sending end current can be expressed as

$$V_S = \mathbf{A} \,.\, V_R + \mathbf{B} \,.\, I_R$$
$$I_S = \mathbf{C} \,.\, V_R + \mathbf{D} \,.\, I_R$$

**In case of short lines**

$$A = 1, B = z, C = 0, D = 1$$

$$AD - BC = 1$$

**Medium Lines**

$$A = D = 1 + \frac{yz}{2}$$

$$B = z\left(1 + \frac{yz}{4}\right)$$

$$C = y$$

$$AD - BC = 1$$

**Long Lines**

$$A = D = \cosh\sqrt{yz}$$

$$B = \sqrt{\frac{z}{y}} \sinh \sqrt{yz}$$

$$C = \sqrt{\frac{y}{z}} \sinh \sqrt{yz}$$

$$AD - BC = 1.$$

**Cables.** A cable consists of one or more conductors with suitable insulation and surrounded by a protecting cover.

**Conductor.** The conductor should be tinned stranded copper or aluminium of high conductivity. Stranding is done so that conductor is flexible.

**Insulation.** Each core or conductor is provided with a suitable thickness of insulation, the thickness is decided by the voltage. Impregnated paper, varnished cambric, rubber are the commonly used insulations.

**Metallic sheath** is provided to protect the cable from moisture, gases or other damaging liquids in the soil and atmosphere. This is generally made of aluminium or lead.

**Bedding.** This is a layer of fibrous material like jute or hessian tape, provided over sheath, to protect the sheath against corrosion as well as mechanical injuries due to armouring.

**Armouring** is provided over the bedding. This consists of one or two layers of galvanished steel wire or steel tape. This protects the cable against any mechanical injury during laying.

For high voltage applications oil filled cables and gas pressure cables are used.

## OBJECTIVE TYPE QUESTIONS

1. As compared to ac transmission, dc transmission lines are free from

   (*a*) inductance (*b*) capacitance

   (*c*) phase displacement (*d*) all of the above.

2. As compared to ac transmission, dc transmission lines are free from
(*a*) skin effect (*b*) corona effect (*c*) resistance drop (*d*) all of the above.

3. For the transmission of same power
(*a*) dc line requires more copper than a ac line
(*b*) ac line requires more copper than a dc line
(*c*) both ac as well as dc line need same copper.

4. DC power is transmitted through cable system. For the same quantity of power to be transmitted, which system will need volume of conductor material more than the remaining systems?
(*a*) DC two wire system (*b*) DC two wire system with mid-point earthed
(*c*) Three wire DC system.

5. If $K$ be the volume of the cable conductor material required to transmit power, then for the transmission of same power, the volume of cable conductor required for single phase 2 wire ac system will be
(*a*) $2K$ (*b*) $K \cos \phi$ (*c*) $\dfrac{K}{\cos^2 \phi}$ (*d*) $\dfrac{2K}{\cos^2 \phi}$.

6. For the transmission of same power, which ac transmission system will need maximum volume of conductor material?
(*a*) Single phase–2 wire system with mid-point earthed
(*b*) Single phase–3 wire system
(*c*) 3 phase–3 wire system
(*d*) 3 phase-4 wire system.

7. If $K$ be the volume of conductor material required for two wire dc system with one conductor earthed, then the volume of conductor material required for the transmission of same power in two wire dc system with mid-point earthed will be
(*a*) $4K$ (*b*) $2K$ (*c*) $K$ (*d*) $\dfrac{K}{4}$.

8. If $K$ be the volume of conductor material required for two wire dc system with one conductor earthed, then the volume of conductor required for the transmission of same power in three wire dc system will be
(*a*) $3K$ (*b*) $\dfrac{K}{3}$ (*c*) $\dfrac{5}{16}K$ (*d*) $\dfrac{13}{16}K$.

9. If $K$ be the volume of conductor material required for two wire dc system with one conductor earthed, then the volume of conductor material required for the transmission of same power in single phase, 2 wire ac system with one conductor earthed will be
(*a*) $2K$ (*b*) $\dfrac{2K}{\cos \phi}$ (*c*) $\dfrac{2K}{\cos^2 \phi}$ (*d*) $2K \cos^2 \phi$.

10. If $K$ be the volume of conductor material required for two wire dc system with one conductor earthed, then the volume of conductor material required for the transmission of same power in single phase three wire system will be

(*a*) $\frac{K}{3}\cos\phi$ (*b*) $5K\cos^2\phi$ (*c*) $\frac{K}{5\cos^2\phi}$ (*d*) $\frac{5K}{8\cos^2\phi}$

**11.** For the same power to be transmitted, which cable system will need more volume of cable conductor then the remaining?

(*a*) single phase 2 wire ac system (*b*) single phase 2 wire with mid-point earthed

(*c*) single phase three wire system (*d*) two phase 4 wire system.

**12.** A dc 2 wire system is to be converted into ac 3 phase 3 wire system by the addition of 3 conductors of the same cross-section as the existing two conductors. If the voltage between the wires, the percentage loss remains unchanged and power factor is unity, the additional load that the system can now carry will be

(*a*) 200% (*b*) 100% (*c*) 50% (*d*) 25%.

**13.** The inductance of a single phase two wire line is given by (*D* is distance between conductors and 2*r* is diameter of conductor)

(*a*) $0.4 \log_e \frac{D}{r}$ mH/km (*b*) $0.55 \log_e \frac{D}{r}$ mH/km

(*c*) $0.4 \log_e \left(\frac{r}{D}\right)$ mH/km (*d*) $0.55 \log_e \left(\frac{r}{D}\right)$ mH/km.

**14.** For the same power to be transmitted through cables, which system will need more volume of cable conductor as compared to the remaining systems?

(*a*) single phase two wire system

(*b*) single phase two wire system with mid-point earthed

(*c*) two phase four wire system

(*d*) all will need same volume of cable conductor.

**15.** For the transmission of the same power, by 2 wire dc system, if the transmission voltage is raised from 200 V to 400 V, the saving in conductor material will be nearly

(*a*) 10% (*b*) 25% (*c*) 50% (*d*) 75%.

**16.** In a transmission system a 33 kV/0.4 kV transformer will be installed on

(*a*) primary transmission side (*b*) secondary transmission side

(*c*) secondary distribution side (*d*) any of the above.

**17.** A dc 3 wire system is designated for 100 kW load. If this system is now converted into 3 phase-4 wire system by adding a fourth wire equal in cross-section to each outer of the dc system (with percentage power loss and voltage at consumer ends remaining same), the power than can be transmitted now will be

(*a*) 200 kW (*b*) 150 kW (*c*) 100 kW (*d*) 50 kW.

**18.** Which of the following is not a standard transmission voltage in India?

(*a*) 33 kV (*b*) 66 kV (*c*) 99 kV (*d*) 220 kV.

**19.** 3 phase 4 wire system is commonly used on

(*a*) primary transmission (*b*) secondary transmission

(*c*) primary distribution (*d*) secondary distribution.

**20.** A single phase ac system is capable of transmitting 400 kW load. This system is now converted to 3 phase, 3 wire ac system by running a third similar conductor by maintaining same transmission efficiency and unity power factor. The three phase load that the system can carry now will be

(*a*) 100 kW (*b*) 200 kW (*c*) 400 kW (*d*) 800 kW.

**21.** The variation of capital cost with transmission voltage is represented by

(*a*) curve *A*

(*b*) curve *B*

(*c*) curve *C*

(*d*) curve *D*.

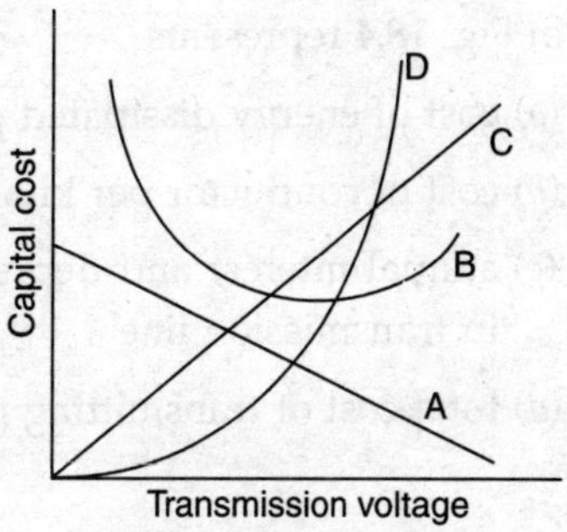


*FIGURE 18.3*

**22.** Highest transmission voltage used in India is

(*a*) 440 V (*b*) 11 kV (*c*) 132 kV (*d*) 400 kV.

**23.** When two conductors each of radius *r* are at a distance *D*, the capacitance between the two will be proportional to

(*a*) $\log_e\left(\frac{D}{r}\right)$ (*b*) $\log_e\left(\frac{r}{d}\right)$ (*c*) $\frac{1}{\log_e\left(\frac{D}{r}\right)}$ (*d*) $\frac{1}{\log_e\left(\frac{r}{D}\right)}$.

**24.** Due to skin effect

(*a*) portion of the conductor near the surface carries more current and core of the conductor carries less current .

(*b*) portion of the conductor near the surface carries less current and core of the conductor carries more current

(*c*) current flows through half cross section of the conductor

(*d*) none of the above.

**25.** While skin effect result in

(*a*) uniformity of current through conductor, proximity effect results in non-uniformity of current

(*b*) non uniformity of current through conductor, proximity effect result in uniformity of current

(*c*) non-uniformity of current through conductor , proximity also results in non-uniformity of current

(*d*) none of the above.

**26.** According to Kelvin's law, most economical area of cross- section of conductor is given by the condition (*a*) voltage drop cost of get erating voltage

(*b*) cost of transmission line = cost of energy transmitted

(*c*) annual cost of transmission line = annual cost of energy transmitted

(*d*) none of the above.

**27.** For determination of economical size of conductor, curve *A* in Fig. 18.4 represents

(*a*) cost of energy dissipated per annum

(*b*) cost of conductor per kilometre length

(*c*) annual interest and depreciation on capital investment in transmission line

(*d*) total cost of transmitting power.

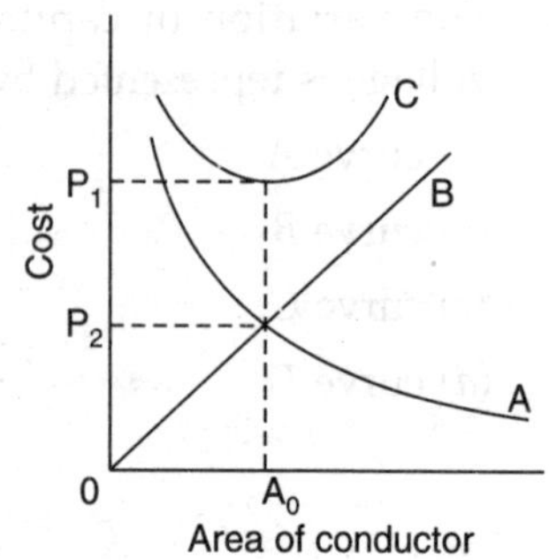


***FIGURE 18.4***

**28.** In Fig. 18.4 curve *B* represents

(*a*) annual cost of energy dissipated per annum

(*b*) total cost of transmitting power

(*c*) cost of conductor per kilometre length

(*d*) annual interest and depreciation on capital investment in transmission line.

**29.** Transmission voltage of 11kV is normally used for distances upto

(*a*) 1–2 km (*b*) 20–25 km (*c*) 50–60 km (*d*) 100–150 km.

**30.** When power is to be transmitted over a distance of 200 km, the transmission voltage should be

(*a*) 11 kV (*b* ) 33 kV (*c*) 66 kV (*d*) 132 kV.

**31.** 66 kV transmission voltage is usually restricted upto distances of

(*a*) 50 km (*b*) 80 km (*c*) 150 km (*d*) 180 km.

**32.** Which of the following material is used for overhead power transmission lines?

(*a*) Galvanised steel (*b*) Cadmium copper

(*c*) Steel cored aluminium (*d*) Any of the above.

**33.** Which material has relatively high scrap value as compared to its initial cost?

(*a*) Steel (*b*) Copper (*c*) Aluminium (*d*) Cast cron.

**34.** Which of the following is not favourable factor for the use of copper as a conductor for power transmission?

(*a*) High current density (*b*) High electrical conductivity

(*c*) High cost (*d*) High tensile strength.

**35.** The electrical conductivity of aluminium is aproximately what percent of that of copper?

(*a*) 140% (*b*) 98% (*c*) 60% (*d*) 33.3%.

36. For the same resistance

$$\frac{\text{Diameter of aluminium conductor}}{\text{Diameter of copper conductor}}$$

will be nearly

(*a*) 0.5 (*b*) 0.8 (*c*) 1.25 (*d*) 1.75.

37. The specific gravity of copper is

(*a*) 12.2 (*b*) 10.2 (*c*) 8.9 (*d*) 2.9.

38. The specific gravity of aluminium is

(*a*) 11 (*b*) 2.7 (*c*) 3.9 (*d*) 7.9.

39. All the following factors have higher value for aluminium as compared to those for copper, EXCEPT

(*a*) co-efficient of linear expansion (*b*) specific volume

(*c*) resistance per unit length for same cross-section

(*d*) electrical conductivity.

40. Steel core is provided on aluminium conductors to

(*a*) increase the tensile strength (*b*) reduce line inductance

(*c*) compensate for skin effect (*d*) neutralise proximity effect.

41. The core steel for aluminium conductors is usually

(*a*) painted blue (*b*) anodized (*c*) galvanised (*d*) nickel plated.

42. The steel used in steel cored conductors is usually

(*a*) mild steel (*b*) stainless steel (*c*) high speed steel (*d*) alloy steel.

43. By alloying copper with cadmium

(*a*) electrical conductivity is reduced but tensile strength is increased

(*b*) electrical conductivity is increased but tensile strength is reduced

(*c*) both electrical conductivity as well as tensile strength are reduced

(*d*) both electrical conductivity as well as tensile strength are increased.

44. Line supports for overhead transmission lines should

(*a*) have high mechanical strength (*b*) be cheap

(*c*) have longer life (*d*) all of the above.

45. The life span of wooden poles for transmission lines is expected to be

(*a*) 1–2 years (*b*) 2–5 years (*c*) 10–20 years (*d*) 25–50 years.

46. Pin type insulators are generally not used for voltages beyond

(*a*) 1000 V (*b*) 10 kV (*c*) 33 kV (*d*) 66 kV.

47. Insulator are generally made of

(*a*) glass (*b*) porcelain (*c*) RCC (*d*) cadmium.

48. Which of the following is not a constituent for making porcelain insulators?

(*a*) Silica (*b*) Kaolin (*c*) Felspar (*d*) Quartz.

49. String efficiency is given by

(*a*) $\dfrac{\text{voltage across the string}}{\text{number of discs on the string} \times \text{voltage across disc nearest to the conductor}}$

(*b*) $\dfrac{\text{number of discs in the string} \times \text{voltage across the string}}{\text{voltage across disc nearest to the conductor}}$

(*c*) $\dfrac{\text{number of discs in the string} \times \text{voltage across disc nearest to the conductor}}{\text{voltage across the string}}$

(*d*) $\dfrac{\text{voltage across disc nearest to the conductor}}{\text{number of discs in the string} \times \text{voltage across the string}}$

50. String efficiency can be improved by

(*a*) using longer cross arm (*b*) grading the insulator

(*c*) using a guard ring (*d*) any to the above.

51. Without changing the phase displacement between any two stations, the transfer of power is possible by boosting

(*a*) the in-phase voltage of the station to which the power is to be fed

(*b*) the in-phase voltage of the station having surplus capacity

(*c*) the quadrature voltage of the station to which the power is to be fed

(*d*) the quadrature voltage of the station with surplus capacity.

52. Quadrature voltage boost is of great help in affecting

(*a*) transfer of real power between stations

(*b*) transfer of wattless power between stations

(*c*) both (*a*) and (*b*) above (*d*) none of the above.

53. In-phase voltage control enables transfer of

(*a*) wattless currents (*b*) real power

(*c*) both (*a*) and (*b*) above (*d*) none of the above.

54. The transfer of power between two stations is maximum when the phase displacement between the voltages of two stations is

(*a*) zero (*b*) 90° (*c*) 120° (*d*) 180°.

55. A 100 MW alternator is connected to infinite busbar and its excitation is increaded, then the terminal voltage of the alternator will

(*a*) rise (*b*) fall (*c*) remain unaltered.

56. Synchronous generators can

(*a*) generate reactive power only

(*b*) absorb reactive power only

(*c*) both generate or absorb reactive power

(*d*) neither generate nor absorb reactive power.

**57.** Transformers can

(*a*) generate reactive power only

(*b*) absorb reactive power only

(*c*) both generate or absorb reactive power

(*d*) neither generate nor absorb reactive power.

**58.** Transmission lines

(*a*) absorb reactive power when fully loaded and generate reactive power on light loads

(*b*) generate reactive power when fully loaded and absorb reactive power on light loads

(*c*) generate reactive power on light loads as well as full load

(*d*) absorb reactive power on light loads as well as full load.

**59.** A transmission line consists of which of the following distributed constants?

(*a*) resistance and inductance only

(*b*) inductance and shunt capacitance only

(*c*) resistance and series capacitance only

(*d*) resistance, inductance, series and shunt capacitance.

**60.** Cables

(*a*) generate reactive power at light loads and absorb reactive power at full load

(*b*) generate reactive power at light loads as well as full load

(*c*) absorb reactive power at light loads as well as full load

(*d*) absorb reactive power at light loads and generate reactive power at full load.

**61.** A load power factor of 0.95 lagging implies reactive power demand of

(*a*) 0.05 kW per kVAR (*b*) 0.10 kW per kVAR

(*c*) 0.33 kW per kVAR (*d*) 0.95 kW per kVAR.

**62.** Due to low supply voltage for a 40 Hp, three phase induction motor, which of the following equipment can be used to boost the voltage?

(*a*) synchronous condenser (*b*) additional tap changing transformer

(*c*) series capacitors (*d*) shunt capacitors.

**63.** Series capacitors on transmission lines are of little use when

(*a*) the load VAR requirement is small

(*b*) the load VAR requirement is large

(*c*) the load VAR requirement is fluctuating

(*d*) series capacitors are never used on transmission lines.

**64.** In a tap changing transformer, the tappings are provided on

(*a*) high voltage winding (*b*) primary winding

(*c*) secondary winding (*d*) any of the above.

65. A synchronous generator generates reactive power while working at
(*a*) zero power factor (*b*) lagging power factor
(*c*) leading power factor (*d*) any of the above.

66. Electro mechanical voltage regulators generally find applications on
(*a*) generators (*b*) transformers (*c*) reactors (*d*) all of the above.

67. In magnetic amplifier type voltage regulator, the voltage is regulated by
(*a*) varying the resistance (*b*) varying the reactance
(*c*) variable transformer (*d*) electromagnetic induction.

68. For regulating the voltage in distribution feeder, which of the following equipment will be most economical?
(*a*) Booster transformer (*b*) Tap changing transformer
(*c*) Synchronous condenser (*d*) Static condenser.

69. In case the interconnector between two stations has large reactance
(*a*) the transfer of power will take place with least loss
(*b*) the transfer of power will take place with voltage fluctuation and noise
(*c*) although the stations will work in synchronism, however, the transfer of power between the two stations will be least
(*d*) the stations will fall out of step because of large angular displacement between the stations.

70. The angular displacement between two interconnected stations is mainly due to
(*a*) reactance of the interconnector (*b*) synchronous reactance of both the alternators
(*c*) armature reactance of both alternators (*d*) all of the above.

71. When two alternators are running in parallel, if the excitation of one of the alternator is increased
(*a*) the power factor of other alternator will become leading
(*b*) the power factor of both alternators will change
(*c*) the working components of both alternators will change
(*d*) none of the above.

72. Whenever an alternator connected to infinite busbar is shut down, the busbar voltage will
(*a*) rise (*b*) fall
(*c*) remain unaltered (*d*) change if the capacity of alternator is large.

73. The effect of dead zone is found in which of the following voltage regulator?
(*a*) Electronic type using $IC_s$ (*b*) Magnetic amplifier
(*c*) Electromagnetic type (*d*) All of the above.

74. Automatic flat tie load control system is used for frequency regulation of
(*a*) small systems (*b*) large systems
(*c*) generators (*d*) transformers.

75. In case of generators the frequency of voltage generated can be increased by
(a) reducing the terminal voltage (b) increasing the load
(c) using reactors (d) adjusting the governor.

## CABLES

76. In cables the thickness of the layer of insulation on the conductor depends upon
(a) current carrying capacity (b) voltage
(c) power factor (d) reactive power.

77. The insulation used on cables is
(a) paper (b) rubber
(c) varnished cambric (d) any of the above.

78. The metallic sheath is usually made of
(a) lead (b) aluminium (c) lead or aluminium (d) none of the above.

79. In a cable immediately above metallic sheath
(a) bedding is provided (b) armouring is providing
(c) earthing connection is provided
(d) none of the above is provided.

80. The bedding on a cable consists of
(a) jute (b) hessian cloth (c) any of the above (d) none of the above.

81. A cable is protected against any mechanical injury by
(a) armouring (b) bedding (c) sheath (d) none of the above.

82. The material for armouring on cable is usually
(a) galvanised steel wire (b) steel tape
(c) any of the above (d) none of the above.

83. In single core cables armouring is not done to
(a) reduce the coal (b) make it flexible
(c) avoid excessive sheath losses (d) all of the above.

84. If the length of a cable is doubled, its capacitance
(a) becomes half (b) becomes one-fourth
(c) remains unaltered (d) become double.

85. A certain cable has an insulation of relative permittivity 2. If the insulation is replaced by one of relative permittivity 4, the capacitance of the cable will become
(a) one-fourth (b) half (c) double (d) four times.

86. The charging current in cables
(a) lags the voltage by 90° (b) lags the voltage by 180°
(c) leads the voltage by 90° (d) leads the voltage by 180°.

**87.** The insulating material for a cable should have

(*a*) high dielectric strength (*b*) high mechanical strength

(*c*) low cost (*d*) all of the above.

**88.** The insulating material for cables should

(*a*) be non-hygroscopic (*b*) be non-inflammable

(*c*) be acid proof (*d*) have all above properties.

**89.** Dielectric strength of rubber is around

(*a*) 3 kV/mm (*b*) 10 kV/mm (*c*) 30 kV/mm (*d*) 300 kV/mm.

**90.** The relative permittivity of rubber is

(*a*) 1 (*b*) between 2 and 3 (*c*) between 6 and 7 (*d*) between 10 and 12.

**91.** Empire tape is

(*a*) impregnated paper (*b*) vulcanised rubber

(*c*) varnished cambric (*d*) none of the above.

**92.** PVC stands for

(*a*) Pressed and varnished cloth (*b*) Positive voltage conductor

(*c*) Poly vinyl chloride (*d*) Post varnish conductor.

**93.** Low tension cables are generally used upto

(*a*) 220 V (*b*) 400 V (*c*) 660 V (*d*) 1000 V.

**94.** High tension cables are generally used up to

(*a*) 11 kV (*b*) 33 kV (*c*) 66 kV (*d*) 132 kV.

**95.** Which of the following cable is used for 132 kV lines?

(*a*) Super tension cables (*b*) Extra super voltage cables

(*c*) Extra high tension cables (*d*) High tension cables.

**96.** Pressure cables are generally not used beyond

(*a*) 11 kV (*b*) 33 kV (*c*) 66 kV (*d*) 132 kV.

**97.** Under operating conditions the maximum stress in a cable is at

(*a*) conductor surface (*b*) sheath

(*c*) armour (*d*) insulation layer.

**98.** Beyond 66 kV solid type cables are considered unreliable because

(*a*) skin effect dominates on the conductor

(*b*) corona loss between conductor and sheath material

(*c*) there is danger of break down of insulation due to the presence of voids

(*d*) insulation may melt due to higher temperatures.

**99.** The minimum dielectric stress in a cable is at

(*a*) conductor surface (*b*) lead sheath

(*c*) armour (*d*) bedding.

**100.** If a cable of homogeneous insulation has a maximum stress of 10 kV/mm, then the dielectric strength of insulation should be

(*a*) (10 π) kV/mm (*b*) 15 kV/mm (*c*) 10 kV/mm (*d*) 5 kV/mm.

**101.** In capacitance grading of cables we use a

(*a*) hygroscopic dielectric (*b*) porous dielectric

(*c*) homogeneous dielectric (*d*) composite dielectric.

**102.** Cables generally used beyond 66 kV are

(*a*) belted (*b*) armoured (*c*) S.L. type (*d*) oil filled.

**103.** The colour of the neutral in case of three core flexible cable is

(*a*) Brown (*b*) Black

(*c*) Blue (*d*) Neutral is not provided in 3 core cable.

**104.** The surge resistance of cable is

(*a*) 10 ohms (*b*) 30 ohms (*c*) 50 ohms (*d*) 80 ohms.

**105.** Conduit pipes are normally used to protect

(*a*) armoured cables (*b*) PVC sheathed cables

(*c*) unsheathed cables (*d*) all cables.

**106.** The location of fault in the cable is usually found out by

(*a*) comparing the resistance of conductor

(*b*) comparing the inductance of conductors

(*c*) comparing the capacitances of insulated conductors

(*d*) comparing all above parameters.

**107.** The current carrying capacity of cables in dc is more than that in ac mainly due to

(*a*) smaller dielectric loss (*b*) non-existance of any stability limit

(*c*) absence of harmonics (*d*) absence of ripples.

## CORONA

**108.** Corona effect can be identified by

(*a*) bushy sparks (*b*) faint violet glow

(*c*) red light (*d*) arcing between conductors are earth.

**109.** The phenomenon of corona is generally accompanied by

(*a*) a bang (*b*) a hissing sound (*c*) magnetic hum (*d*) all of the above.

**110.** Corona results in

(*a*) improvement in power factor

(*b*) increased capacitive reactance of transmission lines

(*c*) radio interference

(*d*) better regulation.

**111.** Corona loss is more in

(*a*) solid conductors (*b*) ACSR conductors

(*c*) stranded conductors

(*d*) conductors with rough surface.

**112.** The corona effect is accompanied by all of the following EXCEPT

(*a*) hissing sound (*b*) leading power factor

(*c*) production of ozone (*d*) radio interference.

**113.** During corona effect on dc transmission lines

(*a*) positive wire has uniform glow while negative conductor has spotty glow

(*b*) positive wire has spotty glow while negative conductor has uniform glow

(*c*) both positive as well as negative conductors have spotty glow

(*d*) both positive as well as negative conductors have uniform glow.

**114.** If the conductors are polished and smooth

(*a*) the corona glow will not occur

(*b*) power loss due to corona will be least

(*c*) hissing sound will be more intense during corona

(*d*) corona glow will be uniform along the length of the conductor.

**115.** The tendency for corona will be more under all the following conditions EXCEPT

(*a*) rough surface of conductor (*b*) stormy weather

(*c*) larger spacing between conductors (*d*) higher line voltage.

**116.** Under normal atmospheric conditions the breakdown voltage for air is

(*a*) 1.1 kV/mm (*b*) 11 kV/cm (*c*) 22 kV/cm (*d*) 33 kV/cm.

**117.** Power loss due to corona is not directly proportional to

(*a*) supply voltage frequency (*b*) phase-neutral voltage

(*c*) spacing between the conductors (*d*) all of the above.

**118.** Which of the following has least effect on corona?

(*a*) Size and charge per ion (*b*) Number of ions

(*c*) Mean free path (*d*) Atmospheric temperature.

## MECHANICAL DESIGN

**119.** When the conductors are suspended between two supports at the same level, it takes the shape of a

(*a*) cycloid (*b*) involute (*c*) parabola (*d*) catenary.

**120.** When a conductor is suspended between two supports

(*a*) the tension at any point acts vertically

(*b*) the vertical component of the tension remains constant throughout the length of the conductor

(c) the horizontal component of the tension remains constant throughout the length of the conductor

(d) the tension at the supports is negligible.

121. The potential across the various discs of suspension string is different because of

(a) shunt capacitance (b) series capacitance
(c) leakage current (d) eddy currents.

122. A string efficiency of 100% implies that

(a) shunt capacitance is 1 MF (b) potential across each disc is the same
(c) potential across each disc is zero (d) one of the insulator disc is shorted.

123. Suspension type insulators are used for voltages beyond

(a) 11 kV (b) 33 kV (c) 66 kV (d) 132 kV.

124. The effect of ice on transmission line conductors is to

(a) increase resistance to flow of current
(b) increase tendency for corona
(c) increase the weight of the conductor
(d) increase transmission losses.

125. Maximum span in case of wooden poles in generally restricted to

(a) 150 metres (b) 300 metres (c) 500 metres (d) 750 metres.

126. Poles which carry transformer are usually

(a) A type (b) I type (c) H type (d) circular.

127. 70/6 ACSR conductor has

(a) 70 aluminium conductors and 6 steel conductors
(b) 70 steel conductors and 6 aluminium conductors
(c) 76 aluminium conductors and 64 steel conductors
(d) 35 aluminium conductors and 3 steel conductors.

128. Which type of poles will be used for a 220 kV transmission line?

(a) Wooden poles (b) Concrete poles (c) Transmission (d) Tubular steel poles.

129. Which type of poles are bulky?

(a) Wooden poles (b) Steel poles (c) Concrete poles (d) Transmission towers.

## DISTRIBUTION

130. The underground system has

(a) low initial cost and more flexibility as compared to over-head system
(b) high initial cost and less flexibility as compared to over-head system
(c) high initial cost and more flexibility as compared to over-head system
(d) low initial cost and less flexibility as compared to over-head system.

131. The service mains connect

(*a*) distributor and consumer terminals (*b*) distributor and transformer

(*c*) transformer and earth (*d*) distributor and relay system.

132. The most common system for secondary distribution is

(*a*) 440 V, single phase, two wire (*b*) 440 V/400 V, three phase, three wire

(*c*) 440/230 V, three phase, four wire (*d*) 440 V three phase, four wire.

133. The size of a feeder is based on

(*a*) voltage (*b*) voltage drop

(*c*) current carrying capacity (*d*) frequency.

## Questions 134 and 135 Refer to Data Given Below

A two wire dc distributor cable 1000 metres long is loaded with 0.5 A/metre. Resistance of each conductor is 0.05 Ω/km. The distributor is fed from both ends with equal voltages of 220 V.

134. The maximum voltage drop will be

(*a*) 2.25 V (*b*) 5 V (*c*) 6.25 V (*d*) 12.5 V.

135. The value of minimum voltage will be

(*a*) 220 V (*b*) 217.5 V (*c*) 213.75 V (*d*) 207.50 V.

136. A distributor arranged to form a closed loop and fed at one or more points is called a

(*a*) closed distributor (*b*) open distributor

(*c*) ring distributor (*d*) short distributor

137. The major consideration in the design of a distributor is

(*a*) frequency (*b*) current carrying capacity

(*c*) voltage drop (*d*) kVA of system.

138. In a 3 wire dc system, the load on positive side is 300 A and on the negative side is 200 A. The current in the neutral wire will be

(*a*) 100 A (*b*) 200 A (*c*) 250 A (*d*) 300 A.

139. A booster is used to

(*a*) reduce voltage drop (*b*) compensate for voltage drop

(*c*) increase current (*d*) reduce current.

140. The major cost on distribution system is that of

(*a*) distribution transformer (*b*) conductors

(*c*) metres (*d*) earthing system.

141. Which of the following is not a transmission voltage in India?

(*a*) 132 kV (*b*) 220 kV (*c*) 330 kV (*d*) 400 kV.

**142.** In a uniformly loaded distributor for dc power, which is fed at both ends, the minimum voltage will occur at

(*a*) one of the ends (*b*) nearer to one of the ends

(*c*) at the mid-point (*d*) at the heaviest load.

**143.** In the above case if *I* be the total current fed to the distributor from both ends and *R* be the total resistance of the distributor, then maximum voltage drop will be

(*a*) $IR$ (*b*) $\frac{IR}{2}$ (*c*) $\frac{IR}{4}$ (*d*) $\frac{IR}{8}$.

**144.** A booster is connected

(*a*) in series with the feeder (*b*) in series with earth connection

(*c*) in parallel with the feeder (*d*) in parallel with earth connection.

**145.** For exact compensation of voltage drop in the feeder

(*a*) the booster must work on line voltage

(*b*) the booster must be earthed

(*c*) the booster must work on linear portion of its V–I characteristics

(*d*) the booster must work on non-linear portion of its V–I characteristic.

**146.** Guard ring transmission is used to

(*a*) reduce transmission losses (*b*) improve power factor

(*c*) reduce the earth capacitance of the lowest unit

(*d*) increase the earth-capacitance of the lowest unit.

### Questions 147 to 150 Refer to Data Given Below

A 2 wire dc distributor AB 500 metres long is fed from both ends and is loaded uniformly at the rate of 1.0 A/metre. At feeding point A, the voltage is maintained at 255 V and at B at 250 V. The resistance of each conductor is 0.1 Ω per kilometre.

**147.** The minimum voltage will occur at

(*a*) 300 metres from A (*b*) 300 metres from B

(*c*) 250 metres from B (*d*) 100 metres from A.

**148.** The value of minimum voltage is

(*a*) 255 V (*b*) 246 V (*c*) 216 V (*d*) 125.5 V.

**149.** The current supplied from A will be

(*a*) 100 A (*b*) 200 A (*c*) 300 A (*d*) 400 A.

**150.** The current supplied from B will be

(*a*) 100 A (*b*) 200 A (*c*) 300 A (*d*) 400 A.

**151.** In a balanced 3 wire dc system, the potential of the neutral is

(*a*) same as that of positive wire (*b*) same as that of negative wire

(*c*) mid-way between that of outers (*d*) zero.

**152.** In a balanced 3 wire dc system, if voltage across the outers is 400 V, then voltage between the outer and the neutral will be

(*a*) zero (*b*) 200 V (*c*) 200 $\sqrt{2}$ V (*d*) 400 V.

**153.** In a 3 wire dc system the area of cross-section of neutral is generally

(*a*) double of the area of cross-section of main conductors

(*b*) same as the area of cross-section of main conductor

(*c*) half the area of cross-section of main conductor

(*d*) one-fourth the area of cross section of main conductor.

**154.** The voltage variations in a Brown Boveri regulator never exceeds

(*a*) ± 20% (*b*) ± 10% (*c*) ± 5% (*d*) ± 1%.

**155.** Induction regulators are used for voltage control in

(*a*) primary distribution (*b*) secondary distribution

(*c*) alternators (*d*) none of the abvoe.

**156.** A synchronous condenser is generally installed at the

(*a*) sending end of the transmission line

(*b*) receiving end of the transmission line

(*c*) middle of the transmission line

(*d*) point of least resistance along the transmission line.

**157.** For voltage control, which of the following method is directly connected with alternator?

(*a*) Booster (*b*) Induction regulator

(*c*) Teraril regulator (*d*) Synchronous condenser.

**158.** A terril regulator cuts in or cuts out resistance out of the

(*a*) generator armature coils (*b*) generator field magnets

(*c*) exciter field current (*d*) none of the above.

**159.** Excitation control method is satisfactory for

(*a*) short lines (*b*) long lines

(*c*) low voltages lines (*d*) high voltage lines.

**160.** In a 3 phase, 4 wire ac system, the current in the neutral wire will be zero when

(*a*) the supply voltage is 230 V (*b*) the supply voltage is sinusoidal

(*c*) the supply voltage frequency is 50 Hz (*d*) the loads are balanced

**161.** 3 phase, three wire ac distribution system is used for

(*a*) arc furnaces (*b*) resistive loads

(*c*) unity power factor loads (*d*) balanced loads.

**162.** In a balance 3 phase, 4 wire ac system, the phase sequence is RYB. If the voltage of *R* phase is 230 0 ∠0° V, then for ∠Y phase will be

(*a*) 230 ∠0° V (*b*) 230° ∠60° V (*c*) 130° ∠90° V (*d*) 230° V ∠120° V.

**163.** In a terril regulator, capacitor is provided across the relay contacts to
(*a*) improve power factor (*b*) compensate for reactive kVA
(*c*) reduce time of relay contact opening
(*d*) reduce arcing at the time relay contacts are opened.

**Questions 164 to 166 Refer to Fig. 18.5**

**164.** A part of key diagram for a substation is shown in Fig. 18.5. In this diagram *P* represents
(*a*) double pole double throw switch
(*b*) relay
(*c*) isolator
(*d*) circuit breaker.

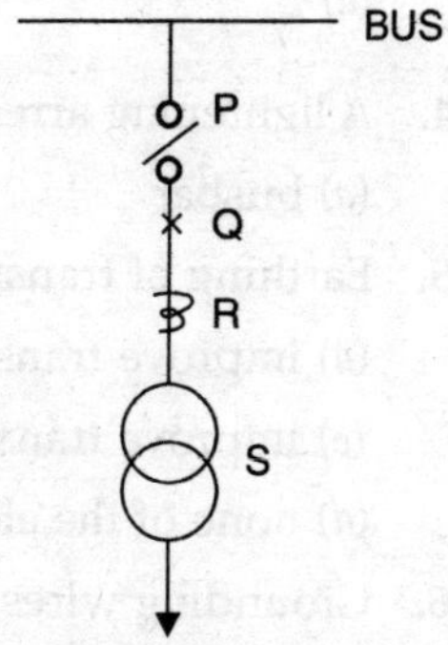

*FIGURE 18.5*

**165.** The circuit breaker is represented by
(*a*) *P* (*b*) *Q* (*c*) *R* (*d*) *S*.

**166.** *R* represents
(*a*) voltage regulator (*b*) relay
(*c*) earthing connection (*d*) current transformer.

**167.** For a short transmission line
(*a*) resistance is zero (*b*) capacitance is zero
(*c*) capacitance is very high (*d*) earth current is heavy.

**168.** For short transmission lines
(*a*) $A = 1, B = z, C = 0, D = 1$ (*b*) $A = 0, B = z, C = 1, D = 0$
(*c*) $A = 0, B = z, C = 1, D = 1$ (*d*) $A = 1, B = z, C = 1, D = 0$.

**169.** A transmission line of 50 km length will be considered as
(*a*) short transmission line (*b*) medium transmission line
(*c*) long transmission line (*d*) extra long transmission line.

**170.** A transmission line of 150 km is termed as
(*a*) short transmission lines (*b*) medium transmission line
(*c*) long transmission line (*d*) extra long transmission line.

**171.** Shunt capacitance is negligible in case of
(*a*) short transmission lines (*b*) medium transmission lines
(*c*) long transmission lines (*d*) all transmission lines.

**172.** In case of medium transmission lines

(*a*) $AB = 0, CD = 1$ (*b*) $-BC + AD = 1$

(*c*) $BC - AD = 1$ (*d*) $AB - CD = -1$.

**173.** When Z is the series impedance and Y is the shunt admittance, the surge impedance is given by

(*a*) $\frac{Z}{Y}$ (*b*) $\frac{Y}{Z}$ (*c*) $\sqrt{\frac{Z}{Y}}$ (*d*) $\sqrt{\frac{Y}{Z}}$.

**174.** A lightening arrester is usually located nearer to

(*a*) busbar (*b*) transformer (*c*) circuit breaker (*d*) isolator.

**175.** Earthing of transformer neutral is usually done through reactance to

(*a*) improve transient stability (*b*) improve steady state stability

(*c*) improve transient as well steady state stability

(*d*) none of the above.

**176.** Grounding wires are usualy mode of

(*a*) stainless steel (*b*) aluminium (*c*) galvanised steel (*d*) cast iron.

**177.** Which of the following is pilotless protection for feeder lines?

(*a*) Time grade protection (*b*) Carrier current protection

(*c*) interlock protection (*d*) None of the above.

**178.** The capacitance in the equivalent circuit of a transmission line is mainly due to

(*a*) current in the line (*b*) presence of magnetic flux

(*c*) potential difference (*d*) leakage of current.

**179.** The presence of earth

(*a*) reduces the inductance (*b*) increases the effective capacitance

(*c*) reduces capacitance increases inductance

(*d*) neither changes inductance nor capacitance.

**180.** Which parameters have zero value in case of a lossless transmission line?

(*a*) R and C (*b*) R and G (*c*) L and C (*d*) G and C.

**181.** A distortionless transmisson line is obtained when

(*a*) $RLGC = -1$ (*b*) $\frac{R}{L} = \frac{C}{G}$ (*c*) $\frac{R}{G} = \frac{C}{L}$ (*d*) $\frac{R}{G} = \frac{L}{C}$.

**182.** Which of the following are frequency dependent parameters?

(*a*) L only (*b*) G only (*c*) G, L and C only (*d*) R, L, G and C.

**183.** The reflection co-efficient for an open circuit transmission line is

(*a*) 1 (*b*) – 1 (*c*) zero (*d*) infinite.

**184.** The reflection co-efficient for a short circuit transmission line is

(*a*) 1 (*b*) – 1 (*c*) zero (*d*) infinite.

**185.** The reflection co-efficient for the voltage wave in case of over-head lines is given by

(*a*) $\frac{R_L - R_0}{R_L + R_0}$ (*b*) $\frac{R_L + R_0}{R_L - R_0}$ (*c*) $\frac{R_L + R_0}{R_L R_0}$ (*d*) $\frac{R_L R_0}{R_L - R_0}$.

**186.** The surge resistance for overhead lines is

(*a*) 50 ohms (*b*) 100 ohms

(*c*) 500 ohms (*d*) 1000 ohms.

**187.** Surge absorbers are mainly used for protection against

(*a*) low voltage high frequency oscillations

(*b*) high voltage low frequency oscillations

(*c*) high voltage high frequency oscillations

(*d*) none of the above.

**188.** Travelling wave in a transmission line moves

(*a*) from one end to another (*b*) from one end to earth

(*c*) from centre to both ends (*d*) from both ends to centre.

**189.** Which of the following condition is unlikely to cause transient disturbance?

(*a*) Switching operation (*b*) Load more that steady state value

(*c*) Load less than steady state value (*d*) Sudden variations in load.

**190.** For transmission line conductors, stock bridge dampers are provided

(*a*) to damp out harmonics (*b*) to reduce conductor swinging

(*c*) to damp out vibrations (*d*) to protect against high lightening.

**191.** In a long transmission line constant

(*a*) $A = B = 0$ (*b*) $A = B = \sqrt{Y/Z}$

(*c*) $A - D = \cosh \sqrt{YZ}$ (*d*) $A = \frac{Z}{Y}$.

**192.** The surge impedance of a transmission line can be expressed as

(*a*) $\frac{B}{C}$ (*b*) $\frac{B}{D}$ (*c*) $\frac{A}{D}$ (*d*) $\frac{C}{D}$.

**193.** When the load at the receiving end of a long transmission line is removed, the sending end voltage is less than the receiving end voltage. This effect is known as

(*a*) Proximity effect (*b*) Kelvin effect (*c*) Ferranti effect (*d*) Faraday effect.

**194.** Skin effect is noticeable only at

(*a*) low frequencies (*b*) high frequencies

(*c*) audio frequencies (*d*) all frequencies

**195.** Skin effect exists only in

(*a*) low voltage dc overhead transmission

(*b*) high voltage dc overhead transmission

(c) cables carrying dc current

(d) ac transmission.

196. For reducing the inductive interference with parallel communication circuits

(a) the stranded conductors should be used

(b) the conductors should be insulated

(c) the conductors should be transposed at regular intervals

(d) thin conductors should be used.

197. If the transmission voltage is increased $k$ times, the size of conductor will

(a) also increase by $k$ times (b) increase by $\sqrt{k}$ times

(c) reduce to $\frac{1}{k}$ times (d) reduce to $\frac{1}{k^2}$ times.

198. The distribution of electrostatic stress is uniform in

(a) Shackle insulators (b) Hewlett type suspension insulators

(c) Pin type insulator (d) Cemented cap type suspension insulator.

199. Power system stability is least affected by

(a) input torque (b) losses

(c) reactance of generator (d) reactance of transmission line.

200. A metallic sheath is provided on cables to protect against

(a) moisture (b) gases

(c) any damaging liquids in the soil or atmosphere

(d) all of the above.

## ANSWERS

| | | | | | | |
|---|---|---|---|---|---|---|
| 1. (d) | 2. (a) | 3. (b) | 4. (c) | 5. (d) | 6. (b) | 7. (d) |
| 8. (c) | 9. (c) | 10. (d) | 11. (c) | 12. (b) | 13. (a) | 14. (d) |
| 15. (d) | 16. (c) | 17. (b) | 18. (c) | 19. (d) | 20. (d) | 21. (b) |
| 22. (d) | 23. (c) | 24. (a) | 25. (c) | 26. (c) | 27. (a) | 28. (d) |
| 29. (b) | 30. (d) | 31. (b) | 32. (d) | 33. (b) | 34. (c) | 35. (c) |
| 36. (c) | 37. (c) | 38. (b) | 39. (d) | 40. (a) | 41. (c) | 42. (a) |
| 43. (a) | 44. (d) | 45. (c) | 46. (c) | 47. (b) | 48. (a) | 49. (a) |
| 50. (d) | 51. (d) | 52. (a) | 53. (a) | 54. (b) | 55. (c) | 56. (c) |
| 57. (b) | 58. (a) | 59. (d) | 60. (b) | 61. (d) | 62. (d) | 63. (a) |
| 64. (c) | 65. (c) | 66. (a) | 67. (b) | 68. (a) | 69. (d) | 70. (a) |
| 71. (c) | 72. (c) | 73. (c) | 74. (a) | 75. (d) | 76. (b) | 77. (d) |
| 78. (c) | 79. (a) | 80. (c) | 81. (a) | 82. (c) | 83. (c) | 84. (a) |

| | | | | | | |
|---|---|---|---|---|---|---|
| **85.** (*c*) | **86.** (*c*) | **87.** (*d*) | **88.** (*d*) | **89.** (*c*) | **90.** (*b*) | **91.** (*c*) |
| **92.** (*c*) | **93.** (*d*) | **94.** (*a*) | **95.** (*b*) | **96.** (*c*) | **97.** (*a*) | **98.** (*c*) |
| **99.** (*b*) | **100.** (*c*) | **101.** (*d*) | **102.** (*d*) | **103.** (*c*) | **104.** (*c*) | **105.** (*c*) |
| **106.** (*c*) | **107.** (*a*) | **108.** (*b*) | **109.** (*b*) | **110.** (*c*) | **111.** (*a*) | **112.** (*b*) |
| **113.** (*a*) | **114.** (*d*) | **115.** (*c*) | **116.** (*d*) | **117.** (*c*) | **118.** (*d*) | **119.** (*d*) |
| **120.** (*c*) | **121.** (*a*) | **122.** (*b*) | **123.** (*b*) | **124.** (*c*) | **125.** (*a*) | **126.** (*c*) |
| **127.** (*a*) | **128.** (*c*) | **129.** (*c*) | **130.** (*b*) | **131.** (*a*) | **132.** (*c*) | **133.** (*c*) |
| **134.** (*c*) | **135.** (*c*) | **136.** (*c*) | **137.** (*c*) | **138.** (*a*) | **139.** (*b*) | **140.** (*a*) |
| **141.** (*c*) | **142.** (*c*) | **143.** (*d*) | **144.** (*a*) | **145.** (*c*) | **146.** (*c*) | **147.** (*a*) |
| **148.** (*b*) | **149.** (*c*) | **150.** (*b*) | **151.** (*c*) | **152.** (*b*) | **153.** (*c*) | **154.** (*d*) |
| **155.** (*a*) | **156.** (*b*) | **157.** (*c*) | **158.** (*c*) | **159.** (*a*) | **160.** (*d*) | **161.** (*d*) |
| **162.** (*d*) | **163.** (*d*) | **164.** (*c*) | **165.** (*b*) | **166.** (*d*) | **167.** (*b*) | **168.** (*a*) |
| **169.** (*a*) | **170.** (*b*) | **171.** (*a*) | **172.** (*b*) | **173.** (*c*) | **174.** (*b*) | **175.** (*a*) |
| **176.** (*c*) | **177.** (*b*) | **178.** (*c*) | **179.** (*b*) | **180.** (*b*) | **181.** (*d*) | **182.** (*d*) |
| **183.** (*a*) | **184.** (*b*) | **185.** (*a*) | **186.** (*c*) | **187.** (*a*) | **188.** (*c*) | **189.** (*c*) |
| **190.** (*c*) | **191.** (*c*) | **192.** (*a*) | **193.** (*c*) | **194.** (*b*) | **195.** (*d*) | **196.** (*c*) |
| **197.** (*d*) | **198.** (*d*) | **199.** (*b*) | **200.** (*a*). | | | |

Chapter **19**

# Switchgear and Protection

Faults in a power system may occur due to any of the following reasons.

1. Overvoltages caused by switching surges or lightening strokes.
2. Broken insulators or conductors..
3. Insulation failure.

Faults may cause heavy short circuit current to flow in the power system. For this reason suitable protective devices must be incorporated within the system so that the damage resulting from these faults is minimum.

**Circuit breakers.** A circuit breaker is a device which can

1. make or break a circuit either manually or by remote control under normal conditions.
2. break the connected circuit automatically under the conditions of fault.
3. make the circuit either manually or by remote control under no fault conditions.

A circuit breaker has manual (or remote control) as well as automatic control for switching functions. The latter control employs relays which operate whenever fault occurs in the circuit.

A circuit breaker has fixed and moving contacts. These contacts remain in contact whenever circuit breaker is in normal operation. Whenever fault occurs the contacts separate out resulting in arc between the electrodes. The production of arc not only delays the current interruption process but it also generates enormous heat which may cause damage to the system or the breaker itself. Thus the main problem is to extinguish the arc whenever it occurs in circuit breaker.

The voltage across an arc is given approximately by the relation

$$e_a = A + \frac{B}{\sqrt{i}}$$

where $e_a$ = arc voltage

$i$ = arc current

$A$ and $B$ are constants.

**Arc interruption.** There are two methods of interrupting the arc current:

1. **High resistance interruption.** By increasing the effective resistance with time current can be reduced to a value insufficient to maintain it. The resistance can be increased by

   (*i*) Lengthening the arc (*ii*) Cooling the arc

   (*iii*) Splitting the arc (*iv*) Constraining the arc.

2. **Low resistance interruption.** In this the arc resistance is kept low until the current zero so that arc extinguishes automatically and is prevented from restriking inspite of high restriking voltage.

This can be achieved by

(*i*) Lengthening of the gap (*ii*) Cooling

(*iii*) Blast effect.

**Recovery voltage.** The normal frequency rms voltage that appears across the breaker poles after final arc extinction has occurred, is called the recovery voltage. The instantaneous value of the recovery voltage at the instant of arc extinction is called the active recovery voltage.

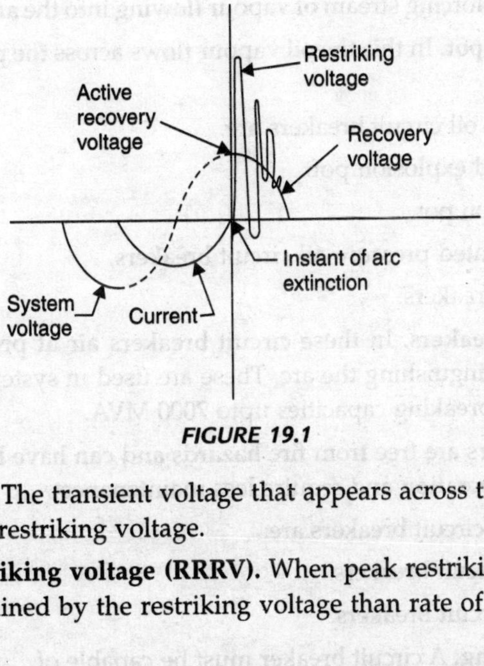


***FIGURE 19.1***

**Restriking voltage.** The transient voltage that appears across the contacts at the instant of arc extinction is called the restriking voltage.

**Rate of rise of restriking voltage (RRRV).** When peak restriking voltage is the maximum instantaneous voltage obtained by the restriking voltage than rate of rise of restriking voltage is given by

$$\text{RRRV} = \frac{\text{Peak restriking voltage}}{\text{Time between voltage zero and peak voltage}}$$

RRRV depends on

1. active recovery voltage
2. natural frequency of oscillation.

**Classification of circuit breakers.** Circuit breakers may be classified as

1. Oil circuit breakers
2. Oilless circuit breakers.

The advantages of oil circuit breakers include absorption of arc energy in decomposing oil, good cooling properties of the gas formed, ability of cooling oil of high dielectric strength of flow into the arc space after the current zero. The disadvantages include inflammability of oil which may form an explosive mixture.

**Plain oil circuit breaker.** In this moving and fixed contacts are immersed in oil. Whenever arc occurs, oil gets vapourised and the extinguishing action of the arc occurs in gaseous medium. These circuit breakers cannot be used for short circuit rating above 150 MVA and are generally used on low voltage installations (less than 11 kV).

**Self generated pressure oil circuit breaker.** In this type of circuit breaker, the pressure generated due to arc is used in speeding up the movement of oil into the contact space immediately after the current zero.

**Plain explosion pot.** In this the fixed and the moving contacts are enclosed in a strong shell of insulating material. Whenever the contacts part, the gas generated by the arc produces high pressure in the pot thereby forcing stream of vapour flowing into the arc, resulting in arc extinction.

**Cross jet explosion pot.** In this the oil vapour flows across the path of the arc, thus splitting the arc.

Some other types of oil circuit breakers are:

1. Self compensated explosion pot.
2. Oil blast explosion pot.
3. Externally generated pressure oil circuit breakers.
4. Low-oil circuit breakers.

**Air blast circuit breakers.** In these circuit breakers air at pressures ranging from 18 to 20 kg/cm$^2$ are used for extinguishing the arc. These are used in systems operating at 132 kV and above (upto 400 kV) with breaking capacities upto 7000 MVA.

These circuit breakers are free from fire hazards and can have high speed operation. These are suitable for frequent operation and require less maintenance.

Various types of air circuit breakers are:

1. Cross blast air circuit breakers
2. Axial blast air circuit breakers.

**Circuit breaker rating.** A circuit breaker must be capable of

1. opening on the occurrence of fault and of clearing the fault.
2. being closed on to a fault.
3. carrying fault current for short time.

The capacity of a circuit breaker is expressed in terms of

1. Breaking capacity
2. Making capacity
3. Short time capacity.

Clearing and closing time of a circuit breaker is illustrated in Fig. 19.2.

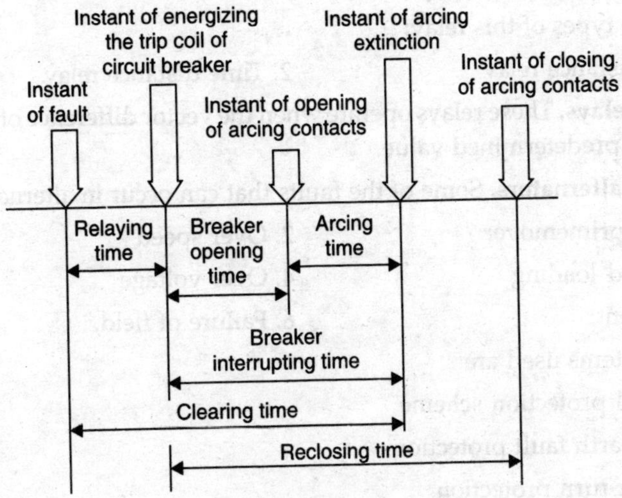


*FIGURE 19.2*

**Protective relays.** A protective relay is a device that detects the fault and initiates the operation of the circuit breaker to isolate the defective element from the rest of the system. The relays detect the abnormal conditions in the electrical circuits by constantly measuring the electrical quantities which are different under normal and fault conditions. The electrical quantities that change under fault condition are

1. Current
2. Voltage
3. Frequency
4. Phase angle.

For the satisfactory performance of a relay it should have the following qualities:

1. **Selectivity.** Ability to detect faulty part of the system, disconnect the same without disturbing the rest of the system.
2. **Speed.** Ability of relay system to quickly disconnect the faulty section so that damage may be minimized.
3. **Sensitivity.** Ability of relay system to operate with low value of actuating quantity.
4. **Simplicity.** Simple construction to facilitate operation and maintenance.

**Reliability.** Ability to operate under predetermined conditions. Failure to operate, when fault occurs, may result in colossal damage to the system.

**Economy.** Low initial cost as well as low cost of maintenance.

**Type of Relays**

1. **Induction type overcurrent relays.** This relay operates on the principle of electro-magnetic induction. The relay operates on actuating current supplied through a current transformer. These are used on ac circuits and can operate for fault flow in either direction.
2. **Induction type directional power relay.** This relay operates by the interaction of magnetic field. This is a directional relay.

3. **Distance or impedance relay.** This relay operates whenever the impedance of the protected zone falls below a predetermined value.

There are two types of this relay:

1. Definite distance relay
2. Time distance relay.

**Differential relays.** These relays operate when the vector difference of two or more electrical quantities exceeds a predetermined value.

**Protection of alternators.** Some of the faults that can occur in alternators are:

1. Failure of primemover
2. Over speed
3. Unbalanced loading
4. Over voltage
5. Over current
6. Failure of field.

Protection systems used are:

1. Differential protection scheme
2. Balanced earth fault protection
3. Stator inter-turn protection.

**Protection of transformers.** Common faults occurring with transformers are:

1. Open circuits
2. Over heating
3. Winding short circuits resulting from
   (*i*) earth faults
   (*ii*) phase-to-phase faults
   (*iii*) inter-turn faults.

The protection scheme for transformers includes the use of following relays:

1. Buchholz relay
2. Earth fault relay
3. Over current relay.

**Over voltages.** A sudden rise in voltage for a very short duration on a power system is known as voltage surge or transient voltage surge may occur due to any of the following causes:

**1. Internal Causes**

(*a*) **Switching surges.** In a transmission line travelling waves are set up, whenever a switch is closed. This results in over voltages.

(*b*) **Insulation failure.** The grounding of a conductor is equivalent to insulation failure.

(*c*) **Arcing ground.** The phenomenon of intermittent arc taking place in line-to-ground fault of a 3 phase system with consequent production of transients is called arcing grounds.

(*d*) **Resonance.** When inductive reactance of the circuit becomes equal to capacitive reactance, resonance occurs. Resonance causes high voltages in the system.

**Lightening** is the electrical discharge between cloud and earth, between clouds of between the charge centres of the same cloud. Lightening may strike a power system in two ways.

(*i*) **Direct stroke.** In which there is direct discharge from the cloud to the object which may be building, transmission line or tower. Direct strokes are rare and usually occur on tall buildings.

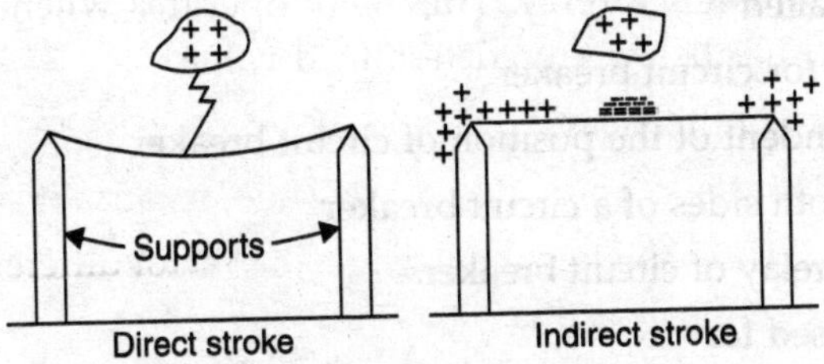


*FIGURE 19.3*

(*ii*) **Indirect stroke** results from the electrostatically induced charges on the conductor, due to the presence of charged clouds. Majority of the surges in a transmission line are caused by indirect lightening strokes.

Protection against lightening may be provided by any of the following methods:

1. Earthing screen
2. Overhead ground wires
3. Lightening arresters or surge diverters.

Lightening arresters may be of any of the following types:

1. Rod gap arrester
2. Horn gap arrester
3. Multi gap arrester
4. Thyrite arrester
5. Expulsion type arrester.

## OBJECTIVE TYPE QUESTIONS

1. A switchgear is a device for
   (*a*) interrupting an electrical circuit
   (*b*) switching an electrical circuit
   (*c*) switching and controlling an electrical circuit
   (*d*) switching, controlling and protecting the electrical circuit and equipment.
2. Arcing horns are
   (*a*) pieces of metal between which arc is formed during opening operation
   (*b*) devices for reducing circuit transients during breaking of a circuit
   (*c*) devices for absorbing heat produced during breaking of a circuit
   (*d*) any of the above.
3. An isolator is
   (*a*) an automatic device for breaking a circuit
   (*b*) a relay operated device for breaking a circuit
   (*c*) opens a circuit under no load.
   (*d*) opens a circuit under load.

4. An isolator is installed
   (*a*) as a substitute for circuit breaker
   (*b*) always independent of the position of circuit breaker
   (*c*) generally on both sides of a circuit breaker
   (*d*) to operate the relay of circuit breaker.
5. Oil switches are used for
   (*a*) circuits of low voltages (*b*) circuits of low currents
   (*c*) circuits of high voltages and large currents
   (*d*) all circuits.
6. The main purpose of providing oil in oil switches is
   (*a*) to provide insulation effect (*b*) to avoid ionisation of surrounding air
   (*c*) to facilitate low temperature operation
   (*d*) to cool and quench the arc that tends to form when the circuit is opened.
7. A fuse blows off by
   (*a*) melting (*b*) burning (*c*) arcing (*d*) none of the above.
8. A fuse operates on which effect of electrical current?
   (*a*) Magnetic effect (*b*) Electrostatic effect (*c*) Heating effect (*d*) Photoelectric effect.
9. The fuse rating is expressed in terms of
   (*a*) voltage (*b*) current (*c*) kVA (*d*) VAR.
10. The fusing current of a fuse is proportional to ($D$ being diameter of fuse wire)
   (*a*) $D$ (*b*) $D^{3/2}$ (*c*) $\frac{1}{D^{3/2}}$ (*d*) $D^2$.
11. If a existing fuse wire is to be replaced by a fuse wire of lower specific resistance
   (*a*) fuse wire of lesser diameter should be used
   (*b*) fuse wire of larger diameter should be used
   (*c*) fuse wire of reduced current rating should be used
   (*d*) fuse wire of higher voltage rating should be used.
12. If a fuse wire is replaced by another fuse wire of same current rating but higher specific heat and higher resistance temperature co-efficient, then it will become
   (*a*) spontaneous in operation (*b*) inspontaneous in operation
   (*c*) sluggish or dull in operation (*d*) unreliable.
13. Two stranded fuse wires are twisted together. The fusing current for this twisted wire will be
   (*a*) equal to fusing current of one of the stranded wires
   (*b*) more than the sum of fusing current for two wires
   (*c*) equal to the sum of fusing currents for two wires
   (*d*) less than the sum of fusing currents for two wires.

**14.** A delay fuse is generally used for the protection of

(*a*) three phase motors (*b*) GLS lamp circuits

(*c*) fluorescent tube circuits (*d*) all of the above.

**15.** A fuse is connected

(*a*) in series with circuit (*b*) in parallel with circuit

(*c*) either series of parallel with circuit

(*d*) series or parallel location of fuse depends on line current.

**16.** Fuse wires are usually made of

(*a*) copper-aluminium alloy (*b*) tin-load alloy

(*c*) lead-zinc alloy (*d*) stainless steel.

**17.** The current rating of a normal rewirable fuse

(*a*) increases with increase in diameter and decreases with increase in length of wire

(*b*) increases with increase in diameter and increases with increase in length of wire

(*c*) decreases with increase in diameter and increases with increase in length of wire

(*d*) decreases with increase in diameter and decreases with increase in length of wire.

**18.** As compared to rewirable fuse, a HRC fuse has

(*a*) high rupturing capacity (*b*) high speed of operation

(*c*) no aging effect (*d*) all of the above.

**19.** Which of the following provides relatively steady load?

(*a*) Heating load (*b*) Transformer

(*c*) Single phase motor load (*d*) Capacitor load.

**20.** A fuse wire made of which material will have relatively sharp melting action

(*a*) Copper (*b*) Silver (*c*) Tin-lead alloy (*d*) Aluminium.

**21.** A 10 HP, 3 phase, 400 V, 50 Hz induction motor takes 15 A on full load. If the motor is to be started by using a direct-on-line starter, the current rating of the fuse will be

(*a*) 15 A (*b*) 30 A (*c*) 45 A (*d*) 60 A.

**22.** For a given size of motor, the fuse rating of a induction motor depends on

(*a*) shaft diameter of motor (*b*) nature of load to be applied on motor

(*c*) rpm of motor (*d*) method of starting the motor.

**23.** The starting current for a 3 phase, 400 V, 50 Hz induction motor will be least, when the motor is started by

(*a*) directly switching on the line (*b*) using DOL starter

(*c*) using autotransformer with 65% lagging

(*d*) using star-delta starter.

**24.** A fuse in a motor circuit provides protection against

(*a*) overload (*b*) short circuit and overload

(*c*) open circuit short circuit and overload

(*d*) none of the above.

25. For dc circuits, the fuse wire is inserted in
(*a*) positive circuit only (*b*) negative circuit only
(*c*) both positive as well as negative circuit
(*d*) either positive or negative circuit.

26. A fuse wire should have
(*a*) low resistance and high melting point
(*b*) low resistance and low melting point
(*c*) high resistance and high melting point
(*d*) high resistance and low melting point.

27. The rating of fuse in the circuit of a three phase induction motor will be highest when the motor is to be started by
(*a*) star delta starter—air break type (*b*) star delta starter—oil filled type
(*c*) autotransformer starter (*d*) DOL starter.

28. In a HRC fuse major portion of the heat generated is dissipated by
(*a*) conduction (*b*) convection
(*c*) radiation (*d*) convection and radiation.

29. Which fuse material is more susceptible to oxidation as compared to the remaining?
(*a*) Tin lead alloy (*b*) Copper (*c*) Silver.

30. In case of HRC fuses, HRC stands for
(*a*) Highly Regulated Current (*b*) Hot Resever Capacity
(*c*) High Resistance Circuit (*d*) High Rupturing Capacity.

31. Protection by fuses is generally not used beyond
(*a*) 10 A (*b*) 25 A (*c*) 100 A (*d*) 250 A.

32. Which of the following is used in liquid fuses?
(*a*) Carbon tetra chloride (*b*) Sulphur hexafluoride
(*c*) Distilled water (*d*) Transformer oil.

33. A fuse in never inserted in
(*a*) phase line (*b*) positive of dc circuit
(*c*) negative of dc circuit (*d*) neutral wire.

34. HRC fuses possess
(*a*) linear time — current characteristic
(*b*) inverse time — current characteristic
(*c*) parabolic time — current characteristic
(*d*) none of the above.

35. HRC fuses can be used upto
(*a*) 100 A (*b*) 1000 A (*c*) 33 kA (*d*) 330 kA.

36. Under fault conditions, the information to the circuit breaker is provided by
(*a*) rewirable fuse (*b*) HRC fuse (*c*) relay (*d*) all of the above.

37. In a single busbar system there will be complete shut down when
(*a*) fault occurs on neutral line
(*b*) fault occurs with respect to earthing
(*c*) two or more faults occur simultaneously
(*d*) fault occurs on the bus itself.

38. Outdoor type switchgear is generally used for voltages beyond
(*a*) 11 kV (*b*) 33 kV (*c*) 66 kV (*d*) 132 kV.

39. An isolator is usually a
(*a*) manually operated air switch (*b*) manually operated oil switch
(*c*) relay operated oil switch (*d*) relay controlled air switch.

40. A circuit breaker is an equipment which can open or close a circuit under
(*a*) no load conditions (*b*) full load conditions
(*c*) fault conditions (*d*) all above conditions.

41. A circuit beaker can be operated
(*a*) manually (*b*) automatically (*c*) by remote control (*d*) any of the above.

42. A short circuit is identified by
(*a*) voltage drop (*b*) voltage rise (*c*) heavy current flow (*d*) no current flow.

43. For a circuit breaker the contacts
(*a*) remain open during normal conditions and closed under fault conditions
(*b*) remain closed during normal conditions and open under fault conditions
(*c*) remain open during normal conditions as well as fault conditions
(*d*) closed during normal conditions as well as fault conditions.

44. If the percentage reactance of an element is 20% and the full load current is 50 A, then short circuit current will be
(*a*) 10 A (*b*) 50 A (*c*) 100 A (*d*) 250 A.

45. The most serious result of a major uncleared short circuit fault could be
(*a*) deterioration of insulation (*b*) heavy current in neutral
(*c*) blowing off of fuses (*d*) fire.

46. Which is the worst fault for a power system?
(*a*) Single line to ground fault (*b*) Line-to-line fault
(*c*) Double line to ground fault (*d*) Symmetrical fault.

47. In order to limit short circuit current in a power system
(*a*) isolators are used (*b*) reactors are used
(*c*) HRC fuses are used (*d*) earth wires are used.

48. The rating of a circuit breaker is generally determined on the basis of

(*a*) single line to ground fault current (*b*) line-to-line fault current

(*c*) double line to ground fault current (*d*) symmetrical fault current.

49. A 1000 kVA transformer with 5% reactance will have a reactance of

(*a*) 2.5% at 2000 kVA base (*b*) 5% at 2000 kVA base

(*c*) 10% at 2000 kVA base (*d*) 25% at 2000 kVA base.

50. Which of the following gives rise to symmetrical fault?

(*a*) Single phase to ground (*b*) Phase-to-phase

(*c*) Double phase to ground (*d*) Three phase short circuit.

51. The percentage reactance of a system upto the fault point is 25% and the base kVA is 4000. The short circuit kVA will be

(*a*) 1000 kVA (*b*) 4000 kVA (*c*) 8000 kVA (*d*) 16000 kVA.

52. A balanced 3 phase system consists of

(*a*) positive sequence components only (*b*) zero sequence currents only

(*c*) negative and zero sequence currents

(*d*) zero, negative and positive sequence currents.

53. Which of the following statement is incorrect?

(*a*) The presence of negative or zero sequence currents in a 3 phase system introduces unsymmetry

(*b*) The presence of negative or zero sequence currents in a 3 phase system indicates an abnormal condition in the circuit

(*c*) The vector sum of the positive and negative sequence currents of an unbalanced 3 phase system is non-zero

(*d*) In a 3 phase, 4 wire unbalanced system, the magnitude of zero sequence component is one-third of the component in the neutral wire.

54. Which of the following statement is incorrect?

(*a*) In a 3 phase unbalanced system the magnitude of negative sequence cannot exceed that of positive sequence component

(*b*) In a 3 phase unbalanced system, if the negative sequence components is greater than the positive sequence component, then the phase sequence of the resultant system would be reversed.

(*c*) The current of a single phase load drawn from a 3 phase system comprises equal positive, negative and zero sequence components

(*d*) In a 3 phase, 4 wire unbalanced system, the magnitude of zero sequence component is one fourth of the current in the neutral wire.

55. The current from neutral to ground connection is 24 A. The zero sequence current in each phase will be

(*a*) 24 A (*b*) 8 A (*c*) $8\sqrt{3}$ A (*d*) 4 A.

**56.** The positive sequence impedance of a transmission line is
(*a*) always zero (*b*) one-third of negative sequence impedance
(*c*) equal to negative sequence current (*d*) three times the negative sequence current.

**57.** When '*a*' is the operator, then $(a - a^2)$ is
(*a*) zero (*b*) 1 (*c*) $j\sqrt{3}$ (*d*) $j\,3$.

**58.** When '*a*' is the operator, then $1 + a + a^2 =$
(*a*) 1 (*b*) 0 (*c*) – 1 (*d*) $\sqrt{3}$.

**59.** When '*a*' is the operator, then $a^2 =$
(*a*) 0 (*b*) $-0.5 - j\,0.866$ (*c*) $0.5 - j\,0.866$ (*d*) $0.5 + j\,0.866$.

**60.** When '*a*' is the operator, then $a^3 =$
(*a*) 0 (*b*) – 1 (*c*) 1 (*d*) $j\sqrt{3}$.

**61.** A high speed circuit breaker can complete its operation in
(*a*) 1 micro second (*b*) 1 milli second (*c*) 2 – 3 cycles (*d*) 4 – 8 cycles.

**62.** The power factor of arc in a circuit breaker is
(*a*) zero leading (*b*) zero lagging (*c*) 0.8 lagging (*d*) unity.

**63.** In air blast circuit breakers the pressure of air is around
(*a*) 10 to 15 cm of water (*b*) 10 to 15 cm of mercury
(*c*) 2 to 5 kg/cm$^2$ (*d*) 20 to 25 kg/cm$^2$.

**64.** Which figure represents the symbol for a lightening arrester?
(*a*) figure *A* (*b*) figure *B* (*c*) figure *C* (*d*) figure *D*.

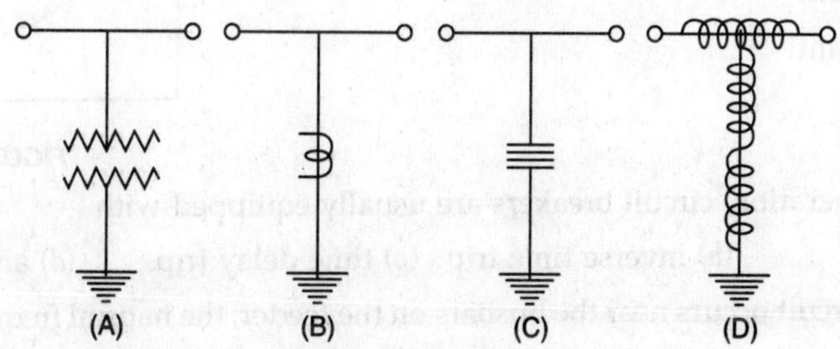


***FIGURE 19.4***

**65.** In case of circuit breakers the break time is
(*a*) the same as opening time (*b*) the same as arc duration
(*c*) the sum of opening time and arc duration
(*d*) the sum of opening time, arc duration and resistor current duration.

**66.** The time instant from the instant of opening of the contacts to the instant of extinction of arc is known as
(*a*) arc clearance time (*b*) break time
(*c*) arc duration (*d*) none of the above.

**67.** The time interval from the instant of fault to the instant of energizing the trip coil is known as

(*a*) arcing time (*b*) operating time (*c*) breaking time (*d*) protection time.

**68.** The time interval from the instant of energizing the trip coil to the instant of opening of the contact is known as

(*a*) arcing time (*b*) operating time (*c*) breaking time (*d*) total clearing time.

**69.** The time interval from the instant of fault to the instant of extinction of arc is known as

(*a*) clearing time (*b*) break time (*c*) make time (*d*) none of the above.

**70.** The braking capacity of a 3 phase circuit breaker is given by

(*a*) service voltage × rated symmetrical current

(*b*) 3 × service voltage × rated symmetrical current

(*c*) $\sqrt{3}$ × service voltage × rated symmetrical current

(*d*) none of the above.

**71.** The advantage of using oil as arc quenching medium in circuit breakers is

(*a*) absorption of arc energy (*b*) oil acts as insulator

(*c*) oil has high dielectric strength (*d*) all of the above.

**72.** In Fig. 19.5 if equation for curve *A* is $I^2.t$ = constant, then equation for curve *B* could be

(*a*) $I^0t$ = constant

(*b*) $It$ = constant

(*c*) $I^2t$ = constant

(*d*) $I^4t$ = constant.

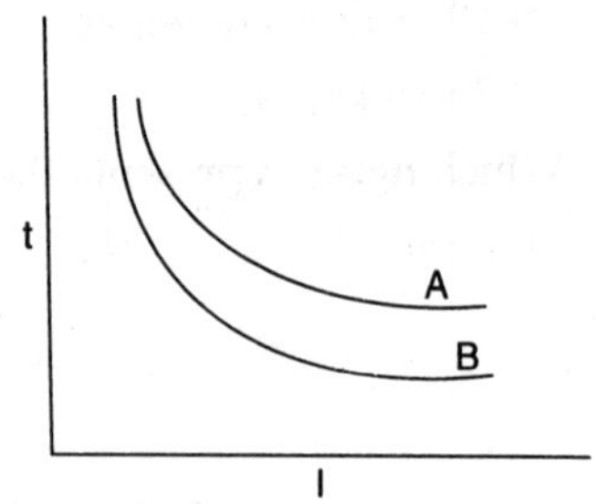


***FIGURE 19.5***

**73.** For remote operation, circuit breakers are usually equipped with

(*a*) shunt trip (*b*) inverse time trip (*c*) time delay trip (*d*) any of the above.

**74.** When short circuit occurs near the busbars on the feeder, the natural frequency of oscillatory circuit will be

(*a*) $2\pi\sqrt{LC}$ (*b*) $\frac{2\pi}{\sqrt{LC}}$ (*c*) $\frac{2\pi}{\sqrt{L/C}}$ (*d*) $\frac{1}{2\pi\sqrt{LC}}$.

**75.** Which of the following is not a part of circuit breaker?

(*a*) Explosion pot (*b*) Current transformer

(*c*) Fixed and moving contacts (*d*) Operating mechanism.

**76.** The active recovery voltage will be maximum when

(*a*) power factor is zero (*b*) power factor is lagging

(*c*) power factor is leading (*d*) power factor is unity.

**77.** In the event of a fault on connected circuit, a circuit breaker will operate
(*a*) manually (*b*) manually through control switch
(*c*) automatically (*d*) depends on the design.

**78.** The material used for circuit breaker contacts should have
(*a*) high electrical conductivity and low thermal conductivity
(*b*) high electrical as well as thermal conductivity
(*c*) low electrical conductivity and high thermal conductivity
(*d*) low electrical as well as thermal conductivity.

**79.** In which of the following circuit breaker the arc energy produced will be least?
(*a*) Air blast circuit breaker (*b*) Minimum oil circuit breaker
(*c*) Plain oil break circuit breaker (*d*) Self generated pressure oil circuit breaker.

**80.** Air blast circuit breakers are used above
(*a*) 11 kV (*b*) 33 kV (*c*) 66 kV (*d*) 132 kV.

**81.** In case of $SF_6$ circuit breakers the time of interruption is nearly
(*a*) 1 micro second (*b*) 3 milli seconds (*c*) 3 seconds (*d*) 3 cycles.

**82.** Which of the following is not true for $SF_6$ gas?
(*a*) $SF_6$ gas has insulating properties (*b*) $SF_6$ gas is non-inflammable
(*c*) $SF_6$ gas is poisonous (*d*) $SF_6$ gas is heavier than air.

**83.** The dielectric strength of $SF_6$ gas at atmospheric pressure is nearly
(*a*) same as that of air (*b*) less than that of air
(*c*) 2.5 times that of air (*d*) 10 to 15 times that of air.

**84.** In $SF_6$ circuit breakers, during tripping operation the gas at
(*a*) 5 $kg/cm^2$ is fed to the interrupting nozzle
(*b*) 15 $kg/cm^2$ is fed to the interrupting nozzle
(*c*) 50 $kg/cm^2$ is fed to the interrupting nozzle
(*d*) 100 $kg/cm^2$ is fed to the interrupting nozzle.

**85.** In air blast circuit breaker to interrupt the arc, compressed air is used at a pressure of about
(*a*) 1 $kg/cm^2$ (*b*) 2 to 3 $kg/cm^2$ (*c*) 5 to 8 $kg/cm^2$ (*d*) 18 to 20 $kg/cm^2$.

**86.** In circuit breakers, the material for arcing contacts is usually
(*a*) copper (*b*) copper-tungsten alloy
(*c*) *Y*-alloy (*d*) mica.

**87.** The circuit breaker preferred for extra high tension application is
(*a*) air blast circuit (*b*) $SF_6$ circuit breaker
(*c*) minimum oil circuit breaker (*d*) none of the above.

**88.** A circuit breaker perferred for railway work is
(*a*) $SF_6$ circuit breaker (*b*) minimum oil circuit breaker
(*c*) bulk oil circuit breaker (*d*) air break circuit breaker.

**89.** An arc in a circuit behaves like

(*a*) infinite resistance

(*b*) a resistance decreasing with voltage rise across the arc

(*c*) an inductive reactance (*d*) a capacitive reactance.

**90.** A magnetic circuit breaker has

(*a*) instantaneous trip action while a thermal circuit breaker has delayed trip action

(*b*) instantaneous trip action and a thermal circuit breaker has also instantaneous trip action

(*c*) delayed trip action as in case of a thermal circuit breaker

(*d*) delayed trip action while a thermal circuit breaker has instantaneous trip action.

**91.** Relays can be designed to respond to changes in

(*a*) voltage (*b*) current (*c*) light intensity (*d*) any of the above.

**92.** The power factor of an arc is

(*a*) zero leading (*b*) zero lagging (*c*) 0.5 (*d*) unity.

**93.** The speed of operation of a relay depends on all of the following factors EXCEPT

(*a*) spring tension (*b*) atmospheric temperature

(*c*) armature core air gap (*d*) rate of flux built up.

**94.** The operating time for a high speed relay is usually

(*a*) 1 second (*b*) 1 milli second (*c*) 1 to 2 cycles (*d*) 10 to 15 cycles.

**95.** The operation of a distance relay is based on the measurement of any of the following EXCEPT

(*a*) admittance (*b*) reactance (*c*) impedance (*d*) frequency.

**96.** A DC shunt relay consists of

(*a*) few turns of thick wire (*b*) several turns of thick wire

(*c*) few turns of thin wire (*d*) several turns of thin wire.

**97.** Copper oxide is

(*a*) bad conductor of electricity while silver oxide is good conductor of electricity

(*b*) bad conductor of electricity just like silver oxide

(*c*) good conductor of electricity while silver oxide is bad conductor of electricity

(*d*) good conductor of electricity just like silver oxide.

**98.** If relay contacts get overheated the probable cause may be

(*a*) some foreign matter on the contact surface

(*b*) low contact pressure (*c*) slow opening of contacts

(*d*) any of the above.

**99.** In Fig. 19.6 which curve represents the inverse time characteristics

(*a*) curve *A*

(*b*) curve *B*

(*c*) curve *C*

(*d*) curve *D*.

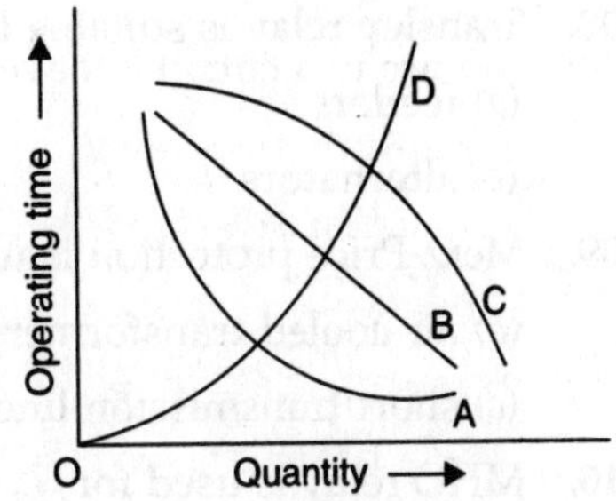


FIGURE 19.6

**100.** The operating time of a relay with inverse-time characteristics could be

(*a*) 2 – 3 cycles (*b*) 2 – 3 micro seconds

(*c*) 5 – 10 milli seconds (*d*) 5 – 10 seconds.

**101.** A single phasing relay is used on

(*a*) single phase motors (*b*) universal motors

(*c*) three phase motors (*d*) three phase transformers.

**102.** A differential relay responds to

(*a*) algebraic difference between two voltages

(*b*) algebraic difference between two currents

(*c*) algebraic difference between two electrical quantities

(*d*) vector difference between two electrical quantities.

**103.** Generally the temperature rise in electrical machines follows

(*a*) exponential law (*b*) quadratic law (*c*) square law (*d*) linear law.

**104.** In distance relay the operation depends upon

(*a*) voltage (*b*) current

(*c*) voltage to current ratio (*d*) none of the above.

**105.** Buchholz's relay is

(*a*) located on the top of conservator tank

(*b*) is installed inside the breather

(*c*) is connected in the pipe connecting main tank of transformer and conservator

(*d*) part of circuit breaker installed near the transformer.

**106.** Buchholz's relay provides protection to the transformer from

(*a*) lightening faults (*b*) dielectric breakdown

(*c*) internal faults (*d*) external faults.

**107.** Which relay is used for protecting a transformer against frame leakage?

(*a*) directional relay

(*b*) instantaneous earth fault relay

(*c*) instantaneous over voltage relay

(*d*) instantaneous over current relay.

**108.** Transley relay is suitable for

(*a*) feeders (*b*) welding transformers

(*c*) alternators (*d*) none of the above.

**109.** Merz-Price protection is used for

(*a*) air cooled transformers (*b*) alternator

(*c*) short transmission lines (*d*) long transmission lines.

**110.** MHO relay is used for

(*a*) transmission lines (*b*) feeders

(*c*) alternators (*d*) transformers.

**111.** The pilot relay is provided to obtain

(*a*) preset tripping (*b*) delayed tripping

(*c*) high speed tripping (*d*) none of the above.

**112.** For the protection of radial feeders, the relay used must have

(*a*) linear characteristics (*b*) inverse time characteristics

(*c*) exponential time characteristics (*d*) none of the above.

**113.** The time $V_s$ impedance characteristic of definite distance impedance is represented in Fig. 19.7 by

(*a*) curve *A*

(*b*) curve *B*

(*c*) curve *C*

(*d*) curve *D*.

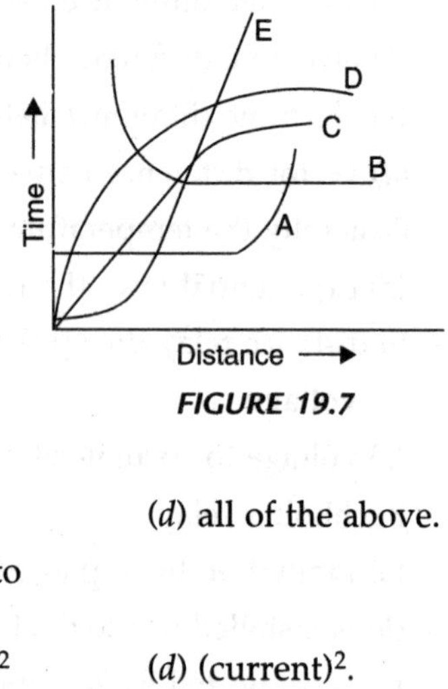


*FIGURE 19.7*

**114.** Reactance relays are used for phase fault in

(*a*) long line (*b*) medium line (*c*) short line (*d*) all of the above.

**115.** In induction type relays, torque is directly proportional to

(*a*) current (*b*) $\dfrac{1}{\text{current}}$ (*c*) $(\text{current})^{1/2}$ (*d*) $(\text{current})^2$.

**116.** The relay preferred for phase fault along long transmission line is

(*a*) reactance relay (*b*) impedance relay (*c*) MHO relay (*d*) Buchholz relay.

**117.** The relay preferred for phase fault along medium transmission line is

(*a*) MHO relay (*b*) Buchholz relay (*c*) impedance relay (*d*) reactance relay.

**118.** Which of following relay has inherent directional characteristics?

(*a*) Impedance relay (*b*) Reactance relay (*c*) MHO relay (*d*) None of the above.

**119.** Relay generally used for phase fault is

(*a*) distance relay (*b*) induction relay (*c*) over current relay (*d*) MHO relay.

**120.** Buchholz's relay is invariably used on

(*a*) long transmission lines (*b*) short transmission lines

(*c*) transformers (*d*) alternators.

**121.** Which of the following is not a distance relay?

(*a*) Reactance relay (*b*) Induction relay (*c*) Buchholz relay (*d*) MHO relay.

**122.** Earth fault relays are invariably

(*a*) short-operating time relays (*b*) long-operating time relays

(*c*) directional relays (*d*) non-directional relays.

**123.** Which of the following is a distance relay?

(*a*) MHO relay (*b*) Buchholz relay (*c*) Directional relay (*d*) Induction relay.

**124.** In case the fault occurs near the impedance relay

(*a*) the $V/I$ ratio will be zero

(*b*) the $V/I$ ratio will remain unaffected

(*c*) the $V/I$ ratio will be higher than its value in case the fault occurred away from the relay

(*d*) the $V/I$ ratio will be lower than its value in case the fault occured away from the relay.

**125.** A thermal protection circuit provides protection against

(*a*) over load (*b*) short circuit (*c*) over voltage (*d*) temperature rise.

**126.** The general characteristic of a relay is given by

$$I^n \cdot t = \text{constant}$$

the relation for definite time relay will be

(*a*) $I^0 \cdot t$ = constant (*b*) $I \cdot t$ = constant (*c*) $I^2 \cdot t$ = constant (*d*) $I^8 \cdot t$ = constant.

**127.** Which of the following relation represents the operating characteristic of an impedance relay?

(*a*) $t = aI - bV$ (*b*) $t = aI^2 - bV^2$ (*c*) $t = aI + bV^2$ (*d*) $t = aI^2 + bV$.

**128.** The purpose of using reactors in electrical circuits is to

(*a*) absorb reactive power in transmission lines

(*b*) limit the magnitude of short-circuit current

(*c*) suppress arcing grounds

(*d*) any of the above.

**129.** The use of air cored reactors is generally restricted upto

(*a*) 1 kV (*b*) 11 kV (*c*) 33 kV (*d*) 132 kV.

**130.** The use of oil immersed reactors is generally restricted upto

(*a*) 1 kV (*b*) 11 kV

(*c*) 33 kV (*d*) there is no restriction of voltage normally.

**131.** The magnitude of short circuit current does not depend on

(*a*) voltage of the system at the point of fault

(*b*) size of conductor

(*c*) reactance between the generator and the fault

(*d*) generating capacity feeding the fault.

**132.** Reactors for the protection of transmission lines are connected in

(*a*) series (*b*) parallel (*c*) delta (*d*) star.

**133.** Any voltage surge travelling on the transmission line will first enter

(*a*) switchgear (*b*) current transformer

(*c*) circuit breaker (*d*) lightening arrester.

**134.** A lightening arrester is always located

(*a*) before the isolator (*b*) before the switchgear

(*c*) before the main transformer (*d*) before the current transformer.

**135.** An earth conductor provided on the top of the transmission line provides protection against

(*a*) travelling waves (*b*) direct lightening stroke

(*c*) electrostatically induced voltage due to a charged cloud

(*d*) all of the above.

**136.** A lightening arrester provides

(*a*) a high resistance path between line and ground, during operation

(*b*) a low resistance path between line and ground, during operation

(*c*) a low impedance path between line and ground, during operation

(*d*) a high impedance path between line and ground, during operation.

**137.** The main purpose of providing lightening arrester in a substation is to protect

(*a*) circuit breakers (*b*) relays (*c*) transformers (*d*) civil structure.

**138.** Basically a lightening arrester is a

(*a*) surge alternator (*b*) surge reflector

(*c*) surge absorber (*d*) surge diverter.

**139.** Expulsion type lightening arresters are used with systems having operating voltages upto

(*a*) 1000 volts (*b*) 2 kV (*c*) 11 kV (*d*) 33 kV.

**140.** Which of the following needs to be protected against direct stroke due to lightening?

(*a*) Power station (*b*) Substation

(*c*) Overhead transmission lines (*d*) All of the above.

**141.** Which of the following system suffers most from arcing grounds?

(*a*) Solid ground neutral system (*b*) Insulated neutral system

(*c*) Resistance grounded neutral system (*d*) Reactance grounded neutral system.

142. Which system will need the lightening arrester of least voltage rating?
(*a*) Insulated neutral system (*b*) Solid ground neutral system
(*c*) Reactance grounded neutral system (*d*) Resistance grounded neutral system.

143. Which section of a power system generally suffers from more faults as compared to the remaining?
(*a*) Generators (*b*) Transformers
(*c*) Transmission lines (*d*) Underground cables.

144. Least number of faults are generally reported for
(*a*) cables (*b*) transformers (*c*) switchgears (*d*) transmission lines.

145. Batteries are provided at substations mainly for
(*a*) emergency lighting (*b*) voltage regulation
(*c*) trip circuits (*d*) meter movements.

146. The gap length of rod type surge diverter used for 132 kV line should be
(*a*) 10 cm (*b*) 15 cm (*c*) 50 cm (*d*) 65 cm.

147. Wave trap is used to trap waves of
(*a*) low frequency (*b*) power frequency
(*c*) higher frequency (*d*) microwave frequency.

148. When a transmission line is energized, the wave that propagates on it is
(*a*) current wave only (*b*) voltage wave only
(*c*) current and voltage waves (*d*) current, voltage and power factor waves.

149. Transient state stability is generally improved by
(*a*) using high speed governors on machines
(*b*) using low inertia machines
(*c*) dispensing with neutral grounding
(*d*) any of the above.

150. A three phase delta-star connected transformer is provided with Merz price percentage differential protection. This windings of current transformers will be connected in
(*a*) star-delta (*b*) star-star (*c*) delta-star (*d*) delta-delta.

151. Buchholz relay cannot be used on
(*a*) 500 kV transformers (*b*) 1000 kV transformers
(*c*) three phase transformers (*d*) air-cooled transformers.

152. A three phase power transformer is provided with star-delta connections. In order to protect against fault, the connections for current transformer should be in
(*a*) star-star (*b*) delta-star (*c*) delta-star (*d*) star-delta.

## ANSWERS

| | | | | | | |
|---|---|---|---|---|---|---|
| **1.** (*d*) | **2.** (*a*) | **3.** (*c*) | **4.** (*c*) | **5.** (*c*) | **6.** (*d*) | **7.** (*a*) |
| **8.** (*c*) | **9.** (*b*) | **10.** (*b*) | **11.** (*a*) | **12.** (*a*) | **13.** (*d*) | **14.** (*a*) |
| **15.** (*a*) | **16.** (*b*) | **17.** (*a*) | **18.** (*d*) | **19.** (*a*) | **20.** (*b*) | **21.** (*c*) |
| **22.** (*d*) | **23.** (*d*) | **24.** (*b*) | **25.** (*c*) | **26.** (*d*) | **27.** (*d*) | **28.** (*a*) |
| **29.** (*b*) | **30.** (*d*) | **31.** (*c*) | **32.** (*a*) | **33.** (*d*) | **34.** (*b*) | **35.** (*c*) |
| **36.** (*c*) | **37.** (*d*) | **38.** (*c*) | **39.** (*a*) | **40.** (*d*) | **41.** (*d*) | **42.** (*c*) |
| **43.** (*b*) | **44.** (*d*) | **45.** (*d*) | **46.** (*d*) | **47.** (*b*) | **48.** (*d*) | **49.** (*c*) |
| **50.** (*d*) | **51.** (*d*) | **52.** (*a*) | **53.** (*d*) | **54.** (*d*) | **55.** (*b*) | **56.** (*c*) |
| **57.** (*c*) | **58.** (*b*) | **59.** (*b*) | **60.** (*c*) | **61.** (*d*) | **62.** (*d*) | **63.** (*d*) |
| **64.** (*c*) | **65.** (*d*) | **66.** (*a*) | **67.** (*d*) | **68.** (*b*) | **69.** (*a*) | **70.** (*c*) |
| **71.** (*a*) | **72.** (*d*) | **73.** (*a*) | **74.** (*d*) | **75.** (*b*) | **76.** (*a*) | **77.** (*c*) |
| **78.** (*b*) | **79.** (*a*) | **80.** (*d*) | **81.** (*d*) | **82.** (*c*) | **83.** (*c*) | **84.** (*b*) |
| **85.** (*d*) | **86.** (*b*) | **87.** (*b*) | **88.** (*d*) | **89.** (*b*) | **90.** (*a*) | **91.** (*d*) |
| **92.** (*d*) | **93.** (*b*) | **94.** (*c*) | **95.** (*d*) | **96.** (*d*) | **97.** (*a*) | **98.** (*d*) |
| **99.** (*a*) | **100.** (*d*) | **101.** (*c*) | **102.** (*d*) | **103.** (*a*) | **104.** (*c*) | **105.** (*c*) |
| **106.** (*c*) | **107.** (*b*) | **108.** (*a*) | **109.** (*b*) | **110.** (*a*) | **111.** (*c*) | **112.** (*b*) |
| **113.** (*a*) | **114.** (*c*) | **115.** (*d*) | **116.** (*c*) | **117.** (*c*) | **118.** (*a*) | **119.** (*a*) |
| **120.** (*c*) | **121.** (*b*) | **122.** (*c*) | **123.** (*a*) | **124.** (*c*) | **125.** (*a*) | **126.** (*a*) |
| **127.** (*b*) | **128.** (*d*) | **129.** (*c*) | **130.** (*d*) | **131.** (*b*) | **132.** (*a*) | **133.** (*d*) |
| **134.** (*c*) | **135.** (*d*) | **136.** (*c*) | **137.** (*c*) | **138.** (*d*) | **139.** (*d*) | **140.** (*d*) |
| **141.** (*b*) | **142.** (*b*) | **143.** (*c*) | **144.** (*a*) | **145.** (*c*) | **146.** (*d*) | **147.** (*c*) |
| **148.** (*c*) | **149.** (*a*) | **150.** (*a*) | **151.** (*d*) | **152.** (*c*). | | |

# Chapter 20 UTILISATION

**Electric Drives.** Electric motors for industrial applications are being widely used these days. Both ac and dc motors are being used. The selection of an electric motor for any application depends on the following factors.

### Electrical Characteristics

1. Starting characteristics
2. Running characteristics
3. Speed control
4. Braking.

### Mechanical Characteristics

1. Transmission of power
2. Cooling
3. Noise
4. Type of enclosure.

### Size and Rating of Motors

1. Load cycle:
   (*i*) Continuous
   (*ii*) Intermittent
   (*iii*) Variable.

2. Overload capacity.

**Cost**

1. Initial cost
2. Cost of control gear
3. Running cost.

Characteristics of electric motors are given in table below:

| *Motor* | *Voltage limit* | *HP limit* | $\frac{I_S}{I_{FL}}$ | $\frac{T_S}{I_{FL}}$ | *Method of starting* | *Method of speed control* |
|---|---|---|---|---|---|---|
| **DC Motors** | | | | | | |
| 1. Series | 1.5 kV | 3000 | 2 | 3 | 1. Series resistance<br>2. Series parallel method | 1. By tapping the field<br>2. By field diverter<br>3. Variable resistance in series<br>4. Series parallel control |
| 2. Shunt | 3 kV | 25,000 | 2 | 2 | 1. Series resistance in armature | 1. Variable resistance in field circuit<br>2. Variable resistance in armature circuit |
| 3. Compound | 1.5 kV | 3000 | 2 | 2 – 3 | 1. Series resistance in armature | 1. Variable resistance in shunt field<br>2. Series field diverter |
| **AC Motors** | | | | | | |
| 4. Single phase series motor | 500 V | 3000 | 2 | 3 | 1. Variable voltage supply | 1. Voltage variation |
| 5. Slip-ring induction motor | 11 kV | 11000 | 2 | 2 | 1. Resistance in rotor circuit | 1. Variation in applied voltage<br>2. Cascading<br>3. Pole changing<br>4. Variation of frequency<br>5. Rotor rheostat control<br>6. Injecting emf in rotor |

| Motor | Voltage limit | HP limit | $\frac{I_S}{I_{FL}}$ | $\frac{T_S}{I_{FL}}$ | Method of starting | Method of speed control |
|---|---|---|---|---|---|---|
| 6. Squirrel cage induction motor | 11 kV | 300 | 1 | 1 | 1. Star delta starter<br>2. Auto-transformer<br>3. Resistance in stator circuit | 1. Pole changing<br>2. Variation of frequency |
| 7. Synchronous motor | 15 kV | 10,000 | 2 | 3 | 1. Damper winding<br>2. Prony brake | 1. Fixed speed |
| 8. Pole-changing induction motor | 11 kV | 300 | 6 | 1 | 1. Direct switching<br>2. Star-delta starter<br>3. Auto-transformer | 1. Pole changing<br>2. Variation of frequency |
| 9. Schrage motor | 1 kV | 1000 | 2 | 2.6 | 1. Low voltage application<br>2. Direct switching | 1. Shifting the brushes<br>2. Inserting impedance in secondary winding |
| 10. Single phase induction motor | 250 kV | 1 | 2 | 1.5 – 3.0 | 1. Repulsion start<br>2. Pole shading<br>3. Phase splitting by C, L or C and R | Voltage drop in series impedance |

**Illumination.** Some definitions are given below:

**Luminous Flux** is the rate of energy radiation in the form of light waves. It is the energy per second. The unit is lumen.

**Luminuous Intensity** in a given direction is the luminous flux per unit solid angle.

**Lumen** is defined as the luminous flux emitted in a unit solid angle by a source of one candle power. Thus

**Lumens = candle power × solid angle.**

**Candle Power** is the number of lumens emitted by a source in a unit solid angle in a given direction.

**Illumination** is the luminous flux received by a surface per unit area. The unit is lux *i.e.,* metre-candle. Thus

$$\text{Illumination} = \frac{\text{Flux}}{\text{Area}} = \frac{\text{Lumens}}{\text{Area}}$$

$$= \frac{\text{Candle power} \times \text{Solid angle}}{\text{Area}}$$

**Brightness** of a surface is the luminous intensity per unit projected area of the surface in a given direction.

The candle power of a surface will be different in different directions.

**Mean Horizontal Candle Power** (M.H.C.P.) is the mean of the candle powers in all directions in the horizontal plane containing the source of light.

**Mean Spherical Candle Power** (M.S.C.P.) is the mean or average of the candle power in all planes.

**Mean Hemi-spherical Candle Power** (M.H.S.C.P.) is the mean or average of the candle powers in all directions below the horizontal.

**Electric Lamps.** There are three types of electric lamps in use.

**Filament Lamps** also known as general light service (GLS) lamps use tungsten filament which illuminates when current passes through it. The hot resistance of tungsten filament is 15 to 18 times the cold resistance. For two filaments of the same material, at the same temperature the diameter $d$ is proportional to $I^{2/3}$. Most filaments radiate only 10 to 15 percent of input energy as light. Arc lamps operate at unity power factor. Inside the bulb either there is vacuum or inert gases like nitrogen and argon may be field. The filament may be used either as plane coil or as coiled coil.

**Carbon Arc Lamps.** These lamps use hard carbon electrodes and arc is sustained between them. The voltage necessary to sustain the arc is supplied and distance between the electrodes adjusted to main luminosity. The electrode is consumed while producing arc. The arc has negative resistance characteristic, therefore a stablising resistor has to be placed in series with the arc.

**Electric Discharge Lamps.** The operation of these lamps is based on the phenomenon of excitation and ionisation in a gas or vapour. The colours given by various vapours are given below :

| *Vapour or gas* | *Colour* | *Lumens/Watt* | |
|---|---|---|---|
| | | *Theoretical* | *Actual* |
| Neon | Red | 190 | 15–40 |
| Sodium vapour | Yellow | 475 | 40–50 |
| Mercury vapour (at low pressure) | Bluish green | 245 | 15–20 |
| Mercury vapour (at high pressure) | Bluish white | 300 | 20–30 |

**Fluorescent Lamps.** Various phosphors have been developed which when coated on the lamp surface, give different colours:

| *Phosphor* | *Colour* |
|---|---|
| Zinc Silicate | Green |
| Calcium tungstate | Blue |
| Magnesium tungstate | Bluish white |
| Cadmium Borate | Pink |
| Calcium halo phosphate | White |

**Electric Traction.** Various systems for traction in India are given below:

1. **Steam Locomotive.** This consists of a steam boiler which produces superheated steam at a pressure of 10 to 15 kg/cm$^2$. This steam drives two double acting, non-condensing steam engines which provide the motive power for the train. The speed is controlled by regulating the flow of steam to the engine.

   These engines gave now become obsolete and are being gradually withdrawn from services.

2. **Diesel Engines.** In these engines a multicylinder diesel engine is coupled to a dc generator which supplies power to the dc traction motors. These engines are available in 450 HP (shunter) to 2500 HP range. These are manufactured in India at Diesel Locomotive Workshops, Varanasi.

   These engines can be easily started from cold conditions. Their availability is higher as compared to a steam engine. Also these engines have better overall efficiency (around 25%) as compared to about 5 to 10 percent for steam locomotives.

   However these engines have shorter service life span and cost of maintenance is high.

3. **Electric Traction.** In this the power is supplied to the vehicle from an overhead wire suspended above the track. In dc system voltages of 1500 V and 3000 V are used. These are used on suburban trains for relatively shorter distances. AC systems use single phase voltage of 11 kV to 16 kV.

The motors used are dc series motors which have high starting torque. Single phase series motors are used on ac traction.

For dc motors the braking systems are:

1. Plugging
2. Rheostatic braking
3. Regenerative braking.

For single phase series motors, rheostatic braking and regenerative braking is used.

**Speed Time Curves.** These curves are graphical representation of actual run of a locomotive. A typical speed-time curve is shown in Fig. 20.1. In this *OA* represents the period during which the traction motors accelerate from rest, the current taken by the motors and the tractive effort are practically constant. This period is also known as notching up.

During *AB* the train continues to accelerate. During this period the torque gradually falls and speed increases. Finally the train reaches a speed at which the tractive effort is equal to the resistance. During *BC* the train runs at constant speed.

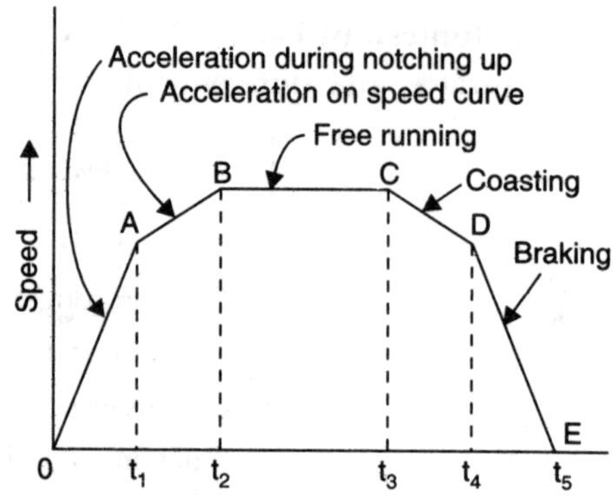


*FIGURE 20.1*

During *CD* the power supply to the train is cut off and the train moves due to its kinetic energy as at *C*. Naturally the speed will drop due to friction offered by the track and wind resistance. This is known as coasting. During *DE* the brakes are applied to bring the train to halt at desired place. The shape of the speed-time curve will be different for suburban service and main line service.

**Electric Heating.** Heating by the use of resistance wires is often preferred due to cleanliness, absence of polluting gases, accurate control of heating, quicker operation, higher efficiency and low maintenance cost. Electric heater is the well known appliance used for heating at home. Convective heaters, immersion heaters, cooking range are some of the appliances often used for heating purposes at home.

For industrial heating resistance ovens are used. Nichrome heating element is often used in heating appliances. In a resistance heating oven, number of resistances are used. The temperature inside the furnace may be controlled by any one of the following means:

1. Control on operating voltage by auto-transformer.
2. Control of current by
   (*i*) Cutting off some of the parallel resistances.
   (*ii*) Using star-delta connections for resistance bank.
   (*iii*) Use of external resistance to dissipate part of energy.
3. Automatic temperature control—in this the supply is cut off frequently so that desired temperature inside the furnace is maintained.

**High Frequency Induction Heating.** In this case heat is produced due to eddy currents produced within the material to be heated. The heat is generated near the surface of the work piece. The area of the surface over which heat is produced can be accurately controlled. The amount of heat produced can also be accurately controlled by suitable timing devices. This method can be used for production of heat in vacuum or any other inert atmosphere.

This method of heating is used primarily for surface hardening of steels. The surface to be heated is surrounded by a copper tube coil which produces a high frequency (400 kHz) a magnetic field. The coil is water cooled. The depth of penetration of heat is controlled by frequency and time. Crank bearings, geartooth, cams, blades of saws, etc. need surface hardening.

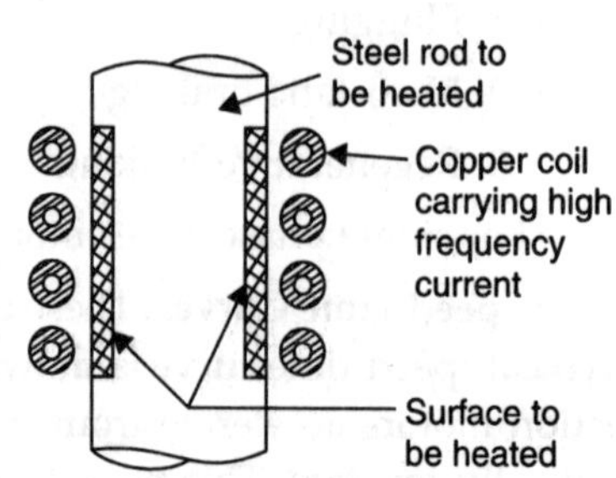


*FIGURE 20.2*

**Dielectric Heating.** Non-metals like wood, plastics etc. can be heated by using high frequency current (10 MHz to 30 MHz) and using higher voltages (upto 30 kV). This method is used for drying of rayon cakes in textile units, dehydration of food stuffs, gluing of laminated glass, rubber vulcanizing, drying of explosives, etc.

**Electric Welding.** One common method of electric welding is use of electric arc for producing heat. The arc is produced by using ac or dc current. For ac current, generally a transformer is used. This gives low voltage high current output. Arc is struck between welding electrode (often coated with flux) and the metal to be welded. Depending upon the size of the job the size of electrode and current rating is fixed. For dc arc welding motor generator sets are used.

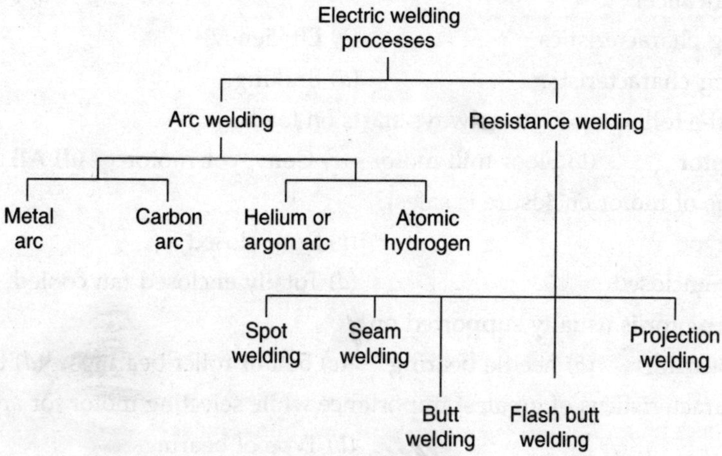


***FORURE 20.3*** Various methods for electric welding.

**Resistance Welding.** In this case a heavy current is passed through the joint required to be welded. The heat is produced due to the resistance of the joint and is proportional to $I^2R$. As a result of heat produced at the joint the fusion takes place and the welded joint is produced. No separate electrode is needed in this case. Various methods of resistance welding use this principle and are only modifications to suit specific requirements of the welding.

## OBJECTIVE TYPE QUESTIONS

### Electric Drives

1. Which of the following process essentially needs direct current?
   (*a*) Welding (*b*) Heating (*c*) Refrigeration (*d*) Electroplating.
2. For which of the following application dc motors are still preferred?
   (*a*) High starting torque (*b*) Variable speed drive
   (*c*) Reversibility (*d*) High efficiency operation.
3. For traction work which motor is preferred?
   (*a*) Synchronous motor (*b*) Three phase induction motor
   (*c*) DC series motor (*d*) Universal motor.

4. Which one of the following will be cheaper?
(*a*) A 200 HP ac three phase motor (*b*) Eight motors of 25 HP each
(*c*) Twenty motors of 10 HP each
(*d*) Five motors of 20 HP each and ten motors of 10 HP each.
5. In a paper mill, where constant speed is required
(*a*) group drive is preferred (*b*) individual drive is preferred
(*c*) ac motors are preferred (*d*) synchronous motors are preferred.
6. While selecting an electric motor for a floor mill, which electrical characteristic will be of least significance?
(*a*) Starting characteristics (*b*) Efficiency
(*c*) Running characteristics (*d*) Braking.
7. Which of the following motor always starts on load?
(*a*) Fan motor (*b*) Floor mill motor (*c*) Conveyor motor (*d*) All of the above.
8. Which type of motor enclosure is safest?
(*a*) Open type (*b*) Semi closed
(*c*) Totally enclosed (*d*) Totally enclosed fan cooled.
9. Rotor of a motor is usually supported on
(*a*) bush bearings (*b*) needle bearing (*c*) ball or roller bearings (*d*) thrust bearings.
10. Which characteristic is of greatest importance while selecting motor for an air conditioner?
(*a*) Noise (*b*) Type of bearings
(*c*) Type of enclosure (*d*) Arrangement for transmission or power.
11. Which part of motor needs maximum attention for maintenance?
(*a*) Rotor winding (*b*) Stator winding (*c*) Bearings (*d*) Frame.
12. The load cycle for a motor driving a power press will be
(*a*) continuous (*b*) variable load
(*c*) intermittent and variable load (*d*) continuous but periodical.
13. Which accessary is essentially needed while selecting a motor?
(*a*) Bearings (*b*) Starter (*c*) Foundation pedestal (*d*) Pulley.
14. The diameter of the rotor shaft for an electric motor depends on
(*a*) horse power only (*b*) rpm only
(*c*) horse power and rpm (*d*) horse power, rpm and power factor.
15. Which of the following is not a part of ball bearing?
(*a*) Cage (*b*) Bush (*c*) Inner race (*d*) Outer race.
16. While selecting a motor for centrifugal pump, which feature will be of least significance?
(*a*) Horse power (*b*) Operating speed
(*c*) Speed control (*d*) Starting characteristics.
17. Motor for which application has to start frequently with high acceleration?
(*a*) Lifts and hoists (*b*) Floor mill (*c*) Centrifugal pump (*d*) Oil expeller.

**18.** Which of the following application needs frequent starting and stopping of electric motor?
(*a*) Air conditioner (*b*) Lifts and hoists (*c*) Grinding mill (*d*) Paper mill.

**19.** The starting torque of a dc motor is independent of
(*a*) speed
(*b*) flux
(*c*) armature current
(*d*) flux and armature current.

**20.** Which of the following is not true for a dc shunt motor?
(*a*) Torque varies as armature current
(*b*) Torque-armature current curve is a straight line
(*c*) Torque is zero for zero armature current
(*d*) This motor is not suitable for heavy duty staring.

**21.** Which method of starting a three phase induction motor needs six terminals?
(*a*) Auto-transformer
(*b*) Star delta
(*c*) Resistance starting
(*d*) None of the above.

**22.** While using star-delta starter
(*a*) starting current is reduced
(*b*) starting voltage is reduced
(*c*) both starting current as well as voltage are reduced
(*d*) neither starting current nor starting voltage is reduced.

**23.** In which method of starting three phase induction motors, the starting voltage is not reduced?
(*a*) Star-delta starter
(*b*) Slip ring starter
(*c*) Auto-transformer starter
(*d*) None of the above.

**24.** During star-delta starting of three phase induction motor, the factor by which the line voltage impressed on each phase is reduced during starting is
(*a*) $\frac{1}{6}$ (*b*) $\frac{1}{3}$ (*c*) $\frac{1}{\sqrt{3}}$ (*d*) $\frac{1}{3\sqrt{3}}$.

**25.** The cost of an induction motor will increase as
(*a*) horsepower rating and operating speed increases
(*b*) horsepower rating and operating speed decreases
(*c*) horsepower rating increases but rpm decreases
(*d*) horsepower rating decreases but rpm increases.

## Questions 26 to 29 Refer to Data Given Below:

A 30 HP, 440 V, 4 pole, 50 Hz cage type induction motor with a mesh connected stator takes a full-load current of 33 A and has a slip of 4%. The impedance per phase is 3.5 ohms.

**26.** In which case the starting torque will be least?
(*a*) direct switching
(*b*) star-delta starting
(*c*) starting with a star connected auto-transformer with 60% tap
(*d*) starting with a series-parallel switch which connects the sections of the winding in series at starting and in parallel when running.

27. The starting torque will be maximum when a motor is started by
    (*a*) direct switching (*b*) star-delta starting
    (*c*) starting with a start connected auto-transformer with 60% tap
    (*d*) starting with a series-parallel switch which connects the sections of the winding in series at starting and in parallel when running.
28. The starting current will be least when started by
    (*a*) direct switching (*b*) star-delta starter (*c*) auto-transformer (*d*) series parallel switch.
29. In which two methods of starting the starting current will be the same?
    (*a*) Direct switching and star-delta starting
    (*b*) Star delta starting and series parallel switch starting
    (*c*) Series parallel switching and direct switching
    (*d*) Star delta starting and auto transformer starting.
30. Which of the following is a constant speed motor?
    (*a*) Schrage motor (*b*) Induction motor (*c*) Universal motor (*d*) Synchronous motor.
31. Which of the following application essentially needs constant speed operation?
    (*a*) Refrigerator (*b*) Water pump (*c*) Clock timer (*d*) Ceiling fan.
32. Which of the following motor has series characteristics?
    (*a*) Repulsion motor (*b*) Shaded pole motor
    (*c*) Capacitor start motor (*d*) None of the above.
33. In which motor the speed can be varied by changing the position of brushes?
    (*a*) Induction motor (*b*) Schrage motor (*c*) Slip ring motor (*d*) Repulsion motor.
34. In which of the following application, variable speed operation is preferred?
    (*a*) Water pump (*b*) Ceiling fan (*c*) Refrigerator (*d*) Exhaust fan.
35. Which of the following method can be used for changing the flux per pole in series motors ?
    (*a*) Diverter field control (*b*) Tapped field control
    (*c*) Series-parallel control (*d*) Any of the above.

**Questions 36 to 39 Refer to Fig. 20.4:**

Speed-torque characteristics for motors are shown in Fig. 20.4.

36. Which curve represents the speed-torque characteristics for dc shunt motor?
    (*a*) curve *A*
    (*b*) curve *B*
    (*c*) curve *C*
    (*d*) curve *D*.

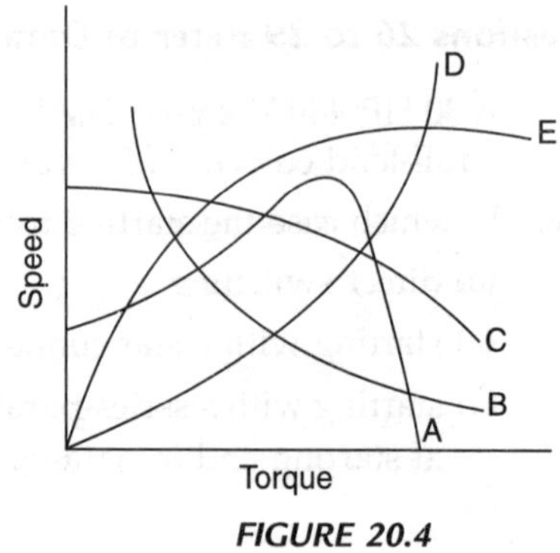


***FIGURE 20.4***

37. Which curve represents the torque-speed characteristics for a dc series motor?
(*a*) curve *A* (*b*) curve *B* (*c*) curve *C* (*d*) curve *E*.

38. Which curve represents the torque-speed characteristics for slip-ring induction motor?
(*a*) curve *A* (*b*) curve *B* (*c*) curve *C* (*d*) curve *E*.

39. Which curve represents the torque-speed characteristics for a synchronous motor?
(*a*) curve *D* (*b*) curve *E* (*c*) curve *B* (*d*) none of the curves.

40. A pony motor is used for the starting of a
(*a*) schrage motor (*b*) synchronous motor
(*c*) squirrel cage induction motor (*d*) none of the above.

41. Which of the following motor has relatively least range of speed control?
(*a*) DC shunt motor (*b*) Schrage motor
(*c*) Sychronous motor (*d*) Slip ring induction motor.

42. Which of the following pair of motors have almost identical torque-speed characteristics?
(*a*) Schrage motor and dc shunt motor
(*b*) Synchronous motor and slip-ring induction motor
(*c*) Single phase induction motor and dc compound motor
(*d*) DC series motor and squirrel cage induction motor.

43. The ratio of starting current to full load current can be highest in case of
(*a*) pole changing induction motor (*b*) schrage motor
(*c*) slip ring induction motor (*d*) dc shunt motor.

44. Speed control by injecting emf in the rotor circuit is possible in case of
(*a*) synchronous motor (*b*) slip-ring inducting motor
(*c*) schrage motor (*d*) dc shunt motor.

45. Which of the following motor will be preferred for motor generator set?
(*a*) slip ring induction motor (*b*) squirrel cage induction motor
(*c*) pole changing induction motor (*d*) synchronous motor.

46. Which motor has widest variety of methods for speed control?
(*a*) DC shunt motor (*b*) synchronous motor
(*c*) slip-ring induction motor (*d*) schrage motor.

47. Which of the following motor has relatively wider range of speed control?
(*a*) Squirrel cage induction motor (*b*) Slip ring induction motor
(*c*) Synchronous motor (*d*) DC shunt motor.

48. Which of the following motor has least value of starting torque to full load torque ratio?
(*a*) Squirrel cage induction motor (*b*) Slip ring induction motor
(*c*) DC series motor (*d*) DC shunt motor.

**49.** Which of the following method of starting cannot be used in case of squirrel cage conduction motors?

(*a*) Star-delta starting (*b*) Auto-transformer starting

(*c*) Resistance in rotor circuit (*d*) Resistance in stator circuit.

**Questions 50 to 53 Refer to Fig. 20.5:**

Speed torque characteristic for electric motors are shown in Fig. 20.5.

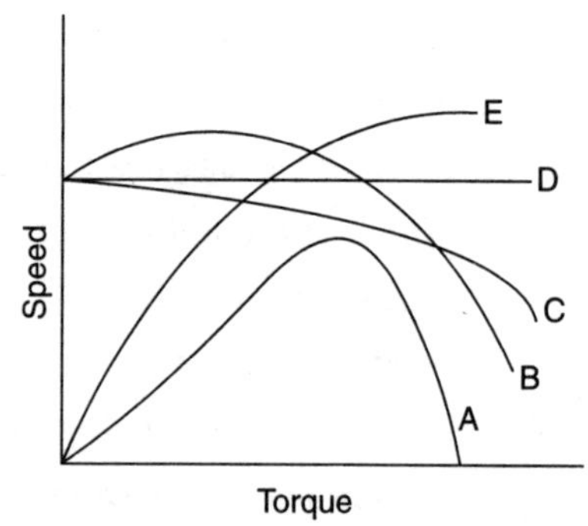


*FIGURE 20.5*

**50.** Which curve represents speed-torque characteristics for synchronous motor?

(*a*) curve *A*

(*b*) curve *B*

(*c*) curve *C*

(*d*) curve *D*.

**51.** The speed-torque characteristic for single phase induction motor is represented by

(*a*) curve *A* (*b*) curve *B* (*c*) curve *C* (*d*) curve *E*.

**52.** Speed torque characteristics for schrage motor is represented by

(*a*) curve *B* (*b*) curve *C* (*c*) curve *E* (*d*) none of the curves.

**53.** Speed-torque characteristics for squirrel cage induction motor is represented by

(*a*) curve *A* (*b*) curve *B* (*c*) curve *E* (*d*) none of the curves.

**54.** Flame proof motors are used in

(*a*) moist atmospheres (*b*) explosive atmospheres

(*c*) steel mills (*d*) paper mills.

**55.** For under water movement of submarine, power is provided by

(*a*) diesel engines (*b*) steam accumulators

(*c*) batteries (*d*) solar cells.

**56.** For a ship, power required for propeller is proportional to

(*a*) $\frac{1}{\text{speed}}$ (*b*) speed (*c*) $(\text{speed})^2$ (*d*) $(\text{speed})^3$.

**57.** In which of the following applications the load on motor changes in a cyclic order?

(*a*) Rolling mills (*b*) Cranes (*c*) Electric shovels (*d*) All of the above.

**58.** A motor has to perform the following duty cycle:

100 kW for 10 minutes, no load for 5 minutes, 60 kW for 8 minutes, no load for 4 minutes, which is repeated indefinitely. The suitable size of the motor will be

(*a*) 100 kW (*b*) 80 kW (*c*) 60 kW (*d*) 25 kW.

59. Which of the following machine has heavy fluctuation of load?
(*a*) Lathe (*b*) Planner
(*c*) Punching machine (*d*) Printing machine.

60. Battery operated scooter for braking uses
(*a*) regenerative braking (*b*) reheostatic braking
(*c*) plugging (*d*) mechanical braking.

## Illumination

61. The velocity of light is
(*a*) $10^8$ cm/sec (*b*) $10^8$ m/sec (*c*) $3 \times 10^8$ cm/sec (*d*) $3 \times 10^{10}$ cm/sec.

62. The range of wavelengths that can produce the sensation of sight is
(*a*) $4 \times 10^{-5}$ to $7.5 \times 10^{-5}$ cm (*b*) $4 \times 10^{-6}$ to $7.5 \times 10^{-6}$ cm
(*c*) $4 \times 10^{-7}$ to $7.5 \times 10^{-7}$ cm (*d*) $4 \times 10^{-8}$ to $7.5 \times 10^{-7}$ cm.

63. When $v$ is the velocity of propagation, $\lambda$ is the wavelength and $f$ is the frequency, then
(*a*) $f = \frac{\lambda}{v}$ (*b*) $fv = \lambda$ (*c*) $f\lambda = \frac{1}{v}$ (*d*) $v = f\lambda$.

64. 1 Angstrom unit is equal to
(*a*) $1^{-8}$ cm (*b*) $10^{-10}$ cm (*c*) $10^{-8}$ m (*d*) $10^{-8}$ mm.

65. The wavelength of yellow light is around
(*a*) 4000 A.u. (*b*) 4500 A.u. (*c*) 5000 A.u. (*d*) 6000 A.u.

66. A solid angle is expressed in terms of
(*a*) radians (*b*) radians/m (*c*) steradians (*d*) grads.

67. Luminous flux is
(*a*) the rate of energy radiation in the form of light waves
(*b*) the light energy, radiated by sun
(*c*) the part of light energy, radiated by sun, which is received on earth
(*d*) none of the earth.

68. The unit of luminous flux is
(*a*) watt (*b*) watt/$m^2$ (*c*) lumen (*d*) lumen/$m^2$.

69. The luminous flux received by a surface per unit area is called
(*a*) candle power (*b*) illumination
(*c*) brightness (*d*) brightness co-efficient.

70. According to Lambert's law, the illumination on a surface is proportional to
(*a*) $\cos^2 \theta$ (*b*) $\cos^3 \theta$ (*c*) $\frac{1}{\cos^3 \theta}$ (*d*) $\frac{1}{\cos^2 \theta}$.

**71.** One lux is equal to

(*a*) one lumen per square millimetre (*b*) one lumen per square centimetre
(*c*) one lumen per square metre (*d*) one lumen per square foot.

**72.** What percentage of the input energy is radiated by filament lamps?

(*a*) 10 to 15 percent (*b*) 25 to 40 percent (*c*) 50 to 70 percent (*d*) 75 to 95 percent.

**73.** The hot resistance of tungsten is

(*a*) $\frac{1}{15}$ to $\frac{1}{20}$ of cold resistance (*b*) $\frac{1}{5}$ to $\frac{1}{10}$ of cold resistance
(*c*) 5 to 10 times the cold resistance (*d*) 15 to 20 times the cold resistance.

**74.** The filament of a *GLS* lamp is made of

(*a*) carbon (*b*) copper (*c*) aluminium (*d*) tungsten.

**75.** Fine diameter tungsten wires are made by

(*a*) compressing (*b*) swaging (*c*) wire drawing (*d*) turning.

**76.** Which gas can be filled in *GLS* lamps?

(*a*) Oxygen (*b*) Xenon (*c*) Carbon dioxide (*d*) Any intent gas.

**77.** Which of the following is not the standard rating of *GLS* lamps?

(*a*) 15 W (*b*) 40 W (*c*) 75 W (*d*) 100 W.

**78.** Increase in voltage for a *GLS* lamp reduces

(*a*) efficiency (*b*) wattage (*b*) light output (*d*) life.

**79.** Filament lamps operate normally at a power factor of

(*a*) 0.5 leading (*b*) 0.5 lagging (*c*) unity (*d*) 0.8 lagging.

**80.** The gas filled in vacuum filament lamps is

(*a*) air (*b*) nitrogen (*c*) argon (*d*) none.

**Questions 81 to 83 Refer to Fig. 20.6:**

The characteristic curves for tungsten filament lamps are shown in Fig. 20.6.

**81.** Which curve represents life of the lamp?

(*a*) curve *A*
(*b*) curve *B*
(*c*) curve *C*
(*d*) curve *D*.

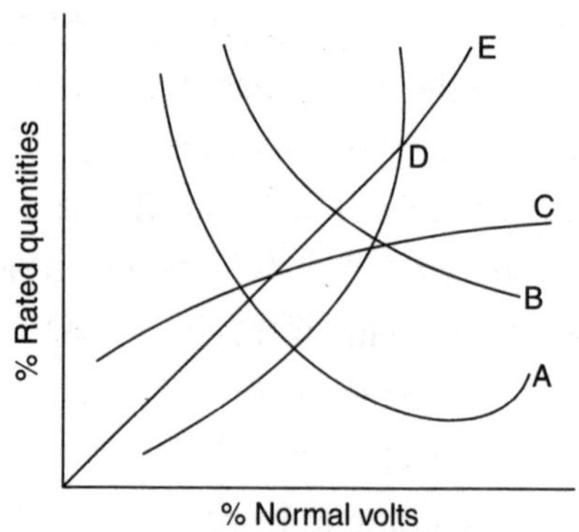


*FIGURE 20.6*

**82.** Which curve represents resistance in ohms?

(*a*) curve *A* (*b*) curve *B* (*c*) curve *C* (*d*) curve *D*.

**83.** Which curve represents variation of lumens ?

(*a*) curve *A* (*b*) curve *B* (*c*) curve *C* (*d*) curve *D*.

**84.** Carbon arc lamps are commonly used in

(*a*) domestic lighting (*b*) street lighting

(*c*) cinema projectors (*d*) photography.

**85.** For magnetic arc the electrodes are made of magnetic oxide of iron and

(*a*) carbon (*b*) copper (*c*) tungsten (*d*) aluminium.

**86.** Light is produced in electric discharge lamps by

(*a*) heating effect of current (*b*) magnetic effect of current

(*c*) ionisation in a gas or vapour (*d*) cathode ray emission.

**87.** Electrodes for electric discharge lamps use

(*a*) tungsten electrodes (*b*) copper electrodes

(*c*) barium and strontium oxide coated electrodes

(*d*) carbon electrodes.

**88.** The colour of sodium vapour discharge lamp is

(*a*) yellow (*b*) red (*c*) bluish green (*d*) pink.

**89.** In electric discharge lamp, neon gas gives

(*a*) yellow colour (*b*) violet colour (*c*) red colour (*d*) green colour.

**90.** In electric discharge lamps, bluish white colour is given by

(*a*) sodium vapour (*b*) mercury vapour at high pressure

(*c*) mercury vapour at low pressure (*d*) chlorine gas.

**91.** In electric discharge lamps, bluish green colour is given by

(*a*) mercury vapour at low pressure (*b*) chlorine gas

(*c*) flourine gas (*d*) neon gas at low pressure.

**92.** Which of the following electric discharge lamp gives highest lumens per watt ?

(*a*) Sodium vapour lamp (*b*) Neon lamp

(*c*) Mercury vapour at low pressure (*d*) Mercury vapour at high pressure.

**93.** Which of the following phosphor produce green colour ?

(*a*) Zinc silicate (*b*) Cadmium borate

(*c*) Magnesium tungstate (*d*) Calcium halo phosphate.

**94.** Blue colour is given by which of the following phosphor ?

(*a*) Calcium tungstate (*b*) Cadmium borate

(*c*) Phosphorus pentoxide (*d*) None of the above.

**95.** Pink colour is given by which of the following phosphor ?

(*a*) Magnesium tungstate (*b*) Calcium halo phosphate

(*c*) Cadmium borate (*d*) Zinc silicate.

96. Magnesium tungstate gives
(*a*) bluish white colour
(*b*) yellow colour
(*c*) green colour
(*d*) pink colour.

97. The colour of light depends on
(*a*) wavelength
(*b*) frequency
(*c*) wavelength and frequency
(*d*) wavelength, frequency, speed and intensity.

98. A substance which changes its electrical resistance when illuminated by light is called
(*a*) photovoltaic
(*b*) photoelectric
(*c*) photoconductive
(*d*) none of the above.

99. The mercury vapour discharge tube used for domestic lighting
(*a*) has no filament
(*b*) has one filament
(*c*) has two filaments
(*d*) has three filaments.

100. Least illumination level is required in case of
(*a*) hospital wards
(*b*) drawing offices
(*c*) fine machine work
(*d*) skilled bench work.

101. The illumination level required for precision work is of the order of
(*a*) 10 – 50 lumens/$m^2$
(*b*) 50 – 100 lumens/$m^2$
(*c*) 100 – 500 lumens/$m^2$
(*d*) 500 – 1000 lumens/$m^2$.

102. The illumination level required for railway platforms and corridors is
(*a*) 5 to 10 lumens/$m^2$
(*b*) 10 to 50 lumens/$m^2$
(*c*) 50 to 100 lumens/$m^2$
(*d*) 100 to 500 lumens/$m^2$.

103. The illumination level required for normal reading is around
(*a*) 500 lumens/$m^2$
(*b*) 300 – 400 lumens/$m^2$
(*c*) 100 – 250 lumens/$m^2$
(*d*) 60 – 100 lumens/$m^2$.

104. As per *ISS* the expected life of a *GLS* lamp is
(*a*) 100 hours
(*b*) 200 hours
(*c*) 1000 hours
(*d*) 10000 hours.

105. Which of the following is a cold cathode lamp ?
(*a*) Mercury vapour lamp
(*b*) Neon lamp
(*c*) Sodium vapour lamp
(*d*) Fluorescent lamp.

106. In electric discharge lamps for stabilising the arc
(*a*) a condenser is connected in parallel to the supply
(*b*) a condenser is connected in series to the supply
(*c*) a variable resistor is connected in the circuit
(*d*) a reactive choke is connected in series with the supply.

107. The lamp used in cinema projectors is
(*a*) neon lamp
(*b*) tungsten filament lamp
(*c*) mercury vapour lamp
(*d*) carbon arc lamp.

**108.** The function of choke in the fluorescent tube circuit is to
(*a*) eliminate radio interference (*b*) reduce noise
(*c*) reduce flicker (*d*) initiate the arc and stablies it.

**109.** A condenser is provided in tube light circuit to
(*a*) facilitate starting (*b*) improve power factor
(*c*) reduce noise (*d*) avoid radio interference.

**110.** Which of the following lamp has least capacity to sustain voltage fluctuations ?
(*a*) sodium vapour lamp (*b*) fluorescent lamp
(*c*) incandescent lamp (*d*) mercury vapour lamp.

## Traction

**111.** In case of locomotives the tractive power is provided by
(*a*) single stage steam turbine
(*b*) single cylinder double acting steam engine
(*c*) double cylinder, single acting steam engine
(*d*) double cylinder, double acting steam engine.

**112.** In case of steam engines the steam pressure is
(*a*) 2 to 5 kg/cm$^2$ (*b*) 5 to 8 kg/cm$^2$ (*c*) 10 to 15 kg/cm$^2$ (*d*) 20 to 30 kg/cm$^2$.

**113.** The steam engine provided on steam locomotives is
(*a*) single acting, condensing type (*b*) double acting, condensing type
(*c*) double acting, non-condensing type (*d*) single acting, non-condensing type.

**114.** Electric power for illumination in locomotive is provided by
(*a*) main steam engine (*b*) small turbo generator
(*c*) battery (*d*) solar cells.

**115.** The speed of a locomotive is controlled by
(*a*) gearbox (*b*) flywheel
(*c*) regulating steam flow to engine (*d*) applying brakes.

**116.** The overall efficiency of steam locomotive is around
(*a*) 5 to 10 percent (*b*) 15 to 25 percent (*c*) 30 to 50 percent (*d*) 60 to 85 percent.

**117.** Which of the following engine has relatively shorter life span ?
(*a*) Steam engine (*b*) Diesel engine
(*b*) Electric engine (*d*) All have same life span.

**118.** The advantage of electric traction over other methods of traction is
(*a*) no pollution problems (*b*) faster acceleration
(*c*) better braking action (*d*) all of the above.

**119.** The horsepower of steam locomotives is
(*a*) upto 1500 (*b*) 1500 to 2500 (*c*) 2500 to 4000 (*d*) more than 4000.

**120.** The horsepower of diesel locomotives is

(*a*) 1000 to 1500 (*b*) 1500 to 2500 (*c*) 2500 to 5000 (*d*) 5000 to 10000.

**121.** Which is the least efficient method of speed control for 3 phase induction motors ?

(*a*) Rheostatic control (*b*) Cascade control

(*c*) Pole changing (*d*) Combination of cascade and poles changing.

**122.** The number of passenger coaches that can be attached to a diesel engine locomotive on broad gauge is usually restricted to

(*a*) 10 (*b*) 12 (*c*) 15 (*d*) 17.

**123.** The voltage for single phase ac system is

(*a*) 440 V (*b*) 1100 V (*c*) 110 V (*d*) 230 V.

**124.** Diesel locomotives in India are manufactured at

(*a*) Bhopal (*b*) Varanasi (*c*) Perambur (*d*) Ajmer.

**125.** Electric locomotives in India are manufactured at

(*a*) Chittranjan (*b*) Gorakhpur (*c*) Bangalore (*d*) Jamalpur.

**126.** The distance between the rails for metre gauge track is

(*a*) 2 ft $6\frac{3}{8}$ inches (*b*) 3 ft $4\frac{3}{4}$ inches (*c*) 4 ft $6\frac{1}{2}$ inches (*d*) one metre.

**127.** Which of the following state capital is not on broadgauge track ?

(*a*) Bhopal (*b*) Chandigarh (*c*) Jaipur (*d*) Lucknow.

**128.** The wheels of a train, engine as well as bogies, are slightly tapered to

(*a*) facilitate in taking turns (*b*) facilitate braking

(*c*) increase friction (*d*) reduce friction.

**129.** The speed of a superfast train is

(*a*) 50 kmph (*b*) 75 kmph (*c*) 100 kmph (*d*) more than 100 kmph.

**130.** Automatic signalling is used for

(*a*) superfast trains (*b*) mail and express trains

(*c*) suburban and urban electric trains (*d*) all trains.

**131.** For suburban trains in dc system, the voltage is usually

(*a*) 12 V (*b*) 24 V (*c*) 220 V (*d*) 600 to 750 V.

**132.** The voltage of long distance dc system is

(*a*) 200 V ro 250 V (*b*) 600 V ro 750 V (*c*) 1500 V or 3000 V (*d*) 15000 V.

**133.** The normal value of adhesion friction is

(*a*) 0.15 (*b*) 0.25 (*c*) 0.50 (*d*) 0.66.

**134.** The value of co-efficient of adhesion will be highest when

(*a*) rails are cleaned with sand (*b*) rails are wet

(*c*) rails are sprayed with oil (*d*) rails are greased.

135. The pulsating torque exerted by steam locomotives causes

(a) hammer blow (b) pitching

(c) jolting and skidding (d) all of the above.

136. The braking system used on steam locomotives is

(a) pneumatic system (b) hydraulic system

(c) vacuum system (d) none of the above.

137. Vacuum is created by

(a) ejector (b) vacuum pump (c) any of the above (d) none of the above.

**Questions 138 and 139 Refer to Data Given Below :**

A suburban electric train has a maximum speed of 65 km per hour. The schedule speed including the station stop of 30 seconds is 43.5 km per hour. The acceleration is 1.3 km per hour per second and the average distance between the stops is 3 km.

138. The time for actual run is

(a) 278 sec (b) 248 sec (c) 218 sed (d) 208 sec.

139. The retardation is

(a) 2.416 km phps (b) 1.208 km phps (c) 0.604 km phps (d) 0.302 km phps.

**Questions 140 to 143 Refer to Fig. 20.7 :**

Speed — time curve for train movement is shown in Fig. 20.7.

140. OA represents

(a) acceleration during notching up

(b) acceleration of speed curve

(c) free running

(d) coasting.

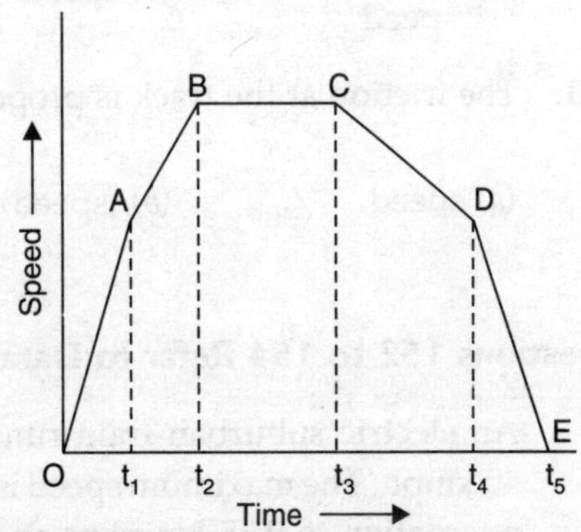


*FIGURE 20.7*

141. BC represents

(a) acceleration (b) free running (c) braking (d) coasting.

142. Coasting or running with power cut off is represented by

(a) OA (b) BC (c) CD (d) DE.

143. DE represents

(a) braking retardation (b) breakdown

(c) regeneration (d) none of the above.

**144.** The rate of acceleration on suburban or urban services is restricted by the consideration of
(*a*) track size (*b*) track curves
(*c*) passenger discomfort (*d*) engine power.

**145.** The rate of acceleration on suburbon or urban services is in the range
(*a*) 0.1 to 0.4 km per hour per second (*b*) 1.6 to 4 km per hour per second
(*c*) 4 to 8 km per hour per second (*d*) 10 to 40 km per hour per second.

**146.** The coasting retardation is around
(*a*) 0.16 km phps (*b*) 1.6 km phps (*c*) 16 km phps (*d*) 321 km phps.

**147.** The braking retardation is usually in the range
(*a*) 0.16 to 0.3 km phps (*b*) 0.3 to 0.5 km phps
(*c*) 0.5 to 2.5 km phps (*d*) 3 to 5 km phps.

**148.** The specific energy consumption of a train depends on
(*a*) gradient (*b*) distance covered
(*c*) acceleration and retardation (*d*) all of the above.

**149.** The resistance encountered by a train in motion is on account of
(*a*) friction at various parts of the rolling stock
(*b*) friction at the track
(*c*) resistance offered by air (*d*) all of the above.

**150.** The air resistance to the movement of train is proportional to
(*a*) $\frac{1}{\text{speed}}$ (*b*) speed (*c*) $(\text{speed})^2$ (*d*) $(\text{speed})^3$.

**151.** The friction at the track is proportional to
(*a*) speed (*b*) $(\text{speed})^2$ (*c*) $\frac{1}{\text{speed}}$ (*d*) $\frac{1}{(\text{speed})^2}$.

## Questions 152 to 154 Refer to Data Given Below :

An electric suburban train runs between two stations 2 km apart. The average speed is 40 kmph. The maximum speed is to be limited to 60 kmph, acceleration to 2 km phps, coasting retardation to 0.15 km phps and braking retardation to 3 km phps. The run is to be made according to simplified quadrilateral speed time curve.

**152.** The time for acceleration is
(*a*) 15 seconds (*b*) 20 seconds (*c*) 25 seconds (*d*) 30 seconds.

**153.** The coasting period is nearly
(*a*) 67 seconds (*b*) 84 seconds (*c*) 137 seconds (*d*) 248 seconds.

**154.** The braking period is nearly
(*a*) 5 seconds (*b*) 13 seconds (*c*) 22 seconds (*d*) 33 seconds.

**155.** The method of braking in which motor armature remains connected to the supply and draws power from it producing torque opposite to the direction of motion, is known as
(*a*) plugging (*b*) rheostatic braking
(*c*) regenerative braking (*d*) none of the above.

**156.** Tram service in India exists in
(*a*) Delhi (*b*) Mumbai (*c*) Kolkata (*d*) Chennai.

**157.** Battery operated trucks are used in
(*a*) narrow gauge traction (*b*) factories for material transportation
(*c*) power stations (*d*) steel mills.

**158.** The traction collection gear used in 25 kV, 50 Hz ac traction is
(*a*) Faivelay type pantograph (*b*) Diamond type pantograph
(*c*) Pole collector (*d*) Bow collector.

**159.** Which of the following track is electrified ?
(*a*) Delhi — Ahmedabad (*b*) Delhi — Chennai
(*c*) Delhi — Howrah (*d*) Delhi — Mumbai.

**160.** Which method of braking can bring the locomotive to dead stop ?
(*a*) Regenerative braking (*b*) Rheostatic braking
(*c*) Plugging braking (*d*) All of the above.

## Welding

**161.** Type of welding used for manufacturing *ERW* pipes is
(*a*) gas welding (*b*) thermal welding
(*c*) electrical resistance welding (*d*) induction welding.

**162.** Which of the following method is normally not preferred for welding of chromium molybdenum steels ?
(*a*) Thermit welding (*b*) Submerged arc welding
(*c*) Oxyacetylene welding (*d*) Resistance welding.

**163.** Gray iron is usually welded by
(*a*) arc welding (*b*) gas welding (*c*) resistance welding (*d*) *MIG* welding.

**164.** The process used for the welding of car body is
(*a*) gas welding (*b*) spot welding
(*c*) carbon arc welding (*d*) atomic hydrogen welding.

**165.** Which of the following metal cannot be flush welded ?
(*a*) Lead (*b*) Zinc (*c*) Antimony (*d*) All of the above.

**166.** Which of the following is not a resistance welding process ?
(*a*) Flush welding (*b*) Seam welding
(*c*) Carbon arc welding (*d*) Projection welding.

**167.** Heat produced at the joint during resistance welding is proportional to
(*a*) voltage (*b*) current (*c*) kVA (*d*) $I^2R$.

**168.** For spot welding the tipe of the electrodes are made of
(*a*) carbon (*b*) copper alloy (*c*) mica (*d*) porcelain.

**169.** During spot welding the current may be
(*a*) 1 kA to 10 kA (*b*) 10 kA to 30 kA (*c*) 40 kA to 60 kA (*d*) 60 kA to 100 kA.

**170.** The time for which the current flows during spot welding is
(*a*) few micro seconds
(*b*) few milli seconds
(*c*) fraction of a second to several seconds
(*d*) fraction of a minute to several minutes.

**171.** In electrical resistance welding the greatest resistance is offered by
(*a*) metal surface
(*b*) contact layer of metals to be welded
(*c*) contact point of electrode with metal top
(*d*) contact point of electrode with metal bottom.

**172.** For electrical resistance welding, the metal surfaces
(*a*) must be cleaned (*b*) must be moistened
(*c*) must be rough (*d*) must be lubricated.

**173.** The power factor of a spot welding machine is expected to be around
(*a*) 0.3 to 0.5 leading (*b*) unity (*c*) 0.8 lagging (*d*) 0.3 to 0.5 lagging.

**174.** The problem with welding of aluminium alloy is
(*a*) low specific resistance requiring heavy current
(*b*) high thermal conductivity — quickly dissipating heat
(*c*) on melting becomes thin liquid-tendency to run off.
(*d*) all of the above.

**175.** In which of the following method of welding the molten metal is poured for joining the metals ?
(*a*) *TIG* welding (*b*) Gas welding (*c*) Arc welding (*d*) Thermit welding.

**176.** During carbon arc welding
(*a*) electrode is negative with respect to the work if dc is used
(*b*) electrode is positive with respect to the work if dc is used
(*c*) electrode is connected to neutral if ac is used
(*d*) electrode is not connected to any voltage source when ac is used.

**177.** During carbon arc welding if electrode is connected to positive
(*a*) arc will not strike (*b*) arc will be dull
(*c*) metal will not melt
(*d*) carbon will have tendency to go into the weld joint.

**178.** During atomic hydrogen welding the electrode is made of
(*a*) mild steel (*b*) carbon (*c*) tungsten (*d*) graphite.

**179.** Weld spatter defect in welding is generally the result of
(*a*) Low voltage during welding
(*b*) low current during welding
(*c*) too high current during welding
(*d*) too high voltage during welding.

**180.** Poor fusion in a welded joint may be due to
(*a*) uncleaned metal surface (*b*) lack of flux
(*c*) improper current (*d*) high welding speed.

**181.** The porosity of welded joint may be caused by
(*a*) poor base metal (*b*) low welding current
(*c*) incorrect size of electrode (*d*) any of the above.

**182.** Arc blow is a welding defect, generally encountered in
(*a*) thermit welding (*b*) gas welding
(*c*) arc welding using dc current (*d*) arc welding using ac current.

**183.** In helium arc welding, the electrode is made of
(*a*) copper (*b*) carbon (*c*) mildsteel (*d*) tungsten.

**184.** 'Sweating' is
(*a*) tinning of two surfaces to be joined, applying flux and heating
(*b*) heating two surfaces to be joined, to a high temperature in the presence of flux
(*c*) heating two surfaces at low temperature
(*d*) joining two surfaces without use of flux.

**185.** The range of open circuit voltage for arc welding is generally
(*a*) 20 – 40 V (*b*) 40 – 90 V (*c*) 100 – 200 V (*d*) 200 – 250 V.

**186.** During argon arc welding argon is used as a
(*a*) source of heat (*b*) agent for heat transfer
(*c*) shield to protect the work from oxidation
(*d*) flux.

**187.** Although open circuit voltage for arc welding may be around 80 V, however, once the arc is struck, the voltage required to maintain the arc is
(*a*) 1 to 2 volts (*b*) 2 to 10 volts (*c*) 20 to 30 volts (*d*) 50 to 70 volts.

**188.** DC for arc welding is produced by
(*a*) transformer (*b*) regulator
(*c*) motor generator set (*d*) none of the above.

**189.** For ultra sonic welding the frequency range is generally

(*a*) 4 kHz to 20 kHz (*b*) 20 kHz to 40 kHz

(*c*) 40 kHz to 80 kHz (*d*) 1 MHz to 10 MHz.

**190.** Which of the following welding process uses consumable electrodes ?

(*a*) *MIG* welding (*b*) *TIG* welding (*c*) Laser welding (*d*) All of the above.

**191.** Plain and butt welds may be used on materials upto thickness of nearly

(*a*) 25 mm (*b*) 50 mm (*c*) 75 mm (*d*) 100 mm.

**192.** During arc welding as the thickness of metal to be welded increases

(*a*) current should increase, voltage remaining the same

(*b*) current should increase, voltage should decrease

(*c*) voltage should increase, current should remain the same

(*d*) current should decrease, voltage should increases.

**193.** Motor generator set for dc arc welding has generator of

(*a*) shunt type (*b*) series type

(*c*) level compound type (*d*) differentially compound type.

**194.** For ac arc welding the equipment used in generally

(*a*) transformer (*b*) single phase alternator

(*c*) two phase alternator (*d*) three phase alternator.

**195.** In dc arc welding

(*a*) electrode is made negative and workpiece positive

(*b*) electrode is made positive and workpiece negative

(*c*) both electrode as well as workpiece are made negative

(*d*) both electrode as well as workpiece are made positive.

**196.** Argon

(*a*) is heavier than air while helium is lighter than air

(*b*) is lighter than air while helium is heavier than air

(*c*) and helium both are lighter than air

(*d*) and helium both are heavier than air.

**197.** Arc sun burns caused by

(*a*) infrared rays (*b*) ultra violet rays (*c*) alpha rays (*d*) cathode rays.

**198.** A seamless pipe has

(*a*) spot welded joint (*b*) seam welded joint

(*c*) arc welded joint (*d*) no joint.

**199.** A solder is made of

(*a*) lead and zinc (*b*) tin and lead

(*c*) copper and tin (*d*) aluminium and copper.

**200.** Galvanising is the process of applying a layer of

(*a*) copper (*b*) lead (*c*) zinc (*d*) aluminium.

**201.** Pipes are galvanised

(*a*) from inside only (*b*) from outside only

(*c*) both from inside and outside.

**202.** Spot welding is used for

(*a*) castings only (*b*) thick sections

(*c*) rough and irregular surfaces (*d*) thin metal sheets.

**203.** Which of the following is not an inert gas ?

(*a*) Helium (*b*) Argon

(*c*) Carbon dioxide (*d*) All of the above.

**204.** Current range for arc welding usually

(*a*) 1 to 5 A (*b*) 5 to 15 A (*c*) 15 to 75 A (*d*) 100 to 350 A.

**205.** Air craft body is

(*a*) gas welded (*b*) spot welded (*c*) seam welded (*d*) rivetted.

**206.** The purpose of coating on arc welding electrodes is to

(*a*) provide a protecting atmosphere (*b*) provide slag to protect the molten metal

(*c*) stablise the arc (*d*) all of the above.

**207.** Electronic components are joined by

(*a*) spot welding (*b*) seam welding (*c*) soldering (*d*) brazing.

**208.** For varying the current for welding in a tap changing welding transformer the tap changes occur on

(*a*) high voltage side (*b*) primary side (*c*) secondary side (*d*) either of the above.

**209.** 40% duty cycle of a welding machine means

(*a*) machine efficiency is 40%

(*b*) machine works on 40% output

(*c*) machine input is 40% of rated input

(*d*) machine works for 4 minutes in a duration of 10 minutes.

**210.** Resistance welding cannot be used for

(*a*) ferrous materials (*b*) non-ferrous metals

(*c*) copper and aluminium (*d*) dielectrics.

## Heating

**211.** The advantage of heating by electricity is

(*a*) absence of flue gases (*b*) quicker operation

(*c*) higher efficiency (*d*) all of the above.

212. Which law is applicable to heat transfer by radiation ?

(*a*) Newtons low of cooling (*b*) Fourier's law

(*c*) Stefan Boltzmann law (*d*) Faraday's law.

213. Radiative heat transfer is proportional to ($T$ = absolute temperature)

(*a*) $T$ (*b*) $T^2$ (*c*) $T^4$ (*d*) $T^8$.

214. Body $P$ is kept in contact with body $Q$. Heat will flow from $P$ to $Q$ if

(*a*) heat content of $P$ is greater than that of $Q$

(*b*) temperature of $P$ is greater than that of $Q$

(*c*) specific heat of $P$ is greater than that of $Q$

(*d*) all above are valid.

215. Which of the following is not a unit of thermal conductivity ?

(*a*) Watt/m$^2$k (*b*) Kcal/m-hr-°C (*c*) BTU/ft-hr-°C (*d*) Cal/cm-sec-°K.

216. Which of the following material has highest value of thermal conductivity ?

(*a*) Steel (*b*) Aluminium (*c*) Brass (*d*) Copper.

217. Heat transfer by conduction will not occur when

(*a*) bodies are immersed in water

(*b*) bodies are kept in vacuum

(*c*) temperatures of the two bodies are identical

(*d*) bodies are exposed to thermal radiations.

218. Heat is transferred by all the three modes, *viz.* conduction convection and radiation, in case of

(*a*) insulated pipes carrying steam (*b*) refrigerator freezer coils

(*c*) steam condensers (*d*) boiler furnaces.

219. A relationship between the temperature of a black body and the wavelength at which the maximum value of monochromatic emission power occurs, is established under

(*a*) Planck's law (*b*) Kirchhoff's law

(*c*) Wein's displacement law (*d*) D'Alambert's rule.

220. If a body reflects entire radiation incident on it, then it is known as

(*a*) black body (*b*) gray body (*c*) transparent body (*d*) white body.

221. A perfect black body is one that

(*a*) absorbs all incident radiations

(*b*) reflects all incident radiations

(*c*) transmits all incident radiations

(*d*) absorbs, reflects and transmits all incident radiations.

222. A non dimensional number generally associated with natural convection heat transfer is

(*a*) Pecelet number (*b*) Nusselt number (*c*) Grashoff number (*d*) Prandtl number.

223. Which one of the following approximates closely to the black body conditions ?
(*a*) Ice (*b*) Water (*c*) Black painted lamp (*d*) All of the above.

224. What should be the shape of an ideal thermometer ?
(*a*) Rectangular (*b*) Cylindrical (*c*) Spherical (*d*) Square.

225. The process of heat transfer during the re entry of satellites and missiles, at very high speeds, into the earth's atmosphere is known as
(*a*) radiation (*b*) ablation (*c*) irradiation (*d*) viscous dissipation.

226. The statement that the emissivity and absorptivity of a surface is surrounded by its own temperature are the same for both monochromatic and total radiation, is known as
(*a*) Kirchhoff's law (*b*) Stefan Boltzmann's law
(*c*) D'Alambert's rule (*d*) Law of emissivity.

227. The numerical value of which property of air generally drops with rising temperature ?
(*a*) Density (*b*) Dynamic viscosity
(*c*) Kinematic viscosity (*d*) Thermal diffusivity.

228. Three rods made of glass, aluminium and wrought iron have identical dimensions *i.e.*, 20 mm diameter, 600 mm length. If all the three rods are heated and maintained at one end at a temperature of 200°C, the lowest temperature at free end will occur in case of
(*a*) aluminium rod (*b*) wrought iron rod
(*c*) glass rod (*d*) all the rods will have the same temperature.

229. A body cools from 100°C to 90°C in 15 minutes. The time taken to cool further from 90°C to 80°C will be
(*a*) 20 minutes (*b*) 15 minutes
(*c*) less than 15 minutes (*d*) more than 15 minutes.

230. A thermister is a
(*a*) thermocouple element (*b*) thermal insulation
(*c*) miniature resistance (*d*) thermal relay.

231. The temperature inside the furnace is usually measured by
(*a*) mercury thermometer (*b*) optical pyrometer
(*c*) alcohol thermometer (*d*) any of the above.

232. The highest value of thermal conductivity can be expected for
(*a*) solid ice (*b*) melting ice (*c*) water (*d*) steam.

233. Which variety of steel may be expected to have lowest value of thermal conductivity ?
(*a*) High carbon steel (0.83% C) (*b*) Low nickel steel (1.37% Ni)
(*c*) High chromium steel (0.26% Cr) (*d*) Chromium nickel steel (9% Ni, 18% Cr).

234. Which insulating material is suitable for low temperature applications ?
(*a*) cork (*b*) asbestos paper (*c*) diatomaceous earth (*d*) 85% magnesia.

**235.** The efficiency of electric oven is generally in the range

(*a*) 10% to 25% (*b*) 30% to 50% (*c*) 60% to 80% (*d*) 80% to 95%.

**236.** If the thickness of refractory wall of furnace is increased

(*a*) temperature inside the furnace will fall

(*b*) energy consumption will increase

(*c*) heat loss through furnace walls will increase

(*d*) temperature on the outer surface of furnace walls will drop.

**237.** Nichrome can be used for furnace temperatures upto

(*a*) 200°C (*b*) 400°C (*c*) 800°C (*d*) 1000°C.

**238.** The material of the heating element for a furnace should have

(*a*) high specific resistance (*b*) lower melting point

(*c*) high temperature co-efficient (*d*) all of the above.

**239.** Cremation furnaces are invariably

(*a*) oil fired (*b*) gas fired (*c*) wood fired (*d*) resistance furnaces.

**240.** The atmosphere in a resistance furnace is

(*a*) deoxidising (*b*) neutral (*c*) oxidising (*d*) reducing.

**241.** Non-metallic resistors are used in

(*a*) low temperature furnaces (*b*) variable temperature furnaces

(*c*) high temperature furnaces (*d*) cryogenic range of temperatures.

**242.** The temperature inside a resistance furnace can be varied by

(*a*) varying the operating voltage

(*b*) varying the current through heating elements

(*c*) disconnecting some of the heating elements

(*d*) any of the above.

**243.** Which of the following is used for automatic temperature control of resistance furnaces ?

(*a*) Circuit breaker (*b*) Double pole, double throw switch

(*c*) Thermostat (*d*) Pyrometer.

**244.** For radiant heating around 2250°C, the heating elements are made of

(*a*) tungsten alloy (*b*) stainless steel alloy

(*c*) carbon (*d*) copper alloy.

**245.** For induction heating, which of the following is abnormally high ?

(*a*) Voltage (*b*) Current (*c*) Frequency (*d*) Phase angle.

**246.** High frequency power supply for induction furnaces can be obtained from

(*a*) current transformer (*b*) multi phase transformer

(*c*) coreless transformers (*d*) motor generator set.

**247.** A valve oscillator is used for frequencies around

(*a*) 1 kHz (*b*) 10 kHz (*c*) 100 kHz (*d*) 1 MHz.

**248.** Induction furnaces are used for

(*a*) melting aluminium (*b*) heat treatment of castings

(*b*) heating of insulators (*d*) none of the above.

**249.** The power factor of an induction motor is

(*a*) low — lagging (*b*) low — leading (*c*) around unity (*d*) around 0.8 lagging.

**250.** During eddy current heating

(*a*) articles are uniformly heated

(*b*) heat is generated near the surface of the material core

(*c*) heat is generated in the core of the material

(*d*) heat generated depends on composition of the material.

**251.** The advantage of eddy current heating is

(*a*) heat at very high rate can be generated

(*b*) the area of the surface over which heat is produced can be accurately controlled

(*c*) the amount of heat generated can be controlled accurately

(*d*) all of the above.

**252.** Non-metallic substances like wood and plastics can be heated by

(*a*) gas firing (*b*) liquid bath heating

(*c*) resistance heating (*d*) induction heating.

**253.** In direct arc furnace the electrode is made of

(*a*) graphite (*b*) copper (*c*) silver (*d*) tungsten.

**254.** In direct arc furnaces arc is produced between

(*a*) two electrodes (*b*) electrode and furnace wall

(*c*) electrode and charge (*d*) charge and furnace wall.

**255.** Direct arc furnaces are commonly used for

(*a*) heat treatment of castings (*b*) melting of metals

(*c*) annealing of castings (*d*) heating of dielectrics.

**256.** The power factor of direct arc furnaces is

(*a*) low, lagging (*b*) low, leading (*c*) unity (*d*) around 0.8 lagging.

**257.** In direct arc furnace, which of the following has high value ?

(*a*) current (*b*) voltage (*c*) power factor (*d*) all of the above.

**258.** The electrodes for induction furnace should have

(*a*) high conduction (*b*) good mechanical strength

(*c*) low thermal conductivity (*d*) all of the above.

**259.** The consumption of electrode per 100 kWh will be around

(*a*) 1 kg (*b*) 5 kg (*c*) 20 kg (*d*) 100 kg.

260. Dielectric heating can be used for
(*a*) dehydration of food stuffs
(*b*) gluing of laminated glass
(*c*) drying of explosives
(*d*) any of the above.

## Rectification

261. The advantage of motor generator set is
(*a*) dc output voltage is practically constant
(*b*) dc output can be controlled by adjusting shunt field regulator
(*c*) the unit is self starting
(*d*) all of the above.

### Questions 262 to 264 Refer to Data Given Below :

In a 500 kW motor-generator set the induction motor has six poles and the dc generator has 8 poles. The supply frequency is 50 Hz.

262. The cascade speed of the set will be
(*a*) 1000 rpm (*b*) 1500 rpm (*c*) 750 rpm (*d*) 428.6 rpm.

263. The frequency of currents in generator will be
(*a*) 14/3 Hz (*b*) $16\frac{2}{3}$ Hz (*c*) 25 Hz (*d*) 50 Hz.

264. The power transmitted through shaft will be
(*a*) 500 kW (*b*) 375.5 kW (*c*) 215 kW (*d*) 147 kW.

265. A synchronous convertor can be used as
(*a*) ac to dc or dc to ac convertor
(*b*) synchronous motor or dc motor
(*c*) dc generator or ac generator
(*d*) any of the above.

266. The number of slip rings on a 3-phase synchronous convertor will be
(*a*) 1 (*b*) 3 (*c*) 4 (*d*) 6.

267. The cathode of a mercury arc rectifier is in the form of
(*a*) cup shape (*b*) pointed electrode (*c*) pool of mercury (*d*) spherical bowl.

268. A mercury arc rectifier works on the principle of
(*a*) ionization by collision
(*b*) directional movement of electrons
(*c*) release of electrons by thermal excitation
(*d*) none of the above.

269. In mercury arc rectifiers the arc voltage drop at the anode is about
(*a*) 1 to 2 volts (*b*) 6 to 8 volts (*c*) 12 to 20 volts (*d*) 25 to 50 volts.

270. In mercury arc rectifiers the arc voltage drop at the cathode is about
(*a*) 0.1 to 0.5 volts (*b*) 0.5 to 1 volt (*c*) 1 to 2 volts (*d*) 5 to 6 volts.

**271.** In mercury arc rectifiers the voltage drop in arc is nearly
(*a*) 0.1 volt per centimetre of arc length (*b*) 1 volt per centimetre of arc length
(*c*) 10 volts per centimetre of arc length (*d*) 12 to 15 volts per centimetre of arc length.

**272.** In mercury arc rectifiers, mercury is selected as cathode because
(*a*) its ionization potential is low (*b*) it has low specific heat
(*c*) it has small latent heat of vapourisation
(*d*) all of the above.

**273.** A rectifier is a
(*a*) bilateral device (*b*) linear device (*c*) non-linear device (*d*) passive device.

**274.** A rectifier has
(*a*) low resistance in forward direction and high resistance in reverse direction
(*b*) high resistance in forward direction and no resistance in reverse direction
(*c*) same resistance in forward as well as reverse direction
(*d*) resistance always inversely proportional to load.

**275.** An ideal rectifier having a linear characteristic in forward direction and an infinite resistance in the reverse direction, can be represented by
(*a*) figure *A* (*b*) figure *B*
(*c*) figure C (*d*) none of the figures shown.

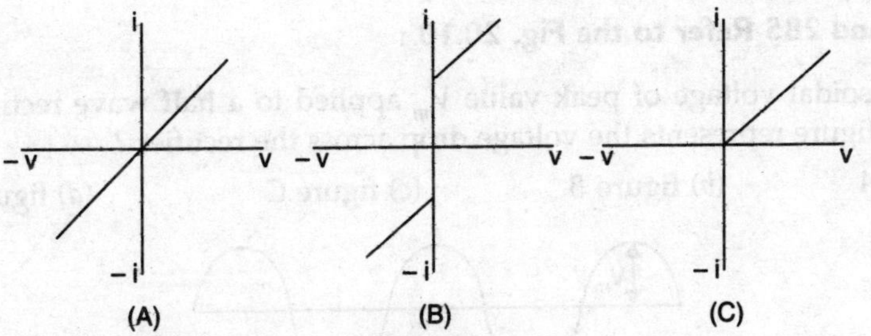


*FIGURE 20.8*

**Questions 276 to 280 Refer to Full Wave Rectified Sine Wave :**

**276.** The rms value of the voltage will be
(*a*) 0.707 $v_m$
(*b*) 0.636 $v_m$
(*c*) 0.5 $v_m$
(*d*) 0.318 $v_m$.

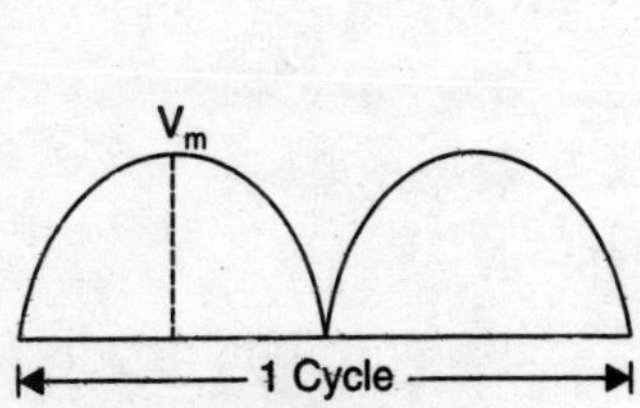


*FIGURE 20.9*

**277.** For the same peak value, the rms values of which of the following waves will be identical ?
(*a*) Sine wave and square wave (*b*) Sine wave and triangular wave
(*c*) Full wave rectified sine wave and half wave rectified sine wave
(*d*) Sine wave and full wave rectified sine wave.

278. The mean value of the voltage will be

(a) 0.707 $v_m$ (b) 0.636 $v_m$ (c) 0.5 $v_m$ (d) 0.414 $v_m$.

279. The form factor of the wave is

(a) 1.57 (b) 1.41 (c) 1.11 (d) 1.0.

280. The peak factor of the wave is

(a) 1.41 (b) 1.33 (c) 1.11 (d) 1.00.

**Questions 281 to 283 Refer to Data Given Below :**

A half wave rectifier is used to charge a 40 V battery from an ac supply of 200 V (rms). The forward resistance of the circuit is 200 ohms and the reverse resistance is very high. The average battery voltage during the period is 40 V.

281. The angle at which charging commences is

(a) $\sin^{-1}\frac{40}{283}$ (b) $\sin^{-1}\frac{283}{40}$ (c) $\tan^{-1}\frac{283}{40}$ (d) $\cos^{-1}\frac{40}{283}$.

282. Average charging current during the half cycle will be

(a) 0.713 A (b) 0.515 A (c) 0.357 A (d) 0.115 A.

283. Charge per 24 hours will be

(a) 2.25 Ah (b) 4.50 Ah (c) 8.57 Ah (d) 12.22 Ah.

**Questions 284 and 285 Refer to the Fig. 20.10 :**

284. For a sinusoidal voltage of peak value $V_m$ applied to a half wave rectifier, which of the following figure represents the voltage drop across the rectifier ?

(a) figure *A* (b) figure *B* (c) figure *C* (d) figure *D*.

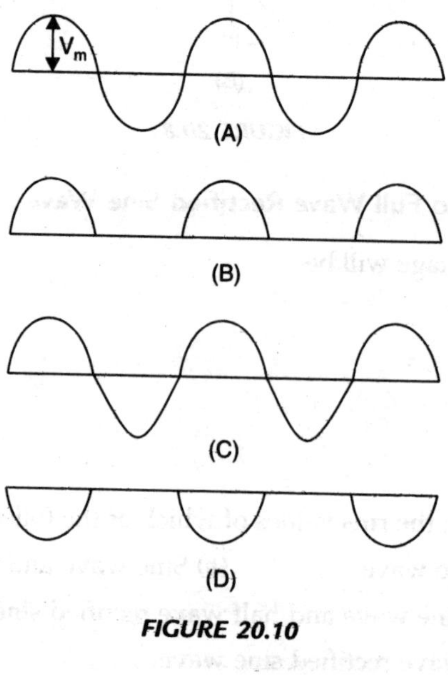


*FIGURE 20.10*

285. Which of the above figure represents the voltage drop across load
(*a*) figure *A* (*b*) figure *B* (*c*) figure *C* (*d*) figure *D*.

286. Peak reverse voltage for a diode is the
(*a*) voltage corresponding to rated maximum voltage
(*b*) maximum voltage that can be applied across the diode in the conducting direction
(*c*) maximum voltage that can be applied across the diode in the non-conducting direction
(*d*) none of the above.

287. When voltage applied to a diode is more than the peak reverse voltage, it is likely to result in
(*a*) more distortion on output side (*b*) poor regulation
(*c*) conduction in both directions (*d*) breakdown at the junction.

288. Peak reverse voltage for a diode is usually
(*a*) half the dc voltage (*b*) same as dc voltage
(*c*) twice the dc output voltage (*d*) hundred times the dc output voltage.

289. A silicon diode has low internal voltage drop of the order of
(*a*) 1.01 V (*b*) 0.1 V (*c*) 1 V (*d*) 10 V.

290. The internal voltage drop in case of vacuum tubes may be around
(*a*) 1 V (*b*) 2 to 5 V (*c*) 18 V (*d*) 50 V.

291. Which of the following is known as metal rectifier ?
(*a*) Selenium disc rectifier (*b*) Copper oxide rectifier
(*c*) Gas tube diode (*d*) All of the above.

292. As compared to selenium rectifier, a gas tube diode has
(*a*) low internal voltage drop (*b*) high current rating
(*c*) low current rating (*d*) less distortion.

293. A thyratron is
(*a*) a semiconductor (*b*) a cold cathode vacuum device
(*c*) a hot cathode gas filled device (*d*) none of the above.

294. The gas filled in a thyratron is
(*a*) marcury vapour (*b*) inert gas
(*c*) any of the above (*d*) none of the above.

295. In a mercury vapour filled thyratron
(*a*) vapour pressure is a function of temperature
(*b*) rectified voltage has rectangular wave form
(*c*) critical grid voltage is low
(*d*) most of the energy is consumed in mercury vapour ionisation.

**296.** For a argon gas thyratron, the ionisation time is of the order of

(*a*) few seconds (*b*) hundredth of a second

(*c*) milli second (*d*) $10^{-5}$ sec.

**Questions 297 to 300 Refers to Fig. 20.11 :**

**297.** For a half-wave rectified sine wave, the rms value of the current will be

(*a*) 0.875 $v_m$

(*b*) 0.5 $v_m$

(*c*) 0.3535 $v_m$

(*d*) 0.318 $v_m$.

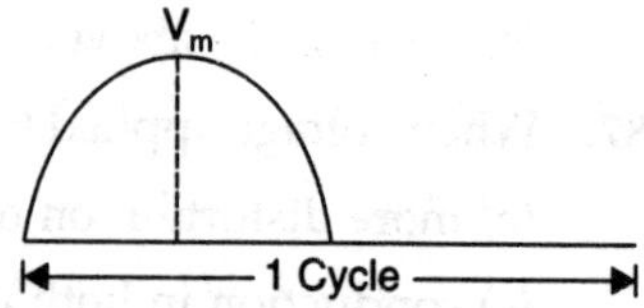


*FIGURE 20.11*

**298.** The mean value will be

(*a*) 0.5 $v_m$ (*b*) 0.3535 $v_m$ (*c*) 0.318 $v_m$ (*d*) 0.25 $v_m$.

**299.** The form factor of the wave will be

(*a*) 1.0 (*b*) 1.11 (*c*) 1.41 (*d*) 1.57.

**300.** The peak factor of the wave is

(*a*) 1.57 (*b*) 2.0 (*c*) 2.25 (*d*) infinite.

**301.** Which of the following rectifier will not have the ripple frequency double the supply frequency ?

(*a*) Full wave rectifier (*b*) Full wave bridge rectifier

(*c*) Cascade doubler (*d*) None of the above.

**302.** In a half-wave rectifier dc output voltage is

(*a*) half the rms ac input (*b*) nearly 0.85 times the rms input

(*c*) same as rms input (*d*) nearly 1.2 times rms input.

**303.** In a rectifier the rms ripple voltage is 20 mV for a 15 V dc output. The ripple factor will be

(*a*) 0.13 percent (*b*) 0.26 percent (*c*) 33 percent (*d*) 25 percent.

**304.** The numbered statements at the left are to be matched with lettered statements on the right

1. Uses four diodes (*a*) no $V^+$ output

2. Removes ac ripple (*b*) bridge rectifier

3. Open filter choke (*c*) filter capacitor

4. Full wave rectifier (*d*) double the supply frequency.

Which of the following represents the correct pairing ?

(*a*) 1 – *d* ; 2 – *c* ; 3 – *b* ; 4 – *a* (*b*) 1 – *a* ; 2 – *b* ; 3 – *c* ; 4 – *d*

(*c*) 1 – *c* ; 2 – *a* ; 3 – *d* ; 4 – *b* (*d*) 1 – *b* ; 2 – *c* ; 3 – *a* ; 4 – *d*.

**305.** In the semiconductor diode symbol, the arrow head is

(*a*) cathode (*b*) anode (*c*) base (*d*) any of the above.

306. In a semiconductor diode the forward voltage at the anode is
(a) positive (b) negative
(c) either positive or negative (d) zero.

307. The ripple frequency of half wave rectifier is
(a) same as supply frequency (b) half the supply frequency
(c) double the supply frequency (d) $\sqrt{2}$ times the supply frequency.

308. The ripple frequency of full-wave rectifier is
(a) same as supply frequency (b) half the supply frequency
(c) double the supply frequency (d) $\sqrt{2}$ times the supply frequency.

309. A semiconductor diode tested with a ohmmeter shows low resistance in both forward as well as reverse direction. It can be concluded that
(a) the diode is capable of conducting in both directions
(b) the diode will conduct only when low voltage is applied
(c) the diode is open
(d) the diode is shorted.

310. A semiconductor diode tested with a ohmmeter shows high resistance in both forward as well as reverse direction. It can be concluded that the diode
(a) have low peak reverse voltage (b) has high current density
(c) is open (d) is shorted.

311. A zener diode needs
(a) forward voltage (b) reverse voltage
(c) both forward and reverse voltage (d) none of the above.

312. A zener diode is used as a
(a) shunt regulator (b) series regulator
(c) either shunt as series regulator (d) none of the above.

313. If $V_N$ is no load voltage and $V_L$ is load voltage, then voltage regulation is given by

(a) $\frac{V_N - V_L}{V_L}$ (b) $\frac{V_L - V_N}{V_L}$ (c) $\frac{V_L - V_N}{V_N}$ (d) $\frac{V_N - V_L}{V_N}$.

314. A power supply with a dc output of 140 V has a 50 Hz ripple of 2.8 V. The percentage ripple is
(a) 0.1 (b) 0.5 (c) 1 (d) 2.

315. In a power supply which of the following symptoms can cause vary high current in the rectifier and power transformer, which can even burnout
(a) Shorted input filter capacitor (b) Open output filter capacitor
(c) Open filter choke (d) Deflective rectifier.

316. The filters used in power supplies are
(a) paper capacitor
(b) mica capacitors
(c) electrolytic capacitors
(d) any of the above.

317. In a power supply incorporating two rectifiers, if one of the rectifier is defective, this is likely to result in
(a) excessive hum
(b) reduced output voltage
(c) burning out of other rectifier
(d) reduced ripple.

318. Which of the following rectifier needs a transformer with centre-tap ?
(a) Half wave rectifier
(b) Full wave rectifier
(c) Full wave bridge rectifier
(d) Cascade doubler.

319. Which of the following rectifier will not have ripple frequency same as supply frequency ?
(a) Cascade doubler
(b) Half waves rectifier
(c) Full wave bridge rectifier
(d) None of the above.

320. A semiconductor rectifier has dc output voltage of 40 V at full load and 41 V without any load current. The percentage load regulation factor will be
(a) 0.024 percent
(b) 0.24 percent
(c) 2.4 percent
(d) 24 percent.

321. In a power supply excess hum is caused by
(a) choke
(b) capacitor
(c) resistor
(d) all of the above.

322. Under normal power supply of there is no $V^+$ when measured with a dc voltmeter, then all of the following could be the reasons for this EXCEPT
(a) defective rectifier
(b) open heater in series string
(c) shorted input filter capacitor
(d) open output filter capacitor.

**Questions 323 to 326 Refer to Fig. 20.12 :**

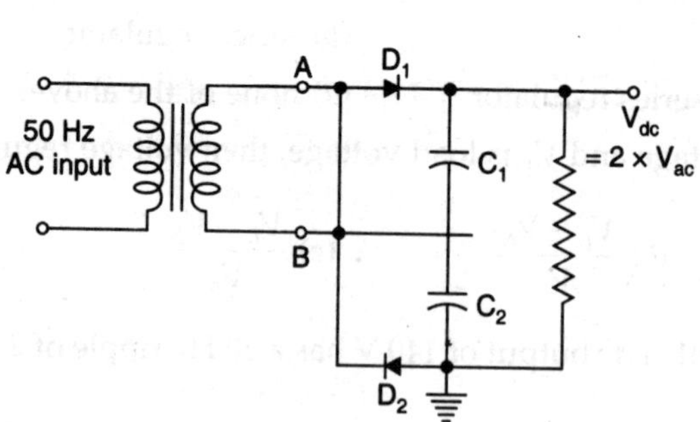


*FIGURE 20.12*

323. The circuit shown is that of a
(a) half wave rectifier
(b) full wave rectifier
(c) voltage doubler
(d) voltage quadrupler.

**324.** The ripple frequency will be

(*a*) 25 Hz (*b*) 50 Hz (*c*) 100 Hz (*d*) 200 Hz.

**325.** The diode $D_1$ charges

(*a*) $C_1$ (*b*) $C_2$ (*c*) both $C_1$ and $C_2$ (*d*) none.

**326.** How many diodes are needed in a bridge rectifier ?

(*a*) two (*b*) three (*c*) four (*d*) six.

**Questions 327 to 329 Refer to Fig. 20.13 :**

In the half wave rectifier the generator has internal resistance or 100 ohms as shown in Fig. 20.13. The diode forward resistance is 200 ohms and the reverse resistance is 200 kΩ.

**327.** The instantaneous output voltage $v_0$ when 10 sin ωt ≥ 0 will be $v_0$ =

(*a*) 9.7 sin ωt

(*b*) 9.7 cos ωt

(*c*) 0.97 sin ωt

(*d*) 0.97 cos ωt.

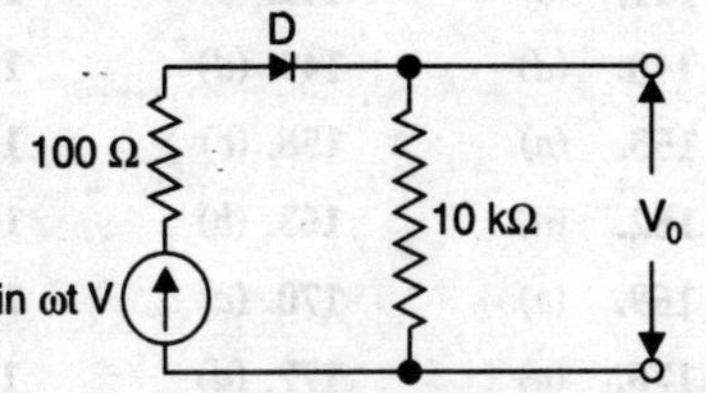


*FIGURE 20.13*

**328.** The instantaneous voltage for 10 sin ωt < 0 will be $v_0$ =

(*a*) 0.475 sin ωt volts (*b*) 4.75 sin ωt volts

(*c*) 9.5 sin ωt volts (*d*) 11.5 sin ωt volts.

**329.** The instantaneous current *i* when 10 sin ωt ≥ 0 will be

(*a*) 10 sin ωt (*b*) $\frac{10}{10000}$ sin ωt (*c*) $\frac{10}{10300}$ sin ωt (*d*) $\frac{10}{20300}$ sin ωt .

**330.** The instantaneous current *i* when 10 sin ωt < 0 will be

(*a*) 10 sin ωt (*b*) $\frac{10}{10,000}$ sin ωt (*c*) $\frac{10}{20,000}$ sin ωt (*d*) $\frac{10}{210,000}$ sin ωt .

## ANSWERS

| | | | | | | |
|---|---|---|---|---|---|---|
| 1. (*d*) | 2. (*b*) | 3. (*c*) | 4. (*a*) | 5. (*b*) | 6. (*d*) | 7. (*d*) |
| 8. (*d*) | 9. (*c*) | 10. (*a*) | 11. (*c*) | 12. (*c*) | 13. (*b*) | 14. (*c*) |
| 15. (*b*) | 16. (*c*) | 17. (*a*) | 18. (*b*) | 19. (*a*) | 20. (*d*) | 21. (*b*) |
| 22. (*c*) | 23. (*b*) | 24. (*c*) | 25. (*c*) | 26. (*b*) | 27. (*d*) | 28. (*b*) |
| 29. (*c*) | 30. (*d*) | 31. (*c*) | 32. (*a*) | 33. (*b*) | 34. (*b*) | 35. (*d*) |
| 36. (*c*) | 37. (*b*) | 38. (*a*) | 39. (*d*) | 40. (*b*) | 41. (*c*) | 42. (*a*) |
| 43. (*a*) | 44. (*b*) | 45. (*d*) | 46. (*c*) | 47. (*d*) | 48. (*a*) | 49. (*c*) |
| 50. (*d*) | 51. (*a*) | 52. (*b*) | 53. (*d*) | 54. (*b*) | 55. (*c*) | 56. (*d*) |
| 57. (*d*) | 58. (*d*) | 59. (*c*) | 60. (*d*) | 61. (*d*) | 62. (*a*) | 63. (*d*) |

**64.** (*a*) **65.** (*d*) **66.** (*c*) **67.** (*a*) **68.** (*c*) **69.** (*b*) **70.** (*b*)
**71.** (*c*) **72.** (*a*) **73.** (*d*) **74.** (*d*) **75.** (*c*) **76.** (*d*) **77.** (*c*)
**78.** (*d*) **79.** (*c*) **80.** (*d*) **81.** (*b*) **82.** (*c*) **83.** (*d*) **84.** (*c*)
**85.** (*b*) **86.** (*c*) **87.** (*c*) **88.** (*a*) **89.** (*c*) **90.** (*b*) **91.** (*a*)
**92.** (*a*) **93.** (*a*) **94.** (*a*) **95.** (*c*) **96.** (*a*) **97.** (*c*) **98.** (*c*)
**99.** (*c*) **100.** (*a*) **101.** (*d*) **102.** (*a*) **103.** (*d*) **104.** (*c*) **105.** (*b*)
**106.** (*d*) **107.** (*d*) **108.** (*d*) **109.** (*a*) **110.** (*c*) **111.** (*d*) **112.** (*c*)
**113.** (*c*) **114.** (*b*) **115.** (*c*) **116.** (*a*) **117.** (*b*) **118.** (*d*) **119.** (*a*)
**120.** (*b*) **121.** (*a*) **122.** (*d*) **123.** (*d*) **124.** (*b*) **125.** (*a*) **126.** (*d*)
**127.** (*c*) **128.** (*a*) **129.** (*d*) **130.** (*c*) **131.** (*d*) **132.** (*c*) **133.** (*b*)
**134.** (*a*) **135.** (*c*) **136.** (*c*) **137.** (*c*) **138.** (*c*) **139.** (*b*) **140.** (*a*)
**141.** (*b*) **142.** (*c*) **143.** (*a*) **144.** (*c*) **145.** (*b*) **146.** (*a*) **147.** (*d*)
**148.** (*d*) **149.** (*d*) **150.** (*c*) **151.** (*a*) **152.** (*d*) **153.** (*c*) **154.** (*b*)
**155.** (*a*) **156.** (*c*) **157.** (*b*) **158.** (*a*) **159.** (*c*) **160.** (*c*) **161.** (*c*)
**162.** (*d*) **163.** (*b*) **164.** (*b*) **165.** (*d*) **166.** (*c*) **167.** (*d*) **168.** (*b*)
**169.** (*a*) **170.** (*c*) **171.** (*b*) **172.** (*a*) **173.** (*d*) **174.** (*d*) **175.** (*d*)
**176.** (*a*) **177.** (*d*) **178.** (*c*) **179.** (*c*) **180.** (*c*) **181.** (*a*) **182.** (*c*)
**183.** (*d*) **184.** (*a*) **185.** (*b*) **186.** (*c*) **187.** (*c*) **188.** (*c*) **189.** (*a*)
**190.** (*a*) **191.** (*a*) **192.** (*a*) **193.** (*c*) **194.** (*a*) **195.** (*a*) **196.** (*a*)
**197.** (*b*) **198.** (*d*) **199.** (*b*) **200.** (*c*) **201.** (*c*) **202.** (*d*) **203.** (*c*)
**204.** (*d*) **205.** (*d*) **206.** (*d*) **207.** (*c*) **208.** (*c*) **209.** (*d*) **210.** (*d*)
**211.** (*d*) **212.** (*c*) **213.** (*c*) **214.** (*b*) **215.** (*a*) **216.** (*d*) **217.** (*c*)
**218.** (*d*) **219.** (*c*) **220.** (*d*) **221.** (*a*) **222.** (*c*) **223.** (*d*) **224.** (*c*)
**225.** (*b*) **226.** (*a*) **227.** (*a*) **228.** (*c*) **229.** (*d*) **230.** (*c*) **231.** (*b*)
**232.** (*a*) **233.** (*d*) **234.** (*c*) **235.** (*c*) **236.** (*d*) **237.** (*d*) **238.** (*a*)
**239.** (*d*) **240.** (*c*) **241.** (*d*) **242.** (*d*) **243.** (*c*) **244.** (*a*) **245.** (*c*)
**246.** (*d*) **247.** (*d*) **248.** (*b*) **249.** (*a*) **250.** (*b*) **251.** (*d*) **252.** (*d*)
**253.** (*a*) **254.** (*c*) **255.** (*b*) **256.** (*a*) **257.** (*a*) **258.** (*d*) **259.** (*c*)
**260.** (*d*) **261.** (*d*) **262.** (*d*) **263.** (*a*) **264.** (*c*) **265.** (*d*) **266.** (*b*)
**267.** (*c*) **268.** (*a*) **269.** (*b*) **270.** (*d*) **271.** (*a*) **272.** (*d*) **273.** (*c*)
**274.** (*a*) **275.** (*c*) **276.** (*a*) **277.** (*d*) **278.** (*b*) **279.** (*c*) **280.** (*a*)
**281.** (*a*) **282.** (*a*) **283.** (*c*) **284.** (*c*) **285.** (*b*) **286.** (*c*) **287.** (*d*)
**288.** (*c*) **289.** (*c*) **290.** (*c*) **291.** (*b*) **292.** (*b*) **293.** (*c*) **294.** (*c*)
**295.** (*a*) **296.** (*d*) **297.** (*b*) **298.** (*c*) **299.** (*d*) **300.** (*b*) **301.** (*d*)
**302.** (*d*) **303.** (*a*) **304.** (*d*) **305.** (*b*) **306.** (*a*) **307.** (*a*) **308.** (*c*)
**309.** (*d*) **310.** (*c*) **311.** (*b*) **312.** (*a*) **313.** (*d*) **314.** (*d*) **315.** (*a*)
**316.** (*c*) **317.** (*b*) **318.** (*b*) **319.** (*c*) **320.** (*c*) **321.** (*b*) **322.** (*d*)
**323.** (*c*) **324.** (*c*) **325.** (*a*) **326.** (*c*) **327.** (*a*) **328.** (*a*) **329.** (*c*)
**330.** (*d*).

Chapter **21**

# INSTRUMENTS AND MEASUREMENTS

Electrical instruments may be divided into two classes.

1. **Absolute Instruments.** Those which measure electrical quantities in terms of the constants of instrument and its deflection.
2. **Secondary Instruments.** Those in which the value of the electrical quantity to be measured can be determined from the deflection of the instruments only when they have been pre-calibrated by comparison with an absolute instrument. Of course without calibration, the deflection of such instruments is meaningless.

Instruments may also be classified as

(*i*) **Indicating Instruments.** Those which indicate the instantaneous values of electrical quantity being measured. A pointer moving over a calibrated scale gives the reading. Ordinary voltmeters, ammeters etc., are indicating instruments.

(*ii*) **Recording Instruments.** Those which give a continuous record of the variations of an electrical quantity with respect to time usually a graph paper is used in these instruments and a pointer registers the reading on the graph paper.

(*iii*) **Integrating Instruments.** Those which measure and register by a set of dials and pointers. A energy meter installed at the residence is an integrating instrument.

**Effects Utilised in Measurements.** All instruments operate utilizing one or more of the following effects:

| *Effect utilised* | *Category of instruments* |
|---|---|
| 1. Magnetic effect | Voltmeters, ammeters |
| 2. Thermal effect | Voltmeters, ammeters |
| 3. Electrodynamic effect | Voltmeters, ammeters, wattmeters |
| 4. Electromagnetic effect | Voltmeters, ammeters, wattmeters, watthour meters |
| 5. Chemical effect | DC ampere hour meters |
| 6. Electrostatic effect | Voltmeters only |

**Torques on Moving Systems.** The moving system of an instrument is subjected to following torques:

1. **Deflecting Torque.** Produced due to one or more effects, mentioned above. This torque causes the moving system and hence pointer to move from initial 'zero' condition.
2. **Controlling Torque.** Which opposes the deflecting torque and increases with deflection of the moving system. The points comes to rest at a position where these torques are equal (and opposite). Controlling torque may be provided by

   (*i*) Spring control (*ii*) Gravity control.
3. **Damping Torque.** Which acts only when the instrument system is moving. This torque tries to stabilize the moving system quickly so that pointer comes to rest at the point of reading without oscillating around. Damping force can be produced by

   (*i*) Air friction (*ii*) Eddy currents

   (*iii*) Fluid friction.

## Classification of Instruments

**1. Ammeters and Voltmeters**

(*a*) **Moving iron type** (both AC and DC)

(*i*) Attraction type (*ii*) Repulsion type.

(*b*) **Moving coil type**

(*i*) Permanent magnet type (for DC only)

(*ii*) Dynamometer type (for AC/DC).

(*c*) **Induction type** (for AC only)

(*i*) Split phase type (*ii*) Shaded pole type.

(*d*) **Electrostatic type** (for AC/DC both)

(*i*) for voltmeters only.

(*e*) **Hot wire type** (for AC/DC both)

## Wattmeters

**For AC/DC both**

(*i*) Induction type (*ii*) Dynamometer type.

**For AC only** (*i*) Induction type.

## kWh meters

(*a*) Electrolytic type (for DC only).

(*b*) Motor meters.

(*i*) Induction type (for AC only).

(*ii*) Mercury motor meters (for DC only).

(*iii*) Commutator motor meter (for AC and DC both).

(*c*) Clock meters

## AC Bridges

**Maxwell's Inductance Bridge.** This bridge can be used for medium inductances and can be used for fair accuracies. In this $R_2$ and $R_3$ are two pure resistances. $Z_1$ is the coil of unknown impedance. $Z_4$ is the known impedance. The positive phase angle of $Z_1$ can be compensated in two ways.

1. A known impedance with an equal positive phase angle may be used so that $\phi_1 = \phi_4$ and $\phi_2 = \phi_3 = 0$. Such a bridge is known as Maxwell's ac bridge.
2. An impedance with equal negative phase angle may be used in opposite arm. Such an arrangement is called Maxwell-Wein bridge or Maxwell *L*/*C* bridge, as shown in Fig. 21.1.

Under the conditions of balance

$$Z_1Z_3 = Z_2Z_4$$

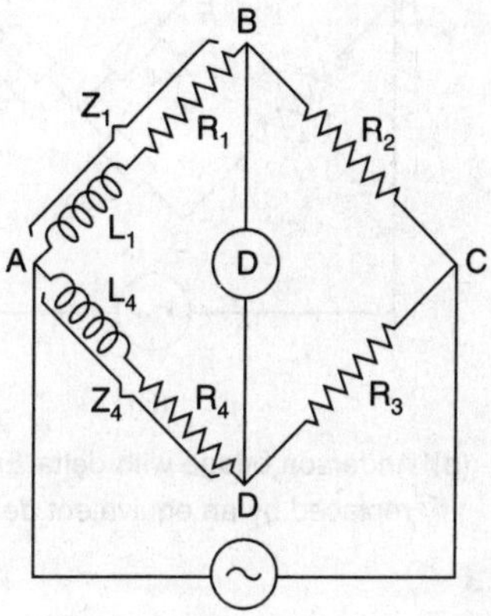


*FIGURE 21.1*

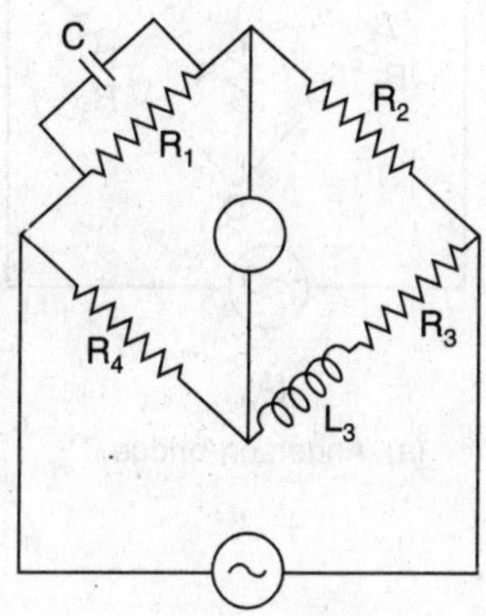


*FIGURE 21.2*

or $$(R_1 + j\omega L_1)R_3 = (R_4 + j\omega L_4)R_2$$

∴ $$L_1 = L_4\frac{R_2}{R_3}$$

In case of Maxwell Weins bridge,

$$Z_1 = \frac{R_1}{1 + j\omega CR_1}$$

$$Z_2 = R_3 + j\omega L_3$$

under the conditions of balance

$$Z_1Z_3 = Z_2Z_4$$

Solving $$L_3 = CR_2R_4.$$

**Anderson Bridge.** In this bridge the unknown inductance is measured in terms of a known capacitance and resistance. Under the conditions of balance, *C*, $R_5$ and $R_3$ forming delta may be replaced by equivalent star. In that case

$$Z_{OD} = \frac{R_3 R_5}{R_3 + R_5 + \dfrac{1}{j\omega C}}$$

$$Z_{OC} = \frac{R_3 j\omega C}{R_3 + R_5 + \dfrac{1}{j\omega C}} = Z_3.$$

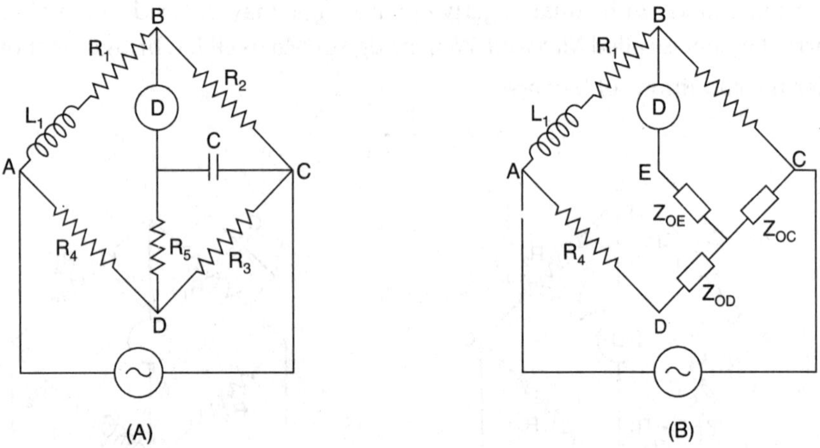


(a) Anderson bridge

(b) Anderson bridge with delta *ECD* of Fig. 21.3 (A) replaced by an equivalent delta.

*FIGURE 21.3*

The bridge can be solved for $L_1$ the value of which is given by

$$L_1 = CR_2\left(R_4 + R_5 + \frac{R_4 R_5}{R_3}\right)$$

This method can be used for precise measurement of $L_1$ over a wide range of values extending from few micro-henrys to several henries.

**Hay's Bridge.** This bridge is used for measurement of inductances particularly when the phase angle of the inductive impedance is large.

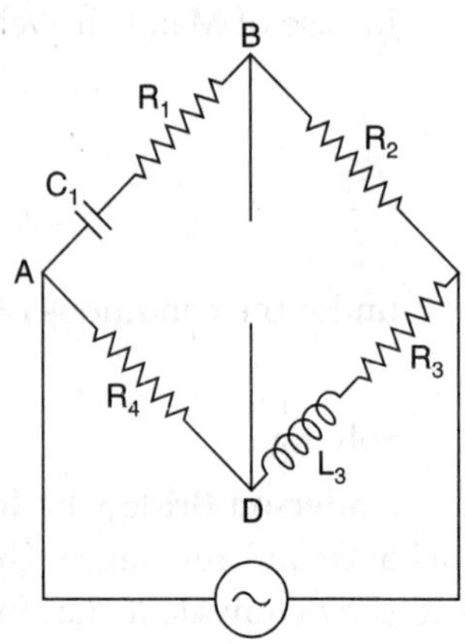


*FIGURE 21.4*

In this case

$$Z_1 = R_1 - \frac{j}{\omega C_1}$$

$$Z_2 = R_2$$

$$Z_3 = R_3 + j\omega L_3$$

$$Z_4 = R_4$$

The value of $L_3$ is given by

$$L_3 = \frac{C_1 R_2 R_4}{1 + \omega^2 R_1^2 C_1^2}$$

and

$$R_3 = \frac{\omega^2 C_1^2 R_1 R_2 R_4}{1 + \omega^2 R_1^2 C_1{}^2}$$

**De Sauty Bridge.** In this $C_2$ is the capacitance the value of which is to be determined. $C_3$ is a standard capacitor and $R_1$, $R_2$ are non-inductive resistors. For obtaining the balance either $R_1$ or $R_2$ is varied. The value of $C_2$ is given by

$$C_2 = C_3 \frac{R_1}{R_2}$$

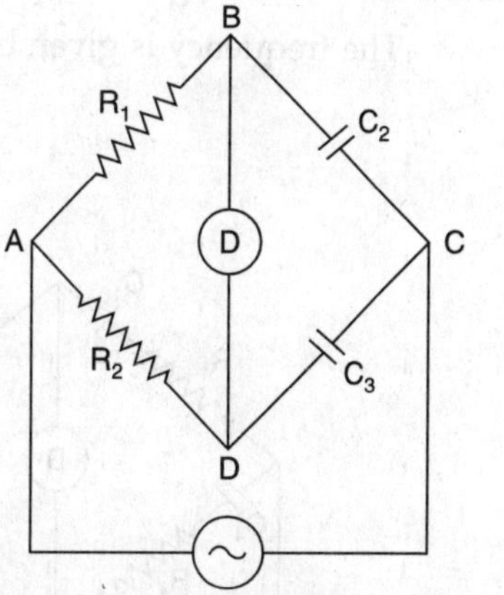

*FIGURE 21.5*

It is difficult to get balance in case capacitors are not free from the dielectric loss. The bridge has maximum sensitivity when $C_2 = C_3$.

**Schering Bridge.** This bridge can be used for measuring the capacitance and dielectric loss of a capacitor. This bridge is also used for comparing imperfect capacitor $C_2$ in terms of loss free capacitor $C_1$. In Fig. 21.6 *(B)* the imperfect capacitor has been replaced by a loss free capacitor $C_2$ and a series resistance *r*.

Under the conditions of balance

$$C_2 = C_1\left(\frac{R_4}{R_3}\right)$$

and

$$r = R_3\left(\frac{C_4}{C_1}\right)$$

$$\begin{aligned}\text{Dissipation factor} &= \omega r C_2\\ &= \omega C_4 R_4\\ &= \text{Power factor.}\end{aligned}$$

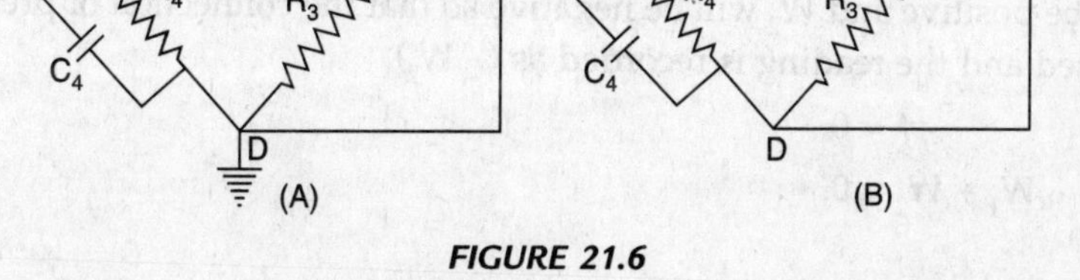

*FIGURE 21.6*

**Wein Series Bridge.** This bridge is mainly used for audio frequency measurement of capacitors over a wide range. Under the conditions of balance.

$$C_1 = C_4\left(\frac{R_3}{R_2}\right)$$

**Wein Parallel Bridge.** This bridge can be used for measuring audio frequencies. In the analysis of this bridge the capacitors $C_1$ and $C_2$ are assumed to be loss-free.

The frequency is given by

$$f = \frac{1}{2\pi\sqrt{R_1R_2C_1C_2}} \text{ Hz.}$$

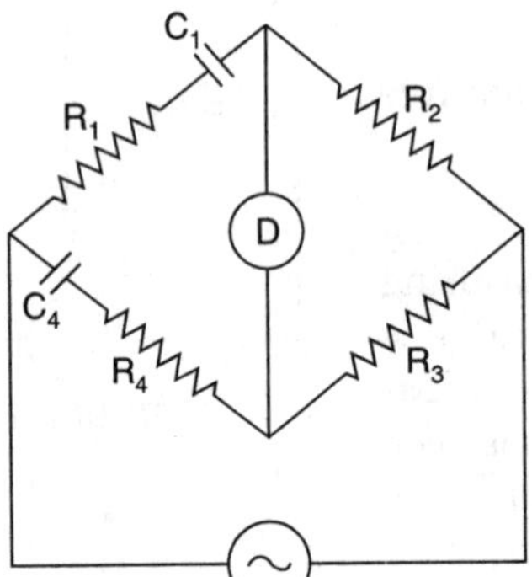


*FIGURE 21.7.* Wien series bridge.

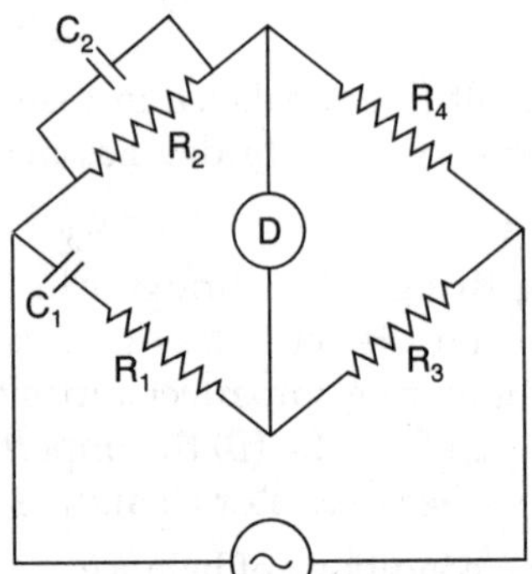


*FIGURE 21.8.* Wien parallel bridge.

**Power Measurement in Three Phase Circuits.** Following methods can be used for power measurement in three phase circuits:

1. **Three Wattmeter Method.** In this one wattmeter is introduced in each phase and the total power is the algebraic sum of the three wattmeter readings.
2. **Two Wattmeter Method.** The method of connecting wattmeters in series and delta circuits is shown in Fig. 21.9. Total power is given by the sum of two wattmeter readings.

However, following points may be noted:

(1) When power factor is unity $W_1 = W_2$.

(2) When power factor is 0.5 (lagging).

One of the wattmeter will read zero and the total power is given by the reading of other wattmeter.

(3) When $0.5 > pf > 0$

$W_1$ will be positive and $W_2$ will be negative so that the connection of pressure coil of $W_2$ is reversed and the reading is recorded as $(-W_2)$.

(4) When $pf = 0$

$$W_1 + W_2 = 0.$$

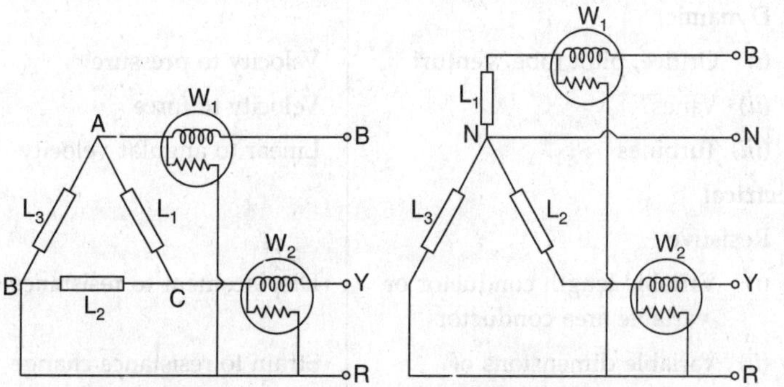


Delta connected load　　Star connected load

*FIGURE 21.9*

**Balanced Load.** In this case power factor angle is given by

$$\tan \phi = -\frac{\sqrt{3}\,(W_1 - W_2)}{W_1 + W_2}.$$

**Transducers.** A transducer is a device that converts energy from one form to another. Thus springs, bellows, rods, tubes, etc. can be used as transducers.

Those transducers which directly sense the physical phenomenon are known as **primary transducers.** Sometimes the primary signals are to be converted into electrical signals using what is known as **secondary transducers.**

Those transducers which generate their own voltage are known as active transducers. Thus piezoelectric transducers are active transducers. On the other hand some transducers need external power for their action. They may absorb some energy from the physical phenomenon under study. Such transducers are known as passive transducers. Some transducers and their operations are given below:

| *Type of transducer* | *Operation* |
|---|---|
| **1. Mechanical** | |
| (*a*) Thermal | |
| (*i*) Thermocouple | Temperature to current |
| (*ii*) Bimetals | Temperature to displacement |
| (*iii*) Temp-stik | Temperature to phase |
| (*b*) Elastic member | |
| (*i*) Bellows, diaphragm, Bourdon tubes | Pressure to displacement |
| (*ii*) Proving ring and springs | Force to displacement |
| (*c*) Hydro pneumatic | |
| (*i*) Float | Fluid level to displacement |
| (*ii*) Hydrometer | Specific gravity to displacement |

| | |
|---|---|
| (*d*) Dynamic | |
| (*i*) Orifice, pitot tube, venturi | Velocity to pressure |
| (*ii*) Vanes | Velocity to force |
| (*iii*) Turbines | Linear to angular velocity |
| **2. Electrical** | |
| (*a*) Resistive | |
| (*i*) Variable length conductor or variable area conductor | Displacement to resistance change |
| (*ii*) Variable dimensions of conductor | Strain to resistance change |
| (*b*) Inductance | |
| (*i*) Variable coil dimensions, variable air gap, change of core material change of coil positions, change of core positions | Displacement to inductance change |
| (*ii*) Movement of coil, movement of permanent magnet, movement of core | Velocity to inductance change |
| (*c*) Capacitance | |
| Change of air gap, plate area, dielectric | Displacement to capacitance change |
| (*d*) Piezoelectric | Displacement to voltage |
| (*e*) Photoelectric | Light intensity to voltage. |

## OBJECTIVE TYPE QUESTIONS

1. The resolution of an instrument is
   (*a*) the minimum input increment that gives definite numerical change in the output
   (*b*) the extent to which the indicated output is different from the actual output
   (*c*) the ratio of output to input in a identical units
   (*d*) the ratio of the incremental output to the incremental input.
2. If the input to an instrument is increased very slowly from zero value there will be some minimum value of input below which no output can be detected. This minimum value of input is known as
   (*a*) sensitivity ratio of the instrument
   (*b*) static sensitivity of the instrument
   (*c*) threshold of the instrument
   (*d*) drift of the instrument.

3. Two resistors $R_1 = 38.5\ \Omega$ and $R_2 = 45.62\ \Omega$ with an uncertainty of one unit in the last digit of each number are connected in series. The total resistance will be
(*a*) 84.12 Ω (*b*) 84.14 to 84.11 Ω (*c*) 84.1 Ω (*d*) 84 Ω.

4. In the above case if resistors are connected in parallel, the equivalent resistance will be
(*a*) 20.8 Ω (*b*) 20.88 Ω (*c*) 20.881 Ω (*d*) 20 Ω.

**Questions 5 and 6 Refer to Data Given Below:**

5. A 0 – 50 V voltmeter has 100 scale divisions that can be read to half a division. The resolution of the meter will be
(*a*) 0.025 V (*b*) 0.25 V (*c*) 0.5 V (*d*) 2.5 V.

6. The resolution in percentage will be
(*a*) 0.05% (*b*) 0.01% (*c*) 0.05% (*d*) 0.5%.

7. In case the pointer of an instrument is not fine and straight, then the error in reading is termed as
(*a*) systematic error (*b*) personal error
(*c*) accidental error (*d*) random error.

**Questions 8 and 9 Refer to Data Given Below:**

A 0 – 150 V voltmeter having a sensitivity of 1000 Ω/V reads 100 V across a resistance. A milliammeter connected in series with the resistance reads 50 mA.

8. The apparent value of the unknown resistor will be
(*a*) 1 kΩ (*b*) 2 kΩ (*c*) 20 kΩ (*d*) 20.2 kΩ.

9. The error due to loading effect of the voltmeter will be
(*a*) 0.0133% (*b*) 0.133% (*c*) 1.33% (*d*) 13.3%.

**Questions 10 and 11 Refer to Data Given Below:**

The solution for the unknown resistance in a Wheatstone bridge is given by

$$P = \frac{Q}{R} S$$

where $Q = 100 \pm 0.5\ \%$
$R = 50 \pm 0.5\%$
$S = 75.5 \pm 0.5\%$.

10. The unknown resistance is
(*a*) 3 Ω (*b*) 150 Ω (*c*) 151 Ω (*d*) 445 Ω.

11. The maximum possible error in the unknown resistance $P$ will be
(*a*) 2.5% (*b*) 1.5% (*c*) 1.0% (*d*) 0.5%.

**Questions 12 and 13 Refer to Data Given Below:**

Three resistors have the following values

$$R_1 = 40 \pm 5\%$$
$$R_2 = 50 \pm 5\%$$
$$R_3 = 80 \pm 5\%.$$

12. If these resistors are connected in series the maximum possible error will be
(*a*) 25% (*b*) 15% (*c*) 5% (*d*) 1%.

13. If these resistors are connected in parallel, the maximum possible error could be
(*a*) 25% (*b*) 20% (*c*) 15% (*d*) 5%.

14. In e.s.u. system, the dimensions of charge $Q$ will be
(*a*) $[\varepsilon MLT^{-1}]$ (*b*) $[\varepsilon^{1/2} MLT^{-1}]$
(*c*) $[\varepsilon^{1/2} M^{1/2}L^{1/2}T^{-1/2}]$ (*d*) $[\varepsilon^{1/2}M^{1/2}L^{3/2}T^{-1}]$.

15. In e.s.u. system, the dimensions of current $I$ will be
(*a*) $[\varepsilon^{1/2}M^{1/2}L^{1/2}T^{1/2}]$ (*b*) $[\varepsilon^{1/2}M^{1/2}L^{1/2}T^{-1/2}]$
(*c*) $[\varepsilon^{1/2}M^{1/2}L^{3/2}T^{-2}]$ (*d*) $[\varepsilon^{1/2}M^{3/2}L^{3/2}T^{-3/2}]$.

16. In e.s.u. system, the dimensions of potential difference will be
(*a*) $[\varepsilon^{1/2}M^{1/2}L^{1/2}T^{-1}]$ (*b*) $[\varepsilon^{-1/2}M^{1/2}L^{1/2}T^{-1}]$
(*c*) $[\varepsilon^{-1/2}M^{-1/2}L^{2}T^{-1}]$ (*d*) $[\varepsilon^{-1/2}M^{-1/2}L^{-3/2}T^{-1/2}]$.

17. In e.s.u. system, $[\varepsilon L]$ could be the dimension of
(*a*) Resistance (*b*) Flux (*c*) Resistance (*d*) Capacitance.

18. In e.s.u. system, the dimension of resistance will be
(*a*) $\varepsilon LT$ (*b*) $\varepsilon^{-1}L^{-1}T$ (*c*) $\varepsilon^{-1}L^{-1/2}T^{1/2}$ (*d*) $\varepsilon^{-1}L^{1/2}T^{3/2}$.

19. In e.s.u. system, the dimension of inductance will be
(*a*) $\varepsilon^{-1}L^{-1}T^{2}$ (*b*) $\varepsilon^{-1}LT^{-2}$ (*c*) $\varepsilon^{-1}LT^{-1}$ (*d*) $\varepsilon^{-1}LT^{2}$.

20. In e.m.u. system, the dimension of current will be
(*a*) $[\mu^{-1/2}M^{1/2}L^{1/2}T^{-1}]$ (*b*) $[\mu^{1/2}M^{1/2}L^{-1/2}T^{1}]$
(*c*) $[\mu^{1/2}M^{1/2}L^{-3/2}T^{-1/2}]$ (*d*) $[\mu^{1/2}M^{1/2}L^{3/2}T^{1/2}]$.

21. In e.m.u. system, the dimensions of charge will be
(*a*) $[\mu^{-1/2}M^{1/2}L^{1/2}]$ (*b*) $[\mu^{1/2}M^{1/2}L^{1/2}T^{1/2}]$
(*c*) $[\mu^{1/2}M^{1/2}L^{-1/2}T^{-1}]$ (*d*) $[\mu^{-1/2}M^{-1/2}L^{-1/2}]$.

22. In e.m.u. system $[\mu L]$ could be the dimension of
(*a*) inductance (*b*) capacitance (*c*) resistance (*d*) none of the above.

23. In e.m.u. system $(\mu LT^{-1})$ could be the dimensions of
(*a*) Resistance (*b*) Flux density (*c*) Potential difference (*d*) Power.

24. In e.m.u. system, the dimensions of capacitance could be
(*a*) $[\mu LT]$ (*b*) $[\mu LT^{-1}]$ (*c*) $[\mu L^{-1}T^{-1}]$ (*d*) $[\mu^{-1}L^{-1}T]$.

25. The number of e.m. units and the number of e.s. units is the same for which of the following practical unit?
(*a*) Coulomb (*b*) Ohm (*c*) Henry (*d*) Watt.

26. SI unit for luminous intensity is
(*a*) lumen (*b*) lux (*c*) candela (*d*) none of the above.

27. The prefix atto is used for
(*a*) $10^{12}$ (*b*) $10^{18}$ (*c*) $10^{-18}$ (*d*) $10^{-12}$.

28. The prefix femto is used for
(*a*) $10^{-15}$ (*b*) $10^{-12}$ (*c*) $10^{6}$ (*d*) $10^{24}$.

29. In *LTVI* system, the dimensions of charge will be
(*a*) $TI$ (*b*) $LTI$ (*c*) $LT^{-1/2}I$ (*d*) $L^{-12}TV$.

30. In *LTVI* system, the dimensions of magnetic flux density will be
(*a*) $LTV$ (*b*) $L^2TV$ (*c*) $L^{-2}TV$ (*d*) $L^{-2}TV^{-1}$.

31. In *LTVI* system, the dimensions of permittivity are
(*a*) $LTVI$ (*b*) $LTVI^{-1}$ (*c*) $L^{-1}TV^{-1}I$ (*d*) $L^{-1}T^{-1}V^{-1}I^{-1}$.

32. In *LTVI* system, *TV* could be the dimensions of
(*a*) Permeability (*b*) Magnetic pole (*c*) Resistivity (*d*) None of the above.

33. Platinum is used as a resistance material in
(*a*) precision resistors (*b*) high wattage resistors
(*c*) thermometry (*d*) rheostats.

34. Which resistance material is preferred for standards?
(*a*) Manganin (*b*) Platinum (*c*) Nichrome (*d*) German silver.

35. Which instrument is used in instruments for the measurement of electrical quantities?
(*a*) Heating effect (*b*) Chemical effect
(*c*) Electrostatic effect (*d*) Any of the above.

36. Which of the following instrument cannot be made using heating effect?
(*a*) Ammeter (*b*) Voltmeter (*c*) Frequency meter (*d*) All of the above.

37. Which of the following instrument can be made using magnetic effect?
(*a*) Energy meters (*b*) Wattmeters (*c*) Voltmeters (*d*) All of the above.

38. Which of the following instrument can be made using electromagnetic induction effect?
(*a*) Energy meters (*b*) AC ammeters (*c*) Wattmeters (*d*) All of the above.

39. Energy meter installed at the residence of electric power consumers is
(*a*) indicating type instrument
(*b*) recording type instrument
(*c*) indicating as well as recording type instrument
(*d*) integrating type instrument.

**40.** The term 'artificial aging' in instruments is associated with

(*a*) springs (*b*) permanent magnets (*c*) controlling torque (*d*) damping.

**41.** Artificial aging of permanent magnets is done by

(*a*) putting the magnet in strong magnetic field

(*b*) rubbing a magnet with other magnet of higher intensity

(*c*) placing the magnet in a weak alternating magnetic field or by heating

(*d*) any of the above.

**42.** Which of the following material is preferred for permanent magnets?

(*a*) Alnico (*b*) Y-alloy (*c*) Silicon steel (*d*) Wrought iron.

**43.** A Ballistic galvanometer is

(*a*) moving coil type galvanometer (*b*) moving magnet type galvanometer

(*c*) electro-magnet type galvanometer (*d*) none of the above.

**44.** Which of the following instrument cannot be used for the measurement of ac as well as dc quantities?

(*a*) Hot wire instruments (*b*) Moving iron instruments

(*c*) Induction type instruments (*d*) Electrostatic instruments.

**45.** Which of the following error in moving iron instruments affects the instrument on ac as well as dc circuits?

(*a*) Hysteresis error (*b*) Effect of stray magnetic fields

(*c*) Temperature error (*d*) All of the above.

**46.** Which of the following error in moving iron instruments affects the instrument on ac circuits only?

(*a*) Temperature error (*b*) Hysteresis error (*c*) Waveform error (*d*) Frictional error.

**47.** Which of the following is not a constituent of permanent magnet material 'Alnico'?

(*a*) Nickel (*b*) Silicon (*c*) Cobalt (*d*) Aluminium.

**48.** Which of the following material for permanent magnets has the maximum value of coercive force in $AT/m$?

(*a*) Cobalt steel (35% cobalt) (*b*) Cobalt steel (9% cobalt)

(*c*) Alcomax (*d*) Alnico.

**49.** Which type of instrument will draw least current from the circuit in which it is incorporated?

(*a*) Electrostatic instruments (*b*) Hot wire instruments

(*c*) Rectifier instruments (*d*) Thermocouple instruments.

**50.** Which of the following instrument is free from hysteresis and eddy current errors?

(*a*) Moving iron instrument (*b*) Electrostatic instrument

(*c*) Moving coil-permanent magnet type instrument

(*d*) Moving coil dynamometer type instrument.

51. Two voltmeters of 0–300 V range are connected in parallel to a ac circuit. One voltmeter is moving iron type and reads 200 V. If the other instrument is moving coil type, its reading will be

(a) 200 $\sqrt{3}$ volts (b) 200 × 1.41 volts (c) slightly less than 200 V (d) zero.

52. For getting best results, indicating instruments are

(a) undamped (b) overdamped

(c) critically damped

(d) damped slightly less than the critical value.

53. In a critically damped instrument

(a) the pointer will slowly move from zero to the actual value

(b) the pointer will move in number of steps stopping momentarily at various stages

(c) quickly move to final position with large oscillations

(d) quickly move to final position without oscillations.

54. Moving coil instruments can be used on

(a) dc only (b) sinusoidal ac only (c) all ac wave forms (d) ac and dc both.

55. Permanent magnet moving coil instruments can be used on

(a) dc only (b) sinusoidal ac only

(c) all ac wave forms and frequencies (d) ac and dc both.

56. Moving coil and moving iron type instruments can be distinguished by looking at

(a) their range (b) pointer (c) scale (d) size of terminals.

57. Instruments measuring values indicate

(a) peak values (b) rms values (c) average values (d) mean values.

58. The scale of moving iron type instruments is

(a) uniform (b) cramped (c) logarithmic (d) none of the above.

59. If a instrument is installed in vertical position, which type of damping it can have?

(a) Fluid friction type (b) Gravity (c) Eddy current type (d) Any of the above.

60. Bearing used for supporting pointer is

(a) needle bearing (b) ball bearing (c) roller bearing (d) jewel bearing.

61. Which of the following instrument has identical calibration for ac as well as dc values?

(a) Induction type (b) Hot wire type (c) Moving iron type (d) Moving coil type.

62. Which type of wattmeter is widely used?

(a) Induction type (b) Dynamometer type (c) Moving iron type (d) Electrostatic type.

63. A dynamometer type wattmeter can be used on

(a) dc only (b) ac only (c) rectified ac only (d) ac as well as dc.

64. A dynamometer type wattmeter has

(a) uniform scale (b) square law scale (c) logarithmic scale (d) non-linear scale.

65. Damping generally used in dynamometer type wattmeters is
(a) eddy current damping
(b) air friction damping
(c) fluid friction damping
(d) no damping is required in these instruments.

66. For shielding a dynamometer type wattmeter against stary fields
(a) meter is housed in a soft iron case
(b) meter is earthed
(c) neutral wire connection is provided
(d) meter components are made of non-magnet materials.

67. The readings of a dynamometer type wattmeter can be highly erratic at
(a) high voltages (b) fluctuating loads (c) low power factors (d) low frequencies.

68. The meter constant of single phase energy meter is expressed in terms of
(a) volts/kWh (b) kW/kWh (c) amps/kW (d) revolutions/kWh.

69. In dynamometer type instruments used for the measurement of ac values, the coil is generally air cored to minimise
(a) temperature errors
(b) eddy current error
(c) hysteresis error
(d) eddy current and hysteresis error.

70. Multimeters generally do not have facility for the measurement of
(a) current (b) voltage (c) resistance (d) frequency.

71. A multimeter can be used for measuring
(a) dc quantities only
(b) ac quantities
(c) dc as well ac quantities.

72. A multimeter indicates less reading as compared to the actual reading. The probable reason could be
(a) poor damping (b) non-linear scale (c) pointer bend (d) battery weak.

73. Which of the following instrument is an integrating type instrument?
(a) Wattmeter (b) Energy meter (c) Power factor meter (d) None of the above.

74. Which instrument is least affected by stray fields?
(a) Moving iron type
(b) Induction type
(c) Hot wire type
(d) Moving iron type.

75. Which of the following instrument can have full scale deflection of 300 degrees?
(a) Hot wire type (b) Rectifier type (c) Induction type (d) None of the above.

76. For the same range which voltmeter will have least power consumption?
(a) Induction type (b) Hot wire type (c) Electrostatic type (d) Moving iron type.

77. In dynamometer type moving coil instruments, a swamping resistance is connected in series with the moving coil mainly to
(a) equalise time constants of fixed and moving coils
(b) reduce damping force required for the instrument
(c) reduce the flow of current through the coil
(d) all of the above.

**78.** In a Ferraris induction instrument of the current to be measured is halved, the mean torque will

(*a*) remain unaffected (*b*) reduce marginally
(*c*) reduce to half value (*d*) reduce to one fourth value.

**79.** In loss of charge method for measuring the change in voltage, which of the following instrument is used

(*a*) *VTVM* (*b*) D'Arsonnal galvanometer
(*c*) Ballistic galvanometer (*d*) Digital voltmeter.

**80.** A potentiometer works on

(*a*) magnetic effect (*b*) electromagnetic induction
(*c*) heating effect (*d*) none of the above.

**81.** Which method will be suitable for the measurement of a resistance of expected value less than one ohm?

(*a*) Wheatstone bridge method (*b*) Kelvin's double bridge method
(*c*) Loss of charge method (*d*) Substitution method.

**82.** Schering bridge is used to measure

(*a*) variable resistances (*b*) dielectric loss and power factor
(*c*) frequency and capacitance (*d*) all of the above.

**83.** Which of the following ac bridge is preferred for the measurement of inductance having high *Q*-factor?

(*a*) Hay bridge (*b*) Maxwell bridge (*c*) Desauty bridge (*d*) Owen bridge.

**84.** A signal of 10 mV at 75 MHz is to be measured. Which of the following instrument can be used?

(*a*) *VTVM* (*b*) Cathode ray oscilloscope
(*c*) Moving iron voltmeter (*d*) Digital multimeter.

**85.** For measuring small voltages of the order of 10 micro volts, which voltmeter would you select?

(*a*) *CRO* (*b*) Peak reading diode *VTVM*
(*c*) Amplifier—rectifier *VTVM* (*d*) Electrostatic voltmeter.

**Questions 86 to 88 Refer to Fig. 21.10:**

**86.** The bridge shown in Fig. 21.10 is

(*a*) Anderson bridge
(*b*) Maxwell bridge
(*c*) Heaviside campbell bridge
(*d*) Desauty bridge.

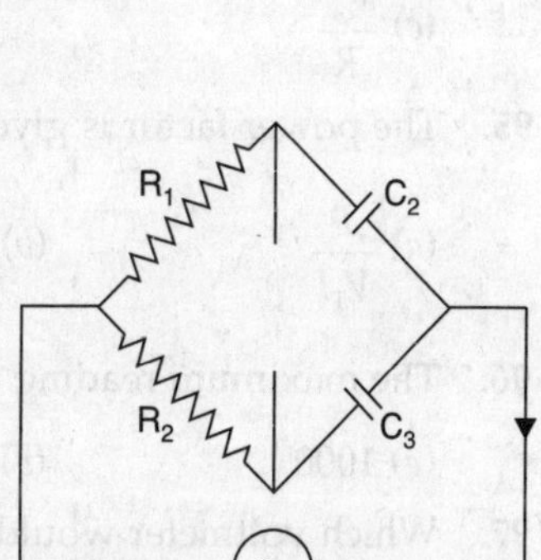


*FIGURE 21.10*

**87.** When the bridge is balanced, the value of $C_2$ is given by

(*a*) $C_2 = C_3 \frac{R_3}{R_1}$ (*b*) $C_2 = C_3 \frac{R_1}{R_3}$ (*c*) $C_2 = \frac{R_1}{C_3 R_3}$ (*d*) $C_2 = \frac{R_3}{R_1 C_3}$.

**88.** The bridge will have maximum sensitivity when

(*a*) $C_3 = 0$ (*b*) $C_2 = C_3$ (*c*) $R_2 = R_1$ (*d*) $R_1 R_2 = 1$.

**89.** Which of the following bridge is used for the precise measurement of inductance values ?

(*a*) Maxwell bridge (*b*) Schering bridge

(*c*) Anderson bridge (*d*) Heaviside-campbell bridge.

**90.** Hay bridge is used for

(*a*) small inductances (*b*) variable inductances

(*c*) air cored inductances (*d*) inductance having large time constant.

**91.** The dissipation factor of a capacitor can be measured with

(*a*) Potentiometer (*b*) Galvanometer (*c*) Campbell bridge (*d*) Schering bridge.

**92.** Which detector is generally used in ac potentiometer work ?

(*a*) D'Arsonvel galvanometer (*b*) Ballistic galvanometer

(*c*) Vibration galvanometer (*d*) None of the above.

**93.** A wattmeter is measuring the power supplied to a circuit whose power factor is 0.7. The frequency is 50 Hz. The wattmeter has a voltage—coil circuit of resistance 1000 ohms and inductance 0.4 H. The percentage error will be nearly

(*a*) 1% (*b*) 2% (*c*) 10% (*d*) 20%.

**94.** A consuming device Z, is connected in series with a non-inductive resistor *R*, the three voltage drops $V_1$, $V_2$ and $V_3$ are measured by voltmeter.

The power *P* of the device is given by

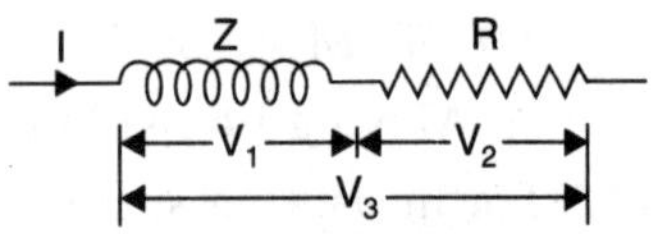

***FIGURE 21.11***

(*a*) $\frac{V_1^2 + V_2^2 + V_3^2}{R}$ (*b*) $\frac{V_1^{\ 2}}{R}$

(*c*) $\frac{V_3^2}{R}$ (*d*) $\frac{V_3^2 - V_1^2 - V_2^2}{2R}$.

**95.** The power factor is given by

(*a*) $\frac{P}{V_1 I}$ (*b*) $\frac{P}{V_2 I}$ (*c*) $\frac{P}{V_3 I}$ (*d*) none of the above.

**96.** The maximum reading of a $3\frac{1}{2}$ digit voltmeter will be

(*a*) 1000 (*b*) 999 (*c*) 1999 (*d*) 9999.

**97.** Which voltmeter would you select for measuring 50,000 V direct current ?

(*a*) Moving coil voltmeter (*b*) Hot wire voltmeter

(*c*) Electrodynamometer voltmeter (*d*) Electrostatic voltmeter.

98. A 0–100 V voltmeter is guaranteed for 1% accuracy when full scale reading is 80 V. The percentage limiting error will be

(a) 0.8% (b) 1% (c) 1.25% (d) 5%.

99. In case of first grade instruments the relative error is generally restricted to

(a) 10% (b) 5% (c) 1% (d) 0.1%.

100. A wattmeter will be free from the effects of power factor and frequency variations in case

(a) damping is not provided

(b) voltage coil resistance is zero

(c) pressure coil inductance is zero

(d) a capacitance is connected in parallel to pressure coil.

101. In case of current transformers the burden is expressed in terms of

(a) volts (b) volt-amperes (c) kWh (d) watt-hours.

102. Mumetal alloy is used for current transformer because it has

(a) high permeability and low retentivity (b) high permeability and high retentivity

(c) low permeability and high retentivity (d) low permeability and low retentivity.

103. A Q-meter measures

(a) frequency (b) losses in a capacitor

(c) properties of the coils (d) accurate values of electrical quantities.

104. An integrating digital voltmeter measures

(a) rms volues (b) true average values (c) peak to peak values (d) all of the above.

105. Which of the following instrument will always indicate true rms values irrespective of the waveform ?

(a) Thermocouple meter (b) Digital voltmeter

(c) Moving iron meter (d) None of the above.

106. In the measurement of power by two wattmeter method in three phase circuit, when one of the wattmeter shows zero reading, it can be concluded that

(a) one of the phase is earthed (b) power factor is unity

(c) power factor is leading (d) power factor is 0.5 lagging.

107. In the measurement of power by two wattmeter method, when the readings of the two wattmeters are equal and opposite, it can be concluded that

(a) power factor is zero (b) power factor is 0.5 leading

(c) power factor is unity (d) power factor is 0.5 lagging.

108. In the measurement of power by two wattmeter method, when the power factor is unity

(a) one of the wattmeters will read zero

(b) both wattmeters will read zero

(c) both wattmeters will have equal but opposite readings

(d) both wattmeters will have equal and positive reading.

109. In the measurement of power by two wattmeter method, the connections of one of the wattmeter will have to be reversed when

(a) p.f. is zero (b) $0.5 > p.f. > 0$ (c) $0.5 < p.f. < 0$ (d) p.f. is 0.5 leading.

110. While measuring power in a three phase balanced circuit by two wattmeter method, the readings are – 2 kW and + 7 kW. The power angle will be

(a) $\tan^{-1} \frac{9\sqrt{3}}{5}$ (b) $\tan^{-1} \frac{5}{9}\sqrt{3}$ (c) $\tan^{-1} \sqrt{3}$ (d) none of the above.

111. The cathode of a Cathode Ray Oscilloscope is usually coated with

(a) Capper oxide (b) Tungsten or thorium oxide

(c) Barium or strontium oxide (d) Alkali metals.

112. Cathode rays can be deflected by

(a) electric field (b) magnetic field

(c) both (a) and (b) above (d) neither of above.

113. The input impedance of a CRO is nearly

(a) zero (b) around 10 ohms (c) around 100 ohms (d) around 1 MΩ.

114. A 100 MHz oscilloscope has

(a) 50 MHz sweep (b) 50 MHz vertical oscillator

(c) 50 MHz horizontal oscillator (d) 50 MHz supply voltage frequency.

115. A cathode ray oscilloscope can be used to measure

(a) ac voltages only (b) dc voltages only (c) frequency (d) any of the above.

116. Green screen for cathode ray tubes uses which of the following phosphor ?

(a) $P_1$ (b) $P_4$ (c) $P_{22}$ (d) $P_{33}$.

117. Which phosphor number is used for television screens ?

(a) $P_1$ (b) $P_2$ (c) $P_3$ (d) $P_4$.

118. Phosphor number $P_{19}$ is generally used for

(a) low frequency oscilloscopes (b) radar indicators

(c) fast photographic oscillography (d) all of the above.

119. Which of the following colour of screen has least persistence ?

(a) Bluish purple (b) Orange (c) Blue green (d) Blue white.

120. Phosphor coating for cathode ray tubes is provided on

(a) inside surfaces only (b) outside surfaces only

(c) both the surfaces (d) within the glass.

121. With a sweep time of one millisecond across the screen, the approximate horizontal saw tooth frequency will be

(a) 1 kHz (b) 1 MHz (c) 1 GHz (d) none of the above.

**122.** On a cathode ray screen, the brightest spot occurs at
(*a*) the outer periphery (*b*) the centre
(*c*) midway between centre and outer periphery of screen
(*d*) brightness is same all over the screen.

**123.** The secondary emission of electrons is due to
(*a*) thermal excitation (*b*) solar radiation
(*c*) electric field (*d*) kinetic energy of striking electrons.

**124.** Secondary electrons inside *CRT* are collected by
(*a*) aquadag (*b*) boron (*c*) glass (*d*) silver coating.

**125.** In *CRO* astigmatism is
(*a*) a source of generating fast electrons
(*b*) media for absorbing secondary emission electrons
(*c*) an additional focussing contral
(*d*) internal saw tooth voltage generator.

**126.** A double beam oscilloscope has
(*a*) two screens (*b*) two electron guns
(*c*) two different phosphor coatings (*d*) one waveform divided into two parts.

**Questions 127 to 131 Rrefer to Fig. 21.12 :**

Various lissajous patterns on *CRO* screen are shown in Fig. 21.12. These patterns have been obtained for two sinewave voltages of the same frequency.

**127.** A phase difference of zero degree is represented by
(*a*) figure *A* (*b*) figure *B* (*c*) figure *C* (*d*) figure *D*.

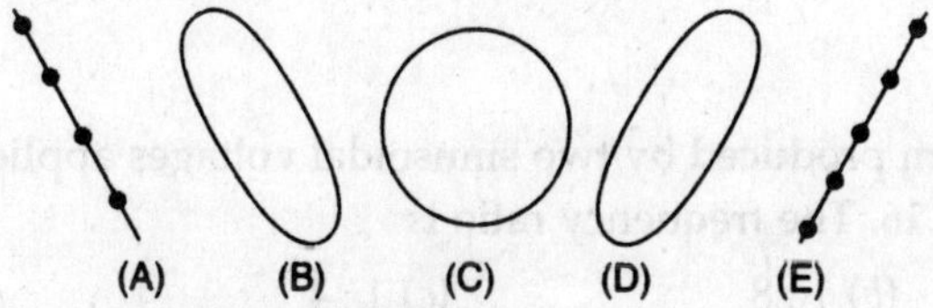


***FIGURE 21.12***

**128.** A phase difference of 90° is represented by
(*a*) figure *A* (*b*) figure *B* (*c*) figure *C* (*d*) figure *D*.

**129.** Figure *A* represents the phase difference of
(*a*) 360° (*b*) 270° (*c*) 180° (*d*) 90°.

**130.** Figure *D* represents the phase difference of
(*a*) 30° (*b*) 45° (*c*) 60° (*d*) 120°.

**131.** Which figure represents phase difference of 30° ?
(*a*) figure *A* (*b*) figure *B* (*c*) figure *C* (*d*) none of the figures.

132. If the lissajous pattern obtained on the screen is as shown in Fig. 21.13 the phase difference between two sine wave voltages is given by

(a) $\sin^{-1} \frac{X}{Y}$

(b) $\sin^{-1} \frac{Y}{X}$

(c) $\tan^{-1} \frac{X}{Y}$

(d) $\tan^{-1} \frac{Y}{X}$.

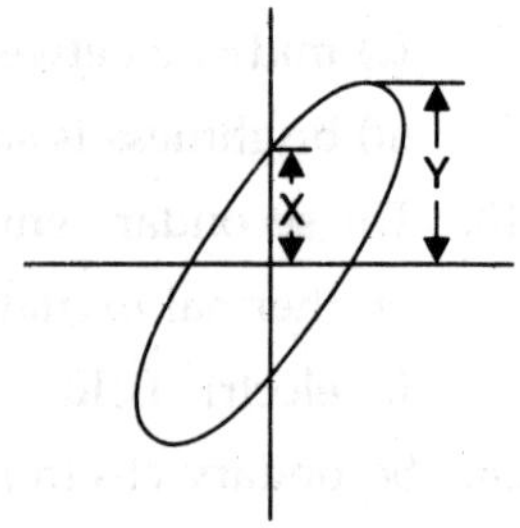


FIGURE 21.13

133. The lissajous pattern on CRO screen is as shown in Fig. 21.14. If the frequency applied to horizontal plate is 100 Hz, the frequency of the vertical plate will be

(a) 200 Hz
(b) 100 Hz
(c) 50 Hz
(d) 33.3 Hz.

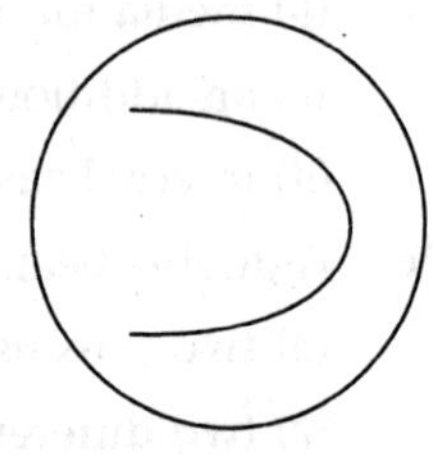

FIGURE 21.14

134. The lissajous pattern on CRO screen is as shown in Fig. 21.15. If the frequency applied to horizontal plate is 100 Hz the frequency of vertical plate will be

(a) 200 Hz
(d) 100 Hz
(c) 50 Hz
(d) 25 Hz.

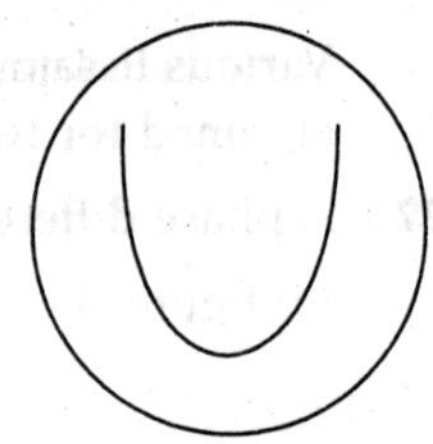

FIGURE 21.15

135. The lessajous pattern produced by two sinusoidal voltages applied to X and Y terminals is as shown in Fig. 21.16. The frequency ratio is

(a) 1 : 16 (b) 1 : 8 (c) 1 : 4 (d) 3 : 4.

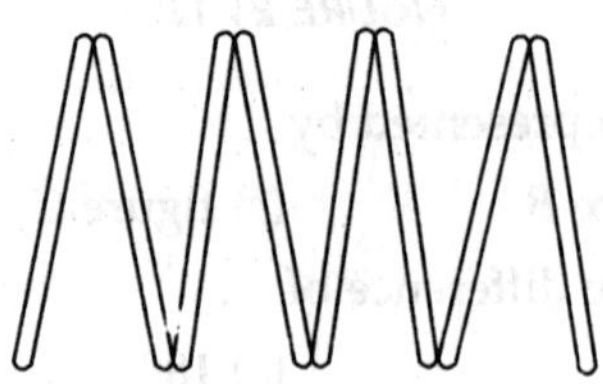

FIGURE 21.16

136. In a *CRO* a sine wave signal of unknown frequency is connected to the vertical input terminals and a sine wave signal of 50 Hz frequency is connected to horizontal input. The lissajous pattern is as shown in Fig. 21.17. The unknown frequency will be

(*a*) 200 Hz

(*b*) 100 Hz

(*c*) 50 Hz

(*d*) 25 Hz.

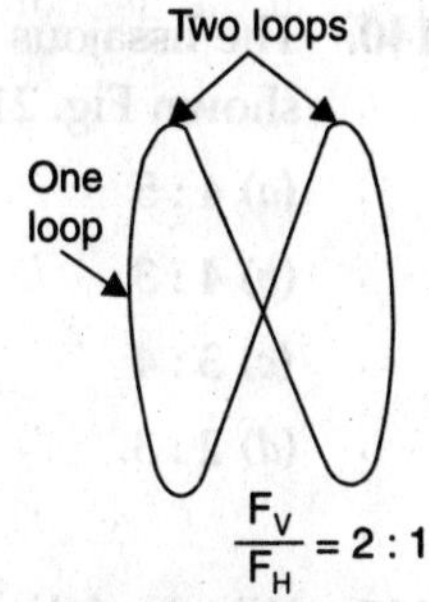


*FIGURE 21.17*

137. In a *CRO* a sine wave signal if unknown frequency is connected to the vertical terminals and a sine wave signal of 50 Hz frequency is connected to horizontal input. The lissajous pattern is as shown in Fig. 21.18. The unknown frequency will be

(*a*) $66\frac{2}{3}$

(*b*) $33\frac{1}{3}$

(*c*) 25

(*d*) $16\frac{2}{3}$.

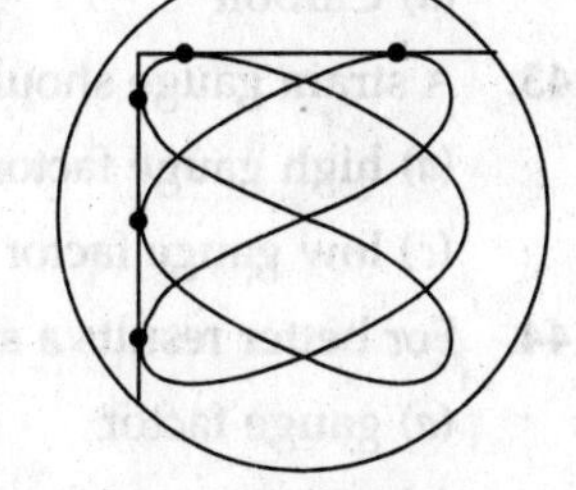

*FIGURE 21.18*

138. In a *CRO* a sine wave signal of unknown frequency is connected to the vertical input terminals and sine wave signal of 50 Hz frequency is connected to horizontal input. The lissajous pattern is as shown in Fig. 21.19. The unknown frequency will be

(*a*) $16\frac{2}{3}$ Hz (*b*) 25 Hz

(*c*) $66\frac{2}{3}$ Hz (*d*) 150 Hz.

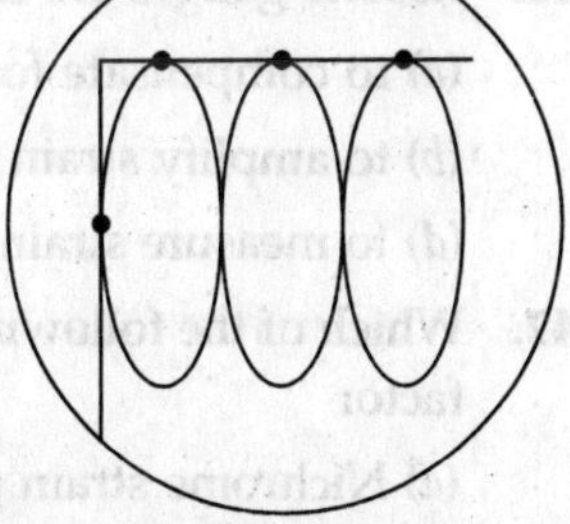

*FIGURE 21.19*

139. In a *CRO* a sine wave signal of unknown frequency is connected to the vertical input terminals and a sine wave signal of 50 Hz is connected to horizontal input. The lissajous pattern is as shown in Fig. 21.20. The unknown frequency will be

(*a*) 100 Hz

(*b*) 75 Hz

(*c*) 25 Hz

(*d*) $12\frac{1}{2}$ Hz.

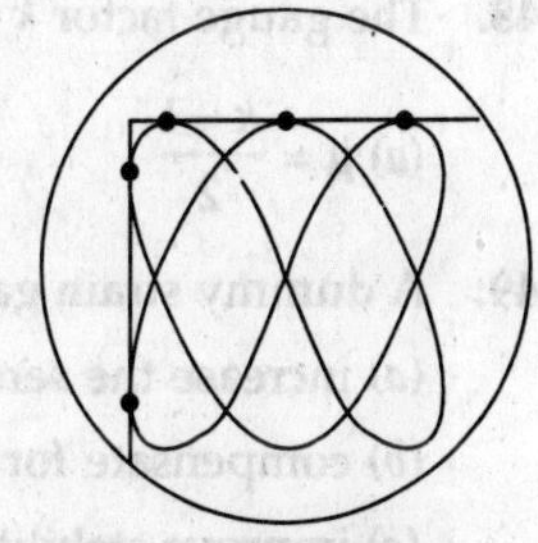

*FIGURE 21.20*

**140.** The lissajous pattern produced by two sinusoidal voltages applied to $X$ and $Y$ plates is as shown Fig. 21.21. The frequency ratio is

(*a*) 4 : 5

(*b*) 4 : 3

(*c*) 3 : 4

(*d*) 2 : 3.

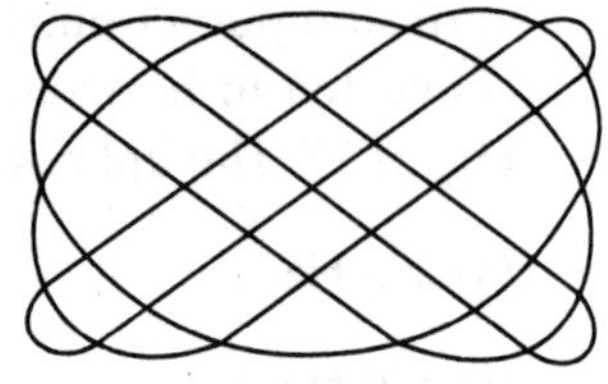

*FIGURE 21.21*

**141.** Which of the following is not a source of power ?

(*a*) Solar cell (*b*) Thermocouple (*c*) Photoelectric cell (*d*) Photovoltaic cell.

**142.** Which of the following strain gauge material has highest gauge factor ?

(*a*) Carbon (*b*) Platinum (*c*) Nichrome (*d*) Soft iron.

**143.** A strain gauge should have

(*a*) high gauge factor and low sensitivity (*b*) high gauge factor and high sensitivity

(*c*) low gauge factor and low sensitivity (*d*) low gauge factor and high sensitivity.

**144.** For better results a strain gauge should have low

(*a*) gauge factor (*b*) resistance value

(*c*) resistance temperature co-efficient (*d*) all of the above.

**145.** Which of the following strain gauge material is an alloy of Nickel and Copper ?

(*a*) Nichrome (*b*) Isoelastic (*c*) Constantan (*d*) None of the above.

**146.** Rosette gauges are used

(*a*) to compensate for temperature variations

(*b*) to amplify strain values (*c*) to measure variable strain

(*d*) to measure strain in more than one direction.

**147.** Which of the following strain gauge transducer has relatively highest sensitivity and gauge factor

(*a*) Nichrome strain gauge (*b*) Semiconductor strain gauge

(*c*) Stabiloy and dinaloy strain gauge transducer

(*d*) Platinum-tunsducer.

**148.** The gauge factor $k$ is related to Poisson's ratio as

(*a*) $\mu = \frac{k-1}{2}$ (*b*) $\mu = \frac{1-2k}{2}$ (*c*) $\mu = \frac{1}{2}(2-k)$ (*d*) $\mu = \frac{2k}{k-1}$.

**149.** A dummy strain gauge is used to

(*a*) increase the sensitivity of measuring system

(*b*) compensate for temperature variations

(*c*) improve stability of the system

(*d*) none of the above.

150. The gauge factor of semi-conductor strain gauges is of the order of
(a) ± 20 (b) ± 50 (c) ± 80 (d) ± 130.

151. Which of the following is not an advantage of semi-conductor strain gauges as compared to conventional strain gauges ?
(a) Excellent hysteresis characteristics (b) Least sensitive to temperature changes
(c) High fatigue life (d) Smaller size.

**Questions 152 and 153 Refer to Fig. 21.22 :**

152. Which curve represents temperature resistance characteristics for a thermistor ?
(a) curve A (b) curve B (c) curve C (d) curve E.

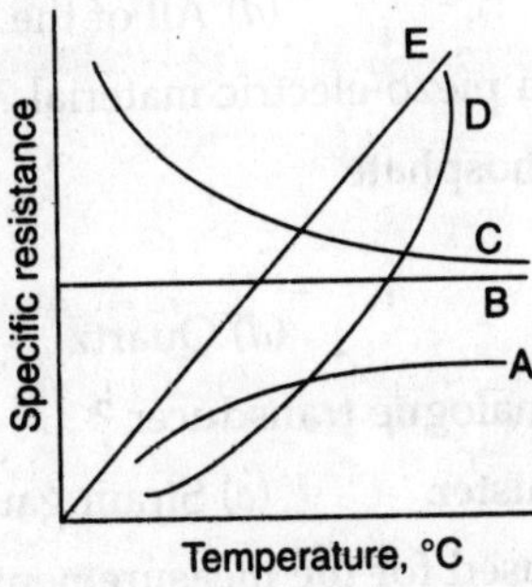


*FIGURE 21.22*

153. Which curve represents temperature-resistance characteristics for platinum ?
(a) curve A (b) curve B (c) curve D (d) curve E.

154. Which of the following is a digital transducer ?
(a) Piezo electric transducer (b) Thermocouple
(c) Encoder (d) Photo voltaic cell.

155. Which of the following is passive transducer ?
(a) Thermocouple (b) Strain gauge
(c) Photo voltaic cell (d) Piezo-electric pick-up.

156. In which of the following transducer, the variation of inductance is used for measurement purposes ?
(a) Differential transformer (b) Strain gauge
(c) Encoders (d) Photo voltaic cell.

157. *LVDT* is a
(a) resistance transducer (b) inductive transducer
(c) eddy current transducer (d) magnetostriction transducer.

158. The output of *LVDT* is in the form of
(a) linear displacement of core (b) rotary movement of core
(c) pulses (d) high frequency signals.

**159.** The core of *LVDT* is made of

(*a*) carbon (*b*) copper (*c*) soft iron (*d*) nichrome.

**160.** The advantage of *LVDT* is

(*a*) high linearity of output (*b*) infinite resolution

(*c*) excellent repeatability (*d*) all of the above.

**161.** A synchro is a

(*a*) linear output transducer (*b*) angular position transducer

(*c*) variable capacitance transducer (*d*) variable speed transducer.

**162.** Which of the following is piezo-electric salt ?

(*a*) Rochelle salt (*b*) Sodium thiosulphate

(*c*) Dichloro difluoromethane (*d*) All of the above.

**163.** Which of the following is not a piezo-electric material ?

(*a*) Ammonium dihydrogen phosphate

(*b*) Lithium sulphate

(*c*) Copper sulphate (*d*) Quartz.

**164.** Which of the following is an analogue transducer ?

(*a*) *LVDT* (*b*) Thermister (*c*) Strain gauge (*d*) All of the above.

**165.** A Hall's effect pick up can be used for the measurement of

(*a*) magnetic flux (*b*) pressure (*c*) relative humidity (*d*) temperature change.

**166.** Which of the following is a self generating transducer ?

(*a*) Thermopile (*b*) Photoemissive cell

(*c*) Magnetostriction gauge (*d*) Capacitor microphone.

**167.** Which of the following is a passive transducer ?

(*a*) Photoconductive cell (*b*) Photovoltaic cell

(*c*) Thermister (*d*) Capacitor microphone.

**168.** A Pirani gauge is used for measurement of

(*a*) velocity of gases (*b*) pressure of fluids in motor

(*c*) low vacuum (*d*) low temperature.

**169.** A Pirani can be used for the measurement of vacuum upto

(*a*) 500 mm Hg (*b*) 100 mm Hg (*c*) 50 mm Hg (*d*) 1 mm Hg.

**170.** A Pirani gauge works on the principle of variation of

(*a*) thermal conductivity of gas with pressure

(*b*) ionisation of air at low pressures

(*c*) reduced oxidation of metals at low pressures

(*d*) difference in kinetic energy of oxygen and nitrogen molecules at low pressures.

171. A compound pressure is used to measure
(a) vacuum pressures (b) high pressures
(c) fluctuating pressures
(d) both pressure above and below atmospheric pressure.

172. Which instrument can be used for the measurement of temperature of a oil fired furnace ?
(a) Gas thermometer (b) Mercury thermometer
(c) Optical pyrometer (d) Any of the above.

173. In optical pyrometer temperature is measured by
(a) thermocouple effect (b) photo cell principle
(c) comparing the brightness of the source with the brightness of a standard source
(d) none of the above.

174. In optical pyrometer for comparing the brightness
(a) current through standard filament is varied
(b) voltage across standard filament is varied
(c) resistance of filament is varied
(d) VI through filament is varied.

175. Optical pyrometers can be used for measurement of temperatures upto
(a) 1000°C (b) 1500°C (c) 2000°C (d) 35000°C.

176. Which thermocouple material can be used for measurement of temperatures above 1000°C ?
(a) copper—constantan (b) silver—constantan (c) alumel—chromel (d) all of the above.

177. Copper—constantan thermocouples can be used upto
(a) 600°C (b) 1000°C (c) 1200°C (d) 1500°C.

178. The range for alumel – chromel thermocouple is
(a) 600°C (b) 800°C (c) 1200°C (d) 2000°C.

179. Which of the following materials for resistance thermometers, has the highest melting point ?
(a) Platinum (b) Copper (c) Nickel (d) Tungsten.

180. Which of the following materials for resistance thermometers, has the lowest melting point ?
(a) Copper (b) Nickel (c) Tungsten (d) Platinum.

181. Which of the following thermometer can be used for measurement of low temperatures in cryogenic region ?
(a) Platinum thermometer (b) Semiconductor thermometer
(c) Mercury thermometer (d) Any of the above.

182. Which of the following is a low cost thermocouple ?
(a) Chromel—constantan (b) Iron—constantan
(c) Rhodium—Indium (d) Tungsten—rhenium.

183. A hot wire anemometer is used for the measurement of
(a) flow of fluids (b) temperature of gases
(c) ultra violet radiations (d) solar temperature.

**184.** A hygrometer is used for the measurement of

(*a*) specific gravity of liquids (*b*) specific gravity of gases in motion

(*c*) moisture content of foodgrains (*d*) relative humidity.

**185.** Which of the following substance exhibits change in resistivity with humidity ?

(*a*) Copper sulphate (*b*) Rochelle salt (*c*) Lithium chloride (*d*) Barium oxide.

**186.** A pH value of 6 indicates

(*a*) acidic solution (*b*) neutral solution

(*c*) alkaline solution (*d*) fully ionised solution.

**187.** The range of pH scale is

(*a*) 0–100 (*b*) 0–20 (*c*) 0–14 (*d*) 0–10.

**188.** Neutral solution has a pH value of

(*a*) 0 (*b*) 1 (*c*) 7 (*d*) 14.

**189.** In pH meters, reference electrode

(*a*) always remains at constant voltage (*b*) is made of copper

(*c*) is made of non-conducting material (*d*) has four leads.

**190.** In pH meters, the pH value is given by the

(*a*) current between the electrodes (*b*) potential difference between the electrodes

(*c*) resistance between electrodes (*d*) none of the above.

**191.** Neper is

(*a*) larger than a decibel (*b*) smaller than a decibel

(*c*) same as a decibel (*d*) half of a decibel.

**192.** The number of base units in SI system is

(*a*) 5 (*b*) 6 (*c*) 7 (*d*) 9.

**193.** A voltmeter has full scale reading of 100 volts. The accuracy is guaranted as 2%. At the reading of 50 volts, the chances of error are upto

(*a*) 1% (*b*) 2% (*c*) 4% (*d*) 5%.

**194.** A varactor diode can be used as variable

(*a*) current source (*b*) capacitor (*c*) inductor (*d*) resistor.

**195.** For sweep frequency plots of impedance and phase angle versus frequency, which of the following meter can be used ?

(*a*) Magnetic potentiometer (*b*) Varactor diode

(*c*) Maxwell's bridge (*d*) Vactor impedance meter.

**196.** A potentiometer recorder is used for

(*a*) ac signal only (*b*) dc signals only (*c*) harmonics only (*d*) all of the above.

**197.** Relative permittivity can be measured by which of the following bridge ?

(*a*) Schering bridge (*b*) Hay's bridge (*c*) De sauty bridge (*d*) Weins bridge.

**198.** Which of the following instrument cannot be used for the measurement of flux density ?

(*a*) Gauss meter (*b*) Ballistic galvanometer

(*c*) Mumetal wire (*d*) Bismuth spiral.

**199.** The maximum reading on a four digit voltmeter can be

(*a*) 10000 (*b*) 9999 (*c*) 9000 (*d*) 1000.

**200.** Which of the following cannot be used for the measurement of alpha, beta and gamma ray intensity ?

(*a*) *Q*-meter (*b*) Ionisation chamber

(*c*) Geiger-muller tube (*d*) Scintillation counter.

**201.** A Westorn frequency meter is

(*a*) moving coil instruments (*b*) moving iron instruments

(*c*) dynamometer instrument (*d*) none of the above.

**202.** In a vibrating reed frequency meter the neutral frequencies of two adjacent reeds have a difference of

(*a*) 0.1 Hz (*b*) 0.25 Hz (*c*) 0.5 Hz (*d*) 1.5 Hz.

**203.** A ............ device prevents the vibration of moving system and enables the latter to reach its final position quickly

(*a*) deflecting (*b*) controlling (*c*) damping (*d*) any of the above.

**204.** Which of the following can be easily melted in a resistance furnace

(*a*) lead (*b*) high speed steel (*c*) zinc (*d*) tin.

**205.** Which of the following is the main disadvantage of using resistors as voltage divider

(*a*) high voltage of the battery (*b*) small current in the load

(*c*) complicated calculations (*d*) large power dissipation.

**206.** A Weston Synchronoscope is

(*a*) moving coil instruments (*b*) moving iron instruments

(*c*) dynamometer instrument (*d*) none of the above.

**207.** A dual-beam CRO has which of the following

(*a*) two vertical amplifiers (*b*) two horizontal amplifiers

(*c*) two trigger circuits (*d*) all of the above.

**208.** Which of the following signal can be displayed by a CRO

(*a*) d.c. signals (*b*) time variant signals

(*c*) a.c. signals (*d*) a.c. as well as d.c. signals.

**209.** The use of instruments is merely confined within laboratories as standardizing instrument

(*a*) absolute (*b*) indicating (*c*) recording (*d*) integrating

**210.** The first stage of vertical amplifier of CRO consists of

(*a*) Triac (*b*) SCR (*c*) FET (*d*) BJT.

211. In a CRO the colour of the spot on the screen is a characteristic of
(*a*) the velocity of the electrons striking the screen
(*b*) the coating material of the screen
(*c*) electrons gun in a CRT
(*d*) the type of the waveform being observed.

212. Primary and Secondary winding of LVDT are wound on a tube of
(*a*) ferrite (*b*) copper
(*c*) laminated sheets of electric steel
(*d*) an insulating material like paper board.

213. In a hot wire instruments the sensing wire is made up of
(*a*) copper-nickel (*b*) silver (*c*) copper (*d*) platinum-irridium.

214. Hot wire instruments read
(*a*) average value (*b*) r.m.s. value (*c*) peak value (*d*) none of these.

215. The chemical effect of current is used in
(*a*) d.c. ammeter hour meter (*b*) d.c. ammeter
(*c*) d.c. energy meter (*d*) none of these.

216. Which of the following instruments indicate the instantaneous value of the electrical quantity being measured at the time at which it is being measured
(*a*) absolute (*b*) indicating (*c*) recording (*d*) integrating.

217. In a portable instruments , the controlling torque provided by
(*a*) spring (*b*) gravity (*c*) eddy currents (*d*) all of the above.

218. In order to achieve high accuracy, the slide wire of a potentiometer should be
(*a*) as long as possible (*b*) as short as possible
(*c*) neither too small nor too large (*d*) very thick.

219. The pointer of an indicating instruments should be
(*a*) very light (*b*) very heavy (*c*) either (*a*) or (*b*) (*d*) neither (*a*) nor (*b*).

220. In case of strain gauge is the gauge factor '*k*' is related to poission's ratio μ by the relation
(*a*) $k = 1+2\mu$ (*b*) $k = 1-\mu$ (*c*) $\mu = (k-1)/2$ (*d*) $\mu = (k+1)/2$.

221. Wein's bride is used ROE the measurement of frequency in the applied voltag waveform is
(*a*) sinusoidal (*b*) square (*c*) rectangular (*d*) triangular.

222. Strain gauge is
(*a*) not a transducer (*b*) an active transducer
(*c*) not an electronic instrument (*d*) none of these.

223. A high *Q* coil has
(*a*) large bandwidth (*b*) high losses (*c*) low losses (*d*) flat response.

**224.** For a pulse transformer, the material used for its core and the turn ratio from primary to secondary winding are

(*a*) ferrite, 20 : 1 (*b*) laminated core, 1 : 1 (*c*) ferrite, 1 : 1 (*d*) powdered iron, 1 : 1.

**225.** In a full wave rectifier the negative point in a circuit is

(*a*) cathode (*b*) anode

(*c*) the center tap on the high voltage secondary

(*d*) either plate.

**226.** Voltage ratio box is used in conjuction with which of the following

(*a*) potentiometers (*b*) voltmeters (*c*) wheatstone bridge (*d*) ammeters.

**227.** For measuring a very high resistance we should use

(*a*) Kelvin's double bridge (*b*) wheatstone bridge

(*c*) megger (*d*) none of these.

**228.** For strain gauge the gauge factor is generally in the range

(*a*) 0.2 to 0.8 (*b*) 0.8 to 1.5 (*c*) 1.5 to 2.0 (*d*) 2.0 to 4.0.

**229.** A vector voltmeter cannot be used to measure

(*a*) filter transfer function (*b*) *Q*-factor of a circuit

(*c*) Complex insertion loss (*d*) Amplifier gain.

**230.** The sensitivity of wheatstone bridge depends upon

(*a*) the current sensitivity of the galvanometer

(*b*) galvanometer resistance

(*c*) bridge supply voltage

(*d*) all of the above.

**231.** In .................instruments the reflecting torque depend on the frequency

(*a*) induction type (*b*) hot wire (*c*) moving coil (*d*) moving iron.

**232.** ............is commonly used material for thermocouples

(*a*) platinum-rhodium (*b*) chromel-alumel

(*c*) chromel-copal (*d*) any of the above.

**233.** A moving iron instrument can be used for

(*a*) d.c. only (*b*) a.c. only

(*c*) both a.c. and d.c. (*d*) neither a.c. nor d.c.

**234.** Nixie tube display is based on the principle of

(*a*) gaseous glow (*b*) dot matrix

(*c*) seven segment display (*d*) none of the above.

**235.** For measuring current at high frequency we should use

(*a*) moving iron instrument (*b*) electrostatic instrument

(*c*) thermocouple instrument (*d*) none of the above.

**236.** The ratio error of the current transformer is due to

(*a*) exciting current (*b*) stray magnetism (*c*) corona effects (*d*) lagging power factor.

**237.** A voltmeter should have very high internal resistance so that

(*a*) accuracy is high (*b*) range is high

(*c*) loading is high (*d*) minimum current passes through the meter.

**238.** A vector voltmeter cannot be used to measure

(*a*) complex insertion less

(*b*) amplifier gain

(*c*) filter transfer function

(*d*) $Q$ factor of the circuit.

**239.** A power factor meter have

(*a*) only current coil (*b*) only voltage coil

(*c*) both current and voltage coils (*d*) only inductive voltage coil.

**240.** Time response of strain gauge measurement is

(*a*) very poor (*b*) poor (*c*) good (*d*) very good.

**241.** Moving coil and moving iron type instruments can be distinguished by looking at

(*a*) scale (*b*) their range (*c*) pointer (*d*) size of terminals.

**242.** An ammeter is

(*a*) secondary instrument (*b*) absolute instrument

(*c*) recording instrument (*d*) integrating instrument.

**243.** The scale of the rectifier transformer is

(*a*) linear (*b*) non-linear (*c*) either (*a*) or (*b*) (*d*) neither (*a*) nor (*b*).

**244.** Meter accuracy is determined by

(*a*) $1/10^{th}$ of full-scale deflection (*b*) $1/4^{th}$ of full-scale deflection

(*c*) half-scale deflection (*d*) full-scale deflection.

**245.** The device best suited for measurement of temperature of hot bodies radiating energy in the spectrum

(*a*) bolometer (*b*) optical pyrometer (*c*) thermocouple (*d*) thermopile.

**246.** Which of the instrument can be used for the measurement of temperature above 1500 deg. Kelvin

(*a*) mercury thermometer (*b*) gas thermometer

(*c*) thermoelectric pyrometer (*d*) any of the above.

**247.** Maxwell bridge is used for the measurement of

(*a*) resistance (*b*) inductance (*c*) capacitance (*d*) frequency.

**248.** A bridge used for the measurement of capacitance is

(*a*) Wheatstone bridge (*b*) Wein bridge

(*c*) Maxwell bridge (*d*) Schering bridge.

**249.** Voltmeter resistivity is defined as

(*a*) reciprocal of full scale deflection current
(*b*) volts per ohm
(*c*) ohm per volt
(*d*) its degree of sensitivity to impulse change.

**250.** A cadmium sulphide cell is a

(*a*) solar cell (*b*) dry cell
(*c*) photo voltaic cell (*d*) photo conductive cell.

## ANSWERS

| | | | | | | |
|---|---|---|---|---|---|---|
| **1.** (*a*) | **2.** (*c*) | **3.** (*c*) | **4.** (*a*) | **5.** (*b*) | **6.** (*d*) | **7.** (*a*) |
| **8.** (*b*) | **9.** (*c*) | **10.** (*c*) | **11.** (*b*) | **12.** (*c*) | **13.** (*a*) | **14.** (*d*) |
| **15.** (*c*) | **16.** (*b*) | **17.** (*d*) | **18.** (*b*) | **19.** (*a*) | **20.** (*a*) | **21.** (*a*) |
| **22.** (*a*) | **23.** (*a*) | **24.** (*d*) | **25.** (*d*) | **26.** (*c*) | **27.** (*c*) | **28.** (*a*) |
| **29.** (*a*) | **30.** (*c*) | **31.** (*c*) | **32.** (*b*) | **33.** (*c*) | **34.** (*a*) | **35.** (*d*) |
| **36.** (*c*) | **37.** (*d*) | **38.** (*d*) | **39.** (*d*) | **40.** (*b*) | **41.** (*c*) | **42.** (*a*) |
| **43.** (*a*) | **44.** (*c*) | **45.** (*d*) | **46.** (*c*) | **47.** (*b*) | **48.** (*c*) | **49.** (*a*) |
| **50.** (*b*) | **51.** (*d*) | **52.** (*d*) | **53.** (*d*) | **54.** (*d*) | **55.** (*a*) | **56.** (*c*) |
| **57.** (*b*) | **58.** (*b*) | **59.** (*d*) | **60.** (*d*) | **61.** (*b*) | **62.** (*b*) | **63.** (*d*) |
| **64.** (*a*) | **65.** (*b*) | **66.** (*a*) | **67.** (*c*) | **68.** (*d*) | **69.** (*d*) | **70.** (*d*) |
| **71.** (*c*) | **72.** (*d*) | **73.** (*b*) | **74.** (*c*) | **75.** (*c*) | **76.** (*c*) | **77.** (*a*) |
| **78.** (*d*) | **79.** (*c*) | **80.** (*d*) | **81.** (*b*) | **82.** (*b*) | **83.** (*a*) | **84.** (*b*) |
| **85.** (*c*) | **86.** (*d*) | **87.** (*b*) | **88.** (*b*) | **89.** (*c*) | **90.** (*d*) | **91.** (*d*) |
| **92.** (*d*) | **93.** (*c*) | **94.** (*d*) | **95.** (*a*) | **96.** (*c*) | **97.** (*d*) | **98.** (*c*) |
| **99.** (*c*) | **100.** (*c*) | **101.** (*b*) | **102.** (*a*) | **103.** (*c*) | **104.** (*b*) | **105.** (*a*) |
| **106.** (*d*) | **107.** (*a*) | **108.** (*d*) | **109.** (*b*) | **110.** (*a*) | **111.** (*c*) | **112.** (*c*) |
| **113.** (*d*) | **114.** (*a*) | **115.** (*d*) | **116.** (*a*) | **117.** (*d*) | **118.** (*b*) | **119.** (*a*) |
| **120.** (*a*) | **121.** (*a*) | **122.** (*b*) | **123.** (*d*) | **124.** (*a*) | **125.** (*c*) | **126.** (*b*) |
| **127.** (*c*) | **128.** (*c*) | **129.** (*c*) | **130.** (*b*) | **131.** (*d*) | **132.** (*a*) | **133.** (*c*) |
| **134.** (*a*) | **135.** (*b*) | **136.** (*b*) | **137.** (*b*) | **138.** (*d*) | **139.** (*b*) | **140.** (*a*) |
| **141.** (*c*) | **142.** (*a*) | **143.** (*b*) | **144.** (*c*) | **145.** (*c*) | **146.** (*d*) | **147.** (*b*) |
| **148.** (*a*) | **149.** (*a*) | **150.** (*d*) | **151.** (*b*) | **152.** (*c*) | **153.** (*a*) | **154.** (*c*) |
| **155.** (*b*) | **156.** (*a*) | **157.** (*b*) | **158.** (*a*) | **159.** (*c*) | **160.** (*d*) | **161.** (*b*) |
| **162.** (*a*) | **163.** (*c*) | **164.** (*d*) | **165.** (*a*) | **166.** (*a*) | **167.** (*b*) | **168.** (*c*) |
| **169.** (*d*) | **170.** (*a*) | **171.** (*d*) | **172.** (*c*) | **173.** (*c*) | **174.** (*a*) | **175.** (*d*) |
| **176.** (*c*) | **177.** (*a*) | **178.** (*c*) | **179.** (*d*) | **180.** (*a*) | **181.** (*b*) | **182.** (*b*) |

| | | | | | | |
|---|---|---|---|---|---|---|
| **183.** (*a*) | **184.** (*d*) | **185.** (*d*) | **186.** (*a*) | **187.** (*c*) | **188.** (*c*) | **189.** (*a*) |
| **190.** (*b*) | **191.** (*a*) | **192.** (*c*) | **193.** (*c*) | **194.** (*b*) | **195.** (*d*) | **196.** (*b*) |
| **197.** (*a*) | **198.** (*b*) | **199.** (*b*) | **200.** (*a*) | **201.** (*b*) | **202.** (*d*) | **203.** (*c*) |
| **104.** (*b*) | **205.** (*d*) | **206.** (*c*) | **207.** (*a*) | **208.** (*d*) | **209.** (*b*) | **210.** (*c*) |
| **211.** (*b*) | **212.** (*d*) | **213.** (*d*) | **214.** (*b*) | **215.** (*c*) | **216.** (*a*) | **217.** (*a*) |
| **218.** (*a*) | **219.** (*d*) | **220.** (*a*) | **221.** (*a*) | **122.** (*b*) | **223.** (*d*) | **224.** (*c*) |
| **225.** (*c*) | **226.** (*a*) | **227.** (*b*) | **228.** (*b*) | **229.** (*d*) | **230.** (*b*) | **231.** (*d*) |
| **232.** (*a*) | **233.** (*d*) | **234.** (*c*) | **235.** (*d*) | **236.** (*c*) | **237.** (*b*) | **238.** (*d*) |
| **239.** (*a*) | **240.** (*a*) | **241.** (*a*) | **242.** (*a*) | **243.** (*b*) | **244.** (*d*) | **245.** (*b*) |
| **246.** (*c*) | **247.** (*a*) | **248.** (*d*) | **249.** (*c*) | **250.** (*d*). | | |

Chapter 22

# ELECTRICAL ENGINEERING MATERIALS

**The materials used in the construction of electrical machines and associated equipment may be broadly grouped as follows:**

1. Conducting materials
2. Insulating materials
3. Magnetic materials.

Proper selection of machines for various applications is essential. For the selection of proper material one must know physical, mechanical and electrical properties of these materials.

## Mechanical Properties of Materials

1. **Elasticity** is the property of material under which a material regains its original shape after deformation, within elastic limits. Young's modulus of elasticity is a measure of this property.
2. **Plasticity** is the property of material which results in permanent deformation. Thus clay, lead etc. are plastic. Some materials are partly elastic and partly plastic.
3. **Ductility** is the property by which a material can be drawn out or elongated to an appreciable extent before rupture. Ductility is measured in terms of percentage elongation and percentage reduction in area. Some of the materials in order of increasing ductility are given below:

Lead, Tin, Zinc, Nickel, Aluminium, Copper, Iron, Silver, Platinum, Gold.

**Brittleness** is the property in which a material breaks into smaller fragments under the application of load. Thus brittleness is a property opposite to ductility.

**Hardness** is the resistance to abrasion, indentation, machining and scratching. It is expressed in terms of hardness number—brinell hardness number, Rockwell hardness number, Vickers pyramid number, etc.

Some of the materials in increasing order of hardness are given below:

Talc, Gypsum, Calcite, Flourspur, Apatite, Orthoclase, Quartz, Topaz, Corundum, Diamond.

**Toughness** is the property under which a material can be twisted, bent or stretched under a high stress before rupture.

**Stiffness** is the capacity of material to resist deformation.

**Resilience** is the property where in material can store energy and resistance and impact. It is measured by the amount of energy that can be stored per unit volume after being stressed to elastic limit. This property is particularly desirable in materials for springs.

**Malleability** is the property under which a material can undergo change of shape under compressive stresses, without rupture.

**Creep** is the property under which a material undergoes slow but progressive deformation, over a period of time, under constant load.

**Fatigue** is the effect of cyclically repeated stresses on metals. The stresses may be repeated stresses (zero to a certain maximum value), alternating stresses (varying from minimum to maximum value of the opposite nature) and reversed stresses (varying from compressive to tensile values). The fatigue resistance of materials is measured in terms of endurance limit.

### Heat Treatment Operations

1. **Annealing** is the softening process consisting of heating steel to a temperature at or near the critical point, holding there for a proper time and then allowing it to cool slowly in the furnace itself. Annealing is done to
   1. soften the metal
   2. improve machinability
   3. improve mechanical properties, like ductility
   4. refine grain structure
   5. relieve internal stresses
   6. modify electrical and magnetic properties.
2. **Normalising** is the process of heating the steel 40 to 50°C above the upper transformation range, holding there for a specified period and then allowing it to cool in still air at room temperature. This process in normally used with low and medium carbon as well as alloys for any of the following purposes:
   1. improve machinability and increase strength
   2. reduce internal stresses
   3. improve structure of welds.
3. **Tempering** is the process of reheating the hardened steel to some temperature below the cirtical range, followed by any rate of cooling. Tempering is usually done after hardening to modify the properties of the hardened steel (particularly to reduce brittleness). Tempering is done to
   1. reduce hardness, brittleness and tensile strength.
   2. increase ductility and strength.
   3. relieve quenching stresses.

4. **Hardening** is the process in which the steel is heated to a temperature within or above its critical range held at this temperature for a considerable time to ensure thorough penetration of heat inside the component, and then allowed to cool by quenching in water, oil or brine solution. The degree of hardness obtained depends on
   (*a*) carbon content of specimen
   (*b*) quenching rate
   (*c*) work size.

Various methods of hardening are
1. Case hardening
2. Age hardening
3. Induction hardening
4. Flame hardening
5. Cyaniding
6. Nitriding.

**Hot working of metals.** Sometimes required shape can be given to a metal, with lesser force, at elevated temperatures. Some processes for hot working of metals are given below:

1. Rolling
2. Forging
   (*i*) Hammer or smith forging (*ii*) Drop forging
   (*iii*) Upset forging (*iv*) Press forging
   (*v*) Roll forging (*vi*) Swaging.
3. Pipe welding
   (*i*) Butt welding of heated strips
   (*ii*) Butt welding of electrical resistance
   (*iii*) Lap welding
   (*iv*) Hammer welding.
4. Piercing
5. Drawing or cupping
6. Spinning
7. Extruding.

**Cold working of metals.** These operations are carried out relatively at lower temperatures. Some of the cold working operations are given below:

1. Drawing
   (*i*) Blanks (*ii*) Tubes
   (*iii*) Wires (*iv*) Metal spinning
   (*v*) Embossing (*vi*) Stretch forming.
2. Squeezing
   (*i*) Coining (*ii*) Cold rolling
   (*iii*) Sizing (*iv*) Swaging

(*v*) Thread rolling
(*vi*) Knurling
(*vii*) Riveting.

3. Bending
   (*i*) Curling
   (*ii*) Seaming
   (*iii*) Angle bending
   (*iv*) Roll forming
   (*v*) Plate bending.
4. Shearing
   (*i*) Blanking
   (*ii*) Punching
   (*iii*) perforating
   (*iv*) Slitting
   (*v*) Lancing
   (*vi*) Shearing
   (*vii*) Notching
   (*viii*) Trimming
   (*ix*) Cutting off.
5. Extruding
   (*i*) Cold
   (*ii*) Impact
   (*iii*) Shot peening
   (*iv*) Hobbing.

**Processes for joining parts or materials**

1. Welding
2. Soldering
3. Brazing
4. Sintering
5. Pressing
6. Riveting
7. Screw fastening
8. Adhesive joining.

**Processes for surface finish**

1. Polishing
2. Super finishing
3. Electroplating
4. Honing
5. Lapping
6. Abrasive belt grinding
7. Barrel tumbling
8. Metal spraying
9. Inorganic coatings
10. Sheradizing
11. Anodizing
12. Parkerizing.

**Some Alloys and Their Compositions**

| *Alloy* | *Composition (Percentage)* | *Applications* |
|---|---|---|
| Admiralty brass | 70% Cu, 30% Zn | Tubes |
| Alnico | 50% steel, 20% Ni, 20% Al, 10% Co | Permanent magnets |
| Aluminium bronze | 94.8% Cu, 5.2% Al | Coins, frames etc. |
| Brass | 85% Cu, 15% Zn | Gilding metal, cheap jewellery |
| | 75% Cu, 25% Zn | Brazing brass |
| | 70% Cu, 30% Zn | Tubes, wires, pressed components |
| | 60% Cu, 40% Zn | Muntz metal for household fittings etc. |

| | | |
|---|---|---|
| Bronze | 90% Cu, 10% Sn | Utensils, statues |
| Constantan | 50% Cu, 50% Ni | Standard resistance |
| Duralumin | 95% Al, 4% Cu, 0.5% Mn, 0.5% Mg | Light structures, extruded sections, sheets |
| Fusible metal | 50% Bu, 25% Pb, 25% Sn | Fusible plugs |
| Gun metal | 90% Cu, 10% Zn | Balves, water service fittings |
| Germal silver | 50% Cu, 30% Zn, 20% Ni | Household fittings |
| Invar | 64% Fe, 36% Ni | Surveying tapes |
| Mangania | 84% Cu, 12% Mn, 4% Ni | Thermocouple junction metal |
| Nichrome | 80% Ni, 20% Cr | Electrical appliances |
| Perminvar | 45% Ni, 30% Cu, 25% Co | Transformer cores magnetic shield |
| Phosphor bronze | 90% Cu, 9.7% Sn, 0.3% P | Bearings, worm wheels, rods, sheets |
| Silicon bronze | 97.22% Cu, 2.26% Sn, 0.52 % Si | Aerial wires |
| Solder | 66% Pb, 34% Sn | Soldering and tinning |
| White metal | 80% Pb, 20% Bi | Bearing linings |

**Organic Materials**

(*i*) **Polymers.** Mer means a unit. Accordingly a monomer is a small molecule which may combine with others to form a large molecule. A polymer is a large molecule which may be made up of many small suits.

Depending upon their behaviour with respect to change in temperature, the polymers are broadly classified as

1. **Thermoplasts.** Those which have the property of increasing plasticity with increasing temperature. These are long chain polymers helds together by secondary bonds.
2. **Thermosets.** These have three dimensional network of primary bonds as polymerization proceeds in all directions in this case. These are hard and rigid at room temperature and do not soften on heating.

Long chain polymers are further divided as follows:

(*a*) Plastics – randomly oriented molecules

(*b*) Fibres – directional alignment of chain

(*c*) Elastomers – having rubbery behaviour.

**Polymerization** is the process of linking together of monomers. It consists of breaking the double (C = C) bond of monomers. Polymerization mechanism may be

1. **Addition polymerization** is the process of summation of individual molecules into long chains. This mechanism occurs in three steps

   (*i*) Initiation (*ii*) Chain propagation (*iii*) Termination
2. **Condensation polymerization** are reactions between functional molecules leading to the formation of a polymer with elimination of some small molecules usually of water.

**Polymers and Their Uses**

| *Polymers* | *Properties* | *Uses* |
|---|---|---|
| Polyethylene | 1. Water resistant<br>2. Tough<br>3. Low electrical loss factor | 1. Electrical insulation for coaxial cables<br>2. Packaging<br>3. Moisture proofing |
| Polyisobutylene | 1. Flexibility<br>2. Moisture resistance<br>3. Elasticity | 1. Bullet proof tanks<br>2. Cable coatings<br>3. Pour point depressors |
| Polyvinyl chloride | 1. Toughness<br>2. Plasticizer | 1. Cable jackets<br>2. Lead wire insulation<br>3. Rubber substitute<br>4. Fabric coating |
| Polystyrene | 1. High resistivity<br>2. Low dielectric constant<br>3. Moisture resistance | 1. Electrical insulation<br>2. Instrument panals |
| Acrylic polymers | 1. Transparent<br>2. Clarity<br>3. Conforms to shape | 1. Lenses<br>2. Aircraft shields |
| Polyvinyl carbazole | 1. Acid and alkali resistant<br>2. Thermal resistance<br>3. Low power factor | 1. Capacitor material<br>2. Mica bonding<br>3. Electronic components. |

**Plastics** are organic materials which can be easily moulded into shapes. These are cheap, tough, water resistant and have low electrical and thermal conductivity. Plastics may be classified as

(*a*) **Thermosetting Plastics.** These plastics under the action of heat, pressure, etc. undergo chemical changes of condensation and polymerization to form a rigid final shape. These plastics are heat resistant and are widely used for telephone receivers, cabinets, switch boards, electrical outlets, etc.

(*b*) **Thermoplastic Plastics** are formed by substituted derivatives of ethylene which can be made to polymerase under the influence of heat and catalyst. The scrap of these plastics can be recycled. These are widely used for floor tiles, fluorescent light reflectors, leases etc.

**Conducting Materials.** The best conducting material is silver with specific resistance of 1.46 micro-ohm-cm at 0°C. This being costly, cooper is widely used in electrical machines.

Copper when alloyed with silicon, manganese, phosphorus or chromium, gives better machanical properties, Copper is mostly used in the form of wires for windings of machines.

Aluminium thought inferior conductor is light and cheap as compared to copper. It is widely used in the form of conductor.

**Insulating Materials** are used to provide insulation at appropriate places. Besides insulating properties at room temperature, insulating materials must be capable of withstanding temperature rise resulting from constant use of electrical machines. Often the design of electrical machine is restricted due to the considerations of insulating materials. Insulating materials are classified according to their heat resisting properties *i.e.*, permissible temperature rise.

**Classification of Insulating Materials**

| *Insulation* | *Max. Temp. 0°C* | *Materials* |
|---|---|---|
| Y | 90 | Paper, cotton, silk, wood, fibre, cellulose and similar materials neither impregnated nor immersed in oil. |
| A | 105 | All materials of Y class when impregnated with natural resins, cellulose esters, etc. or immersed in a liquid dielectric. |
| E | 120 | Enamelled wire insulations on a base of polyvinyl format, cotton and paper laminates with formaldehyde bonding, epoxy and polyurethane resins, moulded powder, plastics on phenolic formaldehyde, etc. |
| B | 130 | Mica, glass fibre, asbestos and similar organic materials impregnated or glued together with varnishes or compounds. |
| F | 155 | Materials of class B when impregnated or glued together with more thermally resistant varnishes and compounds such as epoxide, polyurethane etc. |
| H | 180 | Built up mica, fibre glass, and asbestos when impregnated with silicons containing no organic fibrous materials. |
| C | > 180 | Class C insulations containing ceramics, quartz, asbestos, glass and mica, which are neither impregnated nor bound with organic substances. |

**Properties of Insulating Materials**

| *Material* | *Dielectric Strength kV/mm* | *Resistivity in ohms-cm* | *Safe Temp. °C* | *Effect of moisture* |
|---|---|---|---|---|
| Cotton | 3–4 | $10^9$–$10^{12}$ | 90° | Absorbent |
| Empire cloth | 12–18 | $10^9$–$10^{12}$ | 90° | Absorbent when varnish layer cracks |
| Paper | 6–10 | $10^{15}$ | 90° | Absorbent |
| Fibre | 5 | $10^9$–$10^{12}$ | 90° | Absorbent when varnish layer cracks |
| Rubber | 15–25 | $10^{15}$–$10^6$ | 40° | Slightly |
| Mica | 40–80 | $5 \times 10^{13}$ | 500 | No |
| Asbestos | 3–4.5 | $10^9$–$10^{15}$ | 500 | Absorbent |
| Bakelite | 20–25 | – | 200 | No |
| Glass | 5–15 | $> 10^{15}$ | Room temp. | No |
| Porcelain | 10–20 | $10^{13}$ | Room temp. | No |
| Polythene | 30 | $10^{16}$ | 70 | No |

**Magnetic Materials** for electrical machines must have the following properties:

1. High magnetic permeability so that even the weakest current flowing through electromagnet should be able to set up large fluxes its core.
2. High electrical resistivity to minimise eddy current losses of the magnetic material. For this purpose core is usually laminated and laminations and insulated from each other by paper or varnish.
3. Hysteresis loop of the magnetic material should be narrow and must have a small area in order to reduce the hysteresis loss. Thus high quality magnetic materials have high saturation limit and low losses at higher flux densities.

## OBJECTIVE TYPE QUESTIONS

1. The nature of atomic bond found in diamond is
   (*a*) ionic (*b*) covalent (*c*) metallic.
2. Which of the following is a ceramic material?
   (*a*) Leather (*b*) MgO (*c*) Invar (*d*) Nylon.
3. 1 atomic mass unit is equal to
   (*a*) $1.6 \times 10^{-10}$ gm (*b*) $1.6 \times 10^{-14}$ gm
   (*c*) $1.6 \times 10^{-20}$ gm (*d*) $1.6 \times 10^{-24}$ gm.
4. The atomic number of sodium is 11 and atomic weight is 23. Therefore it will have total number of electrons as
   (*a*) 3 (*b*) 11 (*c*) 12 (*d*) 23.
5. In the above case, the total number of neutrons will be
   (*a*) 3 (*b*) 11 (*c*) 12 (*d*) 23.
6. Which of the following is non-metal?
   (*a*) Mercury (*b*) Lithium (*c*) Fluorine (*d*) Potassium.
7. The values of spin quantum numbers could be
   (*a*) 1, 2, 3, 4, 5 etc. (*b*) $\frac{1}{2}, \frac{3}{2}, \frac{5}{2}$, ..... etc. (*c*) $\pm \frac{1}{2}$ (*d*) none of the above.
8. Atoms of an element which possess different masses but have the same chemical properties are called
   (*a*) deutrons (*b*) dielectrics (*c*) semi-conductors (*d*) isotopes.
9. The electronic configuration of sodium is
   (*a*) $1s^3\ 2s^2\ 2p^6\ 3s^1$ (*b*) $1s^3\ 2s^2\ 2p^6$ (*c*) $1s^3\ 2s^2\ 3s^1$ (*d*) $1s^3\ 2s^2\ 3s^1\ 4s^2$.
10. $1s^2\ 2s^1$ is the electronic configuration of
    (*a*) Helium (*b*) Lithium (*c*) Silver (*d*) Mercury.
11. The electronic configuration of carbon is
    (*a*) $1s^2\ 2s^2$ (*b*) $1s^2\ 2s^2\ 2p^6\ 2s^2$ (*c*) $1s^2\ 2s^2\ 2p^6$ (*d*) $1s^2\ 2s^2\ 2p^2$.

12. The electronic configuration of an iron atom is
    (a) $1s^2\ 2s^2\ 2p^6\ 3s^2\ 3p^6\ 3d^6 4s^2$ (b) $1s^2\ 2s^2\ 2p^6\ 3s^2\ 3p^6\ 3d^3\ 4s^2$
    (c) $1s^2\ 2s^2\ 2p^6\ 3s^2\ 3p^6\ 3d^8\ 4s^2$ (d) $1s^2\ 2s^2\ 2p^6\ 3s^2\ 3p^6\ 3d^5\ 4s^2$.
13. The atomic number of the element whose outermost electrons fill 3-*p* subshells will be
    (a) 16 (b) 18 (c) 20 (d) 22.
14. The Miller indices of the face ABCD will be
    (a) (0 1 0)
    (b) (0 0 1)
    (c) (1 0 0)
    (d) (1 1 1)

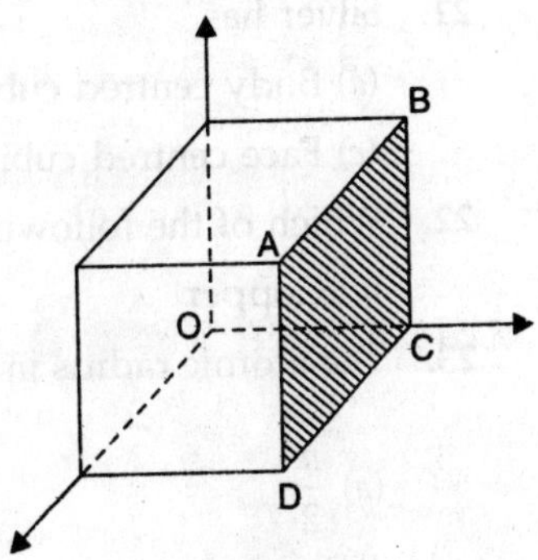

FIGURE 22.1

15. The Miller indices of the diagonal plane of a cube are (Fig. 22.1)
    (a) 1 1 0 (b) 1 1 1 (c) 1 0 0 (d) 0 0 0.
16. The Miller indices of face ABCD in Fig. 22.2 are
    (a) (1 0 0)
    (b) (1 1 1)
    (c) (0 0 1)
    (d) (0 0 0).

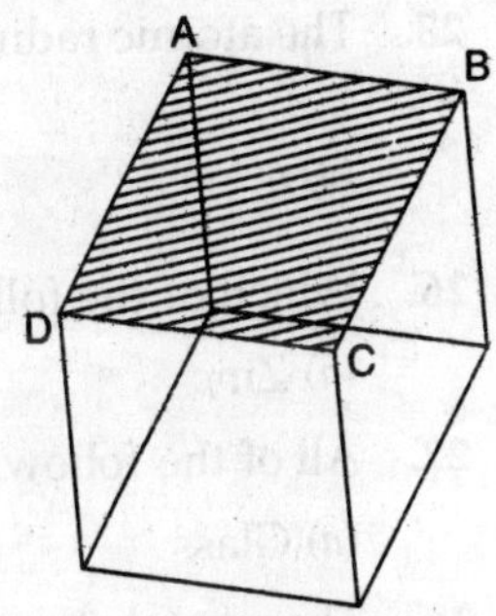

FIGURE 22.2

17. The Miller indices of face ABC in Fig. 22.3 are
    (a) (2 1 1)
    (b) (1 2 1)
    (c) (1 1 2)
    (d) (1 2 2).

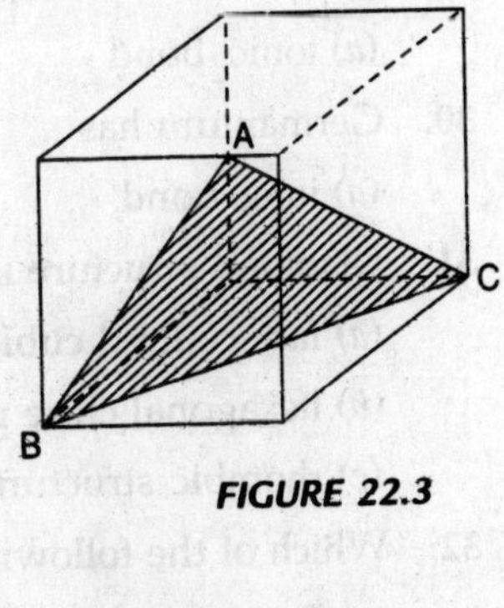

FIGURE 22.3

18. The atomic packing factor for face centred cube is
    (a) 0.74 (b) 0.66 (c) 0.54 (d) 0.44.

19. The atomic packing factor for body centred cube is
(*a*) 0.68 (*b*) 0.55 (*c*) 0.48 (*d*) 0.37.

20. Graphite has
(*a*) Hexagonal close packed structure (*b*) Body centred cubic structure
(*c*) Rhombic structure (*d*) Face centred cubic structure.

21. Silver has
(*a*) Body centred cubic structure (*b*) Rhombic structure
(*c*) Face centred cubic structure (*d*) Hexagonal close packed structure.

22. Which of the following material has body centred cubic structure?
(*a*) Copper (*b*) Lead (*c*) Tungsten (*d*) Nickel.

23. The atomic radius in case of face centred cubic lattice is
(*a*) $\frac{a}{2}$ (*b*) $\frac{a\sqrt{3}}{4}$ (*c*) $\frac{a}{3}$ (*d*) $\frac{a\sqrt{2}}{4}$.

24. The atomic radius in case of body centred cubic lattice is
(*a*) $\frac{a\sqrt{3}}{4}$ (*b*) $\frac{a\sqrt{2}}{3}$ (*c*) $\frac{a\sqrt{2}}{2}$ (*d*) $\frac{a}{2}$.

25. The atomic radius for simple cubic lattice
(*a*) $a$ (*b*) $\frac{a}{2}$ (*c*) $\frac{a}{3}$ (*d*) $\frac{a}{4}$.

26. Which of the following is amorphous solid?
(*a*) Zinc (*b*) Ashphalt (*c*) Lead (*d*) None of the above.

27. All of the following are amorphous except
(*a*) Glass (*b*) Tar (*c*) Unstretched rubber (*d*) Diamond.

28. The crystal structure of most of the common metals is
(*a*) hexagonal (*b*) cubic (*c*) orthorhombic.

29. Sodium chloride has
(*a*) ionic bond (*b*) covalent bond (*c*) metallic bond (*d*) molecular bond.

30. Germanium has
(*a*) ionic bond (*b*) covalent bond (*c*) metallic bond (*d*) molecular bond.

31. Diamond structure is based on
(*a*) face centred cubic lattice
(*b*) hexagonal close packed structure
(*c*) rhombic structure (*d*) body centred cubic structure.

32. Which of the following substances have ionic bond?
(*a*) Potassium oxide (*b*) Lithium hydride (*c*) Molybdenum fluoride (*d*) All of the above.

33. The photoelectric current depends on
(*a*) intensity of incident light (*b*) wavelength of the incident beam
(*c*) both (*a*) and (*b*) above (*d*) none of the above.

34. When the effect of light on a substance is to change its resistance the effect is known as
(*a*) photo conductivity (*b*) photo electric effect
(*c*) photo voltaic effect (*d*) none of the above.

35. The property of photo conducting is also observed in many
(*a*) liquid metals (*b*) fluorides (*c*) semiconductors (*d*) noble metals.

36. Which of the following is photo conductive material?
(*a*) sodium (*b*) selenium (*c*) gold (*d*) copper.

37. When light of suitable wavelength falls on certain metals, the electrons are ejected out from metals. This phenomenon is known as
(*a*) photo-voltaic effect (*b*) photo-electric effect
(*c*) photo-resistivity effect (*d*) none of the above.

38. Cesium under ultraviolet rays demonstrates
(*a*) photo conductive effect (*b*) photo-voltaic effect
(*c*) photo-electric effect (*d*) all of the above.

39. In photo emissive cells which of the following material is used for cathode?
(*a*) Zinc (*b*) Copper (*c*) Gold (*d*) Silver.

40. In photo emissive cells alkali metals are used as cathode materials for
(*a*) infra-red light (*b*) ultra violet light (*c*) visible light (*d*) all of the above.

41. Photo-electric cells can be used for
(*a*) burglar alarms (*b*) fire alarms (*c*) counting machines (*d*) any of the above.

42. Moselay's law is associated with
(*a*) photo-conduction (*b*) X-rays
(*c*) isotopes (*d*) dislocations in metals.

43. The property of the material which enables it to be drawn out or elongated to an appreciable extent, is known as
(*a*) elasticity (*b*) plasticity (*c*) ductility (*d*) hardness.

44. Which of the following material is most ductile?
(*a*) Lead (*b*) Platinum (*c*) Copper (*d*) Gold.

45. Which of the following material is least ductile?
(*a*) Tin (*b*) Zinc (*c*) Silver (*d*) Nickel.

46. The lack of ductility is called
(*a*) brittleness (*b*) hardness (*c*) plasticity (*d*) elasticity.

47. The property of the material which enables it to resist abrasion is known as
(*a*) fatigue (*b*) creep (*c*) hardness (*d*) plasticity.

48. According to Moh's scale, diamond is assigned hardness number of
(a) 0 (b) 1 (c) 10 (d) 100.

49. Which of the following material is least hard?
(a) Topaz (b) Quartz (c) Gypsum (d) Talc.

50. Which of the following represents the increasing order of hardness?
(a) Apatite, Quartz, Topaz (b) Corundum, Calcite, Talc
(c) Quartz, Calcite, Diamond (d) Talc, Topaz, Quartz.

51. For Brinell hardness test
(a) a ball of 10 mm diameter is used (b) a diamond cone of 120° is used
(c) a diamond cone of 136° is used (d) a ball of 1.5875 mm is used.

52. The Brinell hardness number is given by
(a) $\frac{\text{Load}}{\text{Depth of indentation}}$ (b) $\frac{\text{Load}}{\text{Diameter of ball}}$
(c) $\frac{\text{Load}}{\text{Diameter of indentation}}$ (d) $\frac{\text{Load}}{\text{Area of indentation}}$.

53. In hardness test by Brinell hardness tester, the hardness of ball is of the order of
(a) 100 BHN (b) 100 VPN (c) 850 VHN (d) 1000 BHN.

54. Which of the following factors in general increase machinability of metals?
(a) Low hardness (b) Low ductility (c) Low tensile strength (d) All of the above.

55. The property of the material which enables it to store energy and resist shock and impact is known as
(a) toughness (b) stiffness (c) resilience (d) ductility.

56. The phenomenon of slow but progressive deformation of metal under constant load is termed as
(a) fatigue (b) toughness (c) yielding (d) creep.

57. The stresses which vary from one value of compressive stress value to the same value of the opposite nature, is known as
(a) alternating strees (b) reversed strees (c) fluctuating stress (d) repeated stress.

58. During spark test if a material gives small stream of dull red sparks, it can be
(a) medium carbon steel (b) cast iron
(c) mild steel (d) cast steel.

59. Which of the following material has relatively the poorest machinability?
(a) White cast iron (b) Low carbon steel (c) Gun metal (d) Silicon bronze.

60. The Chirpy test of a material measures its
(a) fatigue strength (b) brittleness (c) ductility (d) plasticity.

61. Cementite is
(a) $Fe_2C$ (b) $FeC_2$ (c) $Fe_3C$ (d) $Fe_3C_2$.

62. The carbon percentage is least in
(*a*) cast iron (*b*) wroght iron (*c*) malleable iron (*d*) low carbon steel.

63. The essential constituent of hardened steel is
(*a*) pearlite (*b*) austenite (*c*) martensite (*d*) cementite.

64. The percentage of carbon in mild steel is
(*a*) 0.02 to 0.03 (*b*) 0.2 to 0.3 (*c*) 1.2 to 1.3 (*d*) 2 to 3.5.

65. Fine grain steels are
(*a*) non-magnetic (*b*) corrosion resistant(*c*) ductile (*d*) brittle.

66. The percentage of carbon in cast iron is
(*a*) 0.2 to 0.3 (*b*) 0.7 to 0.8 (*c*) less than 2 (*d*) more than 2.

67. 18-4-1 high speed steel contains
(*a*) 18% cobalt (*b*) 1% chromium (*c*) 4% cobalt (*d*) 0.7% carbon.

68. Which alloying element is added for developing shock resistance in steel?
(*a*) Chromium and Tungsten (*b*) Carbon and Silicon
(*c*) Aluminium and Nickel (*d*) Cobalt and Sulphur.

69. Which of the following alloying element is added to steels for corrosion resistance?
(*a*) Carbon (*b*) Cabalt (*c*) Chromium (*d*) Manganese.

70. Which of the following process is different from the remaining?
(*a*) Tempering (*b*) Sintering (*c*) Annealing (*d*) Cyniding.

71. Annealing is the process of
(*a*) relieving stresses (*b*) reducing grain size
(*c*) inducing hardness (*d*) elevating fatigue limit.

72. Which of the following is not a process for hardness?
(*a*) Nitriding (*b*) Cyniding (*c*) Tumbling (*d*) Case carburising.

73. Which of the following is not a process for getting surface finish?
(*a*) Parkerising (*b*) Lapping (*c*) Sheradizing (*d*) Sintering.

74. Which of the following is not a process for the joining of parts?
(*a*) Sintering (*b*) Brazing (*c*) Tumbling (*d*) Riveting.

75. Which of the following is not a machining process?
(*a*) Routing (*b*) Shot peening
(*c*) Hobbing (*d*) Broaching.

76. Which process is different from the remaining?
(*a*) Flame hardening (*b*) Cyniding (*c*) Electroplating (*d*) Nitriding.

77. Which pressing operation is different from the remaining?
(*a*) Cupping (*b*) Bulging (*c*) Embossing (*d*) Notching.

78. The property of a material by which it can be rolled into sheets is called
(*a*) plasticity (*b*) ductility (*c*) elasticity (*d*) malleability.

79. Bicycle frame pipes are made of
(*a*) cast iron (*b*) cold rolled steel (*c*) hot rolled steel (*d*) carbon-chrome steel.

80. Silicon steel is widely used for
(*a*) transformer cores (*b*) commutators (*c*) bushings (*d*) tank fabrication.

81. The presence of lead in brass
(*a*) reduces ductility (*b*) improves machinability
(*c*) reduces thermal conductivity (*d*) increases electrical resistivity.

82. Constantan is an alloy of
(*a*) copper and nickel (*b*) chromium and zinc
(*c*) copper and aluminium (*d*) chromium vanadium and cobalt.

83. Nichromes are alloys of
(*a*) nickel, chromium and copper (*b*) nickel, chromium and lead
(*c*) nickel, chromium and iron (*d*) nickel, chromium and zinc.

84. Gun metal contains
(*a*) copper and lead (*b*) copper, zinc and tin
(*c*) copper, phosphorus and lead (*d*) copper, nickel and chromium.

85. Phosphor bronze contains
(*a*) phosphorus, lead and zinc (*b*) phosphorus, zinc and aluminium
(*c*) phosphorus, nickel and lead (*d*) phosphorus, copper and tin.

86. German Silver is an alloy of
(*a*) nickel, copper and zinc (*b*) silver and gold
(*c*) silver, copper and zinc (*d*) silver and copper.

87. Brass is an alloy of
(*a*) copper and zinc (*b*) lead and zinc (*c*) zinc and lead (*d*) tin and lead.

88. Solders contain
(*a*) silver and aluminium (*b*) aluminium and lead
(*c*) lead and tin (*d*) copper and nickel.

89. White metal is widely used for
(*a*) soldering (*b*) brazing (*c*) bearing (*d*) painting.

90. The major constituents of white metal are
(*a*) copper and nickel (*b*) aluminium and copper
(*c*) zinc and nickel (*d*) lead, copper and antimony.

91. Which of the following material has highest specific gravity?
(*a*) Aluminium (*b*) High carbon steel (*c*) Leaded brass (*d*) Lead.

92. Which of the following material has lowest melting point?
(*a*) Cast iron (*b*) Copper (*c*) Lead (*d*) Aluminium.

93. Babbit is an alloy of
(*a*) tin, lead and antimony (*b*) tin, nickel and zinc
(*c*) tin, zinc and aluminium (*d*) tin, antimony and copper.

94. The process of making wires of smaller diameters from those of larger diameter is known as
(*a*) trimming (*b*) wire drawing (*c*) rolling (*d*) stretching.

95. Galvanizing is the process of coating materials with a layer of
(*a*) aluminium (*b*) copper (*c*) zinc (*d*) nickel.

96. Dislocations in materials are
(*a*) point defects (*b*) line defects (*c*) planer defects (*d*) none of the above.

97. Alpha brass is a (an)
(*a*) intermediate phase (*b*) interstitial compound
(*c*) substitutional compound.

98. Which of the following material is viscoelastic?
(*a*) Graphite (*b*) Rubber (*c*) Glass (*d*) Silver.

99. Stainless steel is an alloy of
(*a*) iron and nickel (*b*) iron, chromium and nickel
(*c*) iron, chromium and molybdenum (*d*) iron, chromium, molybdenum and nickel.

100. Gun metal is an alloy of
(*a*) 80% Cu + 15% Zn + 5% Sn (*b*) 84% Cu + 11% Zn + 5% Sn
(*c*) 86% Cu + 9% Zn + 5% Sn (*d*) 88% Cu + 2% Zn + 10% Sn.

101. Gamma rays are
(*a*) accelerated helium ions
(*b*) very high frequency electromagnetic radiations
(*c*) accelerated electrons
(*d*) none of the above.

102. Materials having different properties in different directions are known as
(*a*) anisotropic (*b*) ductile (*c*) amorphous (*d*) austenitic.

103. A polymer with high elasticity because of its celled structure is known as
(*a*) dendrite (*b*) copolymer (*c*) elastomer (*d*) ferrite.

104. Fluorescence is
(*a*) luminiscence which occurs immediately after excitation
(*b*) luminiscence which occurs before excitation
(*c*) luminiscence resulting from thermal excitation of electrons
(*d*) luminiscence resulting from magnetic excitation of electrons.

**105.** The main constituent of glass is

(*a*) $SiO_2$ (*b*) $B_2O_3$ (*c*) $Al_2O_3$ (*d*) $CaCO_3$.

**106.** China clay is the major raw material for the production of

(*a*) porcelain (*b*) glass (*c*) fire clay refractory (*d*) all of the above.

**107.** Which of the following is isomorphous system?

(*a*) Copper-gold (*b*) Bismuth-tin (*c*) Iron-vanadium (*d*) None of the above.

**108.** Which variety of glass has least percentage of $SiO_2$?

(*a*) Soda lime glass (*b*) Lead glass (*c*) Borosilicate glass (*d*) High silica glass.

**109.** Polyethylene is produced by

(*a*) condensation polymerization

(*b*) addition polymerization

(*c*) copolymerization of ethylene monomers

(*d*) none of the above.

**110.** Thermoplastic and thermoset polymers differ in

(*a*) glass transition temperature (*b*) thermal behaviour

(*c*) mechanical behaviour (*d*) all of the above.

**111.** Which of the following is not a flexible polymer?

(*a*) Polyvinyl chloride (*b*) Polyethlene

(*c*) Polytetrafluorethylene (*d*) Perspex.

**112.** Which of the following are thermosetting plastics?

(*a*) Silicones (*b*) Acrylics (*c*) Nylon (*d*) Polyvinyls.

**113.** The nature of atomic bond found in diamond is

(*a*) ionic (*b*) covalent

(*c*) metallic (*d*) diamond is amorphorous.

**114.** Neoprene is

(*a*) a variety of rubber (*b*) a plastic

(*c*) a semiconductor (*d*) a rubber like plastic.

**115.** Which of the following polymer is crystalline?

(*a*) Polyethylene (*b*) Polymethyl metacrylate

(*c*) Polyvinyl chloride (*d*) Polyvinylidene chloride.

**116.** Polysterene is a (an)

(*a*) easter (*b*) hydrocarbon (*c*) alkyl halide (*d*) none of the above.

**117.** Plastics are

(*a*) good conductors of heat and bad conductors of electricity

(*b*) bad conductors of heat and good conductors of electricity

(*c*) bad conductors of heat as well as electricity

(*d*) good conductors of heat as well as electricity.

**118.** Inclusions

(*a*) reinforcement material for plastics

(*b*) electrons entering in the vacant spaces in an atom

(*c*) molecules with the same compositions but different structures

(*d*) particles of impurities contained by material.

**119.** Isotope

(*a*) same element with different number of neutrons

(*b*) same element with different number of protons

(*c*) same element with different number of electrons

(*d*) different materials with same number of electrons.

**120.** Magneto striction

(*a*) induced magnetism (*b*) volume changes accompanying magnetization

(*c*) rate of magnetisation (*d*) rate of loss of magnetism.

**121.** Mer

(*a*) a unit of length

(*b*) temporary equilibrium

(*c*) the smallest repetitive unit in a polymer

(*d*) material with weak secondary bonds.

**122.** The resistivity of electrical conductors is most affected by

(*a*) temperature (*b*) pressure (*c*) grain orientation (*d*) composition.

**123.** A randome material should be transparent to

(*a*) *IR* waves (*b*) microwaves (*c*) both (*a*) and (*b*) above

(*d*) none of the above.

**124.** Ferrites are

(*a*) ferromagnetic materials (*b*) ferroelectric materials

(*c*) ferrimagnetic materials (*d*) antiferro magnetic materials.

**125.** Micron

(*a*) $10^6$ cm (*b*) $10^{-12}$ m (*c*) $10^4$ angstrom (*d*) $10^{-18}$ cm.

**126.** Which of the following can be used to form a multiplier?

(*a*) Hall effect (*b*) Piezoelectric effect

(*c*) Zeeman's effect (*d*) Faraday effect.

**127.** Which of the following is piezo-electric material?

(*a*) Corrundum (*b*) Quartz (*c*) Mica (*d*) Stealite.

**128.** Rutile is

(*a*) most common phase of $TiO_2$ (*b*) a polymetic material

(*c*) a discrete unit of energy

(*d*) crystal with three equal axes not at right angles.

129. Zeolite
(*a*) a phase of $ZrSiO_4$ (*b*) silicate with exchangeable ions
(*c*) a cast iron with granite (*d*) a four folded structure.

130. Powder metallurgy technique is widely used for
(*a*) transistors (*b*) ferrite magnets (*c*) cutting tools (*d*) porcelains.

131. Which of the following is not a cold working process?
(*a*) Swaging (*b*) Spinning (*c*) Knurling (*d*) Wire drawing.

132. Which variety of glass can be used for higher temperatures?
(*a*) Soda lime glass (*b*) Lead glass (*c*) Boro silicate glass (*d*) High silica glass.

133. Which of the following is a protective coating for corrosion?
(*a*) Cladding (*b*) Electroplating (*c*) Spraying (*d*) All of the above.

134. Which of the following wire is expected to be strongest in tension?
(*a*) Aluminium wire (*b*) Glass fibre wire (*c*) Piano wire (*d*) Brass wire.

135. Semi-conductors have electrical conductivity of the order of
(*a*) $10^{-30}$ ohm-cm$^{-1}$ (*b*) $10^{-17}$ ohm-cm$^{-1}$ (*c*) $10^{-3}$ ohm-cm$^{-1}$ (*d*) $10^{5}$ ohm-cm$^{-1}$.

136. The behaviour of visco elastic materials is
(*a*) time dependent (*b*) independent of time
(*c*) ductile (*d*) plastic.

137. The phenomenon of age hardening is associated with
(*a*) german silver (*b*) gun metal (*c*) duralumin (*d*) cast iron.

138. For which of the following materials, the magnetization is not linearly related to the applied field?
(*a*) Ferromagnetic material (*b*) Diamagnetic material
(*c*) Paramagnetic material (*d*) None of the above.

139. For which of the following materials the net magnetic moment is zero?
(*a*) Ferromagnetic materials (*b*) Antiferromagnetic materials
(*c*) Ferrimagnetic materials (*d*) All of the above.

140. Ferrites are associated with
(*a*) ferromagnetic materials (*b*) paramagnetic materials
(*c*) diamagnetic materials (*d*) ferrimagnetic materials.

141. A magnetic material above Curie point exists as
(*a*) diamagnetic (*b*) ferromagnetic (*c*) paramagnetic (*d*) ferrite.

142. The magnetic materials, in which the permanent magnetic moment due to electron spin are already aligned due to bonding force is known as
(*a*) ferromagnetic materials (*b*) ferrimagnetic materials
(*c*) paramagnetic materials (*d*) diamagnetic materials.

**143.** In which of the following material the interaction between the neighboring dipoles is negligible ?

(*a*) Ferromagnetic materials (*b*) Ferrimagnetic materials

(*c*) Paramagnetic materials (*d*) Diamagnetic materials.

**144.** Behaviour like semi conductors can be expected from

(*a*) ferromagnetic materials (*b*) ferrimagnetic materials

(*c*) paramagnetic materials (*d*) diamagnetic materials.

**145.** Which of the following material has density closer to that of gold?

(*a*) Lead (*b*) Tungsten (*c*) Chromium (*d*) Nichrome.

**146.** The ratio $\frac{\text{resistivity of manganin}}{\text{resistivity of silver}}$ is nearly

(*a*) 2 (*b*) 12 (*c*) 22 (*d*) 32.

**147.** Copper becomes superconductive

(*a*) when 100% pure (*b*) when electro refined

(*c*) at very high temperatures (*d*) at absolute zero.

**148.** Pure metals generally have

(*a*) high conductivity and low temperature co-efficient

(*b*) high conductivity and large temperature co-efficient

(*c*) low conductivity and zero temperature co-efficient

(*d*) low conductivity and high temperature co-efficient.

**149.** Total number of semi-conductor materials in periodic table is

(*a*) 3 (*b*) 8 (*c*) 13 (*d*) 21.

**150.** Which of the following is not a semi-conductor material?

(*a*) Te (*b*) Ge (*c*) B (*d*) C.

**151.** Sulphur hexafluoride is used for

(*a*) dynamos (*b*) dc series motor (*c*) switch gears (*d*) synchronous motors.

**152.** Ferrites are

(*a*) not easily machinable (*b*) hard materials

(*c*) brittle materials

(*d*) materials possessing all of the above properties.

**153.** Ferrimagnetic material generally find application as

(*a*) Conductors (*b*) Insulators (*c*) Resistors (*d*) Semiconductors.

**154.** Which of the following materials can be used as cathode for vacuum tubes?

(*a*) Thoriated tungsten (*b*) Tungsten

(*c*) Mixture of barium oxide and strontium oxide

(*d*) Any of the above.

155. Superconductivity can be destroyed by

(a) Application of magnetic field (b) Reducing temperature

(c) Adding impurities (d) None of the above.

156. Which of the following troubles are associated with film resistors?

(a) Cracks (b) Spiraling (c) Surface roughness (d) All of the above.

157. Which of the following element has BCC arrangement

(a) Rb (b) Cd (c) Mg (d) Au.

158. Copper is completely miscible with

(a) Nickel (b) Gold (c) Hydrogen (d) Lead.

159. An intrinsic semiconductor behaves as insulator at

(a) 0° K (b) 27° K (c) 300° K (d) 100° K.

160. The materials having permanent electric moment are called

(a) Antiferro-electric materials (b) Dielectrics

(c) Electrolytes (d) Semiconducting materials.

161. Which of the following dielectric material has the lowest permittivity?

(a) Titania (b) Silica (c) Lead glass (d) Ferric oxide.

162. The number of semi-conductors in periodic table is

(a) 3 (b) 5 (c) 7 (d) 13.

163. Which of the following has the highest resistivity?

(a) Mineral oil (b) Paraffin wax (c) Air (d) Mica.

164. Which of the following property of PVC is of prime importance?

(a) Strength (b) Appearance (c) Colour (d) Non-inflammability.

165. Alpha brass is a (an)

(a) Intermediate phase (b) Interstitial compound

(c) Substitutional solid solution (d) None of the above.

166. The relative permeability of superconducting material is

(a) Infinitely large (b) Normal (c) One (d) Zero.

167. Silicon steels are specified as etc. In this figure after E represents

(a) Percentage of silicon in steel (b) Specific loss in steel at 50 Hz

(c) Process of manufacturing (d) Direction of rolling.

168. An H.R.C. fuse is

(a) A ceramic body having metal and caps

(b) A wire of platinum

(c) A heavy cross-section of copper or aluminium

(d) A ceramic tube having carbon rod inside it.

169. The type of bonding that exists in solid Argon is
(*a*) Ionic (*b*) Convalent
(*c*) Van der waals (*d*) A mixture of convalent and Van der waals.

170. High resistivity materials are used in
(*a*) Precision instruments (*b*) Heating elements
(*c*) Motor starters (*d*) Incandescent lamps
(*e*) All of the above.

171. The mean free time of an electron in a metal is
(*a*) Proportional to temperature
(*b*) Inversely proportional to temperature
(*c*) Proportional to the square of the temperature
(*d*) None of the above.

172. Carbon resistors are used extensively because they are
(*a*) Easy to make (*b*) Compact (*c*) Inexpensive (*d*) All of the above.

173. Polyethyelene is produced by
(*a*) Condensation (*b*) Condensation polymerization
(*c*) Addition polymerization (*d*) Copolymerization of ethylene monomers.

174. Polystrene is
(*a*) An ester (*b*) An alcohol (*c*) A hydrocarbon (*d*) An alkyl halide.

175. Effect of moisture on the insulating materials is to
(*a*) Decrease dielectric constant (*b*) Decrease dielectric strength
(*c*) Decrease insulation resistance (*d*) Increase dielectric loss
(*e*) All of the above.

176. Selenium is ……. Semiconductor.
(*a*) Extrinsic (*b*) Intrinsic (*c*) N-type (*d*) P-type.

177. …… has the best damping properties.
(*a*) Diamond (*b*) High speed steel (*c*) Mild steel (*d*) Cast iron.

178. Which photoconductor is widely used?
(*a*) PbS (*b*) CdS (*c*) GaAs (*d*) AsBr.

179. An intrinsic semiconductor at absolute zero temperature behave like
(*a*) An electron (*b*) An insulator (*c*) A conductor (*d*) An ion.

180. Polymers
(*a*) Can be vaporized as well as recycled
(*b*) Can neither be vaporized nor recycled
(*c*) Can be vaporized bur cannot be recycled
(*d*) Can be recycled but cannot be vaporized.

**181.** Which of the following factors affect resistivity of metals?
(*a*) Age hardening (*b*) Alloying (*c*) Temperature (*d*) Cold work
(*e*) All of the above.

**182.** Which allotropic form of iron is magnetic at room-temperature?
(*a*) Alpha iron (*b*) Beta iron (*c*) Delta iron (*d*) Gamma iron.

**183.** Ferrites are ..... materials
(*a*) Ferro-electric (*b*) Ferromagnetic (*c*) Ferromagnetic (*d*) Any of the above.

**184.** Zeolite
(*a*) A phase of $ZrSiO_4$ (*b*) Silicate with exchangeable ions
(*c*) A cast iron with granite (*d*) A four folded structure.

**185.** Spark plug makes use of which of the following materials for insulation?
(*a*) Porcelain (*b*) Slate (*c*) Asbestos (*d*) Glass.

**186.** Super-conductivity results due to
(*a*) Crystal structure having infinite atomic vibrations at 0°K
(*b*) All electrons interacting in the super conducting gate
(*c*) Crystal structure having no atomic vibration at 0°K
(*d*) All electrons having fermi energy at 0°K.

**187.** Carbon bearings are used under which of the following condition?
(*a*) Where lubrication is difficult or impossible
(*b*) Where corrosive chemical action exists
(*c*) Where high temperature exists
(*d*) All of the above.

**188.** Which method is suitable for the measurement of resistivity of good conductors of electricity?
(*a*) Kelvin double bridge method (*b*) Loss of charge method
(*c*) Schering bridge method (*d*) Any of the above.

**189.** The structure of a semi-conductor is like that of a
(*a*) Triangle (*b*) Diamond (*c*) Circle (*d*) Rhombus.

**190.** The common household glass is
(*a*) Soda lime glass (*b*) Bora silicate glass
(*c*) High silica glass (*d*) High leaded glass.

**191.** Above the Curie temperature the ..... have no ferro-electric properties and become ordinary materials
(*a*) Conducting materials (*b*) Semi-conducting materials
(*c*) Insulating materials (*d*) None of the above.

**192.** The materials which exhibit the same elastic properties in all directions are called
(*a*) Isentropic (*b*) Isotropic (*c*) Inelastic (*d*) Visco elastic.

**193.** The manufacturers normally specify that a lead acid battery should not remain discharged for more than

(*a*) 24 hours (*b*) One weak (*c*) One month (*d*) One hour.

**194.** An allotropic material has

(*a*) Different crystal structure at different temperatures

(*b*) Fixed structures at all temperature

(*c*) Atoms in random pattern

(*d*) All of the above.

**195.** Hardness of a material is resistance to

(*a*) Creep (*b*) Scratching (*c*) Machining (*d*) Heating.

**196.** For elastic materials stress and strain are related by

(*a*) Poisson's law (*b*) Gauss's law (*c*) Hooke's law (*d*) Dalton's law.

**197.** The percentage of carbon in cast iron is

(*a*) 0.2 to 0.3 percent (*b*) 2 to 4 percent (*c*) 4 to 10 percent (*d*) 20 to 30 percent.

**198.** The material which has zero temperature coefficient of resistance is

(*a*) Manganin (*b*) Porcelain (*c*) Carbon (*d*) Aluminium.

**199.** The crystal structure of silicon is

(*a*) Simple cubic (*b*) BCC (*c*) FCC (*d*) Diamond.

## ANSWERS

| | | | | | | |
|---|---|---|---|---|---|---|
| **1.** (*b*) | **2.** (*b*) | **3.** (*d*) | **4.** (*b*) | **5.** (*c*) | **6.** (*c*) | **7.** (*c*) |
| **8.** (*d*) | **9.** (*a*) | **10.** (*b*) | **11.** (*d*) | **12.** (*a*) | **13.** (*b*) | **14.** (*c*) |
| **15.** (*a*) | **16.** (*c*) | **17.** (*c*) | **18.** (*a*) | **19.** (*a*) | **20.** (*a*) | **21.** (*c*) |
| **22.** (*c*) | **23.** (*d*) | **24.** (*a*) | **25.** (*b*) | **26.** (*b*) | **27.** (*d*) | **28.** (*b*) |
| **29.** (*a*) | **30.** (*b*) | **31.** (*a*) | **32.** (*d*) | **33.** (*c*) | **34.** (*a*) | **35.** (*c*) |
| **36.** (*b*) | **37.** (*b*) | **38.** (*c*) | **39.** (*a*) | **40.** (*c*) | **41.** (*d*) | **42.** (*b*) |
| **43.** (*c*) | **44.** (*d*) | **45.** (*b*) | **46.** (*a*) | **47.** (*c*) | **48.** (*c*) | **49.** (*d*) |
| **50.** (*a*) | **51.** (*a*) | **52.** (*d*) | **53.** (*c*) | **54.** (*d*) | **55.** (*c*) | **56.** (*d*) |
| **57.** (*b*) | **58.** (*b*) | **59.** (*a*) | **60.** (*b*) | **61.** (*c*) | **62.** (*b*) | **63.** (*c*) |
| **64.** (*b*) | **65.** (*c*) | **66.** (*d*) | **67.** (*d*) | **68.** (*a*) | **69.** (*c*) | **70.** (*b*) |
| **71.** (*a*) | **72.** (*c*) | **73.** (*d*) | **74.** (*c*) | **75.** (*b*) | **76.** (*c*) | **77.** (*d*) |
| **78.** (*b*) | **79.** (*b*) | **80.** (*a*) | **81.** (*b*) | **82.** (*a*) | **83.** (*c*) | **84.** (*b*) |
| **85.** (*d*) | **86.** (*a*) | **87.** (*a*) | **88.** (*c*) | **89.** (*c*) | **90.** (*d*) | **91.** (*d*) |
| **92.** (*d*) | **93.** (*d*) | **94.** (*b*) | **95.** (*c*) | **96.** (*b*) | **97.** (*c*) | **98.** (*b*) |
| **99.** (*b*) | **100.** (*d*) | **101.** (*b*) | **102.** (*a*) | **103.** (*c*) | **104.** (*a*) | **105.** (*a*) |

| | | | | | | |
|---|---|---|---|---|---|---|
| **106.** (*a*) | **107.** (*a*) | **108.** (*b*) | **109.** (*b*) | **110.** (*b*) | **111.** (*d*) | **112.** (*a*) |
| **113.** (*b*) | **114.** (*d*) | **115.** (*a*) | **116.** (*b*) | **117.** (*c*) | **118.** (*d*) | **119.** (*a*) |
| **120.** (*b*) | **121.** (*c*) | **122.** (*a*) | **123.** (*b*) | **124.** (*c*) | **125.** (*c*) | **126.** (*a*) |
| **127.** (*b*) | **128.** (*a*) | **129.** (*b*) | **130.** (*c*) | **131.** (*a*) | **132.** (*d*) | **133.** (*d*) |
| **134.** (*c*) | **135.** (*c*) | **136.** (*a*) | **137.** (*c*) | **138.** (*a*) | **139.** (*b*) | **140.** (*d*) |
| **141.** (*c*) | **142.** (*a*) | **143.** (*c*) | **144.** (*b*) | **145.** (*d*) | **146.** (*d*) | **147.** (*d*) |
| **148.** (*b*) | **149.** (*c*) | **150.** (*d*) | **151.** (*c*) | **152.** (*d*) | **153.** (*d*) | **154.** (*d*) |
| **155.** (*a*) | **156.** (*d*) | **157.** (*a*) | **158.** (*b*) | **159.** (*a*) | **160.** (*c*) | **161.** (*b*) |
| **162.** (*d*) | **163.** (*c*) | **164.** (*d*) | **165.** (*c*) | **166.** (*d*) | **167.** (*a*) | **168.** (*a*) |
| **169.** (*c*) | **170.** (*e*) | **171.** (*b*) | **172.** (*d*) | **173.** (*c*) | **174.** (*c*) | **175.** (*e*) |
| **176.** (*b*) | **177.** (*d*) | **178.** (*b*) | **179.** (*b*) | **180.** (*c*) | **181.** (*c*) | **182.** (*a*) |
| **183.** (*b*) | **184.** (*b*) | **185.** (*a*) | **186.** (*c*) | **187.** (*d*) | **188.** (*a*) | **189.** (*b*) |
| **190.** (*a*) | **191.** (*c*) | **192.** (*b*) | **193.** (*a*) | **194.** (*a*) | **195.** (*b*) | **196.** (*c*) |
| **197.** (*b*) | **198.** (*a*) | **199.** (*d*). | | | | |

Chapter **23** **SEMICONDUCTORS**

**Conducting materials like silver, copper, aluminium are good conductors of electricity. On the other hand we have materials like rubber, wood, porcelain, etc. which are insulator or bad conductors of electricity. Third category of materials exists midway between the conductors and insulators, known as semiconductors for which the electrical conductivity is some where between the two. Germanium and Silicon are well known semiconductors, widely used in electronics.**

**Conductors, Semiconductors and Insulators**

| *Materials* | *Electrical conductivity s/m* | *Resistivity* ($\Omega$ *m*) | *Classification* |
|---|---|---|---|
| Silver | $6.25 \times 10^{7}$ | $1.6 \times 10^{-8}$ | Conductors |
| Copper | $5.88 \times 10^{7}$ | $1.7 \times 10^{-8}$ | |
| Aluminium | $3.85 \times 10^{7}$ | $2.6 \times 10^{-3}$ | |
| Germanium (Pure) | 1.54 | $6.5 \times 10^{-1}$ | Semiconductors |
| Silicon (pure) | $5 \times 10^{-4}$ | $2 \times 10^{3}$ | |
| Porcelain | $3.33 \times 10^{-10}$ | $3.0 \times 10^{9}$ | Insulators |
| Glass | $5.88 \times 10^{-12}$ | $1.7\times10^{11}$ | |
| Hard rubber | $1.0 \times 10^{-16}$ | $1.0 \times 10^{-16}$ | |

**In case of conductors like copper, resistivity increases as temperature increases. However, in case of semiconductors, it is the other way round. As the temperature increases, the resistivity of semiconductor decreases. Thus semiconductors have negative temperature co-efficient of resistance.**

The resistivity of semiconductors can be increased by adding a very small amount of some specific materials (called impurities).

**Intrinsic and Extrinsic Semiconductors.** A semiconductor in pure form (without impurities) is known as intrinsic semiconductor. A semiconductor is considered to be pure when the impurities are less than 1 part in $10^8$ parts of semiconductor.

The process of deliberately adding impurities to a semiconductor material is called "doping". Doping is done after the semiconductor material has been refined to a high degree of purity. A doped semiconductor is known as extrinsic semiconductor.

**N-type and P-type Semiconductors.** The elements used for doping generally have an electron valance of either 5 or 3. As a result, the doped semiconductor has either an excess or shortage of electrons in the covalent bond structure. A semi-conductor with excess electrons is an *N*-type, while a *P*-type has a shortage of electrons. The doping elements are given in table below.

**Doping Elements and Semi-conductors**

| *Element* | *Symbol* | *Atomic Number* | *Valance Electrons* | *Applications* |
|---|---|---|---|---|
| Antimony | Sb | 51 | 5 | Donor impurity elements. Give electrons to form *N*-type semiconductor. As and Sb used for Ge, Phosphorus for Si. |
| Arsenic | As | 33 | 5 | |
| Phosphorus | P | 15 | 5 | |
| Germanium | Ge | 32 | 4 | Intrinsic semiconductors used in pure form to be doped with impurity elements. Doped semiconductors are extrinsic semiconductors. |
| Silicon | Si | 14 | 4 | |
| Aluminium | Al | 13 | 3 | Acceptor impurity elements. Take electrons to form *P*-type semiconductor. Ga and In used for Ge ; Al and B for Si. |
| Boron | B | 5 | 3 | |
| Gallium | Ga | 31 | 3 | |
| Indium | In | 49 | 3 | |

**PN junction.** Fig. 23.1 illustrates a magnified view of the junction of analyze the effect of the *P* and *N* semiconductors. Actually the regular structure of single crystal is the same through the *P* and *N* bulk materials and their junction.

Some electrons in the *N*-side are attracted to the *P*-side, while at the same time an equal number of hole charges move in the opposite direction. These charges produce neutral electron hole pairs. However, this effect at the junction uncovers the charged ions of the impurity atoms. These ions

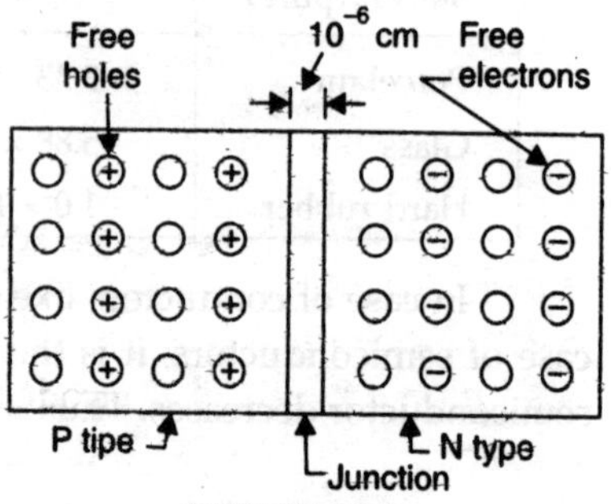


*FIGURE 23.1*

provide an internal barrier potential $V_b$ across the two sides of the junction. $V_b$ has the polarity to prevent any more free electrons or hole charges from crossing the junction. In effect, $V_b$ maintains the free electrons in the *N* semiconductor and the hole charges in the *P* semiconductor, to prevent the opposite sides from becoming neutralized.

**Internal Barrier Potential ($V_b$).** Although it is an internal contact potential that cannot be measured directly, the effect can be overcome by 0.3 V for a Ge junction of 0.7 V for Si. The barrier voltage is more for Si because its lower atomic number allows more stability in the covalent bonds.

**Depletion zone.** Because of its neutral electron-hole pairs, the junction area is the depletion zone, meaning it has no free charges. However, it still has the ion charges anchored in position to produce $V_b$.

**VI Characteristics.** The volt-ampere characteristics for *PN* junction as shown in Fig. 23.2. It offers less resistance in forward direction and high resistance in reverse direction.

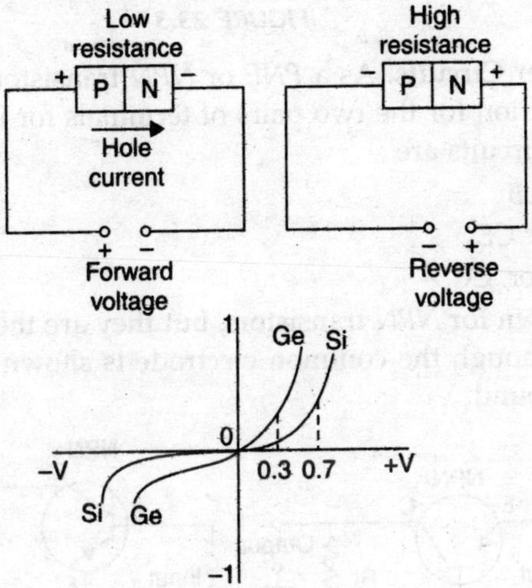


*FIGURE 23.2*

**Diode Rectifiers.** The standard symbol for semiconductor diode is an arrow and bar showing the direction of whole current. Therefore the arrow is the *P* side and the bar is the *N* side. Positive voltage applied to the *P* arrow makes the diode conduct since this side is the anode, while the *N* bar is the cathode.

Semiconductor diode, like other diodes can be used for half wave rectifiers and full wave rectifiers.

**Transistors.** As shown in Fig. 23.3 the transistor consists of a *PN* junction and an *NP* junction, by making either a *P* or *N* semiconductor between opposite types. The purpose is to have the first section supply charges, either holes or electrons, to be collected by the third section, through the middle section. The electrode that supplies charges is the emitter, the electrode at the opposite end to collect the charges is the collector. The base in the middle forms two junctions between emitter and collector, to control the collector current. *PNP* and *NPN* transistors are shown in Fig. 23.3.

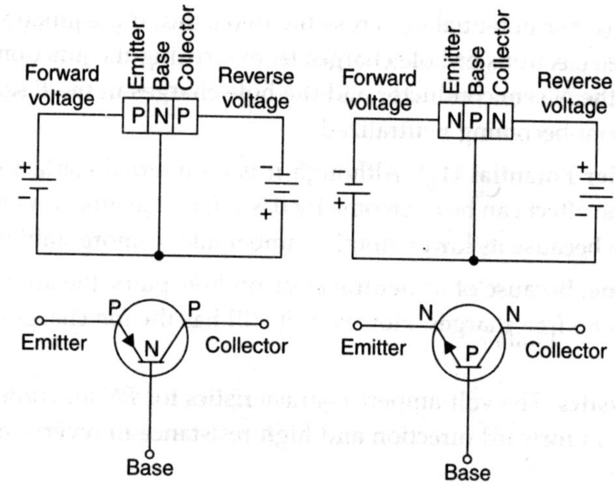


FIGURE 23.3

**Transistor Amplifier Circuits.** As a *PNP* or *NPN* transistor has only three electrodes, one must be a common connection for the two pairs of terminals for input and output signals. Three possibilities for amplifier circuits are :

(1) Common base, *CB*

(2) Common emitter *CE*

(3) Common collector, *CC*.

The circuits are shown for *NPN* transistors, but they are the same for *PNP* transistors with all polarities reversed. Although the common electrode is shown grounded here, it need not be connected to the chasis ground.

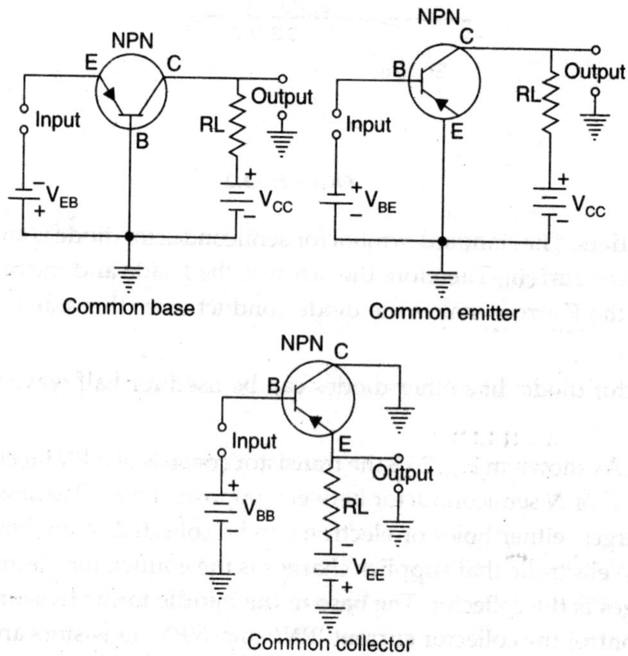


FIGURE 23.4

For all the circuits, the collector has reverse voltage and the emitter base junction requires forward bias. It is marked $V_{BE}$ or $V_{EB}$. This average forward voltage is called bias voltage. The common-emitter circuit is the one generally used for transistor amplifiers because it has the most gain with cascaded stages. The special features for three circuits are given in table below.

**Comparison of Transistor Circuits**

| *Characteristic* | *Common Base* (*CB*) | *Common Emitter* (*CE*) | *Common Collector* (*CC*) |
|---|---|---|---|
| Signal into | Emitter | Base | Base |
| Signal out of | Collector | Collector | Emitter |
| Advantage | Stability | High gain | High $r_i$ |
| Phase inversion | No | Yes | No |
| Similar vacuum tube circuit | Grounded grid | Grounded cathode | Grounded plate (cathode follower) |

## Special Purpose Diodes

**Tunnel Diodes.** These diodes are based on the fact that heavy doping can cause a tunneling effect of charge carriers through the depletion zone at the junction. One important feature of the tunnel diode is its negative resistance for a specific range of forward voltage. In a negative resistance, the current decreases for an increase of applied voltage. Because of the characteristic, the tunnel diode can operate as an amplifier or oscillator. This is a low-power device for microwave frequencies with relative freedom from radiation effects.

**Varistor.** In this two junctions of opposite polarities are used. The varistor is often used as a shunt for the collector to protect against excessive voltage spikes, either positive or negative. Also, a single diode rectifier can be used in series with the collector. This function is to isolate the collector from voltage spikes that have forward polarity.

**Zener Diodes.** These are voltage reference diodes. Basically these are silicon diodes designed for a specific reverse breakdown voltage. When the avalanche current flows, the voltage across the diode remains constant. This characteristic makes the diode useful as a voltage regulator.

**Varactor Diodes.** The barrier voltage at a reverse biased junction enables the junction to serve as a capacitance because of separated charges in the depletion zone. This diode can be used as a voltage-sensitive capacitor. This is used in electronic tuning for television receivers.

**Light Emitting Diodes (LED).** The *PN* junction can radiate light, as energy is released by the recombination of charges. For better efficiency special compounds of gallium are used. A seven segment display is widely used for digital circuitory.

**Photoconductive Diodes.** These are made of a photosensitive material such as cadmium sulphide, in which the resistance decreases with more light. It is widely used in light-control devices.

## OBJECTIVE TYPE QUESTIONS

1. Germanium atom contains
   (*a*) two electron orbits (*b*) two protons
   (*c*) four electrons (*d*) four neutrons.
2. In general the range of conductivity of materials is
   (*a*) $10^3$s/m to $10^{-3}$ s/m (*b*) $10^6$ s/m to $10^{-6}$ s/m
   (*c*) $10^8$ s/m to $10^{-12}$ s/m (*d*) $10^9$ s/m to $10^{-18}$ s/m.
3. The conductivity of silicon can be expected around
   (*a*) $0.5 \times 10^{14}$ s/m (*b*) $0.5 \times 10^{-3}$ s/m (*c*) $0.5 \times 10^3$ s/m (*d*) $0.5 \times 10^{-14}$ s/m.
4. Which of the following is a semiconductor ?
   (*a*) Arsenic (*b*) Gallium arsenide (*c*) Phosphorus (*d*) Diamond.
5. Which of the following is tetravalent ?
   (*a*) Quartz (*b*) Diamond (*c*) Germanium (*d*) Calcium.
6. Which of the following is not a semiconductor ?
   (*a*) Silicon (*b*) Germanium (*c*) Diamond (*d*) Gallium arsenide.
7. Atoms held together by sharing of valance electron
   (*a*) always have tendency to accept electrons
   (*b*) have one or more holes
   (*c*) have a weak bond which tends to break
   (*d*) form a covalent bond.
8. When an electron is removed from an atom it becomes
   (*a*) covalent (*b*) positron (*c*) dielectric (*d*) ionised.
9. An electron in the conduction band
   (*a*) has lower energy than an electron in the valance band
   (*b*) has a higher energy than an electron in the valance band
   (*c*) is always chargeless
   (*d*) has tendency to leave the atom.
10. The outermost orbit in an atom cannot have
    (*a*) more than 8 electrons (*b*) less than four electrons
    (*c*) electrons of other atoms (*d*) all of the above.
11. Atomic number is the
    (*a*) number of electrons in the outermost orbit
    (*b*) number of protons in an atom
    (*c*) total number of protons, neutrons and electrons in an atom
    (*d*) average number of neutrons in an atom.

12. When $n$ is the number of the shell, the maximum number of electrons in the shell can be
(*a*) $n$ (*b*) $2n$ (*c*) $n^2$ (*d*) $2n^2$.

13. Which of the following element belongs to the same group of periodic table as that of Silicon and Germanium ?
(*a*) Boron (*b*) Carbon (*c*) Phosphorus (*d*) Sodium.

14. Germanium and silicon belong to which group of periodic table ?
(*a*) III (*b*) IV (*c*) VI (*d*) VIII.

15. At absolute zero, an intrinsic semiconductor
(*a*) become super conductor (*b*) disintegrates into pieces
(*c*) behaves like an insulator (*d*) becomes extrinsic semiconductor.

16. A semiconductor has
(*a*) zero temperature co-efficient of resistance
(*b*) positive temperature co-efficient of resistance
(*c*) negative temperature co-efficient of resistance
(*d*) none of the above.

17. A silicon atom contains
(*a*) 14 protons (*b*) 14 neutrons (*c*) 14 electrons (*d*) all of the above.

18. The number of electrons in the outermost shell of silicon is
(*a*) 1 (*b*) 2 (*c*) 3 (*d*) 4.

19. The number of valance electrons in phosphorus is
(*a*) 2 (*b*) 3 (*c*) 4 (*d*) 5.

20. The atomic number of Germanium is
(*a*) 16 (*b*) 20 (*c*) 32 (*d*) 48.

21. The number of electrons in the outermost shell of germanium is
(*a*) 1 (*b*) 2 (*c*) 3 (*d*) 4.

22. One electron volts is the same as
(*a*) 1 Joule (*b*) $1.6 \times 10^{-6}$ Joule (*c*) $1.6 \times 10^{-12}$ (*d*) $1.6 \times 10^{-19}$ Joule.

23. An electron will not contribute to electric current when
(*a*) it strikes a positive ion (*b*) it looses its charge
(*c*) it is at higher temperatures (*d*) it is in a completely filled bond.

24. A semiconductor in its purest form is known as
(*a*) intrinsic semiconductor (*b*) extrinsic semiconductor
(*c*) supper conductor (*d*) insulator.

25. When an electron breaks a covalent bond and moves away
(*a*) the semiconductor becomes conductor
(*b*) a vacancy is created in the broken covalent bond
(*c*) the conductivity of the material increases
(*d*) more ions are produced.

26. The concentration of minority carriers, in case of N-type semiconductors mainly depends upon
(*a*) the purity of semiconductor material
(*b*) the doping technique
(*c*) distribution of impurities in semiconductor material
(*d*) the temperature of the material.

27. Which of the following is trivalent ?
(*a*) Boron (*b*) Aluminium (*c*) Indium (*d*) All of the above.

28. A *p*-type semiconductor results when
(*a*) a trivalent impurity is added to an intrinsic semiconductor
(*b*) a pentavalent impurity is added to an intrinsic semiconductor
(*c*) either a trivalent or pentavalent impurity is added to an intrinsic semiconductor
(*d*) any of the above.

29. The number of valance electrons in boron is
(*a*) 1 (*b*) 2 (*c*) 3 (*d*) 4.

30. As the temperature of a semiconductor increases its
(*a*) resistivity increases (*b*) conductivity increases
(*c*) atomic number decreases (*d*) temperature co-efficient becomes zero.

31. The process of deliberately adding impurities to a semiconductor is known as
(*a*) sintering (*b*) intrusion (*c*) drifting (*d*) doping.

32. A doped semiconductor is also known as
(*a*) diffused semiconductor (*b*) intrinsic semiconductor
(*c*) extrinsic semiconductor (*d*) none of the above.

33. At room temperature, when some voltage is applied to an intrinsic semiconductor
(*a*) electrons move to the negative terminal and holes move to the positive terminal
(*b*) electrons move to the positive terminal and holes move to the negative terminal
(*c*) both holes as well as electrons move to the negative terminal
(*d*) both holes as well as electrons move to the positive terminal.

34. As the temperature of an intrinsic semiconductor is increased
(*a*) more holes are created in the conduction band
(*b*) energy of the atoms is increased
(*c*) holes and free electrons fused together
(*d*) none of the above.

35. The movement of a hole results from
(*a*) the vacancy filled by a valance electron from the neighbouring atom
(*b*) excitation due to high temperature
(*c*) change in number of protons in the atom
(*d*) none of the above.

36. The conduction band is

(*a*) the region of free electrons

(*b*) a range of energies corresponding to the energies of the free electrons

(*c*) always above the forbidden energy level

(*d*) concentrates holes for the flow of current.

37. The forbidden energy gap in semiconductors lies

(*a*) between the valance band and conduction band

(*b*) just above the conduction band

(*c*) just below the conduction band

(*d*) either above or below the conduction band.

38. When small amount of antimony is added to germanium

(*a*) the resistance is reduced

(*b*) the resistance is increased

(*c*) there will be more free electrons than holes in the semiconductor

(*d*) there will be more holes than free electrons in the semiconductor.

39. Which of the following has the highest mobility ?

(*a*) Neutron (*b*) Electron (*c*) Positive ions (*d*) Negative ions.

40. The impurity added to extrinsic semiconductors is of the order of

(*a*) 1 in 10 (*b*) 1 in 1000 (*c*) 1 in million (*d*) 1 in billion.

41. The donor type impurity

(*a*) must have only five valance electrons

(*b*) must have only four valance electrons

(*c*) must have only three valance electrons

(*d*) must have only two valance electrons.

42. Which of the following is a trivalent element ?

(*a*) Phosphorus (*b*) Indium (*c*) Antimony (*d*) Arsenic.

43. A *N*-type semiconductor has

(*a*) electrons as majority carriers and holes as minority carriers

(*b*) electrons as minority carriers and holes as majority carriers

(*c*) only electrons as carriers

(*d*) only holes as carriers.

44. In a semiconductor

(*a*) free electrons and holes are generated in pairs due to thermal energy

(*b*) free electrons generated due to thermal energy are always more in number as compared to holes

(*c*) holes generated due to thermal energy are more than the number of electrons

(*d*) the number of holes or electrons generated depends on the rate of heating.

45. Which of the following is trivalent?
(*a*) Phosphorus (*b*) Gallium (*c*) Sulphur (*d*) All of the above.

46. Acceptor-type impurities
(*a*) create excess electrons (*b*) must have five valance electrons
(*c*) must have three valance electrons (*d*) can be added to Silicon but not Germanium.

47. Which of the following material cannot be used to convert an intrinsic semiconductor into *N*-type extrinsic semiconductor?
(*a*) Arsenic (*b*) Antimony (*c*) Phosphorus (*d*) Boron.

48. Thermionic emission of electrons results from
(*a*) strong magnetic fields (*b*) electrostatic fields
(*c*) high temperatures (*d*) photo-voltaic effect.

49. Which of the following cannot exist outside a semiconductor?
(*a*) Electron (*b*) Hole
(*c*) Both hole as well as electrons (*d*) None of the above.

50. The merging of a free electron and a hole is known as
(*a*) recombination (*b*) extrusion (*c*) absorption (*d*) absorption.

51. Which of the following statement is correct?
(*a*) Free electrons are the minority carriers in *N*-type semiconductors and holes are the minority carriers
(*b*) Free electrons are the majority carriers in *N*-type semiconductors and holes are the minority carriers
(*c*) To get excess electrons in an intrinsic semiconductor, we can add trivalent atoms
(*d*) The larger the orbit, the smaller is the energy of the electron.

52. Which of the following impurity is used for silicon?
(*a*) Phosphorus (*b*) Antimony (*c*) Arsenic (*d*) Any of the above.

53. Which of the following acceptor impurity is generally used for silicon?
(*a*) Boron (*b*) Gallium (*c*) Indium (*d*) None of the above.

54. Which of the following donor impurity is used for Germanium?
(*a*) Arsenic (*b*) Antimony (*c*) Any of the above (*d*) None of the above.

55. Which of the following acceptor impurity is used for Germanium?
(*a*) Gallium (*b*) Indium (*c*) Any of the above (*d*) None of the above.

56. A *PN* junction
(*a*) conducts current in forward as well as reverse directions
(*b*) offers high resistance in forward as well as reverse directions
(*c*) conducts current in forward direction and offers high resistance in reverse direction
(*d*) offers high resistance in forward direction and conducts current in reverse direction.

**57.** Which of the following are immobile ?

(*a*) Electrons (*b*) Holes (*c*) Ions (*d*) None of the above.

**58.** In *PN* junction, with no external voltage, the electric field between the acceptor and the donor ions is called a

(*a*) barrier (*b*) threshold (*c*) Peak (*d*) path.

**59.** For a silicon *PN* junction, the barrier potential is about

(*a*) 0.001 V (*b*) 0.07 V (*c*) 0.7 V (*d*) 1.7 V.

**60.** For a germanium *PN*-junction, the barrier potential is nearly

(*a*) 0.003 V (*b*) 0.03 V (*c*) 0.3 V (*d*) 1.3 V.

**61.** At a *PN* junction the potential barrier is due to the charges on either side of the junction. These charges are

(*a*) majority carriers (*b*) minority carriers

(*c*) both majority as well as minority carriers

(*d*) fixed donor and acceptor ions.

**62.** In an unbiased *PN* junction, the junction current at equilibrium is

(*a*) feeble, from *P* to *N* side (*b*) feeble, from *N* to *P* side

(*c*) feeble, localised with *P* and *N* junctions

(*d*) zero.

**63.** In Fig. 23.5 the forward characteristics of a silicon diode is represented by

(*a*) figure *A* (*b*) figure *B* (*c*) figure *C* (*d*) figure *D*.

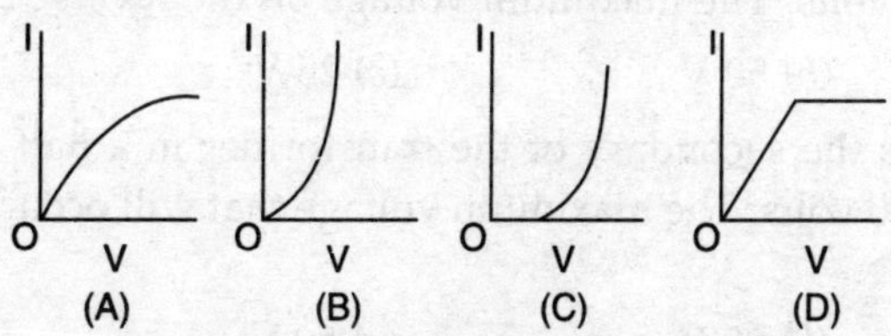


***FIGURE 23.5***

**64.** The reverse bias characteristics of a semiconductor diode is shown in

(*a*) figure *A* (*b*) figure *B* (*c*) figure *C* (*d*) figure *D*.

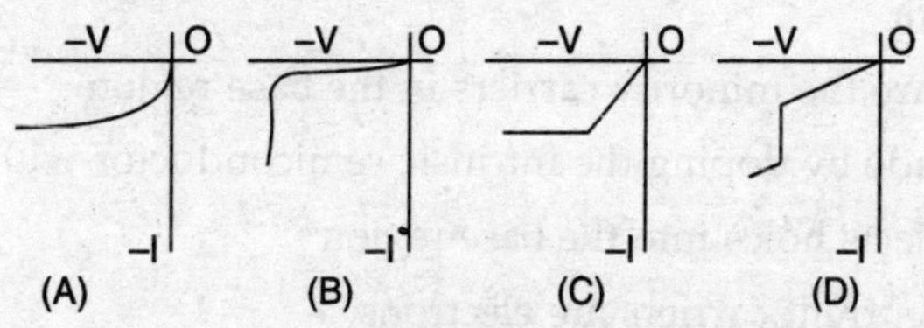


***FIGURE 23.6***

65. Ripple factor is

(a) $\frac{RMS\text{ value}}{DC\text{ value}}$ (b) $\frac{\text{Peak value}}{DC\text{ value}}$ (c) $\frac{\text{Peak value}}{RMS\text{ value}}$ (d) $\frac{RMS\text{ value}}{\text{Peak value}}$.

66. Higher value of ripple factor indicates

(a) poor rectification (b) ideal rectification

(c) rms value equal to peak value (d) none of the above.

67. The best value of rectification efficiency for a half wave rectifier could be around

(a) 99% (b) 80% (c) 60% (d) 40%.

68. The ripple factor for half wave rectified sine wave is

(a) 1.00 (b) 1.21 (c) 1.41 (d) 1.56.

69. The best value of rectification efficiency for a full wave (unfiltered) rectifier could be around

(a) 99% (b) 81% (c) 67% (d) 48.5%.

70. The minimum number of diodes needed for a bridge rectifier is

(a) one (b) two (c) four (d) eight.

71. For a full wave bridge rectifier supplied with 50 Hz ac, the lowest ripple frequency will be

(a) 50 Hz (b) 100 Hz (c) 200 Hz (d) 400 Hz.

72. In a centre tapped full wave rectifier, if 100 V is the peak voltage between the centre-tap and one end of the secondary, then the maximum voltage across the reverse-biased diode will be

(a) 200 V (b) 100 V (c) 88 V (d) 50 V.

73. The voltage across the secondary of the transformer in a halfwave rectifier (without any filter circuit) is 25 volts. The maximum voltage on the reverse biased diode will be

(a) 100 V (b) 50 V (c) 25 V (d) 12.5 V.

74. The voltage across the secondary of the transformer in a half wave rectifier with a shunt capacitor filter is 50 volts. The maximum voltage that will occur on the reverse biased diode will be

(a) 100 V (b) 88 V (c) 50 V (d) 25 V.

75. A zener diode is invariably used with

(a) forward bias (b) reverse bias

(c) either of (a) or (b) above (d) zero bias.

76. In a *PNP* transistor

(a) the electrons are the minority carriers in the base region

(b) the base is made by doping the intrinsic semiconductor with indium

(c) the emitter injects holes into the base region

(d) the principal current carriers are electrons.

77. Which of the following is valid for both *PNP* as well as *NPN* transistors ?

(a) The *EB* junction is forward biased for active operation

(b) The electrons are the minority carriers in the base region

(c) When biased in the active region, current flows into emitter terminal
(d) The emitter injects holes into the base region.

**78.** In a transistor with normal bias
(a) the emitter junction has a low resistance
(b) the emitter junction is reverse biased
(c) the emitter junction offers high resistance
(d) none of the above.

**79.** In the transistor configuration, the arrow head always points in the direction of
(a) positive voltage (b) negative voltage
(c) electron flow in the emitter region (d) hole flow in the emitter region.

**80.** Fig. 23.7 represents a
(a) field effect transistor-*p* channel
(b) *IGFET* with two gates
(c) silicon controlled rectifier
(d) tunnel diode.

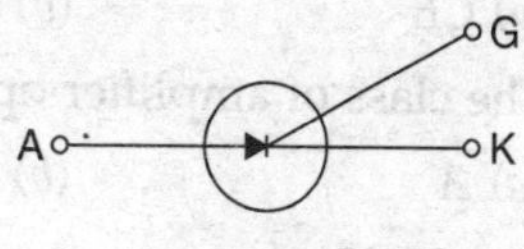

FIGURE 23.7

**81.** Fig. 23.8 represents a
(a) Zener diode
(b) unijunction transistor
(c) diac trigger
(d) none of the above.

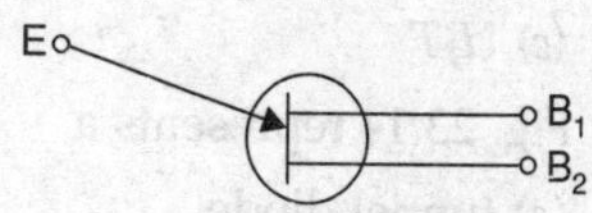

FIGURE 23.8

**82.** Which region of a transistor is lightly doped ?
(a) Base (b) Emitter
(c) Collector (d) All regions are equally doped.

**83.** Which region of a transistor is heavily doped ?
(a) Base (b) Emitter
(c) Collector (d) All regions are equally doped.

**84.** Fig. 23.9 represents a
(a) zener diode
(b) *PNP* transistor
(c) *NPN* transistor
(d) junction field effect transistor.

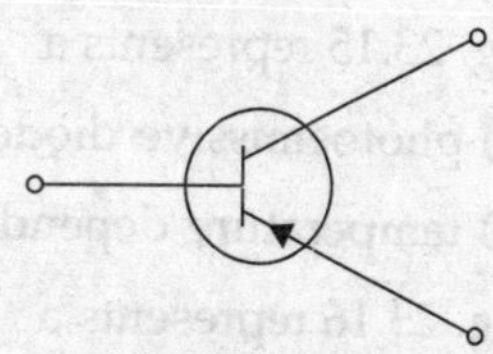
FIGURE 23.9

**85.** Fig. 23.10 represents a
(a) zener diode
(b) silicon controlled rectifier
(c) tunnel diode
(d) junction field effect transistor.

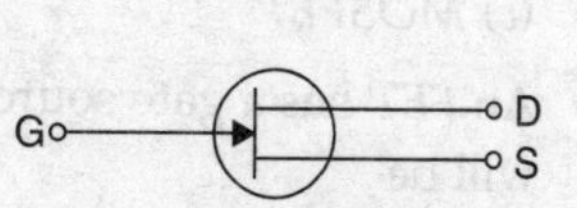

FIGURE 23.10

86. Fig. 23.11 represents a

(a) junction FET
(b) MOSFET
(c) triac
(d) varactor diode.

FIGURE 23.11

87. Fig. 23.12 represents a

(a) zener diode
(b) IGFET
(c) photoconductive cell
(d) photoemissive cell.

FIGURE 23.12

88. The type of circuit that does not need a coupling capacitor is

(a) resistance loaded
(b) transformer coupled
(c) impedance coupled
(d) single-tuned.

89. Which of the amplifier circuit using junction transistors has the best gain ?

(a) CE (b) CC (c) CB (d) all of the above.

90. The class of amplifier operation with least distortion is

(a) A (b) B (c) C (d) $AB_2$.

91. Fig. 23.13 represents a

(a) Triac thyristor
(b) diac triggar diode
(c) UJT
(d) MOSFET.

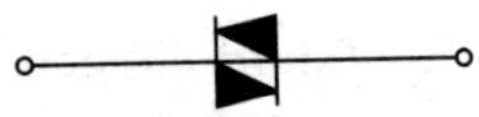

FIGURE 23.13

92. Fig. 23.14 represents a

(a) tunnel diode
(b) temperature dependent diode
(c) capacitive diode
(d) LED.

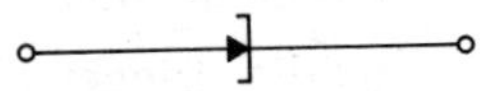

FIGURE 23.14

93. The circuit that would be used for a 455 kHz IF amplifier is

(a) resistance loaded
(b) double tuned transformer
(c) video amplifier
(d) class C.

94. In a RC coupled amplifier, low-frequency response is improved with

(a) lower $R_L$ (b) higher $C_C$ (c) less gaim (d) more bias.

95. Fig. 23.15 represents a

(a) photoemissive diode
(b) photo sensitive diode
(c) temperature dependent diode
(d) capacitive diode.

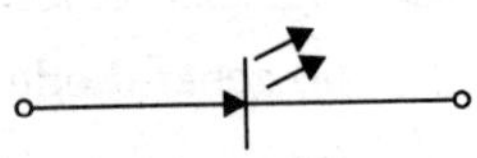

FIGURE 23.15

96. Fig. 23.16 represents a

(a) UJT
(b) photo sensitive diode
(c) MOSFET
(d) none of the above.

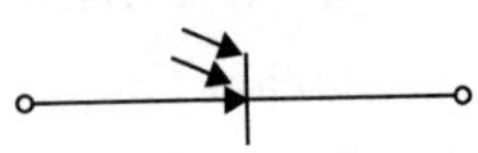

FIGURE 23.16

97. An FET has a gate source bias of – 2 V. The ac input signal is ± 1.2 V. The class of operation will be

(a) A (b) B (c) C (d) $AB_2$.

98. In comparison to *CB* a *CE* amplifier has

(*a*) higher output resistance (*b*) higher current amplification
(*c*) lower current amplification (*d*) lower output resistance.

99. In case of *CE* amplifiers input and output signals are

(*a*) always out of phase (*b*) always in phase
(*c*) always equal in magnitude (*d*) always in phase and equal in magnitude.

100. A transistor in common-base configuration has

(*a*) high input resistance as well as output resistance
(*b*) low input resistance as well as output resistance
(*c*) high input resistance and low output resistance
(*d*) low input resistance and high output resistance.

101. A semiconductor is

(*a*) another name for a transistor or a diode
(*b*) characterized by a very large opposition to the flow of electrical current
(*c*) a material whose conductivity is somewhere between that of a conductor and an insulator
(*d*) a material used for making bipolar transistors, unijunction transistors, field-effect transistors etc.

102. Referring to the atomic structure a semiconductor material has

(*a*) eight electrons in its outermost shell
(*b*) four electrons in its outermost shell
(*c*) two electrons in its outermost shell
(*d*) its outer shell complete

103. Doping of semiconductors is

(*a*) the process of purifying semiconductor materials
(*b*) the process of adding certain impurities to the semiconductor material in controlled amounts
(*c*) the process of converting semiconductor material into some of active device like BJT, FET, UJT etc.
(*d*) one of the steps used in fabrication of ICs.

104. An *N*-type semiconducting material is formed by adding

(*a*) some pentavalent impurity to the semiconductor material
(*b*) some trivalent impurity to the semiconductor material
(*c*) either of the two impurities mentioned above
(*d*) none of these.

105. Most commonly employed trivalent impurities are

(*a*) Boron, Indium (*b*) Arsenic, Antimony
(*c*) Boron, Arsenic (*d*) Indium, Antimony.

**106.** Most commonly employed pentavalent impurities are

(*a*) Boron, Indium (*b*) Aluminium, Antimony

(*c*) Arsenic, Indium (*d*) Arsenic, Antimony

**107.** A *p*-type semiconductor material is formed by adding to the semiconductor controlled amounts of

(*a*) some trivalent impurity (*b*) some pentavalent impurity

(*c*) arsenic (*d*) none of these.

**108.** Referring to the energy level diagram of semiconductor materials, the width of forbidden gap is

(*a*) About 10 eV (*b*) about 100 eV (*c*) about 1 eV (*d*) about 0.1 eV.

**109.** A *p*-type semiconductor material as a whole is

(*a*) positively charged (*b*) negatively charged

(*c*) electrically neutral (*d*) none of these.

**110.** An *N*-type semiconductor material as a whole is

(*a*) electrically neutral (*b*) positively charged

(*c*) negatively charged (*d*) none of these.

**111.** One of the following statements justifies the extensive use of semiconductor materials

(*a*) It is because of their low forbidden energy gap

(*b*) It is because of their resistance value which lies between that of good conductors and insulators

(*c*) It is because of ease of fabrication of semiconductor material into practical active and passive devices

(*d*) It is because of the fact that they exhibit some wide ranging characteristics when certain specific impurities are added to them in controlled amounts.

**112.** Pick the odd-man out (referring to the material used for their fabrication)

(*a*) fixed resistors (*b*) thermistors (*c*) sensistors (*d*) varistors.

**113.** A semiconductor is said to be in thermal equilibrium with its environment if

(*a*) its characteristics remain unaltered with charge in device temperature

(*b*) its characteristics do not change with temperature

(*c*) for any process, there is a reverse process that occurs at the same rate

(*d*) the device temperature is equal to environmental temperature.

**114.** Typical value of forbidden energy gap in germanium is

(*a*) 1 eV (*b*) 1.4 eV (*c*) 10 eV (*d*) 0.67 eV.

**115.** One of the following is not a semiconductor

(*a*) Ga As (Gallium Arsenide) (*b*) Indium

(*c*) Germanium (Ge) (*d*) Silicon (Si).

**116.** Out of the three most commonly employed semiconductor materials

(*a*) Ge (*b*) Si (*c*) GaAs (*d*) None of the above.

**117.** Indicate the false statement

(*a*) The resistivity of a semiconductor material is of the order of $10^{-3}$ ohm cm

(*b*) A semiconductor is any solid or liquid substance whose resistivity lies between $10^{-2}$ ohm-cm and $10^{9}$ ohm-cm

(*c*) Indium is an acceptor impurity

(*d*) Arsenic is a donor impurity.

**118.** An intrinsic semiconductor

(*a*) has a lower value of resistivity among other semiconductors of the same material

(*b*) is one whose electrical properties are essentially those of pure crystal

(*c*) is one whose electrical properties depend upon the nature of impurities added

(*d*) none of these.

**119.** An extrinsic semiconductor is one

(*a*) whose electrical characteristics depend upon nature of impurities added

(*b*) whose electrical characteristics are those of ideal or pure crystal

(*c*) with a higher resistivity than intrinsic of the same material

(*d*) none of these.

**120.** A defect semiconductor

(*a*) is another name for N-type semiconductors

(*b*) is the one suffering from some inherent defects

(*c*) is one whose electrical properties depend upon the departure of the material from stoichiometric structure

(*d*) is one which has a deficiency of charge carriers.

**121.** Semiconductor generation rate is a term

(*a*) used in semiconductor industry indicating the number of semiconductors produced everyday

(*b*) used to indicate time rate of creation of electron-hole pairs in semiconductors

(*c*) used to indicate rise in current in a semiconductor as the voltage is increased

(*d*) none of these.

**122.** Semiconductor junction is

(*a*) a transition between two semiconducting regions

(*b*) produced when a p-type material is joined to an N-type material or vice-versa

(*c*) a region of transition between semiconducting regions of different electrical properties

(*d*) none of these.

**123.** A junction that is formed by adding controlled amounts of an impurity to the melt during crystal growth is termed as

(*a*) doped junction (*b*) alloy junction (*c*) fused junction (*d*) none of these.

124. A junction which is produced by recrystallization on a base crystal from a liquid phase of one or more components and the semiconductor is called

(*a*) doped junction (*b*) alloy junction (*c*) fused junction (*d*) none of these.

125. A grown junction is the one which is formed

(*a*) During growth of a crystal from a melt

(*b*) By the addition of an impurity to the melt during crystal growth

(*c*) By recrystallization technique

(*d*) None of these.

126. An *N-N* junction is a transition

(*a*) between two semiconducting materials having different electrical properties

(*b*) region between two regions having different properties in a *N*-type semiconducting material

(*c*) between two semiconducting materials having identical electrical properties

(*d*) none of these.

127. A *P-P* junction is a transition

(*a*) between two semiconducting materials having different electrical properties

(*b*) between two semiconducting materials having identical electrical properties

(*c*) region between two regions having differential properties in a P-type semiconducting material

(*d*) none of these.

128. In the forward bias condition, the potential of *P*-side with respect to *N*-side is

(*a*) positive (*b*) negative

(*c*) either positive or negative (*d*) none of these.

129. In a semiconductor diode, *P*-side is grounded and *N*-side applied a potential of –2 V. The diode shall

(*a*) conduct (*b*) not conduct

(*c*) conduct partially (*d*) none of these.

130. *P*-side of a semiconductor diode is applied a potential of 2.5 V whereas *N*-side is applied a potential of –5.0 V. The diode shall

(*a*) conduct (*b*) not conduct

(*c*) conduct partially (*d*) breakdown.

131. In a semiconductor diode, *V-I* relationship is such that

(*a*) current varies linearly with voltage

(*b*) current increases exponentially with voltage

(*c*) current varies inversely with voltage

(*d*) none of these.

**132.** Cut-in voltage of a silicon diode is

(*a*) 0.1 V (*b*) 0.2 V (*c*) 0.6 V (*d*) none of these.

**133.** Cut-in voltage of a germanium diode is

(*a*) 0.1 V (*b*) 0.2 V (*c*) 0.6 V (*d*) none of these.

**134.** Peak-inverse-voltage of a semiconductor diode is

(*a*) the maximum voltage a diode can withstand in the forward bias condition

(*b*) the maximum voltage diode can withstand in the reverse bias condition

(*c*) the voltage across the diode in reverse bias condition

(*d*) the voltage across the diode in forward bias condition.

**135.** The static resistance of a diode is

(*a*) its DC opposition to the current flow

(*b*) its opposition to AC flow

(*c*) resistance of the diode when forward biased

(*d*) none of these.

**136.** The dynamic resistance of diode is

(*a*) the resistance of diode when forward biased

(*b*) the resistance of diode when reverse biased

(*c*) the AC opposition to the current flow

(*d*) none of these.

**137.** A semiconductor transistor operates in the active region only when

(*a*) the emitter junction is forward biased and collector junction is reverse biased

(*b*) both emitter and collector junctions are reverse biased

(*c*) both emitter and collector junctions are reverse biased

(*d*) the emitter junction is reverse biased and collector junction is forward biased.

**138.** The capacitance appearing across a reverse biased semiconductor junction

(*a*) increases with increase in the bias voltage

(*b*) decreases with increase in the bias voltage

(*c*) is independent of bias voltage

(*d*) none of these.

**139.** A semiconductor diode used as a rectifier must be operated

(*a*) in the non-linear region

(*b*) in the linear region

(*c*) in the cut-off region

(*d*) none of these.

**140.** A zener diode is generally operated

(*a*) in a reverse-biased mode

(*b*) in a forward-biased mode

(*c*) with a very large value of reverse bias

(*d*) none of these.

**141.** One of the following diodes will produce the highest pulsed power output

(*a*) Varactor (*b*) Gunn (*c*) Schottky (*d*) IMPATT.

**142.** Performance of one of the following diodes is not based upon its negative resistance characteristic

(*a*) Gunn (*b*) IMPATT (*c*) Tunnel (*d*) Backward.

**143.** One of the following microwave diodes is suitable for very low power oscillator only. It is

(*a*) Tunnel (*b*) Gunn (*c*) IMPATT (*d*) LSA.

**144.** One of the following is not used as mixer or detector at microwave frequencies

(*a*) Schottky barrier diode (*b*) Crystal diode

(*c*) Backward-diode (*d*) PIN diode.

**145.** The transferred-electron bulk effect occurs in

(*a*) Germanium (*b*) Silicon (*c*) Gallium arsenide (*d*) None of these.

**146.** The biggest advantage of TRAPATT diode over IMPATT diode is its

(*a*) higher efficiency (*b*) higher output

(*c*) lower noise (*d*) capability to operate at higher frequencies.

**147.** A varactor diode may be advantageous at microwave frequencies (indicate false answer)

(*a*) For electronic tuning (*b*) As an oscillator

(*c*) As a parametric amplifier (*d*) For frequency multiplication.

**148.** If high order frequency multiplication is required from a diode multiplier

(*a*) the resistive cut-off frequency must be high

(*b*) a small value of base resistance is required

(*c*) a step recovery diode must be used

(*d*) none of these.

**149.** Negative resistance is obtained with a Gunn diode because of

(*a*) electron transfer to a less mobile energy level

(*b*) avalanche breakdown with the high voltage gradient

(*c*) tunneling across the junction

(*d*) electron domains forming at the junction.

**150.** For making Gunn diodes, preferred semiconductor material is

(*a*) Silicon (*b*) Germanium (*c*) Gallium arsenide (*d*) None of these.

151. The LSA diode differs from the Gunn diode in that the former
   (*a*) is a true diode
   (*b*) does not require negative resistance
   (*c*) need not be made from GaAs as it is in Gunn diode
   (*d*) prevents the formation of domains.

152. Pick the odd-man out
   (*a*) Backward diode   (*b*) Schottky-barrier diode
   (*c*) Hot-carrier diode   (*d*) Laser diode.

153. The biggest advantage of gallium antimonide over germanium for tunnel diode use is that the former has
   (*a*) lower noise   (*b*) higher iron mobility
   (*c*) larger voltage swing   (*d*) none of these.

154. For Gunn diodes, gallium arsenide is preferred to silicon because
   (*a*) of higher ion mobility
   (*b*) of lower noise at even highest frequencies
   (*c*) of capability to handle higher power densities
   (*d*) none of these.

155. The tunnel diode
   (*a*) has a tiny hole through its centre to facilitate tunneling
   (*b*) is a point contact diode with a very high value of reverse resistance
   (*c*) uses a high doping level to provide a narrow junction
   (*d*) none of these.

## ANSWERS

| | | | | | | |
|---|---|---|---|---|---|---|
| 1. (*a*) | 2. (*d*) | 3. (*b*) | 4. (*b*) | 5. (*c*) | 6. (*c*) | 7. (*d*) |
| 8. (*d*) | 9. (*b*) | 10. (*a*) | 11. (*b*) | 12. (*d*) | 13. (*b*) | 14. (*b*) |
| 15. (*c*) | 16. (*c*) | 17. (*d*) | 18. (*b*) | 19. (*d*) | 20. (*c*) | 21. (*d*) |
| 22. (*d*) | 23. (*d*) | 24. (*a*) | 25. (*b*) | 26. (*d*) | 27. (*d*) | 28. (*a*) |
| 29. (*c*) | 30. (*b*) | 31. (*d*) | 32. (*c*) | 33. (*b*) | 34. (*b*) | 35. (*a*) |
| 36. (*b*) | 37. (*a*) | 38. (*c*) | 39. (*b*) | 40. (*d*) | 41. (*a*) | 42. (*b*) |
| 43. (*a*) | 44. (*a*) | 45. (*b*) | 46. (*c*) | 47. (*d*) | 48. (*c*) | 49. (*b*) |
| 50. (*a*) | 51. (*b*) | 52. (*a*) | 53. (*a*) | 54. (*c*) | 55. (*c*) | 56. (*c*) |
| 57. (*c*) | 58. (*a*) | 59. (*c*) | 60. (*c*) | 61. (*d*) | 62. (*d*) | 63. (*c*) |
| 64. (*b*) | 65. (*a*) | 66. (*a*) | 67. (*d*) | 68. (*b*) | 69. (*b*) | 70. (*c*) |
| 71. (*b*) | 72. (*d*) | 73. (*c*) | 74. (*a*) | 75. (*b*) | 76. (*c*) | 77. (*a*) |

| | | | | | | |
|---|---|---|---|---|---|---|
| **78.** (*a*) | **79.** (*d*) | **80.** (*c*) | **81.** (*b*) | **82.** (*a*) | **83.** (*b*) | **84.** (*b*) |
| **85.** (*d*) | **86.** (*d*) | **87.** (*a*) | **88.** (*b*) | **89.** (*a*) | **90.** (*a*) | **91.** (*b*) |
| **92.** (*a*) | **93.** (*b*) | **94.** (*b*) | **95.** (*a*) | **96.** (*b*) | **97.** (*a*) | **98.** (*b*) |
| **99.** (*a*) | **100.** (*d*) | **101.** (*c*) | **102.** (*b*) | **103.** (*b*) | **104.** (*a*) | **105.** (*a*) |
| **106.** (*d*) | **107.** (*a*) | **108.** (*c*) | **109.** (*c*) | **110.** (*a*) | **111.** (*d*) | **112.** (*a*) |
| **113.** (*c*) | **114.** (*d*) | **115.** (*b*) | **116.** (*c*) | **117.** (*c*) | **118.** (*b*) | **119.** (*a*) |
| **120.** (*c*) | **121.** (*b*) | **122.** (*c*) | **123.** (*a*) | **124.** (*c*) | **125.** (*a*) | **126.** (*b*) |
| **127.** (*c*) | **128.** (*a*) | **129.** (*a*) | **130.** (*a*) | **131.** (*b*) | **132.** (*c*) | **133.** (*b*) |
| **134.** (*b*) | **135.** (*a*) | **136.** (*c*) | **137.** (*a*) | **138.** (*b*) | **139.** (*a*) | **140.** (*a*) |
| **141.** (*d*) | **142.** (*d*) | **143.** (*a*) | **144.** (*d*) | **145.** (*c*) | **146.** (*a*) | **147.** (*b*) |
| **148.** (*c*) | **149.** (*a*) | **150.** (*c*) | **151.** (*d*) | **152.** (*a*) | **153.** (*c*) | **154.** (*d*) |
| **155.** (*c*). | | | | | | |

Chapter 24

# AMPLIFIERS

Amplification means increasing the amplitude of a desired ac signal voltage or current such as an audio signal for sound or a video signal for a television picture. The amplifier allows a small input signal to control a larger amount of power in the output circuit. Although the output signal is a copy of the input signal, but it has more amplitude.

Transistors are used as the amplifiers in most circuits. In addition, resistors, inductors, and capacitors are required to form complete amplifier circuits. They provide paths for the input and output signals.

## 1. Resistance Loaded Amplifier

This amplifier is often used for audio circuits. In this $R_L$ allows the same amount of gain for a wide range of audio frequencies. Since $R_L$ has the same resistance at all audio frequencies, the amplifier has uniform gain. However, $R_L$ results in high dc voltage drop.

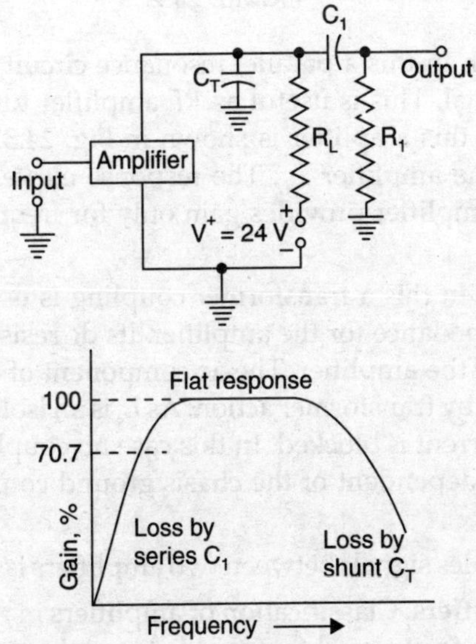


*FIGURE 24.1*

At low frequencies the output is reduced because of the reactance of $C_1$. In this $X_c$ increases as the frequency decreases. More of the amplifier output voltage is across $C_1$. At a frequency, where $X_c = R$ the gain is 70.7% of the flat response gain.

At high frequencies amplifier gain decreases because of the shunting effect of the stray, distributed capacitances. The total parallel capacitance $C_T$ is typically 10 to 40 picofarads. Even that small value of $C$ can bypass high frequencies around a high value of $R_L$

**Impedance Coupling.** In this an inductance $L$ instead of $R_L$ has been used for load impedance of the amplifier. In this again $R_1C_1$ coupling circuit is still needed to block $V^+$ from the next circuit. An iron core audio choke can be used for an *AF* amplifier, an air core choke can be used for an *RF* amplifier.

Choke offers the advantage of low dc resistance with high ac impedance. Low resistance means a small *IR* drop, which allows most of $V^+$ to be available at the amplifier. The high ac impedance for the signal allows high gain. Since $Z_L$ varies with frequency, this amplifier does not have as uniform a frequency response.

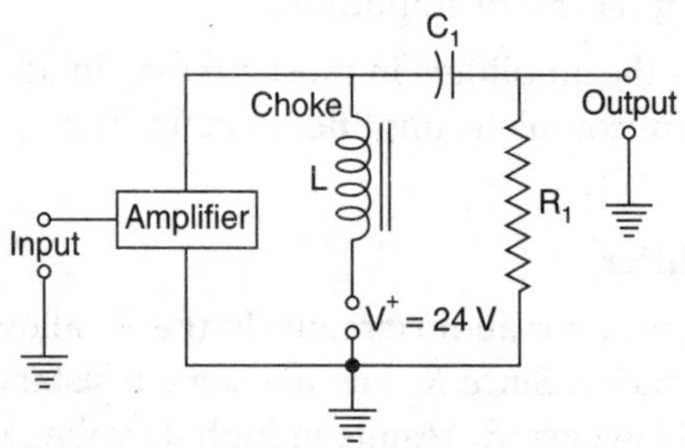


*FIGURE 24.2*

**Single tuned amplifier.** In this a parallel resonance circuit is used to provide the required high impedance for an ac signal. This is useful as *RF* amplifier tuned to a specific frequency. The frequency response curve for this amplifier is shown in Fig. 24.3. In this the voltage gain of the amplifier is proportional to the amplifier $Z_L$. The response curve of this amplifier resembles the resonance curve. The tuned amplifier provides gain only for frequencies at or near the resonance frequencies.

**Transformer coupling.** In this a transformer coupling is used to drive the loudspeaker. $L_D$ provides the primary load impedance for the amplifier. Its dc resistance is low, which allows most of the $V^+$ supply to be used at the amplifier. The ac component of the signal current in $L_D$ induces the desired signal voltage in $L_s$ by transformer action. As $L_s$ is an isolated winding, the dc component of the primary voltage and current is blocked. In this case no coupling capacitor is needed. Also ac voltage in the secondary is independent of the chasis ground connections. One side of $L_s$ may or may not be grounded.

A transformer that couples signals between two amplifiers is called an interstage transformer.

**Class A, B and C amplifiers.** Classification of amplifiers in *A*, *B* and *C* categories is based on the percentage of the cycle of input signal that is able to produce output current.

**Class A.** In class *A* amplifiers, the output current $I_0$ flows for full cycle of 360° of the input signal. An audio amplifier stage operates this way to follow the signal variations without too much distortion. In this distortion is less (5 to 10%) but efficiency is also low (20 to 40%).

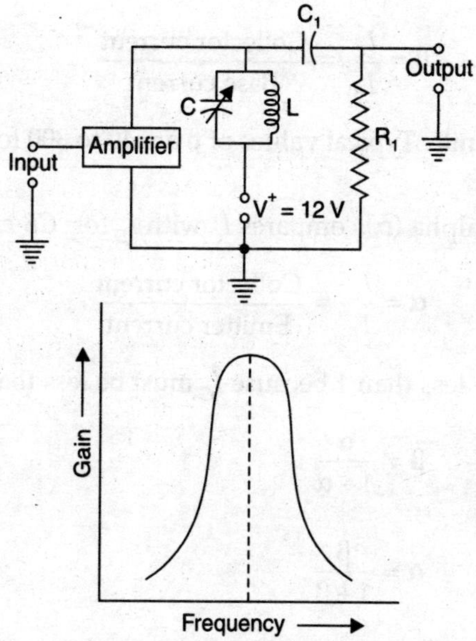


*FIGURE 24.3*

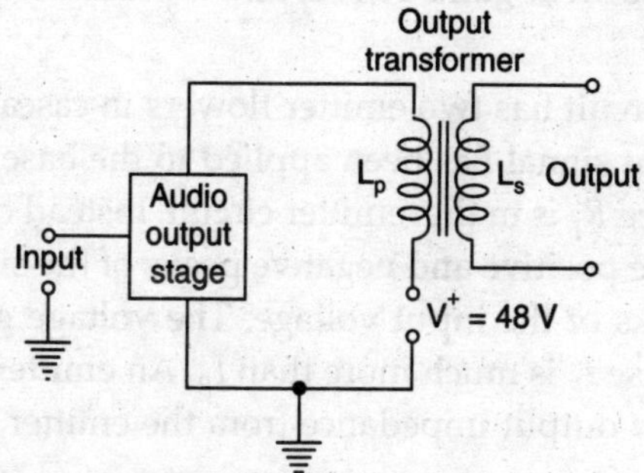


*FIGURE 24.4*

**Class B.** In this the output current flows for 180° or approximately one half of the input cycle. The dc bias is at or near the cut-off value. Class *B* operation requires more dc bias and more ac signal drive than class *A*. This has distortion and efficiency values between those for class *A* and *C*.

**Class C.** In this the output current flows for less than onehalf the input cycle. This may be for 120° or so during the positive half cycle of input. Because of its high efficiency, class C operation is used mainly for tuned *RF* power amplifiers. In this efficiency is highest (about 80%) and this allows the greatest ac power output but with the most distortion. These are generally used as *RF* amplifiers with tuned circuit in output.

**Class AB.** This offers a compromise between the low distortion of class *A* and the higher power of class *B*. It is generally used for push pull audio amplifiers.

**Characteristics.** The beta and alpha values determine how good a junction transistor is as an amplifier. β signifies how the base current controls the amount of collector current. Thus

$$\beta = \frac{I_C}{I_B} = \frac{\text{Collector current}}{\text{Base current}}.$$

The value of β has no units. Typical values of β are 40 to 300 for small signal transistors and 10 to 30 for power transistors.

Another characteristic alpha (α) compares $I_C$ with $I_E$ for *CB* circuit. Thus

$$\alpha = \frac{I_C}{I_E} = \frac{\text{Collector current}}{\text{Emitter current}}.$$

The value of α must be less than 1 because $I_C$ must be less than $I_E$, α and β are related as

$$\beta = \frac{\alpha}{1-\alpha}$$

or
$$\alpha = \frac{\beta}{1+\beta}.$$

**Amplifiers in cascade.** When there are junction transistors in the *CE* circuit, the amplified output from the collector drives the base electrode of the next stage. This is known as cascading of amplifier. Thus is amplifiers with gains $A_1$, $A_2$ and $A_3$ are cascaded, the overall gain will be $(A_1 \times A_2 \times A_3)$.

**Darlington pair.** This circuit has two emitter flowers in cascade. One stage of *CC* circuit is shown in Fig. 24.5, where input signal has been applied to the base and the output is taken from the emitter. The load impedance $R_L$ is in the emitter circuit, instead of the collector circuit. These is no phase inversion, because the positive and negative peaks of the signal occur at the same time as the positive and negative peaks of the input voltage. The voltage gain is slightly less than 1 but current gain can be high because $I_E$ is much more than $I_B$. An emitter follower provides high input impedance at the base and low output impedance from the emitter circuit.

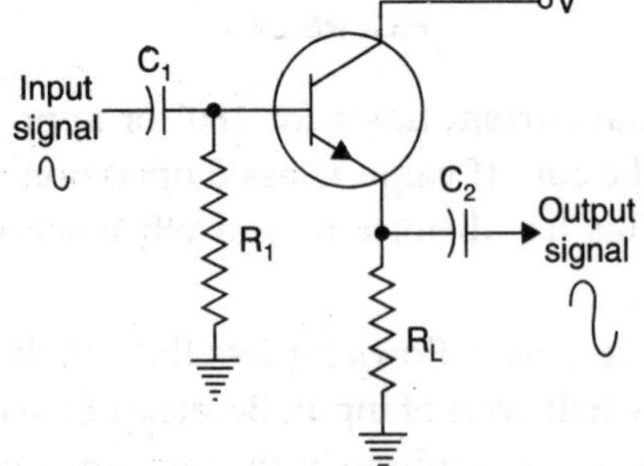


*FIGURE 24.5*

The advantage of Darlington pair is higher input impedance, lower output impedance and more current gain as compared to single emitter follower.

**Push pull amplifier.** A push pull amplifier I uses two stages to operate on opposite half cycles of the signal. The input signals are equal but 180° out of phase. The amplified output signals are also 180° out of phase.

In complementary symmetry, *PNP* and *NPN* transistors are used for push pull amplification, but with a single ended input and output.

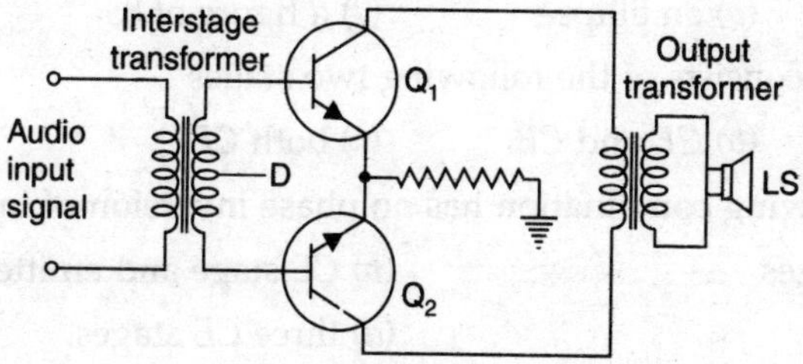


*FIGURE 24.6*

Quasi-complementary symmetry means the driver stages are *PNP* and *NPN* but both output stages are *NPN*.

## OBJECTIVE TYPE QUESTIONS

**1.** In case of *CE* amplifiers, the output voltage wave form is
(*a*) in phase with input voltage wave form
(*b*) always lagging by 90° with respect to input voltage wave form
(*c*) always lagging by 180° with respect to input voltage wave form
(*d*) always leading by 90° with respect to input voltage wave form.

**2.** Which of the following is the typical value of input impedance for *CE* amplifier?
(*a*) 10 Ω (*b*) 150 Ω (*c*) 50 k Ω (*d*) 1 M Ω.

**3.** How many cascaded stages of *CE* amplifiers will result in polarity inversion of the input signal?
(*a*) two (*b*) three
(*c*) any of the above
(*d*) polarity inversion is not possible in *CE* amplifiers.

**4.** In an amplifier the purpose of coupling capacitor is to
(*a*) control the bandwidth (*b*) prevent dc mixing with input
(*c*) control the output (*d*) match the impedance.

**5.** In an amplifier if the output current flows for 120° of the input cycle, the class of amplifier will be
(*a*) *A* (*b*) *B* (*c*) *C* (*d*) *AB*.

**6.** A silicon *NPN* transistor has cut-off voltage of 0.5 V when bias is + 0.6 V. If the ac signal is ± 50 mV, the class of amplifier will be
(*a*) *A* (*b*) *B* (*c*) *C* (*d*) *AB*.

**7.** A transistor amplifier with 85% efficiency is expected to be
(*a*) class *A* (*b*) class *B* (*c*) class *C* (*d*) class *AB*.

8. In case the input and output signals of a *CE* amplifier are fed to an oscilloscope, the lissajous figure will be
(*a*) a circle (*b*) an ellipse (*c*) a figure of 8 (*d*) a straight line.

9. A Darlington pair consists of the following two stages
(*a*) *CE* and *CC* (*b*) *CE* and *CB* (*c*) both *CE* (*d*) both *CC*.

10. Which of the following combination has no phase inversion of input signal?
(*a*) *CE* and *CC* stages (*b*) *CE* stage and emitter follower
(*c*) two *CE* stages (*d*) three *CE* stages.

11. For a good voltage amplifier, its input impedance should be
(*a*) low as compared to the resistance of the source
(*b*) high as compared to the resistance of the source
(*c*) always inductive
(*d*) always capacitive.

12. Class *A* amplifiers are preferred when
(*a*) minimum distortion is required (*b*) maximum voltage gain is required
(*c*) output signal is desired in phase with input signal
(*d*) signals having ac super imposed on dc are to be amplified.

13. An ideal voltage amplifier should have
(*a*) low input impedance and high output impedance
(*b*) high input impedance and low output impedance
(*c*) high input impedance and high output impedance
(*d*) low input impedance and low output impedance.

14. If three cascaded stages of amplifier have gains 10, 20 and 30, the overall gain will be
(*a*) 30 (*b*) 60 (*c*) 600 (*d*) 6000.

15. An amplifier circuit with gain of 200 has output voltage as 2 V. The input voltage will be
(*a*) 400 V (*b*) 2 V (*c*) 20 mV (*d*) 10 mV.

16. In class *AB* amplifiers, the current flows through the active device for
(*a*) full duration of input cycle (*b*) half duration of input cycle
(*c*) more than half but less than full cycle duration
(*d*) less than half of the duration of input cycle.

17. In an amplifier for maximum power transfer, the load resistance should be
(*a*) as small as possible (*b*) as large as possible
(*c*) equal to half the value of output resistance of amplifier
(*d*) equal to the value of output resistance of amplifier.

18. The power gain of an amplifier is *P* and the voltage gain is *Q*. The current gain will be
(*a*) $PQ$ (*b*) $P + Q$ (*c*) $\frac{P}{Q}$ (*d*) $\frac{Q}{P}$.

19. When the power level of an amplifier is changed from 10 W to 20 W, the equivalent dB gain will be

(*a*) 3 dB (*b*) 10 dB (*c*) 20 dB (*d*) 30 dB.

20. Which of the following configuration has the highest current gain?

(*a*) Common base (*b*) Common emitter

(*c*) Common collector (*d*) All configurations have identical gain.

21. One octave above 500 Hz will be

(*a*) 501 Hz (*b*) $e^{506}$ Hz (*c*) $\log_e$ (500) dB (*d*) 1000 Hz.

22. The range of audio frequencies is

(*a*) 1 Hz to $10^6$ Hz (*b*) 16 Hz to 16 kHz

(*c*) 16 kHz to 16 MHz (*d*) 160 kHz to 160 MHz.

23. A woman's speaking voice produces sound with frequencies around

(*a*) 25 Hz (*b*) 250 Hz (*c*) 25 kHz (*d*) 250 kHz.

24. A third harmonic of 400 Hz is

(*a*) 300 Hz (*b*) 400 Hz (*c*) 600 Hz (*d*) 1200 Hz.

25. Treble tone corresponds to

(*a*) high frequencies (*b*) low frequencies (*c*) high amplitudes (*d*) low amplitudes.

26. Typical loudspeaker impedance rating is

(*a*) 0.1 Ω (*b*) 8 Ω (*c*) 120 Ω (*d*) 5000 Ω.

27. Typical output from a magnetic microphone is

(*a*) one microvolt (*b*) one millivolt (*c*) 1 volt (*d*) 1 kV.

28. Which of the following would apply to the audio power output stage?

(*a*) no distortion (*b*) class C operation

(*c*) low level signal (*d*) complementary symmetry.

29. The maximum efficiency a class C amplifier is

(*a*) 100% (*b*) 80% (*c*) 60% (*d*) 43%.

30. Which of the following will be preferred for high gain?

(*a*) Darlington amplifier (*b*) Push pull amplifier

(*c*) Cascode amplifier (*d*) Cascade amplifier.

31. In a *RC* coupled amplifier the component producing distortion is

(*a*) power supply (*b*) resistor (*c*) capacitor (*d*) transistor itself.

32. Which capacitor is preferred for coupling capacitor in a *RC* coupled amplifier using triodes ?

(*a*) Electrolytic (*b*) Mica

(*c*) Paper (*d*) Any of the above.

33. The advantage of push pull amplifier is
(*a*) cancellation of even harmonics
(*b*) can be used for class *B* or class *AB* operation
(*c*) the magnetic saturation of the core by dc current does not occur
(*d*) all of the above.

34. The gain of an amplifier with feed back is given by
(*a*) $\frac{A}{1+\beta A}$ (*b*) $\frac{A}{1-\beta A}$ (*c*) $\frac{\beta}{1-\beta A}$ (*d*) $\frac{\beta}{1+\beta A}$.

35. What is the dB gain for an increase in power level from 12 to 24 W?
(*a*) 2 (*b*) 3 (*c*) 20 (*d*) 231.

36. A resistive attenuator reduces a 10 mW input to 5 mW. The dB loss will be
(*a*) 3 dB (*b*) 2 dB (*c*) – 3 dB (*d*) – 30 dB.

37. A circuit consists of two transistors connected in series with the dc supply voltage. This circuit is known as
(*a*) Push pull amplifier (*b*) Differential pair
(*c*) Complementary symmetry (*d*) Stacked $V^+$.

38. Which two transistors are used for complementary symmetry?
(*a*) *PNP* and *NPN* (*b*) both *PNP* (*c*) both *NPN* (*d*) both *FET*.

39. Negative feedback is used in audio amplifiers to
(*a*) increase gain (*b*) increase distortion
(*c*) for class C operation (*d*) reduce distortion.

40. The advantage of negative feedback is
(*a*) reduction in noise (*b*) extended bond width
(*c*) reduced harmonic distortion (*d*) all of the above.

41. Which of the following configuration can be used as a buffer?
(*a*) *CC* (*b*) *CB* (*c*) *CE* (*d*) all of the above.

42. Which configuration is preferred as a constant current source in sweep circuits?
(*a*) *CC* (*b*) *CB* (*c*) *CE* (*d*) any of the above.

43. The stability factor *S* for *CE* amplifier using a fixed bias is
(*a*) $\beta(1+\beta)$ (*b*) $\frac{\beta}{1+\beta}$ (*c*) $\frac{\beta}{1-\beta}$ (*d*) $\frac{1}{\beta}-\frac{1}{\beta^2}$.

44. In general conversion efficiency of an amplifier lies between
(*a*) 90% and 99% (*b*) 75% and 99% (*c*) 25% and 90% (*d*) 10% and 35%.

**Questions 45 to 48 refer to data given below:**

A sinusoidal signal $V_s = 1.75 \sin 600t$ is fed to an amplifier. The resulting output current is $i_0 = 15 \sin 600t + 1.5 \sin 1200t + 1.2 \sin 1800t + 0.5 \sin 2400t$.

**45.** The second harmonic distortion is

(*a*) 3% (*b*) 5% (*c*) 8% (*d*) 10%.

**46.** The third harmonic distortion is

(*a*) 12.5% (*b*) 8% (*c*) 4% (*d*) 3.33%.

**47.** The fourth harmonic distortion is

(*a*) 1% (*b*) 2.5% (*c*) 3.33% (*d*) 5%.

**48.** The percentage increase in power because of distortion is

(*a*) 1.75% (*b*) 3.33% (*c*) 6.66% (*d*) 12.5%.

**49.** Which of the following is a unipolar diode?

(*a*) Zener diode (*b*) Step recovery diode

(*c*) Schottky diode (*d*) All of the above.

**50.** Which of the following is a bipolar device?

(*a*) Tunnel diode (*b*) Gunn diode (*c*) *FET* (*d*) *UJT*.

**51.** Which of the following diode has negative resistance region?

(*a*) Tunnel diode (*b*) *PIN* diode (*c*) Zener diode (*d*) GoAs as diode.

**52.** Which of the following diode does not show negative resistance region?

(*a*) Gunn diode (*b*) Zener diode (*c*) Tunnel diode (*d*) All of the above.

**53.** The maximum efficiency of class *B* amplifier is

(*a*) 78.5% (*b*) 66% (*c*) 50% (*d*) 40%.

**54.** The maximum efficiency of class *A* amplifier is

(*a*) 100% (*b*) 81.8% (*c*) 50% (*d*) 40%.

**55.** Negative feedback generally results in

(*a*) decrease in bandwidth (*b*) increase in noise

(*c*) increase in harmonic distortion (*d*) decrease in harmonic distortion.

**56.** The main source of harmonic distortion in amplifiers is

(*a*) positive feedback (*b*) noisy transformer

(*c*) defective resistors (*d*) non-linearity in active device.

**57.** Which harmonics are balanced out in push-pull amplifiers?

(*a*) Odd harmonics (*b*) Even harmonics

(*c*) Both odd as well as even harmonics (*d*) Neither odd nor even harmonics.

**58.** A video amplifier is expected to amplify frequencies in the range

(*a*) 1 Hz to 5 kHz (*b*) 1 Hz to 5 MHz

(*c*) 5 kHz to 5 MHz (*d*) 5 MHz to 5 GHz.

**59.** Which of the following statement is false?

(*a*) Emitter following is the same as a common base amplifier circuit

(*b*) Negative feedback in an amplifier increases the stability of its voltage gain

(*c*) By introducing negative feedback the upper cut off frequency of an amplifier is increased

(*d*) None of the above.

**60.** Which of the following is a unipolar device

(*a*) *FET* (*b*) Zener diode (*c*) *PNP* transistor (*d*) All of the above.

**61.** *AF* amplifier can amplify frequencies in the range

(*a*) 15 Hz to 15 kHz (*b*) 15 kHz to 150 kHz

(*c*) 150 kHz to 15 MHz (*d*) 15 MHz to 15 GHz.

**62.** The slope of the skirts on an *RF* response curve determines

(*a*) sensitivity (*b*) selectivity

(*c*) both (*a*) and (*b*) above (*d*) none of the above.

**63.** An *RF* amplifier has a gain of 10 at 100 per cent response. What will be the gain at edge frequencies – 3 dB down?

(*a*) 2.02 (*b*) 3.03 (*c*) 5.05 (*d*) 7.07.

**64.** The phenomenon of thermal run away is associated with

(*a*) Vacuum tubes (*b*) Transistors (*c*) Gas diode (*d*) All of the above.

**65.** Tuned voltage amplifiers are generally not used in

(*a*) TV (*b*) radio receivers

(*c*) public address system (*d*) all of the above.

**66.** Which of the following statement is false?

In *RF* amplifiers

(*a*) A crystal filter has sharper skirt selectivity than a resonant *LC* circuit

(*b*) A transducer can convert mechanical energy into electrical energy

(*c*) A monolithic crystal filter is smaller than a filter with discrete components

(*d*) None of the above.

**67.** The gain of an *RF* amplifier with a tuned *LC* circuit for the collector is maximum at the resonant frequency because of

(*a*) low *Q* (*b*) high emitter resistance

(*c*) series resonance (*d*) parallel resonance.

**68.** Which of the following circuit can operate class *AB* for audio power output?

(*a*) Darlington pair (*b*) Push pull

(*c*) Emitter follower (*d*) Cascade.

**69.** A radio frequency signal contains frequencies of 880 Hz, 870 Hz and 860 Hz. For amplifying this signal, which amplifier will be preferred?

(a) Push pull amplifier (b) AF amplifier
(c) RC coupled amplifier (d) Tuned voltage amplifier.

**70.** The Q of a single tuned LC circuit is lower when its
(a) capacitance is reduced (b) series $r_s$ is reduced
(c) shunt $r_d$ is reduced (d) none of the above.

**71.** The response curve of a double-tuned circuit shows two peaks when the transformer has
(a) loose coupling (b) critical coupling
(c) light coupling (d) excessive damping.

**72.** Which of the following amplifier has the highest input impedance?
(a) Cascade amplifier (b) Darlington pair amplifier
(c) Cascaded amplifier (d) Boot strap darlington amplifier.

**73.** The negative feedback in an amplifier
(a) increases voltage gain (b) reduces voltage gain
(c) does not affect voltage gain.

**74.** Which of the following method cannot be used to increase the bandwidth for cascaded amplifier stages?
(a) A shunt damping resistor across each tuned circuit
(b) Stagger tuning
(c) Double tuned transformers with tighter coupling
(d) Wave traps in each stage.

**75.** Which of the following circuit has the greatest bandwidth?
(a) frequency = 1 MHz; Q = 10
(b) frequency = 455 kHz; Q = 100
(c) frequency = 1 MHz; Q = 100
(d) frequency = 50 MHz; Q = 50.

**76.** The transferred-electron bulk effect occurs in
(a) Silicon (b) Germanium
(c) Gallium arsenide (d) Metal semiconductor junction.

**77.** An ideal power supply has
(a) very small output resistance (b) very large output resistance
(c) very large input resistance (d) zero internal resistance.

**78.** Bleeder resistance is often connected
(a) across the input of power supply
(b) across the output of power supply
(c) in series with the output of power supply
(d) in the mid way.

79. Which of the following statement is false?

(a) A capacitor – input filter has a higher dc voltage output than a choke-input filter

(b) Excessive 50 Hz ripple in the $V^+$ supply for an audio amplifier can cause hum in the output

(c) A typical value of ripple factor for a *H* type filter is 40%

(d) None of the above.

80. If the dc output voltage of a power supply drops from 48 V with no load to 46 V at full load, the percentage regulation is

(a) 2% (b) 4% (c) 4.2% (d) 8.4%.

81. Excessive hum is caused by

(a) an open filter choke (b) capacitor

(c) wire wound resistor (d) all of the above.

82. If a power supply gives no $V^+$ output even with full ac input, the defect could be

(a) defective rectifier (b) open output filter capacitor

(c) shorted output filter capacitor (d) open filter choke.

83. Which of the following type of noise does not occur in transistors?

(a) Shot noise (b) Flicker noise

(c) Partition noise (d) Resistance noise.

84. Indicate the noise whose source is in a category different from that of the other three

(a) Solar noise (b) Galactic noise

(c) Cosmic noise (d) Atmospheric noise.

85. Which of the following configuration can be used for matching low impedance source to high impedance load?

(a) *CE* (b) *CC* (c) *CB* (d) any of the above.

86. Heat sinks are used in power amplifier circuits primarily to increase

(a) the output power (b) the voltage gain

(c) collector dissipation rating of the transistor

(d) dissipation of energy of free electrons.

87. When both the primary and secondary resonant circuits of a double tuned transformer have a *Q* of 50, the value of critical coupling will be

(a) 50 (b) 5 (c) 0.5 (d) 0.2.

88. Which of the following transistors can be used in enhancement mode?

(a) *NPN* (b) *UJT* (c) *JFET* (d) *MOSFET*.

89. In high frequency region, a amplifier behaves like a

(a) high pass filter (b) band pass filter (c) low pass filter (d) none of the above.

90. A *IGFET* is a

(a) linear device (b) logarithmic device

(c) square law device (d) half power device.

91. Staggered tuned amplifiers generally use
(a) CC configuration (b) CB configuration
(c) CE configuration (d) JEET.

92. Which of the following amplifier has output closer to the outs put of a half wave rectifier?
(a) Class A amplifier (b) Class B amplifier
(c) Class C amplifier (d) None of the above.

93. The bandwidth of an amplifier can be reduced by
(a) Miller effect compensation (b) Pole zero compensation
(c) Dominant pole compensation (d) Lead compensation.

94. In a full wave voltage doubler circuit, the output has a ripple frequency
(a) 50 Hz (b) half the ripple frequency of input
(c) same as the ripple frequency of input (d) double the ripple frequency of input.

95. Which of the following noise becomes of great importance at high frequencies?
(a) Shot noise (b) Agitation noise (c) Flicker noise (d) Transit-time noise.

96. Which of the following statement is false?
(a) Pentodes are generally noisier than triodes in the same connection
(b) Partition noise voltage is independent of the bandwidth
(c) Thermal noise is independent of the frequency at which it is measured
(d) Man made noise is usually of the impulse type.

97. If the plate supply voltage for a plate modulated class C amplifier is E volts the maximum plate-cathode voltage could be almost as high as
(a) $E$ (b) $\frac{E}{2}$ (c) $2E$ (d) $4E$.

98. Leak type bias is used in a plate modulated class C amplifier to
(a) increase the bandwidth (b) prevent tuned circuit damping
(c) prevent excessive grid current (d) none of the above.

99. Which of the following diode will produce the highest pulsed power output?
(a) Varactor (b) Gunn (c) RIMPATT (d) Schottky barrier.

100. A diac is a
(a) 2 terminal unidirectional switch (b) 2 terminal bidirectional switch
(c) 3 terminal bidirectional switch (d) 4 terminal bidirectional switch.

## ANSWERS

| | | | | | | |
|---|---|---|---|---|---|---|
| 1. (c) | 2. (c) | 3. (b) | 4. (b) | 5. (c) | 6. (a) | 7. (c) |
| 8. (d) | 9. (d) | 10. (c) | 11. (b) | 12. (a) | 13. (b) | 14. (d) |

15. (*d*) 16. (*c*) 17. (*d*) 18. (*c*) 19. (*a*) 20. (*c*) 21. (*d*)
22. (*b*) 23. (*b*) 24. (*d*) 25. (*a*) 26. (*b*) 27. (*b*) 28. (*d*)
29. (*a*) 30. (*d*) 31. (*d*) 32. (*c*) 33. (*d*) 34. (*b*) 35. (*b*)
36. (*c*) 37. (*d*) 38. (*a*) 39. (*d*) 40. (*d*) 41. (*a*) 42. (*b*)
43. (*a*) 44. (*c*) 45. (*d*) 46. (*b*) 47. (*c*) 48. (*a*) 49. (*c*)
50. (*a*) 51. (*a*) 52. (*b*) 53. (*a*) 54. (*c*) 55. (*d*) 56. (*d*)
57. (*a*) 58. (*b*) 59. (*a*) 60. (*a*) 61. (*a*) 62. (*b*) 63. (*d*)
64. (*b*) 65. (*c*) 66. (*d*) 67. (*d*) 68. (*b*) 69. (*d*) 70. (*c*)
71. (*b*) 72. (*d*) 73. (*b*) 74. (*d*) 75. (*d*) 76. (*c*) 77. (*d*)
78. (*d*) 79. (*c*) 80. (*c*) 81. (*b*) 82. (*a*) 83. (*c*) 84. (*d*)
85. (*c*) 86. (*c*) 87. (*d*) 88. (*d*) 89. (*c*) 90. (*c*) 91. (*b*)
92. (*b*) 93. (*c*) 94. (*d*) 95. (*d*) 96. (*b*) 97. (*d*) 98. (*c*)
99. (*c*) 100. (*b*).

Chapter **25**

# COMMUNICATION ENGINEERING

A transmitter sends out an *RF* signal to the receiver. Two features of the transmitter are appreciable *RF* power output and a specific frequency of the transmitted *RF* signal. The more is the power in the radiated signal, the greater is the transmission distance. Further more, the frequency of the *RF* signal is in an assigned band.

The heart of any transmitter is a tuned *RF* oscillator stage that generates the *RF* carrier wave. In addition, transmitters usually include amplifiers for the oscillator output and circuits for modulating the *RF* wave.

**Modulation.** Modulation is the process of varying the *RF* output in step with a lower-frequency signal such as an audio signal. The most common modulating methods are amplitude modulation (A.M) and frequency modulation (F.M). In A.M., the instantaneous amplitude of the *RF* carrier varies with the modulating signal. To recover the modulation, The A.M. detector circuit in the receiver must extract the amplitude variations of the *RF* carrier signal.

In *FM* the varying amplitudes of the modulating signal are made to change the instantaneous frequency of the *RF* carrier wave. In effect, the frequency of the *RF* carrier is wobbled around its centre value, or resting frequency. However, the average frequency is still the assigned carrier frequency. In order to recover the modulation, the *FM* detector circuit in the receiver must convert the frequency variations into corresponding amplitude variations and then extract the modulating signal.

In pulse modulation (P.M.) the *RF* carrier wave is transmitted in pulses *AM* and *FM* are widely used in radio and television communication systems.

**Amplitude Modulation.** Let the carrier voltage and the modulating voltage, $e_c$ and $e_m$, respectively, be represented by

$$e_c = E_c \sin \omega_c t$$
$$e_m = E_m \sin \omega_m t$$

The instantaneous voltage of the resulting amplitude – modulated wave will be

$$e = E_c (1 + m \sin \omega_m t) \sin \omega_c t$$

**Major Radio Frequency Bands**

| *Name* | *Abbreviation* | *Frequencies* | *Remarks* |
|---|---|---|---|
| Very Low Frequency | *VLF* | Below 30 kHz | Radio location equipment |
| Low Frequency | *LF* | 30 kHz to 300 kHz | Martime radio navigation |
| Medium Frequency | *MF* | 300 kHz to 3 MHz | Includes *AM* radio broadcast band |
| High Frequency | *HF* | 3 MHz to 30 MHz | Radio |
| Very High Frequency | *VHF* | 30 MHz to 300 MHz | Includes *FM* broadcast band and television *VHF* channels |
| Ultra High Frequency | *UHF* | 300 MHz to 3 GHz | Includes television *UHF* channels |
| Super High Frequency | *SHF* | 3 GHz to 30 GHz | Satellite communications |
| Extremely High Frequency | *EHF* | 30 GHz to 300 GHz | Satellite communications |

where $$m = \frac{E_m}{E_c} = \text{modulation index.}$$

The modulation index *m*, is a number lying between 0 and 1, and it is very often expressed as a percentage and called the percentage modulation.

Total power in the modulated wave will be

$$P_i = \frac{E_{carr}^2}{R} + \frac{E_{LSB}^2}{R} + \frac{E_{USB}^2}{R}$$

where all three voltages are rms values and *R* is the resistance (antenna resistance) in which the power is dissipated.

The first term in the above equation is the unmodulated carrier power and is given by

$$P_c = \frac{(E_{carr})^2}{R} = \frac{(E_c/\sqrt{2})^2}{R}$$

$$= \frac{E_c^2}{2R}$$

Also $$P_{LSB} = P_{USB} = \frac{(E_{SB})^2}{R}$$

$$= \left(\frac{m\,E_c/2}{\sqrt{2}}\right)^2 \div R$$

$$= \frac{m^2}{4}\frac{E_c^2}{2R}$$

Hence $$P_t = \frac{E_c^2}{2R} + \frac{m^2}{4}\frac{E_c^2}{2R} + \frac{m^2}{4}\frac{E_c^2}{2R}$$

or $$\frac{P_t}{P_c} = \left[1 + \frac{m^2}{2}\right]$$

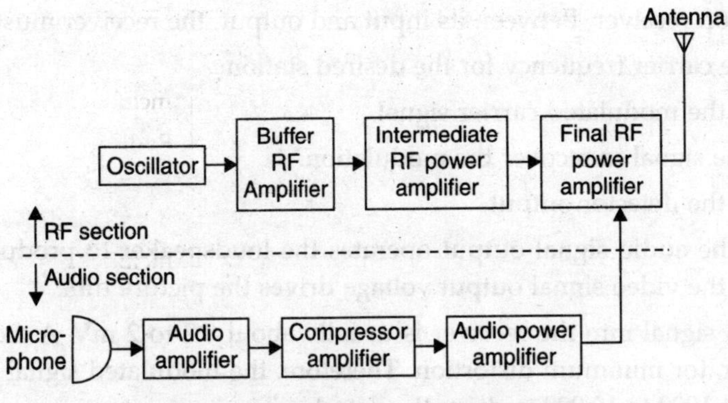


***FIGURE 25.1*** Block diagram for *AM* transmitters.

From the above relation it can be seen that the maximum power in *AM* wave is $P_t = 1.5\ P_c$, when $m = 1$. A relation for current can be shown as

$$\frac{I_t}{I_c} = \sqrt{1 + \frac{m^2}{2}}.$$

**Frequency Modulation.** The general equation for unmodulated wave of carrier may be written as

$$x = A \sin (\omega t + \phi)$$

where $x$ = instantaneous value (of voltage or current)

$A$ = (maximum) amplitude

$\omega$ = angular velocity (rad/s)

$\phi$ = phase angle, rad.

The instantaneous frequency $f$ of the frequency modulated wave is given by

$$f = f_c\ (1 + k\ E_m \cos \omega_m\ t)$$

where $f_c$ = unmodulated (or average) carrier frequency

$k$ = proportionality constant

$E_m \cos \omega_m\ t$ = instantaneous modulating voltage

The modulation index for *FM*, $m_f$ is defined as

$$m_f = \frac{\text{(Maximum) Frequency deviation}}{\text{Modulating frequency}} = \frac{\delta}{f_m}$$

hence instantaneous value of *FM* voltage is given by

$$e = A \sin (\omega_c t + m_f \sin \omega_m t)$$

As modulating frequency decreases and the modulating voltage amplitude remains constant, the modulation index increases.

**Radio Receivers.** The receiver input is a modulated *RF* carrier signal from the transmitter. The output is the desired modulated information, such as an audio signal for a radio or the video signal for a television receiver. Between its input and output, the receiver must

(1) Select the carrier frequency for the desired station.

(2) Amplify the modulated carrier signal.

(3) Detect the signal to recover the modulation.

(4) Amplify the detector output.

In a radio, the audio signal output operates the loudspeaker to produce the sound. In a television receiver, the video signal output voltage drives the picture tube.

The antenna signal into the receiver is usually about 0.2 to 2 mV. A diode detector needs about 2 V, however, for minimum distortion. Therefore, the modulated signal must be amplified with again of about 1000 to 10,000 to drive the detector.

Only two types of radio receivers have commercial significance.

(1) Tuned Radio Frequency Receiver.

(2) Superheterodyne Receiver.

Only superheterodyne receivers are widely used these days.

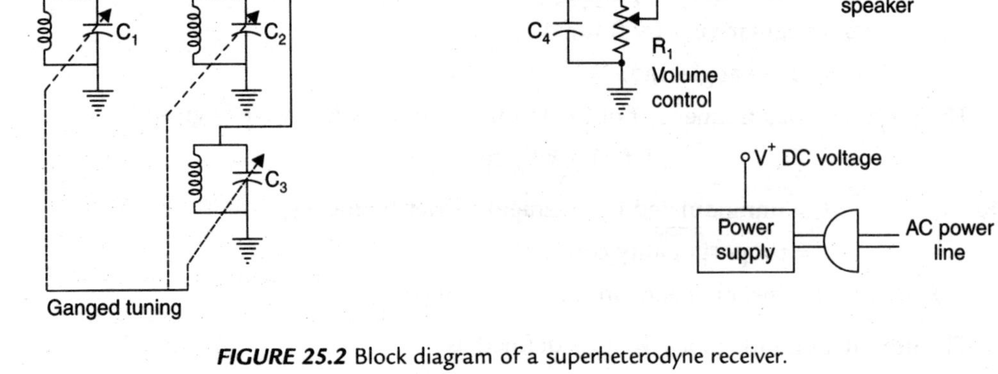


*FIGURE 25.2* Block diagram of a superheterodyne receiver.

**Image Frequencies.** The main advantage of the superheterodyne circuit is that any signal at the intermediate frequency will be amplified in the *IF* section, regardless of whether the signal is the tuned station or an interfering signal. Any false signal is called spurious response. The most important spurious response is the image of the desired signal. An image frequency differs from the oscillator frequency by the amount of the *IF*. The image frequency is higher instead of lower than the oscillator frequency.

$$\text{Image Frequency} = RF + (2 \times IF)$$

The problem of image-frequency interference is minimized by making the $H^-$ value as high as possible. That puts the image frequencies out side the *RF* band of receivers.

**Television.** A television system is required to reproduce faithfully the following:

1. The shape of each object, or structural content.
2. The relative brightness of each object.
3. Motion or kinematic content.
4. Sound.
5. Colour or chromatic content.
6. Perspective or stereoscopic content.

Various television systems in use are given below:

| *System* | *Places where used* |
|---|---|
| American System | North and South America (except Argentina and Venezuela), Phillipines and Japan |
| *SECAM* | France, Belgium |
| *PAL* | U.K., Japan, India |
| European System | Europe and most of the other countries |

In India sound for television is frequency modulated and video is amplitude modulated.

**Frequency allocation for TV in India**

| *Channel* | *Frequency Range (MHz)* |
|---|---|
| 1 | 40–47 |
| 2 | 47–54 |
| 3 | 54–61 |
| 4 | 61–68 |
| 5 | 81–88 |
| 6 | 174–181 |
| 7 | 181–188 |
| 8 | 188–195 |
| 9 | 195–202 |
| 10 | 202–209 |
| 11 | 209–216 |
| 12 | 216–223 |
| 13 | 223–230 |
| 14 | 230–237 |

For channel range of (*A*) to (*B*) MHz, the frequency of

Picture carrier ⟶ $(A + 1.25)$ MHz

Sound carrier ⟶ $(B - 0.25)$ MHz

For colour transmission the primary colours are red, blue and green. Only these colours need to be produced on screen.

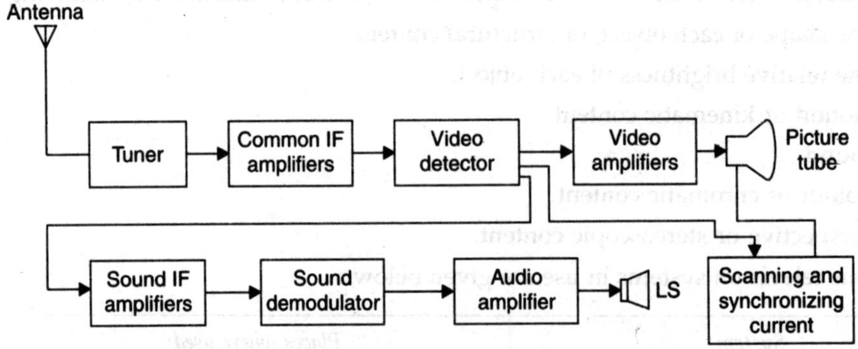


***FIGURE 25.3*** Block diagram for monochrome television receiver.

## OBJECTIVE TYPE QUESTIONS

1. The modulation index of an amplitude modulated wave is changed from 0 to 1. The transmitted power is

   (*a*) unchanged (*b*) halved (*c*) increased by 50% (*d*) doubled.

2. A carrier is simultaneously modulated by two sine waves with modulation indices of 0.3 and 0.4. The total modulation index will be

   (*a*) 0.5 (*b*) 0.7

   (*c*) 1

   (*d*) cannot be ascertained unless phase relations are known.

3. In case the carrier of a 100 percent modulated *AM* wave is suppressed, the percentage power saving will be

   (*a*) 33.33 (*b*) 66.66 (*c*) 100 (*d*) 150.

4. In a low level amplitude modulation system, the amplifier following the modulated stage must be

   (*a*) a linear device (*b*) a non-linear device

   (*c*) a harmonic device (*d*) a class C amplifier.

5. The purpose of using leak-type bias in a plate modulated class C amplifier is to

   (*a*) increase the band width (*b*) prevent over modulation

   (*c*) prevent tuned circuit damping (*d*) prevent excessive grid current.

6. For a signal amplitude modulated to a depth of 100% by a sinusoidal signal, the power is
   (*a*) same as the power of unmodulated signal
   (*b*) twice as the power of unmodulated signal
   (*c*) four times the power of unmodulated signal
   (*d*) one and a half times the power of unmodulated carrier.
7. Which one of the following is the advantage of base modulation over collector modulation of a transistor class *C* amplifier is
   (*a*) better efficiency (*b*) better linearity
   (*c*) higher power output per transistor (*d*) lower modulating power required.
8. The output stage of a television transmitter is most likely to be a
   (*a*) grid-modulated class *A* amplifier (*b*) screen-modulated class *C* amplifier
   (*c*) plate-modulated class *C* normal (*d*) grid-modulated class *C* amplifier.

**Questions 9 and 10 refer to data given below:**

9. The tuned circuit of the oscillator in a simple *AM* transmitter uses a 50 μH coil and a 1 *nF* capacitor. If the oscillator output is modulated by audio frequencies upto 10 kHz, the frequency of the carrier will be
   (*a*) 356 kHz (*b*) 356 MHz (*c*) 712 kHz (*d*) 712 MHz.
10. The frequency range occupied by the side bands will be
    (*a*) 713 kHz to 711 kHz (*b*) 715 kHz to 709 kHz
    (*c*) 712 kHz to 722 kHz (*d*) 702 kHz to 722 kHz.
11. A 200 W carrier is modulated to a depth of 75 percent. The total power of the modulated wave is given by
    (*a*) 128 W (*b*) 156.3 W (*c*) 256.25 W (*d*) 288.33 W.
12. A broadcast radio transmitter radiates 20 kW when the modulation percentage is 60. How much of this is carrier power?
    (*a*) 4.23 W (*b*) 8.47 W (*c*) 12.7 W (*d*) 16.94 W.
13. When $E_1$, $E_2$, $E_3$ and $E_4$ are the simultaneous modulating voltages, then the total modulating voltage will be
    (*a*) $E_1 + E_2 + E_3 + E_4$ (*b*) $\frac{E_1 + E_2 + E_3 + E_4}{4}$
    (*c*) $\sqrt{E_1E_2 + E_2E_4}$ (*d*) $\sqrt{E_1^2 + E_2^2 + E_3^2 + E_4^2}$.
14. A 1000 kHz carrier is simultaneously modulated with 300 Hz, 800 Hz and 2 kHz audio sine waves. The frequencies present in the output will be
    (*a*) 1000 kHz, 300 Hz, 800 Hz and 2 kHz
    (*b*) 1800 kHz, 1300 kHz and 1002 kHz
    (*c*) 200 kHz, 1800 kHz, 700 kHz, 1300 kHz, 1002 kHz, 998 kHz and 1000 kHz
    (*d*) 998.0 kHz, 999.2 kHz, 999.7 kHz, 1000.3 kHz, 1000.8 kHz and 1002.0 kHz.

**Questions 15 to 17 refer to data given below:**

A transistor class *C* amplifier has a maximum permissible collector dissipation of 20 W and a collector efficiency of 75 percent. It is to be collector modulated to a depth of 90 percent.

**15.** The maximum unmodulated carrier power will be

(*a*) 22.7 W (*b*) 27.2 W (*c*) 42.7 W (*d*) 66.3 W.

**16.** The side band power generated will be

(*a*) 17.3 W (*b*) 21.3 W (*c*) 29.6 W (*d*) 66.3 W.

**17.** If the maximum depth of modulation is restricted to 70%, the new maximum side band power generated will be

(*a*) 11.9 W (*b*) 21.1 W (*c*) 33.3 W (*d*) None of the above.

**18.** Which one of the following is an indirect way of generating *FM*?

(*a*) Armstrong modulator (*b*) Reactance bipolar transistor modulator

(*c*) Reactance *FET* modulator (*d*) Varactor diode modulator.

**Questions 19 to 22 refer to data given below:**

**19.** In an *FM* system, when the audio frequency is 500 Hz and *AF* voltage is 2.4 V, the deviation is 4.8 kHz. If the *AF* voltage is now increased to 7.2 V, the new deviation will be

(*a*) 10.2 kHz (*b*) 14.4 kHz (*c*) 16.6 kHz (*d*) 18.8 kHz.

**20.** In the above case the modulation index will be

(*a*) 9.6 (*b*) 28.8

(*c*) 43.2 (*d*) 57.6.

**21.** In case the *AF* voltage is raised to 10 V while the *AF* is dropped to 200 Hz, the deviation will be

(*a*) 10 kHz (*b*) 15 kHz (*c*) 16 kHz (*d*) 20 kHz.

**22.** Under the above conditions, the modulation index will be

(*a*) 20 (*b*) 40 (*c*) 80 (*d*) 100.

**23.** Which one of the following is not necessarily an advantage of *FM* over *AM*?

(*a*) Lower band width is required (*b*) Less modulating power is required

(*c*) Better noise immunity is provided (*d*) The transmitted power is more useful.

**Questions 24 to 26 refer to data given below:**

A 25 MHz carrier is modulated by a 400 Hz audio sine wave. The carrier voltage is 4 V and the maximum deviation is 10 kHz.

**24.** The modulation index will be

(*a*) 2.5 (*b*) 5 (*c*) 15 (*d*) 25.

**25.** The equation of the modulated wave for *FM* will be

(*a*) 10 sin $(1.57 \times 10^6 t)$ (*b*) 10 sin $(1.57 \times 10^6 t + 25 \sin 2513\, t)$

(*c*) 4 sin $(1.57 \times 10^6 t)$ (*d*) 4 sin $(1.57 \times 10^6 t + 25 \sin 2513\, t)$.

**26.** The equation of the modulated wave for *PM* will be

(*a*) $4 \sin (1.57 \times 10^6 t)$ (*b*) $4 \sin 800 t + 4 \sin 50 \times 10^6 t$

(*c*) $4 \sin (1.57 \times 10^6 t + 25 \sin 2513 t)$ (*d*) $4 \sin (1.57 \times 10^6 t - 25 \sin 2513 t)$.

**27.** The difference between *PM* and *FM* is

(*a*) in the poorer audio response of phase modulation

(*b*) purely theoretical otherwise the two are the same

(*c*) merely in the different modulation indices

(*d*) too great to make the two systems compatible.

**28.** In frequency modulation for a given frequency deviation, the modulation index varies

(*a*) inversely as the modulating frequency

(*b*) directly as the modulating frequency

(*c*) independent of the changes in modulating frequency.

**29.** An *FM* signal with a deviation '*d*' passes through a mixer as a result of which its frequency reduces five folds. The deviation in the output of the mixer will be

(*a*) $\sqrt{5d}$ (*b*) $5d$ (*c*) $\frac{d}{5}$ (*d*) $d$.

**30.** In the spectrum of a frequency modulated wave

(*a*) the carrier frequency cannot disappear

(*b*) the total number of side bands depends on the modulation index

(*c*) the carrier frequency disappears when the modulation index is large

(*d*) the amplitude of any side band depends on the modulation index.

**31.** In *FM*, the frequency deviation is

(*a*) proportional to modulating frequency

(*b*) proportional to amplitude of modulating signal

(*c*) constant

(*d*) zero.

**32.** A pre-emphasis circuit provides extra noise immunity by

(*a*) converting the phase modulation to frequency modulation

(*b*) preamplifying the whole audio band

(*c*) boosting the bass frequencies

(*d*) amplifying the higher audio frequencies.

**33.** In the stablized reactance modulator *AFC* system,

(*a*) the phase modulation is converted into frequency modulation by the equalizer circuit

(*b*) the discriminator must have a fast time constant to prevent demodulation

(*c*) the discriminator frequency must not be too low, or the system will fail

(*d*) the higher the discriminator frequency, the better the oscillator frequency stability.

34. In which of the following modulation system when the modulating frequency is doubled, the modulation index reduces to half while modulating voltage remains constant?

(*a*) Phase modulation (*b*) Amplitude modulation

(*c*) Frequency modulation (*d*) None of the above.

35. Which of the following is the advantage of frequency modulation over amplitude modulation ?

(*a*) Noise can be reduced by increasing deviation

(*b*) the amplitude of the *FM* wave in *FM* is independent of the depth of modulation, whereas in *AM* it is dependent on this parameter

(*c*) There is a large increase in noise to signal ratio in *FM*

(*d*) All of the above.

36. Which of the following statement is not valid for Armstrong modulation system?

(*a*) the system is basically *PM* not *FM*

(*b*) *AFC* is not needed, as a crystal oscillator is used

(*c*) Frequency multiplication must be used

(*d*) Equalization is unnecessary.

37. In an *FM* signal, the power

(*a*) increases as the modulation index increases

(*b*) reduces as the modulation index increases

(*c*) increases as the modulation index decreases

(*d*) remains constant when the modulation index increases.

38. It is known that noise phase modulates the *FM* wave. As the noise side band frequency approaches the carrier frequency, the noise amplitude

(*a*) will remain constant (*b*) will decrease

(*c*) will increase (*d*) will reduce to negligible value.

39. Which of the following is the disadvantage of *FM* over *AM*?

(*a*) a much wider channel is required by *FM* as compared to that for *AM*

(*b*) the area of reception is limited

(*c*) the modulating and demodulating equipment is complex

(*d*) all of the above.

40. A *FM* signal with modulation index $M_f$ is passed through a frequency doubler. The wave in the output of the doubler will have a modulation index of

(*a*) $M_f$ (*b*) $\frac{M_f}{2}$ (*c*) $2M_f$ (*d*) $4M_f$.

**Questions 41 and 42 refer to data given below:**

A 500 Hz modulating voltage fed into an *FM* generator produces a frequency deviation of 2.25 kHz

**41.** The modulation index will be

(*a*) 2.25 (*b*) 4.5 (*c*) 6.77 (*d*) 8.00.

**42.** If the amplitude of the voltage is kept constant, but its frequency is raised to 6 kHz, the new deviation will be

(*a*) 13.5 kHz (*b*) 27 kHz (*c*) 54 kHz (*d*) 108 kHz.

**43.** In television transmission in India,

(*a*) picture is *AM* and sound is *FM* (*b*) picture is *FM* and sound is *AM*

(*c*) both picture as well as sound are *FM* (*d*) both picture as well as sound are *AM*.

**44.** The signal to quantisation noise ratio in a *PCM* system depends upon

(*a*) sampling rate (*b*) number of quantisation levels

(*c*) message signal bandwidth (*d*) none of the above.

**45.** The frequency deviation in phase modulated signal is proportional to

(*a*) amplitude only (*b*) frequency only

(*c*) amplitude and frequency of modulating signal.

**46.** *PCM* transmission requires

(*a*) more bandwidth than amplitude modulated transmission

(*b*) less bandwidth than amplitude modulated transmission

(*c*) same bandwidth as in amplitude modulated transmission

(*d*) None of these.

**47.** For the transmission of normal speech signal, the *PCM* channel needs a bandwidth of

(*a*) 64 kHz (*b*) 16 kHz (*c*) 8 kHz (*d*) 4 kHz.

**48.** The main advantage of *PCM* system is

(*a*) lower bandwidth (*b*) lower power

(*c*) lower noise (*d*) none of the above.

**49.** A *TDM* system

(*a*) needs lower bandwidth

(*b*) gives low signal to noise ratio

(*c*) uses simpler circuits as compared to *FDM* system

(*d*) needs higher bandwidth.

**50.** To transmit *N* signals each band limited to $f_m$ hertz by *TDM* will require a minimum bandwidth of

(*a*) $f_m$ (*b*) $2f_m$ (*c*) $Nf_m$ (*d*) $\sqrt{2Nf_m}$.

51. When power ratios are expressed in $dB_m$, the reference power is
(*a*) 1 W (*b*) 1 mW (*c*) 1 μW (*d*) none of the above.

52. Frequency frogging is used in carrier system to
(*a*) conserve frequencies (*b*) reduce distortion
(*c*) reduce cross talk (*d*) reduce noise.

53. To provide two or more voice circuits with the same carrier, it is necessary to use
(*a*) *ISB* (*b*) Lincompex
(*c*) *SSB* with pilot carrier (*d*) carrier reinsertion.

54. *PPM* is
(*a*) linear modulation technique (*b*) analogue modulation technique
(*c*) digital modulation technique (*d*) not a modulation technique.

55. Intersymbol interference is a problem in
(*a*) *AM* transmission (*b*) *FM* transmission
(*c*) *PCM* transmission (*d*) All above transmissions.

56. L.O.S. microwave analogue radio-relay links normally use a modulation format which is
(*a*) *TDM — PPM* (*b*) *FDM — FM* (*c*) *FDM — SSB* (*d*) Any of the above.

57. An amplitude modulated wave is
(*a*) the sum of the carrier and the modulating wave
(*b*) the difference between the carrier and the modulating wave
(*c*) the product of the carrier and the modulating wave
(*d*) sum of the carrier and its product with modulating signal.

58. In an amplitude-modulated waveform, the amplitude of side bands is
(*a*) independent of the carrier amplitude
(*b*) independent of the modulation index
(*c*) carrier amplitude × modulation index
(*d*) $\frac{1}{2}$ carrier amplitude × modulation index.

59. Which one of the following cannot be used to demodulate *SSB*?
(*a*) Complete phase shift generator (*b*) Product detector
(*c*) Diode balanced modulator (*d*) Bipolar transistor balanced modulator.

60. In an *FM* waveform, the side bands are spaced at intervals equal to
(*a*) four times the modulating frequency (*b*) twice the modulating frequency
(*c*) half the modulating frequency (*d*) equal to the modulating frequency.

61. In a superhet broadcast receiver, the frequency of local oscillator is
(*a*) higher than the incoming signal frequency
(*b*) lower than the incoming signal frequency
(*c*) equal to the incoming signal frequency
(*d*) none of these.

62. A heterodyne frequency changer is called a
(*a*) modulator (*b*) mixer (*c*) demodulator (*d*) frequency translator.

63. The noise generated by a resistor depends upon
(*a*) its resistance value
(*b*) its operating temperature
(*c*) both its resistance value and operating temperature
(*d*) none of the above.

64. A carbon microphone is a
(*a*) variable resistance device (*b*) variable inductance device
(*c*) variable capacitance device (*d*) variable *LC* device.

65. In a carbon microphone, the electrical power output is
(*a*) more than the incident acoustical power
(*b*) less than the incident acoustical power
(*c*) equal to the incident acoustical power.

66. High intermediate frequency in a superhet receiver
(*a*) increases tracking problems (*b*) improves selectivity
(*c*) reduces adjacent channel rejection (*d*) decreases tracking problems.

67. If a signal band limited to *fm* is sampled at a rate less than 2 *fm*, the reconstructed signal
(*a*) will be smaller in amplitude (*b*) will have higher frequencies suppressed
(*c*) will be distorted (*d*) will not be distorted.

68. Which of the following statement about the advantages of the phase discriminator over the slope detector is false?
(*a*) Fewer tuned circuits (*b*) Much easier alignment
(*c*) Better linearity (*d*) Greater limiting.

69. For broadcasting purposes, full *AM* signal is preferred to *SSB* signal because
(*a*) it requires large bandwidth (*b*) it requires small bandwidth
(*c*) generation of full *AM* is easier (*d*) detection of full *AM* is simpler.

70. Which of the following statement about the amplitude limiter is false?
(*a*) Leak type bias must be used
(*b*) The output must be tuned
(*c*) The circuit is always biased in class *C*, by virtue of the leak type bias
(*d*) When the input increases past the threshold of limiting, the gain decreases to keep the output constant.

71. Three point tracking is achieved with
(*a*) variable selectivity (*b*) the padder capacitor
(*c*) double spotting (*d*) double conversion.

72. Neutralization is used in *RF* amplifiers to
    (*a*) stop oscillations (*b*) increase bandwidth
    (*c*) improve selectivity (*d*) all of the above.
73. In a radio receiver
    (*a*) all stages contribute equally to noise
    (*b*) *RF* stage has no effect on *S/N* ratio
    (*c*) Mixer stage contributes most of the noise generated
    (*d*) *RF* stage increases *S/N* ratio.
74. In a radio receiver with simple *AGC*
    (*a*) the highest *AGC* voltage is produced between stations
    (*b*) an increase in signal strength produces more *AGC*
    (*c*) the audio stage gain is normally controlled by the *AGC*
    (*d*) the faster the *AGC* time constant, the more accurate the output.
75. A receiver has poor *IF* selectivity. It implies that it will also have
    (*a*) sensitivity (*b*) blocking (*c*) double-spotting (*d*) diversity reception.
76. In superheterodyne reception, the local oscillator is used to
    (*a*) detect the modulating signal
    (*b*) amplify the received modulated carrier
    (*c*) shift the frequency of the received modulated carrier to the *IF* band.
77. The household radio receiver uses
    (*a*) synchronous detector (*b*) envelope detector
    (*c*) radio detector (*d*) none of the above.
78. *AGC* is used in a radio receiver to
    (*a*) maintain the carrier level at the second detector input constant
    (*b*) maintain the carrier level at the second detector input variable
    (*c*) improve selectivity of the receiver
    (*d*) improve noise figure of the receiver.
79. A superheterodyne receiver with an *IF* of 450 kHz is tuned to a signal at 1400 kHz. The image frequency will be
    (*a*) 900 kHz (*b*) 1125 kHz (*c*) 1625 kHz (*d*) 2300 kHz.
80. Power spectral density of white noise
    (*a*) is constant with frequency (*b*) increases with frequency
    (*c*) decreases with frequency (*d*) none of the above.
81. In a radio receiver, the maximum contribution to noise is made by
    (*a*) power supply (*b*) power amplifier stage
    (*c*) mixer stage (*d*) all above.

**82.** In case the *IF* of a receiver is very high, then when of the following will not be true?

(*a*) Selectivity will be poor
(*b*) Tracking will be improved
(*c*) The local oscillator need not be extremely stable
(*d*) Image frequency rejection will be better.

**83.** Most of the amplification of the received signal is obtained in a superheterodyne receiver from the

(*a*) *RF* stage (*b*) *IF* stage
(*c*) Power amplifier stage (*d*) None of the above stages.

**84.** In order to prevent over loading of the last *IF* amplifier in a receiver, one should use

(*a*) squelch (*b*) variable sensitivity
(*c*) variable selectivity (*d*) double conversion.

**85.** When an *FET* is used as the first stage amplifier in a transistor receiver, this is likely to result in

(*a*) improved selectivity of receiver
(*b*) improved effectiveness of *AGC*
(*c*) reduced negative-peak clipping
(*d*) reduced effect of noise at low-modulation depths.

**86.** In which of the following frequency bands are the standard *AM* radio broadcast stations?

(*a*) *MF* (*b*) *HF* (*c*) *VHF* (*d*) *UHF*.

**87.** A frequency multiplier circuit

(*a*) operates class *A* (*b*) is tuned to a harmonic of the input signal
(*c*) needs parasitic oscillations (*d*) is usually pulse modulated.

**88.** A negative feedback amplifier will tend to oscillate if

(*a*) $A\beta$ is zero (*b*) $A\beta$ is unity (*c*) $A = \beta = 0$ (*d*) $A\beta$ is infinite.

**89.** Oscillator crystals are made of

(*a*) Silicon (*b*) Germanium (*c*) Quartz (*d*) Tungsten.

**90.** An oscillator with best frequency stability is

(*a*) multivibrator (*b*) clapp oscillator
(*c*) phase shift circuit (*d*) crystal oscillator.

**91.** In an oscillator crystal, the parallel resonant frequency is

(*a*) equal to series resonant frequency
(*b*) higher than series resonant frequency
(*c*) half of series resonant frequency
(*d*) one quarter of series resonant frequency.

**92.** The local oscillator of a broadcast receiver is tuned to a frequency higher than the incoming frequency

(*a*) to allow adequate frequency coverage without switching
(*b*) to permit easier tracking
(*c*) to help the image frequency rejection
(*d*) none of the above.

93. The vertical synchronizing pulses in television occur after every
(*a*) scanning line (*b*) field (*c*) frame (*d*) none of the above.

94. Increasing the height of the radiating antenna, the coverage area of a TV station would
(*a*) increase (*b*) decrease (*c*) not change.

95. In television, equalizing pulses are used for
(*a*) horizontal synchronization
(*b*) vertical synchronization
(*c*) both horizontal as well as vertical synchronization
(*d*) neither horizontal nor vertical synchronization.

96. The output of the vertical amplifier, applied to the yoke in a TV receiver, consists of
(*a*) direct current (*b*) amplified vertical sync pulses
(*c*) a sawtooth voltage (*d*) a sawtooth current.

97. Horizontal retrace in a TV receiver is also known as
(*a*) fly back (*b*) damper (*c*) burst (*d*) ringing.

98. Interlacing is used in television to
(*a*) produce the illusion of motion
(*b*) ensure that all the lines of the screen are scanned, not merely the alternate one
(*c*) simplify the vertical sync pulse train
(*d*) avoid flicker.

99. Sync separator in a TV receiver
(*a*) separates horizontal and vertical sync pulses
(*b*) separates sync pulses (horizontal and vertical) from the composite video signal
(*c*) separates only horizontal sync pulses
(*d*) separates only vertical sync pulses.

100. The output of the vertical amplifier, applied to the yoke in a TV receiver consists of
(*a*) direct current (*b*) amplified vertical sync pulses
(*c*) a sawtooth voltage (*d*) a sawtooth current.

101. An odd number of lines per frame forms part of every one of the world 's TV systems. This is
(*a*) done to assist interlace (*b*) purely an accident
(*c*) to ensure that line and frame frequencies can be obtained from the same original source
(*d*) done to minimize interference with the chroma subcarrier.

**102.** The baseband bandwidth of television transmission is
(*a*) 1 MHz (*b*) 5 MHz (*c*) 10 MHz (*d*) 30 MHz.

**103.** In a TV receiver, the circuit that separates sync pluses from the composite video waveform is
(*a*) the keyed *AGC* amplifier (*b*) a clipper
(*c*) an integrator (*d*) a differentiator.

**104.** Which of the following voltage is not found in the output of a normal monochrome receiver video detector?
(*a*) Sync (*b*) Video (*c*) Sweep (*d*) Sound.

**105.** The service area of a TV transmitter for satisfactory reception is determined by
(*a*) line-of sight zone (*b*) diffraction zone
(*c*) tropospheric zone (*d*) regional tropospheric zone.

**106.** The separation of sound and picture carriers in Indian TV system is
(*a*) 1 MHz (*b*) 5.5 MHz (*c*) 6.0 MHz (*d*) 7.5 MHz.

**107.** The number of frames per second in Indian TV system is
(*a*) 24 (*b*) 25
(*c*) 48 (*d*) 50.

**108.** The maximum swing of the sound carrier in Indian TV system is
(*a*) ± 5 kHz (*b*) ± 10 kHz (*c*) ± 15 kHz (*d*) ± 25 kHz.

**109.** The line frequency of TV system in India is
(*a*) 50 Hz (*b*) 625 Hz (*c*) 15,625 Hz (*d*) 25,625 Hz.

**110.** Equalizing pulses in TV are sent during
(*a*) horizontal blanking (*b*) vertical blanking
(*c*) the serrations (*d*) the horizontal retrace.

**111.** In television, 4: 3 represents the
(*a*) interlace ratio (*b*) maximum horizontal deflection
(*c*) aspect ratio (*d*) ratio of the two diagonals.

**112.** When the size of TV picture tube is expressed as 51 cm, it represents
(*a*) width of screen (*b*) height of screen
(*c*) average of width and height of screen (*d*) diagonal of screen.

**113.** The video voltage applied to the picture tube of a television receiver is fed in
(*a*) between grid and ground (*b*) to the yoke
(*c*) to the anode (*d*) between grid and cathode.

**114.** Interlaced scanning is used in television mainly to
(*a*) reduce large area flicker effect (*b*) reduce bendwidth requirements
(*c*) both (*a*) and (*b*) above (*d*) none of the above.

**115.** The demodulation of the sound signals in a TV receiver is accomplished by

(*a*) linear detector (*b*) product detector (*c*) discriminator (*d*) envelope detector.

**116.** The *HV* anode supply for the picture tube of a TV receiver is generated in the

(*a*) mains transformer (*b*) vertical output stage

(*c*) horizontal output stage (*d*) horizontal deflection oscillator.

**117.** The signals sent by the TV transmitter to ensure correct scanning in the receiver are called

(*a*) sync (*b*) chroma (*c*) luminance (*d*) video.

**118.** The base bandwidth of the TV signal for 625 line scanning is

(*a*) 1 MHz (*b*) 4.5 MHz (*c*) 5.5 MHz (*d*) 10 MHz.

**119.** Which of the following frequencies is not found in the output of a normal TV receiver tuner ?

(*a*) 45.75 MHz (*b*) 41.25 MHz (*c*) 42.17 MHz (*d*) 4.5 MHz.

**120.** In TV broadcast, the sound signal is modulated in

(*a*) *SSB* (*b*) *VSB* (*c*) *FM* (*d*) *AM*.

**121.** The function of serrations in the composite video waveform is to

(*a*) help horizontal synchronization

(*b*) help vertical synchronization

(*c*) simplify the generation of the vertical sync pulse

(*d*) equalize the charge in the integrator before the start of vertical retrace.

**122.** The three primary colours are

(*a*) red, yellow and purple (*b*) black, white and gray

(*c*) violet, red and green (*d*) red, blue and green.

**123.** The shadow mask in colour TV picture tube is used to

(*a*) reduce X-ray emission (*b*) reduce gamma-ray emission

(*c*) increase screen brightness (*d*) ensure that each beam hits its own target.

**124.** In a colour TV receiver, the colour killer

(*a*) cuts off the chroma stages during monochrome reception

(*b*) prevents colour over loading

(*c*) makes sure that the colour burst is not mistaken for sync pulses, by cutting off reception during the back porch.

(*d*) ensures that no colour is transmitted to monochrome receivers.

**125.** Which of the following signal is not transmitted in colour TV?

(*a*) *Y* (*b*) *Q* (*c*) *R* (*d*) *I*.

**126.** Due to which of the following reasons radar detection is limited to line of sight

(*a*) Short wavelength are used (*b*) Long wavelength are used

(*c*) Curvature of the earth (*d*) The waves are not reflected by the ionosphere.

**127.** The channel capacity is exactly equal to which of the following

(a) Noise rate in the channel (b) BW of demand

(c) Amount of information per second (d) None of above.

**128.** By using which of the following reliable ship-to-ship communication may be secured with little fading

(a) more directional antenna (b) frequency diversity reception

(c) broadband antenna (d) space diversity reception.

**129.** The term "delayed *AGC*" applies application of *AGC*

(a) only when signal strength has increased beyond a specified value

(b) to the last stage of receiver

(c) after switch off-on-off

(d) after sometime lag where $I_p$ = dc diode current.

**130.** The pilot carrier in *SSB* is used for

(a) lower power consumption (b) providing better noise immunity

(c) frequency stability purpose (d) none of the above.

**131.** A low ratio of the *AC* to the *DC* load impedance of a diode detector results in

(a) poor *AF* response (b) diagonal clipping

(c) negative peak clipping (d) poor *AGC* operation.

**132.** The characteristic impedance $Z_O$ equals

(a) $\sqrt{Z_{SC}\ Z_{OC}}$ (b) $\sqrt{Z_{SC}/Z_{OC}}$ (c) $Z_{OC} / Z_{SC}$ (d) $Z_{SC} / Z_{OC}$

**133.** ........... is the main advantage of *PCM* system

(a) Lower noise (b) Lower *BW* (c) Lower power (d) Any of the above.

**134.** To modulate *PCM*, it is first converted into

(a) *PDM* (b) *PPM* (c) *PWM* (d) *PAM*.

**135.** For transmission of normal speech signal, the *PCM* channel needs a bandwidth of

(a) 16 Hz (b) 32 Hz (c) 64 Hz (d) 96 Hz

**136.** Telephone traffic is measured

(a) with echo cancellers (b) in terms of grade of service

(c) in Erlanges (d) none of the above.

**137.** Which of the following microphones is used in an ordinary telephone?

(a) Pressure microphone (b) Crystal microphone

(c) Carbon microphone (d) Piezoelectric microphone.

**138.** In the phase shift method of producing *SSB*, the number of balanced modulators and phase shift circuits required respectively are

(a) 1, 1 (b) 2, 1 (c) 2, 2 (d) 1, 2.

**139.** In a radio detector

(*a*) the linearity is worse than in a phase discrimination

(*b*) the output is twice that obtainable from a similar phase discrimination

(*c*) stabilization against signal strength variations is provided

(*d*) the circuit is the same as in a discriminator, except that the diode are reversed.

**140.** In an *AM* system, the modulated basically

(*a*) causes the carrier envelope to vary in accordance with the modulating signal

(*b*) provides the power required by the transmission system

(*c*) multiplies the modulating signal and the carrier signal

(*d*) (*a*) and (*b*) both.

**141.** In *PCM* if the transmission path is very long

(*a*) repeaters stations are used (*b*) pulse width may be increased

(*c*) pulse amplitude is increased (*d*) pulse spacing is reduced

**142.** A source generates four messages. The entropy of the source will be maximum, when

(*a*) all probabilities are equal

(*b*) one of the probability is 1.0 and other are zero

(*c*) the probabilities are 1/2, 1/4, 1/8 and 1/9

(*d*) two of the probabilities are 1/2 each and others are zero.

**143.** *ARF* carrier of 12 kV at 1 kHz signals of 6 kV peak. The modulation pattern is obtained on a *CRO* calibrated suitably. The voltage indicated will be

(*a*) 18 kV (*b*) 1.001 kV (*c*) 18 V (*d*) 36 kV.

**144.** A radio receiver is tuned to 750 kHz and its oscillator frequency is 925 kHz. The image frequency would be

(*a*) 900 kHz (*b*) 1000 kHz (*c*) 1100 kHz (*d*) 1150 kHz.

**145.** A communication channel disturbed by additive white gaussien noise has a bandwidth of 4 kHz and SNR of 15. The highest transmission rate that such a channel can support (in *k*-bits/sec.) is

(*a*) 16 (*b*) 1.6 (*c*) 3.2 (*d*) 60.

**146.** The value of a resistor creating thermal noise is doubled. The noise power generated will be

(*a*) unchanged (*b*) halved (*c*) doubled (*d*) quadrupled.

**147.** The discrete zero memory information source has 40 symbols and each symbol is equally likely. The minimum number of bits required to code the source with uniform length code and the entropy of the source are resp.

(*a*) 5, 5.03 (*b*) 6, 5.83 (*c*) 5, 6.64 (*d*) 6, 5.32.

**148.** Master oscillator in a transmitter commonly uses crystal oscillator because

(*a*) it provides high power output (*b*) no buffer stage is required

(*c*) tuned circuit is not required (*d*) it has high frequency stability.

**149.** Which of the following is most likely to be found in an *SSB* transistor

(*a*) class *A* , *RF* output amplifier (*b*) class *B* , *RF* amplifier

(*c*) tuned modulator (*d*) class *C* , audio amplifier.

**150.** Term neutralisation in radio transmitter generally refers to

(*a*) Simultaneous modulation and demodulation

(*b*) Modulation at low frequencies

(*c*) Eliminating spurious oscillation due to unwanted feedback

(*d*) Cancelling the modulation caused in modulator.

**151.** Which of the following should be used to prevent loading in the last I.F. amplifier in a receiver

(*a*) variable sensitivity (*b*) variable selectivity

(*c*) double conversion (*d*) squelch circuit.

**152.** Channel capacity is exactly equals to

(*a*) BW of demand (*b*) amount of information per second

(*c*) noise rate in demand (*d*) none of the above.

**153.** In a superhet receiver, the three-point tracking is achieved with the help of which of the following

(*a*) double conversion (*b*) band spread control

(*c*) phase discrimination (*d*) a patter capacity.

**154.** In a synchronous satellite link the signal propagation time between two ground stations is about

(*a*) 125 mts (*b*) 200 mts (*c*) 270 mts (*d*) 370 mts.

**155.** By the use of which of the following filter a *PAM* signal can be demodulated

(*a*) a high pass filter (*b*) a low pass filter

(*c*) a band pass filter (*d*) none of the above.

**156.** Quadrature mux is a form of

(*a*) *TDM* (*b*) *FDM*

(*c*) combined *TDM* and *FDM* (*d*) none of the above.

**157.** While selecting a satellite system which of the following is the first determining factor

(*a*) antenna size (*b*) eirp (*c*) antenna gain (*d*) coverage area

**158.** The channel capacity *C* of a band limited Gaussian channel equals

(*a*) $B \log_2 (1 + s/n)$ (*b*) $B \log_2 (s/n)$ (*c*) $(1/B) \log_2 (1 + s/n)$ (*d*) none of the above.

where 'B' is the channel *BW*.

**159.** To transmit '*n*' signals each band limited to $f_m$ Hz by *TDM* will require a minimum *BW* of

(*a*) $f_m$ (*b*) $2f_m$ (*c*) $N f_m$ (*d*) $10 f_m$.

**160.** For a Gaussian process , auto-correlation also implies
(*a*) statistical dependence (*b*) statistical independence
(*c*) Ergodic process (*d*) none of the above.

**161.** The complete *BW* of a typical multipurpose satellite is
(*a*) 5 kHz (*b*) 5 MHz (*c*) 50 MHz (*d*) 500 MHz.

**162.** Linear velocity of the satellite in a circular orbit is
(*a*) independent of its mass
(*b*) directly proportional to its mass
(*c*) directly proportional to square of its mass
(*d*) directly proportional to the square root of its mass.

**163.** The amplifier following the modulator stage in low level *AM* must be
(*a*) linear device (*b*) harmonic device
(*c*) class-C amplifier (*d*) none of the above.

**164.** From which of the following places APPLE was shot into orbit by European Space Agency (on 19/06/1981)
(*a*) Bangalore in south India (*b*) Cape Canaveral in Florida
(*c*) Kourou in French Guyana (*d*) Carnarvon in western Australia.

**165.** For wnich of the following purposes the candascent beam pattern is essentially used
(*a*) detecting complicated objects
(*b*) direction finding
(*c*) providing a constant vertical signal returns over a range of angles in the vertical pattern
(*d*) none of the above.

**166.** In TV transmission, the type of modulation for picture is
(*a*) frequency modulation (*b*) vestigial sideband amplitude
(*c*) DSB amplitude modulation (*d*) none of the above.

**167.** In ............. microphone background noise is present
(*a*) crystal (*b*) ribbon (*c*) carbon (*d*) condenser.

**168.** In colour TV system interleaving implies
(*a*) converting black and white signals into colour signals
(*b*) mixing colour information with the black and white signals
(*c*) extracting black and white signals from colour signals
(*d*) none of the above.

**169.** The number of spectral component of a sine wave is
(*a*) zero (*b*) one (*c*) two (*d*) four.

**170.** The input of the mixer stage is
(*a*) very small carrier power (*b*) RF signals only
(*c*) local oscillator signals only (*d*) small carrier power.

**171.** A telegraph sounder is basically a ............. device
(*a*) electronics (*b*) mechanical (*c*) electrostatics (*d*) electromagnetic.

**172.** Under simplex telegraph system
(*a*) signals can neither be sent nor received
(*b*) signals can only be sent
(*c*) signals can only be received
(*d*) signals can either be sent or received from another station at a time.

**173.** Which of the following is the most modern telephone exchange system?
(*a*) Electronics system (*b*) Cross bar system
(*c*) Strowger system (*d*) None of the above.

**174.** In communication receivers the fidelity is provided by
(*a*) audio stage (*b*) detector stage (*c*) mixer stage (*d*) all of the above.

**175.** ............. microphone does not require polarizing current
(*a*) Carbon (*b*) Crystal (*c*) Condenser (*d*) All of the above.

**176.** If the output power of radio receiver is doubled, volume is increased by..........
(*a*) – 3 db (*b*) 1 db (*c*) 2 db (*d*) 3 db.

**177.** A stable multivibrator can be used as squaring circuit
(*a*) squaring circuit (*b*) comparator circuit
(*c*) $V/F$ converter circuit (*d*) $F/V$ converter circuit.

**178.** The *RF* amplifier in TV receiver is of ............... configuration
(*a*) differential (*b*) cascade (*c*) common base (*d*) none of the above.

**179.** In carrier system frequency frogging is used to
(*a*) reduce cross talk (*b*) reduce distortion
(*c*) conserve frequencies (*d*) all of the above.

**180.** Which of the following is the main advantage of *PCM* system?
(*a*) Lower noise (*b*) Lower power (*c*) Lower BW (*d*) None of the above.

**181.** The Emphasis circuit is employed for improving $S/N$ ratio at .......... frequencies
(*a*) higher (*b*) middle (*c*) lower (*d*) any of the above.

**182.** The term noise temperature is used for which of the following
(*a*) Thermal noise source only (*b*) White noise source only
(*c*) Non-thermal noise source only (*d*) All of the above.

**183.** If the value of the resistor is doubled, the thermal noise power generated
(*a*) Gets halved (*b*) Gets doubled
(*c*) Gets quadrupled (*d*) Remains unchanged.

**184.** The auto-correction function of the white noise is
(*a*) impulse function (*b*) step function
(*c*) constant (*d*) none of the above.

**185.** The photo masking process during fabrication of *ICs*

(*a*) is used to remove selected regions of $SiO_2$

(*b*) is used to remove photo resist

(*c*) control the depth of diffusion

(*d*) forms an insulating layer.

**186.** With reference to advantage of *SSB* over double side band full carrier *AM*, which of the statement is correct

(*a*) transmitter circuit are more stable giving better reception

(*b*) more channel space is available

(*c*) much less power is needed for the same signal strength

(*d*) signal is more noise resist.

**187.** Armstrong modulator generates .............. signals

(*a*) frequency modulated (*b*) phase modulated

(*c*) *AM* (*d*) both (*a*) and (*b*).

**188.** Quantization noise occurs in

(*a*) *TDM* (*b*) *PCM* (*c*) *FDM* (*d*) None of the above.

**189.** Broad band long distance communication was originally made possible by the advent of

(*a*) telegraph cable (*b*) geostationary satellite

(*c*) HF radio system (*d*) repeaters amplifier.

**190.** With reference to 'noise temperature' which of the following statement is true

(*a*) It is measure of noise figure of a receiver

(*b*) It is the same as physical temperature of the body

(*c*) It is the measure of the available noise power for non-thermal sources

(*d*) None of the above.

**191.** Pick odd one out

(*a*) Cosmic noise (*b*) Galactic noise

(*c*) Atmospheric noise (*d*) Polar noise.

**192.** In communication system, redundancy

(*a*) helps to correct error (*b*) helps to detect error

(*c*) reduces efficiency of communication (*d*) all of the above.

**193.** Which of the instrument is used in an aeroplane to measure its height above ground?

(*a*) Radar altimeter (*b*) *MTI*

(*c*) *CW* radar (*d*) Doppler radar

**194.** For direct reception of TV signal from a communication satellite, the type of antenna required is

(*a*) loop antenna (*b*) horn antenna

(*c*) chicken-mesh antenna (*d*) none of the above.

**195.** For a periodic function , the spectral density and the auto-correlation function are

(*a*) Fourier transform pair (*b*) Laplace transform pair

(*c*) One and the same thing (*d*) None of the above.

**196.** The sensitivity of a receiver is an important quality because

(*a*) It defines the receiver's ability to respond efficiently to a weak signal

(*b*) It defines usable receiver *BW*

(*c*) It defines the *IF BW* of the receiver

(*d*) None of the above.

**197.** For suppressing unwanted side bands in SSB which of the following method can't be used?

(*a*) Filter method (*b*) Phase shift method

(*c*) Third method (*d*) Balanced modulator.

**198.** A *TDM* system

(*a*) needs lower *BW*

(*b*) gives low *SNR*

(*c*) uses simpler circuits as compared to *FDM* system

(*d*) gives high *SNR*.

**199.** The auto-correlation function of signal ($t$) at $t_0 = 0$ is equals to

(*a*) zero (*b*) infinite

(*c*) average voltage of the signal (*d*) average power of the signal.

**200.** Synchronous *TDM* is called Synchronous because

(*a*) synchronous transmission is used

(*b*) time slots are pre-assigned to sources and fixed

(*c*) none of these.

**201.** A random process is defined as ergodic if

(*a*) all type of ensembles averages are not changed

(*b*) all type of ensembles averages are constant

(*c*) all type of ensembles averages are interchangeable

(*d*) in all cases.

**202.** The *D*-layer in the ionosphere, during the day time

(*a*) split into two layers (*b*) disappears

(*c*) exists (*d*) none of these .

**203.** In *AM* transmitters, the function of buffer amplifier is to

(*a*) isolate *RF* stages (*b*) multiply frequency

(*c*) amplify audio signal (*d*) none of these.

**204.** Which of the following factors does not contribute to the drift of a geostationary satellite from its stationary position in space?

(*a*) Oblate ness of the earth (*b*) Weight of the satellite
(*c*) Pressure of the solar radiation (*d*) Gravitational changes due to sun and moon.

**205.** The critical frequency of an ionospheric layer during the day time is …….. that for night time
(*a*) less than (*b*) same as (*c*) more than (*d*) negligible.

**206.** In *AM* transmitter, the type of the modulated amplifier used is
(*a*) Class *A* (*b*) Class *B*
(*c*) Class *AB* (*d*) Class C with '–' ve feedback.

**207.** The horizontal retrace in a TV receiver is also called
(*a*) Flyback (*b*) Blanker (*c*) Equalizer (*d*) Burst.

**208.** Interlacing in TV
(*a*) Increase frame frequency (*b*) Avoids flicker
(*c*) Provides vertical sync pulse position (*d*) Avoid picture noise.

**209.** Which of the following is not the function of detector circuit?
(*a*) Eliminate corner (*b*) Eliminate unwanted signals
(*c*) Rectify signal (*d*) None of these.

**210.** In case of movies the number of still pictures displayed in the second is
(*a*) 12 (*b*) 16 (*c*) 24 (*d*) 32.

**211.** Which of the following is an advantage of DME in navigation?
(*a*) It displays the location of the aircraft
(*b*) It employs low frequencies
(*c*) It measures distance using one range stations only
(*d*) It measures distance accurately by using two range station

**212.** Which of the following has the least wavelength?
(*a*) *UHF* (*b*) X-ray (*c*) Infra-red (*d*) Ultra-voilet.

**213.** In simplex telegraph working
(*a*) a signal can either be sent or received from another station at a time
(*b*) two signals can be transmitted
(*c*) two signals can be received
(*d*) none of these.

**214.** Three point tracking is achieved with which of the following
(*a*) Double conversion (*b*) Variable selectivity
(*c*) The paddar capacitor (*d*) Double spotting.

**215.** Which of the following is correct regarding a radio receiver?
(*a*) Mixer stage contributes most of the noise generated
(*b*) All stages contribution equally to noise
(*c*) *RF* stage has no effect on *SNR*.

**216.** Auto-correlation, for a Gaussian process, also implies
(*a*) Ergodic process (*b*) Statistical dependence
(*c*) Statistical independence (*d*) None of these.

**217.** In TV interlacing is used
(*a*) to avoid flickers (*b*) to produce the illusion of motion
(*c*) to simplify the vertical sync pulse train
(*d*) to ensure that all the lines of the screen are scanned.

**218.** In TV, 4 : 3 represents which of the following?
(*a*) Aspect ratio (*b*) Interlace ratio
(*c*) Ratio of the diagonals (*d*) Maximum horizontal deflection.

**219.** In order to connect the output of a source to a telephone line, it is necessary to have
(*a*) *PSK* (*b*) a modem (*c*) a lease for the line (*d*) none of these.

**220.** In the Indian TV system the no. of lines per field is
(*a*) 625/2 (*b*) 625 (*c*) 825 (*d*) 1025.

**221.** In case the carrier of a 100% modulated *AM* wave is suppressed, the percentage saving will be
(*a*) 30 (*b*) 66.66 (*c*) 75 (*d*) 125.

**222.** In a *PCM* system the signal to quantization noise ratio depends upon
(*a*) message signal BW (*b*) sampling rate
(*c*) no. of quantization levels (*d*) none of these.

**223.** Spectral density of white noise
(*a*) varies with frequency (*b*) is constant
(*c*) varies with BW (*d*) none of these.

**224.** In a *TDM* system , different signals are distinguished from each other
(*a*) only in time (*b*) only in frequency
(*c*) both in time and frequency (*d*) in amplitude.

**225.** The auto-correlation function is
(*a*) an odd function in $t_0$ (*b*) an even function in $t_0$
(*c*) an exponential function in $t_0$ (*d*) none of these.

## ANSWERS

| | | | | | | |
|---|---|---|---|---|---|---|
| **1.** (*a*) | **2.** (*a*) | **3.** (*b*) | **4.** (*a*) | **5.** (*d*) | **6.** (*d*) | **7.** (*a*) |
| **8.** (*d*) | **9.** (*c*) | **10.** (*d*) | **11.** (*c*) | **12.** (*d*) | **13.** (*d*) | **14.** (*d*) |
| **15.** (*c*) | **16.** (*a*) | **17.** (*a*) | **18.** (*a*) | **19.** (*b*) | **20.** (*b*) | **21.** (*d*) |

22. (*d*) 23. (*a*) 24. (*d*) 25. (*d*) 26. (*c*) 27. (*c*) 28. (*a*)
29. (*d*) 30. (*d*) 31. (*b*) 32. (*d*) 33. (*c*) 34. (*c*) 35. (*d*)
36. (*d*) 37. (*d*) 38. (*b*) 39. (*d*) 40. (*c*) 41. (*b*) 42. (*c*)
43. (*a*) 44. (*c*) 45. (*a*) 46. (*a*) 47. (*d*) 48. (*c*) 49. (*b*)
50. (*c*) 51. (*b*) 52. (*c*) 53. (*a*) 54. (*c*) 55. (*c*) 56. (*b*)
57. (*a*) 58. (*d*) 59. (*d*) 60. (*c*) 61. (*a*) 62. (*b*) 63. (*b*)
64. (*a*) 65. (*b*) 66. (*a*) 67. (*c*) 68. (*d*) 69. (*d*) 70. (*a*)
71. (*b*) 72. (*a*) 73. (*c*) 74. (*b*) 75. (*b*) 76. (*c*) 77. (*b*)
78. (*a*) 79. (*d*) 80. (*a*) 81. (*c*) 82. (*b*) 83. (*b*) 84. (*b*)
85. (*d*) 86. (*a*) 87. (*b*) 88. (*b*) 89. (*c*) 90. (*d*) 91. (*a*)
92. (*a*) 93. (*c*) 94. (*a*) 95. (*c*) 96. (*c*) 97. (*a*) 98. (*d*)
99. (*b*) 100. (*d*) 101. (*a*) 102. (*b*) 103. (*b*) 104. (*b*) 105. (*a*)
106. (*b*) 107. (*b*) 108. (*d*) 109. (*c*) 110. (*b*) 111. (*c*) 112. (*d*)
113. (*d*) 114. (*a*) 115. (*c*) 116. (*c*) 117. (*a*) 118. (*b*) 119. (*d*)
120. (*c*) 121. (*a*) 122. (*d*) 123. (*d*) 124. (*a*) 125. (*c*) 126. (*d*)
127. (*b*) 128. (*c*) 129. (*c*) 130. (*c*) 131. (*d*) 132. (*c*) 133. (*c*)
134. (*b*) 135. (*d*) 136. (*c*) 137. (*a*) 138. (*c*) 139. (*b*) 140. (*d*)
141. (*b*) 142. (*c*) 143. (*b*) 144. (*c*) 145. (*d*) 146. (*d*) 147. (*a*)
148. (*d*) 149. (*c*) 150. (*c*) 151. (*d*) 152. (*b*) 153. (*a*) 154. (*c*)
155. (*d*) 156. (*b*) 157. (*d*) 158. (*c*) 159. (*b*) 160. (*a*) 161. (*b*)
162. (*a*) 163. (*b*) 164. (*b*) 165. (*d*) 166. (*d*) 167. (*d*) 168. (*a*)
169. (*d*) 170. (*c*) 171. (*b*) 172. (*a*) 173. (*b*) 174. (*c*) 175. (*d*)
176. (*c*) 177. (*b*) 178. (*c*) 179. (*a*) 180. (*a*) 181. (*c*) 182. (*a*)
183. (*b*) 184. (*a*) 185. (*c*) 186. (*a*) 187. (*b*) 188. (*a*) 189. (*d*)
190. (*d*) 191. (*a*) 192. (*a*) 193. (*c*) 194. (*a*) 195. (*d*) 196. (*a*)
197. (*a*) 198. (*a*) 199. (*b*) 200. (*c*) 201. (*c*) 202. (*b*) 203. (*a*)
204. (*d*) 205. (*c*) 206. (*b*) 207. (*d*) 208. (*b*) 209. (*a*) 210. (*d*)
211. (*a*) 212. (*c*) 213. (*a*) 214. (*d*) 215. (*d*) 216. (*d*) 217. (*b*)
218. (*a*) 219. (*a*) 220. (*d*) 221. (*c*) 122. (*d*) 223. (*a*) 224. (*c*)
225. (*d*).

Chapter 26

# ANTENNA, WAVE PROPAGATION AND MICROWAVES

An antenna is just a metal conductor, usually either a length of wire or hollow tubing. A conductor must be used for the antenna so that the current can flow in it. At the transmitter the antenna current produces electromagnetic radio waves. The wave consists of varying electric and magnetic fields that move out into space from the antenna. At the receiver, radio waves induce a current in the antenna. This current is the input signal to a receiver.

**Velocity of Radio Waves.** In general, an electromagnetic wave is a form of radiation that transmits energy through space. Light radiation, heat radiation, X-rays and radio waves are examples of electromagnetic waves. When $\lambda$ is the wavelength, $f$ the frequency and $c$ the velocity of light, then

$$\lambda = \frac{c}{f}$$

The value of $c$ is known to be $3 \times 10^{10}$ cm/s.

**Half Wave Dipole Antenna.** This is the basic type of resonant antenna. Each pole of the antenna is quarter wavelength long. The insulating space between the two poles is negligible The dipole can be mounted either horizontally or vertically. A half wave dipole is also called a Hertz antenna. The current and voltage distribution is also shown in Fig. 26.1. The ends of the antenna are points of maximum voltage. The waveforms for the values of $I$ and $V$ along the antenna are known as *standing waves.* Points of minimum $I$ or $V$ are *nodes*. The maximum points are *loops*. In a half-wave antenna, successive loops or nodes for $V$ and $I$ are a half wavelength apart.

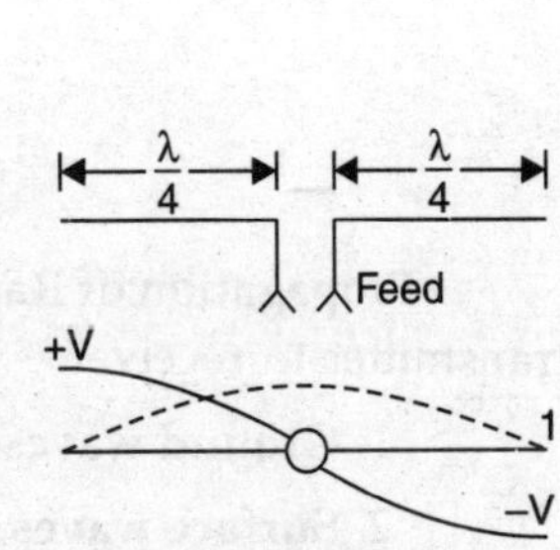


***FIGURE 26.1*** Half wave dipole antenna

**Standing Wave Ratio** (SWR) compares the voltage at the maximum point to that at a minimum point. A resonant antenna has a high SWR.

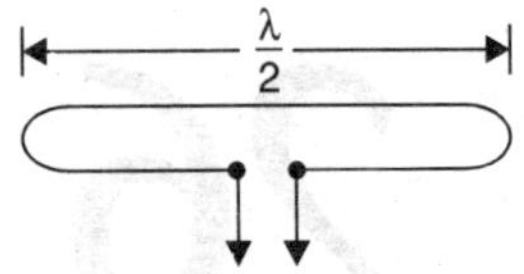


*FIGURE 26.2* Folded dipole antenna.

**Folded Dipole Antenna** is constructed of two half wave conductors joined at the ends but with one conductor open at the centre where it connects to the transmission line. The spacing between the conductors is small compared with a half wavelength. The antenna characteristics are essentially the same as those of a straight dipole except that $R$ is four times as compared to folded dipole antenna.

**Antenna Loading.** When the antenna is not long enough for resonance, a small inductance can be added in series with the antenna. The inductance is called a *loading coil* because it increases antenna load current by making the antenna resonant. A loading coil makes the antenna electrically longer by adding the inductance that would be there if the antenna were of the right length. The reason for using an antenna shorter than necessary is that the required physical length may not be practical for mounting.

Another way to make an antenna electrically longer is to use capacitive top loading. A metal cylinder, disc or spoked wheel at the top end of the antenna conductor adds shunt capacitance. In case the effective length of antenna is to be reduced, a series capacitance may be added.

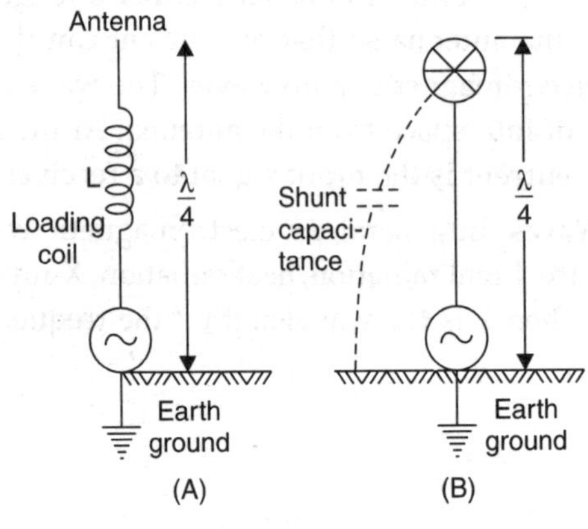


*FIGURE 26.3* (A) Antenna loading with inductance.
(B) Top loading to increase shunt capacitance.

**Propagation of Radio Waves.** There are several ways by which radio waves can travel from transmitter to receiver.

1. **Ground waves.** In this the waves travel close to the earth and in it.
2. **Surface waves.** In this the waves travel just above the earth's surface, mainly in a direct line to the receiving antenna.
3. **Sky waves.** These waves travel in the sky. Part of the energy is lost, but some of the sky wave returns to the earth from the ionosphere by reflection and refraction.

In general, high frequency waves (above 30 MHz) are propagated mainly by surface waves for line-of-sight transmission to the receiving antenna (as in case of television). Long distance transmission is possible through sky waves only.

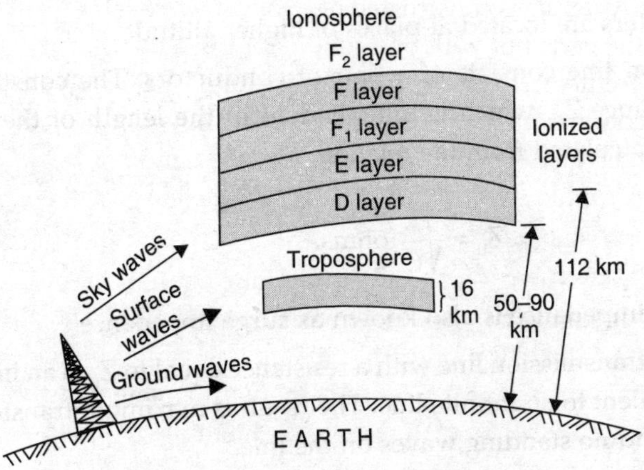


***FIGURE 26.4*** Propagation of radiowaves.

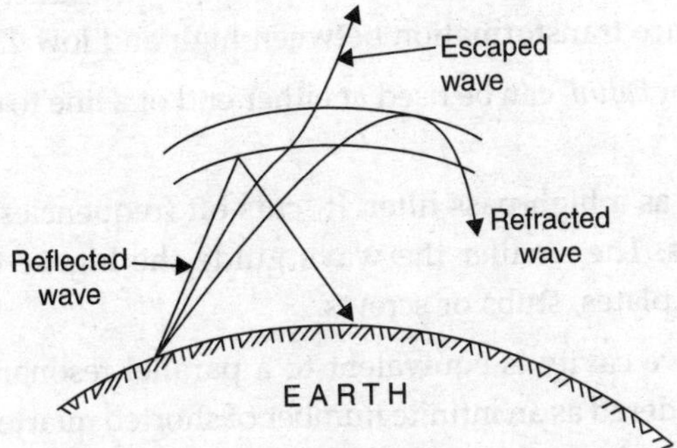


***FIGURE 26.5*** Reflection and refraction of waves.

**Ionosphere.** Ultraviolet rays and other radiations from the sun result in ionization in the upper part of the earth's atmosphere. The charged particles consist of free electrons and ions in a rarefied gas. Ionosphere is the layer containing such charged particles. This extends from 112 km to 360 km. Depending upon the charge density, the ionosphere layers are designated as $F_2$, $F$, $F_1$ $E$ and $D$. Radio waves upto 30 MHz get reflected or refracted through ionosphere layers. At higher frequencies they escape through ionosphere layers and pass into space until all energy is dissipated. That is why for satellite communication higher frequencies are used.

The highest frequency that can be bent back from a specific layer of the ionosphere is called the *critical* frequency.

The total distance from the transmitting antenna to the point where the sky wave returns to the earth is known as 'skip distance'.

**Line of sight transmission.** In this the surface wave, like light waves, is propagated in a straight line to the horizon. It generally applies to FM radio broadcast band of 88 to 108 MHz and television broadcasting in VHF and UHF bands.

In this transmission, the height of the antenna is very important for good reception. The purpose is to overcome the shadowing effect of the curvature of the earth. For this reason

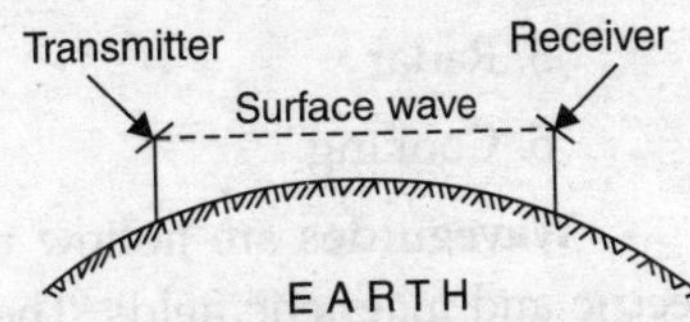


***FIGURE 26.6.*** Line of sight transmission.

TV antenna transmitters are located at places of higher altitude.

A transmission line consists of a pair of conductors. The constant spacing provides a characteristic impedance $Z_0$, which is independent of the length of the line. The characteristic impendance can be calculated from the relation.

$$Z_0 = \sqrt{\frac{L}{C}} \text{ ohms.}$$

Characteristic impedance is also known as surge impedance.

Terminating a transmission line with a resistance equal to $Z_0$ is an impedance matching that makes the line equivalent to an infinite line. The result is maximum transfer of power from line to load, no reflections and no standing waves on the line.

A resonant line section or stub is a $\frac{\lambda}{2}$ or $\frac{\lambda}{4}$ length either open or shorted at the end. The $\frac{\lambda}{4}$ stub provides an impedance transformation between high and low $Z$ at opposite ends.

A balancing unite or *balun*, can be used at either end of a line to convert between unbalanced $Z$ and balanced $Z$.

A wave guide acts as a high-pass filter. It cuts off frequencies below the cut off value but passes higher frequencies. The smaller the wave guide the higher the cut off frequency. Wave guides can be tuned with plates, stubs or screws.

**Cavity.** A microwave cavity is equivalent to a parallel resonant L–C circuit. The reason is that the cavity can be considered as an infinite number of shorted quarter wave stubs joined together. The smaller is the cavity the higher its resonant frequency. The resonant cavity is used as a band pass or band stop filter and as the tuned circuit for oscillators.

Probe and aperture coupling are used to transfer energy in or out of the cavity. A probe is a small metal tip inerted in the field. An aperture is a small opening.

**Microwaves.** The term microwaves refers to high frequencies (above 300 MHz) and short wavelengths less than 1 m. The microwave frequencies span three major bands at the highest end of *RF* spectrum — UHF, SHF and EHF. At such high frequencies, the components depend on the varying electromagnetic field rather than on the current in a wire conductor or the voltage across two points. Instead of resonant L–C circuits and conventional wire conductors, therefore, resonant cavities and wave guides are often at microwave frequencies.

Major application of microwaves is in the following field:

1. Telecommunications
2. World wide television
3. Radar
4. Industrial heating
5. Cooking
6. Research.

**Waveguides** are hollow metal tubes used to propagate microwave energy in the form of electric and magnetic fields. The mode describes the method of using the electric and magnetic fields for propagation of the microwave energy. In the transverse electric (TE) mode, the electric field $E$ is transverse to the long axis. In the transverse magnetic (TM) mode, the magnetic field is transverse to the long axis.

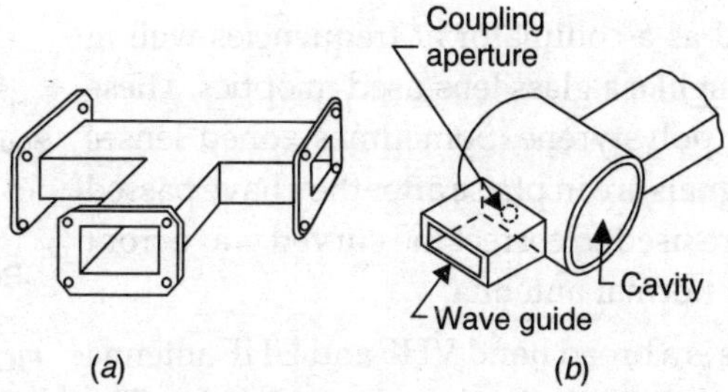


*FIGURE 26.7* (*a*) Wave guide junction – *H* plane tee.
(*b*) Microwave cavity used as a resonant band pass element.

**Microstrip.** A microstrip is a flat co-axial transmission line made as a printed circuit for application in an amplifier or oscillator in which microwave diodes and transistors are used. The characteristic impedance for microstrip is determined by the thickness of the substrate the dielectric constant of the substrate, and the width of the conductor strip.

**Microwave Antennas.** Transmitting and receiving antennas for use in the microwave spectrum tend to be directive. High gain may often be the reason, but other requirements are:

(1) Microwave antennas need not be omnidirectional.

(2) The signal at the input terminals of the receiver should be as large as possible.

(3) Power gain should be high.

(4) Directional application is often required.

(5) Need not be physically large in size.

**Antenna with paraboloidal reflectors.** The directional pattern of this antenna has a very sharp main lobe surrounded by a number of miner lobes. In these antennas all the waves are in phase as a result, radiation is very strong and concentrated. The gain is influenced by the aperture ratio and the uniformity of illumination.

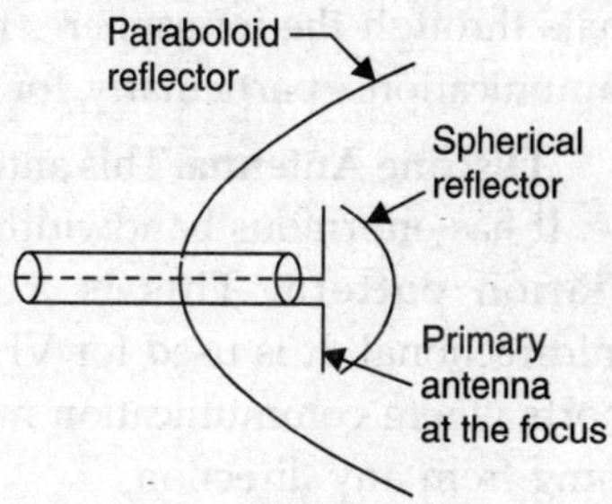


*FIGURE 26.8*

**Horn Antennas.** The born antenna is just an open section of a rectangular waveguide. It is flared out to match the characteristic impedance $Z_0$. If the flare angle of the horn is too small, resulting in a shallow horn, the wave front leaving the horn will be spherical rather than plane and the radiated beam will not be directive. This angle usually is between 40° and 66°. This antenna has fairly good directivity, an adequate bandwidth and simple mechanical construction.

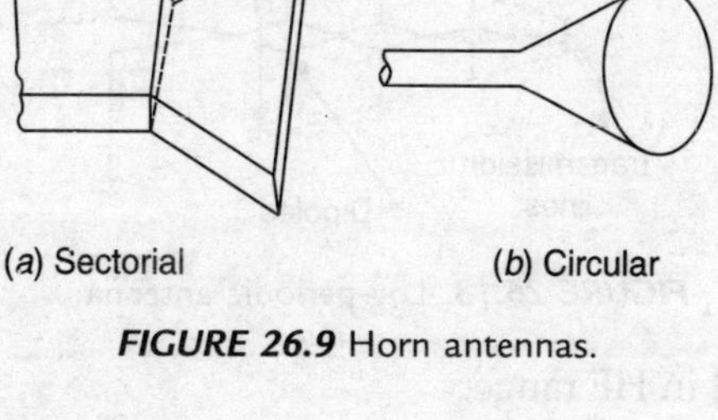

(*a*) Sectorial (*b*) Circular

*FIGURE 26.9* Horn antennas.

**Lens Antenna** is used as a collimator at frequencies well in excess of 3 GHz and work just like a glass lens used in optics. These antennas are often made of polystyrene. Sometimes zoned lenses are used to ensure that the signals are in phase after they have passed through it. These antennas are used to correct the curved wave front from a shallow horn, or as a normal antenna.

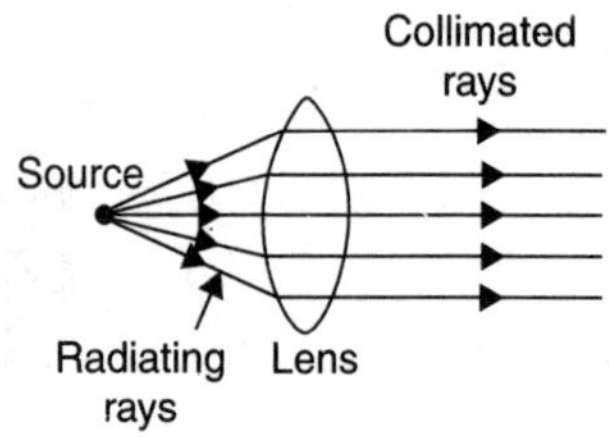


*FIGURE 26.10* Lens antenna.

**Helical Antenna.** This is a broad hand VHF and UHF antenna which is used to provide circular polarization characteristics. The antenna consists of a loosely wound helix backed up by a ground plane which is simply a screen made of "chicken" wire. There are two modes of radiation:

(*i*) **Normal** is which the radiation is in a direction perpendicular to the axis of the helix.

(*ii*) **Axial** produces a broad band, fairly directional radiation in the axial direction. When the proportions of this antenna are as shown in Fig. 26.11, it has typical values of directivity close to 25, beam width of 90° between nulls and frequency range of about 20% on either side of the centre frequency.

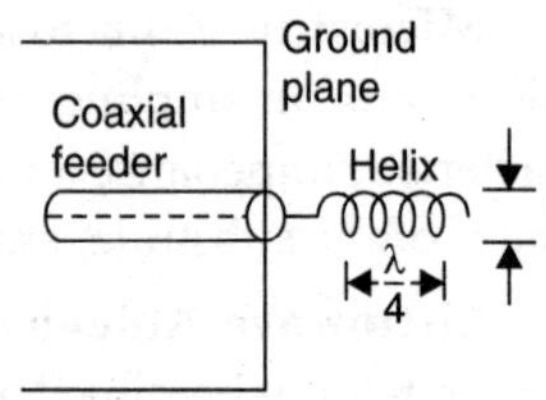


*FIGURE 26.11.* Helical antenna.

This antenna is used either singly or in an array for transmission and reception of VHF signals through the ionosphere. It is used for satellite and probe communications, particularly for radiotelemetry.

**Discone Antenna.** This antenna is a combination of disc and cone. It has enormous bandwidth for both input impedance and radiation pattern. This is a low gain antenna, but it is omnidirectional. It is used for VHF and UHF signals, specially at airports where communication must be maintained with air craft coming from any direction.

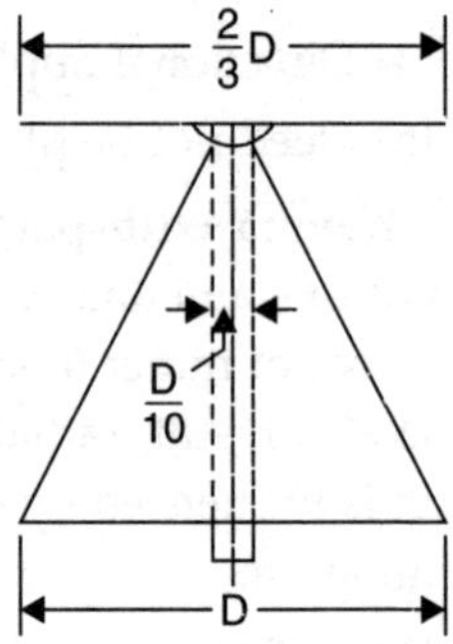


*FIGURE 26.12.* Discone antenna.

**Log Periodic Antenna.** These antennas can have bandwidths of 10 : 1, the directive gains are low to moderate and the radiation pattern may be unidirectional or bidirectional. In these antennas

$$\frac{R_1}{l_1} = \frac{R_2}{l_2} = \frac{R_3}{l_3} = \frac{R_4}{l_4} = \text{constant}$$

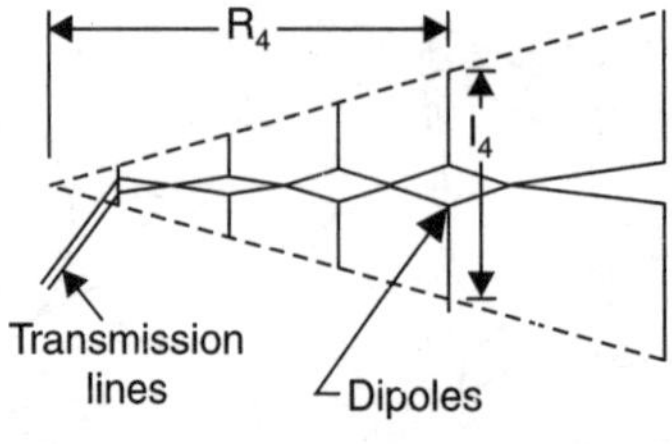


*FIGURE 26.13.* Log-periodic antenna.

These antennas are used in HF range.

## OBJECTIVE TYPE QUESTIONS

1. 10–30 kHz frequency range is categorized as
   (*a*) VLF (*b*) LF (*c*) MF (*d*) HF.
2. The frequency range for *EHF* is
   (*a*) 100 to 1000 kHz (*b*) 300 to 3000 MHz
   (*c*) 3000 to 30,000 MHz (*d*) 30,000 to 300,000 MHz.
3. The wavelength of 10 to 1 metres occurs in
   (*a*) VLF (*b*) VHF (*c*) UHF (*d*) SHF.
4. Which of the following will have the least range of wave lengths?
   (*a*) LF (*b*) VHF (*c*) SHF (*d*) VLF.
5. The frequency rang for *UHF* is
   (*a*) 300–3000 kHz (*b*) 300–3000 MHz
   (*c*) 300–3000 GHz (*d*) none of the above.
6. *EHF* is mainly used for
   (*a*) radio transmission (*b*) telephony
   (*c*) satellite communication (*d*) television.
7. Which of the following frequency range is used for maritime radio navigation?
   (*a*) VLF (*b*) LF (*c*) MF (*d*) HF.
8. The velocity of electromagnetic waves in free space
   (*a*) increases as frequency increases (*b*) is independent of frequency
   (*c*) decreases as frequency increases.
9. The intrinsic impedance of free space is
   (*a*) 1 ohm (*b*) 4 $\pi$ ohm
   (*c*) 120 $\pi$ ohm (*d*) $3 \times 10^8$ ohm.
10. A plane electromagnetic wave travels in dielectric medium of relative permittivity 9. Relative to free space, the velocity of propagation in the dielectric is
    (*a*) increased by a factor of 9 (*b*) increased by a factor of 3
    (*c*) unchanged (*d*) reduced by a factor of $\frac{1}{2}$.
11. An electromagnetic wave is incident normally on a dielectric boundary. It will be
    (*a*) totally reflected (*b*) totally refracted
    (*c*) totally absorbed (*d*) partly reflected and partly refracted.
12. Electromagnetic waves are reflected when they
    (*a*) pass through a small slot in a conducting plane
    (*b*) encounter a perfectly conducting surface
    (*c*) pass into a medium of different dielectric constant
    (*d*) are polarized at right angles to the direction of propagation.

13. The velocity of electromagnetic waves in free space is
(a) $10^{10}$ metres per second (b) $3 \times 10^8$ metres per second
(c) $3 \times 10^{10}$ metres per second (d) $3 \times 10^{12}$ metres per second.

14. Frequencies in UHF range propagate by means of
(a) space waves (b) surface waves (c) ground waves (d) sky waves.

15. The electromagnetic waves travelling in free space can be only
(a) absorbed (b) attenuated (c) refracted (d) reflected.

16. The electromagnetic waves reflected by the ionosphere due to their interaction with
(a) electrons (b) protons (c) ultraviolet rays (d) infrared rays.

17. Diffraction of electromagnetic waves is caused by
(a) the edges of sharp obstacles (b) the reflections from the ground
(c) the spherical wave fronts (d) the waves passing through a slot.

18. The polarisation of electromagnetic waves
(a) is always vertical in an isotropic medium
(b) is caused by reflection
(c) is due to the transverse nature of the waves
(d) results from the longitudinal nature of the waves.

19. The ionosphere consists of
(a) negative charge layer (b) positive charge layer
(c) neutral.

20. The intrinsic impedance of free space is given by

(a) $\sqrt{\dfrac{\mu_0}{\varepsilon_0}}$ (b) $\sqrt{\dfrac{\varepsilon_0}{\mu_0}}$ (c) $\sqrt{\mu_0 \varepsilon_0}$ (d) $\dfrac{1}{\sqrt{\mu_0 \varepsilon_0}}$.

21. The ionosphere plays a significant role in radio wave propagation at
(a) high frequencies (b) ultra-high frequencies
(c) microwave frequencies (d) optical frequencies.

22. The wavelength $\lambda$, frequency $f$ and the velocity of light are related to each other by the relation

(a) $\lambda = \dfrac{f}{c}$ (b) $\lambda fc = 1$ (c) $\lambda f = c$ (d) $\lambda c = f$.

23. For a radio station broadcasting at 1 MHz the wavelength of electromagnetic waves will be
(a) 3 m (b) 30 m (c) 300 m (d) 3000 m.

24. Virtual height of an ionospheric layer is
(a) more as the height a wave actually penetrates
(b) less as the height a wave actually penetrates
(c) same as the height a wave actually penetrates.

25. *D*-layer of ionosphere occurs only during
(*a*) day time (*b*) night time (*c*) summer time (*d*) moonlit nights.

26. The absorption of radio waves by the atmosphere depends primarily on
(*a*) their frequency (*b*) the polarisation of the waves
(*c*) the absorption of the waves (*d*) their distance from the transmitter.

27. The intrinsic impedance of free space
(*a*) increases with increase of frequency
(*b*) decreases with increase of frequency
(*c*) is independent of frequency.

28. Which of the following statement is false?
(*a*) sky waves are reflected from the ionosphere
(*b*) the skip distance is equal to the zone of silence
(*c*) surface waves are propagated through the troposphere
(*d*) none of the above.

29. Which of the following remains unaffected by changes on account of day and night?
(*a*) Ground wave (*b*) Sky wave
(*c*) Tropospheric wave (*d*) None of the above.

30. *MUF* is given by
(*a*) $f_c \sin \theta$ (*b*) $f_c \cos \theta$ (*c*) $f_c \tan \theta$ (*d*) $f_c \sec \theta$.

31. For a wave of critical frequency 25 MHz departing at an angle 60°, *MUF* will be
(*a*) 12.5 MHz (*b*) 30 MHz (*c*) 45 MHz (*d*) 50 MHz.

32. The critical frequency for ionospheric propagation is proportional to
(*a*) the maximum electron density
(*b*) square of the maximum electron density
(*c*) square root of the maximum electron density
(*d*) logarithm of the maximum electron density.

33. Tropospheric line-of-sight propagation mode is used for communications at carrier frequencies
(*a*) less than 1 kHz (*b*) less than 30 kHz
(*c*) less than 1600 kHz (*d*) greater than 30 MHz.

34. Which of the following frequencies cannot be used for reliable beyond—the—horizon terrestrial communication without the use of repeaters?
(*a*) 15 kHz (*b*) 20 MHz (*c*) 775 MHz (*d*) 12 GHz.

35. During day time, the *D* layer in the ionosphere
(*a*) disappears quickly (*b*) disappears slowly
(*c*) exists (*d*) splits into two layers.

**36.** Fading in ship to ship communication can be reduced by using

(*a*) frequency diversity (*b*) space diversity

(*c*) a broad band antenna (*d*) a more directional antenna.

**37.** High frequency waves are

(*a*) absorbed by the $F_2$ layer (*b*) reflected by the *D* layer

(*c*) affected by the solar cycle

(*d*) capable of use for long distance communications on the mars.

**38.** During day time, the ionosphere consists of

(*a*) *D*, *E*, *F* layers (*b*) *D*, *E*, $F_1$ layers

(*c*) *D*, *E*, $F_2$ layers (*d*) *D*, *E*, $F_1$ $F_2$ layers.

**39.** The critical frequency of an ionospheric layer during the day time is

(*a*) more than that for night time

(*b*) less than that for night time

(*c*) same as that for night time.

**40.** Distances near skip distance should be used for sky-wave propagation

(*a*) to avoid tilting

(*b*) to prevent sky-wave and upper ray interference

(*c*) to avoid the Faraday effect

(*d*) to not to exceed the critical frequency.

**41.** The effective refractive index of the ionosphere for radiowaves is

(*a*) more than unity (*b*) less than unity

(*c*) equal to unity

(*d*) either more or less than unity depending on the electron concentration.

**42.** Anomalous propagation is caused by

(*a*) super-refractive duct (*b*) meteoreological factors

(*c*) troposcatter (*d*) none of the above.

**43.** Short-term fading is commonly observed in

(*a*) ground wave propagation (*b*) satellite communication

(*c*) troposcatter communication (*d*) all of the above.

**44.** Which of the following term applies to troposcatter propagation?

(*a*) Faraday rotation (*b*) Fading

(*c*) SIDs (*d*) Atmospheric storms.

**45.** Tropospheric scatter is used with frequencies in the range of

(*a*) VLF (*b*) UHF (*c*) VHF (*d*) HF.

## Antenna

**46.** A duplexer is a device used to

(*a*) feed more than one receiver from a single antenna

(*b*) connect two transmitters to the same antenna

(*c*) connect a receiver and a transmitter to the same antenna

(*d*) none of the above.

**47.** For a satellite tracking at *VHF*, helical antenna is often used because of

(*a*) troposcatter (*b*) super refraction

(*c*) ionospheric refraction (*d*) Faraday effect.

**48.** Broadcasting antennas are generally

(*a*) omnidirectional type (*b*) vertical type

(*c*) horizontal type (*d*) any of the above.

**49.** In a turnstile antenna the crossed dipoles are excited with voltages

(*a*) in phase with each other (*b*) 90° out of phase with each other

(*c*) 120° out of phase with each other (*d*) 180° out of phase with each other.

**50.** Consider an isotropic radiator radiating power of 1 watt. At a distance of 1 metre, the power per unit area will be

(*a*) 1 watt/m$^2$ (*b*) $4\pi$ watt/m$^2$ (*c*) $\frac{1}{4\pi}$ watt/m$^2$ (*d*) $2\pi$ watt/m$^2$.

**51.** For which of the following antenna, the length is $\frac{\lambda}{2}$?

(*a*) Discone (*b*) Monopole (*c*) Hertz (*d*) All of the above.

**52.** For which of the following antenna, the length is $\frac{\lambda}{4}$?

(*a*) Marconi (*b*) Dipole (*c*) Discone (*d*) Hertz.

**53.** Which of the following statement is false?

A parabolic reflector is

(*a*) omnidirectional antenna

(*b*) pencil beam antenna

(*c*) popularly used antenna in terrestial line-of-sight communication

(*d*) not a broadcast antenna.

**54.** The bandwidth of an antenna is proportional to

(*a*) $Q$ (*b*) $Q^2$ (*c*) $\frac{1}{Q}$ (*d*) $\frac{1}{Q^2}$.

55. PCM channels can be engineered on

(*a*) wire circuits only (*b*) radio circuits only

(*c*) both radio and wire circuits (*d*) neither radio nor wire circuits.

56. In a pulse transmission system, the error rate

(*a*) increases as S/N ratio increases

(*b*) increases as logarithm of S/N ratio increases

(*c*) decreases as S/N ratio increases

(*d*) is independent of noise at S/N ratio above 20 dB.

57. The beamwidth of a parabolic dish antenna

(*a*) decreases with increase in the dish diameter

(*b*) decreases with decreases in the dish diameter

(*c*) is independent of the dish diameter.

58. Parabolic and lens antenna are extensively used for

(*a*) microwaves (*b*) short waves

(*c*) medium frequency waves (*d*) all of the above.

59. The term "critical frequency" denotes

(*a*) the lowest frequency that is reflected by ionosphere

(*b*) the highest frequency that is reflected by the ionosphere at vertical incidence

(*c*) the lowest frequency by which communication is possible between two given points

(*d*) none of the above.

60. A parabolic antenna is commonly used at

(*a*) 50 kHz (*b*) 500 kHz (*c*) 500 MHz (*d*) 5000 MHz.

61. A loop antenna is commonly used for

(*a*) radar (*b*) direction finding

(*c*) satellite communication (*d*) none of the above.

62. The radiation pattern of parabolic antenna is

(*a*) omni-directional (*b*) a figure of eight

(*c*) a figure of *S* (*d*) highly directional.

63. The random pattern of loop antenna is

(*a*) circle (*b*) semi-circle (*c*) cardioid (*d*) none of the above.

64. A half wave dipole at a frequency of 100 MHz has a length of

(*a*) 0.75 metre (*b*) 1.5 metre (*c*) 3 metres (*d*) 100 metres.

65. The power gain of a half wave dipole with respect to an isotropic radiator is

(*a*) 2.15 dB (*b*) 3 dB (*c*) 4.5 dB (*d*) 6 dB.

66. The power gain of a half wave dipole with respect to an isotropic radiator is

(*a*) 2.15 dB (*b*) 3 dB (*c*) 4.5 dB (*d*) 6 dB.

67. Which of the following antenna is independent of frequency?
(a) Yagi-uda (b) Log periodic antenna
(c) Folded dipole (d) All of the above.

68. The current distribution in half wave dipole is
(a) uniform (b) sinusoidal (c) triangular (d) sawtooth.

69. A Hertzian dipole has a radiation pattern with
(a) a null in the direction of dipole (b) a null in the plane perpendicular to the dipole
(c) a peak on the direction of the dipole (d) none of the above features.

70. The radiation pattern of Hertzian dipole in the plane perpendicular to the dipole is a
(a) null (b) circle (c) figure of eight (d) none of the above.

71. The electric field intensity of a Hertzian dipole at a remote point varies as
(a) $\frac{1}{r}$ (b) $\frac{1}{r^2}$ (c) $\frac{1}{r^3}$ (d) $\frac{1}{r^4}$.

72. In order to receive a vertically polarized waves, the conductor of the dipole
(a) should be mounted at an angle of 30° (b) should be mounted at an angle of 45°
(c) should be mounted at an angle of 60° (d) should be mounted vertically.

73. The input impedance of a shorted lossless line of length $\frac{\lambda}{4}$ is
(a) zero (b) $z_0$ (c) $\frac{z_0}{2}$ (d) infinity.

74. For a transmission line with $wL \geq R$ and $wC \leq G$, $\alpha$ is
(a) proportional to frequency (b) proportional to the square of frequency
(c) inversely proportional to frequency (d) constant.

75. An antenna above its critical frequency is
(a) inductive (b) capacitive (c) resistive (d) either of the above.

76. On a transmission line with standing waves the distance between a voltage maximum and adjacent current maximum is
(a) $\frac{\lambda}{8}$ (b) $\frac{\lambda}{6}$ (c) $\frac{\lambda}{4}$ (d) $\frac{\lambda}{2}$.

77. In a transmission line terminated with a load equal to the characteristic impedance, the reflection co-efficient is
(a) zero (b) plus one (c) minus one (d) infinities.

78. The television receiver antenna are usually
(a) helical antennas (b) parabolic dish type
(c) dipoles (d) none of the above.

**79.** A matching stub should be

(*a*) nearest to load (*b*) nearest to transmitter

(*c*) mid-way between the load and the transmitter.

**80.** Stub-matching eliminates standing waves on

(*a*) load side of stub (*b*) transmitter side of stub

(*c*) both sides of stub.

**81.** Which of the following antenna gets circularly polarized?

(*a*) Dipole (*b*) Helical (*c*) Parabolic (*d*) Yegi-uda.

**82.** If the spacing between the wires of a transmission line is increased, the characteristic impedance will

(*a*) increase (*b*) decrease (*c*) reduce to zero (*d*) remain antlered.

**83.** Antenna reciprocity is put to use in

(*a*) television broadcast systems (*b*) commercial radio broadcast systems

(*c*) radar system (*d*) none of these.

**84.** The reduction resistance of a half-wave dipole is

(*a*) 73 ohm (*b*) 300 ohm (*c*) 50 ohm (*d*) none of these.

**85.** Directivity of an antenna is

(*a*) same as its directive gain (*b*) directive gain expressed in decibels

(*c*) the max. value of its directive gain (*d*) none of these.

**86.** A antenna behave as a resonant circuit only when its length

(*a*) equal($\lambda/4$) (*b*) equals($\lambda/2$)

(*c*) equals($\lambda/2$) or its integral multiple (*d*) none of these.

**87.** A half wave dipole produces

(*a*) a figure-*O* directional pattern (*b*) a broadside directional pattern

(*c*) an endfire directional pattern (*d*) none of these.

**88.** An antenna having higher '*Q*' has

(*a*) a higher BW (*b*) a lower BW (*c*) a flat response (*d*) none of these.

**89.** An antenna placed in horizontal plane produces

(*a*) horizontally polarized electromagnetic waves

(*b*) vertically polarized electromagnetic waves

(*c*) elliptically polarized electromagnetic waves

(*d*) none of these.

**90.** An antenna placed in vertical plane produces

(*a*) horizontally polarized electromagnetic waves

(*b*) vertically polarized electromagnetic waves

(*c*) secularly polarized electromagnetic waves

(*d*) none of these.

**91.** A yagi antenna in the horizontal plane produces

(*a*) an end fire directional pattern

(*b*) a broadside directional pattern

(*c*) a figure eight directional pattern

(*d*) none of these.

**92.** Most commonly used antenna type with TV receivers is the

(*a*) *V*-antenna (*b*) rhombic antenna (*c*) loop antenna (*d*) yagi antenna.

**93.** Electrical length or effective length of the antenna is

(*a*) same as its physical length (*b*) greater than its physical length

(*c*) smaller than its physical length (*d*) none of these.

**94.** Addition of more and more directors in a yagi structure

(*a*) increases the directivity and *BW* of the antenna

(*b*) decreases the directivity and increases the *BW*

(*c*) increases the beam width of the antenna

(*d*) none of these.

**95.** An additional capacitance

(*a*) increases the effective length of the antenna

(*b*) decreases the effective length of the antenna

(*c*) has no effect on antenna's length

(*d*) none of these.

**96.** An additional inductance

(*a*) increases the resonant frequency of the antenna

(*b*) decreases the resonant frequency of the antenna

(*c*) compensates for parasitic capacitances

(*d*) none of these.

**97.** One of the following antenna type is broadband

(*a*) parabolic reflector (*b*) rhombic antenna

(*c*) pyramidal horns (*d*) *v*-antenna.

**98.** A rhombic antenna in a horizontal plane produces

(*a*) vertically polarized waves (*b*) an end fire pattern

(*c*) a broadside pattern (*d*) none of these.

**99.** A non-resonant rhombic antenna produces

(*a*) a unidirectional pattern (*b*) vertically polarized waves

(*c*) horizontally polarized waves (*d*) an endfire pattern.

**100.** A rhombic antenna comprises of

(*a*) 4 half wave dipoles (*b*) 4 quarter wave dipoles

(*c*) 2 quarter wave dipoles (*d*) None of these.

**101.** A *v*-antenna consists of

(*a*) two wires with an apex angle of 45 degree

(*b*) two ($\lambda/4$) long wires with an apex angle of 45 degree

(*c*) two ($\lambda/2$) long wires with an apex angle of 45 degree

(*d*) none of these.

**102.** One of the following is a non-resonant antenna

(*a*) the folded dipole (*b*) the endfire array

(*c*) the broadside array (*d*) the rhombic antenna.

**103.** One of the antenna types is the best excited from a waveguide

(*a*) horn antenna (*b*) helical antenna

(*c*) biconical antenna (*d*) log-periodic antenna.

**104.** One of the antenna types is circularly polarized

(*a*) parabolic reflector (*b*) yagi uda

(*c*) helical (*d*) small circular loop.

**105.** Top loading is used with antennas to

(*a*) increase the effective height (*b*) reduce the effective height

(*c*) increase the *BW* (*d*) increase the input capacitance.

**106.** Of the following antenna types, one is not an omni directional antenna

(*a*) log periodic antenna (*b*) half wave dipole

(*c*) Marconi antenna (*d*) Discone antenna.

**107.** A helical antenna is used for satellite tracking because of its

(*a*) large *BW* (*b*) large gain

(*c*) circular polarization (*d*) maneuverability.

**108.** Identify wideband antenna

(*a*) Marconi antenna (*b*) Discone antenna

(*c*) Folded dipole antenna (*d*) Both (*b*) and (*c*).

**109.** Directive gain of elementary doublet as compared to an isotropic radiation is

(*a*) 2 (*b*) 1.5 (*c*) 3.5 (*d*) none of these.

**110.** The radiation resistance of a ($\lambda/16$) wire dipole in free space is approximately

(*a*) 1 ohm (*b*) 3 ohm (*c*) 73 ohm (*d*) 300 ohm.

**111.** One of the following antenna can be used for directive finding

(*a*) Half wave dipole (*b*) Loop antenna

(*c*) Broadside antenna array (*d*) End fire antenna array.

**112.** Microwave antenna has

(*a*) high gain (*b*) high front to back ratio

(*c*) high input impedance (*d*) low input impedance.

**113.** If the diameter of the parabolic reflector of a microwave antenna is doubled, its gain will increase by

(*a*) 0 db (*b*) 6 db (*c*) 3 db (*d*) None of these.

**114.** A higher directivity is specified by

(*a*) high gain and low BW (*b*) high gain and large BW

(*c*) low gain and low BW (*d*) none of these.

**115.** A antenna is synonymous to a

(*a*) generator (*b*) reflector (*c*) transducer (*d*) regulator.

**116.** A yagi-uda antenna does not have

(*a*) high BW (*b*) high gain

(*c*) a parasitic director (*d*) parasitic reflector.

**117.** A resonant antenna is a system with

(*a*) zero SWR (*b*) a very high SWR

(*c*) a very low SWR (*d*) none of these.

**118.** Most of the man made noise has

(*a*) circular polarization (*b*) horizontal polarization

(*c*) vertical polarization (*d*) elliptic polarization.

**119.** The Marconi antenna produces

(*a*) figure eight directional pattern (*b*) pencil beam direction pattern

(*c*) circular directional pattern (*d*) none of these.

**120.** The radiation resistance of a long wire resonant antenna (5 λ) long will be

(*a*) 73 ohm (*b*) 142 ohm (*c*) 365 ohm (*d*) 730 ohm.

**121.** One of the following antenna types is also referred to as a frequency independent antenna

(*a*) Log periodic antenna (*b*) Dipole antenna

(*c*) Marconi antenna (*d*) Rhombic antenna.

**122.** That portion of the plane near the antenna, perpendicular to the direction of radiation, through which major part of radiation passes is called

(*a*) near field region (*b*) antenna aperture

(*c*) effective antenna diameter (*d*) effective radiating area.

**123.** The region of the field of antenna where the angular field distribution is essentially independent of radial distance from the antenna is called the

(*a*) near field region (*b*) induction field region

(*c*) crossover region (*d*) far field region.

**124.** The ratio of the directive of an antenna to its directive gain in a specified direction towards the back is called

(*a*) side lobe suppression (*b*) front-to-back ratio

(*c*) directivity (*d*) none of these.

**125.** A hypothetical lossless antenna having equal radiation intensity in all direction is the

(*a*) isotropic radiator (*b*) half wave dipole

(*c*) elementary doublet (*d*) wire antenna.

**126.** The ratio of the total power radiated by an antenna to the net power accepted by the antenna is the

(*a*) directivity (*b*) directive gain

(*c*) effective radiated power (ERP) (*d*) radiation efficiency.

**127.** If the maximum linear dimension of an antenna ($D$) is not large as compared to (?), then one of the following regions may not exist

(*a*) far field region (*b*) radiating near field region

(*c*) induction field region (*d*) none of these.

**128.** If the maximum linear dimension of an antenna ($D$) is large as compared to ($\lambda$), then the far field region can be taken to exist at a distance

(*a*) greater than ($2D^2/\lambda$) (*b*) less than ($2D^2/\lambda$)

(*c*) greater than ($2D/\lambda$) (*d*) greater than ($\lambda/2D^2$).

**129.** Higher value of SNR

(*a*) higher ohmic losses

(*b*) higher ohmic losses and possibility of insulation breakdown

(*c*) higher radiation resistance

(*d*) none of these.

**130.** A transmission line has a VSWR of 3 when terminated in certain load. The reflection co-efficient is

(*a*) 1/3 (*b*) 1/2 (*c*) 3 (*d*) 1/3.

**131.** A non-resonant transmission line has no reflections irrespective of the length because

(*a*) its characteristics impedance is very small

(*b*) its characteristics impedance is very large

(*c*) it is always terminated in its characteristics impedance

(*d*) none of these.

**132.** Which of the following applies to 300 ohm twin lead transmission line?

(*a*) Low loss, shielded, unbalanced (*b*) High loss, shielded, unbalanced

(*c*) Low loss, unshielded, balanced (*d*) None of these.

## Microwaves

**133.** Microwaves refer to high frequencies above

(*a*) 1 MHz (*b*) 30 MHz (*c*) 300 MHz (*d*) 3000 MHz.

**134.** The wavelength of microwaves ranges from

(*a*) 100 m to 10 m (*b*) 10 m to 1 m (*c*) 1 m to 1 mm (*d*) 1 mm to 1 μm.

**135.** Microwave frequencies span over which three major frequency bands?

(*a*) LF, MF and HF (*b*) MF, HF and VHF

(*c*) VHF, UHF and SHF (*d*) UHF, SHF and EHF.

**136.** How many megahertz are there in one giga hertz?

(*a*) 10 (*b*) 100 (*c*) 1000 (*d*) 10,000.

**137.** The wavelength of microwaves at 100 GHz will be

(*a*) 3 m (*b*) 30 cm (*c*) 3 cm (*d*) 0.3 cm.

**138.** Antennas commonly used for microwave links are

(*a*) loop antennas (*b*) log-periodic antennas

(*c*) paraboloidal dishes (*d*) rhombic antennas.

**139.** A range of microwave frequencies more easily passed by the atmosphere than are the others is called

(*a*) a gyro-frequency range (*b*) a window

(*c*) a critical frequency (*d*) a resonance in the atmosphere.

**140.** A duplexer is used

(*a*) to couple two different antennas to transmitter without mutual interference

(*b*) to allow the one antenna to be used without .mutual interference

(*c*) to prevent interference between two antennas when they are connected to a receiver

(*d*) to increase the speed of the pulses in pulsed radar.

**141.** Which of the following microwave tube amplifier uses an axial magnetic field and a radial electric field?

(*a*) travelling wave magnetron (*b*) CFA

(*c*) coaxial magnetron (*d*) none of the above.

**142.** The characteristic impedance of a transmission line of infinite length is

(*a*) $\dfrac{R + j\omega L}{G + j\omega C}$ (*b*) $\sqrt{\dfrac{R + j\omega L}{G + j\omega C}}$

(*c*) $\sqrt{(R + j\omega L)\,(G + j\omega C)}$ (*d*) none of the above.

**143.** Which of the following is unlikely to be used as a pulsed device?

(*a*) TWT (*b*) CFA (*c*) BWO (*d*) all of the above.

**144.** When a microwave signal follows the curvature of the earth this is known as

(*a*) ionospheric reflection (*b*) tropospheric scatter

(*c*) the Faraday effect (*d*) ducting.

**145.** Which of the following statement is false?

(*a*) waveguides are used to minimize skin effect at microwave frequencies

(*b*) a rectangular waveguide serves as a high pass filter

(*c*) $TE_0$ is the dominant mode for coaxial cable

(*d*) none of the above.

**146.** The resonant frequency of a magnetron oscillator is determined by

(*a*) the cathode (*b*) the cavities

(*c*) both (*a*) and (*b*) above (*d*) none of the above.

**147.** VSWR is defined as the ratio of

(*a*) $VSWR = \frac{V_{min.}}{V_{max.}}$ (*b*) $VSWR = \frac{V_{max.}}{V_{min.}}$

(*c*) $VSWR = \frac{V_{max.} - V_{min.}}{V_{max.} + V_{min.}}$ (*d*) $VSWR = \frac{V_{max.} + V_{min.}}{V_{max.} - V_{min.}}$.

**148.** A quarterwave monopole at a frequency of one MHz has a length of

(*a*) one metre (*b*) 75 metres (*c*) 150 metres (*d*) 300 metres.

**149.** The main reasons one to which vacuum tubes are not used at microwave frequencies is

(*a*) too low series inductive reactance at microwave frequencies

(*b*) too large value of shunt capacitive reactance at microwave frequencies

(*c*) too high noise figure

(*d*) availability of too short transit time.

**150.** Which of the following statement is false?

(*a*) The GASFET is a field effect transistor used for microwave amplifier

(*b*) Generally the power ratings for the GASFET are less than 1 W

(*c*) A microstrip is similar to a flat coaxial cable but is in printed circuit form

(*d*) A microstrip is too lossy for use in microwave oscillator circuits.

**151.** VSWR is related with the reflection co-efficient *k* as

(*a*) $VSWR = \frac{1+|k|}{|k|}$ (*b*) $VSWR = \frac{1-|k|}{|k|}$

(*c*) $VSWR = \frac{1+|k|}{1-|k|}$ (*d*) $VSWR = \frac{1-|k|}{1+|k|}$.

**152.** The major source of thermal noise in a microwave system is

(*a*) waveguide feeder (*b*) receiver mixer

(*c*) TWT amplifier in transmitter (*d*) FM demodulator.

**153.** Which of the following diode is not used as a microwave mixer or detector?

(*a*) PIN diode (*b*) Crystal diode

(*c*) Schottky-barrier diode (*d*) Backward diode.

**154.** When a microwave signal follow the curvature of the earth, this is known as

(*a*) the Faraday effect (*b*) ducting

(*c*) tropospheric scatter (*d*) ionospheric reflection.

**155.** The normal fade margin in a line of sight microwave system is

(*a*) 10 dB (*b*) 25 dB (*c*) 40 dB (*d*) 60 dB.

**156.** The magnese ferrite may be used as a

(*a*) circulator (*b*) garnet (*c*) phase shifter (*d*) isolator.

**157.** One of the following is not a reason for the use of antenna coupler

(*a*) to make the antenna link resistive

(*b*) to provide the output amplifier with the correct load impedances

(*c*) to discriminate against harmonics

(*d*) to prevent reradiation of the local oscillator.

**158.** For the same size of antenna dish more gain will be provided at

(*a*) 6 MHz (*b*) 8 MHz (*c*) 10 MHz (*d*) 12 MHz.

**159.** A PIN diode is

(*a*) a microwave miser diode

(*b*) a device used for microwave detection

(*c*) a metal semi-conductor point-contact diode

(*d*) generally used as a microwave switch.

**160.** Which of the following antenna is best excited from a wave guide?

(*a*) Biconical (*b*) Helical (*c*) Horn (*d*) Discone.

**161.** A helical antenna is used for satellite tracking because of its

(*a*) broad bandwidth (*b*) maneuverability

(*c*) circular polarization (*d*) good front-to-back ratio.

**162.** Which of the following is digital system?

(*a*) PCM (*b*) PPM (*c*) PWM (*d*) all of the above.

**163.** The largest application of microwaves is in

(*a*) industrial heating (*b*) cooking

(*c*) communications (*d*) research.

**164.** The angle between electric and magnetic fields in a waveguide is

(*a*) 0° (*b*) 90° (*c*) 180° (*d*) 360°.

**165.** The primary purpose of the helix in a travelling-wave tube is to

(*a*) prevent the electron beam from spreading in the long tube

(*b*) reduce the axial velocity of the RF field

(*c*) ensure broad band operation

(*d*) reduce the noise figure.

**166.** Waveguide feeders are pressurised to

(*a*) reduce loss (*b*) reduce noise

(*c*) prevent ingress of moisture (*d*) detect faults.

**167.** Which of the following is preferred for coupling a coaxial line to a parallel line?

(*a*) Slotted line (*b*) Directional coupler

(*c*) Balun (*d*) Quarter wave transformer.

**168.** The dominant mode for rectangular waveguides is

(*a*) TEM (*b*) $TE_{10}$ (*c*) $TE_{11}$ (*d*) $TM_{11}$.

**169.** The dB loss of a 500 W signal attenuated to $1.26 \times 10^{-16}$ W will be

(*a*) 196 dB (*b*) 88 dB (*c*) – 88 dB (*d*) – 196 dB.

**170.** The attenuation in waveguides near the cutt off frequency is

(*a*) very high (*b*) very low (*c*) zero (*d*) infinite.

**171.** The reflex klystron is often preferred to the two cavity klystron oscillators because the former

(*a*) is more efficient (*b*) has higher number of modes

(*c*) is easier to tune (*d*) produces a higher output power.

**172.** The length of a quarter wave stub in free space, at 4 GHz will be

(*a*) 1.875 m (*b*) 18.75 cm (*c*) 1.875 cm (*d*) 0.1875 cm.

**173.** For satellite communication, the frequency should be

(*a*) less than the critical frequency of ionosphere

(*b*) equal to the critical frequency of ionosphere

(*c*) more than the critical frequency of ionosphere.

**174.** The uplink and downlink frequencies are made different in satellite links in order to

(*a*) reduce transmitter power

(*b*) increase the isolation between the satellite transmitter and receiver

(*c*) reduce antenna size (*d*) none of the above.

**175.** The signal propagation time between two ground stations in a synchronous satellite link is about

(*a*) 25 mts (*b*) 50 mts (*c*) 135 mts (*d*) 270 mts.

## ANSWERS

| | | | | | | |
|---|---|---|---|---|---|---|
| **1.** (*a*) | **2.** (*d*) | **3.** (*b*) | **4.** (*c*) | **5.** (*b*) | **6.** (*c*) | **7.** (*b*) |
| **8.** (*b*) | **9.** (*c*) | **10.** (*d*) | **11.** (*d*) | **12.** (*c*) | **13.** (*b*) | **14.** (*a*) |
| **15.** (*b*) | **16.** (*a*) | **17.** (*a*) | **18.** (*c*) | **19.** (*a*) | **20.** (*a*) | **21.** (*a*) |
| **22.** (*c*) | **23.** (*c*) | **24.** (*a*) | **25.** (*a*) | **26.** (*a*) | **27.** (*c*) | **28.** (*b*) |
| **29.** (*a*) | **30.** (*d*) | **31.** (*d*) | **32.** (*c*) | **33.** (*d*) | **34.** (*d*) | **35.** (*c*) |
| **36.** (*a*) | **37.** (*c*) | **38.** (*b*) | **39.** (*a*) | **40.** (*b*) | **41.** (*b*) | **42.** (*b*) |
| **43.** (*c*) | **44.** (*b*) | **45.** (*b*) | **46.** (*c*) | **47.** (*d*) | **48.** (*b*) | **49.** (*b*) |
| **50.** (*c*) | **51.** (*c*) | **52.** (*a*) | **53.** (*a*) | **54.** (*c*) | **55.** (*c*) | **56.** (*d*) |
| **57.** (*a*) | **58.** (*a*) | **59.** (*b*) | **60.** (*d*) | **61.** (*b*) | **62.** (*d*) | **63.** (*c*) |

| | | | | | | |
|---|---|---|---|---|---|---|
| 64. (*b*) | 65. (*a*) | 66. (*a*) | 67. (*b*) | 68. (*b*) | 69. (*a*) | 70. (*c*) |
| 71. (*a*) | 72. (*d*) | 73. (*d*) | 74. (*d*) | 75. (*a*) | 76. (*c*) | 77. (*a*) |
| 78. (*c*) | 79. (*a*) | 80. (*c*) | 81. (*b*) | 82. (*a*) | 83. (*c*) | 84. (*c*) |
| 85. (*d*) | 86. (*c*) | 87. (*d*) | 88. (*c*) | 89. (*b*) | 90. (*a*) | 91. (*b*) |
| 92. (*b*) | 93. (*b*) | 94. (*d*) | 95. (*c*) | 96. (*b*) | 97. (*b*) | 98. (*b*) |
| 99. (*c*) | 100. (*d*) | 101. (*c*) | 102. (*c*) | 103. (*b*) | 104. (*b*) | 105. (*b*) |
| 106. (*b*) | 107. (*d*) | 108. (*d*) | 109. (*d*) | 110. (*b*) | 111. (*c*) | 112. (*c*) |
| 113. (*c*) | 114. (*b*) | 115. (*b*) | 116. (*b*) | 117. (*c*) | 118. (*b*) | 119. (*d*) |
| 120. (*c*) | 121. (*c*) | 122. (*c*) | 123. (*c*) | 124. (*b*) | 125. (*d*) | 126. (*a*) |
| 127. (*a*) | 128. (*c*) | 129. (*c*) | 130. (*b*) | 131. (*b*) | 132. (*a*) | 133. (*b*) |
| 134. (*b*) | 135. (*b*) | 136. (*d*) | 137. (*d*) | 138. (*d*) | 139. (*b*) | 140. (*a*) |
| 141. (*c*) | 142. (*b*) | 143. (*a*) | 144. (*c*) | 145. (*d*) | 146. (*d*) | 147. (*a*) |
| 148. (*c*) | 149. (*a*) | 150. (*a*) | 151. (*c*) | 152. (*b*) | 153. (*d*) | 154. (*b*) |
| 155. (*b*) | 156. (*b*) | 157. (*b*) | 158. (*a*) | 159. (*c*) | 160. (*a*) | 161. (*b*) |
| 162. (*c*) | 163. (*c*) | 164. (*b*) | 165. (*a*) | 166. (*b*) | 167. (*d*) | 168. (*b*) |
| 169. (*a*) | 170. (*d*) | 171. (*b*) | 172. (*a*) | 173. (*b*) | 174. (*b*) | 175. (*c*). |

Chapter 27

# MICROPROCESSOR AND DIGITAL ELECTRONICS

In the decimal number system of counting of 10s the digits 0, 1, 2, 3, 4, 5, 6, 7, 8 and 9 are used. Thus there are 10 digits including 0. The system is known as decimal system. It has a radix of 10.

In digital system, however, only the digits 0 and 1 are used. This system is also known as binary system. It has a radix of 2. The reason for using binary system is that digital circuits operate with square wave pulse waveforms which have only two amplitudes —HIGH and LOW or ON and OFF.

There are many applications of digital electronics in addition to the natural function of counting pulses. Even audio and video information can be converted into digital form by pulse sampling. The advantage of digital circuits is the excellent signal to noise ratio.

**Decimal to Binary Conversion.** The divide by two method for the conversion of decimal number to digital number is illustrated in Fig. 27.1. The method is self explanatory. The representation of number is expressed as

$$23_{10} = 10111_2$$

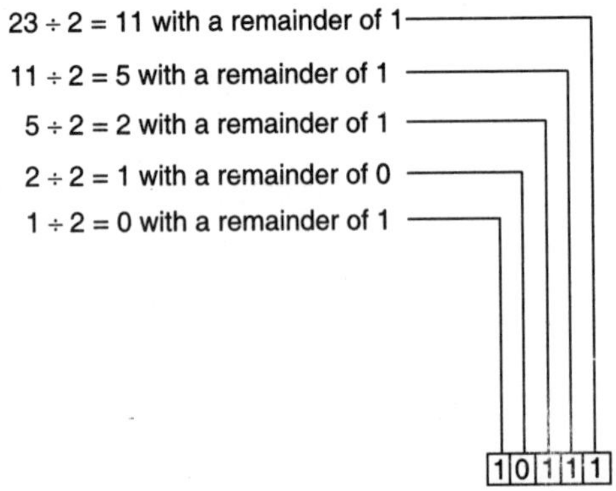


*FIGURE 27.1*

Subscript 10 stands for decimal number and subscript 2 for binary numbers.

**Binary Equivalents of Decimal Numbers**

| *Decimal Count* | *Binary Count* | | | | |
|---|---|---|---|---|---|
| | $2^4$ *Sixteenths* | $2^3$ *Eights* | $2^2$ *Fours* | $2^1$ *Twos* | $2^0$ *Ones* |
| 0 | 0 | 0 | 0 | 0 | 0 |
| 1 | 0 | 0 | 0 | 0 | 1 |
| 2 | 0 | 0 | 0 | 1 | 0 |
| 3 | 0 | 0 | 0 | 1 | 1 |
| 4 | 0 | 0 | 1 | 0 | 0 |
| 5 | 0 | 0 | 1 | 0 | 1 |
| 6 | 0 | 0 | 1 | 1 | 0 |
| 7 | 0 | 0 | 1 | 1 | 1 |
| 8 | 0 | 1 | 0 | 0 | 0 |
| 9 | 0 | 1 | 0 | 0 | 1 |
| 10 | 0 | 1 | 0 | 1 | 0 |
| 11 | 0 | 1 | 0 | 1 | 1 |
| 12 | 0 | 1 | 1 | 0 | 0 |
| 13 | 0 | 1 | 1 | 0 | 1 |
| 14 | 0 | 1 | 1 | 1 | 0 |
| 15 | 0 | 1 | 1 | 1 | 1 |
| 16 | 1 | 0 | 0 | 0 | 0 |
| 17 | 1 | 0 | 0 | 0 | 1 |
| 18 | 1 | 0 | 0 | 1 | 0 |
| 19 | 1 | 0 | 0 | 1 | 1 |
| 20 | 1 | 0 | 1 | 0 | 0 |

**Binary to Decimal Conversion.** The process of converting from binary to decimal just involves counting the place values for base 2. From right to left, they are 1, 2, 4, 8, 16, 32, 64, 128, etc, all in powers of 2.

**Binary Logic Gates.** Logic gates are circuits the output voltage of which can be predicted from the conditions at the input. Various gates are given in table below.

## Logic Gates

| *Gate* | *Truth table* | | | *Boolean Equation* |
|---|---|---|---|---|
| *AND* | *Inputs* | | *Output* | |
| | *A* | *B* | *Y* | $A \, . \, B = Y$ |
| | 0 | 0 | 0 | |
| | 0 | 1 | 0 | |
| | 1 | 0 | 0 | |
| | 1 | 1 | 1 | |
| *OR* | *Inputs* | | *output* | |
| | *A* | *B* | *Y* | $A + B = Y$ |
| | 0 | 0 | 0 | |
| | 0 | 1 | 1 | |
| | 1 | 0 | 1 | |
| | 1 | 1 | 1 | |
| *INVER NAND* | *Inputs* | | *Output* | $A = \overline{A}$ |
| | *A* | *B* | *Y* | $\overline{AB} = Y$ |
| | 0 | 0 | 1 | |
| | 0 | 1 | 1 | |
| | 1 | 0 | 1 | |
| | 1 | 1 | 0 | |
| *NOR* | *Inputs* | | *Output* | |
| | *A* | *B* | *Y* | $\overline{A + B} = Y$ |
| | 0 | 0 | 1 | |
| | 0 | 1 | 0 | |
| | 1 | 0 | 0 | |
| | 1 | 1 | 0 | |
| *XOR* | *Inputs* | | *Output* | |
| | *A* | *B* | *Y* | $A \oplus B = Y$ |
| | 0 | 0 | 0 | |
| | 0 | 1 | 1 | |
| | 1 | 0 | 1 | |
| | 1 | 1 | 0 | |
| *XNOR* | *Inputs* | | *Output* | |
| | *A* | *B* | *Y* | $\overline{A \oplus B} = Y$ |
| | 0 | 0 | 1 | |
| | 0 | 1 | 0 | |
| | 1 | 0 | 0 | |
| | 1 | 1 | 1 | |

**Electronic Adders.** The rules of binary addition are

$$0 + 0 = 0$$
$$0 + 1 = 1$$
$$1 + 0 = 1$$

We cannot have 1 + 1 = 2 in binary system. It is 1 + 1 = 0 plus a carry of 1 to the next position towards left. This addition with a carry in can occur for any column except the 1s place.

**Half Adders.** Electronic adders are constructed from logic gates to add binary numbers. A half adder combines, 0 and 1 with a carry of 1 to the next place if needed, but is has no terminal for carry in. A circuit for half adder using *XOR* and *AND* gate is shown in Fig. 27.2. The truth table is given below.

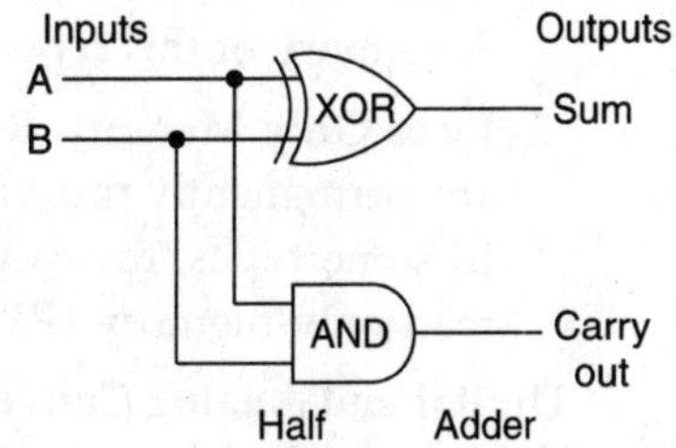


***FIGURE 27.2***

**Truth Table for Half Adder**

| *Inputs* | | *Outputs* | |
|---|---|---|---|
| *B* | *A* | *Sum* | *Carry* |
| 0 | 0 | 0 | 0 |
| 0 | 1 | 1 | 0 |
| 1 | 0 | 1 | 0 |
| 1 | 1 | 0 | 1 |
| Digits to be Added | | *XOR* | *AND* |

**The Full Adders.** A full adder has the provision for carry in and carry out. A full adder is formed by using two half-adders and an *OR* gate. The output of the *OR* gate forms the carry-out output. The full adder is for binary addition in all places except the 1s place.

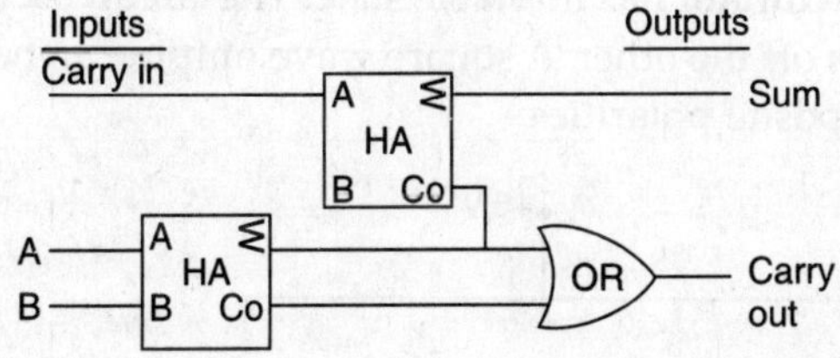


***FIGURE 27.3*** Full adder. *HA* represents half adder.

**Flip Flop.** The flip-flop (*FF*) is a basic digital circuit with a memory characteristic that can be used to store information. The circuit is a bistable multivibrator that stays in one state until switched to the opposite state. The two outputs $Q$ and $\overline{Q}$ are always opposite as HIGH and LOW logic levels.

The main types of flip flops are the *R* – *S*, with set and reset input terminals, *D* and *J* – *K*. Clock input pulses to time the switching make the flip flop synchronous. The clocked *J* – *K* flip-flop is often used in the toggle mode, in which the $Q$ and $\overline{Q}$ outputs are reversed by each clock input pulses.

**Semi-conductor Memories.** Two types of semi-conductor memories are in use. These are

1. **Random Access Memory (RAM).** The function of memory is the storage of digital information as 1*s* and 0*s*. *RAM* is also known as **read-write-memory** because it can provide both functions. However, this is a temporary type of storage because the information is destroyed when the power is turned off. For this reason it is also called volatile memory. A memory of this type can use the flip-flop circuit for storage.
2. **Read Only Memory, ROM.** This is a permanent memory. It contains memory cells that are permanently programmed by the manufacturer with a specific pattern of 1*s* and 0*s*. In some types, the *ROM* can be programmed by the user. It is then a programmable read-only-memory (*PROM*).

**Digital and Analog Converters.** These converters change information from one form to the other. Thus a digital multimeter takes an analog input and gives a digital output. Thus it is analog to digital, *A*/*D* converter. We can also have devices that take digital input and give out analog output. Such converters are known as digital to analog or *D*/*A* converters.

The circuit for *A*/*D* converter includes an op amp as a comparator to indicate when the analog signal is higher or lower than a reference voltage.

The circuit for *D*/*A* converter includes an op amp as a summing amplifier to form an analog voltage from the bits of digital information.

**Multiplexers and Demultiplexers.** An electronic multiplexer or demultiplexer operates like a single pole many position rotary switch. The multiplexer selects data one at a time for serial output from the parallel line at the input. The multiplexer does the opposite; it changes serial data from the input to the parallel form at the output.

**Clock Timer.** Most digital systems need a clock to time the operation of the various system parts. An electronic clock is a timing device that generates a continuous series of pulses. The pulses provide a digital waveform in which HIGH and LOW are repeated at a specific frequency.

**Free Running Multivibrator.** Fig. 27.4 shows a free running multivibrator using transistors $Q_1$ and $Q_2$. An a stable multivibrator has no stable state. The circuit oscillates continuously because conduction in one stage cuts off the other. A square wave output can be taken from the collector of either $Q_1$ or $Q_2$ but with opposite polarities.

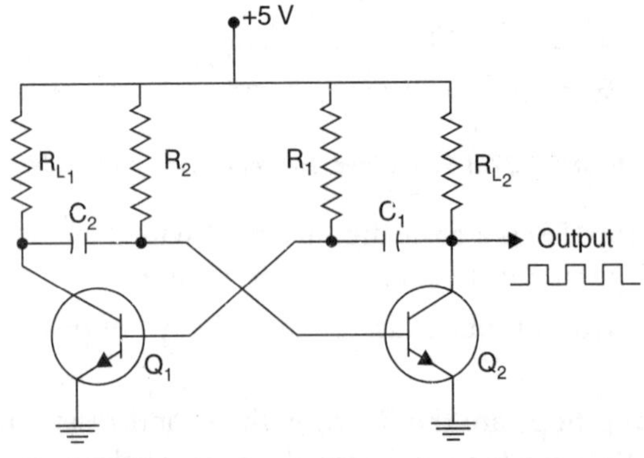


*FIGURE 27.4* Free running multivibrator.

The pulse output may be symmetrical or not, depending on the *R* and *C* values in each stage. When one stage is cut off longer than the other, the output is asymmetrical.

### Microprocessor Architecture

The microprocessor is the central processing unit (*CPU*) of a micro computer. It is the heart of the microcomputer.

**Intel 8085.** It is an 8 bit, *N MOS* microprocessor. It is a 40 pin *IC* package fabricate on a single *LSI* chip. The intel 8085 uses a single +5V dc supply for its operation. Its clock speed is about 3 MHz. the clock cycle is of 320 ns. The time for the clock cycle of the intel 8085 is 200 ns. It has 80 basic instructions and 246 opcodes. Intel 8085 consists of three main sections, arithmetic and logic unit, timing and control unit and several registers.

**Arithmetic Logic Unit.** The arithmetic and logic unit (*ALU*) performs the following the arithmetic and logic operations.

1. Addition
2. Subtraction
3. Logical AND
4. Logical OR
5. Logical exclusive OR
6. Complement
7. Increment
8. Decrement
9. Left shift
10. Clear

**Timing and Control Unit.** It generates timing and control signals which are necessary for the execution of instructions. It controls data flow between *CPU* and peripherals including memory. It provides status, control and timing signals which are required for the operation of memory and *I/O* device. It controls the entire operation of the microprocessor and peripherals connected to it. Thus it is seen that control unit of the *CPU* acts as the brain of the computer system.

**Registers.** Registers are used by the microprocessor for temporary storage and manipulation of data and instruction. Data remain in the register till they are sent to the memory or *I/O* device.

Intel 8085 microprocessor has following registers.

1. One 8- bit accumulator- register *A*
2. Six 8- bit general purpose registers. These are *B, C, D, E, H* and *L*.
3. One 16-bit stack pointer, *SP*.
4. One 16 bit program counter, *PC*.
5. Instruction register
6. Status register
7. Temporary register.

Intel 8085 microprocessor also contains address buffer and data/address buffer.

The program counter *PC* contains the address of the next instructions. The *CPU* fetches an instruction from the memory, executes it and increments the contents of the program counter.

Thus in the next instruction cycle it will fetch the next instruction of the program. Instructions are executed sequentially unless an instruction changes the contents of the program counter. The instruction register holds the instruction until it is decoded. This can not be accessed by the programmer.

The stack pointer *SP*, holds the address of the address of the stack top. The stack is the sequence of memory locations defined by the programmer. The stack is used to save the contents of the register. During the execution of the program, the last memory location of the occupied portion of the stack is called stack top.

These are six 8-bit registers which are used for general purposes as desired by the programmer. There are; *A, B, C, D, E, H* and *L*. To handle 16-bit data two 8-bit registers can be combined. The combination of two 8-bit registers is called a register pair. The valid register pair of intel 8085 are B-C, D-E, and H-L. The *HL* pair is used to address memories. The register A in accumulator in intel 8085. This is used during the execution of a program for temporary storage. There are five flip-flops, which acts as status flags. Each of these flip-flops holds 1-bit flag that indicates certain conditions which arises during arithmetic and logic operation. The following status flags have been provided in intel 8085.

1. CARRY (*CS*),
2. ZERO (*Z*),
3. SIGN (*S*),
4. PARITY (*P*) and
5. AUXILIARY CARRY (*AC*).

**CARRY (*CS*):** It holds carry out of the most significant bit resulting from the execution of an arithmetic operation. If there a carry from the addition or borrow from subtraction or comparison the carry flag(S is set to 1) otherwise 0.

**ZERO (*Z*):** The zero status flag *Z* is set to 1 if the result of arithmetic or logical operation is zero. For non zero result, it is set to zero.

**SIGN (*S*):** The sign status flag is set to 1. If the most significant bit of the result of an arithmetic or logical operations is 1, otherwise 0.

**PARITY (*P*):** The parity status flag is set to 1, if the most significant bit of the result of arithmetic or logical operations contains even no. of 1s. It is set to zero when there is odd no. of 1s.

**AUXILIARY CARRY (*AC*):** The auxiliary carry status flags hold carry out of bit 3 to 4, resulting from the execution of an arithmetic operation.

**DATA and ADDRESS bus:** The intel 8085 is an eight bit microprocessor. Its data bus is 8-bit wide and hence 8-bit data can be transmitted in parallel from or to the microprocessor. The 8085 microprocessor requires 16 –bit wide address bus as the memory addresses are of 16-bits. The most significant bits of the address are transmitted by the address bus, A-bus. The 8 least significant bits of the addresses/data bus transmits data and addressee at different moments. At a particular moment it transmits either data or address. Thus the *AD* bus operates in time shared mode. This technique is known as multiplexing.

**Addressing modes:** Each instruction requires certain data on which it has to operate. Various techniques to specify data for instruction is called addressing mode. Intel 8085 uses the following addressing modes.

1. Direct addressing
2. Register addressing
3. Register indirect addressing
4. Immediate addressing

**Instruction cycle:** It is a command given to a computer to perform a specified operation on given data. Program and data store in the memory. The *CPU* fetches one instruction from the memory at a time and executes it. The necessary steps that a *CPU* carries out to fetch an instructions and necessary data form the memory and to execute it constitutes and instruction cycle. An instruction cycle consists of a fetch cycle and executes cycle. In fetch cycle a *CPU* fetches opcodes from the memory. The necessary steps which are carried out to fetch an opcode from the memory constitutes the fetch cycle. The necessary steps which are carried out to get data, if any from the memory and to perform the specified operation specified in an instruction constitutes an execute cycle. The total time required to execute an instruction is given by

$$IC = FC + EC$$

**Fetch operation:** 1st byte of an instruction is its opcodes and instruction may be more than one byte long. The other bytes are the data or operand address. The program counter (*PC*) keeps the memory address of the next instruction to be executed. In the beginning of a fetch cycle contain of the program counter, where is the address of the memory location where the opcode is available, is sent to the memory. The memory places the opcode on the data bus so as to transfer it to *CPU*.

The entire operation of the fetching an opcode takes three clock cycle.

**Execute operation:** The opcode fetched from the memory goes to data register, *DR* (data/address buffer in 8085) and then to instruction register, *IR*. From the instruction register it goes to the decoder circuitry which decodes the instructions. The decoder circuitry is with in the microprocessor. After the instruction is decoded execution begins. If the operand is in general purpose registers, execution is immediately performed. The time taken in decoding and execution is one clock cycle. If an instruction contains data or operand and address which are still in memory, the *CPU* has performed some read operation to get the desired data. After receiving the data it performs executes operation. A read cycle is similar to a fetch cycle. In some instruction write operation is performed. In write cycle data send from the *CPU* to the memory or an output device.

**Machine cycle and state:** The necessary step is carried out to perform the operation of accessing either memory or *I/O* device, constitute a machine cycle. In other words necessary steps carried out to perform fetch, a read or a write operation constitutes a machine cycle. In machine cycle one basic operation such as opcode fetch, memory read, memory write, *I/O* read or write operation is performed. An instruction cycle consists of several machine cycles.

## OBJECTIVE TYPE QUESTIONS

**1.** A pure sine wave is

(*a*) a digital signal
(*b*) a digital signal at higher frequencies
(*c*) an analog signal
(*d*) neither digital nor analog signal.

2. Which of the following can provide a digital signal?
   (*a*) A cosine wave (*b*) Slow turning of a potentiometer
   (*c*) Variable resistor (*d*) A square wave.
3. The number of levels in a digital signal is
   (*a*) one (*b*) two (*c*) always odd (*d*) ten.
4. Decimal 1 in binary system can be witten as
   (*a*) 1000 (*b*) 0100 (*c*) 0010 (*d*) 0001.
5. Binary 111 represents
   (*a*) decimal 222 (*b*) decimal 8 (*c*) decimal 7 (*d*) decimal 4.
6. Binary 1111 represents
   (*a*) decimal 2222 (*b*) decimal 100 (*c*) decimal 25 (*d*) decimal 15.
7. If decimal 10 in binary is 1010, then decimal 100 in binary will be
   (*a*) 1000100 (*b*) 1100100 (*c*) 1110100 (*d*) 1111111.
8. Binary 111111 represents
   (*a*) decimal 99 (*b*) decimal 87 (*c*) decimal 63 (*d*) decimal 51.
9. Binary 1111 when added to binary 1111 the result is decimal
   (*a*) 10000 (*b*) 10010 (*c*) 10110 (*d*) 11110.
10. Binary 11111 when added to binary 1111, the result will be binary
    (*a*) 100010 (*b*) 100110 (*c*) 101110 (*d*) 111110.
11. Binary 1000 when subtracted from binary 11112, the result will be
    (*a*) 111 (*b*) 1110 (*c*) 1010 (*d*) 1000.
12. Binary 1111 when subtracted from binary 11111, the result will be binary
    (*a*) 1000 (*b*) 10000 (*c*) 11111 (*d*) 1111.
13. Binary 1000 when multiplied by binary 11112, the result will be binary
    (*a*) 1111111 (*b*) 1111100 (*c*) 1111000 (*d*) 1110000.
14. Binary 1000 multiplied by binary 1000 gives.
    (*a*) 10000 (*b*) 100000 (*c*) 1000000 (*d*) 10000000.
15. Which of the following binary relation is invalid?
    (*a*) $0 \times 0 = 0$ (*b*) $1 + 1 = 10$ (*c*) $1 \times 1 = 1$ (*d*) $0 \div 1 = 1$.
16. Decimal 0.875 is represented in binary system by
    (*a*) .0101 (*b*) 0.011 (*c*) 0.111 (*d*) 0.001.
17. Decimal 10.75 is represented in binary system by
    (*a*) 10.1010 (*b*) 101.1110 (*c*) 111.1111 (*d*) 1010.11.
18. Binary 1111.1111 is equivalent to decimal
    (*a*) 15.9375 (*b*) 10.9375 (*c*) 10.8750 (*d*) 15.96875.

**19.** Which of the following binary addition is incorrect?

(*a*) 1000 + 0110 = 1111 (*b*) 11111 + 10001 = 110000

(*c*) 11111 + 10000 = 1111111 (*d*) 11111 + 1111 = 101110.

**20.** Binary 1000 will be the result of which of the following subtraction in binary system?

(*a*) 1111 – 111 (*b*) 1010 – 101 (*c*) 1011 – 1110 (*d*) 11111 – 1110.

**21.** The truth table shown is valid for

(*a*) *NOR* gate

(*b*) *NAND* gate

(*c*) *AND* gate

(*d*) *OR* gate.

| *Input* | | *Output* |
|---|---|---|
| A | B | Y |
| 0 | 0 | 0 |
| 0 | 1 | 0 |
| 1 | 0 | 0 |
| 1 | 1 | 1 |

**22.** The truth table shown is valid for

(*a*) *NAND* gate

(*b*) *OR* gate

(*c*) *NOR* gate

(*d*) *XNOR* gate.

| *Input* | | *Output* |
|---|---|---|
| A | B | Y |
| 0 | 0 | 0 |
| 0 | 1 | 1 |
| 1 | 0 | 1 |
| 1 | 1 | 1 |

**23.** The truth table shown is valid for

(*a*) *NAND* gate

(*b*) *AND* gate

(*c*) *OR* gate

(*d*) *NOR* gate.

| *Input* | | *Output* |
|---|---|---|
| A | B | Y |
| 0 | 0 | 1 |
| 0 | 1 | 0 |
| 1 | 0 | 0 |
| 1 | 1 | 0 |

**24.** The Boolean expression for AND gate is

(*a*) $A + B = Y$ (*b*) $A \cdot B = Y$ (*c*) $A = B$ (*d*) $A - B = Y$.

**25.** $A + B = Y$ is the Boolean expression for

(*a*) *AND* gate (*b*) *XNOR* gate (*c*) *NOR* gate (*d*) None of the above.

**26.** $\overline{AB} = Y$ is the Boolean expression for

(*a*) *OR* gate (*b*) *NOR* gate (*c*) *NAND* gate (*d*) None of the above.

**27.** $A = \overline{A}$ is the Boolean expression for

(*a*) adder (*b*) subtractor (*c*) inverter (*d*) multiplier.

**28.** $\overline{A + B} = Y$ is the Boolean expression for

(*a*) *XNOR* gate (*b*) *XOR* gate (*c*) *AND* gate (*d*) *NOR* gate.

**29.** The Boolean expression for the truth table shown below will be

| *Inputs* | | | *Output* |
|---|---|---|---|
| *C* | *B* | *A* | *Y* |
| 0 | 0 | 0 | 0 |
| 0 | 0 | 1 | 0 |
| 0 | 1 | 0 | 0 |
| 0 | 1 | 1 | 0 |
| 1 | 0 | 0 | 0 |
| 1 | 0 | 1 | 1 |
| 1 | 1 | 0 | 1 |
| 1 | 1 | 1 | 0 |

(*a*) $CBA + \overline{CBA} = Y$ (*b*) $\overline{CBA} + C\overline{BA} = Y$ (*c*) $CB\overline{A} + \overline{CBA} = Y$ (*d*) $C\overline{B}A + CB\overline{A} = Y$.

**30.** Which gate is formed by inversion of the output of the AND gate?

(*a*) *NAND* gate (*b*) *NOR* gate (*c*) *OR* gate (*d*) *XNOR* gate.

**31.** The symbol shown in Fig. 27.5 represents

(*a*) inverter

(*b*) flip flop

(*c*) *AND* gate

(*d*) *NAND* gate.

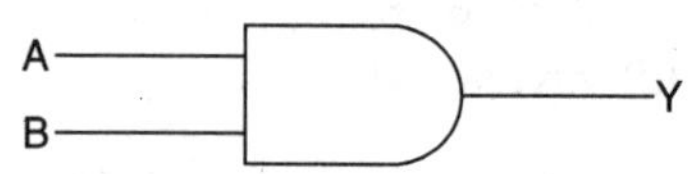


***FIGURE 27.5***

**32.** Which gate corresponds to the action of parallel switches?

(*a*) *AND* gate (*b*) *OR* gate (*c*) *NOR* gate (*d*) *NAND* gate.

**33.** The symbol shown in Fig. 27.6 represents

(*a*) *NOR* gate

(*b*) *OR* gate

(*c*) *XOR* gate

(*d*) *XNOR* gate.

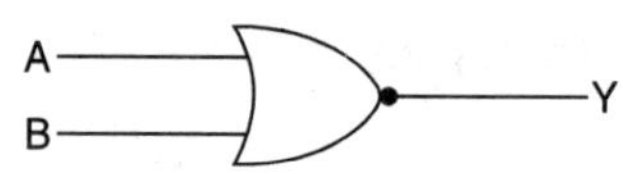


***FIGURE 27.6***

**34.** The symbol shown in Fig. 27.7 represents

(*a*) invert

(*b*) *XNOR* gate

(*c*) *XOR* gate

(*d*) none of the above.

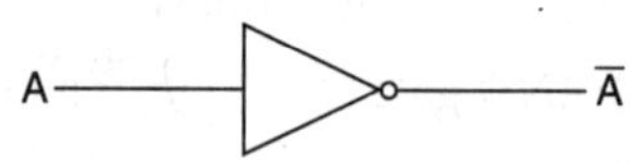


***FIGURE 27.7***

**35.** In case of *OR* gate, no matter what the number of inputs, a

(*a*) 0 at any input causes the output to be at logic 1

(*b*) 0 at any input causes the output to be at logic 0

(*c*) 1 at any input causes the output to be at logic 0

(*d*) 1 at any input causes the output to be at logic 1.

36. The symbol shown in Fig 27.8 represents
(*a*) *OR* gate
(*b*) *XOR* gate
(*c*) *NOR* gate
(*d*) *XNOR* gate.

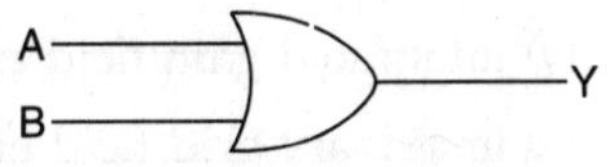


*FIGURE 27.8*

37. The symbol shown in Fig. 27.9 represents
(*a*) *OR* gate
(*b*) *NOR* gate
(*c*) *XNOR* gate
(*d*) *NOR* gate.

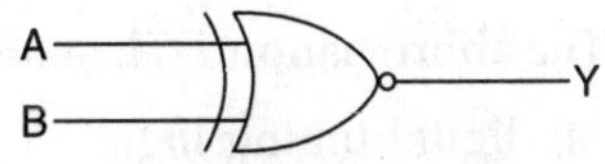


*FIGURE 27.9*

38. A combination of *AND* function and *NOT* function will result in
(*a*) *AND* gate (*b*) *NAND* gate (*c*) *NOR* gate (*d*) *XNOR* gate.

39. Which gate is identical to the action of two series switches?
(*a*) *AND* (*b*) *NAND* (*c*) *OR* (*d*) *NOR*.

40. Which logic function has the output low only when both inputs are high?
(*a*) *AND* (*b*) *NAND* (*c*) *OR* (*d*) *NOR*.

41. The radix for binary system is
(*a*) 0 (*b*) 1 (*c*) 2 (*d*) 10.

42. The radix for decimal system is
(*a*) 0 (*b*) 1 (*c*) 10 (*d*) $\log_e 10$.

43. The radix for hexadecimal system is
(*a*) 1 (*b*) $\log_e 6$ (*c*) 6 (*d*) 8.

44. In boolean algebra, $A + A + A + A + ..... + A$ is the same as
(*a*) $nA$ (*b*) $A^n$ (*c*) zero (*d*) $A$.

45. In Boolean algebra, $1 + A + B + C$ is equal to
(*a*) $A$ (*b*) 1 (*c*) $1 + A$ (*d*) $1 + 3A$.

46. According to Boolean algebra, if $A = 0$, $B = 0$, then $\overline{A}\,\overline{B}$ is
(*a*) 0 (*b*) 1 (*c*) 10 (*d*) none of the above.

47. One's complement of a binary number can be found by
(*a*) changing all zeros to one's
(*b*) changing all ones to zeros
(*c*) changing all ones to zeros and all zeros to ones
(*d*) changing all zeros to ones in case of odd decimal numbers and changing all ones to zeros in case of even decimal numbers.

48. Integrated circuit logic gates contain the properties of
(*a*) resistors (*b*) diodes
(*c*) bipolar junction transistors (*d*) all of the above.

**49.** *IGFET* is

(*a*) integrated gain field effect transistor (*b*) infinite gain field effect transistor

(*c*) imaginary grid field effect transistor (*d*) insulated gate field effect transistor.

**50.** *MOSFET* is

(*a*) metal oxide semi-conductor *FET* (*b*) mixed oscillator series *FET*

(*c*) mono overload silicon *FET* (*d*) multiband oscillator static *FET*.

**51.** The abbreviation *DTL* stands for

(*a*) digital timing logic (*b*) diode transistor logic

(*c*) dynamic transient logic (*d*) delayed tracking logic.

**52.** The abbreviation *TTL* stands for

(*a*) transmitter transistor load (*b*) tuned transistor loop

(*c*) transistor — transistor logic (*d*) transistor — transformer logic.

**53.** The logic represented by the abbreviation *ECL* is

(*a*) emitter coupled logic (*b*) electron carrier logic

(*c*) encoding clock logic (*d*) electrostatic channel logic.

**54.** In *DTL* the junction transistor serves as

(*a*) inverter (*b*) buffer amplifier

(*c*) any of (*a*) and (*b*) above (*d*) none of the above.

**55.** The basic *DTL* configuration is

(*a*) *NAND* gate (*b*) *NOR* gate (*c*) *OR* gate (*d*) *NOR* gate.

**56.** The output of basic *DTL* configuration is

(*a*) low when one of the input is high (*b*) low when all inputs are high

(*c*) high when all inputs are high (*d*) high when all inputs are low.

**57.** The basic circuit configuration for *TTL* resembles that of a

(*a*) *AND* gate (*b*) *NAND* gate (*c*) *NOR* gate (*d*) *OR* gate.

**58.** Typical switching time for *ECL* is

(*a*) 5 nanoseconds (*b*) 5 micro seconds (*c*) 5 milli seconds (*d*) 5 seconds.

**59.** In Schottky *TTL*, a Schottky diode is used for

(*a*) forming the gate (*b*) connecting the resistor

(*c*) clamping of the base-collector junction

(*d*) none of the above.

**60.** In Schottky *TTL*, a Schottky diode is used primarily to

(*a*) prevent saturation of the transistor (*b*) saturate the transistor

(*c*) act as a switch (*d*) act as a controlling switch.

61. A Schottky diode has
(*a*) no minority carriers and very low voltage drop in forward direction
(*b*) no minority carriers and very high voltage drop in forward direction
(*c*) large number of minority carriers and very high voltage drop in forward direction
(*d*) large number of minority carriers and very low voltage drop in forward direction.

62. Which of the following binary addition is incorrect?
(*a*) 0 + 0 = 0 (*b*) 0 + 1 = 1 (*c*) 1 + 0 = 1 (*d*) 1 + 1 = 0.

63. The truth table shown in for a
(*a*) CMO (*b*) counter (*c*) flip flop (*d*) half adder.

| *Inputs* | | *Outputs* | |
|---|---|---|---|
| *B* | *A* | *Sum* | *Carry* |
| 0 | 0 | 0 | 0 |
| 0 | 1 | 1 | 0 |
| 1 | 0 | 1 | 0 |
| 1 | 1 | 0 | 1 |
| Digits to be added | | *XOR* | *AND* |

64. The logic diagram shown in Fig. 27.10 represents
(*a*) elip flop
(*b*) half adder
(*c*) parallel
(*d*) none of the above.

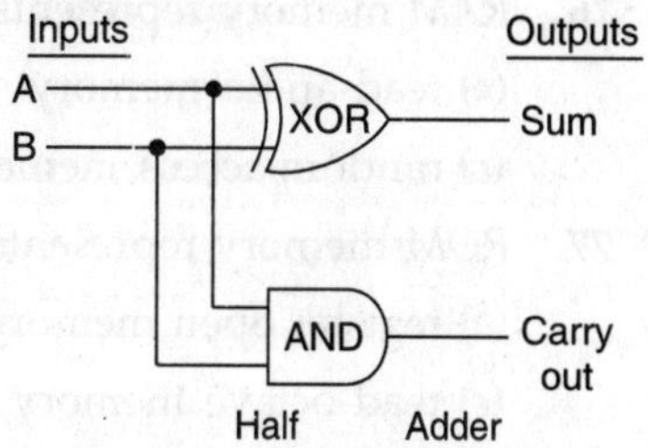


*FIGURE 27.10*

65. The binary sum of $111010_2$ and $11011_2$ is
(*a*) $1010101_2$ (*b*) $1010111_2$ (*c*) $1011111_2$ (*d*) $1111111_2$.

66. The number of full adders in a 4-bit parallel adder will be
(*a*) two (*b*) three (*c*) four (*d*) six.

67. A flip flop circuit is
(*a*) monostable (*b*) bistable (*c*) multistable (*d*) unstable.

68. The two outputs of *RS* flip flop are
(*a*) always low (*b*) always high
(*c*) either low or high (*d*) always complementary.

69. In a *RS* flip flop no change occurs during
(*a*) prohibited mode (*b*) set mode
(*c*) reset mode (*d*) disabled mode.

**70.** A *RS* flip flop during 'disabled' mode will do nothing because

(*a*) both inputs are at low (*b*) only one input is at low

(*c*) both inputs are at high (*d*) only one input is at high.

**71.** Which of the following is 'synchronous'?

(*a*) Half adder (*b*) Full adder

(*c*) *R-S* flip flop (*d*) Clocked *R-S* flip flop.

**72.** A *D*-flip flop is

(*a*) dial type flip flop (*b*) delayed flip flop

(*c*) differential flip flop (*d*) digital flip flop.

**73.** A digital counter consists of a group of

(*a*) half adders (*b*) full adders (*c*) flip flops (*d*) none of the above.

**74.** In a digital counter the number of flip flops is

(*a*) always 2 (*b*) always even

(*c*) always odd

(*d*) equal to the number of bits required in the final binary count.

**75.** A 4-bit counter with four flip flops will count upto decimal

(*a*) 8 (*b*) 15 (*c*) 31 (*d*) 63.

**76.** *RAM* memory represents

(*a*) read-audio memory (*b*) radio-amplified memory

(*c*) random-access memory (*d*) record-attenuated memory.

**77.** *ROM* memory represents

(*a*) register open memory (*b*) read-out memory

(*c*) read-octave memory (*d*) record-oscillation memory.

**78.** Which of the following is a temporary memory?

(*a*) *RAM* only (*b*) *ROM* only

(*c*) Both *RAM* and *ROM* (*d*) Neither *RAM* nor *ROM*.

**79.** A temporary memory is

(*a*) destroyed in few milli seconds (*b*) destroyed in few seconds

(*c*) destroyed in few minutes (*d*) destroyed when power is switched off.

**80.** Which of the following is a volatile memory?

(*a*) *RAM* (*b*) *ROM*

(*c*) Both *RAM* and *ROM* (*d*) Neither *RAM* nor *ROM*.

**81.** Which of the following is a permanent memory?

(*a*) *RAM* only (*b*) *ROM* only

(*c*) Both *RAM* and *ROM* (*d*) Neither *RAM* nor *ROM*.

**82.** *PROM* is

(*a*) permanent read only memory

(*b*) polarised read only memory

(*c*) positive read only memory

(*d*) programmable read only memory.

**83.** Which of the following memory has both read and write capabilities?

(*a*) *RAM* only (*b*) *ROM* only

(*c*) Both *RAM* and *ROM* (*d*) None of the above.

**84.** The input to a multiplexer is a

(*a*) serial data and output from a demultiplexer is a parallel data

(*b*) parallel data and output from a demultiplexer is a serial data

(*c*) parallel data and output from a demultiplexer is also a parallel data

(*d*) serial data and output from a demultiplexer is also a serial data.

**85.** A digital voltmeter has

(*a*) digital input and digital output

(*b*) digital input and analog output

(*c*) analog input and analog output

(*d*) analog input and digital output.

**86.** An *A*/*D* converter uses for reference purposes

(*a*) dc voltage (*b*) a saw tooth generator

(*c*) set of keys (*d*) a flip flop.

**87.** Which of the following has a binary input?

(*a*) *A*/*D* converter (*b*) *D*/*A* converter

(*c*) Both (*a*) and (*b*) above (*d*) None of the above.

**88.** Which of the following can be used to change analog voltage to binary data?

(*a*) *A*/*D* converter (*b*) *D*/*A* converter

(*c*) Both (*a*) and (*b*) above (*d*) None of the above.

**89.** For a *D*/*A* converter, inputs and outputs are given:
The output corresponding to digital input 0100 will be

| *Digital inputs* | *Analog outputs* |
|---|---|
| 0000 | 0.0 V |
| 0001 | 0.5 V |
| 0010 | 1.0 V |
| 0011 | 1.5 V |

(*a*) 2.0 V

(*b*) 2.5 V

(*c*) 3.0 V

(*d*) 4.0 V.

**90.** *LED* display represents

(*a*) logic electrostatic diode (*b*) loop emitter decoder
(*c*) long emission diac (*d*) light emitting diode.

**91.** *LCD* display represents

(*a*) lead cascade demonstrator (*b*) linear digital demonstrator
(*c*) liquid crystal display (*d*) light decoding device.

**92.** The display that emits a reddish glow is usually

(*a*) *LED* type (*b*) *LCD* type (*c*) fluorescent type (*d*) any of the above.

**93.** Which material is generally used for *LED*?

(*a*) Compounds of silica (*b*) Compounds of gallium
(*c*) Compounds of phosphorus (*d*) Compounds of sulphur.

**94.** Which of the following display consumes least amount of power?

(*a*) *LCD* (*b*) *LED*
(*c*) Fluorescent display (*d*) All displays consume same power.

**95.** Which liquid crystal is used for *LCD* display?

(*a*) Mercury (*b*) Nemanic fluid (*c*) Aqua regia (*d*) Liquid boron.

**96.** Which multivibrator can be used as a clock timer?

(*a*) Astable multivibrator (*b*) Bistable multivibrator
(*c*) Any of the above (*d*) None of the above.

**97.** When all the seven segments of a display are energized, the number shown is

(*a*) 0 (*b*) 1 (*c*) 7 (*d*) 8.

**98.** The *OP* amp is used in

(*a*) *A/D* converters (*b*) *D/A* converters
(*c*) both (*a*) and (*b*) above (*d*) shift registers.

**99.** Which family of logic circuits uses field effect transistors?

(*a*) *TTL* (*b*) *CMOS*
(*c*) Both *TTL* and *CMOS* (*d*) None of the above.

**100.** Which of the following needs dc forward voltage to emit light?

(*a*) *LED* (*b*) *LCD*
(*c*) Both *LED* as well as *LCD* (*d*) None of the above.

**101.** In a digital computer

(*a*) only numerical data can be memorized indefinitely
(*b*) only non-numerical data can be memorized indefinitely
(*c*) both numerical and non-numerical data can be memorized indefinitely
(*d*) neither numerical nor non-numerical data can be memorized indefinitely.

**102.** Program counter in a digital computer

(*a*) counts the number of programs run in the machine

(*b*) counts the number of times a subroutine is called

(*c*) counts the number of times the loops are executed

(*d*) points the memory address of the current or the next instruction.

**103.** An index register in a digital computer is used for

(*a*) address modification (*b*) for indirect addressing

(*c*) storing one of the operands (*d*) pointing to the stack address.

**104.** Subroutines are used in larger programs not

(*a*) to increase the programming ease

(*b*) to reduce storage equipment

(*c*) to reduce program extension time

(*d*) for ease of program testing at the program development time.

**105.** In digital computer programming, subroutines are used

(*a*) to reduce program execution time at the expense of more memory

(*b*) to reduce storage requirements

(*c*) to increase programming ease and reduce storage requirements

(*d*) because most of the functions are same.

**106.** A semi-conductor Read-Only-Memory basically is

(*a*) a combinational logic circuit

(*b*) a set of flip-flop memory elements

(*c*) a sequential circuit with flip-flops and gates

(*d*) none of the above.

**107.** Out of the following memory types, one that is volatile is

(*a*) ferrite core (*b*) semi-conductor *RAM*

(*c*) semi-conductor *ROM* (*d*) magnetic disc.

**108.** Introducing parity bit for error detection does not imply

(*a*) increase in the length of the code (*b*) increase in the hardware in the system

(*c*) automatic error correction (*d*) odd number of error detection.

**109.** "Microprogramming" is a technique

(*a*) for programming the microprocessors

(*b*) for writing small programs efficiently

(*c*) for programming the control steps of a computer

(*d*) for programming output/input.

**110.** An operational amplifier is

(*a*) a low gain amplifier without feedback

(*b*) a high gain amplifier indirectly coupled without feedback (*c*) a very-high-gain direct coupled amplifier with feedback

(*d*) none of the above.

**111.** The advantage of solid state amplifiers over the amplifiers using electron tubes is

(*a*) smaller size (*b*) low power requirements

(*c*) better reliability (*d*) all of the above.

**112.** An ideal operational amplifier has

(*a*) infinite input impedance and zero output impedance

(*b*) infinite input impedance as well as output impedance

(*c*) zero input impedance and infinite output impedance

(*d*) zero input impedance as well as output impedance.

**113.** An operational amplifier has

(*a*) low gain at low frequencies and high gain at high frequencies

(*b*) high gain at low frequencies and low gain at high frequencies

(*c*) high gain at all frequencies

(*d*) low gain at all frequencies.

**114.** Ideal operational amplifiers have

(*a*) wide phase shift at all frequencies and narrow bandwidth

(*b*) zero phase shift at all frequencies and infinite bandwidth

(*c*) zero phase shift at all frequencies and narrow bandwidth

(*d*) phase shift and bandwidth proportional to frequencies.

**115.** The gain of an operational amplifier will be maximum at

(*a*) direct current (*b*) 1 Hz (*c*) 50 Hz (*d*) 100 Hz.

**116.** Practically the gain of an operational amplifier at dc could be in the range

(*a*) $10^6$ to $10^8$ (*b*) $10^3$ to $10^4$ (*c*) 100 to 999 (*d*) 10 to 99.

**117.** In an ideal operational amplifier the output impedance should be zero. In practice it is in the range of

(*a*) few milli ohms to one ohm (*b*) few ohms to a few hundred ohms

(*c*) few hundred ohms to few kilo ohms (*d*) few kilo ohms to few mega ohms.

**118.** The noise in an operational amplifier can be reduced by

(*a*) shielding

(*b*) use of low pass filters on incoming power leads

(*c*) attenuation grounding

(*d*) all of the above.

**119.** Noise in op amp may be due to

(*a*) semiconductors (*b*) discrete components

(*c*) pick up from external source (*d*) any of the above.

**120.** In general the operating temperature range for operational amplifiers is

(*a*) 0°C (*b*) 0°C to 25°C (*c*) 0°C to 70°C (*d*) 0°C to 100°C.

**121.** An operational amplifier can be used for implementing which of the following mathematical operations?

(*a*) integration and differentiation (*b*) scale changing and sign reversal

(*c*) addition and subtraction (*d*) any of the above.

**122.** The operational amplifier shown in Fig. 27.11 will act as

(*a*) integrator

(*b*) scale changer

(*c*) sine changer

(*d*) differentiator.

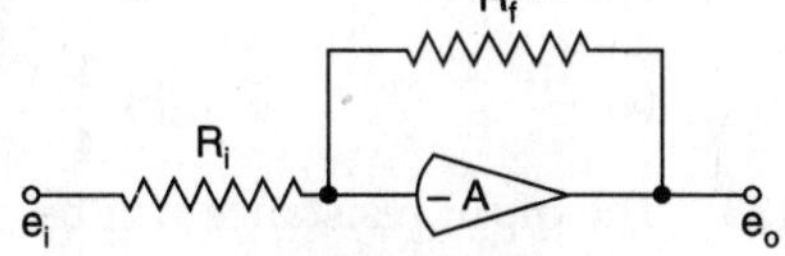


*FIGURE 27.11*

**123.** If $R_f$ = 5 MΩ and $R_i$ = 500 kΩ, the output to input ratio will be

(*a*) 1000 (*b*) 100

(*c*) – 100 (*d*) – 10.

**124.** In which of the following case the output to input ratio will be – 100?

(*a*) $R_f$ = 100,000,000 Ω, $R_i$ = 100,000 (*b*) $R_f$ = 10,000,000 Ω, $R_i$ = 100,000 Ω

(*c*) $R_f$ = 10,000 Ω, $R_i$ = 10,000 Ω (*d*) $R_f$ = 10,000 Ω, $R_i$ = 1000 Ω.

**125.** In an op amp the loop gain of the amplifier can be changed by

(*a*) adding dc voltage source (*b*) adding ac voltage source

(*c*) external feedback registers (*d*) none of the above.

**126.** The output for the operational amplifier shown in Fig. 27.12 will be

(*a*) – X

(*b*) – 10X

(*c*) – 100X

(*d*) – 1000X.

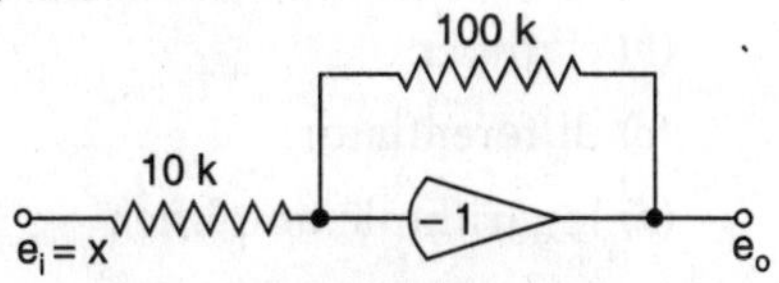


*FIGURE 27.12*

**127.** In op amp the ratio of the voltage, that is feedback, to the input is known as

(*a*) gain (*b*) loop gain (*c*) feedback factor (*d*) transfer function.

**Questions 128 and 129 refer to data given below:**

A scale changer is constructed with an amplifier whose gain is – 250. If the input resistance is 5 MΩ and $R_f$ = 500 kΩ.

**128.** The feedback factor will be

(*a*) 0.999 (*b*) 0.909 (*c*) 0.818 (*d*) 0.717.

**129.** The loop gain will be

(*a*) 250 (*b*) 247.7 (*c*) 227.25 (*d*) 199.99.

**130.** In op amp for minimum percentage error

(*a*) open loop gain should be high but feedback factor should be low

(*b*) open loop gain should be low but feedback factor should be high (*c*) open loop gain as well as feedback factor should be high

(*d*) open loop gain as well as feedback factor should be low.

**131.** In the circuit for scale changer, the gain of the amplifier is 10,000. The percentage error is 0.05%. If the input resistance is 100 ohms, the feedback resistor will be

(*a*) 500 Ω (*b*) 400 Ω (*c*) 250 Ω (*d*) 125 Ω.

**132.** In op amp of the percentage error is 0.10%, amplifier gain is – 5000 and feedback resistance is 100 kΩ

The feedback factor will be

(*a*) $\frac{1}{5}$ (*b*) $\frac{1}{4}$ (*c*) $\frac{1}{3}$ (*d*) $\frac{1}{2}$.

**133.** The input resistance will be

(*a*) 100 kΩ (*b*) 50 kΩ (*c*) 25 kΩ (*d*) 10 kΩ.

**134.** Fig. 27.13 represents a

(*a*) high gain amplifier

(*b*) summer

(*c*) subtractor

(*d*) multiplier.

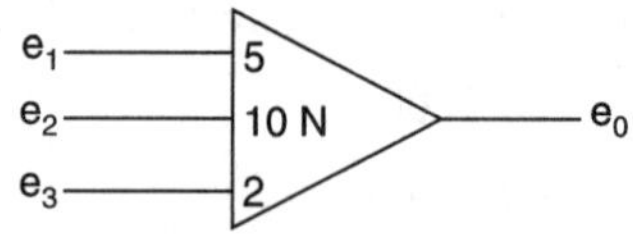


*FIGURE 27.13*

**135.** Which of the following system is non-linear?

(*a*) current to voltage converter (*b*) voltage to current converter

(*c*) logarithmic amplifiers (*d*) none of the above.

**136.** The circuit shown in Fig. 27.14 represents a

(*a*) integrator

(*b*) clamper

(*c*) differentiator

(*d*) logarithmic amplifier.

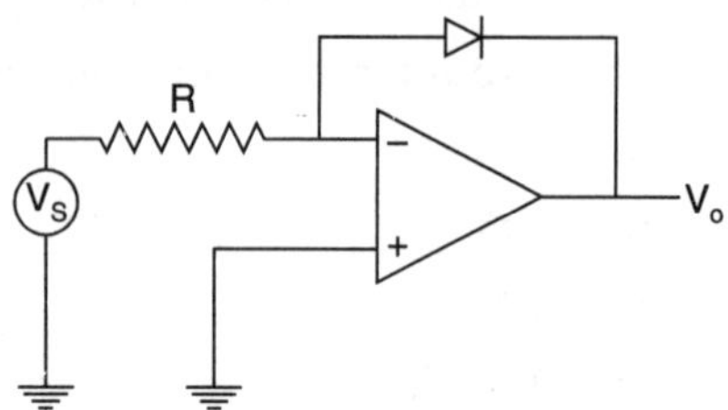


*FIGURE 27.14*

**137.** Log amplifiers generally find application in

(*a*) dividers

(*b*) dividers and multipliers

(*c*) dividers, multipliers and differentiators

(*d*) dividers, multipliers, differentiators and integrators.

**138.** The circuit shown in Fig. 27.15 represents a

(*a*) log amplifier

(*b*) clamper

(*c*) voltage to current converter

(*d*) current to voltage converter.

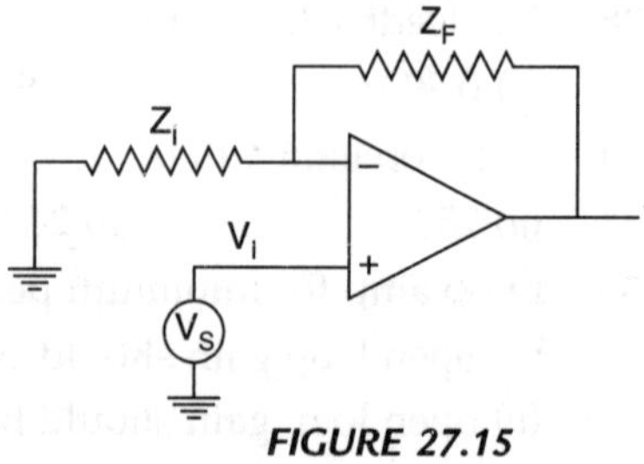


*FIGURE 27.15*

**139.** The circuit shown in Fig. 27.16 represents a

(*a*) subtractor

(*b*) integrator

(*c*) differentiator

(*d*) none of the above.

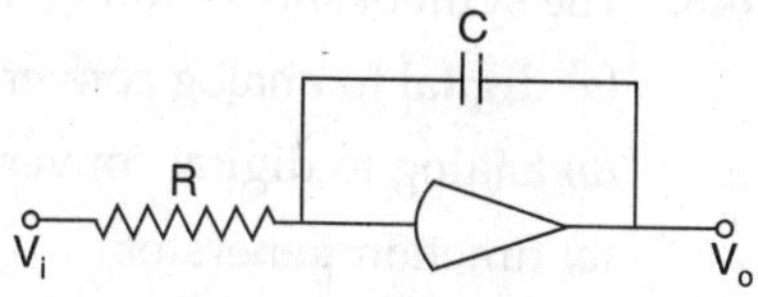


FIGURE 27.16

**140.** Amplitude scaling is done in analog computers

(*a*) for highest precision (*b*) to reduce noise

(*c*) to reduce voltage drift (*d*) all of the above.

**141.** The circuit shown in Fig. 27.17 represents a

(*a*) sign changer

(*b*) differentiator

(*c*) integrator

(*d*) subtractor.

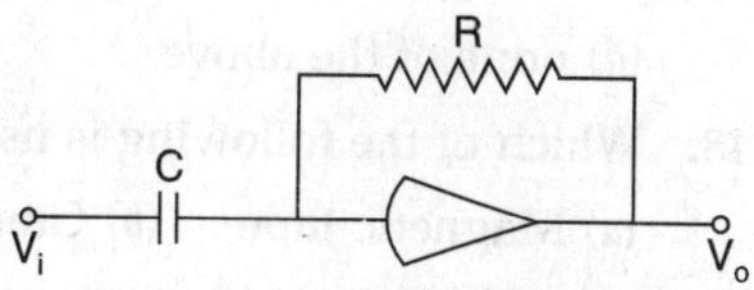


FIGURE 27.17

**142.** Common Mode Rejection Ratio (CMRR) for a differential amplifier should be

(*a*) large (*b*) small (*c*) unity (*d*) zero.

**143.** The equation for the output of diagram shown in Fig. 27.18 will be

(*a*) $-\left[10\int_0^t e_1 dt + 20\int_0^t e_2 dt + 50\int_0^t e_3 dt\right]$

(*b*) $-\left[\frac{1}{10}\int_0^t e_1 dt + \frac{1}{20}\int_0^t e_2 dt + \frac{1}{50}\int_0^t e_3 dt\right]$

(*c*) $-\left[10\int_0^t e_1 dt + 5\int_0^t e_2 dt + 2\int_0^t e_3 dt\right]$

(*d*) $-\left[\frac{1}{10}\int_0^t e_1 dt + \frac{1}{5}\int_0^t e_2 dt + \frac{1}{2}\int_0^t e_3 dt\right]$

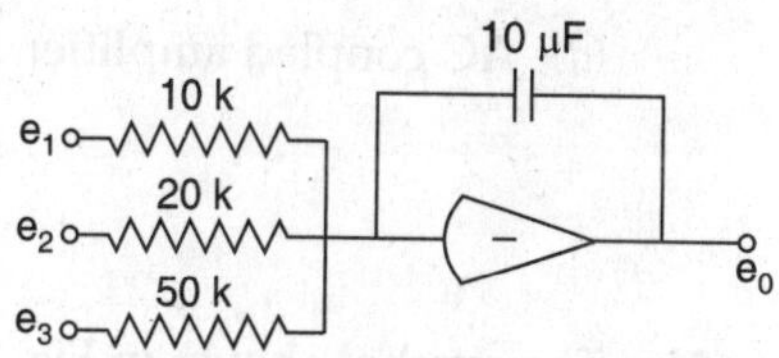


FIGURE 27.18

**144.** A sine wave $v_i$ = 10 sin 200 *t* volts is applied to the circuit shown in Fig. 27.19. The output voltage will be

(*a*) – 10 sin 100 *t*

(*b*) – 10 sin 200 *t*

(*c*) cos 200 *t*

(*d*) cos 100 *t*.

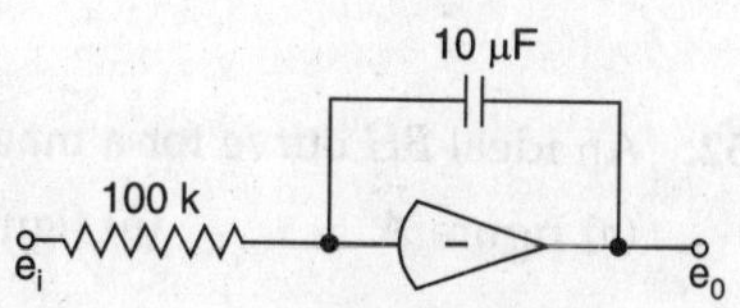


FIGURE 27.19

**145.** A sine wave $e_i$ = 100 sin 200 *t* volts is applied to the circuit shown in Fig. 27.20. The equation for the output voltage $e_0$ will be

(*a*) – 2 cos 200 *t* (*b*) – 20 cos 100 *t*

(*c*) 20 cos 200 *t* (*d*) 2 cos 100 *t*.

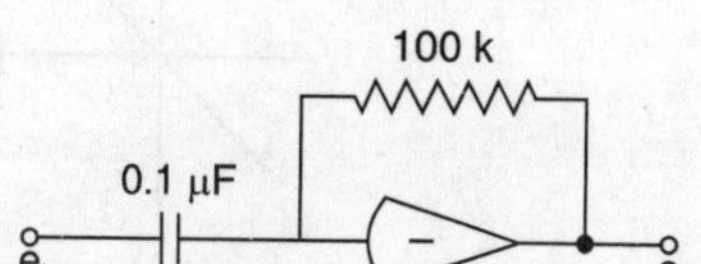


FIGURE 27.20

146. The symbol shown in Fig. 27.21 represents
(*a*) digital to analog converter
(*b*) analog to digital converter
(*c*) function generator
(*d*) multiplier.

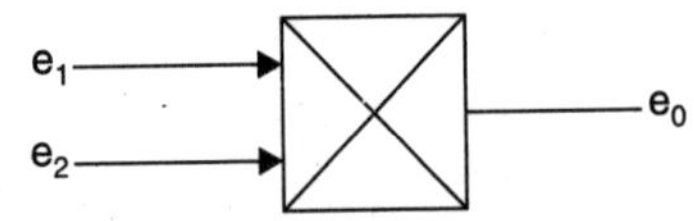


*FIGURE 27.21*

147. The symbol shown in Fig. 27.22 represents
(*a*) track store unit
(*b*) reducer
(*c*) comparator
(*d*) none of the above.

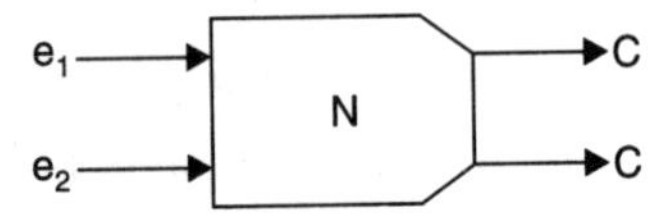


*FIGURE 27.22*

148. Which of the following is usually the output device for an analog computer?
(*a*) Magnetic tape (*b*) Graphic display (*c*) Line printer (*d*) *X-Y* plotter.

149. The use of *OPAMP* is generally not preferred as
(*a*) subtractor (*b*) divider (*c*) differentiator (*d*) integrator.

150. The circuit shown in Fig. 27.23 represents
(*a*) clipper
(*b*) AC voltage follower
(*c*) DC coupled amplifier
(*d*) AC coupled amplifier.

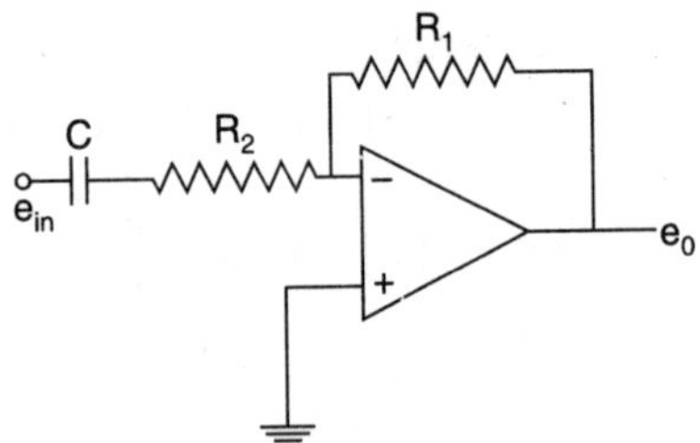


*FIGURE 27.23*

151. The symbol shown in Fig. 27.24 represents
(*a*) DC voltage follower
(*b*) AC voltage follower
(*c*) adder
(*d*) clamper.

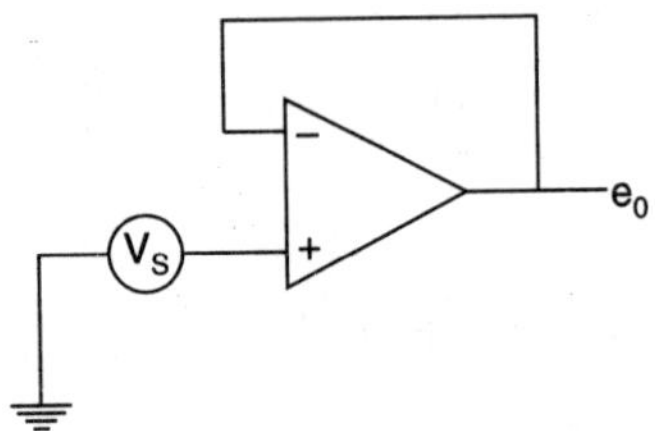


*FIGURE 27.24*

152. An ideal *BH* curve for a magnetic amplifier is shown in
(*a*) figure *A* (*b*) figure *B* (*c*) figure *C* (*d*) figure *D*.

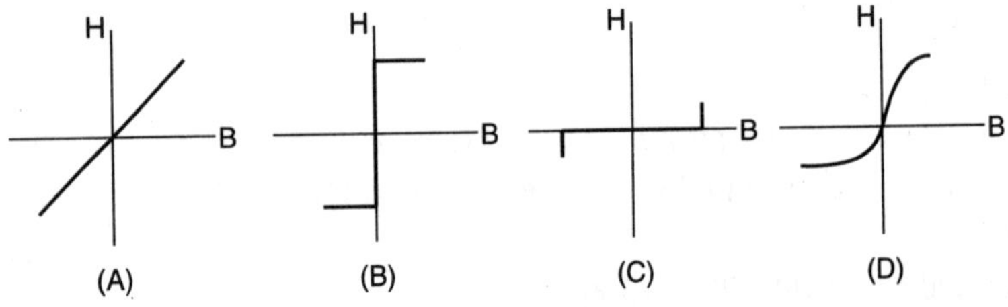


*FIGURE 27.25*

**153.** Feedback in an amplifier

(*a*) reduces sensitivity as well as gain (*b*) increases sensitivity as well as gain

(*c*) increases sensitivity and reduces gain (*d*) reduces sensitivity and increases gain.

**154.** The differential equation for the circuit shown in Fig. 27.26 will be

(*a*) $\frac{d^2x}{dt^2} - 0.7\frac{dx}{dt} + 0.5 = 0$

(*b*) $7\frac{d^2x}{dt^2} - 5\frac{dx}{dt} = 0$

(*c*) $\frac{d^2x}{dt^2} + 7\frac{dx}{dt} + 5 = 0$

(*d*) $\frac{d^2x}{dt^2} + \frac{dx}{dt} = 5.$

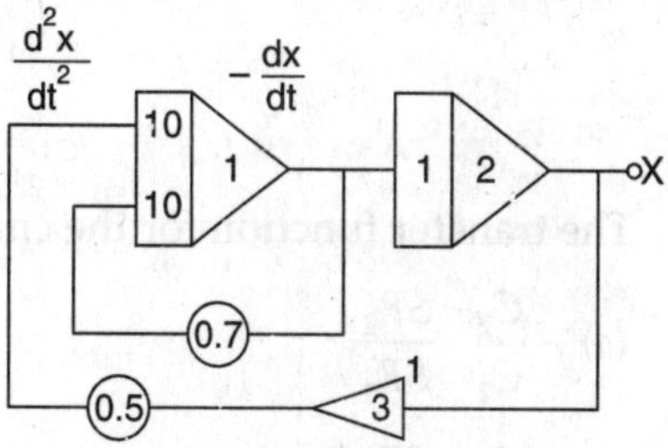


FIGURE 27.26

**155.** The initial condition in above is

(*a*) none (*b*) $y = 0.5$ (*c*) $y = 0.7$ (*d*) $y = 0.5, y' = 0.7.$

**156.** The differential equation for the computer set up shown in Fig. 27.27 is

(*a*) $x'' + x' + 4 = 0$

(*b*) $x'' + 4x' = 0$

(*c*) $x'' + 4x = 0$

(*d*) $x'' + 4 = 0.$

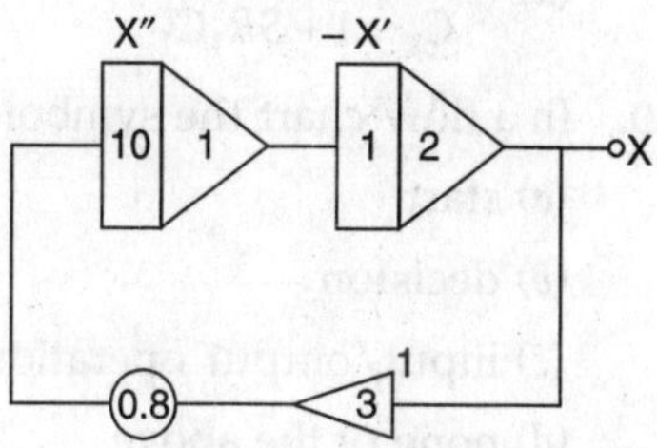


FIGURE 27.27

**157.** The differential equation for the computer set up shown in Fig. 27.28 is

(*a*) $y'' + y' + y = 2$

(*b*) $y'' + 5y' + y = 5$

(*c*) $5y'' + 5y' + 5y = 0$

(*d*) $y'' - 5y' + 5y = 0.$

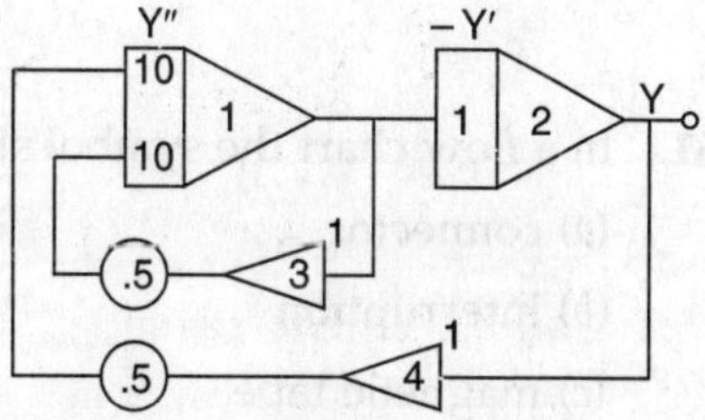


FIGURE 27.28

**158.** The transfer function for the circuit diagram shown in Fig. 27.29 will be

(*a*) $-\frac{R_2}{R_1}\cdot\frac{1+SR_1C_1}{1+SR_2C_2}$ (*b*) $-\frac{R_1}{R_2}\cdot\frac{1+SR_2C_2}{1+SR_1C_1}$

(*c*) $-\frac{R_2C_2}{R_1C_1}\cdot\frac{1+SR_1R_2}{1+SC_1C_2}$ (*d*) $-\frac{R_1C_1}{R_2C_2}\cdot\frac{1+SC_1C_2}{1+SR_1R_1}.$

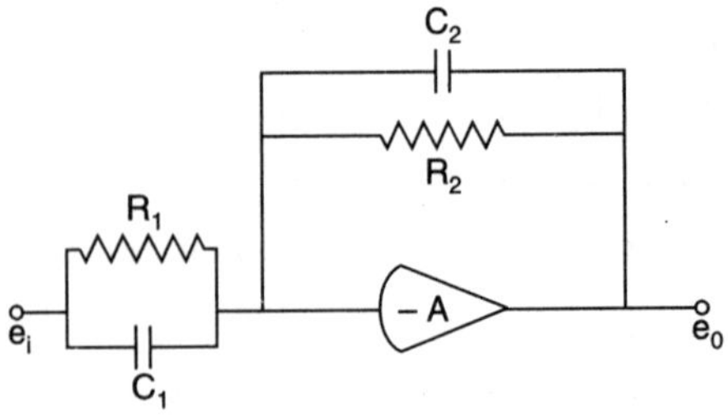


FIGURE 27.29

**159.** The transfer function for the circuit shown in Fig. 27.30 will be

(*a*) $-\frac{C_2}{C_1} \cdot \frac{SR_2}{SR_1}$

(*b*) $-\frac{C_1}{C_2} \cdot \frac{SR_2C_2}{SR_1C_1}$

(*c*) $-\frac{C_2}{C_1} \cdot \frac{SR_2C_2}{SR_1C_1}$

(*d*) $-\frac{C_1}{C_1} \cdot \frac{1+SR_2C_2}{1+SR_1C_1}$.

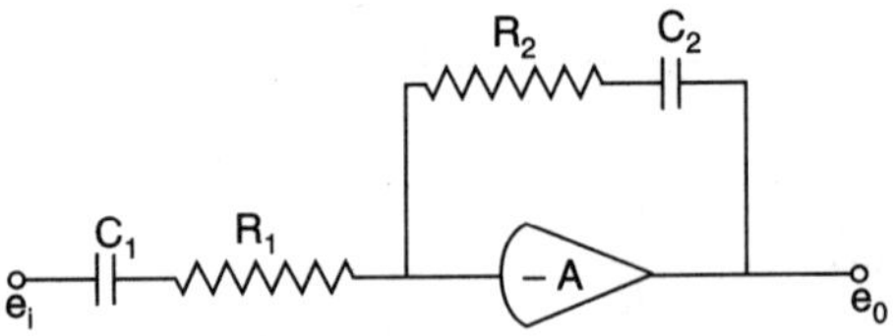


FIGURE 27.30

**160.** In a flow chart the symbol shown in Fig. 27.31 represents

(*a*) start

(*b*) decision

(*c*) input/output operation

(*d*) none of the above.

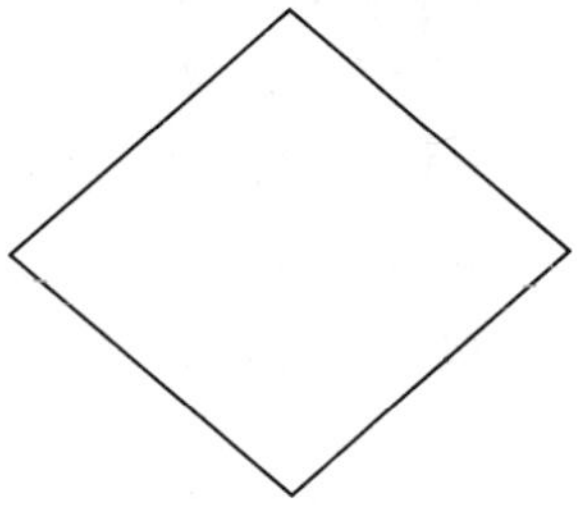

FIGURE 27.31

**161.** In a flow chart the symbol shown in Fig. 27.32 represents

(*a*) connector

(*b*) interruption

(*c*) magnetic tape

(*d*) paper document

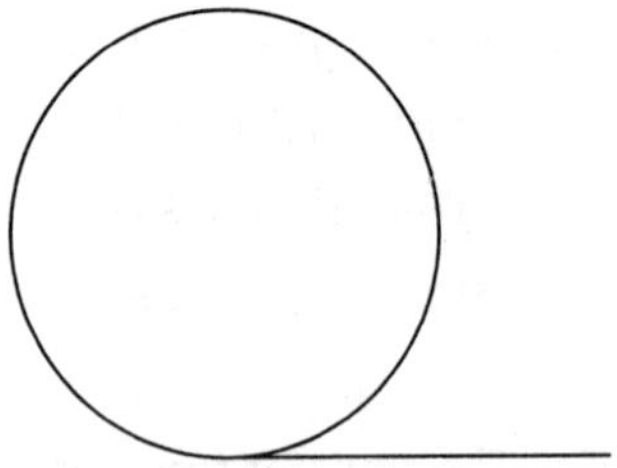

FIGURE 27.32

**162.** In a flow chart the symbol shown in Fig. 27.33 represents

(*a*) manual task (*b*) end

(*c*) general processing (*d*) punched card.

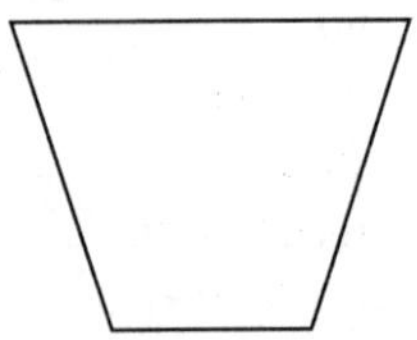

FIGURE 27.33

**163.** In a flow chart the symbol shown in Fig. 27.34 represents

(*a*) input/output operation

(*b*) addition

(*c*) stop

(*d*) punched paper tape

*FIGURE 27.34*

**164.** In a flow chart the symbol shown in Fig. 27.35 represents

(*a*) input/output operation

(*b*) end

(*c*) interruption

(*d*) decision.

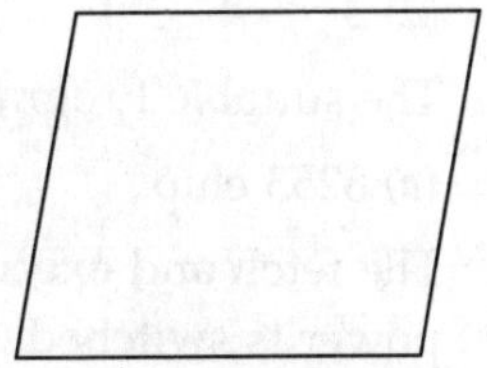

*FIGURE 27.35*

**165.** In a flow chart the symbol shown in Fig. 27.36 represents

(*a*) manual task

(*b*) connector

(*c*) general processing

(*d*) none of the above.

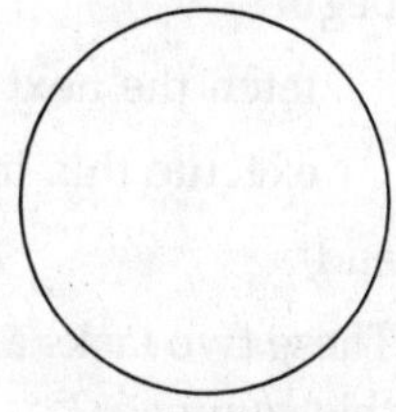

*FIGURE 27.36*

**166.** An instruction or signal in a digital computer, which conditionally or unconditionally specifies and directs the computer to the next instruction is known as

(*a*) jump (*b*) loop (*c*) level (*d*) none of the above.

**167.** A name or number that designates the locations of information in a storage or memory device is known as

(*a*) address (*b*) code (*c*) channel (*d*) memory.

**168.** In a computer an isolating circuit used to avoid any reaction of a driven circuit load on the corresponding driving circuit, is known as

(*a*) matrix (*b*) jump (*c*) buffer (*d*) cell.

**169.** Match List I (chips related to microprocessors and controllers) with list II (their function) and select the correct answer using the codes given below the list:

| *List I* | *List II* |
|---|---|
| A – 8255 *A* | 1 – Interrupt controller |
| B – 8254 | 2 – Microcontroller |
| C – 8259 | 3 – Counter |
| D – *i* 8096 | 4 – Peripheral interface |

**Codes:**

| | A | B | C | D |
|---|---|---|---|---|
| (a) | 3 | 4 | 2 | 1 |
| (b) | 4 | 3 | 2 | 1 |
| (c) | 4 | 3 | 1 | 2 |
| (d) | 3 | 4 | 1 | 2 |

**170.** The suitable Programmable counter for 8086 microprocessors is

(*a*) 8253 chip (*b*) 8254 chip (*c*) 8359 *A*. chip (*d*) 8251 chip.

**171.** The fetch and execute instructions are repeatedly executed by the microprocessor unit the power is switched off. The microprocessor into O loop described by

while power is ON do

begin

fetch the next instruction

execute this instruction

end

These two tasks are sequential. Which one of the following microprocessors does not follow this sequence?

(*a*) 8085 (*b*) Z 80 (*c*) 8086 (*d*) 68000.

**172.** A 4-bit modulo 16 ripple counter uses JK flip-flop. If the propagation delay of each flip-flop is 50 ns, the maximum clock frequency that can be used is

(*a*) 20 MHz (*b*) 10 MHz (*c*) 5 MHz (*d*) 4 MHz.

**173.** In a JK flip-flop, J is connected to $\overline{Q}$ and K is connected to $Q$ outputs. The JK flip-flop converts into a

(*a*) $R - S$ flip-flop (*b*) *D* flip-flop (*c*) *T* flip-flop (*d*) clocked *RS* flip-flop.

**174.** A system which accepts on M-bit word and establishes the state '1' on one and only one of $2^M$ output lines is called.

(*a*) decoder (*b*) demultiplexer (*c*) multiplexer (*d*) encoder.

**175.** In 8085 microprocessor whenever the POPH instruction is executed

(*a*) data bytes in the *HL* pair are stored in the stack

(*b*) two data bytes at the top of the stack are transferred to the *HL* pair.

(*c*) two data bytes at the top of the stack are transferred to the *PC*.

(*d*) two data bytes from the *HL* registers that were previously stored on the stack are transferred back to the *HL* pair.

**176.** The instruction queue length in INTEL 8086 and 8088 are

(*a*) 6 bytes in both processors (*b*) 4 bytes in both processors

(*c*) 4 bytes in 8086 and 6 bytes in 8088 (*d*) 6 bytes in 8086 and 4 bytes in 8088.

**177.** A program accesses ten continuous memory locations. The CPU will take less time if the addressing mode used is

(*a*) direct (*b*) implicit (*c*) register indirect (*d*) immediate.

**178.** If a mod-6 counter is constructed using 3 flip-flops, the counter will skip

(*a*) 4 counts (*b*) 3 counts (*c*) 2 counts (*d*) none of the counts.

**179.** Which one of the following is an example of non-volatile memory?

(*a*) Static *RAM* (*b*) Dynamic *RAM* (*c*) *ROM* (*d*) Cache memory.

**180.** In a 8085 microprocessor, the following sequence of instructions is executed:

*STC*

*CMC*

*MOVE A, B*

*RAL*

*MOVE B, A*

After the last instruction, the output will

(*a*) rotate the contents of the accumulator and store it in *B*

(*b*) get the contents of *B* register into accumulator and rotate it to left by one bit

(*c*) double contents of *B* register

(*d*) manipulate carry in *A* and *B*.

**181.** The highest priority in 3085 microprocessor system is

(*a*) *RST* 7.5 (*b*) *RST* 6.5 (*c*) *INTR* (*d*) *TRAP*.

**182.** In 8085 microprocessor system, the direct addressing instruction is

(*a*) MOV *A, B* (*b*) MOV *B, OAH* (*c*) MOV, *C, M* (*d*) *STA* addr.

**183.** A fragment of assembly code is given below for 8085

(*a*) *EL* (*b*) *MV* 109

(*c*) *SIM* (*d*) This sequence of instruction.

**184.** Consider the following assembly language program

*MVIA*, 30 H

*ACI* 30 H

*XRA*, A

*POP* H

After the execution of the above program the contents of the accumulator will be

(*a*) 30 H (*b*) 60 H (*c*) 00 H (*d*) contents of stack.

**185.** After execution of following statements

X = 5

A = 7

If (A . GT. 5) Then

If (A. GT. 10) then $x = 10$

ELSE $x = 15$

The value assigned to $x$ will be

(*a*) 5 (*b*) 7

(*c*) 10 (*d*) 15.

**186.** Three open collector *TTL* gates connected in wired *AND* configuration. Drive the inputs of five *TTL* gates what would be a suitable for external pull-up resistor connected to the wired-*AND* circuit, if the following specifications applied to *TTL* gates?

Low level input current = 1.6 mA

Low level output current = 16 mA

Low level input voltage = 0.8 V

Low level input voltage = 0.4 V

(*a*) 600 Ω (*b*) 300 Ω (*c*) 60 Ω (*d*) 30 Ω.

**187.** If the number of bits in input and output codes is 4 and 8 respectively for a *ROM*, then the memory of this chip is equal to

(*a*) 12 bits (*b*) 32 bits (*c*) 128 bits (*d*) 256 bits.

**188.** The voltage levels for a negative logic system

(*a*) must necessarily be negative (*b*) could be negative or positive

(*c*) must necessarily be positive (*d*) must necessarily be either zero or – 5 V.

**189.** Match list I (Type of instruction) with list II (Instruction) and select the correct answer using the codes given below the lists:

| *List I* | *List II* |
|---|---|
| A – One byte instruction | 1 – MOV *A*, data |
| B – Two bytes instruction | 2 – MOV *A*, *B* |
| C – Three bytes instruction | 3 – MOV *A*, *M* |
| D – Register indirect addressing | 4 – *JMP* addr. |

**Codes:**

| | *A* | *B* | *C* | *D* |
|---|---|---|---|---|
| (*a*) | 1 | 2 | 4 | 3 |
| (*b*) | 2 | 1 | 4 | 3 |
| (*c*) | 2 | 1 | 3 | 4 |
| (*d*) | 1 | 2 | 3 | 4. |

**190.** A 4-input multiplexer can be used to implement

(*a*) four combinational functions of 2-variables each

(*b*) two combinational functions of 4-variables each

(*c*) one combinational functions of 4-variables

(*d*) one combinational functions of 3-variables.

**191.** In a microprocessor system, the stack is used for storing

(*a*) the program return address whenever a subroutine jump instruction is executed

(*b*) all important CPU register contents whenever an interrupt is to be serviced

(*c*) program instructions for interrupt serive routine

(*d*) all of these.

**192.** For the circuit shown below, output

(*a*) $F = 1$ (*b*) $F = 0$

(*c*) $F = X$ (*d*) $F = \overline{X}$.

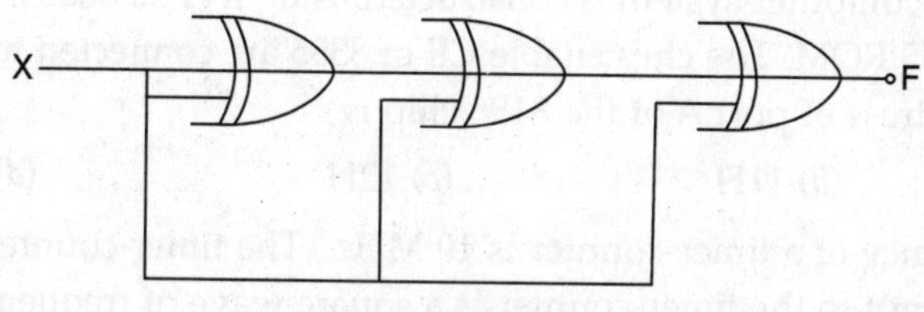


***FIGURE 27.37***

**193.** In register index addressing mode, the effective address is given by the

(*a*) index register value

(*b*) sum of the index register value and the operand

(*c*) operand

(*d*) difference of the index register value and the operand

**194.** For a $\mu P$ system using *IO* mapped *IO*, which of the following statement (*S*) is true?

(*a*) Memory space available is greater

(*b*) Not all data transfer instructions are available

(*c*) *IO* and memory address spaces are distinct

(*d*) All of these.

**195.** When a *FF* is used with asynchronous inputs *Pr*, *clr*, their levels four synchronous inputs should be

(*a*) *LL* (*b*) *LH* (*c*) *HL* (*d*) *HH*.

**196.** The logic circuitry in *ALU* is

(*a*) entirely combinational (*b*) entirely sequential

(*c*) combinational cum sequential (*d*) none of these.

**197.** The 8085 assembly language instruction that stores the content of H and L registers into the memory locations $2050_H$ and $2051_H$, respectively, is

(*a*) SPHL $2050_H$ (*b*) SPHL $2051_H$

(*c*) SHLD $2050_H$ (*d*) STAX $2050_H$.

**198.** A digital-to-analog converter with a full-scale output voltage of 3.5 V has a resolution close to 14 mV. Its bit size is

(*a*) 4 (*b*) 8 (*c*) 16 (*d*) 32.

**199.** If the following program is executed in a microprocessor, the number of instruction cycles it will take from START to HALT is

START MVI A, 14H; Move 14H to register A

SHIFT RLC ; Rotate left without carry

JNZ SHIFT; Jump on non-zero to SHIFT

HALT

(*a*) 4 (*b*) 8 (*c*) 13 (*d*) 16.

**200.** A minimal microcomputer system is constructed using INTEL 8085 microprocessor, an 8156 RAM and an 8355 ROM. The chip enable CE of 8355 are connected to the address line A12. of 8085. The address of port A of the 8156 chip is

(*a*) 21H (*b*) 11H (*c*) 12H (*d*) 20H.

**201.** The clock frequency of a timer-counter is 10 MHz. The timer-counter is used in the period mode and the input to the timer-counter is a square wave of frequency 2 kHz. The display of the timer-counter will show a value

(*a*) 200 (*b*) 5000 (*c*) 50000 (*d*) 2000.

**202.** In an INTEL 8085 microprocessor the ADDRESS-DATA bus and the DATA bus are

(*a*) non multiplexed (*b*) multiplexed

(*c*) duplicated (*d*) same as CONTROL bus.

**203.** In 8085 microprocessor, CY flag may be set by the instruction

(*a*) *SUB* (*b*) *INX* (*c*) *CMA* (*d*) *ANA*.

## ANSWERS

| | | | | | | |
|---|---|---|---|---|---|---|
| **1.** (*c*) | **2.** (*d*) | **3.** (*b*) | **4.** (*d*) | **5.** (*c*) | **6.** (*d*) | **7.** (*b*) |
| **8.** (*c*) | **9.** (*d*) | **10.** (*c*) | **11.** (*a*) | **12.** (*b*) | **13.** (*c*) | **14.** (*c*) |
| **15.** (*d*) | **16.** (*c*) | **17.** (*d*) | **18.** (*a*) | **19.** (*c*) | **20.** (*a*) | **21.** (*c*) |
| **22.** (*b*) | **23.** (*d*) | **24.** (*b*) | **25.** (*d*) | **26.** (*c*) | **27.** (*c*) | **28.** (*d*) |
| **29.** (*d*) | **30.** (*a*) | **31.** (*c*) | **32.** (*b*) | **33.** (*a*) | **34.** (*a*) | **35.** (*d*) |
| **36.** (*a*) | **37.** (*c*) | **38.** (*b*) | **39.** (*a*) | **40.** (*b*) | **41.** (*c*) | **42.** (*c*) |
| **43.** (*d*) | **44.** (*d*) | **45.** (*b*) | **46.** (*a*) | **47.** (*c*) | **48.** (*d*) | **49.** (*d*) |
| **50.** (*a*) | **51.** (*b*) | **52.** (*c*) | **53.** (*a*) | **54.** (*c*) | **55.** (*a*) | **56.** (*b*) |
| **57.** (*b*) | **58.** (*a*) | **59.** (*c*) | **60.** (*a*) | **61.** (*a*) | **62.** (*d*) | **63.** (*d*) |
| **64.** (*b*) | **65.** (*a*) | **66.** (*b*) | **67.** (*b*) | **68.** (*d*) | **69.** (*d*) | **70.** (*c*) |
| **71.** (*d*) | **72.** (*b*) | **73.** (*c*) | **74.** (*d*) | **75.** (*b*) | **76.** (*c*) | **77.** (*b*) |
| **78.** (*a*) | **79.** (*d*) | **80.** (*a*) | **81.** (*b*) | **82.** (*d*) | **83.** (*a*) | **84.** (*c*) |

| | | | | | | |
|---|---|---|---|---|---|---|
| **85.** (*d*) | **86.** (*b*) | **87.** (*b*) | **88.** (*a*) | **89.** (*a*) | **90.** (*d*) | **91.** (*c*) |
| **92.** (*a*) | **93.** (*b*) | **94.** (*a*) | **95.** (*b*) | **96.** (*a*) | **97.** (*d*) | **98.** (*c*) |
| **99.** (*b*) | **100.** (*a*) | **101.** (*c*) | **102.** (*d*) | **103.** (*a*) | **104.** (*d*) | **105.** (*c*) |
| **106.** (*a*) | **107.** (*c*) | **108.** (*c*) | **109.** (*c*) | **110.** (*c*) | **111.** (*d*) | **112.** (*a*) |
| **113.** (*c*) | **114.** (*b*) | **115.** (*a*) | **116.** (*a*) | **117.** (*b*) | **118.** (*d*) | **119.** (*d*) |
| **120.** (*c*) | **121.** (*d*) | **122.** (*b*) | **123.** (*d*) | **124.** (*b*) | **125.** (*c*) | **126.** (*b*) |
| **127.** (*c*) | **128.** (*b*) | **129.** (*c*) | **130.** (*c*) | **131.** (*b*) | **132.** (*a*) | **133.** (*c*) |
| **134.** (*b*) | **135.** (*c*) | **136.** (*d*) | **137.** (*b*) | **138.** (*c*) | **139.** (*b*) | **140.** (*d*) |
| **141.** (*b*) | **142.** (*a*) | **143.** (*c*) | **144.** (*c*) | **145.** (*a*) | **146.** (*d*) | **147.** (*c*) |
| **148.** (*d*) | **149.** (*c*) | **150.** (*d*) | **151.** (*b*) | **152.** (*b*) | **153.** (*a*) | **154.** (*c*) |
| **155.** (*a*) | **156.** (*c*) | **157.** (*d*) | **158.** (*a*) | **159.** (*d*) | **160.** (*b*) | **161.** (*c*) |
| **162.** (*a*) | **163.** (*d*) | **164.** (*a*) | **165.** (*b*) | **166.** (*a*) | **167.** (*a*) | **168.** (*c*) |
| **169.** (*c*) | **170.** (*c*) | **171.** (*c*) | **172.** (*c*) | **173.** (*c*) | **174.** (*a*) | **175.** (*b*) |
| **176.** (*b*) | **177.** (*a*) | **178.** (*c*) | **179.** (*c*) | **180.** (*a*) | **181.** (*d*) | **182.** (*d*) |
| **183.** (*a*) | **184.** (*c*) | **185.** (*a*) | **186.** (*b*) | **187.** (*b*) | **188.** (*b*) | **189.** (*c*) |
| **190.** (*c*) | **191.** (*a*) | **192.** (*b*) | **193.** (*b*) | **194.** (*d*) | **195.** (*d*) | **196.** (*a*) |
| **197.** (*c*) | **198.** (*b*) | **199.** (*c*) | **200.** (*d*) | **201.** (*b*) | **202.** (*b*) | **203.** (*a*). |

# Chapter 28 POWER ELECTRONICS

## Thyristors

Thyristor is a general name given to all four layer *PNPN* devices including the commonly used silicon controlled rectifier called *SCR*. The diac, the triac and the *SCS* are the other popular devices belonging to the family of Thyristor.

An *SCR* is a four layer *PNPN* device with three semiconductor junctions shown in figure below. An *SCS* also works like an *SCR* with the difference that if it has relatively smaller current and voltage specification and is usually a four terminal device.

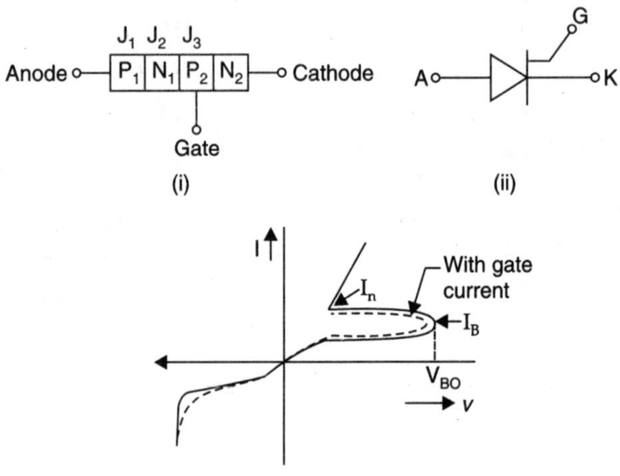


SCR Characteristics

*FIGURE 28.1*

Unlike a conventional junction transistor which conducts only as long as there is a derive at the base, the *SCR* and the *SCS* remains in the conducting state even after the removal of the gate trigger pulse.

A triac is a bidirectional Thyristor that can be considered equivalent to two *SCR*s connected in inverse parallel. While the three terminals of *SCR* are designed as anode, cathode and gate, the terminal of the triac are called *MT*-1, *MT*-2 and gate. The terminals of an *SCS* are anode, cathode anode gate and cathode gate. The characteristics are also shown in the figure. The *SCR* is a current controlled device and its *I-V* characteristics can be expressed by a single valued function of the current.

The some parameters related with the Thyristor are explained here.

### Repetitive Peak Reverse Voltage

It is defined for reverse blocking mode type of Thyristors like *SCRs* for the operation in the third quadrant. It is the maximum repetitive reverse voltage that can be applied to the device safely and still keeping it in the blocked state. The current that flows through the device in this state is referred to as the Repetitive Peak Reverse Current.

### Non-Repetitive Peak Reverse Voltage

It is also defined in the similar way. For obvious reason, this rating is little higher than the one discussed above. If the application is such that the reverse voltage does not appear across the device respectively, may be because the device is being operated in the forward off-state, and if it is to be protected against any occasional reverse transients, then non-repetitive peak reverse voltage and not the repetitive peak reverse voltage is the rating we should look for. Normally, if we have not allowed any reverse voltage to exceed repetitive peak reverse voltage, we are very safe.

### Repetitive Peak Off-state Voltage

It is the maximum forward offset voltage that an be allowed to appear across the device when it is being operated in the first quadrant. This voltage is much less than the actual break-over voltage of the device which is the basically a characteristics of the construction of the thyristor. The maximum value of the current that can flow in the forward blocking state is called Repetitive Peak Off-State Current.

### Holding Current

It is the minimum ON state current required to keep the thyristor in the conducting state once it has been triggered to that state. In order to turn the thyristor off, the anode current must be brought below the holding current value. The voltage corresponding to the holding current is called Holding Voltage.

### Latching Current

It is little more than the holding current and an ON-state current equal to or more than the latching current ensures that a switched on device remains in the ON-state even after removal of gate signal.

### Peak off state voltage or break over voltage

It is the maximum off state voltage. A forward voltage greater than this breaks the junction and switches the device to the ON-state even in the absence of any gate signal. As mentioned, it does not depend upon the voltage grades of different thyristor in the same family.

### Critical Rate of rise of on state current $\left(\frac{di}{dt}\right)$

It tells about the maximum rate of change of ON-state current the device can handle safely. If this rating is exceeded during device switch on, may be due to a faster switch ON, there could be a development of localised hot spots in the pellet at the time of switch ON and this could result in the device getting damaged. The typical value of $\frac{di}{dt}$ is 100 amp per micro sec.

### Critical Rate of rise of on state current $\left(\frac{dv}{dt}\right)$

It determines the maximum allowable rate of change of applied forward off-state voltage. When the applied forward voltage is a time varying one, a current flows through a reverse biased junction capacitance of the thyristor. This value of the current is more than a certain value then a premature firing of the thyristor may cause due to decrease in the break-over voltage. The $\frac{dv}{dt}$ rating can be enhanced by connecting a low resistance between gate and cathode thus providing a low resistance shunt path for the changing current mentioned above by passing a gate cathode junction. Majority of high $\frac{dv}{dt}$ rating thyristor use what is called a Shorted Emitter Construction which is nothing but an internally created low value resistance across gate cathode junction.

### Amperes Squared Seconds

This rating of a thyristor tells about the surge current handling capability of the device for sub-cycle time periods when the device is being used as a rectifier. This rating is usually specify or it can be calculated from the maximum peak surge current that the device can sustain fir a given time period which is usually one full cycle of a applied 50 Hz or 60 Hz waveform. Rating of the thyristor can used to calculate the surge current capability when the thyristor conducts for only a part of the full cycle time.

## OBJECTIVE TYPE QUESTIONS

1. From the Fig. 28.2 the *VI* characteristics of an ideal diode are
   (*a*) curve *a* (*b*) curve *b* (*c*) curve *c* (*d*) curve *d*.

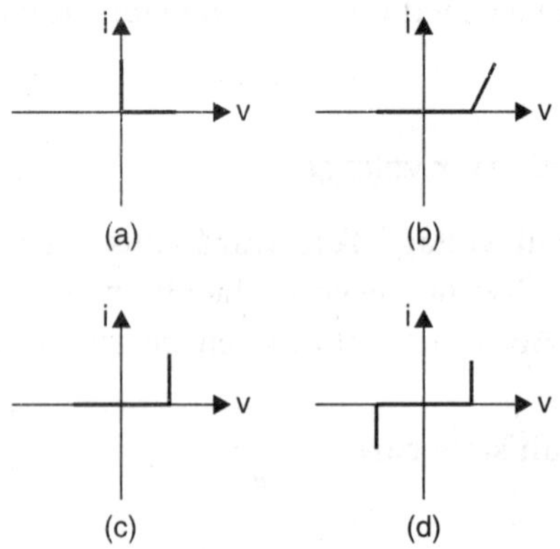


*FIGURE 28.2*

2. Choose the major advantage of bridge rectifier
   (*a*) No centre tap transformer is required
   (*b*) The output peak inverse voltage is higher
   (*c*) Peak inverse voltage of each diode is half of that for a full wave rectifier.
   (*d*) The output waveform is ripple free.
3. The ripple factor is defined as

   (*a*) $\dfrac{\text{average output level}}{\text{rms of the ripple}}$ (*b*) $\dfrac{\text{rms of the ripple}}{\text{average output level}}$

   (*c*) $\dfrac{\text{average output level}}{\text{peak of the ripple}}$ (*d*) $\dfrac{\text{peak output level}}{\text{rms of the ripple}}$.
4. If the value of ripple factor is low, it implies
   (*a*) output voltage will have less ripple
   (*b*) output is low
   (*c*) rms value will be more than average value
   (*d*) none of these.
5. The circuit that can not be operated directly from the mains,
   (*a*) centre tap full wave rectifier (*b*) potentiometer
   (*c*) bridge rectifier (*d*) half wave rectifier.
6. In the voltage regulator shown below, if the current through the load decreases,
   (*a*) The current through $R_1$ will increase
   (*b*) The current through $R_1$ will decrease
   (*c*) Zener diode current will decrease
   (*d*) Zener diode current will increase.

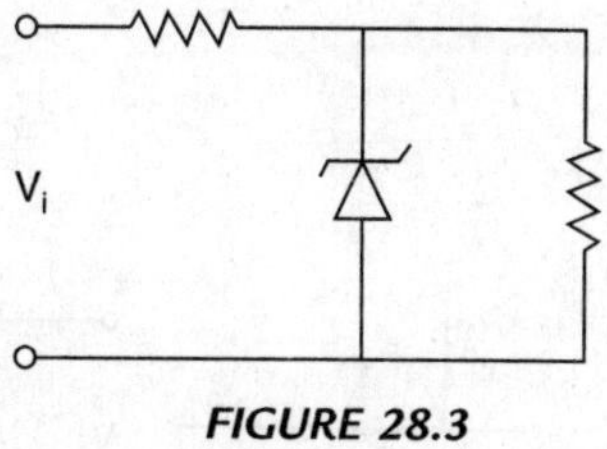

*FIGURE 28.3*

7. In the last problem, if the input voltage decreases
   (*a*) Load current will increase (*b*) Load current will increase
   (*c*) Zener diode current will decrease (*d*) Zener diode current will increase.
8. In a three phase half wave rectifier, each diode conducts for a duration of
   (*a*) 180 degree (*b*) 30 degree (*c*) 60 degree (*d*) 45 degree.
9. In a three phase half wave rectifier, the output voltage is equal to
   (*a*) the most positive input phase voltage at any instant
   (*b*) the difference of most positive and most negative input phases at any instant
   (*c*) the average value of the three phases
   (*d*) the differences of the two positive phase voltage.

10. In a three phase half wave rectifier, if the input phase voltage is 200 V, the *PIV* required for each diode is
    (*a*) 400 V (*b*) 346 V (*c*) 370 V (*d*) 200 V.
11. The rectifier which requires minimum amount of filtering is
    (*a*) Half wave rectifier (*b*) Full wave rectifier
    (*c*) Voltage doubler circuit (*d*) *SCR* half wave rectifier.
12. The circuit shown below is
    (*a*) integrator
    (*b*) low pass filter
    (*c*) high pass filter
    (*d*) band pass filter.
13. The circuit shown below is a
    (*a*) integrator
    (*b*) low pass filter
    (*c*) high pass filter
    (*d*) band pass filter.

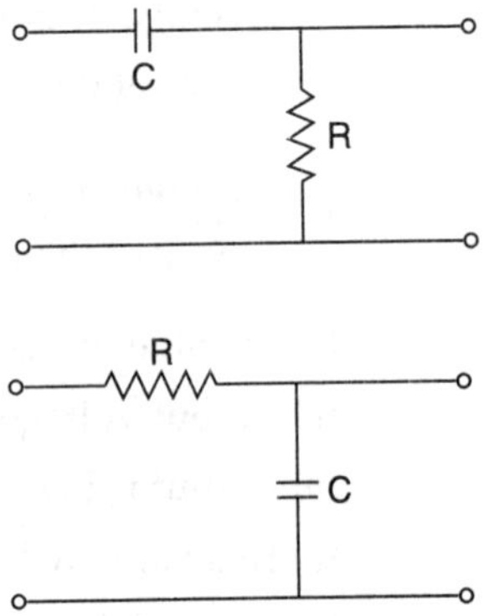


***FIGURE 28.4***

14. The output of the circuit shown for the given input is shown in
    (*a*) curve *A* (*b*) curve *B* (*c*) curve *C* (*d*) curve *D*.

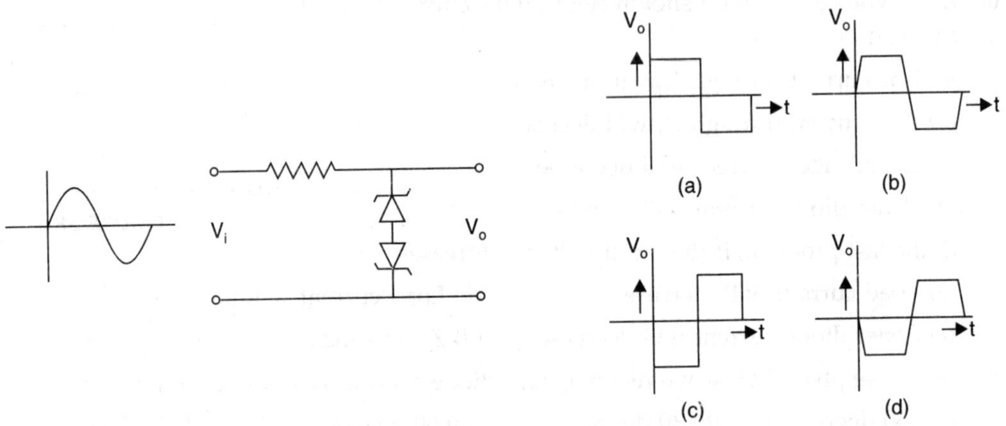


***FIGURE 28.5***

15. The output of the circuit shown for the given input is shown in
    (*a*) curve *A* (*b*) curve *B*
    (*c*) curve *C* (*d*) curve *D*.

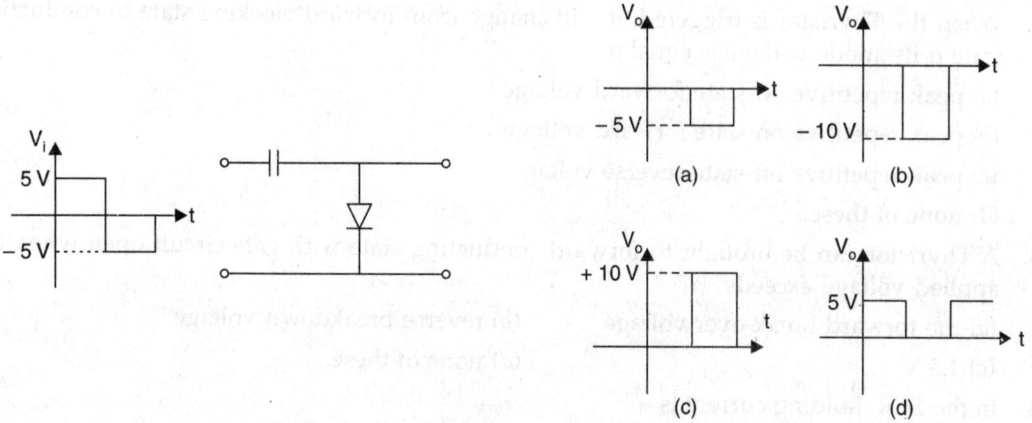


*FIGURE 28.6*

**16.** The output of the circuit shown for the given input is shown in

(*a*) curve *A* (*b*) curve *B* (*c*) curve *C* (*d*) curve *D*.

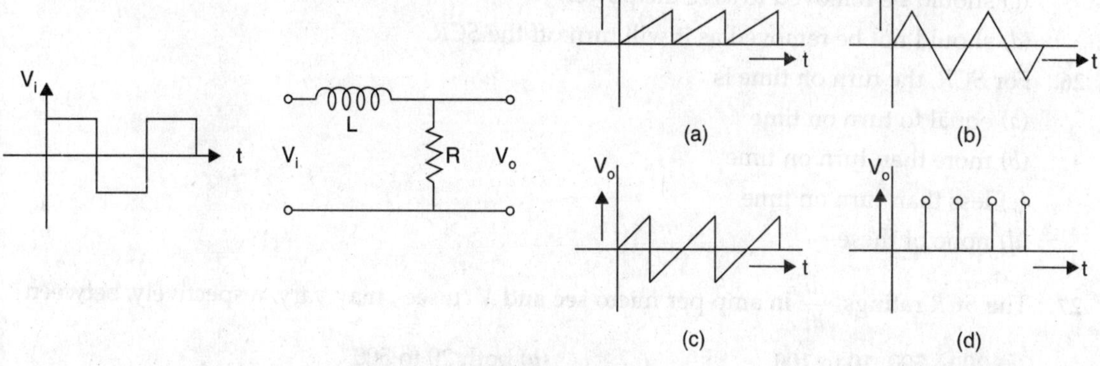


*FIGURE 28.7*

**17.** Thyristor is basically

(*a*) *SCR* (*b*) Triac

(*c*) all *PNPN* devices (*d*) both *SCR* and triac.

**18.** The number of *p-n* junctions in a Thyristor

(*a*) 3 (*b*) 4 (*c*) 1 (*d*) 2.

**19.** When anode is positive with respect to cathode in *SCR*, the number of reversed *p-n* junction

(*a*) 1 (*b*) 2 (*c*) 3 (*d*) 0.

**20.** When anode is negative with respect to cathode in *SCR*, the number of reversed *p-n* junction

(*a*) 1 (*b*) 2 (*c*) 3 (*d*) 0.

**21.** In the Thyristor, the anode current is respond by

(*a*) electrons (*b*) holes (*c*) both (*d*) none.

22. When the Thyristor is triggered, it will change from forward blocking state to conduction state if its anode voltage is equal to
    (*a*) peak repetitive off-state forward voltage
    (*b*) peak repetitive on-state forward voltage
    (*c*) peak repetitive off-state reverse voltage
    (*d*) none of these.
23. A Thyristor can be brought to forward conducting state with gate-circuit open when the applied voltage exceeds
    (*a*) the forward break-over voltage
    (*b*) reverse breakdown voltage
    (*c*) 1.5 V
    (*d*) none of these.
24. In the *SCR*, holding current is
    (*a*) more than latching current
    (*b*) less than latching current
    (*c*) equal to latching current
    (*d*) very small.
25. When a Thyristor gets turned on, the gate drive
    (*a*) should be removed in order to avoid increased losses and higher junction temperatures
    (*b*) may or may not be removed
    (*c*) should be removed to save the power
    (*d*) should not be removed as it will turn off the *SCR*.
26. For *SCR*, the turn on time is
    (*a*) equal to turn on time
    (*b*) more than turn on time
    (*c*) less than turn on time
    (*d*) none of these.
27. The *SCR* ratings, $\frac{di}{dt}$ in amp per micro sec and $V/\mu$ sec, may vary, respectively, between
    (*a*) 20 to 500, 10 to 100
    (*b*) both 20 to 500
    (*c*) both 10 to 100
    (*d*) 50 to 300, 20 to 500.
28. In a thyristor
    (*a*) holding current is associated with turn off process and latching current with turn on process
    (*b*) latching current is associated with turn off process of holding current with turn on process
    (*c*) both latching current and holding current are associated with turn off process
    (*d*) both latching current and holding current are associated with turn on process.
29. Turn off time of an *SCR* is measured from the instant
    (*a*) anode current becomes zero
    (*b*) anode voltage and anode current become zero at the same time
    (*c*) anode voltage becomes zero
    (*d*) gate current becomes zero.

**30.** Turn on time for an *SCR* is 10 μ sec. If an inductance is inserted in anode circuit. Then the turn on time will be

(*a*) about 10 μ sec (*b*) 10 μ sec

(*c*) less than 10 μ sec (*d*) more than 10 μ sec.

**31.** In an *SCR*, anode current flows over a narrow region near the gate during

(*a*) delay time (*b*) rise time and spread time

(*c*) delay time and peak time (*d*) delay time and rise time.

**32.** The gate characteristics of a thyristor

(*a*) is a straight line passing through origin

(*b*) has a spread between two curves of gate voltage and gate current

(*c*) is straight line with positive slope with positive interception on *y*-axis

(*d*) is a curve between gate voltage and gate current.

**33.** During forward blocking state, a Thyristor is associated with

(*a*) large current, low voltage (*b*) low current, large voltage

(*c*) medium current, large voltage (*d*) none of these.

**34.** In an *SCR*, the ratio of latching current to holding current is

(*a*) 0.4 (*b*) 1 (*c*) 2.5 (*d*) none of these.

**35.** On state voltage drop across a Thyristor used in a 250 V supply system is of the order of

(*a*) 100–110 V (*b*) 240–250 V (*c*) 1–1.5 V (*d*) none of these.

**36.** Once *SCR* starts conducting a forward current, its gate looses control over

(*a*) anode circuit voltage only

(*b*) anode circuit current only

(*c*) anode circuit voltage and anode circuit current

(*d*) none of these.

**37.** For an *SCR*, $dV/dt$ protection is achieved through the use of

(*a*) *RL* in series with *SCR* (*b*) *RL* across *SCR*

(*c*) *L* in series with *SCR* (*d*) none of these.

**38.** The $\frac{di}{dt}$ rating of an *SCr* is specified for its

(*a*) rising anode current (*b*) decaying anode current

(*c*) decaying gate current (*d*) rising gate current.

**39.** Surge current rating of an *SCR* specifies the maximum

(*a*) repetitive current with sin wave

(*b*) non-repetitive current with rectangular wave

(*c*) non-repetitive current with sin wave

(*d*) repetitive current with triangular wave.

40. Specification sheet for an *SCR* gives its maximum rms on-state current as 35 A. This rms rating for a conduction angle of 120 degree would be

(*a*) more than 35 A (*b*) less than 35 A (*c*) 35 A (*d*) none of these.

41. The average on-state current for an *SCR* is 20 A for a resistive load. If an inductance of 5 mH is included in the load then average on-state current would be

(*a*) 20 A (*b*) 15 A (*c*) less than 20 A (*d*) more than 20 A.

42. In a Thyristor the magnitude of anode current will

(*a*) increase if gate current is increased

(*b*) not change with an variation of gate current

(*c*) decrease if gate current is decreased

(*d*) increase if gate current is decreased.

43. Turn on time of an *SCR* can be reduce by using a

(*a*) rectangular pulse of low amplitude and narrow width

(*b*) rectangular pulse of high amplitude and narrow width

(*c*) triangular pulse

(*d*) trapezoidal pulse.

44. Turn on time of *SCR* in series with *RL* circuit can be reduced by

(*a*) increasing circuit resistance *R* (*b*) decreasing *R*

(*c*) increasing circuit inductance *L* (*d*) decreasing *L*.

45. For an *SCR* with turn on time of 5 μ sec and ideal trigger pulse should have

(*a*) short rise time with pulse width 3 μ sec

(*b*) long rise time with pulse width 6 μ sec

(*c*) short rise time with pulse width 6 μ sec

(*d*) long rise time with pulse width 3 μ sec.

46. A forward voltage can be applied to an *SCR* after its

(*a*) anode current is reduced to zero (*b*) gate recovery time

(*c*) reverse recovery time (*d*) anode voltage reduced to zero.

47. For series connected *SCRs*, dynamic equalizing circuit consists of

(*a*) resistor *R* and capacitor *C* in series but with a diode *D* across capacitor *C*

(*b*) resistor *R* and diode *D* in series but with a capacitor *C* across resistance *R*

(*c*) resistor *R* and capacitor *C* in series but with a diode *D* across capacitor *C*

(*d*) none of these.

48. Practical way of obtaining static voltage equalization in series connected *SCRs* is by the use of

(*a*) one resistor across the string

(*b*) resistors of different values across each *SCR*

(*c*) resistor of the same value across each *SCR*

(*d*) none of these.

**49.** Thermal resistance in *SCRs* as the units of degree C per *W* and heat sink are made from

(*a*) copper (*b*) steel (*c*) aluminium (*d*) none of these.

**50.** A single-phase half-controlled rectifier is driving a separately excited dc motor. The dc motor has a back emf constant of 0.5 V/rpm. The armature current is 5 A without any ripple. The armature resistance is 2 Ω. The converter is working from a 280 V, single phase ac source with a firing angle of 80°. Under this operating condition, the speed of the motor will be

(*a*) 339 rpm (*b*) 359 rpm (*c*) 366 rpm (*d*) 386 rpm.

**51.** The conduction loss versus device current characteristic of a power *MOSFET* is best approximated by

(*a*) a parabola (*b*) straight line

(*c*) a rectangular hyperbola (*d*) an exponentially decaying function.

**52.** A bipolar junction transistor (*BJT*) is used as a power control switch by biasing it in the cut-off region (OFF state) or in the saturation region (ON state). In the ON state, for the *BJT*

(*a*) both the base-emitter and base-collector junctions are reverse biased

(*b*) the base-emitter junction is reverse biased, and the base-collector junction is forward biased

(*c*) the base-emitter junction is forward biased, and the base-collector junction is reverse biased

(*d*) both the base-emitter and base-collector junctions are forward biased.

**53.** The output voltage waveform of a three-phase square-wave inverter contains

(*a*) only even harmonics (*b*) only odd harmonics

(*c*) both odd and even harmonics (*d*) only triplen harmonics.

**54.** In a thyristor *di*/*dt* protection is achieved through the use of

(*a*) *L* in series with thyristor (*b*) *R* in series with thyristor

(*c*) *RC* in series with thyristor (*d*) *RL* in series with thyristor.

**55.** In a 3-ϕ half wave rectifier circuit, each diode is subjected to a *PIV* of

(*a*) $V_m$ (*b*) $\sqrt{2V_m}$ (*c*) $\sqrt{3}V_m$ (*d*) $2V_m$.

**56.** For a 3-phase, pulse diode rectifier, average output voltage is

(*a*) $\dfrac{3V_m}{\pi}$ (*b*) $\dfrac{\sqrt{3}}{\pi}V_m$ (*c*) $\dfrac{3\sqrt{3}}{\pi}V_m$ (*d*) $\dfrac{3\sqrt{3}}{2\pi}V_m$.

**57.** The following figure represents

(*a*) triac (*b*) diac trigger (*c*) capacity diode (*d*) none of these.

**58.** An UPS is an example of

(*a*) A DC *SMPS* (*b*) An AC *SMPS*

(*c*) AC to DC converter (*d*) A voltage regulation device.

**59.** An *N*-channel enhancement *MOSFET*

(*a*) has a physical channel

(*b*) has no physical channel

(*c*) is similar in construction to a *P*-type depletion mosfet

(*d*) none of these.

**60.** A 1-phase voltage regulator feeds power to a resistance of 10 ohm. The source voltage is 200 V for a firing angle of 90°, the *RMS* value of THYRISTOR current is

(*a*) 10 A (*b*) 22 A (*c*) 16 A (*d*) 4 A.

**61.** In case of an AC *SMPS*

(*a*) the input side inverter uses *PWM* control at a high frequency

(*b*) the output side inverter does not operate at output frequency

(*c*) there is no need for a static switch

(*d*) none of these.

**62.** In a thyristor, *dv*/*dt* protection is achieved through the use of

(*a*) *L* across thyristor (*b*) *R* across thyristor

(*c*) *RC* across thyristor (*d*) *RL* across thyristor.

**63.** Chopper control for DC motor provides variation in

(*a*) input voltage (*b*) frequency (*c*) current (*d*) none of these.

**64.** The ward leonard system is uses for controlling the speed of

(*a*) DC motor (*b*) 1-ϕ AC motor (*c*) 3-ϕ motor (*d*) universal motor.

**65.** Inverter converts

(*a*) DC to AC (*b*) AC to DC (*c*) DC to DC (*d*) AC to AC.

**66.** In a 3-ϕ half wave diode rectifier, the ratio of average output voltage to per phase maximum AC voltage is

(*a*) 1 (*b*) 1.169 (*c*) 0.827 (*d*) 1.571.

**67.** In a 3-ϕ half wave diode rectifier, each diode conducts for a duration of

(*a*) 180° (*b*) 120° (*c*) 90° (*d*) 60°.

**68.** A device can not be triggered with voltage of either polarity is

(*a*) *DIAC* (*b*) *TRIAC* (*c*) *SCS* (*d*) all of the above.

**69.** Which of the following finds applications in speed control of DC motor?

(*a*) *FET* (*b*) *NPN* transistor (*c*) *SCR* (*d*) None of these.

**70.** The capacitive current flowing through the junction of a *SCR* is

(*a*) $Ic = C\ (dv/dt)$ (*b*) $Ic = (1/C)dv/dt$ (*c*) $Ic = C\ (di/dt)$ (*d*) $Ic = C^2\ (di/dt)$.

**71.** When a voltage is applied to a thyristor

(*a*) Holes moves to positive electrode (*b*) Electrons moves to negative electrode

(*c*) Holes moves to negative and electrons to positive electrode

(*d*) None of these.

**72.** There are four configuration of a *PWM* inverter. Three of them are push-pull, half bridge and full bridge. The fourth one is

(*a*) Start-up (*b*) Fly-back (*c*) Start-down (*d*) Fly-kick.

**73.** In a 3-phase voltage regulator

(*a*) 3 *SCR* combination are reqd. per phase, each combination having 4 *SCR*

(*b*) 6 *SCR* combination are reqd. per phase, each combination having 5 *SCR*

(*c*) 3 *SCR* combination are reqd. per phase, each combination having 7 *SCR*

(*d*) none of these.

**74.** A 1-phase voltage controller is use for controlling the flow of power from 260 V, 50 Hz source in to a load consisting of $R = 5$ ohm and $X = 12$ ohm. Find the value of maximum *RMS* load current and the firing angle are respectively.

(*a*) 20 A, 90° (*b*) 25 A, 45° (*c*) 20 A, 0° (*d*) 25 A, 90°.

**75.** A 3-phase half wave controlled rectifier has a supply of 150 V per phase. Find the average load voltage for a firing angle of 60°. Assume a thyristor voltage drop of 1.5 V

(*a*) 46.2 V (*b*) 86.2 V (*c*) 75.4 V (*d*) 80.2 V.

**76.** A resistance of 10 ohm is supplied through a single phase voltage controller from a source of 200 sin 314 *t*. For a firing angle of 90°, the power delivered to the load is

(*a*) 2.0 kW (*b*) 0.5 kW (*c*) 2.5 kW (*d*) 1.0 kW.

**77.** The input impedance of an *IGBT* is

(*a*) low (*b*) moderate (*c*) high (*d*) none of these.

**78.** Refer to the Fig. 28.8 given below

This figure shows the load voltage $V_L$ and load current $I_L$ for an single phase bridge inverter. The thyristor are being fired at continuous train of gate pulses of 180°. The output voltage is square wave. Can you guess the type of load?

(*a*) Resistive (*b*) Capacitive

(*c*) Inductive (*d*) None of these.

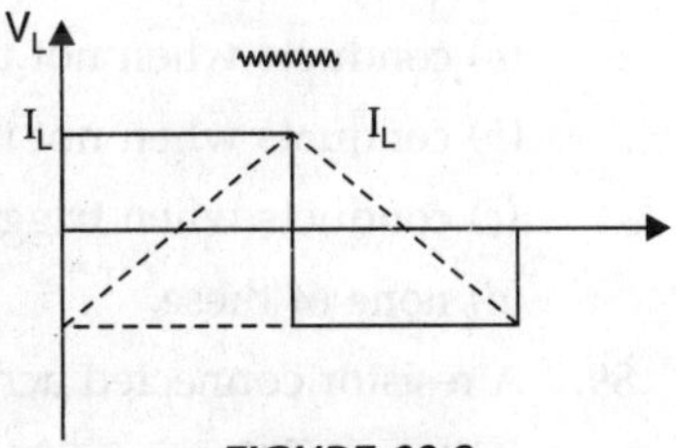


**FIGURE 28.8**

**79.** A series inverter circuit has an inductor of 10 mH, a capacitor of 47 μF, connected in series with a load resistance of 5 ohm. What is the resonant frequency?

(*a*) 1292 Hz (*b*) 1392 Hz (*c*) 1492 Hz (*d*) 1592 Hz.

**80.** In a dual single phase DC to AC converter

(*a*) two full converter are connected in parallel

(*b*) two full converter are connected back to back

(*c*) there is only one full converter

(*d*) all of the above could be true.

**81.** In a 3-phase full converter the 6 thyristor are fired at an interval of

(*a*) 30° (*b*) 60° (*c*) 90° (*d*) 120°.

**82.** A 3-phase full wave converter has an average output of 220 V for 0° firing angle and for resistive load. For a firing angle of 90° the output voltage would be

(*a*) zero (*b*) 50 V (*c*) 100 V (*d*) 26.8 V.

**83.** In the context of inverters, forced commutation would imply that

(*a*) the inverter would be cut-off from the load

(*b*) the inverter would have to be force fully turned off through some external circuit

(*c*) the inverter would be turned off automatically by DC supply

(*d*) all of the above.

**84.** In a 3-phase semi-converter, for a firing angle of 120° and extinction angle of 110°, the freewheeling diode conducts for

(*a*) 50° (*b*) 110° (*c*) 30° (*d*) none of these.

**85.** The major part of an inverter is its

(*a*) step up transformer (*b*) oscillating circuitry

(*c*) filter (*d*) DC source.

**86.** In a 3-phase semi-converter, for firing angle less than or equals to 60°, freewheeling diode conducts for

(*a*) 0° (*b*) 30° (*c*) 45° (*d*) 75°.

**87.** In a half wave controlled rectifier if input voltage is 400 sin 314 $t$, the average output voltage for a firing angle of 60° is

(*a*) $100/\pi$ (*b*) $200/\pi$ (*c*) $300/\pi$ (*d*) $400/\pi$.

**88.** A triac

(*a*) conducts when not triggered

(*b*) conducts when not triggered in both directions

(*c*) conducts when triggered in one direction

(*d*) none of these.

**89.** A resistor connected across the gate and cathode of a thyristor increases its

(*a*) turn off time (*b*) $(di/dt)$ rating

(*c*) noise immunity and holding current (*d*) none of these.

**90.** In a thyristor

(*a*) holding current is greater than latching current

(*b*) the two currents are equal

(*c*) the latching current is about 3 times the holding current

(*d*) none of these.

**91.** Comparing *SCR* and triac

(*a*) Both are unidirectional devices

(*b*) Triac requires more current for turn on than *SCR* at a particular voltage

(*c*) A triac has less time or turn off than *SCR*

(*d*) Both (*b*) and (*c*).

**92.** A device that does not exhibit negative resistance characteristics is
(*a*) *FET* (*b*) *UJT* (*c*) tunnel diode (*d*) *SCR*.

**93.** *UJT* when used for triggering *SCR* has waveform
(*a*) Sine wave (*b*) Square wave
(*c*) Saw-tooth wave (*d*) Trapezoidal wave.

**94.** Each diode in 3-phase half wave diode rectifier conducts for
(*a*) 130° (*b*) 60° (*c*) 90° (*d*) 120°.

**95.** A *SCR* is a
(*a*) unijunction device (*b*) device with 3 junctions
(*c*) device with 4 junctions (*d*) none of these.

**96.** Which of these find applications in speed control of DC motor
(*a*) *FET* (*b*) *NPN* transistor (*c*) *SCR* (*d*) none of these.

**97.** A chopper can be used to step up a DC voltage
(*a*) True (*b*) False (*c*) Partially true (*d*) Not certain.

**98.** For current commutated chopper, output voltage waveform is

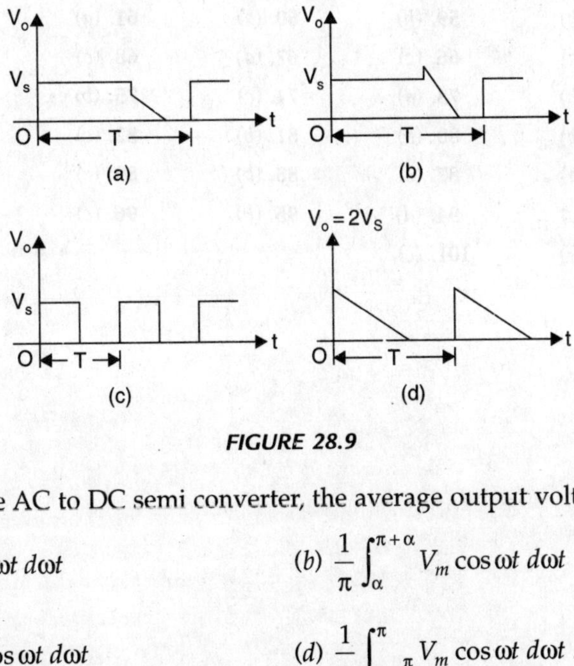


*FIGURE 28.9*

**99.** In a single phase AC to DC semi converter, the average output voltage is given by

(*a*) $\frac{1}{\pi}\int_{\alpha}^{\pi} V_m \cos \omega t \, d\omega t$ (*b*) $\frac{1}{\pi}\int_{\alpha}^{\pi+\alpha} V_m \cos \omega t \, d\omega t$

(*c*) $\frac{1}{\pi}\int_{\alpha-\frac{\pi}{2}}^{\pi+\frac{\pi}{2}} V_m \cos \omega t \, d\omega t$ (*d*) $\frac{1}{\pi}\int_{\alpha-\frac{\pi}{2}}^{\pi} V_m \cos \omega t \, d\omega t$ .

**100.** The following figure is the transfer characteristics of
(*a*) *n*-channel enhancement *MOSFET*
(*b*) *n*-channel depletion *MOSFET*
(*c*) *p*-channel enhancement *MOSFET*
(*d*) *n*-channel depletion *MOSFET*.

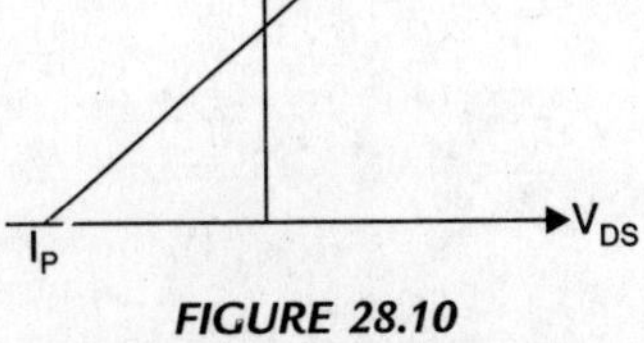


*FIGURE 28.10*

**101.** For a Thyristor shown

(*a*) all the three junctions are forward biased

(*b*) $J_1$ and $J_2$ are forward biased, $J_3$ is reverse biased

(*c*) $J_1$ is forward biased, $J_2$ and $J_3$ are reverse biased

(*d*) $J_1$ and $J_3$ are reverse biased, $J_3$ is forward biased.

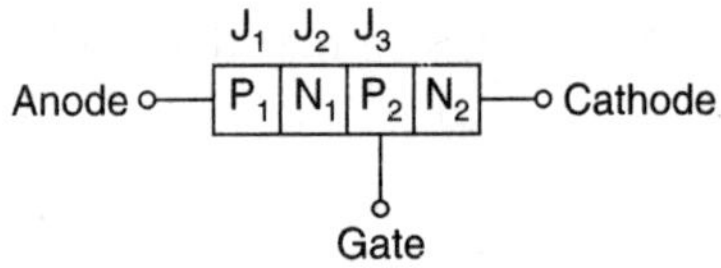


*FIGURE 28.11*

## ANSWERS

**1.** (*a*) **2.** (*c*) **3.** (*b*) **4.** (*b*) **5.** (*a*) **6.** (*d*) **7.** (*c*)
**8.** (*c*) **9.** (*a*) **10.** (*c*) **11.** (*d*) **12.** (*c*) **13.** (*b*) **14.** (*b*)
**15.** (*b*) **16.** (*b*) **17.** (*c*) **18.** (*a*) **19.** (*a*) **20.** (*b*) **21.** (*c*)
**22.** (*a*) **23.** (*a*) **24.** (*b*) **25.** (*a*) **26.** (*a*) **27.** (*b*) **28.** (*a*)
**29.** (*a*) **30.** (*d*) **31.** (*d*) **32.** (*b*) **33.** (*b*) **34.** (*c*) **35.** (*c*)
**36.** (*c*) **37.** (*b*) **38.** (*a*) **39.** (*c*) **40.** (*c*) **41.** (*d*) **42.** (*b*)
**43.** (*b*) **44.** (*d*) **45.** (*c*) **46.** (*b*) **47.** (*c*) **48.** (*c*) **49.** (*a*)
**50.** (*c*) **51.** (*a*) **52.** (*d*) **53.** (*c*) **54.** (*a*) **55.** (*c*) **56.** (*b*)
**57.** (*a*) **58.** (*b*) **59.** (*b*) **60.** (*a*) **61.** (*a*) **62.** (*c*) **63.** (*a*)
**64.** (*a*) **65.** (*a*) **66.** (*c*) **67.** (*d*) **68.** (*c*) **69.** (*c*) **70.** (*a*)
**71.** (*c*) **72.** (*b*) **73.** (*a*) **74.** (*c*) **75.** (*b*) **76.** (*d*) **77.** (*c*)
**78.** (*c*) **79.** (*b*) **80.** (*b*) **81.** (*b*) **82.** (*c*) **83.** (*b*) **84.** (*a*)
**85.** (*b*) **86.** (*a*) **87.** (*c*) **88.** (*b*) **89.** (*c*) **90.** (*c*) **91.** (*d*)
**92.** (*a*) **93.** (*c*) **94.** (*d*) **95.** (*b*) **96.** (*c*) **97.** (*a*) **98.** (*a*)
**99.** (*c*) **100.** (*b*) **101.** (*c*).

Chapter 29

# ELECTROMAGNETIC FIELD THEORY

## 29.1. VECTORS, LINE AND SURFACE INTEGRALS ETC

(*i*) **The coordinate systems used are:**

Cartesian coordinate system $P$ (X, Y, Z)

Cylindrical coordinate system $P$ ($\rho$, $\phi$, $z$)

Spherical coordinate system $P$ ($r$, $\theta$, $\phi$)

Curvilinear coordinate system ($h_1$, $h_2$, $h_3$, $u$, $v$, $w$).

(*ii*) **Differential displacements in:**

Cartesian coordinate system $d\bar{l} = dx\, a_x + dy\, a_y + dz\, a_z$

Cylindrical coordinate system $d\bar{l} = d\rho\, a\rho + \rho d\phi\, ad + dz\, az$

Spherical coordinate systems $d\bar{l} = dr\, a_r + rd\theta\, a_\theta + r \sin\theta\, d\phi\, a_\phi$.

(*iii*) **Differential areas in:**

Cartesian coordinate systems $d\bar{s} = dy\, dz\, a_y, dx\, dz\, a_y, dx\, dy\, a_z$

Cylindrical coordinate systems $d\bar{s} = \rho d\phi\, dz\, a_\rho, d\rho\, dz\, a_\phi, \rho dp\, d\phi\, a_z$

Spherical coordinate systems $d\bar{s} = r^2 \sin\theta\, d\theta\, d\phi\, a_r, r\sin\theta\, dr\, d\phi\, a_\theta, r\, dr\, d\theta\, a_\phi$.

(*iv*) **Differential volumes in:**

Cartesian coordinate systems $dv = dx\, dy\, dz$

Cylindrical coordinate systems $dv = \rho\, d\rho\, d\phi\, dz$

Spherical coordinate system $dv = r^2 \sin\phi\, dr\, d\theta\, d\phi$.

(*v*) **Line Integral of a vector $\overline{A}$ along a path *L*:**

For open path $\int_L \overline{A} \cdot d\overline{l}$

For closed path $\oint_L \overline{A} \cdot d\overline{l}$

Surface integral of a vector $\overline{A}$ across a surface S.

For open path $\int_S \overline{A} \cdot d\overline{s}$ $\int_S \overline{A} \cdot \overline{ds}$

For closed path $\int_S \overline{A} \cdot d\overline{s}$ $\int_S \overline{A} \cdot \overline{ds}$

Volume integral of a scalar ρ over a volume $V = \int_V \rho \, dv$.

(*vi*) **Divergence theorem (*SI* → *VI*)** $\iint_S \overline{A} \cdot d\overline{s} = \iiint_V (\nabla \cdot \overline{A}) \, dv$

Stoke's theorem (*LI* → *SI*) $\oint_C \overline{A} \cdot d\overline{l} = \iint_S (\nabla \times \overline{A}) \, d\overline{s}$.

(*vii*) **A vector field $\overline{A}$ is solenoidal or divergence less if $\nabla \cdot \overline{A} = 0$**

A vector field $\overline{A}$ is Irrotational or Conservative if $\nabla \times \overline{A} = 0$

Scalar field *V* is Harmonic if *V* (Scalar) satisfies Laplace eqn. $\nabla^2 V = 0$.

(*viii*) **Vector identities:**

(*a*) Div curl $\overline{A} = 0$ $\qquad \nabla \cdot (\nabla \times \overline{A}) = 0$

(*b*) Curl grad $\phi = 0$ $\nabla \times (\nabla \phi) = 0$

(*c*) Curl curl $\overline{A}$ = Grad Div $\overline{A}$ – Grad² $\overline{A}$

$$\nabla \times \nabla \times \overline{A} = \nabla (\nabla \cdot \overline{A}) - \nabla^2 \overline{A}$$

$$(A \times B) \cdot C \equiv (B \times C) \cdot A \equiv (C \times A) \cdot B$$

$$A \times (B \times C) \equiv (A \cdot C) B - (A \cdot B) C$$

$$\nabla \cdot (A + B) \equiv \nabla \cdot A + \nabla \cdot B$$

$$\nabla (V + W) = \nabla V + \nabla W$$

$$\nabla (A + B) \equiv \nabla \times A + \nabla + B$$

$$\nabla \cdot (VA) \equiv A \cdot \Delta V + V \nabla \cdot A$$

$$\nabla (VW) \equiv V \nabla W + W \nabla V$$

$$\nabla \times (VA) \equiv \nabla \times A + V \nabla \times A$$

$$\nabla \cdot (A \times B) \equiv \nabla V \times A - A \cdot \nabla \times B$$

$$\nabla (A \cdot B) \equiv (A \cdot \nabla) B + (B \cdot \nabla) A + A \times (\nabla \times B) + B (\nabla \times A)$$

$$\nabla \times (A \times B) \equiv A \nabla \cdot B - B \nabla \cdot A + (B \cdot \nabla) A - (A \cdot \nabla) B$$

$$\nabla \cdot \nabla V \equiv \nabla^2 V.$$

## 29.2 RELATION BETWEEN THE VARIABLES AND COMPONENTS OF THE CARTESIAN AND CYLINDRICAL COORDINATE SYSTEM

| *Cartesian to cylindrical* | *Cylindrical to cartesian* |
|---|---|
| $x = r\cos\phi$ | $r = \sqrt{x^2 + y^2}$ |
| $y = r\sin\phi$ | $\phi = \tan^{-1}\dfrac{y}{x}$ |
| $z = z$ | $z = z$ |
| $A_r = A_x\cos\phi + A_y\sin\phi$ | $A_x = A_r\dfrac{x}{\sqrt{x^2+y^2}} - A_\phi\dfrac{y}{\sqrt{x^2+y^2}}$ |
| $A_\phi = -A_x\sin\phi + A_y\cos\phi$ | $A_y = A_r\dfrac{x}{\sqrt{x^2+y^2}} - A_\phi\dfrac{y}{\sqrt{x^2+x^2}}$ |
| $A_x = A_z$ | $A_z = A_x$ |

### 29.2.1. Relation between the Variables and Components of the Cartesian and Spherical Coordinate System

| *Cartesian to cylindrical* | *Cylindrical to cartesian* |
|---|---|
| $y = r\sin\theta\sin\phi$ | $\theta = \cos^{-1}\dfrac{z}{\sqrt{x^2+y^2+z^2}}$ |
| $z = r\cos\theta$ | $\phi = \tan^{-1}\dfrac{y}{x}$ |
| | $A_x = \dfrac{A_r x}{\sqrt{x^2+y^2+z^2}}$ |
| $A_r = A_x\sin\theta\cos\phi + A_y\sin\theta\sin\phi + A_z\cos\theta$ | $+\dfrac{A_\theta x_z}{\sqrt{(x^2+y^2)(x^2+y^2+z^2)}} - \dfrac{A_\phi y}{\sqrt{x^2+y^2}}$ |
| | $A_y = \dfrac{A_r y}{\sqrt{x^2+y^2+z^2}}$ |
| $A_\theta = A_x\cos\theta\cos\phi + A_y\cos\theta\sin\phi - A_z\sin\theta$ | $+\dfrac{A_\theta y_z}{\sqrt{(x^2+y^2)(x^2+y^2+z^2)}} + \dfrac{A_\phi x}{\sqrt{x^2+y^2}}$ |
| $A_\phi = -A_x\sin\phi + A_y\cos\phi$ | $A_z = \dfrac{A_r z}{\sqrt{x^2+y^2+z^2}} - \dfrac{A_\theta\sqrt{x^2+y^2}}{\sqrt{(x^2+y^2+z^2)}}$ |

## 29.3. MAXWELL'S EQUATIONS

*Differential forms* — *Integral forms*

(*i*) $\nabla \times E = -\frac{\partial \overline{B}}{\partial t}$ $\qquad \oint_C \overline{E} \cdot d\overline{l} = \iint_S \frac{\partial \overline{B}}{\partial t} d\overline{s}$ Modified Faraday's law

(*ii*) $\nabla \times \overline{H} = \overline{J} + \frac{\partial \overline{D}}{\partial t}$ $\qquad \oint_C \overline{H} \cdot d\overline{l} = \iint_S \left( \overline{J} + \frac{\partial \overline{D}}{\partial t} \right) ds$ Modified Amperes law

(*iii*) $\nabla \cdot \overline{D} = \rho$ $\qquad \iint_S D \cdot d\overline{s} = Q = \iiint_V \rho \, dv$ Gauss law of electric field

(*iv*) $\nabla \cdot \overline{B} = 0$ $\qquad \iint_S \overline{B} \cdot d\overline{s} = 0$ Gauss law of magnetic field.

Auxiliary relations are $\overline{D} = \varepsilon \overline{E}$

$\overline{J} = \sigma \overline{E}$

$\overline{B} = \mu \overline{H}$

*Equations of continuity* $\quad \nabla \cdot \overline{J} + \frac{\partial \rho}{\partial t} = 0$

*Displacement current* $\quad I_d = \int \overline{J}\overline{d} \cdot d\overline{s}$, where $\overline{J}_D = \frac{\partial \overline{D}}{\partial t}$

*Displacement current density* $\quad J_D = \frac{I_d}{A} = \varepsilon \frac{\partial E}{\partial t}$

*Conduction current* $\quad I_C = \frac{dQ}{dt}$

*Conduction current density* $\quad J_C = \sigma E$.

Hence the two currents are in time quadrature. The displacement current leads by 90° over the displacement current and is a function of frequency.

Lorentz Gauze condition $\nabla \cdot \overline{A} + \mu\varepsilon \frac{\partial V}{\partial t} = 0$

Instantaneous vector quantity $A(x, y, z, t)$ is related to its phasor form $As(x, y, z)$ as:

$$A(x, y, z, t) = R_c [As(x, y, z) \, e^{j\omega t}]$$

time dependent real $\qquad$ time invariant complex

## 29.4. WAVE EQUATIONS

(*i*) For conducting media $\nabla^2 \overline{E} - \mu\varepsilon \overline{E} - \mu\sigma \overline{E} = 0$

$$\nabla^2 \overline{H} - \mu\varepsilon \overline{H} - \mu\sigma \overline{H} = 0;$$

(*ii*) For free space ($\sigma = 0$) $\nabla^2 \overline{E} = \mu\varepsilon \overline{E}$

$$\nabla^2 \overline{H} = \mu\varepsilon \overline{H};$$

(*iii*) *Helmholtz equations*: Wave equations for time harmonic (sinusoidal) field

$$\nabla^2 \overline{E} = k^2 \overline{E} = 0$$

$$\nabla^2 \overline{H} = k^2 \overline{H} = 0, \text{ where } k^2 = \omega^2\mu\varepsilon.$$

These are second order differential equations which show that electromagnetic radiations propagate with the velocity of light.

Loss tangent $\tan\theta = \dfrac{\sigma}{\omega\varepsilon} = \dfrac{|Js|}{|Jds|} = \dfrac{\varepsilon''}{\varepsilon'}$

$\varepsilon$ complex $= \varepsilon' - J\varepsilon''$

For good dielectric (lossless) tan θ is very small $\tan\theta = \dfrac{\sigma}{\omega\varepsilon} << 1$

For lossy dielectric $\tan\theta = 1 = \dfrac{\sigma}{\omega\varepsilon}$

For good conductors $\tan\theta = \dfrac{\sigma}{\omega\varepsilon} >> 1$

Intrinsic Impedance $\eta = \sqrt{\dfrac{j\,\omega\,\mu}{j + j\,\omega\varepsilon}}$

Attenuation constant $\alpha = \omega\sqrt{\dfrac{\mu\varepsilon}{2}\left[\sqrt{1+\left(\dfrac{\sigma}{\omega\varepsilon}\right)^2} - 1\right]}$

Phase shift $\beta = \omega\sqrt{\dfrac{\mu\varepsilon}{2}\left[\sqrt{1+\left(\dfrac{\sigma}{\omega\varepsilon}\right)^2} + 1\right]}.$

**Skin depth.** It is the distance at which amplitude reduces to $\dfrac{1}{\varepsilon}$ *i.e.*, 37% of its value at the surface

$$\delta = \frac{1}{\alpha} = \sqrt{\frac{2}{\omega\mu\sigma}} = \sqrt{\frac{1}{\pi f \mu\sigma}}.$$

## 29.5. POYNTING VECTOR

It is a power flow vector whose direction is as that wave propagation and magnitude as amount of power flowing through a unit area normal to its direction.

(*i*) $\overline{P} = \overline{E} \times \overline{H}$ power flow per unit area;

(*ii*) Time average power in the wave

$$\overline{P}_{av} = \frac{1}{2} R_c (E \times H^*) = \frac{E_0^2}{2\eta} a_k \text{ watts/m}^2.$$

(*iii*) Given $E$, corresponding magnetic field

$$H = \frac{1}{\mu\,\omega} \overline{k} \times \overline{E} = \frac{a_k \times \overline{E}}{n}$$

$$k = |k| = \omega \sqrt{\mu_0 F_0} = \omega/c$$

$$\eta = 120\,\pi;$$

(*iv*) Average energy density of a plane wave in free space $= \dfrac{\text{Poynting vector}}{\text{Velocity of plane wave}}$.

## 29.6. PLANE WAVES

**Plane wave.** A wave whose phase is constant over a set of planes. It has a plane wave front,

$$E = E(x, y, z) \cos(\omega t - \beta z)\, x.$$

**Uniform plane wave.** A wave whose both phase and magnitude are constant. It is the building block for study of EM waves.

No variation of intensity of plane ⊥ to direction of propagation

$$\frac{\partial E}{\partial x} = \frac{\partial E}{\partial y} = 0$$

$$E = E_0 (z, t)\, x.$$

**Non-uniform plane wave.** A wave whose amplitude may vary (but not phase) within a plane ⊥ to direction of propagation. Hence $\overline{E}$ and $\overline{H}$ are no longer in time phase.

**Travelling waves.** $E = E_0 \sin(\omega t \pm \beta z)$ **or** $E = E_0 e^{-\gamma z}$

–ve sign for wave travelling in + $z$ direction.

+ve sign for wave travelling – $z$ direction.

Here $\quad \omega t - \beta z = \text{constant}$

or $\quad \dfrac{dz}{dt} = \dfrac{\omega}{\beta} = u = \text{velocity of wave.}$

## 29.7. STANDING WAVES

It consists of two travelling waves of equal amplitude but in opposite directions. It stands and does not travel

$$E = E_0\, e^{-\gamma^z} + E_0\, e^{\gamma^z}$$

incident waves reflected waves

For a plane wave in X-direction $\left(\frac{\partial}{\partial y} = \frac{\partial}{\partial z} = 0\right)$ we have

$$\frac{\partial^2 E_x}{\partial x^2} = \mu\varepsilon \frac{\partial^2 E_x}{\partial t^2} \qquad \frac{\partial^2 B_x}{\partial x^2} = \mu\varepsilon \frac{\partial^2 B_x}{\partial t^2}$$

$$\frac{\partial^2 E_y}{\partial x^2} = \mu\varepsilon \frac{\partial^2 E_y}{\partial t^2} \qquad \frac{\partial^2 B_y}{\partial x^2} = \mu\varepsilon \frac{\partial^2 B_y}{\partial t^2}$$

$$\frac{\partial^2 E_z}{\partial x^2} = \mu\varepsilon \frac{\partial^2 E_z}{\partial t^2} \qquad \frac{\partial^2 B_z}{\partial x^2} = \mu\varepsilon \frac{\partial^2 B_z}{\partial t^2}.$$

**Properties of plane waves**

1. It has mutually orthogonal electric and magnetic fields which are also perpendicular to the direction of propagation.
2. No field in the direction of propagation.
3. Electric and magnetic fields are inphase with one another.
4. No variation of field strengths in the plane perpendicular to the direction of propagation.

## 29.8. PROPAGATION THROUGH VARIOUS MEDIA REFLECTION AND REFRACTION

| *Electric field* | *Magnetic field* |
|---|---|
| $E_{t_1} = E_{t_2}$ | $H_{t_1} - H_{t_2} = Js$ |
| $\overline{n} \times (\overline{E}_2 - \overline{E}_1) = 0$ | $\overline{n} \times (\overline{H}_2 - \overline{H}_1) = Js$ |
| | $B_{n_1} = B_{n_2}$ |
| | $H_{n_1} = H_{n_2}$ |
| $D_{n_1} - D_{n_1} = \rho s$ | $\overline{n} \cdot (\overline{B}_2 - \overline{B}_1) = 0$ |
| $\overline{n} \cdot (\overline{D}_2 - \overline{D}_1) = \rho s$ | $\overline{n} \cdot (\overline{H}_2 - \overline{H}_1) = 0$ |
| For perfect conductor (Media 2) | $\overline{n} \times \overline{H}_1 = Js$ |
| $\overline{n} \times \overline{E}_1 = 0\ \overline{n} \cdot \overline{D}_1 = \rho_s$ | $\overline{n} \times \overline{B}_1 = 0$ |
| (1) No field inside region 2 | |
| (2) $Js$ and $\rho_s$ will exist | |

$$\mu_0 = 4\pi \times 10^{-7} \text{ H/m}$$

$$\varepsilon_0 = 8.854 \times 10^{-12} \text{ F/m} = \frac{1}{36\pi \times 10^9} \text{F/m}$$

$$\frac{\overline{E}}{\overline{H}} = 120\pi = 377\ \Omega = \sqrt{\frac{\mu_0}{\varepsilon_0}}$$

$$K = \frac{1}{4\pi\varepsilon_0} = 9 \times 10^9 \text{ m/F.}$$

### 29.8.1. Reflection of Plane Wave by a Perfect Conductor

When an electromagnetic wave travelling in one medium strikes upon a second medium (having a different conductivity), the wave in general will be partially transmitted and partially reflected.

(*i*) When the incidental wave is normal (travelling $x$-direction), the strength of electric field $E(T)\ x$ is given by

$$E_T(x, t) = -\,2E_i \sin \beta x \sin \omega t$$

(Assuming the medium in which the wave is travelling to the perfect, with $P = j\beta$).

Also $$H_r(x, t) = 2\,H_1 \cos \beta_x \cos \omega t$$

(*ii*) When the wave is obliquely incident on the interface between two media, two cases will arise.

(*a*) *Horizontal polarization.* In this case total electric field $E_t$ is given by

$$E_t = 2jE_i \sin \beta_z.z\, e^{-j\beta \sin \theta}$$

where $$\beta_z = \beta \cos \theta.$$

(*b*) *Vertical polarization.* In this case the total $z$-component of electric field strength is given by

$$E_z = 2j\,\eta \sin \theta\, H_i \cos \beta_z.z\, e^{-j\beta_y \gamma}$$

and the total $y$-component of electric field is given by

$$E_y = 2\,j\,\eta \sin \theta\, H_1 \sin \beta_z.z\, e^{-j\beta_y \gamma}$$

### 29.8.2. Snell's Law

According to Snell's law of optics

$$\frac{\sin \theta_1}{\sin \theta_2} = \sqrt{\frac{\varepsilon_2}{\varepsilon_1}}$$

*Normal incidence.* Reflection coefficient $\Gamma = \dfrac{E_{ro}}{E_{io}} = \dfrac{\eta_2 - \eta_2}{\eta_2 + \eta_1}$

## OBJECTIVE TYPE QUESTIONS

1. "The total flux coming out of the closed surface is equal to the net charge within the surface". This statement is the expression of

   (a) Gauss law (b) Divergence theorem

   (c) Faraday's law (d) Maxwell's equation.

2. The electric field intensity magnitude at a point (1, 2, 2) m in Cartesian co-ordinates due to charge (1/9)*nC* located at origin is equal to

   (a) 0.33 mV/m (b) 0.33 V/m (c) 0.33 kV/m (d) zero.

3. The relative permeability is taken as the unity for

   (a) bismuth (b) vacuum (c) air (d) cobalt.

4. "The line integral of the tangential component of the magnetic field strength around the closed path is equal to current enclosed by the path". Statement of

   (a) Gauss law (b) Ampere's law

   (c) Boit-savart law (d) Faraday's law.

5. A beam of protons is deflected sideways

   (a) the deflection is caused by electric field

   (b) deflection is caused by loss of KE

   (c) the deflection is caused by magnetic field

   (d) the deflection is caused by both magnetic field and electric field.

6. In closed surface *S*, enclosing a volume *V* and having current vector *J*, the current *I* emanating from the surface is

   (a) $\int_V J dv$ (b) $\int J$ (c) $\int J ds$ (d) none of these.

7. For a dipole antenna

   (a) the radiation intensity is medium along the normal to the dipole axis

   (b) the current distribution along its length is uniform irrespective of the length

   (c) the effective length equal to its physical length

   (d) the input impedance is independent of the location of the feed-point.

8. The spectral density of pink noise varies as

   (a) $\frac{1}{f^2}$ (b) $f^2$ (c) $\frac{1}{f}$ (d) $f$.

9. The electric flux density (*D*) and the electric field intensity are related as

   (a) $D = \varepsilon.E$ (b) $D = E/\varepsilon$ (c) $D = \varepsilon E^2$ (d) $D = \mu E$.

**10.** A lightening conductor on the top of the building is made into a pointed spike because

(*a*) rain drops may not collect

(*b*) dust particles may not accumulate

(*c*) charge per unit area because very high for lightening to discharge

(*d*) none of these.

**11.** The unit of polarization is same as that of

(*a*) $E$ (*b*) $D$ (*c*) charge (*d*) none of these.

**12.** Only one of the following expressions is correct. The terms have their usual meanings

(*a*) $\nabla \times H = J$ (*b*) $\nabla \times B = H$ (*c*) $\nabla \cdot D = \varepsilon$ (*d*) $\nabla \cdot H = J$.

**13.** The following wave does not exist in a waveguides

(*a*) *TM* wave (*b*) *TE* wave (*c*) *TEM* wave (*d*) *TE* and *TM* wave.

**14.** When $n_1$ and $n_2$ are refractive indices of two media , then according to Snell's law

(*a*) $\frac{n_1}{n_2} = \frac{\sin \theta_r}{\sin \theta_i}$ (*b*) $\frac{n_1}{n_2} = \frac{\sin \theta_i}{\sin \theta_r}$ (*c*) $\frac{n_1}{n_2} = \frac{\cos \theta_i}{\cos \theta_r}$ (*d*) $\frac{n_1}{n_2} = \frac{\cos \theta_r}{\cos \theta_i}$.

**15.** The inconsistency of continuity equation for time varying fields was corrected by Maxwell and correction applied was

(*a*) Ampere's law, $\frac{\partial D}{\partial t}$ (*b*) Gauss's law, $J$

(*c*) Faraday's law, $\frac{\partial B}{\partial t}$ (*d*) Ampere's law, $\frac{\partial \rho}{\partial t}$.

**16.** Surface impedance at high frequencies is a consideration for

(*a*) dielectrics (*b*) semiconductors (*c*) good conductors (*d*) ferrites.

**17.** One of the following is the correct expression for the Coulomb's law of force ($F$) between two charged bodies $Q_1$ and $Q_2$ separated by a distance $d$ in the medium of permittivity $\varepsilon$ in rationalized MKS units.

(*a*) $F = \frac{Q_1 \cdot Q_2}{4\pi\varepsilon d^2}$ (*b*) $F = \frac{Q_1 \cdot Q_2}{4\pi\varepsilon d}$ (*c*) $F = \frac{Q_1 \cdot Q_2 \varepsilon}{4\pi d^2}$ (*d*) $F = \frac{4\pi\varepsilon d^2}{Q_1 Q_2}$.

**18.** The amplitudes and phases of the transmitted and reflected e.m. waves with respect to incident wave are expressed by

(*a*) Maxwell's equation in differential form

(*b*) Snell's law

(*c*) Fresnel's equation

(*d*) Maxwell's equation in integral form.

**19.** Which of the following equations represents the relation between the magnetizing volume current density $J_m$ and the magnetization polarization $M$?

(*a*) $\vec{J}_m = \nabla^2 . \vec{M}$ (*b*) $\vec{J}_m = \nabla . (\nabla \times \vec{M})$ (*c*) $\vec{J}_m = (\nabla \times \vec{M})$ (*d*) $\vec{J}_m = \nabla . \vec{M}$.

20. A body can be charged when it is

1. an insulator
2. an insulated body
3. rubbed in hand
4. charged in humid environment

the false statement are:

(*a*) 1 , 2 (*b*) 3 (*c*) 3 , 4 (*d*) 1, 4.

21. A current is setup in a long copper pipe, choose the correct statement

(*a*) there is a magnetic field inside the pipe
(*b*) there is a magnetic field outside the pipe
(*c*) there is a magnetic field both inside and outside the pipe
(*d*) a back emf is setup in the pipe.

22. The relative permeability is in case of slightly more that 1

(*a*) diamagnetic material (*b*) non magnetic material
(*c*) paramagnetic material (*d*) ferromagnetic material.

23. Sparking occurs when load is switched off because the circuit has high

(*a*) inductance (*b*) capacitance (*c*) resistance (*d*) none of the above.

24. An e.m. wave is incident on a dielectric making an angle θ equal to Brewester angle. Assuming a polarization to be such that the electric vector *E* lies in the plane of reflected wave

(*a*) retraces the incident wave
(*b*) is at angle of 90° with the incident path
(*c*) has the magnitude equal to the incident wave
(*d*) none of these.

25. The average noise temperature of earth, as viewed from space is

(*a*) 254°K (*b*) 303°K (*c*) 100°K (*d*) 500°K.

26. Maxwell's equation of electromagnetics are valid under all conditions except one and that is

(*a*) these do not apply to isotropic media
(*b*) these do not apply to non homogeneous media
(*c*) these do not apply to media which move w.r.t. system of co-ordinates
(*d*) these do not apply to non linear media.

27. One of the following is the scalar quantity

(*a*) electric displacement density (*b*) electric potential
(*c*) electric field strength (*d*) magnetic field strength.

28. Electric displacement current density is measured

(*a*) Coulombs/m (*b*) Coulombs/$m^2$
(*c*) Volts/m (*d*) Amp/m.

29. A charge *Q* is moved a distance *r* on the equipotential line. The work done in moving the charge would be

(*a*) *Q.r* (*b*) *QV*/*r* (*c*) *V*/*r* (*d*) zero.

30. A transmission line whose characteristic is a resistance
(a) must be a lossless line
(b) must be a distortionless line
(c) may not be a lossless line
(d) none of these.

31. For good conductor, the co-efficient of reflection for plane electromagnetic waves is
(a) zero
(b) very small
(c) very large
(d) close to unity.

32. When a particular mode is excited in a wave guide, there appears an extra electric component, in the direction of propagation. The resulting mode is
(a) transverse-electric
(b) transverse-magnetic
(c) longitudinal
(d) transverse-electromagnetic.

33. Which of the following is the ferromagnetic material
(a) copper
(b) palladium
(c) silver
(d) cobalt.

34. Ohm's law when applied to electromagnetics phenomenon gives the expression
(a) $J = \sigma E$
(b) $V = IR$
(c) $J = E/\sigma$
(d) none of these.

35. The value of $\oint I$ along a circle of radius 2 units is
(a) zero
(b) $2\pi$
(c) $4\pi$
(d) $8\pi$.

36. The direction of propagation of electromagnetic waves is given by the direction of
(a) vector $E$
(b) vector $H$
(c) vector $(E \times H)$
(d) none of these.

37. Faraday's law relating the circulation of electric field around the closed path 'C' and magnetic flux density $B$ over the surface $S$ is

(a) $\oint_C \vec{E}.\vec{dl} = +\int_S \vec{B}.\vec{ds}$
(b) $\oint_C \vec{E}.\vec{dl} = -\int_S \vec{B}.\vec{ds}$

(c) $\oint_C \vec{E}.\vec{El} = -\frac{d}{dt}\int_S \vec{B}.\vec{ds}$
(d) $\oint_C \vec{E}.\vec{dl} = \frac{d}{dt}\int_S \vec{B}.\vec{ds}$.

38. In a conductor the relation holds good
(a) $\nabla.D = 0$
(b) $\nabla.D = \rho$
(c) $\nabla \times D = 0$
(d) $\nabla \times D = \rho$.

39. Assuming perfect conductor of a transmission line, pure TEM propagation is not possible in
(a) coaxial cable
(b) air filled cylindrical waveguide
(c) parallel thin wire line in air
(d) semi infinite parallel plate waveguide.

40. Hysteresis losses
(a) are caused by high frequency A.C. current
(b) generally increase with direct current in a coil
(c) can't be produced in an iron core because it is a conductor
(d) none of these.

41. For a plane electromagnetic wave travelling in a linear, isotropic, homogeneous and stationary medium
(a) $\nabla.D = \rho$
(b) $\nabla.D = 0$
(c) $\nabla \times D = \rho$
(d) none of these.

**42.** If the electric field is $\overline{E} = \left[\frac{K}{r}\right] . \overline{a}_r$, in cylindrical co-ordinates, then the work needed to move a point charge $Q$ to a point at twice the radial distance is

(a) $-KQ \ln_2$ (b) $KQ$ (c) $+KQ \ln_2$ (d) $-KQ$.

**43.** According to wave equation for electric field $E$

(a) $\nabla . E = \varepsilon_0 \mu_0 \frac{\partial^2 E}{\partial t^2}$ (b) $\nabla \times E = \varepsilon_0 \mu_0 \frac{\partial^2 E}{\partial t^2}$

(c) $\nabla^2 E = \varepsilon_0 \mu_0 \frac{\partial^2 E}{\partial t^2}$ (d) $\nabla^2 E = \frac{1}{\sqrt{\varepsilon_0 \mu_0}} \frac{\partial^2 E}{\partial t^2}$.

**44.** In free space $\overline{E}$ $(z, t)= 50 \cos(\omega t - \beta z)\ \overline{a}_x$ $(V/m)$, than the average power crossing a circular area of radius 2.5 m in the plane Z = constant

(a) 65.1 W (b) 32.5 W (c) 130.2 W (d) none of these.

**45.** If $R$ and $T$ are the co-efficient of reflection and transmission respectively, then

(a) $R+T = 1$ (b) $R = T$ (c) $R - T = 1$ (d) $R/T$ = constant.

**46.** The current is passed through vertical spring from whose lower end weight is hanging

(a) the weight is pulled down (b) the weight is pulled up

(c) the weight is stationary (d) the weight executes *SHM*.

**47.** Maxwell's curl equation for state magnetic field is given by

(a) $\nabla \times \vec{B} = \mu_o \vec{J}$ (b) $\nabla . \vec{B} = \mu_o \vec{J} = 0$ (c) $\nabla . \vec{B} = \mu_o \vec{J}$ (d) $\nabla . \vec{B} = \mu_o / \vec{J}$ .

**48.** Cross product of two vectors $A$ and $B$ is

(a) $\vec{A} \times \vec{B} = AB \sin \alpha$ (b) $\vec{A} \times \vec{B} = AB \tan \alpha$

(c) $\vec{A} \times \vec{B} = AB \sin \alpha\ \vec{a}_N$ (d) $\vec{A} \times \vec{B} = AB \cos \alpha\ \vec{a}_N$ .

**49.** A spherical conducting shell of radius $a$, centered at the origin, has a potential field

$$V = \begin{cases} V_a & r \le a \\ V_a\, a/r & r < a \end{cases}$$

with the zero reference at infinity. Then the stored energy potential is

(a) $\frac{1}{2}QV$ (b) $QV$ (c) $QV \ln_2$ (d) none of the above.

**50.** The electric field strength at a given point is equal to

(a) the charge gradient at that point

(b) the negative potential gradient at that point

(c) the potential gradient at that point

(d) none of these.

**51.** If the medium of parallel plate capacitor is mica and the air capacitance increases on

(*a*) increasing the area of plates (*b*) increasing the thickness of mica

(*c*) decreasing the thickness of mica (*d*) increasing the air space.

**52.** For an electric medium permeability is given by

(*a*) $\mu = \mu_\theta/\mu_r$ (*b*) $\mu = \mu_\theta.\mu_r$ (*c*) $\mu_\theta = \mu.\mu_r$ (*d*) $\mu_r = \mu_\theta.\mu$.

**53.** The m.k.s. unit of magnetic flux density is

(*a*) Weber's (*b*) Webers/metres$^2$ (*c*) Amperes/metre$^2$ (*d*) Tesla.

**54.** Two isotropic antennas are separated by a distance of two wavelength. If both the antennas are fed with currents of equal phase and magnitude, the number of lobes in radiation pattern in the horizontal plane are

(*a*) 2 (*b*) 4 (*c*) 6 (*d*) 8.

**55.** Fleming's left hand rule can be used to determine

(*a*) the magnitude of the force experienced by a current carrying conductor placed in a magnetic field

(*b*) the direction of force experienced by a current carrying conductor placed in magnetic field

(*c*) the direction of magnetic field intensity *H* at any point due to a current carrying conductor

(*d*) none of these.

**56.** Which of the following statement is true?

(*a*) alnico is commonly used for electromagnets

(*b*) paper can't affect magnetic flux because it is not a magnetic material

(*c*) iron is generally used for permanent magnets

(*d*) all of the above.

**57.** Given vector $A = 4a, + 10a,$ and $B = 2a, + 3a$. The projection of *A* on *B* is

(*a*) $\frac{12}{\sqrt{13}}$ (*b*) $\frac{1}{12.\sqrt{13}}$ (*c*) $\frac{\sqrt{13}}{12}$ (*d*) none of these.

**58.** A radial field $\overline{H} = \frac{2.39 \times 10^6}{r} \cos\phi\, \overline{a}_r$ A/m exists is free space. The magnetic flux $\phi$ crossing the surface defined by $-\frac{\pi}{4} \le \phi \le \frac{\pi}{4}, 0 \le Z \le 1$ m, will be

(*a*) 6.3 Wb (*b*) 8.48 Wb (*c*) 4.24 Wb (*d*) 3.16 Wb.

**59.** The unit vector directed toward the origin from an arbitrary point on the line described by $x = 0, y = 3$

(*a*) $a = \frac{-3a_x - za_y}{\sqrt{9+z^2}}$ (*b*) $a = \frac{-3a_y - za_z}{\sqrt{9+z^2}}$ (*c*) $a = \frac{-3a_x - za_z}{\sqrt{9+z^2}}$ (*d*) none of these.

**60.** The *VSWR* is taken as when the line is
(*a*) open circuited (*b*) short circuited
(*c*) terminated with full load
(*d*) terminated with load equal to the characteristic impedance of the load.

**61.** A parabolic reflector is designed to have a directivity of 30 dB at 300 MHZ. If the aperture efficiency is 55%, then the half-power band width is
(*a*) 4.31° (*b*) 8.62° (*c*) 17.24° (*d*) none of the above.

**62.** The depth of penetration of a wave in a lossy dielectric increase with increasing
(*a*) conductivity (*b*) permeability (*c*) wavelength (*d*) permittivity.

**63.** The maximum usable frequency of an ionospheric layer at 60° incidence and with 8 MHz critical frequency is
(*a*) 16 MHz (*b*) $\frac{16}{\sqrt{3}}$ MHz (*c*) 8 MHz (*d*) 6.93 MHz.

**64.** The perfect conductor is one which has
(*a*) finite conductivity (*b*) infinite conductivity
(*c*) unity conductivity (*d*) zero conductivity.

**65.** The boundary conditions at the interface between two dielectrics in a electric field are
(*a*) $D_{\text{Normal}} = 0; E_{\text{Normal}} = 0$ (*b*) $D_{\text{Normal}} = 0; E_{\text{tangential}} = 0$
(*c*) $D_{\text{tangential}} = 0; E_{\text{tangential}} = 0$ (*d*) $D_{\text{tangential}} = 0; E_{\text{Normal}} = 0$.

**66.** The characteristic impedance of a lossless transmission line is given by
(*a*) $\sqrt{LC}$ (*b*) $\sqrt{\frac{L}{C}}$ (*c*) $1/\sqrt{LC}$ (*d*) $\sqrt{\frac{C}{L}}$.

**67.** White noise is that signal whose frequency spectrum
(*a*) extends over infinite range (*b*) has flat spectral density
(*c*) has spectral density varying as $1/f$ (*d*) has limited no. of frequency components.

**68.** The unit vector directed from (2, – 5, – 2) toward (14, – 5, 3).
(*a*) $a = \frac{13a_x - 5a_z}{12}$ (*b*) $a = \frac{12a_x - 5a_z}{13}$ (*c*) $a = \frac{12a_x - 13a_z}{5}$ (*d*) none of these.

**69.** Intrinsic impedance of free space is given as
(*a*) 75 (*b*) 73 (*c*) 377 (*d*) 300.

**70.** The outward displacement through a closed surface equals
(*a*) 0 (*b*) 1 (*c*) 0 (*d*) none of these.

**71.** A current of $5 \times 10^{-3}$ Amp is passed through a coil of 400 turns an inductance of 8 mH. The magnetic flux through coil is
(*a*) $1 \times 10^{-7}$ Wb (*b*) $1 \times 10^{-7}$ Wb/m$^2$ (*c*) $1 \times 10^{-5}$ Wb (*d*) $1 \times 10^{-6}$ Wb.

**72.** The polarization of dielectric materials results in
(*a*) absorption of electrons (*b*) release of high velocity protons
(*c*) creation of electric dipoles (*d*) production of eddy currents.

**73.** The m.k.s. unit of $\mu$ is

(*a*) farad/metre$^2$ (*b*) farad/metre (*c*) Henry/metre (*d*) henry/metre$^2$.

**74.** The unit of $\mu_o\varepsilon_o$ is

(*a*) farad, henry (*b*) metre$^2$/sec$^2$

(*c*) amper. sec. volt. sec (*d*) netwon metre$^2$/col$^2$.

**75.** The reflection coefficient is generally a

(*a*) real number (*b*) fraction

(*c*) imaginary number (*d*) complex number.

**76.** The relative permeability for iron may be around

(*a*) 1 (*b*) 10 (*c*) 50 (*d*) 5000.

**77.** The inductance per unit length for a solenoid near its center compared to its ends is

(*a*) less (*b*) greater (*c*) same (*d*) none of these.

**78.** The area of the curved surface of a right circular cylinder of radius *a* and height *h*.

(*a*) $2\pi a^2 h$ (*b*) $2\pi ah$ (*c*) $2\pi (a + h)$ (*d*) none of these.

**79.** Two point charges, $Q_1 = 30$ nC and $Q_2 = -100\ \mu$C, are located at (2, 0, 5) *m* and (– 1, 0, – 2) *m*, respectively, then the force on $Q_1$ is

(*a*) $F_1 = (0.465)\left(\dfrac{-3a_x - 7a_y}{\sqrt{58}}\right)$ (*b*) $F_1 = (0.465)\left(\dfrac{-3a_y - 7a_z}{\sqrt{58}}\right)$

(*c*) $F_1 = (0.465)\left(\dfrac{-3a_x - 7a_z}{\sqrt{58}}\right)$ (*d*) none of these.

**80.** The force on charge $Q_2$ in above question is

(*a*) $F_2 = F_1$ (*b*) $F_2 = 3.33\ F_1$ (*c*) $F_2 = -F_1$ (*d*) none of these.

**81.** Ten identical charges of 500 μC each are spaced equally around a circle of radius 2 m. The force on a charge of – 20 μC located on the axis, 2 m from the plane of the circle.

(*a*) $-80\ a_n$ N (*b*) $80\ a_n$ N (*c*) zero (*d*) none of these.

**82.** A one turn coil having a radius of 10 cms is carrying a current of 1 A. The magnetic field intensity *H* in *AT*/m at the center of the coil is

(*a*) 5 (*b*) 10 (*c*) $5/2\pi$ (*d*) $5/\pi$.

**83.** Velocity of plane wave $e^{\sin 2(\omega t - \beta x)}$ is

(*a*) $\dfrac{2\omega}{\beta}$ (*b*) $\dfrac{\omega}{2\beta}$ (*c*) $\dfrac{\omega^2}{\beta^2}$ (*d*) $\dfrac{\omega}{\beta}$.

**84.** A transmission line is distortion less if

(*a*) $\dfrac{L}{C} = \dfrac{G}{R}$ (*b*) $\dfrac{L}{G} = \dfrac{R}{C}$ (*c*) $\dfrac{L}{C} = \dfrac{R}{G}$ (*d*) none of these.

85. The speed of an e.m. wave through PTEE with $\varepsilon_r = 2$ and $\mu_r=1$ is
(*a*) $210 \times 10^6$ Hz (*b*) 0.707 Hz (*c*) $3 \times 10^8$ Hz (*d*) $0.7 \times 10^6$ Hz.

86. A solenoid of 4 cm dia and 2 m long is closely wound with a single layer of 10 gauge (0.10 in dia) copper wire. The inductance per unit length at the center is
(*a*) 2.52 μH (*b*) 0.253 mH/m (*c*) 2.5 H/m (*d*) 2.02 H.

87. Divergence theorem applies to
(*a*) static field only (*b*) time varying field only
(*c*) bath static as well as time varying field
(*d*) magnetic fields only.

88. Mark the incorrect relation
(*a*) $D = \varepsilon E$ (*b*) $B = \mu H$ (*c*) $J = \sigma E$ (*d*) $B = \mu D$.

89. It is given that electric flux density $D$ in a certain region is expressed by $D = (1/r)a_r$ in spherical co-ordinates. The charge density ρ in this region is given by
(*a*) $(1/r^2)$ (*b*) $(1/r)$ (*c*) $(-1/r^2)$ (*d*) $r^2$.

90. The noise temperature of sky is about
(*a*) 100°K (*b*) 273°K (*c*) 0°K (*d*) 30°K.

91. Electric field intensity at any point is
(*a*) negative potential at that point (*b*) positive potential gradient at that point
(*c*) charge at that point (*d*) none of these.

92. For a reflection coefficient $|\rho| =1/2$, *VSWR* is
(*a*) 2 (*b*) 3 (*c*) 4 (*d*) 8.

93. The m.k.s. unit of electric field *E* is
(*a*) volt (*b*) volt per second (*c*) volt per meter (*d*) ampere per meter.

94. Magnetic flux density *B*, magnetic field intensity *H* and the permeability μ are interrelated by
(*a*) $B = H/\mu$ (*b*) $B = \mu H^2$ (*c*) $\mu = B/H$ (*d*) none of these.

95. The m.k.s. unit of magnetic field *H* is
(*a*) ampere *b*) weber
(*c*) weber per square metre (*d*) ampere per metre.

96. Frequencies in UHF range propagate by means of
(*a*) ground waves (*b*) sky waves (*c*) space waves (*d*) surface waves.

97. Transmission of power to a load over a transmission line achieves optimum value when standing-wave ratio (*SWR*) becomes
(*a*) 2 : 1 (*b*) 1 : 2 (*c*) 1 : 1 (*d*) $1 : \sqrt{2}$.

98. The amplitude of *E* (incident) in free space (region 1) at the interface with region 2 is 1V/m. If $H_o{}^r$ (reflected) $= -1.41 \times 10^{-3}$ A/m. $\varepsilon r_2 = 18.5$ and $\sigma_2 = o_p$, then $\mu r_2$ will be
(*a*) 99.2 (*b*) 198.4 (*c*) 123 (*d*) none of the above.

**99.** Wires that carry equal and opposite current are twisted because

(*a*) it is easy to coil

(*b*) while lying the two wires may fall apart

(*c*) it reduces the magnetic field away from it

(*d*) the magnetic field at distant points are reduced by currents in opposite directions and distance from centers is nullified

**100.** The new magnetic field emerging from any closed surface

(*a*) constant (*b*) zero (*c*) unity (*d*) none of these.

**101.** When an electromagnetic wave travelling in one medium strikes upon a second medium having the different conductivity the wave will be

(*a*) totally reflected (*b*) totally transmitted

(*c*) neither transmitted and reflected (*d*) partially transmitted and partially reflected.

**102.** A surface encloses a dipole $\varphi_E$ for the surface is the product of two charges is

(*a*) product of two charges (*b*) sum of two charges

(*c*) zero (*d*) one-half of the surface area.

**103.** Which of the following is the paramagnetic material

(*a*) aluminum (*b*) cobalt (*c*) copper (*d*) silver.

**104.** A short grounded vertical antenna has a length $L$ which is 0.05 $\lambda$ at frequency $f$. If its radiation resistance at $f$ is $R$ ohms, then its radiation resistance at a frequency $2f$ will be

(*a*) $R/2$ ohms (*b*) $R$ ohms (*c*) $2R$ ohms (*d*) $4R$ ohms.

**105.** For a plane e.m. wave travelling in a linear, isotropic, homogeneous and stationary medium,

(*a*) $E$ and $H$ lie in the same plane (*b*) $E \times H = 0$

(*c*) $E$ and $H$ are mutually perpendicular (*d*) none of these.

**106.** When electromagnetics waves are reflected at an angle from a wall, their wavelength along the wall is

(*a*) the same as in free space

(*b*) the same as the wavelength perpendicular to the wall

(*c*) shortened because of Doppler effect

(*d*) greater than in the actual direction of propagation.

**107.** A helical antenna is used for satellite tracking because of its

(*a*) circular polarization (*b*) maneuverability

(*c*) broad bandwidth (*d*) good front-to-back ratio.

**108.** A circular parallel plate capacitor of radius $r$ = 10 cm is charged so that the electric field between the plates are varies at $10^{13}$ V/m sec. The displacement current is

(*a*) 27946 A (*b*) 2.8 A (*c*) 27.946 A (*d*) 0.28 A.

**109.** Volume integral is the

(*a*) integral of volume (*b*) double integral of volume

(*c*) triple integral of volume (*d*) none of these.

**110.** An air filled rectangular wave guide has dimensions 6 cm, 4 cm. The cut-off frequency for $TE_{10}$ is

(a) 2.5 GHz (b) 25 GHz (c) 25 MHz (d) 5 GHz.

**111.** A half-wave dipole is placed at a distance $S = \gamma/2$ from the apex of a 90° corner reflector. Find the radiation intensity in the direction ($\theta = 90°$, $\phi = 0°$), given a feedpoint current of 1 A.

(a) 38.2 W/sr (b) 76.4 W/sr (c) 152.8 W/sr (d) 19.8 W/sr.

**112.** The intrinsic impedance $\eta$ of a conducting medium for which $S$ =58 MS/m, $\mu$ =1 at a frequency $f$ =100 MHZ

(a) $2.14 \times 10^5 \angle 45°\Omega$ (b) $1.84 \times 10^{-3} \angle -45°\Omega$

(c) $3.69 \times 10^{-3} \angle 45°\Omega$ (d) $3.69 \times 10^{-3} \angle -45°\Omega$.

**113.** Impedance inversion may be obtained with

(a) a short circuited stub (b) an open circuited stub

(c) a quarter-wave line (d) a half-wave line

**114.** The surface integral of the electric field intensity

(a) electric charge (b) net flux emanating from the surface

(c) differential of the volume flux (d) none of these.

**115.** The poynting vector for an electromagnetic wave is

(a) $E \times H$ (b) $E \cdot H$ (c) $E/H$ (d) none of these.

**116.** The poynting vector gives

(a) rate of energy flow (b) direction of polarization of e.m. waves

(c) the electric field intensity (d) none of these.

**117.** Magnetic flux density is measured in

(a) Farads/m (b) Webers/m (c) Webers/m$^2$ (d) Amp/m$^2$.

**118.** Equation $\nabla^2 V = -\frac{\rho}{E}$ is called the

(a) Poisson's equation (b) Laplace equation

(c) Continuity equation (d) None of these.

**119.** On doubling the turns of the coil the inductance

(a) remains the same (b) doubles

(c) quadruples (d) becomes one fourth.

**120.** An electron is placed midway between two equal and opposite charges. The electron will

(a) be thrown out of the field (b) spin in the field

(c) remain stationary (d) oscillate between the charges.

**121.** Two waves of same frequency have opposite phases, then the phase angle between them is

(a) 0° (b) 90° (c) 360° (d) $\pi$ rad.

**122.** If a dielectric is placed in an electric field, the field strength

(*a*) increases (*b*) decreases

(*c*) remains unaltered (*d*) becomes zero.

**123.** An electromagnetic wave incident on a perfect dielectric is

(*a*) fully transmitted

(*b*) fully reflected

(*c*) partially transmitted and partially reflected

(*d*) none of these.

**124.** The surface integral of the electric field intensity is:

(*a*) electric charge (*b*) net flux emanating from the surface

(*c*) differential of volume flux (*d*) none of these.

**125.** The electric field intensity (E) and electric potential (*V*) are interrelated as

(*a*) E = – Div *V* (*b*) E = Div *V*

(*c*) E = – Gradient of *V* (*d*) none of these.

**126.** According to Poisson's equation, if *V* is the potential function, then

(*a*) $\nabla^2 V = -\rho/\varepsilon$ (*b*) $\nabla V = -\rho/E$ (*c*) $\nabla^2 V = 0$ (*d*) none of these.

**127.** One of the two main mathematical operators that indicates the manner in which a vector field changes from point to point throughout space is the DIVERGENCE. The other is

(*a*) gradient (*b*) curl (*c*) del (*d*) none of these.

**128.** It is given that electric flux density in a certain region is expressed by $D = (1/r^2)\, a_r$ in spherical coordinate system. $a_r$ here is the unit vector along *r*. The charge density ρ in this region is

(*a*) positive (*b*) negative (*c*) $1/r$ (*d*) zero.

**129.** Four fundamental equations for electromagnetic are grouped under

(*a*) Fleming's laws (*b*) Faraday's laws

(*c*) Lorentz equation (*d*) Maxwell's equations.

**130.** One of the following is the correct expression for one of the four fundamental equations of electromagnetic

(*a*) $\nabla . D = \rho$ (*b*) $\nabla \times D = 0$ (*c*) $\nabla \times B = 0$ (*d*) none of these.

**131.** A particle of charge $-2 \times 10^{-9}$ C is acted on by a downwards force of $3 \times 10^{-6}$ N in an electric field. The strength of the electric field is

(*a*) 6 N (*b*) 6 NC (*c*) 6 N/C (*d*) $2.5 \times 10^3$ N/C.

**132.** The surface integral of field intensity (*E*) is equal to the

(*a*) induced emf (*b*) electric charge

(*c*) electric charge density (*d*) net flux emanating from the surface.

**133.** The function $f(x - v_0 t)$ represents

(*a*) stationary wave (*b*) wave motion in the reverse direction

(*c*) wave motion in the forward direction (*d*) none of these.

**134.** The condition $\sigma >>> \varepsilon\omega$ is valid for a
(*a*) good conducting medium (*b*) good dielectric medium
(*c*) partially conducting medium (*d*) homogeneous and isotropic region.

**135.** A time varying electric field is associated with
(*a*) an electric field only (*b*) a magnetic field
(*c*) both electric and magnetic field (*d*) none of these.

**136.** A time varying magnetic field produces
(*a*) an electric field only (*b*) both electric and magnetic field
(*c*) a magnetic field (*d*) none of these.

**137.** In order to avoid short circuiting of electrical component in case of surface wave the wave should be
(*a*) linearly polarized (*b*) vertical polarized
(*c*) circularly polarized (*d*) elliptically polarized.

**138.** Only one of the following can happen to the electromagnetic waves travelling in free space
(*a*) attenuation (*b*) reflection (*c*) refraction (*d*) absorption.

**139.** Circular polarized waves result when
(*a*) phases are same
(*b*) magnitudes are same and phase difference is 90°
(*c*) magnitudes are same and phase difference is 0°
(*d*) magnitudes are same.

**140.** In a perfect conductor, the incident and the reflected waves combine to produce
(*a*) a stronger incident wave
(*b*) a stronger reflected wave
(*c*) standing wave which is not progressing
(*d*) both (*a*) and (*b*).

**141.** Current changing from 4 to 6 A in 1 sec induces 40 V in a coil, its inductance is
(*a*) 40 mH (*b*) 4 H (*c*) 6 H (*d*) 20 H.

**142.** Which of the following form the highest frequency?
(*a*) $T = 1/6000$ sec (*b*) $T = 1/60$ sec (*c*) $T = 1$ sec (*d*) $T = 2$ sec.

**143.** The wavelength of *X*-rays is 30 pm. The frequency is
(*a*) $10^{16}$ Hz (*b*) $10^{22}$ Hz (*c*) $10^{16}$ Hz (*d*) $10^{19}$ Hz.

**144.** One volts equal
(*a*) one joule (*b*) one joule/coulomb
(*c*) one coulomb/joule (*d*) none of these.

**145.** Two vectors *A* and *B* are equal if their
(*a*) magnitudes are equal (*b*) direction are same
(*c*) either (*a*) or (*b*) (*d*) both (*a*) or (*b*).

146. Which of the following does not characterized ferromagnetic material
(*a*) permittivity (*b*) coercivity (*c*) remanence (*d*) permeability.

147. Air is considered as
(*a*) diamagnetic (*b*) non magnetic (*c*) paramagnetic (*d*) ferromagnetic.

148. The distance between adjacent maxima and minima of a standing wave is
(*a*) $\lambda$ (*b*) $\lambda/2$ (*c*) $\lambda/4$ (*d*) $\lambda/8$.

149. The wave propagation occurs without attenuation in case of a
(*a*) perfect dielectric (*b*) semiconductor
(*c*) conductor (*d*) all of these.

150. Due to law of conservation of charge for a closed surface the equation $[I_c]_s + [I_d]_s$
(*a*) may not necessarily be equal to zero (*b*) should not be equal to zero
(*c*) should be equal to zero (*d*) none of these.

151. Which of the following is incorrect relation?
(*a*) permeability = 1/ reluctance (*b*) resistance × conductance = 1
(*c*) reluctance × mmf = flux (*d*) none of these.

152. Vectors can be added
(*a*) differentially (*b*) commutatively
(*c*) by compounding (*d*) none of these.

153. While comparing a magnetic circuit with an electric circuit, the equivalent of magnetic flux is
(*a*) the voltage of the source (*b*) the current in the lines
(*c*) the conductance of the load (*d*) the conductivity of the lines.

154. A medium is said to be isotropic when
(*a*) $\varepsilon$ is zero (*b*) $\varepsilon$ is a scalar constant
(*c*) $\varepsilon = \mu$ (*d*) $\varepsilon = \mu\sigma$.

155. The electric field strength $E$ of a dipole varies
(*a*) directly as the square of the distance (*b*) inversely as the square of the distance
(*c*) inversely as the distance (*d*) inversely as the cube of the distance.

156. Vacuum is considered as
(*a*) diamagnetic (*b*) non magnetic (*c*) paramagnetic (*d*) ferromagnetic.

157. The dimension of relative permittivity is same as the dimension of
(*a*) $\varepsilon$ (*b*) $\varepsilon_0$ (*c*) $\varepsilon_0/L$ (*d*) none of these.

158. The absorption of radio waves by the atmosphere depends on
(*a*) the polarization of atmosphere (*b*) the polarization of the waves
(*c*) their frequency (*d*) their distance from the transmitter.

**159.** When charges are applied to gold leaf electroscope

(*a*) the leaves converges for positive charges

(*b*) the leaves remain stationary

(*c*) the leaves diverge for negative charges

(*d*) the leaves diverge for both positive and negative charges.

**160.** The unit for electric field strength $E$ is

(*a*) volts/meter (*b*) nt/coul (*c*) joule/coul-meter (*d*) all.

**161.** The force experienced by a small test charge $q$ placed in static electric field $E$ is given by

(*a*) $F = Eq$ (*b*) $F = E/q$ (*c*) $F = E^2/q$ (*d*) $F = Eq^2$.

**162.** A low transmission line has

(*a*) $R = \omega L, G = \omega C$ (*b*) $R <<< \omega L, G <<< \omega C$

(*c*) $R >>> \omega L, G >>> \omega C$ (*d*) none of these.

**163.** The electric flux and field intensity inside a conducting sphere is

(*a*) zero (*b*) uniform (*c*) maximum (*d*) minimum.

**164.** Which of the following can produce the most induced voltage?

(*a*) 1 A d.c. (*b*) 50 A d.c. (*c*) 1 A 60 Hz a.c. (*d*) 1 A 400 Hz a.c.

**165.** The displacement density is measured in

(*a*) volt/meter (*b*) ampere/meter (*c*) coulomb/meter (*d*) coulomb/m$^2$.

**166.** The critical angle for e.m. wave passing from glass to air ($\varepsilon/\varepsilon_0 = 9$) will be nearly

(*a*) 10.5° (*b*) 19.5° (*c*) 26.5° (*d*) 38.5°.

**167.** In good conductor, the velocity of the electromagnetic wave is

(*a*) very low (*b*) high

(*c*) $3 \times 10^8$ m/s (*d*) more than $3 \times 10^8$ m/s.

**168.** Which of the following statement is true?

(*a*) Iron is generally used for permanent magnets

(*b*) Alnico is commonly used for electromagnets

(*c*) Paper can't affect magnetic flux because it is not a magnetic material

(*d*) None of these.

**169.** An e.m. wave incident on perfect conductor is

(*a*) entirely reflected (*b*) fully transmitted

(*c*) partially transmitted (*d*) none of these.

**170.** Which of the following material has maximum magnetic permeability?

(*a*) Copper (*b*) Steel (*c*) Aluminium (*d*) Ebonite.

**171.** Which of the following represents the continuity equation for discrete components?

(*a*) $\nabla \cdot \vec{J} = 0$ (*b*) $\nabla \times \vec{H} = \vec{J}$ (*c*) $\nabla \cdot \vec{J} = \vec{H}$ (*d*) $\nabla \cdot \vec{H} = \vec{J}$.

**172.** The phase velocity $v$ in a transmission line is given by

(*a*) $v = 1/LC$ (*b*) $v = LC$ (*c*) $v = \sqrt{LC}$ (*d*) $v = 1/\sqrt{LC}$.

**173.** The direction of a surface is

(*a*) same as the surface of constant value of the surface

(*b*) normal to the surface of constant value of the surface

(*c*) either (*a*) or (*b*)

(*d*) both (*a*) and (*b*).

**174.** Pick the wrong statement

(*a*) is in the nature of an impedance

(*b*) is an impedance whose value is $\sqrt{\frac{\mu}{\varepsilon}}$

(*c*) is a plane wave

(*d*) is called the intrinsic impedance of free space of value 120π.

**175.** The electric field resulting from electrostatic charge $Q$ is

(*a*) continuous (*b*) isotropic (*c*) irrotational (*d*) none of these.

**176.** An isolated conducting sphere shell carries a positive charge. A negatively charged metal shell is placed inside the shell

(*a*) the object inside will be pushed out

(*b*) the object will be circulating inside the spherical shell

(*c*) the net charge will be zero

(*d*) there will be a net charge depending upon whether positive or negative charge is higher.

**177.** A charged plate is touched by a metal rod standing on wooden platform

(*a*) the plate is discharged completely

(*b*) the charge is unaffected in the plate

(*c*) the charge is transferred to metal rod

(*d*) all of these.

**178.** Electric lines of forces does not cross other because

(*a*) the tangent to the line of force gives the direction of $E$ and if they cross, there will be two tangents for $E$ which is not possible

(*b*) the lines of force mutually repel

(*c*) the lines of force emanate from the point charge

(*d*) none of these.

**179.** The following law relates the forces experienced by two loop of wires carrying current

(*a*) Coulombs law (*b*) Amperes law (*c*) Netwons law (*d*) Maxwell's equation.

**180.** A potential difference of $V$ volts is applied to a copper wire of length $l$. On doubling $v$, $l$ separately, the drift velocity

(*a*) increases, increases (*b*) increases, decreases
(*c*) decreases, increases (*d*) doubled, halved.

**181.** Which of the following terms applies to troposcatter propagation?
(*a*) SIDs (*b*) Faraday rotation (*c*) Fading (*d*) Atmospheric storms.

**182.** As an e.m. wave travels in free space , only one of the following can happen to them
(*a*) reflection (*b*) absorption (*c*) attenuation (*d*) refraction.

**183.** On equipotential surface
(*a*) potential is everywhere same (*b*) potential is everywhere different
(*c*) potential varies radially only (*d*) none of these.

**184.** Inductive reactance applies to sine wave because it
(*a*) increases with lower frequency (*b*) increases with lower inductance
(*c*) increases with higher frequency (*d*) decreases with higher frequency.

**185.** The electric field intensity at a distance '*d*' from a infinitely long charged line will vary as
(*a*) $d^2$ (*b*) $d^4$ (*c*) $1/d$ (*d*) $1/d^2$.

**186.** The relative permeability of a paramagnetic substance will be
(*a*) less than 1 (*b*) 1
(*c*) slightly more than 1 (*d*) around 0.5.

**187.** When the charge moves with a velocity $v$ in an electric field $E$ and magnetic field $B$, the Lorentz force is given by
(*a*) $F = qE + vqB$ (*b*) $F = vqB$ (*c*) $F = qE$ (*d*) $F = qE - vqB$.

**188.** Between a hollow and solid metal sphere, charge resides
(*a*) on the outer surface in both
(*b*) on outer surface in hollow and throughout in solid
(*c*) throughout in both
(*d*) none of these.

**189.** Diffraction of electromagnetic waves
(*a*) is caused by reflection from ground
(*b*) arises only with spherical wave fronts
(*c*) will occur when a wave pass through large slot
(*d*) may occur around the edge of sharp obstacle.

**190.** Between a plate of a capacitor, an aluminium foil is introduced midway and connected to one plate. The capacitance
(*a*) increases (*b*) decreases (*c*) remains same (*d*) increases twice.

**191.** Laplacian of scalar function $V$ is
(*a*) gradient of $V$ (*b*) divergence of $V$
(*c*) gradient of gradient of $V$ (*d*) divergence of gradient of $V$.

**192.** Flux/current could be a dimension of

(*a*) inductance (*b*) charge (*c*) capacitance (*d*) none of these.

**193.** Propagation of frequencies in *UHF* range takes place by means of

(*a*) surface waves (*b*) space wave (*c*) ground wave (*d*) sky wave.

**194.** The product of *RC* an *L*/*R* is

(*a*) ohms farad and henry/ohm (*b*) ohms farad and second

(*c*) both second (*d*) coulomb and ampere.

**195.** The magnetic field intensity emenating from a closed surface

(*a*) depends on dipole moment (*b*) depends on volumetric compaction factor

(*c*) is infinite (*d*) is zero.

**196.** Plasma signifies

(*a*) electronic medium (*b*) solid dielectric

(*c*) liquid dielectric (*d*) gaseous medium.

**197.** $J = \sigma E$ is known as

(*a*) Biot Savart's law (*b*) Ampere's law

(*c*) Coulomb's law (*d*) Ohm's law.

**198.** Permeance is analogous to

(*a*) resistance (*b*) conductance (*c*) reluctance (*d*) capacitance.

**199.** AC current induces voltage because it has

(*a*) constant magnetic field (*b*) varying magnetic field

(*c*) stronger magnetic field (*d*) high peak value.

**200.** When the distance between the two charge is doubled, the force between them will be

(*a*) four times (*b*) double (*c*) half (*d*) one fourth.

## ANSWERS

| | | | | | | |
|---|---|---|---|---|---|---|
| **1.** (*a*) | **2.** (*b*) | **3.** (*b*) | **4.** (*b*) | **5.** (*d*) | **6.** (*c*) | **7.** (*a*) |
| **8.** (*a*) | **9.** (*a*) | **10.** (*c*) | **11.** (*b*) | **12.** (*a*) | **13.** (*c*) | **14.** (*a*) |
| **15.** (*a*) | **16.** (*c*) | **17.** (*a*) | **18.** (*c*) | **19.** (*c*) | **20.** (*c*) | **21.** (*b*) |
| **22.** (*c*) | **23.** (*a*) | **24.** (*d*) | **25.** (*a*) | **26.** (*c*) | **27.** (*b*) | **28.** (*b*) |
| **29.** (*d*) | **30.** (*c*) | **31.** (*c*) | **32.** (*b*) | **33.** (*d*) | **34.** (*a*) | **35.** (*a*) |
| **36.** (*c*) | **37.** (*c*) | **38.** (*c*) | **39.** (*d*) | **40.** (*a*) | **41.** (*b*) | **42.** (*a*) |
| **43.** (*c*) | **44.** (*a*) | **45.** (*a*) | **46.** (*d*) | **47.** (*a*) | **48.** (*c*) | **49.** (*a*) |
| **50.** (*b*) | **51.** (*a*) | **52.** (*b*) | **53.** (*d*) | **54.** (*a*) | **55.** (*b*) | **56.** (*b*) |
| **57.** (*a*) | **58.** (*c*) | **59.** (*b*) | **60.** (*a*) | **61.** (*b*) | **62.** (*c*) | **63.** (*a*) |
| **64.** (*b*) | **65.** (*b*) | **66.** (*b*) | **67.** (*b*) | **68.** (*b*) | **69.** (*c*) | **70.** (*d*) |
| **71.** (*a*) | **72.** (*c*) | **73.** (*c*) | **74.** (*b*) | **75.** (*d*) | **76.** (*d*) | **77.** (*b*) |

| | | | | | | |
|---|---|---|---|---|---|---|
| **78.** (*c*) | **79.** (*c*) | **80.** (*c*) | **81.** (*a*) | **82.** (*a*) | **83.** (*d*) | **84.** (*c*) |
| **85.** (*a*) | **86.** (*b*) | **87.** (*c*) | **88.** (*d*) | **89.** (*a*) | **90.** (*d*) | **91.** (*a*) |
| **92.** (*d*) | **93.** (*d*) | **94.** (*c*) | **95.** (*b*) | **96.** (*c*) | **97.** (*c*) | **98.** (*b*) |
| **99.** (*d*) | **100.** (*b*) | **101.** (*d*) | **102.** (*c*) | **103.** (*a*) | **104.** (*d*) | **105.** (*c*) |
| **106.** (*d*) | **107.** (*a*) | **108.** (*b*) | **109.** (*c*) | **110.** (*a*) | **111.** (*b*) | **112.** (*c*) |
| **113.** (*c*) | **114.** (*b*) | **115.** (*a*) | **116.** (*a*) | **117.** (*c*) | **118.** (*a*) | **119.** (*c*) |
| **120.** (*c*) | **121.** (*d*) | **122.** (*b*) | **123.** (*c*) | **124.** (*b*) | **125.** (*c*) | **126.** (*a*) |
| **127.** (*b*) | **128.** (*d*) | **129.** (*d*) | **130.** (*a*) | **131.** (*d*) | **132.** (*d*) | **133.** (*c*) |
| **134.** (*a*) | **135.** (*c*) | **136.** (*b*) | **137.** (*b*) | **138.** (*a*) | **139.** (*b*) | **140.** (*c*) |
| **141.** (*d*) | **142.** (*a*) | **143.** (*d*) | **144.** (*c*) | **145.** (*d*) | **146.** (*a*) | **147.** (*c*) |
| **148.** (*c*) | **149.** (*a*) | **150.** (*c*) | **151.** (*c*) | **152.** (*b*) | **153.** (*b*) | **154.** (*b*) |
| **155.** (*d*) | **156.** (*b*) | **157.** (*d*) | **158.** (*c*) | **159.** (*d*) | **160.** (*d*) | **161.** (*a*) |
| **162.** (*b*) | **163.** (*a*) | **164.** (*d*) | **165.** (*d*) | **166.** (*b*) | **167.** (*a*) | **168.** (*c*) |
| **169.** (*a*) | **170.** (*d*) | **171.** (*a*) | **172.** (*d*) | **173.** (*b*) | **174.** (*c*) | **175.** (*c*) |
| **176.** (*d*) | **177.** (*c*) | **178.** (*a*) | **179.** (*b*) | **180.** (*d*) | **181.** (*c*) | **182.** (*c*) |
| **183.** (*a*) | **184.** (*c*) | **185.** (*d*) | **186.** (*c*) | **187.** (*a*) | **188.** (*a*) | **189.** (*d*) |
| **190.** (*d*) | **191.** (*d*) | **192.** (*a*) | **193.** (*b*) | **194.** (*c*) | **195.** (*d*) | **196.** (*d*) |
| **197.** (*d*) | **198.** (*d*) | **199.** (*b*) | **200.** (*b*). | | | |